Computer Solutions
in Physics

With Applications in Astrophysics, Biophysics,
Differential Equations, and Engineering

Computer Solutions
in Physics

With Applications in Astrophysics, Biophysics, Differential Equations, and Engineering

Steve VanWyk

Olympic College, USA

 **World Scientific**

NEW JERSEY · LONDON · SINGAPORE · BEIJING · SHANGHAI · HONG KONG · TAIPEI · CHENNAI

Published by

World Scientific Publishing Co. Pte. Ltd.

5 Toh Tuck Link, Singapore 596224

USA office: 27 Warren Street, Suite 401-402, Hackensack, NJ 07601

UK office: 57 Shelton Street, Covent Garden, London WC2H 9HE

British Library Cataloguing-in-Publication Data
A catalogue record for this book is available from the British Library.

COMPUTER SOLUTIONS IN PHYSICS
With Applications in Astrophysics, Biophysics, Differential Equations, and Engineering
(With CD-ROM)

ISBN-13 978-981-270-936-3
ISBN-10 981-270-936-3
ISBN-13 978-981-277-499-6 (pbk)
ISBN-10 981-277-499-8 (pbk)

Printed in Singapore by World Scientific Printers

Introduction

About thirty years ago, David Park wrote a book on Quantum Theory in which he advised the student to determine whether the difficulty in solving a problem was in the Physics or in the Mathematics. With the great amount of progress in numerical methods and with the speed of the modern personal computer, the problem is no longer in the mathematics. Now it is sufficient to set up the correct Physics equations and utilize any of the excellent mathematical softwares to graph and solve the problem.

The computer solutions in this book are primarily written in *Mathematica* because of its reasonably straightforward approach to mathematical problem-solving. It should be noted that the computational powers of *Maple* and *MatLab* are equally good. A comparison of *Mathematica, Maple,* and *MatLab* is given in Appendix A.

The programming with *Mathematica* 5 should present few difficulties. There are 50-plus computer solutions from physics and engineering and 10-plus animations contained on the compact disc. Any of these can be downloaded to your computer and run with *Mathematica* 5 or *Mathematica* 6. Once a program from the CD is on your computer, you may change the conditions or modify the equations to address a problem of your choosing. A listing of all programs is given in Appendix B. If you have never used *Mathematica* before, a capsule summary of *Mathematica* commands is given in Appendix C.

It is important to note that this is not a first-year textbook on Physics. Anyone who wants to program Physics equations should have at least one year of College Physics and a good introduction to the Calculus.

Here is the premise of this text: If you can write down the correct Physics equations, then it is only necessary to program a few lines of code to get the answer. And if the Physics equations are not correct, then the program output will tell you that as well. Either way, you win. Let's get started by setting up some Physics equations ... and let the computer knock them down.

Contents

Chapter 1 Equations of Motion

1.0   Newton's Laws

Some 320 years ago, Isaac Newton wrote down the two basic equations of classical Physics

the basic equation of force $\qquad F = m\dfrac{dv}{dt} = m\,\dot{v}$

and the force of gravity $\qquad m\,\ddot{r} = -\dfrac{mGM}{r^2}$

Let us begin with motion of a planet in a gravity field. We'll use the metric system to describe the motion of the Earth about the Sun.

Cancelling m, $\qquad \ddot{r} = -\dfrac{GM}{r^2} \qquad$ where $M = M_{sun} = 1.989 \times 10^{30}$ kg

$$\text{and} \quad G = 6.673 \times 10^{-11} \ \frac{m^3}{kg \cdot s^2}$$

Notice that if we wish to measure distance in km and time in hours, then we need only change the units of G

$$G = 6.673 \times 10^{-11} \ \frac{m^3}{kg \cdot s^2} = 6.673 \times 10^{-20} \ \frac{km^3}{kg \cdot s^2}$$

$$G = 3600^2 \times 6.673 \times 10^{-20} \ \frac{km^3}{kg \cdot (hr)^2}$$

G is exactly the same. Only the units are changed. Choose the G you need, and express your distance units in m or km and your time in s or hr, but <u>be consistent</u> throughout your equation.

Now, back to the Earth. If we place the Earth at its *average distance* $r = 149.6 \times 10^6$ km away from the Sun, in a circular orbit with velocity v,

$$\frac{mv^2}{r} = \frac{mGM}{r^2}$$

$$v = \sqrt{\frac{GM}{r}} = 29.786 \text{ km/s}$$

Let's now write the physics equations for *circular motion* of the Earth about the Sun. To begin, we'll use x-y coordinates.

$$\ddot{x} = -\frac{GM}{x^2+y^2}\,\mathrm{Cos}\,\theta = -\frac{GM\,x}{(x^2+y^2)^{3/2}} \qquad x_o = 149.6 \times 10^6 \text{ km} \qquad \dot{x}_o = 0$$

$$\ddot{y} = -\frac{GM}{x^2+y^2}\,\mathrm{Sin}\,\theta = -\frac{GM\,y}{(x^2+y^2)^{3/2}} \qquad y_o = 0 \qquad \dot{y}_o \doteq 29.786 \text{ km/s}$$

These two equations describe the motion of the Earth for all time, based on the initial conditions. Let's look at the requirements for finding the speed and location of the Earth at all times after t = 0. First of all these two equations are complicated and they are not linear. We shall find a solution by using numerical methods (**NDSolve** in *Mathematica*). We will solve from t = 0 to t = 9000 hours.

Here is how we do it. To get the time in hours, and the distance from the sun in kilometers, take

$$G = 3600^2 \times 6.673 \times 10^{-20}\,\frac{\text{km}^3}{\text{kg}\cdot(\text{hr})^2} \quad \text{and} \quad M_{sun} = 1.989 \times 10^{30} \text{ kg}$$

Then the x- and y- distance of the Earth from the center of the Sun will be measured in kilometers. Notice that to be consistent we will also have to convert the initial velocity into kilometers per hour.

Here is the *Mathematica* program

```
In[1]:=  G = 3600^2 * 6.673 * 10^-20;  M = 1.989 * 10^30;
         sol = NDSolve[{
             x''[t] == G*M* (-x[t]) / (x[t]^2 + y[t]^2)^1.5,
             y''[t] == G*M* (-y[t]) / (x[t]^2 + y[t]^2)^1.5,
             x[0] == 149.6*10^6, y[0] == 0, x'[0] == 0, y'[0] == 29.786*3600},
             {x, y}, {t, 0, 9000}]
```

When we press **Shift-Enter**, the *Mathematica* program returns

```
Out[1]=  {{x → InterpolatingFunction[{{0., 9000.}}, <>],
          y → InterpolatingFunction[{{0., 9000.}}, <>]}}
```

x and y are given by an <u>Interpolating Function</u>. We can plot x vs y if we first identify an interpolation function in x, and an interpolation function in y and table these values versus time.

```
InterpFunc1 = x /. sol[[1]]; InterpFunc2 = y /. sol[[1]];
InterpFunc3 = x' /. sol[[1]]; InterpFunc4 = y' /. sol[[1]];
tbl = Table[{InterpFunc1[t], InterpFunc2[t]}, {t, 0, 8600, 200}];
```

Now we are free to plot a series of (x, y) points versus time. Either as a series of dots (**ListPlot**) or as a continuous line (**ParametricPlot**).

ListPlot[tbl, AspectRatio → Automatic, Prolog → AbsolutePointSize[3]]

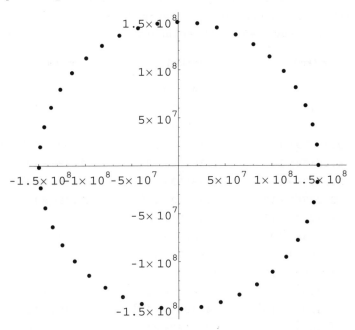

ParametricPlot[{x[t], y[t]} /. sol, {t, 0, 7400}, AspectRatio → Automatic]

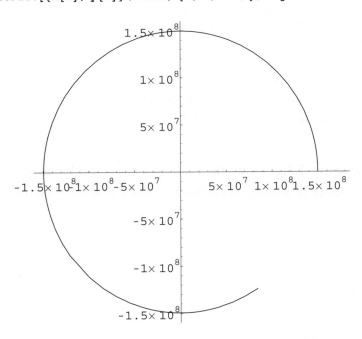

A Parametric Plot of the Earth in a circular orbit.

Let us now **Table** the data from the Interpolation Functions.

$In[191]:=$ `r[t_] =` $\sqrt{\texttt{InterpFunc1[t]^2 + InterpFunc2[t]^2}}$ `;`
         `v[t_] =` $\sqrt{\texttt{InterpFunc3[t]^2 + InterpFunc4[t]^2}}$ `/3600;`
         `Table[{t, InterpFunc1[t], InterpFunc2[t], r[t], v[t]},`
           `{t, 0, 8800, 400}] // TableForm`

t (hr)	x (km)	y (km)	r (km)	v (km/s)
0	1.496×10^8	-1.43599×10^{-21}	1.496×10^8	29.786
400	1.43493×10^8	4.23066×10^7	1.496×10^8	29.786
800	1.25672×10^8	8.11593×10^7	1.496×10^8	29.786
1200	9.75899×10^7	1.13386×10^8	1.496×10^8	29.786
1600	6.15408×10^7	1.36356×10^8	1.496×10^8	29.786
2000	2.04676×10^7	1.48193×10^8	1.496×10^8	29.786
2400	-2.22767×10^7	1.47932×10^8	1.496×10^8	29.786
2800	-6.32023×10^7	1.35594×10^8	1.496×10^8	29.786
3200	-9.8968×10^7	1.12185×10^8	1.496×10^8	29.786
3600	-1.26654×10^8	7.96178×10^7	1.496×10^8	29.786
4000	-1.43999×10^8	4.05503×10^7	1.496×10^8	29.786
4400	-1.49589×10^8	-1.82776×10^6	1.496×10^8	29.786
4800	-1.42966×10^8	-4.40566×10^7	1.496×10^8	29.786
5200	-1.24671×10^8	-8.26886×10^7	1.496×10^8	29.786
5600	-9.61974×10^7	-1.1457×10^8	1.496×10^8	29.786
6000	-5.98704×10^7	-1.37097×10^8	1.496×10^8	29.786
6400	-1.86556×10^7	-1.48432×10^8	1.496×10^8	29.786
6800	2.40823×10^7	-1.47649×10^8	1.496×10^8	29.786
7200	6.48541×10^7	-1.34811×10^8	1.496×10^8	29.786
7600	1.00331×10^8	-1.10968×10^8	1.496×10^8	29.786
8000	1.27617×10^8	-7.80645×10^7	1.496×10^8	29.786
8400	1.44484×10^8	-3.8788×10^7	1.496×10^8	29.786
8800	1.49555×10^8	3.65522×10^6	1.496×10^8	29.786

The above Table shows the Earth in a circular orbit about the Sun. For our choice of starting conditions, r and v remain constant.

Just how accurate is *Mathematica*? Notice that in the above **Table** that we have solved the Differential Equations for a time of one year, and, as expected, the radius of the orbit and the velocity have remained absolutely constant. This is a first test of the *Mathematica* numerical solver, and it is gratifying to see that the LSODA algorithm in **NDSolve** produces good, consistent results. (The Livermore Solver for Ordinary Differential equations Adaptive method was developed at Lawrence Livermore Labs and utilizes both an Adams method and a Gear backward differences method to obtain results of high accuracy.)

In terms of precision, *Mathematica* routinely carries at least 16 places of decimal accuracy. This is far more than what we will need in this text, where we shall report the results of computations to 6 places (1 part per million accuracy) except in those few cases where greater precision is required.

To this point, we have only modeled the Earth as traveling in a circular orbit at one Astronomical Unit (its average distance of 149.6 million kilometers) from the center of the sun. What is necessary next, is to use the computer to find all the parameters of the Earth's *elliptical* orbit about the sun, using only Newton's Law of Gravity, and a starting distance and velocity for the Earth.

1.1 Motion of the Planets about the Sun

Let's program the actual motion of the Earth about the sun. Astronomers tell us that the Earth is closest to the sun every January 3, at a distance of 147.1×10^6 km from the center of the sun, and the velocity of the Earth at that time is 30.288 km/s. Let us program in these two numbers as *initial conditions*, and see if we can find the complete orbit of the Earth about the sun.

From the plot, the orbit appears circular. However, if we **Table** the numbers, we see that the Earth actually slows down a small amount, and moves outward from the sun by a small amount, and halfway through the orbit, speeds up again, as it comes in closer to the sun. The orbit is an ellipse.

We have programmed in a starting velocity a little greater than that required for circular orbit. Therefore our starting point is the <u>perihelion</u> and the point of furthest excursion is the <u>aphelion</u>. What is worth noting is that we only specified the <u>location and velocity</u> of the Earth at ONE POINT.

And this, with Newton's Law of Gravity, is all you need to describe the motion of any planet or comet about the sun, or any satellite about a planet.

Now, the correct equations of motion for the Earth about the sun are

$$\ddot{x} = -\frac{GM}{x^2 + y^2} \cos\theta = -\frac{GM\,x}{(x^2 + y^2)^{3/2}} \qquad x_o = 147.1 \times 10^6 \text{ km} \qquad \dot{x}_o = 0$$

$$\ddot{y} = -\frac{GM}{x^2 + y^2} \sin\theta = -\frac{GM\,y}{(x^2 + y^2)^{3/2}} \qquad y_o = 0 \qquad \dot{y}_o = 30.288 \text{ km/s}$$

```
 G = 3600^2 * 6.673 * 10^-20;  M = 1.989 * 10^30;
sol1 = NDSolve[
   {x''[t] == G*M*(-x[t])/(x[t]^2+y[t]^2)^1.5,
    y''[t] == G*M*(-y[t])/(x[t]^2+y[t]^2)^1.5, x[0] == 147.1*10^6,
    y[0] == 0, x'[0] == 0, y'[0] == 30.288*3600}, {x, y}, {t, 0, 9000}]

InterpFunc1 = x /. sol1[[1]]; InterpFunc2 = y /. sol1[[1]];
tbl = Table[{InterpFunc1[t], InterpFunc2[t]}, {t, 0, 5000, 200}];
ListPlot[tbl, AspectRatio → Automatic, Prolog → AbsolutePointSize[3]]
```

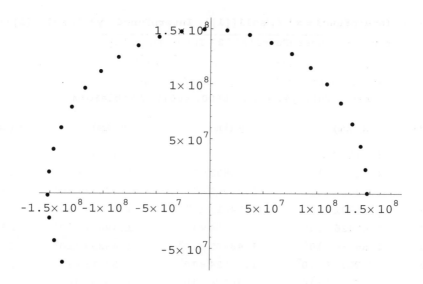

The orbit of the Earth, from January to July.

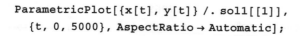

```
ParametricPlot[{x[t], y[t]} /. sol1[[1]],
    {t, 0, 5000}, AspectRatio → Automatic];
```

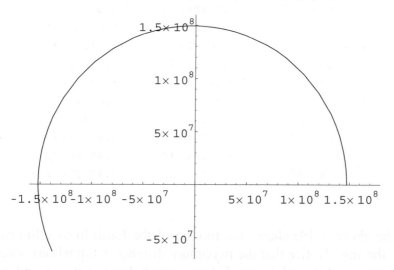

The orbit of the Earth about the sun (at 0, 0). Although the orbit appears circular, the following Table shows that it is an ellipse.

In[210]:= `InterpFunc3 = x' /. sol1[[1]]; InterpFunc4 = y' /. sol1[[1]];`

$$r[t\_] = \sqrt{InterpFunc1[t]\char`\^2 + InterpFunc2[t]\char`\^2}$$

`Vel[t_] := ` $\sqrt{InterpFunc3[t]\char`\^2 + InterpFunc4[t]\char`\^2}$ `/3600`

`Table[{t, InterpFunc1[t], Chop[InterpFunc2[t]],`
`   r[t], Vel[t]}, {t, 0, 8800, 400}] // TableForm`

t (hr)	x (km)	y (km)	r (km)	v (km / s)
0	1.471×10^8	0	1.471×10^8	30.288
400	1.40788×10^8	4.29893×10^7	1.47205×10^8	30.2666
800	1.2242×10^8	8.22999×10^7	1.47512×10^8	30.2046
1200	9.36301×10^7	1.1461×10^8	1.47994×10^8	30.1076
1600	5.69426×10^7	1.37264×10^8	1.48607×10^8	29.9844
2000	1.55064×10^7	1.48492×10^8	1.49299×10^8	29.846
2400	-2.72117×10^7	1.47524×10^8	1.50013×10^8	29.7039
2800	-6.77342×10^7	1.34609×10^8	1.5069×10^8	29.5697
3200	-1.02851×10^8	1.10935×10^8	1.51277×10^8	29.4539
3600	-1.29847×10^8	7.84934×10^7	1.51728×10^8	29.3652
4000	-1.46679×10^8	3.99028×10^7	1.5201×10^8	29.31
4400	-1.52089×10^8	-1.79697×10^6	1.521×10^8	29.2923
4800	-1.45678×10^8	-4.3357×10^7	1.51993×10^8	29.3133
5200	-1.27919×10^8	-8.15376×10^7	1.51696×10^8	29.3715
5600	-1.00141×10^8	-1.13326×10^8	1.51232×10^8	29.4628
6000	-6.44497×10^7	-1.36152×10^8	1.50635×10^8	29.5806
6400	-2.36095×10^7	-1.48083×10^8	1.49953×10^8	29.7158
6800	1.91367×10^7	-1.48007×10^8	1.49239×10^8	29.8581
7200	6.03011×10^7	-1.35761×10^8	1.48551×10^8	29.9957
7600	9.64315×10^7	-1.12202×10^8	1.47947×10^8	30.117
8000	1.24421×10^8	-7.91801×10^7	1.47479×10^8	30.2113
8400	1.41812×10^8	-3.94194×10^7	1.47188×10^8	30.2701
8800	1.47054×10^8	3.71581×10^6	1.47101×10^8	30.2878

The above Table shows the motion of the Earth in its elliptical orbit about the sun. Notice that the maximum distance r (aphelion) is near t = 4400 hours, and the minimum distance (perihelion) at t = 0, and again at t = 8800 hours.

We may zoom in on any part of this Table to get a better read on any part of the orbit. If we wish to examine the point of furthest distance from the sun, we expand the region (t = 4380 to 4390 hours) to find where $y \simeq 0$.

Then if we wish to know the time of orbit completion, we expand (t = 8760 to 8770 hours) to find where $y \simeq 0$.

```
In[214]:= Table[{t, InterpFunc1[t], InterpFunc2[t], r[t], Vel[t]},
              {t, 4380, 4390, 1}] // TableForm
          Table[{t, InterpFunc1[t], InterpFunc2[t], r[t], Vel[t]},
              {t, 8760, 8770, 1}] // TableForm
```

t (hr)	x (km)	y (km)	r (km)	v (km / s)
4380	-1.521×10^8	312033.	1.521×10^8	29.2923
4381	-1.521×10^8	206581.	1.521×10^8	29.2923
4382	-1.521×10^8	101129.	1.521×10^8	29.2923
4383	-1.521×10^8	-4323.56	1.521×10^8	29.2923
4384	-1.521×10^8	-109776.	1.521×10^8	29.2923
4385	-1.521×10^8	-215228.	1.521×10^8	29.2923
4386	-1.521×10^8	-320680.	1.521×10^8	29.2923
4387	-1.521×10^8	-426132.	1.521×10^8	29.2923
4388	-1.52099×10^8	-531584.	1.521×10^8	29.2923
4389	-1.52099×10^8	-637035.	1.521×10^8	29.2923
4390	-1.52098×10^8	-742486.	1.521×10^8	29.2923

t (hr)	x	y	r (km)	v (km / s)
8760	1.47099×10^8	-645273.	1.471×10^8	30.288
8761	1.47099×10^8	-536237.	1.471×10^8	30.288
8762	1.47099×10^8	-427201.	1.471×10^8	30.288
8763	1.471×10^8	-318164.	1.471×10^8	30.288
8764	1.471×10^8	-209128.	1.471×10^8	30.288
8765	1.471×10^8	-100091.	1.471×10^8	30.288
8766	1.471×10^8	8946.02	1.471×10^8	30.288
8767	1.471×10^8	117983.	1.471×10^8	30.288
8768	1.471×10^8	227020.	1.471×10^8	30.288
8769	1.471×10^8	336056.	1.471×10^8	30.288
8770	1.47099×10^8	445093.	1.471×10^8	30.288

From the above two Tables, we find that the maximum excursion from the sun occurs at t = 4383 hours when $y \simeq 0$, $r = 152.1 \times 10^6$ km and $v = 29.292$ km/s.

Then the Earth returns to its starting position at t = 8766 hours when $y \simeq 0$, $x = 147.1 \times 10^6$ km, and $v = 30.288$ km/s.

The time of orbit completion is $T = 8766$ hours $= 365.25$ days, and the average distance of the Earth from the center of the sun is 149.6×10^6 km. Thus we are able to determine all the orbital parameters of the Earth's orbit. Let us now show that these numbers are correct.

From any astronomy text, we may find the following two values:
the *eccentricity* of the Earth's orbit is $\epsilon = 0.01671$ and
the *semi-major axis* of Earth's orbit is $a = 149.6 \times 10^6$ km

Thus the *perihelion distance* is $a(1 - \epsilon) = 147.1 \times 10^6$ km
and the *aphelion distance* is $a(1 + \epsilon) = 152.1 \times 10^6$ km

The *perihelion velocity* is found from $\frac{MG}{r1} - \frac{v^2}{2} = \frac{MG}{2a}$ v1 $= 30.288$ km/s
The *aphelion velocity* is found from $\frac{MG}{r2} - \frac{v^2}{2} = \frac{MG}{2a}$ v2 $= 29.292$ km/s

as is most easily shown using the *Mathematica* Solve command

```
In[1]:= G = 6.673 * 10^-20; M = 1.989 * 10^30;
        a = 1.496 * 10^8; r1 = 147.1 * 10^6; r2 = 152.1 * 10^6;
```

$$
\text{Solve}\left[\frac{GM}{r1} - \frac{v^2}{2} == \frac{GM}{2a}, v \right]
$$

$$
\text{Solve}\left[\frac{GM}{r2} - \frac{v^2}{2} == \frac{GM}{2a}, v \right]
$$

```
Out[2]= {{v → -30.288}, {v → 30.288}}
```

```
Out[3]= {{v → -29.2923}, {v → 29.2923}}
```

The time (in seconds) for the Earth to complete one orbit may be found from

$$
In[8]:= \quad T = \sqrt{\frac{4\pi^2 a^3}{GM}}
$$

```
Out[8]= 3.15573 × 10^7
```

or 365.246 days. Therefore all the *Mathematica* results (of the preceding pages) are in agreement with the classical dynamics of Newton.

Let us now **Animate** the motion of the Earth about the sun.   All we need to do is **Table** the (x,y) values and display them as a data points in a **Do Loop,** incrementing every 200 hours, to show that the Earth moves in a nearly-circular orbit around the sun.

```
In[21]:= Do[ListPlot[
          Table[{InterpFunc1[t] / 10^6, InterpFunc2[t] / 10^6},
          {t, 200, i, 200}], PlotRange → {{-155, 160}, {-155, 160}},
          AspectRatio → Automatic, ImageSize → 4 * 60,
          Prolog → AbsolutePointSize[5]], {i, 200, 8800, 200}];
```

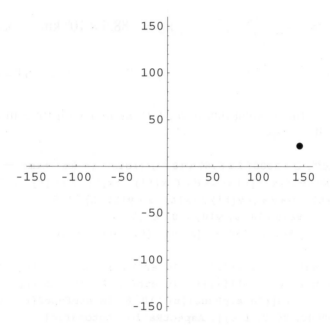

Download the file "**Motion-of-the-Planets**" to your computer and double-click on the figure to see the animation.

For more on the motion of the planets about the sun, see *Feynman's Lost Lecture*.[§] There, Richard Feynman uses arguments from plane geometry to show that the planets move in ellipses about the sun.

[§] *Feynman's Lost Lecture* edited by David and Judith Goodstein, Norton Publications, 1996

1.2 Halley's Comet near the Sun

Every 76 years[§] or so, Halley's comet makes another trip around the sun. The comet approaches to within 0.59 Astronomical Units or 88.5 million kilometers from the sun, and has a velocity of 54.45 km/s at closest approach. If we utilize this information, with Newton's Law of Gravity, we can track Halley's comet in its orbit about the sun. The physics equations are

$$\ddot{x} = -\frac{GM\,x}{(x^2 + y^2)^{3/2}} \qquad x_o = -88.5 \times 10^6 \text{ km} \qquad \dot{x}_o = 0$$

$$\ddot{y} = -\frac{GM\,y}{(x^2 + y^2)^{3/2}} \qquad y_o = 0 \qquad \dot{y}_o = -54.45 \text{ km/s}$$

Let's input these equations into *Mathematica* with time in hours and distance in kilometers.

```
M = 2 * 10^30; G = 3600^2 * 6.67 * 10^-20; (*dist in km   time in hr*)
sol = NDSolve[{x''[t] == G*M* (-x[t]) / (x[t]^2 + y[t]^2)^1.5,
     y''[t] == G*M* (-y[t]) / (x[t]^2 + y[t]^2)^1.5,
     x[0] == -88.5*10^6, y[0] == 0, x'[0] == 0,
     y'[0] == -54.45*3600}, {x, y}, {t, -9600, 9600}]

InterpFunc1 = x /. sol[[1]];   InterpFunc3 = x' /. sol[[1]];
InterpFunc2 = y /. sol[[1]];   InterpFunc4 = y' /. sol[[1]];
ListPlot[Table[{InterpFunc1[t] / 10^6, InterpFunc2[t] / 10^6},
     {t, -9600, 9600, 160}], AspectRatio → Automatic]
```

In solving the physics equations for the comet, notice that we proceed from − 9600 hours to 9600 hours, where the comet makes its closest approach to the sun at time t = 0. It does not matter to the *Mathematica* numerical solver whether we proceed backwards or forwards in time. The comet is shown on the following page from 400 days before to 400 days after its closest approach to the sun.

[§] The period of time for Halley's comet to orbit the sun varies by a few months from century to century. This was explained by Fred Whipple with his "dirty snowball theory". Each time Halley's approaches the sun, it throws off up to 25 metric tons per second of water vapor and volatile gases. This changes the orbital period in an uncertain manner. Nevertheless, Halley's comet will return sometime in 2062.

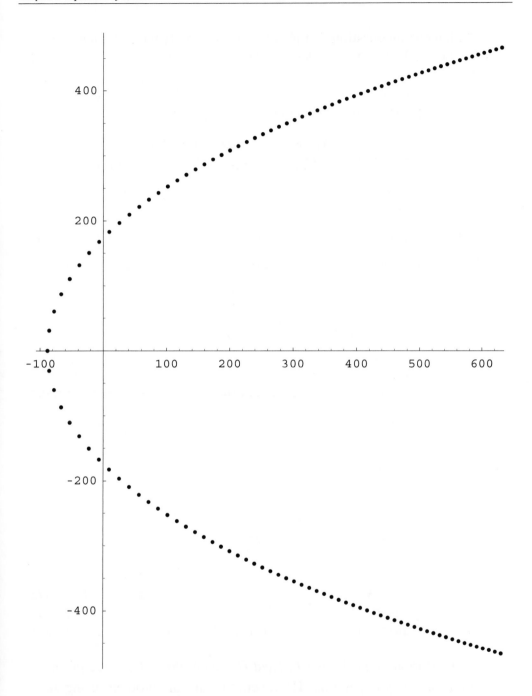

Halley's comet. From near the orbit of Jupiter, sweeping around the sun. Distances shown are in millions of kilometers. Time increment: 7 days between points.

Total time: 400 days from orbit of Jupiter to perihelion, then 400 more days to reach the orbit of Jupiter.

An interesting question that may be solved rapidly with *Mathematica* is the following: Just how long does Halley's comet spend *inside the orbit of the Earth*?   If we **Table** the above results, we can just visually scan the Table and see that the comet is within 1 A.U. of the sun for 40 days coming in, and 40 days going out, or about 80 days altogether.

$In[7]:=$ **Table** $\left[\{t\,/\,24,\;\text{InterpFunc1}[t],\;\text{InterpFunc2}[t],\right.$

$\qquad\sqrt{\text{InterpFunc1}[t]\,\hat{}\,2 + \text{InterpFunc2}[t]\,\hat{}\,2}\;/\;(150*10\,\hat{}\,6),$

$\qquad\sqrt{\text{InterpFunc3}[t]\,\hat{}\,2 + \text{InterpFunc4}[t]\,\hat{}\,2}\;/\;3600\},$

$\qquad\left.\{t,\;-2400,\;2400,\;240\}\right]$ // **TableForm**

t (days)	x (km)	y (km)	r (AU)	v (km / s)
−100	1.16394×10^8	2.61915×10^8	1.91075	29.6814
−90	9.42886×10^7	2.47914×10^8	1.76826	30.9192
−80	7.16813×10^7	2.32585×10^8	1.62253	32.3472
−70	4.85821×10^7	2.1564×10^8	1.47364	34.0163
−60	2.50415×10^7	1.96696×10^8	1.32189	35.9953
−50	1.19482×10^6	1.75222×10^8	1.16818	38.376
−40	-2.26446×10^7	1.50482×10^8	1.01451	41.2716
−30	-4.57804×10^7	1.21465×10^8	0.865371	44.7827
−20	-6.6705×10^7	8.69301×10^7	0.730491	48.8365
−10	-8.24177×10^7	4.59903×10^7	0.629207	52.6968
0	-8.85×10^7	0.	0.59	54.45
10	-8.24177×10^7	-4.59903×10^7	0.629207	52.6968
20	-6.6705×10^7	-8.69301×10^7	0.730491	48.8365
30	-4.57804×10^7	-1.21465×10^8	0.865371	44.7827
40	-2.26446×10^7	-1.50482×10^8	1.01451	41.2716
50	1.19482×10^6	-1.75222×10^8	1.16818	38.376
60	2.50415×10^7	-1.96696×10^8	1.32189	35.9953
70	4.85821×10^7	-2.1564×10^8	1.47364	34.0163
80	7.16813×10^7	-2.32585×10^8	1.62253	32.3472
90	9.42886×10^7	-2.47914×10^8	1.76826	30.9192
100	1.16394×10^8	-2.61915×10^8	1.91075	29.6814

Jerry Marion, in his book *Classical Dynamics* poses the same problem, and finds an analytic solution. However, solving this problem using Jerry Marion's method is a lot more work.

1.3 Voyager at Jupiter

One of the most impressive accomplishments of the United States space program is the odyssey of the Voyager satellite to the outer planets. To get there, NASA used a procedure called "gravity assist" to get the satellite up to speed. This consisted of bringing the satellite in behind Jupiter in its orbit. Then, as Jupiter moved away, the satellite got to keep most of its Kinetic Energy gain from Jupiter's gravitational acceleration.

The mass of Jupiter is 1.9×10^{27} kg, and the giant planet is moving at 13 km/s along $+ x$. At $t = 0$, Jupiter is at $(0, 0)$. At any time t, Jupiter will be at $(13t, 0)$, and the satellite at (x, y). We start the satellite with an initial velocity of 10 km/s on an initial heading toward $(0, 0)$. The initial distance of the satellite will be 10^6 km from Jupiter at an angle of $40°$ below the $+x$ axis.

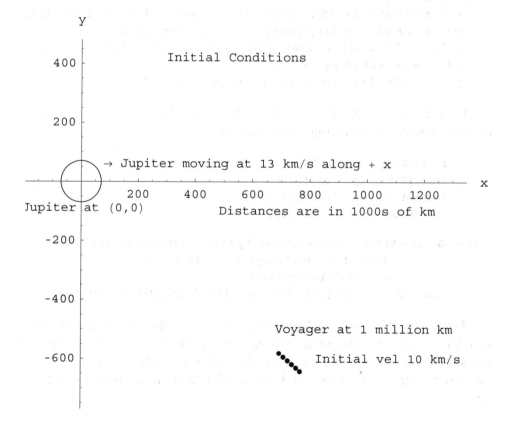

The Gravitational equations of Newton with the initial conditions are:

$$\ddot{x} = -\frac{G M (x - 13 t)}{\left((x - 13 t)^2 + y^2\right)^{3/2}} \quad x_o = 10^6 \text{km} \times \text{Cos } 40° \quad \dot{x}_o = -10 \text{ km/s} \times \text{Cos } 40°$$

$$\ddot{y} = -\frac{G M y}{\left((x - 13 t)^2 + y^2\right)^{3/2}} \quad y_o = -10^6 \text{ km} \times \text{Sin } 40° \quad \dot{y}_o = 10 \text{ km/s} \times \text{Sin } 40°$$

Where the mass of Jupiter and the gravitational constant are given by

$$M = M_{\text{Jupiter}} = 1.9 \times 10^{27} \text{ kg}$$

$$G = 6.67 \times 10^{-11} \frac{\text{m}^3}{\text{kg} \cdot \text{s}^2} = 60^2 \times 6.67 \times 10^{-20} \frac{\text{km}^3}{\text{kg} \cdot \text{min}^2}$$

We will program these equations into *Mathematica* with distance in kilometers, and time in minutes

```
M = 1.9 * 10^27; G = 60^2 * 6.67 * 10^-20;
sol = NDSolve[{
    x''[t] == M * G * (13 * 60 t - x[t]) / ((x[t] - 13 * 60 t)^2 + y[t]^2)^1.5,
    y''[t] == M * G * (-y[t]) / ((x[t] - 13 * 60 t)^2 + y[t]^2)^1.5,
    x[0] == 10^6 * Cos[40 π / 180], y[0] == -10^6 * Sin[40 π / 180],
    x'[0] == -10 * 60 * Cos[40 π / 180],
    y'[0] == 10 * 60 * Sin[40 π / 180]}, {x, y}, {t, 0, 1200}]
```

We may inquire of *Mathematica* as to where the spacecraft is at any time by making the following identifications.

```
In[3]:= InterpFunc1 = x /. sol[[1]];
        InterpFunc2 = y /. sol[[1]];
        InterpFunc3 = x' /. sol[[1]];
        InterpFunc4 = y' /. sol[[1]];
```

```
In[9]:= ListPlot[Table[{InterpFunc1[t] / 1000, InterpFunc2[t] / 1000},
        {t, 0, 1200, 30}], PlotRange → {{0, 900}, {-650, 750}},
        Prolog → AbsolutePointSize[4.0],
        Epilog → Circle[{.78 * 750, 10}, 70], AspectRatio → Automatic]
```

The entire trajectory of Voyager is shown on the next page, with the location of Jupiter at the point of closest approach. The points representing the satellite are at equal time intervals (every 30 minutes) and show the increase in satellite velocity. All distances shown are in thousands of kilometers.

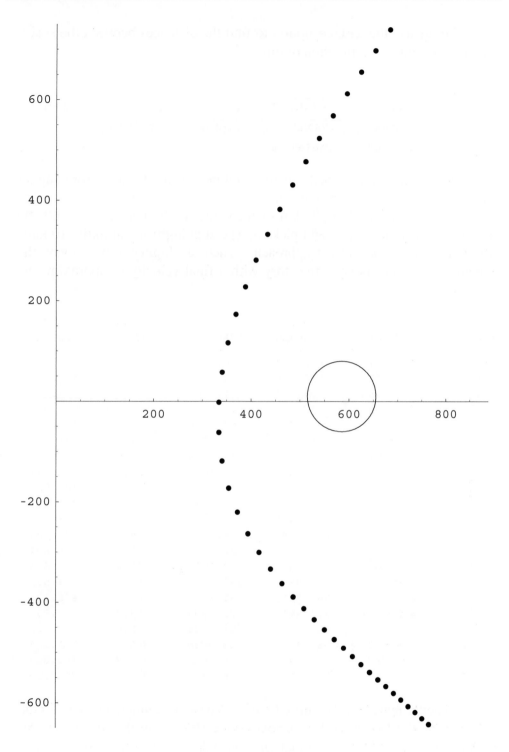

Voyager at Jupiter

Let us use the **Table** option to find the distances between the space craft and Jupiter as a function of time.

```
Table[{t, InterpFunc1[t] / 1000, InterpFunc2[t] / 1000,
    √(InterpFunc3[t]^2 + InterpFunc4[t]^2) / 60, (13.0 * 60 * t / 1000),
    √((InterpFunc1[t] - 780 t)^2 + InterpFunc2[t]^2) / 1000},
    {t, 0, 1200, 60}] // TableForm
```

The location and velocity of the satellite and its distance from Jupiter (in thousands of kilometers) are shown in the following Table. The satellite approaches to within 235,000 km of the center of Jupiter (approx 160,000 km above the cloud-tops), and picks up speed going from an initial 10 km/s to over 33 km/s on closest approach. Then, as Jupiter moves away, the satellite leaves on its new trajectory with a final velocity of approximately 28 km/s.

t (min)	x (1000 km)	y (1000 km)	v (km / s)	x (Jup) (1000 km)	d Jup (1000 km)
0	766.044	-642.788	10.	0	1000.
60	737.812	-619.086	10.4916	46.8	927.774
120	708.155	-594.104	11.0661	93.6	854.774
180	676.879	-567.552	11.7468	140.4	780.977
240	643.75	-539.03	12.5666	187.2	706.393
300	608.49	-507.975	13.5732	234.	631.095
360	570.781	-473.567	14.8379	280.8	555.297
420	530.298	-434.558	16.4692	327.6	479.507
480	486.831	-388.975	18.6327	374.4	404.898
540	440.715	-333.606	21.5597	421.2	334.177
600	394.137	-263.412	25.4325	468.	273.572
660	354.183	-172.54	29.7759	514.8	235.729
720	333.891	-61.4224	32.7412	561.6	235.847
780	340.662	57.6186	33.2933	608.4	273.868
840	368.77	172.88	32.5784	655.2	334.559
900	409.753	280.841	31.6	702.	405.315
960	458.147	381.915	30.6903	748.8	479.936
1020	510.965	477.302	29.9124	795.6	555.728
1080	566.562	568.1	29.2585	842.4	631.525
1140	623.987	655.176	28.7071	889.2	706.819
1200	682.662	739.192	28.238	936.	781.399

To better represent the entire Gravity Assist, an animated series of plots is given by the following **Do Loop**. Three stills from this Animation are shown. To see the entire Animation, you will need to access the program *VoyagerAtJupiter.nb* on the CD, and run it in *Mathematica*.

```
Needs["Graphics`MultipleListPlot`"]
Do[MultipleListPlot[
  Table[{InterpFunc1[t] / 1000, InterpFunc2[t] / 1000}, {t, 0, i, 30}],
  {{0 + .78 * t, 10}} /. {t → i},
  PlotRange → {{0, 1000}, {-680, 800}}, ImageSize → 6 * 72,
  SymbolShape → {PlotSymbol[Star], MakeSymbol[Circle[{0, 0}, 70]]},
  AspectRatio → Automatic], {i, 0, 1200, 30}]
```

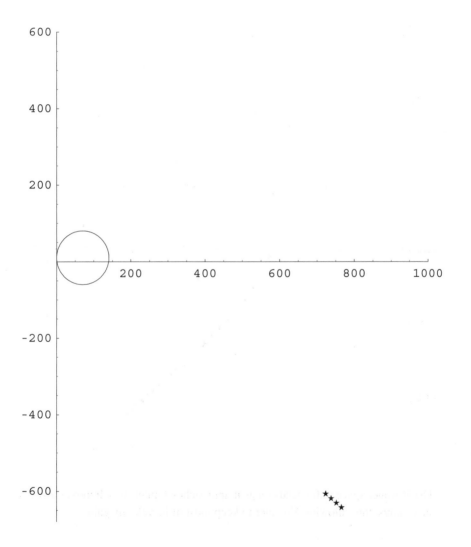

Voyager at the beginning of its approach to Jupiter.

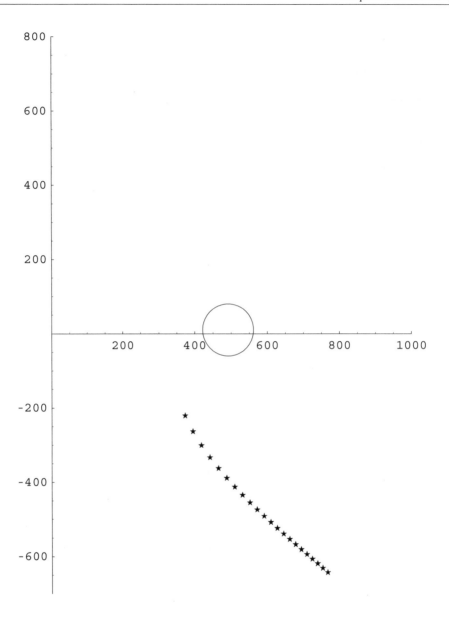

The Voyager spacecraft speeds up as it approaches Jupiter, which moves away at 13 km/s, thus allowing Voyager to keep most of its velocity gain.

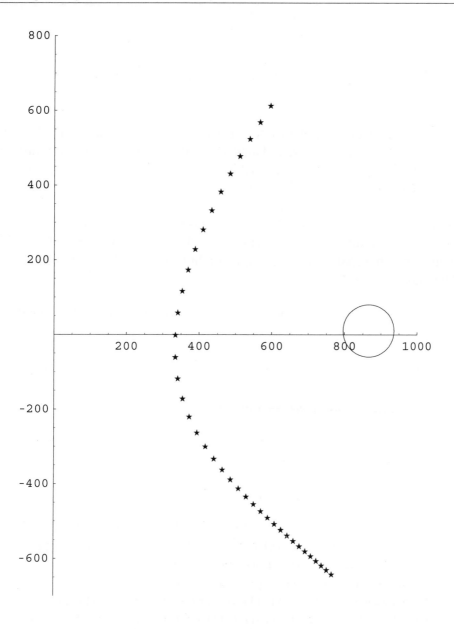

For more on the Gravitational Assist (or Slingshot Effect) see the article "Gravitational Assist in Celestial Mechanics" by James van Allen in the May 2003 *American Journal of Physics*, pages 448-451.

1.4   Baseball with Air Friction

The flight of a baseball once struck by the bat, is determined by the velocity of the ball, gravity, and air-frictional drag. If a bat speed of approximately 90 miles/hour can be achieved (by a very strong major-league batsman), then an analysis of momentum exchange would suggest an initial velocity of approximately 220 ft/s (150 mph) for the baseball. If we assume an initial angle of 30° for the baseball off the bat, then with an assumed constant drag coefficient $C_D = 0.4$, air density $\rho = 0.077$ lb/ft^3, diameter of baseball = 0.238 ft, and weight of the baseball mg = 0.3125 lb. The drag force is $D = \frac{1}{2}\rho A C_D\, v^2$

In units of feet and seconds, the Equations of Motion of the baseball are

$$mx'' = -D * \mathrm{Cos}\,\Theta = -\frac{1}{2}\rho A C_D\, v^2\, \frac{v_x}{v} = -\frac{1}{2}\rho A C_D\, v\, v_x$$

$$my'' = -mg - D * \mathrm{Sin}\,\Theta = -mg - \frac{1}{2}\rho A C_D\, v^2\, \frac{v_y}{v} = -mg - \frac{1}{2}\rho A C_D\, v\, v_y$$

or, in most compact *Mathematica* form

$$x'' = -.00219\, x'\, \sqrt{x'^2 + y'^2} \qquad\qquad x_o = 0 \qquad x'_o = 220 * \mathrm{Cos}\,30°$$

$$y'' = -32 - .00219\, y'\, \sqrt{x'^2 + y'^2} \qquad y_o = 0 \qquad y'_o = 220 * \mathrm{Sin}\,30°$$

Mathematica easily solves these non-linear equations

```
sol = NDSolve[{y''[t] == -32 - .00219 y'[t] * √x'[t]^2+y'[t]^2 ,
    x''[t] == -.00219 x'[t] * √x'[t]^2+y'[t]^2 , x[0] == y[0] == 0,
    x'[0] == 190, y'[0] == 110}, {x, y}, {t, 0, 5.0}]
InterpFunc1 = x /. sol[[1]]; InterpFunc2 = y /. sol[[1]];
InterpFunc3 = x' /. sol[[1]]; InterpFunc4 = y' /. sol[[1]];
tbl = Table[{InterpFunc1[t], InterpFunc2[t]}, {t, 0, 5, .2}];
ListPlot[tbl, Prolog → AbsolutePointSize[4], AspectRatio → 0.3];
```

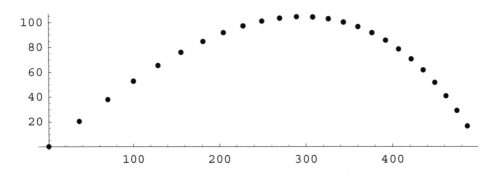

Does this baseball have home-run distance? If we **Table** the results, we will find the distance the baseball travelled $x = 498$ ft, and the time $t = 5$ seconds, for it to return to Earth (when $y \simeq 0$).

```
Table[{t, Chop[InterpFunc1[t]], Chop[InterpFunc2[t]],
    √InterpFunc3[t]^2 + InterpFunc4[t]^2}, {t, 0, 5, .2}] // TableForm
```

The (x,y) coordinates and velocity of the baseball during its 5-second flight

t (s)	x (ft)	y (ft)	v (ft / s)
0	0	0	219.545
0.2	36.2901	20.389	197.42
0.4	69.5674	37.8575	178.966
0.6	100.331	52.774	163.365
0.8	128.967	65.4222	150.043
1.	155.779	76.0249	138.582
1.2	181.01	84.7595	128.679
1.4	204.856	91.7695	120.107
1.6	227.48	97.1723	112.695
1.8	249.012	101.065	106.313
2.	269.564	103.529	100.859
2.2	289.227	104.634	96.2518
2.4	308.078	104.44	92.4257
2.6	326.18	102.999	89.3221
2.8	343.587	100.357	86.8871
3.	360.344	96.5586	85.0684
3.2	376.487	91.6422	83.8133
3.4	392.048	85.6462	83.0677
3.6	407.053	78.6075	82.7761
3.8	421.524	70.5624	82.8825
4.	435.479	61.5473	83.3315
4.2	448.933	51.5985	84.0689
4.4	461.9	40.7528	85.0437
4.6	474.394	29.0473	86.2083
4.8	486.424	16.5197	87.5196
5.	498.002	3.20752	88.9388

We can visualize the flight of the baseball a little better if we **Animate** the plot, and put a center-field wall and an outfielder in the picture. Then, with the wall at 420 ft, and the outfielder running back at 20 ft/s, we will find that the batter is indeed successful in hitting a home-run. To see the entire Animation, access *BaseballwithAirFriction* on the CD, and run it in *Mathematica*.

Notice that to plot multiple objects in the same picture that we need to call "Multiple List Plot". Then the **Do-Loop** programming proceeds as follows

Do-Loop

```
Needs["Graphics`MultipleListPlot`"]
Do[MultipleListPlot[
   Table[{InterpFunc1[t], InterpFunc2[t]}, {t, 0, i, .2}],
   {{300 + 20 * t, 1}} /. {t → i}, Prolog → Rectangle[{420, 0}, {430, 12}],
   PlotRange → {{0, 500}, {0, 210}},
   SymbolShape → {PlotSymbol[Star], PlotSymbol[Box]}], {i, 0, 5}]
```

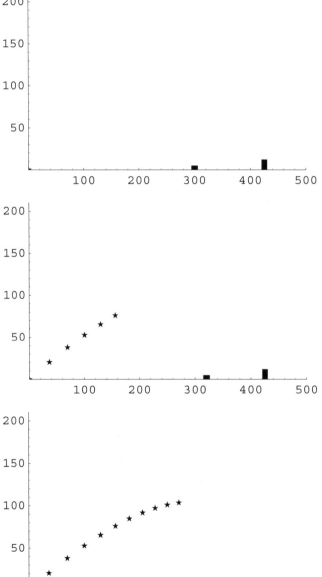

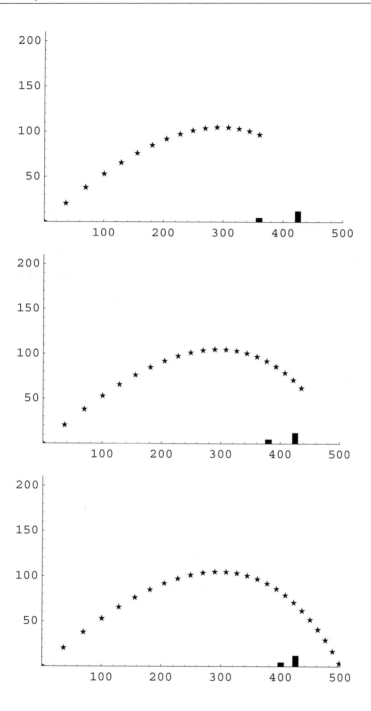

For more on batting, pitching, and fielding, see Robert Adair's 2002 book *The Physics of Baseball*, Harper-Collins, 3rd edition.

1.5 Follow the Bouncing Ball

One of the most famous pictures in Physics is the cover of the PSSC (Physical Science Study Committee) Physics text: the time-lapse photo of the bouncing ball. Let's say we drop a superball with coefficient of restitution $z = 0.9$ from a height of 8 feet. Then it is easy enough to use the energy equation $\frac{1}{2}mv^2 = mgh$ to find the velocity at impact and $h = \frac{1}{2}gt^2$ to find the time of fall.

If we let the floor supply the necessary change in velocity at each impact time, and also give the ball a horizontal velocity of 5 ft/s, so we can visualize the motion, then we have the equations of motion of the superball. (At each impact, the ball springs back with 90% of its vertical velocity. The horizontal velocity stays the same. Note the time between impacts decreases with each bounce.)

The equations of motion are as follows:

$$y' = v_y^* - gt^* \qquad y_0 = 8 \qquad \text{where time and upward velocity}$$

$$\text{are adjusted}$$

$$x' = 5 \qquad\qquad x_0 = 0 \qquad \text{to the start of each bounce}$$

```
sol = NDSolve[
    {y'[t] == 1*Which[t < .707, 0, 0.707 ≤ t ≤ 1.98, 20.36, t > 1.98, 18.32] -
        32*(t - Which[t < .707, 0, 0.707 ≤ t ≤ 1.98, .707, t > 1.98, 1.98]),
    x'[t] == 5, y[0] == 8, x[0] == 0}, {x, y}, {t, 0, 5}];
InterpFunc1 = x /. sol[[1]]; InterpFunc2 = y /. sol[[1]];
InterpFunc3 = y' /. sol[[1]]
ParametricPlot[{x[t], y[t]} /. sol, {t, 0, 3}, AxesLabel → {"x", "y"}];
```

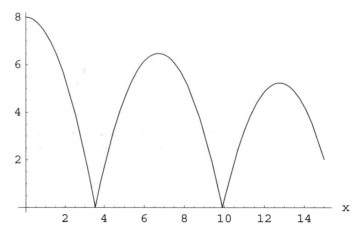

The important thing to notice in the above programming is that the time and upward velocity need to be set to appropriate values at the beginning of each bounce. This can be accomplished in *Mathematica* by using the **Which** command. Then we have a series of parabolas, each 81% as high as the last. Notice that the ball spends less time in the air on each successive bounce.

We have a geometric series which converges in time and horizontal distance travelled.

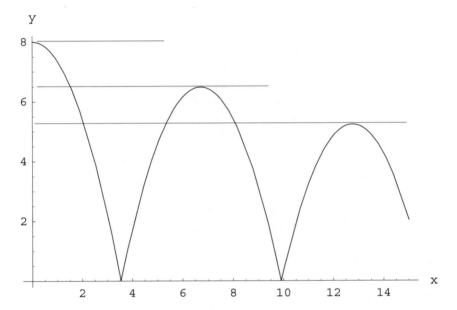

In[46]:= ``tbl = Table[{InterpFunc1[t], InterpFunc2[t]}, {t, .1, 3, .1}];``
          ``ListPlot[tbl, Prolog → AbsolutePointSize[4], AspectRatio → 0.45]``

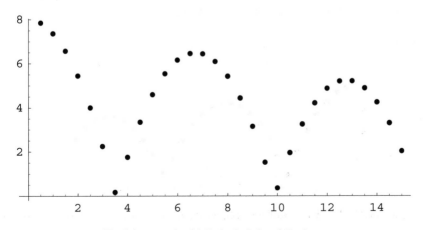

Each bounce is 81% the height of the last

```
Table[{t, InterpFunc1[t], InterpFunc2[t], InterpFunc3[t]}, {t, 0, 3, .1}]
ListPlot[tbl, Prolog → AbsolutePointSize[4], AspectRatio → 0.3]
```

t (s)	x (ft)	y (ft)	y' (ft / s)
0	0.	8.	0
0.1	0.5	7.84	-3.2
0.2	1.	7.36	-6.4
0.3	1.5	6.56	-9.6
0.4	2.	5.44	-12.8
0.5	2.5	4.	-16.
0.6	3.	2.24	-19.2
0.7	3.5	0.16	-22.4
0.8	4.	1.75	17.38
0.9	4.5	3.33	14.18
1.	5.	4.59	10.98
1.1	5.5	5.53	7.784
1.2	6.	6.15	4.584
1.3	6.5	6.44	1.384
1.4	7.	6.42	-1.81
1.5	7.5	6.08	-5.01
1.6	8.	5.42	-8.21
1.7	8.5	4.44	-11.4
1.8	9.	3.14	-14.6
1.9	9.5	1.51	-17.8
2.	10.	0.35	17.68
2.1	10.5	1.96	14.48
2.2	11.	3.24	11.28
2.3	11.5	4.21	8.08
2.4	12.	4.86	4.88
2.5	12.5	5.19	1.68
2.6	13.	5.20	-1.52
2.7	13.5	4.88	-4.72
2.8	14.	4.25	-7.92
2.9	14.5	3.30	-11.1
3.	15.	2.03	-14.3

Technically, the ball has completed 2 bounces when it hits the floor for the 3rd time.

The time to fall from $h_o = 8$ ft to the floor is $t_o = \sqrt{\dfrac{2\,h_o}{g}} = .707$ s

The time to complete the first bounce is $t_1 = 2\,(.9)\,t_o$

The total time for N bounces is $T = t_o + \displaystyle\sum_{N=1}^{N} 2\,(.9^N)\,t_o$

and for 10 bounces,

$In[113]:= \quad T = .707 + \displaystyle\sum_{N=1}^{10} (2*(.9)^N * .707)$

$Out[113]= \quad 8.99572$

Thus, the coefficient of restitution can be found for any ball from a known initial height by using a stopwatch, and finding the time to make N complete bounces.

As an example, if it takes 9 seconds for 10 complete bounces, then we use *Mathematica* to find z, the coefficient of restitution, when the ball is dropped from 8 ft.

$In[2]:= \quad \textbf{FindRoot}\left[9 == .707 + \displaystyle\sum_{N=1}^{10} (2*(z)^N * .707),\ \{z,\ .8\}\right]$

$Out[2]= \quad \{z \to 0.9001\}$

Let's also find the total time for the ball to stop bouncing, and how far it travels along x. Using *Mathematica*, we can evaluate an infinite series

$In[15]:= \quad \textbf{TF} = .707 + \displaystyle\sum_{N=1}^{\infty} (2*(.9)^N * .707);$

```
        XF = 5 * TF;
        Print ["Total Time TF→ ", TF,
            "s   Distance Travelled XF→ ", XF, " ft"]

        Total Time TF→ 13.433s   Distance Travelled XF→ 67.165 ft
```

For an animation of the Bouncing Ball, select the Animation *Bouncing-Ball.nb* on the CD, and run it in *Mathematica*.

1.6   Sky-Diving with a Parachute

An 80-kg skydiver (with parachute !) jumps out of an airplane at several thousand meter altitude, with a 0.8 drag coeff and a 1 m² surface area before parachute deployment, and a 1.33 drag coefficient with a parachute area of 40 m² after deployment. Let us say that the sky-diver gets to enjoy 17 seconds of free-fall, and then pulls the ripcord. The parachute takes 5 seconds to fully deploy. How far has the skydiver fallen in 17 seconds? In 22 seconds? What is the velocity of the skydiver just before parachute deployment? And just after? And what deceleration does the skydiver experience in gees? These are all questions that can be rapidly answered with *Mathematica*.

The Equations of Motion of the sky-diver are

$$mv' = mg - \tfrac{1}{2}\rho A^* C_D{}^* v^2 \qquad v_o = 0 \qquad\qquad \text{*Where A and } C_D \text{ change}$$
$$mx'' = mg - \tfrac{1}{2}\rho A^* C_D{}^* x'^2 \qquad x_o = 0 \qquad x'_o = 0 \qquad \text{value at t = 17 s}$$

m = 80 kg, air density ρ = 1.0 kg / m³, C_D = 0.8 , A = 1 m², g = 9.8 m / s²

Let's first of all look at the approach to terminal velocity during the first 17 seconds. Since the equations are reasonably simple, we will use **DSolve** and look for an analytic solution.

```
DSolve[{80 v'[t] == 80 * 9.8 - .40 v[t] * v[t], v[0] == 0}, v[t], t] ;
Plot[Evaluate[v[t] /. %, {t, 0, 17}], AxesLabel → {"T(s)", "V(m/s)"}];
```

Out [41]=   {{v[t] → 44.2719 Tanh[0. + 0.221359 t]}}

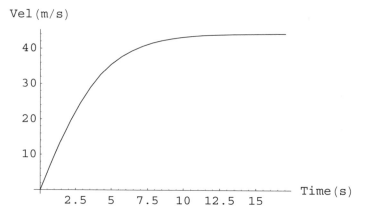

The approach to terminal velocity (44 m/s) in first 17 seconds of free-fall.

Now **DSolve** the second differential equation in x to see how far we've fallen in 17 seconds.

```
sol = DSolve[{80 x''[t] == 80 * 9.8 - .4 x'[t] * x'[t], x[0] == x'[0] == 0}, x, t]
Plot[Evaluate[x[t] /. sol, {t, 0, 17}], AxesLabel → {"Time", "Dist(m)"}]
```

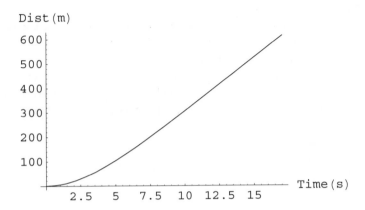

Before deploying the parachute, the skydiver has fallen a distance of six football fields in the first 17 seconds, and now has a velocity of 44 m/s (almost 100 miles per hour).

Let's **Table** the results for distance fallen, velocity, and acceleration.

```
sol = NDSolve[{80 x''[t] == 80 * 9.8 - .4 x'[t] * x'[t],
    x[0] == x'[0] == 0}, x, {t, 0, 17}];
Table[{t, x[t] /. %, x'[t] /. %, x''[t] /. %}, {t, 0, 17, 1}] // TableForm
```

t (s)	x (m)	v (m / s)	a (m / s^2)
0	0	0	9.8
1	4.8605	9.64301	9.33506
2	18.9913	18.4125	8.1049
3	41.1952	25.7251	6.49109
4	69.8903	31.3947	4.87188
5	103.477	35.5471	3.48201
6	140.572	38.4633	2.40286
7	180.095	40.4515	1.61839
8	221.256	41.7796	1.07232
9	263.504	42.6548	0.702818
10	306.465	43.2265	0.457369
11	349.89	43.5976	0.29626
12	393.617	43.8376	0.191325
13	437.537	43.9925	0.123313
14	481.583	44.0922	0.0793827
15	525.71	44.1564	0.0510641
16	569.888	44.1977	0.0328289
17	614.1	44.2242	0.021097

We are now ready to program the entire jump. Here we use the **Which** command in *Mathematica* to tell us the effective area and drag coefficient as the sky-diver proceeds through free-fall, parachute deployment, and final descent to Earth.

```
sol1 = NDSolve[
    {80 x"[t] == 80 * 9.8 - x'[t]^2 * Which[t ≤ 17, 0.40, 17 < t < 22,
        0.4 + 26.2 (t - 17) / 5, t ≥ 22, 26.6], x[0] == x'[0] == 0},
        x, {t, 0, 25}]; InterpFunc4 = x /. sol1[[1]];
Plot[x[t] /. sol1, {t, 0, 25}, AxesLabel → {"Time(s)", "x(m)"}];
Plot[x'[t] /. sol1, {t, 0, 25}, AxesLabel → {"T(s)", "Vel(m/s)"}];
Plot[(x''[t] / 9.8) /. sol1, {t, 0, 25}, AxesLabel → {"Time", "Acc"}]
```

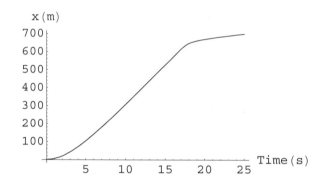

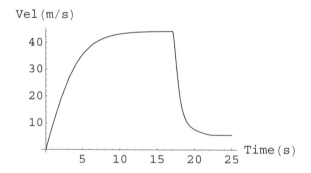

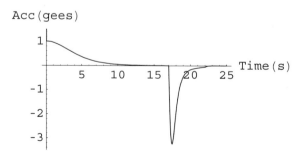

```
sol2 = NDSolve[
    {80 v'[t] == 80 * 9.8 - v[t]^2 * Which[t < 17, 0.4,
        17 ≤ t ≤ 22, 0.4 + 26.2 * (t - 17) / 5, t > 22, 26.6],
    v[0] == 0}, v, {t, 0, 25}];
Acc[t_] = 9.8 - (v[t] /. sol2)^2 * Which[t < 17, 0.4,
        17 ≤ t ≤ 22, 0.4 + 26.2 (t - 17) / 5, t > 22, 26.6] / 80;
Table[{t, x[t] /. sol1, v[t] /. sol2, Acc[t]}, {t, 17, 23, .2}] //
    TableForm
```

t (s)	x (m)	v (m / s)	a (m / s²)
17	614.1	44.2242	0.0210971
17.2	622.784	41.8734	-21.9362
17.4	630.635	36.3202	-31.3578
17.6	637.267	30.0442	-30.1877
17.8	642.705	24.5242	-24.7226
18.	647.155	20.1737	-18.892
18.2	650.847	16.8977	-14.0706
18.4	653.971	14.4657	-10.4352
18.6	656.674	12.6575	-7.79128
18.8	659.064	11.2997	-5.89217
19.	661.216	10.2649	-4.53017
19.2	663.185	9.46227	-3.54959
19.4	665.012	8.8272	-2.83854
19.6	666.724	8.31421	-2.31781
19.8	668.343	7.89114	-1.93168
20.	669.885	7.5352	-1.64102
20.2	671.361	7.23021	-1.41842
20.4	672.78	6.96459	-1.2447
20.6	674.149	6.72998	-1.10645
20.8	675.473	6.52027	-0.994297
21.	676.758	6.33096	-0.901645
21.2	678.007	6.15863	-0.823833
21.4	679.223	6.00066	-0.757531
21.6	680.408	5.85501	-0.700322
21.8	681.565	5.72005	-0.650428
22.	682.697	5.59444	-0.606516
22.2	683.806	5.50872	-0.290053
22.4	684.903	5.46756	-0.139836
22.6	685.994	5.44768	-0.067678
22.8	687.083	5.43805	-0.0328166
23.	688.17	5.43338	-0.015927

A final review of the data shows that the sky-diver was slowed from an initial downward velocity of 44.2 m/s (99 mph) to 5.4 m/s (12 mph) by deploying the parachute. The maximum deceleration experienced by the skydiver during parachute deployment was about 3.2 g.

1.7 Challenge Problems

1. Changing Satellite Speeds

Show that if a satellite is moving at 20 km/s on a straight-line approach to pass in front of Jupiter from 10^6 kilometers out, then after emerging from Jupiter's gravity, the satellite will be slowed. [Start the satellite from the same location as in Section 1.3, but change the satellite's initial velocity to $v_y = 20$ km/s, $v_x = 0$.]

2. Drag Racing

A drag racer accelerates by pushing the huge rear tires against the pavement. Mike Dunn's 1999 record for time and velocity is $t = 4.5$ s, $v = 320$ mph $= 143$ m/s, in traveling the quarter mile ($x = 400$ m). Assuming the dragster maintains the same traction all the way, and air–and–road friction increases as v^2, find the coefficient of friction μ and the terminal velocity v_t that duplicates Mike Dunn's record run, using the equations

$$v' = \mu g - kv^2 \quad \text{and} \quad x'' = \mu g - kx'^2$$

3. Solar Sailing

A novel idea for moving freight from Earth to Mars is the solar sail. The outward velocity of the solar-ship is

$$u = \frac{dr}{dt} = \frac{(2\, r^{-1/2}\, r_o\, (\gamma \cdot \mathrm{Sin}\, x \cdot \mathrm{Cos}^2\, x))}{(a - \gamma \cdot \mathrm{Cos}^3\, x)^{1/2}}$$

For a 1000-kg ship with sail area 20,000 m^2 starting at $r = r_o = 1$ AU $= 1.5 \times 10^{11}$ m from the sun, the acceleration of the ship is $\gamma = 0.00018$ m/s^2, and a = the Sun's gravity = 0.00592 m/s^2.

Find the best angle x for the sail (maximize u). Then find the time required to travel by solar sail from the orbit of Earth (1 AU) to the orbit of Mars (1.5 AU).

4. Lunar Lander

"Lunar Lander" is a popular program for the small computer. The program simulates landing on the moon in a constant gravity field. You are given an initial altitude h , downward velocity v_o and a fuel supply S. By a series of rocket fuel burns you must try to land on the moon's surface at zero velocity before you run out of fuel.

See if you can devise a way to get your 2000 kg craft down safely from an initial height of 10 km if your initial velocity is 100 m/s downwards and you have 200 kg of fuel. Burn rate is proportional to thrust, and burning 1 kg/s of fuel results in an upward thrust of 4000 N.

Your maximum fuel burn rate is 2 kg/s. Assume for simplicity the mass of the landing craft is constant. The moon's gravity near the surface is $g_{moon} = 1.7 \text{ m/s}^2$.

5. Mickey Mantle Home Run

In 1963, Mickey Mantle hit a home run into the third deck at Yankee Stadium. The baseball struck the facade over the third deck some 106 feet above the playing field, and 365 feet from home plate. If the (x, y) coordinates of the ball were (365, 107) then this would have been the first home run ever hit out of Yankee Stadium. How far would this baseball have travelled if it had just cleared the roof of the stadium?   [To match the location of Mantle's home run, adjust the initial angle in Section 1.4.]

6. Satellite Orbits

A satellite with a rocket motor is in a 200-km high orbit about the Earth. If the rocket motor is fired briefly, giving the satellite an 8% increase in forward velocity, what is the new apogee of the orbit? The mass of the Earth is $M_{Earth} = 6 \times 10^{24}$ kg.

7. Jupiter's Moon

Ganymede is 10^6 km from Jupiter. Assume a circular orbit. Write the equations of motion of Ganymede about Jupiter. Find the orbital period and Animate the plot. The mass of Jupiter is $M_{Jupiter} = 1.9 \times 10^{27}$ kg. [Just to make things interesting, put in Jupiter's motion at 13 km/s.]

Chapter 2 Vibrations and Waves

2.1 Damped and Driven Oscillations

Part 1: Natural Response

Let's first of all take an oscillating system $m\,x'' + c\,x' + k\,x = 0$ consisting of a mass on a spring with a damper. We can set this system into motion either by drawing back the mass and releasing, or by starting the mass from $x = 0$ with a velocity, say $x' = 1$ m/s.

For the system $m = 1$ kg, $c = 1$ N·s/m, $k = 16$ N/m, the mass oscillates, but *not* with an angular frequency of 4 rad/s. This is shown by using **DSolve** in *Mathematica*.

```
sol = DSolve[{x''[t] + x'[t] + 16 x[t] == 0, x[0] == 0, x'[0] == 1.0},
    x[t], t] // Chop
Plot[Evaluate [x[t] /. sol, {t, 0, 6}]];
```

Out [9]= $\{\{x[t] \to 0.251976\, e^{-t/2}\, Sin\big[\frac{3\,\sqrt{7}\,t}{2}\big]\}\}$

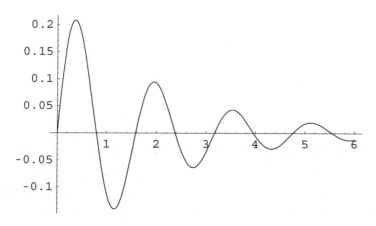

Notice that the angular frequency of vibration is $\omega = \frac{3\,\sqrt{7}}{2} = 3.968$ rad/s. The mass goes from zero to first maximum in a time of 0.364 seconds.

```
x[t_] = .251976 * Exp[-t / 2] * Sin[1.5 √7 t];
Maximize[{x[t], 0 < t < 1}, t]
```

Out [12]= $\{0.208377, \{t \to 0.364224\}\}$

The height at first maximum is 0.2083 meters, as found from

$$In[13]:= \quad \mathtt{x[t\_] = .251976\ Exp[-t/2]\ Sin}\left[\frac{3\sqrt{7}}{2}\ \mathtt{t}\right];$$

$$\mathtt{x[.364224]}$$

$$Out[14]= \quad 0.208377$$

Let's plot the above equation with an exponential envelope

```
In[40]:= Plot[{x[t] /. sol, .251976 Exp[-t/2],
         -.251976 Exp[-t/2]}, {t, 0, 6},
         PlotStyle → {GrayLevel[0], GrayLevel[.5], GrayLevel[.5]}];
```

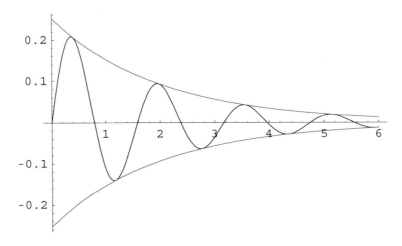

Exponential decay of an underdamped system.

An interesting question is how much does the amplitude decrease *per cycle*? All we have to do is set $\frac{3\sqrt{7}}{2} t = 2\pi$. This gives the time for the mechanical system to go through one cycle. And $t = 1.5832$ s. Therefore the motion decays by a factor of $e^{-t/2} = .453$ from one peak to the next.

If we write this as $e^{-\delta}$ then δ is called the *logarithmic decrement*.

We now know everything about the natural response of an underdamped spring-mass-damper system. The same analysis may be applied to an overdamped or a critically damped system.

We will now use *Mathematica* to examine the steady-state response of a mechanical system to a sinusoidal driving force.

Part 2: Driven Oscillations

For a damped, driven oscillator system,

$$m\ddot{x} + c\dot{x} + kx = F \sin \omega t$$

We will see how easily *Mathematica* handles Complex Algebra.

Set the problem up so that $\gamma = \frac{c}{m}$ and $\omega_o = \sqrt{\frac{k}{m}}$

Then, using Complex Algebra, write the driving force as $F\, e^{i\omega t}$

$$m\ddot{x} + m\gamma\dot{x} + m\omega_o^2 x = F\, e^{i\omega t}$$

We look for solutions of the form $x = A\, e^{i\omega t}$ where A is complex

Using *Mathematica*,

```
In[12]:= Clear[x, A, t]
         x[t_] = A * Exp[I w t];
         sol =
         Solve[m * x''[t] + m g x'[t] + m w0^2 x[t] == F Exp[I w t], {A}]
         A = A /. sol[[1]];
```

$$Out[14]= \left\{\left\{A \to -\frac{F}{m\,(-i\,g\,w + w^2 - w0^2)}\right\}\right\}$$

If we multiply this result times its complex conjugate, we arrive at the steady-state amplitude response to the forcing function $F \sin \omega t$

$$A = \frac{F/m}{\sqrt{(\omega^2 - \omega_o^2)^2 + \gamma^2\,\omega^2}}\,\sin(\omega t - \psi) \quad \text{where } \psi = \mathrm{Tan}^{-1}\left(\frac{\gamma\,\omega}{\omega^2 - \omega_o^2}\right)$$

The amplitude of vibration A for the system $m = 1, \gamma = 0.5, \omega_o = 10, F = 1$ is shown on the next page, as is the phase lag, ψ.

The amplitude at resonance $\omega = \omega_o = 10$ is

$$A = \frac{1/1}{\sqrt{0^2 + .5^2 \cdot 10^2}} = 0.2 \text{ and the phase lag is } \psi = \mathrm{Tan}^{-1}\left(\frac{\gamma\,\omega}{0}\right) = 90°.$$

```
In[16]:=  rule = {m → 1, F → 1, w0 → 10, g → 0.5}
          Plot[Abs[A /. rule], {w, 0, 15}]

Out[16]=  {m → 1, F → 1, w0 → 10, g → 0.5}
```

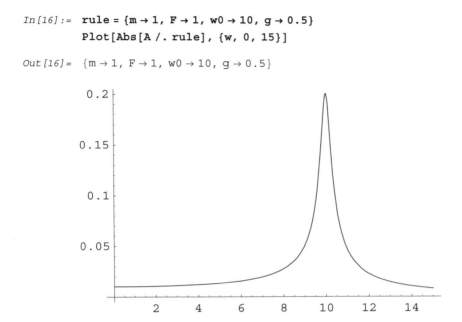

Amplitude response vs driving frequency ω

The Q is the sharpness of the resonance peak. It is numerically equal to $\frac{\omega_o}{\gamma}$
The Q for this curve is 20.

```
rule = {m → 1, F → 1, w0 → 10., g → 0.5};
Plot[-180 / π * ArcTan[Re[A] /. rule, Im[A] /. rule],
  {w, 0, 15}, PlotRange → {0, 180}]
Table[{w, Im[A] /. rule, Re[A] /. rule, Abs[A] /. rule,
    -180 / π * ArcTan[Re[A] /. rule, Im[A] /. rule]}, {w, 0, 15, 1}];
```

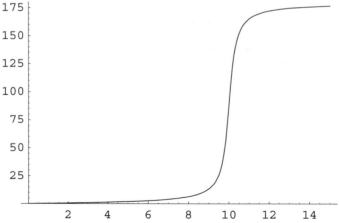

Phase lag in degrees vs driving frequency of force

If we want the *Energy* contained in the vibrating system, we need to look at A^2. Then, we find the *bandwidth* of the system, that is, the width of the curve at half-maximum

$In[10] := $ `F = 1; m = 1; g = 0.5; w0 = 10;`

$$\text{Plot}\left[\frac{(F/m)^2}{(w^2 - w0^2)^2 + g^2 * w^2}, \{w, 0, 15\}\right]$$

P lot of A^2 versus ω . The bandwidth is the width at half-maximum.

To find the *bandwidth*, find the ω values at half maximum

$In[1] := $ `F = 1; m = 1; g = 0.5; w0 = 10;`

$$\text{NSolve}\left[0.02 == \frac{(F/m)^2}{(w^2 - w0^2)^2 + g^2 * w^2}, w, 1\right]$$

$Out[2] = $ $\{\{w \to 10.2409\}, \{w \to -10.2409\}, \{w \to 9.7403\}, \{w \to -9.7403\}\}$

Bandwidth is $\omega 1 - \omega 2$ and note that the Quality factor is $Q = \frac{\omega_o}{\omega_1 - \omega_2}$

$In[1] := $ $Q = \dfrac{10}{(10.24 - 9.74)}$

$Out[1] = $ `20.`

It is worth noting that everything that has been derived for mechanical systems is also good for electrical systems, where Force $\to$ Voltage, mass $\to$ Inductance, damping $\to$ Resistance, and the spring constant $\to$ 1/ Capacitance.

As an example, consider the tuning circuit in a radio.

For AM 1000 kHz, the *bandwidth* is approx 20 kHz, and therefore its $Q \approx 50$.

The value of Complex Analysis is that given a consistent driving force such as $F \sin \omega t$, one may find the long-term or steady-state behavior of a mechanical or electrical system. However if we wish to determine both the initial transient and the steady-state response, then we shall have to model the system with Differential Equations. For the remainder of Chapter 2 and Chapter 3, we shall utilize only Differential Equations to model dynamic systems. We shall return to Complex Analysis in Chapter 4 to see how this very useful tool may be used in Quantum Theory.

2.2 Shock Absorbers

One of the most enjoyable things about physics is to take a physical system out in the country for a test drive. Let's say we have an old Volkswagen with worn shocks and we test-drive our vehicle on an undulating road. Let us say that the road surface rises 0.1 m and then falls below level 0.1 meters in a run of 24 meters. Then the <u>road</u> acts as a driving force on the spring-damper system, and the up-and-down contour of the road is transmitted to the mass of the chassis and the passengers. First, let's look at the road surface as it appears at $v = 24$ m/s (50 mph).

```
In[6]:= Clear[y, t]; L = 24; A = .1 ; v = 24; ω = 2 π * v / L;
        road = Plot[.1 * Sin[ω * t], {t, 0, 2}, AspectRatio → 0.15,
        Prolog → {Text["Time(s)", {1.88, .03}]}]
```

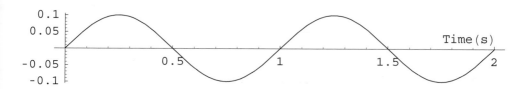

Road surface

On this "washboard" road, the road moves upwards 4 inches and then falls 4 inches below level. The road moves the damper and spring up and down, and this is transmitted to the mass of the chassis. The sum of the forces acting on the damper and spring is $F(t) = k A \sin [\omega t] + \omega c A \cos[\omega t]$.

The differential equation of this "driven" shock-absorber system is

$$m y'' + c y' + k y = k A \sin [\omega t] + \omega c A \cos[\omega t]$$

We will choose values appropriate for a car with worn shocks

$$m = 900 \text{ kg}, \ c = 3394 \text{ N·s/m}, \ k = 80000 \text{ N/m}$$

and now we will ask *Mathematica* to solve the equation of the road-driven system

```
m = 900; c1 = 3394; k = 80000.; A = .1;
v = 24; L = 24.; ω = 2 π * v / L; ω0 = √ k / m ;
 sol1 = NDSolve[
    {900 y''[t] + 3394 y'[t] + 80000 y[t] == 8000 Sin[ω * t] + c1 * A * ω * Cos[ω * t],
    y[0] == 0, y'[0] == 0}, y, {t, 0, 2.0}];
InterpFunc1 = y /. sol1[[1]]; InterpFunc2 = y' /. sol1[[1]];
InterpFunc3 = y'' /. sol1[[1]];
 plot2 = Plot[Evaluate[y[t] /. sol1],
    {t, 0, 2}, PlotStyle → {{RGBColor[1, 0, 0]}},
    AspectRatio → 0.2, AxesLabel → {"time(s)", "y(m)"}];
Show[{road, plot2}, Prolog → {{Text["Road", {0.09, .13}]},
    {Text["▷", {1.3, .178}]}, {Text["Response", {.45, .17}]},
    {Text["Time(s)", {1.75, .05}]}}]
Table[{t, Chop[InterpFunc1[t]], Chop[InterpFunc2[t]], InterpFunc3[t],
    Chop[.1 * Sin[ω t]], Chop[InterpFunc1[t] - .1 Sin[ω t]]},
    {t, 0, 1.6, .04}] // TableForm
```

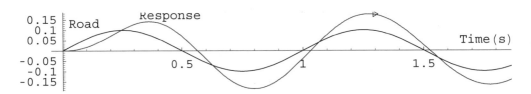

The response of the suspension-system as the automobile
travels the road.

The road "drives" the shock-absorber system, carrying the chassis "over" the top, then allowing almost "free-fall" into the canyon.   What would be the effect of "taking" this series of washboard hills at 24 m/s? If we examine the Table on the next page, we will see an initial acceleration up the first hill of 2 to 4 m/s² followed by an acceleration downward at − 4 to − 7 m/s² and then up-and-down after that as the hills "drive" the suspension system up-and-down. The weakness of the shocks of our chosen automobile allow the still-good springs to "propel" the chassis and the passengers out and over the hill thus allowing a partial "free-fall" of at least the front part of the automobile and the passengers as they descend the hill.

Clearly, this all depends on the shape of the hill, the speed of the automobile, and the strength or weakness of the springs and shocks. Notice also that for this particular automobile, at this particular speed on this particular hill, that there is no danger of "bottoming" the shocks. Most suspension systems have at least 20 cm of clearance upwards and downwards before you experience that dreadful "clunk". The maximum downward deviation z of about 10 cm, occurs at $t = 0.84$ s, on the ascent of the second hill.

Note also that on the particular values of roadway-auto chosen above that we are not even close to resonance. For our choice of m, c, and k, the resonant frequency is $\omega_o = \sqrt{k / m} = 9.43$ whereas the driving frequency of the road is $\omega = 2 \pi v/ L = 6.28$ rad/s. If one were to drive the road at resonance, either $v \to 36$ m/s (75 mph !) on a L = 24 m road, or with v = 24 m/s, L $\to$ 16 m (a gravel road with ruts). Running either of these roads at resonance would lead to a dreadful driving experience, however, by modeling your automobile on computer you can avoid a speeding ticket and/or bottoming-out your shocks.

t(s)	y(m)	Y'(m/s)	acc(m/s^2)	road	z = y-road
0	0	0	2.369	0	0
0.04	0.0023	0.1267	3.819	0.0248	-0.0225
0.08	0.0106	0.2926	4.307	0.0481	-0.0375
0.12	0.0257	0.4579	3.796	0.0684	-0.0427
0.16	0.0467	0.5846	2.411	0.0844	-0.0376
0.2	0.0715	0.6424	0.399	0.0951	-0.0235
0.24	0.0970	0.6126	-1.91	0.0998	-0.0027
0.28	0.1193	0.4901	**-4.17**	0.0982	0.02113
0.32	0.1350	0.2840	**-6.04**	0.0904	0.04461
0.36	0.1412	0.0151	**-7.27**	0.0770	0.06419
0.4	0.1358	-0.286	**-7.68**	0.0587	0.07708
0.44	0.1183	-0.588	**-7.22**	0.0368	0.08149
0.48	0.0892	-0.854	**-5.95**	0.0125	0.07675
0.52	0.0508	-1.055	**-4.00**	-0.012	0.06336
0.56	0.0060	-1.168	-1.60	-0.036	0.04284
0.6	-0.041	-1.181	0.983	-0.058	0.01747
0.64	-0.087	-1.090	3.494	-0.077	-0.0100
0.68	-0.127	-0.905	5.680	-0.090	-0.0368
0.72	-0.158	-0.643	**7.341**	-0.098	-0.0602
0.76	-0.178	-0.327	**8.340**	-0.099	-0.0782
0.8	-0.184	0.0140	**8.613**	-0.095	-0.0892
0.84	-0.176	0.3520	**8.169**	-0.084	-0.0925
0.88	-0.156	0.6589	**7.079**	-0.068	-0.0881
0.92	-0.125	0.9114	5.468	-0.048	-0.0768
0.96	-0.084	1.0916	3.494	-0.024	-0.0598
1.	-0.038	1.1885	1.334	0	-0.0387
1.04	0.0092	1.1982	-0.83	0.0248	-0.0156
1.08	0.0559	1.1237	-2.85	0.0481	0.00777
1.12	0.0981	0.9738	**-4.58**	0.0684	0.02968
1.16	0.1330	0.7622	**-5.92**	0.0844	0.04860
1.2	0.1585	0.5059	**-6.81**	0.0951	0.06341
1.24	0.1731	0.2237	**-7.21**	0.0998	0.07336
1.28	0.1763	-0.064	**-7.14**	0.0982	0.07810
1.32	0.1681	-0.341	**-6.62**	0.0904	0.07764
1.36	0.1493	-0.589	**-5.71**	0.0770	0.07233
1.4	0.1215	-0.794	**-4.49**	0.0587	0.06275
1.44	0.0865	-0.946	-3.05	0.0368	0.04970
1.48	0.0466	-1.037	-1.47	0.0125	0.03410
1.52	0.0044	-1.063	0.154	-0.012	0.01693
1.56	-0.037	-1.025	1.732	-0.036	-0.0007
1.6	-0.076	-0.926	3.181	-0.058	-0.0180

What if the "wavelength" of the road is 32 meters (100 feet) with the same + 4" and – 4" amplitude? We will expect less acceleration up-and-down, because the "driving frequency" ω is now less, and we are moving further away from resonance.

For this second example, $\omega = \frac{2\pi v}{L} = \frac{2\pi(24)}{32} = 4.71$ rad/s whereas $\omega_o = \sqrt{k/m} = 9.43$ rad/s.

```
In[98]:=  Clear[y, t]; (* The road with L=32 m; A=.1 m; v=24 m/s*)
          m = 900; c1 = 3394; k = 80000; L = 32; v = 24; ω = 2 π * v / L; A = .1;
          sol1 =
          NDSolve[{900 y''[t] + 3394 y'[t] + 80000 y[t] == 8000 Sin[ω * t] +
              c1 * A * ω * Cos[ω * t], y[0] == 0, y'[0] == 0}, y, {t, 0, 4.0}]
          road = Plot[.1 * Sin[ω * t], {t, 0, 3.3}, AspectRatio → 0.15]
```

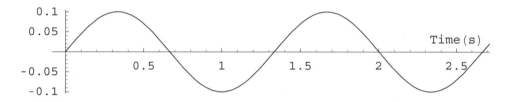

```
In[110]:=  InterpFunc1 = y /. sol1[[1]]; InterpFunc2 = y' /. sol1[[1]];
          InterpFunc3 = y'' /. sol1[[1]];
          plot2 = Plot[Evaluate[y[t] /. sol1], {t, 0, 3.3},
              PlotStyle → {{RGBColor[1, 0, 0]}},
              AspectRatio → 0.3, AxesLabel → {"time(s)", "y(m)"}];
          Show[{road, plot2}, Prolog → {{Text["Road", {.07, .13}]},
              {Text["Response", {.53, .13}]}},
              AxesLabel → {"time(s)", "y(m)"}]
          Table[{t, InterpFunc1[t], InterpFunc2[t], InterpFunc3[t],
              Chop[.1 * Sin[ω * t], InterpFunc1[t] - .1 Sin[ω * t]]},
              {t, 0, 2.25, .05}] // TableForm
```

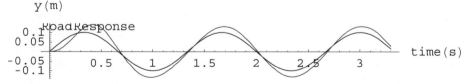

With the increased wavelength L = 32 m road, we find, in the Tabled data, less vertical movement y, and less acceleration up-and-down of the car. The "sailing" over the hill is reduced as is the "free-fall". The "sharper" the hill crest, and the greater the speed, the more pronounced is the "drop" over the hill.

t(s)	y(m)	Y'(m/s)	acc(m/s^2)	road	z = y - road
0	0	0	1.7779	0	0
0.05	0.00286	0.12533	3.0761	0.0233	-0.0204
0.1	0.01319	0.29038	3.3504	0.0453	-0.0322
0.15	0.03170	0.44374	2.6328	0.0649	-0.0332
0.2	0.05662	0.54104	1.1618	0.0809	-0.0242
0.25	0.08438	0.55346	-0.695	0.0923	-0.0080
0.3	0.11039	0.47183	-2.534	0.0987	0.01162
0.35	0.13015	0.30624	**-4.002**	0.0996	0.03046
0.4	0.14003	0.08197	**-4.852**	0.0951	0.04493
0.45	0.13794	-0.1668	**-4.982**	0.0852	0.05267
0.5	0.12353	-0.4047	**-4.426**	0.0707	0.05282
0.55	0.09817	-0.6005	-3.330	0.0522	0.04592
0.6	0.06455	-0.7326	-1.918	0.0309	0.03365
0.65	0.02615	-0.7909	-0.416	0.0078	0.01830
0.7	-0.0133	-0.7762	0.9656	-0.015	0.00232
0.75	-0.0504	-0.6989	2.0749	-0.038	-0.0121
0.8	-0.0824	-0.5747	**2.8318**	-0.058	-0.0236
0.85	-0.1074	-0.4218	**3.2262**	-0.076	**-0.0313**
0.9	-0.1244	-0.2574	**3.3034**	-0.089	**-0.0353**
0.95	-0.1332	-0.0954	**3.1425**	-0.097	**-0.0359**
1.	-0.1341	0.05436	**2.8323**	-0.1	**-0.0341**
1.05	-0.1280	0.18661	2.4514	-0.097	**-0.0308**
1.1	-0.1158	0.29926	2.0547	-0.089	-0.0267
1.15	-0.0984	0.39227	1.6682	-0.076	-0.0224
1.2	-0.0769	0.46627	1.2924	-0.058	-0.0181
1.25	-0.0521	0.52140	0.9098	-0.038	-0.0138
1.3	-0.0251	0.55675	0.4974	-0.015	-0.0094
1.35	0.00315	0.57034	0.0374	0.0078	-0.0046
1.4	0.031511	0.55961	-0.475	0.0309	0.00061
1.45	0.058677	0.52221	-1.025	0.0522	0.00642
1.5	0.083273	0.45695	-1.583	0.0707	0.01256
1.55	0.10391	0.36451	-2.104	0.0852	0.01865
1.6	0.11932	0.24798	**-2.538**	0.0951	0.02421
1.65	0.12840	0.11291	**-2.838**	0.0996	**0.02871**
1.7	0.13043	-0.0330	**-2.967**	0.0987	**0.03166**
1.75	0.12507	-0.1806	**-2.904**	0.0923	**0.03268**
1.8	0.1125	-0.3201	**-2.649**	0.0809	**0.03159**
1.85	0.0933	-0.4422	-2.205	0.0649	**0.02840**
1.9	0.0687	-0.5385	-1.623	0.0453	0.02330
1.95	0.0400	-0.6029	-0.932	0.0233	0.01668
2.	0.0090	-0.6313	-0.195	0	0.00901
2.05	-0.022	-0.6226	0.5445	-0.023	0.00085
2.1	-0.052	-0.5777	1.2394	-0.045	-0.0072
2.15	-0.079	-0.5001	1.8483	-0.064	-0.0147
2.2	-0.102	-0.3949	2.3378	-0.080	-0.0212
2.25	-0.118	-0.2687	2.6863	-0.092	-0.0264

The equations of motion allow predictability in the design of a shock-absorber system. In this section, we have examined the response of a light automobile with worn shocks on a hill.

In designing a shock absorber system for a particular automobile, it is necessary to computer-model the full range of forces and shocks which the automobile will experience. And then compare the actual response of the vehicle to on-road conditions. In automotive design, it will also be necessary to go beyond the relatively simple one-dimensional model presented here, and model all four wheels of the automobile.

2.3 Step Response

Let's say a mechanical system as shown is subject to a step-function driving force which alternates every half-second from + 100 N upwards to – 100 N downwards. What is the response of the 40 kg mass, given that the spring constant is 17720 N/m, and the damping coefficient is 88 N-s/m ? We will need to solve the differential equation

$$40y'' + 88\,y' + 17720y = F(t) \quad \text{where F(t) is the applied force}$$

$$\text{with}\quad y\,(0) = 0 \ \text{ and }\ y'\,(0) = 0$$

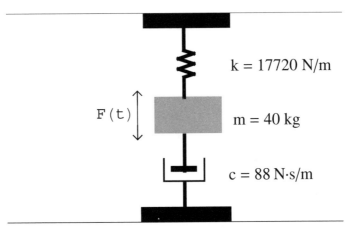

$$k = 17720 \text{ N/m}$$

$$F(t)$$

$$m = 40 \text{ kg}$$

$$c = 88 \text{ N·s/m}$$

A mass-spring-damper system driven by a force F(t)

```
Force =
    Plot[200 UnitStep[Sin[2 π * t]] - 100 UnitStep[t], {t, -.2, 3.6},
        AspectRatio → 0.18, AxesLabel → {"time(s)", "Force(N)"}];
```

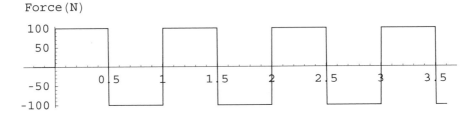

The applied force + 100 N upwards, – 100 N downward

```
In[3]:= Clear[y, t];
        sol = NDSolve[{40 y"[t] + 88 y'[t] + 17720 y[t] ==
            200 * UnitStep[Sin[2 π * t]] - 100 UnitStep[t],
            y[0] == y'[0] == 0}, y, {t, 0, 6}];
        Plot[{y[t] /. sol,
            .025 UnitStep[Sin[2 π * t]] - .0125 * UnitStep[t]}, {t, 1, 4.2},
            PlotStyle → {{RGBColor[0, 0, 0]}, {RGBColor[0, 0, 1]}},
            AspectRatio → 0.33, Prolog → {Text["λ", {2.5, .0185}],
            Text[">", {2.98, .0187}], Text["<", {2.01, .0187}]},
            Epilog → {GrayLevel[0.38], Line[{{2.6, .0188}, {2.95, .0188}}],
            Line[{{2.04, .0188}, {2.4, .0188}}]}]
```

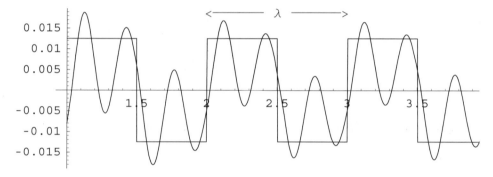

The response of a mechanical system to a square-wave input

Would you have guessed this response to a square-wave input ?

How is it possible that the response of the mechanical system can show *three* up-and-down cycles within one wavelength λ of the driving force ? The answer is very interesting. The square-wave F(t) is not a pure sine-wave or cosine-wave, but rather a composition of many sine waves.

$$F(t) = \frac{400}{\pi} \left(\text{Sin } 2\pi t + \frac{\text{Sin } 6\pi t}{3} + \frac{\text{Sin } 10\pi t}{5} + \frac{\text{Sin } 14\pi t}{7} + \dots \right)$$

The very nice thing about *Mathematica* is that we can isolate the system response to just one of the driving frequencies, and just like a spectrum analyzer, see how much of the system response is due to that frequency. Since there are apparently three cycles per second in the response, let us find the system response to just $f(t) = \frac{400}{3\pi} \text{Sin } (6\pi t)$.

$In[13]:=$ `sol = NDSolve[{40 y''[t] + 88 y'[t] + 17720 y[t] ==` $\dfrac{400}{3\pi}$ `* Sin[6 π * t],`

   `y[0] == y'[0] == 0}, y, {t, 0, 6}];`
   `Plot[{y[t] /. sol,`
      `.025 UnitStep[Sin[2 π * t]] - .0125 * UnitStep[t]}, {t, 1, 4.2},`
      `PlotStyle → {{RGBColor[0, 0, 0]}, {RGBColor[0, 0, 1]}},`
      `AspectRatio → 0.33]`

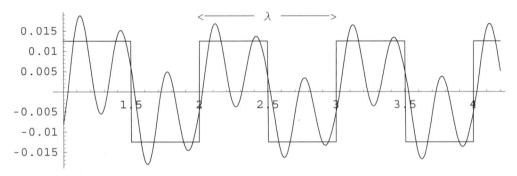

Response of the mechanical system to the complete square-wave driving force

$$F(t) = \frac{400}{\pi}\left(Sin\ 2\pi t + \frac{Sin\ 6\ \pi\ t}{3} + \frac{Sin\ 10\ \pi\ t}{5} + \frac{Sin\ 14\ \pi\ t}{7} + \dots \right)$$

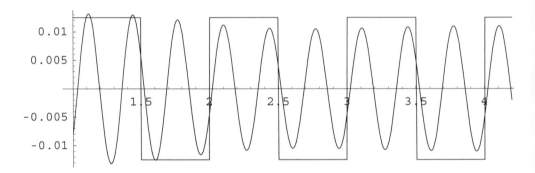

Response of the mechanical system to f(t) = $\frac{400}{3\pi}$ Sin (6π t) only

Notice that in comparing this response to the full square-wave, that almost all of the response is due to the Sin(6πt) term. This is because the resonant frequency of our m = 40, c = 88, k = 17720 system is $\omega_o = \sqrt{\frac{17720}{40}} = 21.04$ rad/s whereas the driving angular frequency is $\omega = 6\pi = 18.85$ rad/s.

Thus the lion's share of the system response is to the second-harmonic 6π term, and the rest of the shape of the response is due to the other terms slightly modulating the 6π response.

2.4 Square Wave – RC Circuit

A very simple and instructive physics lab experiment is to monitor the voltage across a capacitor in an RC-circuit when a square-wave voltage is applied. When the voltage is 1 volt, the capacitor charges. When the applied voltage is 0 volts, the capacitor discharges.

For an RC circuit with $R = 5000 \ \Omega$ and $C = 0.04 \ \mu F$, the time constant is 0.2 milliseconds.

From Kirchoff's Law, one can write the equation for the voltage drops around the circuit. If the voltage drop across the capacitor is V, and the current in the circuit is I,

$$R\,I + V = S(t) \quad \text{where S(t) is the applied voltage}$$

$$R\frac{dQ}{dt} + V = S(t) \quad \text{and since } Q = CV,$$

$$RC\,\frac{dV}{dt} + V = S(t) \quad \text{where } S(t) = UnitStep[Sin\,\pi\,t]$$

$$V(0) = 0$$

```
In[10]:=  voltagePlot =
          Plot[UnitStep[Sin[π*t]], {t, -.8, 3.6}, AspectRatio → 0.28];
          (*Input voltage*)
```

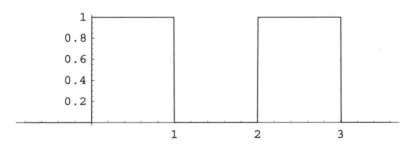

Time in milliseconds

The applied voltage: 1 volt maximum, 0 volts minimum —

```
rcsol = NDSolve[
    {.2 v'[t] + v[t] == UnitStep[Sin[π*t]], v[0] == 0}, v, {t, -1, 4}];
Plot[{UnitStep[Sin[π*t]], v[t] /. rcsol}, {t, -.8, 3.6},
  PlotRange → {0, 1}, AspectRatio → 0.30,
  AxesLabel → {"T(ms)", "Voltage"}, PlotStyle → {{Hue[2/3]}, {Black}}]
```

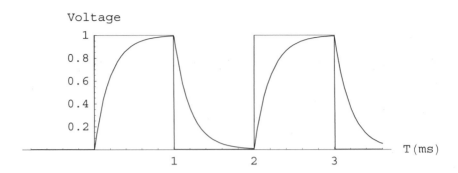

The response of an RC series circuit to a square-wave input

The response of the RC-circuit is this shark-fin charging and discharging as the voltage on the capacitor reaches an exponential charging-maximum, then falls exponentially to zero. The cycle time of S(t), the applied square-wave is 0.002 seconds, corresponding to a frequency of 500 Hz.

If one were to decrease the cycle time, by changing the frequency of the applied voltage to 2000 Hz, then the response of the circuit would be to charge-up only halfway, and give an output waveform of the following shape

```
In[48]:= Clear[v, t];
         sol =
            NDSolve[{.2 v'[t] + v[t] == UnitStep[Sin[4 π * t]], v[0] == 0},
               v, {t, -.2, 1.8}];
         Plot[{UnitStep[Sin[4 π * t]], v[t] /. sol}, {t, -.2, 1.8},
            PlotRange → {0, 1}, AspectRatio → 0.3,
            AxesLabel → {"T (ms)", "Voltage"}]
```

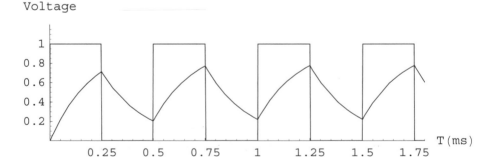

The response of the RC-circuit to a square-wave of higher frequency (2000 Hz).

The capacitor only has time to charge-up part-way before the applied voltage goes to zero.

2.5 Variable Stars

The Physics of Stellar Pulsation

Many giant stars pulsate. Not only does their light output vary in a periodic way, but the outer envelopes of some stars actually move in and out over millions of kilometers over a period of days. One of the most famous of these stars is Delta Cephei, a giant yellow star which oscillates with a time period of approx 5.4 days as the outer layers of the star move outward and then inward a total distance of approx 30 million kilometers. The radial velocity of the envelope exceeds 20 km/s.

It is possible to model the pulsation if we assume that some of the heat and light energy coming up from the stellar interior is absorbed by ionizing hydrogen or helium at a critical depth in the envelope. This heats and drives the envelope outwards. Then, as the hydrogen and helium cool and deionize, the stored-up energy is liberated, causing an increase in the star's luminosity. Now the de-ionized layers fall back, and the process is ready to repeat again. This one-zone model was originally proposed by Sir Arthur Eddington, who unfortunately did not have a computer or mathematical software to fully explore this concept.

We will start by assuming a certain mass of ionizable material, say $m = 10^{26}$ kg at a distance of 1.35×10^7 km from the star's center. Then the forces acting on this mass of material are gravity and pressure, where x is the distance from the stellar center

$$m\ddot{x} = -\frac{GMm}{x^2} + 4\pi x^2 P \qquad x_o = 1.35 \times 10^7 \text{ km} \qquad \dot{x}_o = 0$$

The pressure acting will be assumed to be *adiabatic*. That is, the initial pressure build-up drives the outer layer m outwards and the volume of the star expands from $\frac{4}{3}\pi x_o^3$ to $\frac{4}{3}\pi x^3$. And, from thermodynamics, the pressure driving the expansion falls from its original value of P_o to $P = P_o\left(\frac{x_o^3}{x^3}\right)^{5/3}$ where $P_o = 1.8 \times 10^5$ N/m^2. (These numbers are chosen as representative for the envelope of a giant star). Combining the above equations, we have a mathematical model for a variable star of mass $M = 10^{31}$ kg

$$m\ddot{x} = -\frac{GMm}{x^2} + 4\pi \frac{P_o x_o^5}{x^3} \qquad x_o = 1.35 \times 10^7 \text{ km} \qquad \dot{x}_o = 0$$

In *Mathematica*, with distance in km and time in seconds,

```
In[1]:=  sol =
         NDSolve[{x''[t] == (-6.67 * 10^11) / x[t]^2 + (1 * 10^19) / x[t]^3,
            x[0] == 1.35 * 10^7, x'[0] == 0}, x, {t, 0, 1000000}]
```

We will plot the motion of our variable star for 10^6 seconds, or about 12 days.

```
In[2]:=  InterpFunc1 = x /. sol[[1]];  InterpFunc2 = x' /. sol[[1]];
         Plot[Evaluate[x[t] /. sol], {t, 0, 1000000}, PlotStyle
            → {{RGBColor[1, 0, 0]}}, AspectRatio → 0.45];
```

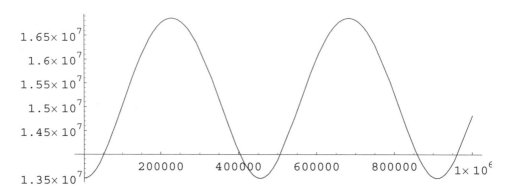

Stellar Radius as a function of time in seconds
The period of oscillation is about 450,000 seconds = 5.2 days

```
In[20]:=  Plot[Evaluate[x'[t] /. sol], {t, 0, 1000000}, AspectRatio → 0.5,
             PlotRange → {-25, 25}, PlotStyle → {{RGBColor[0, 0, 1]}}];
          tbl = Table[{t / 3600., InterpFunc1[t], Chop[InterpFunc2[t]]},
             {t, 0, 720000, 10 * 3600}] // TableForm
```

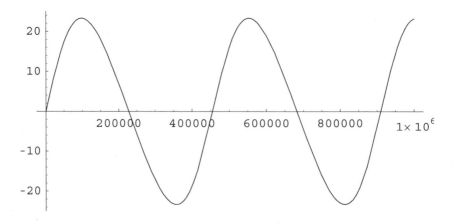

Radial velocity   outward + , inward −
Maximum velocity is approximately 23 km/s

It is reasonably clear that the simple model of Eddington produces a first approximation for stellar oscillation. The pressure that originally comes from the absorption of thermal energy in the ionization zone is converted into mechanical work in pushing the outer layer outward. Then, as the expansion proceeds, the outward pressure weakens, and the star's gravity brings the outer layers back, so the process may begin again. What is amazing, from a Physics standpoint is that the results of this reasonably simple model are fairly close to the oscillation time and expansion velocities of the star Delta Cephei.

Following is a Table of values for Stellar Radius and radial velocity. These are reasonable values and show that the one-zone model provides a good starting point for a more complicated model.

t(hrs)	R(km)	v(km/s)
0	1.35×10^7	0
10.	1.37523×10^7	13.4897
20.	1.44058×10^7	21.679
30.	1.52306×10^7	23.0846
40.	1.60035×10^7	19.1369
50.	1.65689×10^7	11.8766
60.	1.6839×10^7	2.96882
70.	1.67783×10^7	-6.30849
80.	1.63944×10^7	-14.7811
90.	1.57402×10^7	-21.0566
100.	1.49265×10^7	-23.3002
110.	1.41356×10^7	-19.5004
120.	1.36047×10^7	-8.98879
130.	1.35344×10^7	5.23639
140.	1.39542×10^7	17.2465
150.	1.46984×10^7	22.9339
160.	1.55271×10^7	22.1592
170.	1.62388×10^7	16.7849
180.	1.6704×10^7	8.75874
190.	1.68557×10^7	-0.418581
200.	1.66745×10^7	-9.54415

For the full equations of stellar pulsation, and the computer model with many zones from the stellar interior to the ionization zones, see Robert F. Christy *"Variable Stars — RealisticStellarModels"* pp 173-210 in the book Stellar Evolution , Hong-Yee Chiu editor, Plenum Press, 1972).

2.6 Challenge Problems

1. <u>NewYork to Leningrad</u>

A famous physics problem is to consider travel by gravity by tunneling from one city to another over great distances through the Earth. One interesting possibility would be to tunnel from NewYork to Leningrad, a distance of some 12,000 km over the Earth's surface or 108° of a great circle around the Earth.

Such a tunnel would reach a depth of 2640 km and would have to be impervious to the several thousand degree temperatures inside the Earth. However, if such a tunnel could be built, what would be the transit time from NewYork to Leningrad by a train traveling in an evacuated tunnel ? Use the very interesting fact that gravity inside the Earth is nearly constant at 10 m/s^2 to a depth of 3000 km (Adam Dziewonski and Don Anderson (1981) *Physics of Planetary Interiors* pages 297-356, "Preliminary Reference Earth Model" which gives the density, gravity, and sound speed from the Earth's surface to core).

2. <u>Lunar Subway</u>

Assume the moon is of constant density. Find the maximum velocity and transit time from the North pole to the South pole of the moon. The radius of the moon R_{moon} = 1700 km. The surface gravity of the moon g_{moon} = 1.7 m/s^2

3. <u>Speed Bump</u>

Let's say we're cruising along the highway in our favorite automobile when up ahead is a *speed bump*. Should we speed-up or slow down ? You had best slow-down, if you wish to preserve the integrity of your car, and the following problem will show the reason.

Assume the highway engineers have placed a 0.8-meter long, 0.1 meter high sinusoidal *speed bump* in your lane, and you approach the bump at 2 m/s, what is the suspension-system response?

Now take a computer-simulated run at the speed-bump at 4 m/s and at 8 m/s. What is the suspension-system response at each of these velocities? Assume m = 1000 kg, c = 3400 N-s/m, k = 80,000 N/m.

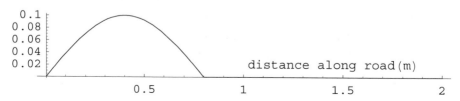

4. RLC Circuit

For an RLC-series circuit, the sum of voltage drops around the loop is equal to the driving voltage. If we have a capacitor of 2-microfarads, a resistor of 100 ohms, and an inductor of 0.5 henries, all connected in series, and an initial voltage on the capacitor of 10 volts, the differential equation of the electrical circuit with no driving voltage is

$$L \frac{dI}{dt} + RI + \frac{1}{C}Q = 0 \quad \text{where Q is the charge on the capacitor.}$$

If we want to know the voltage on the capacitor as a function of time, let's rewrite the equation using $Q = C\,V$. Then,

$$LC \frac{d^2 V}{dt^2} + RC \frac{dV}{dt} + V = 0 \quad \text{where } V_o = 10 \text{ volts and } I_o = 0.$$

Solve the above equation for V for the first 50 milliseconds after the switch is closed. What is the time-constant of this *underdamped* circuit and what is its oscillation frequency?

5. Impulse Response

Let's take a mechanical system ($m = 1, c = 3, k = 2$) at rest at $t = 0$

$$y'' + 3\,y' + 2\,y = F(t) \qquad y_o = 0, \quad y_o' = 0.$$

and give it a unit impulse of 1 force unit for 1 second at $t = 0$.

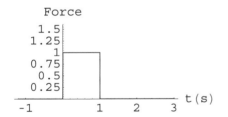

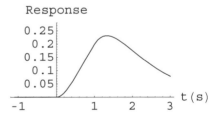

The Impulse   **The Response**

Now let's take the same system and subject it to a unit impulse with a force of 2 for 0.5 seconds. Then take the system and give it a unit impulse with a force of 10 for 0.1 seconds.

It's pretty clear where we're going with this. We will now give the system a unit impulse with a Dirac Delta function where $F \to \infty$ as $\Delta t \to 0$, but $F\Delta t = 1$. Do we need to modify the initial conditions to find the impulse response if the Dirac function acts at $t = 0$? What if the Dirac delta function chimes in at one-one millionth of a second after $t = 0$? Show the system response for each of these unit impulses.

Chapter 3 Building a
Differential Equation

3.1 Modeling Dynamic Systems

<u>Modeling with Differential Equations</u>

The modeling of dynamical systems with differential equations is a little bit of an art, and quite a bit of science. One endeavors to keep the equations as simple as possible while retaining the essential details of the physical system. The essentials of the modeling process are

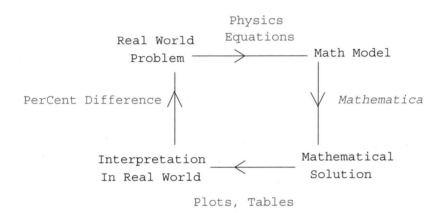

With the advent of modern high-memory personal computers and mathematical software, one is no longer restricted to linear systems. A flow-chart of the modeling process is shown above.

It is now possible to run through the modeling process, complicating the model as necessary, to obtain a reasonable match with the physical system at hand. In the following sections of this chapter, several complicated physical systems will be analyzed.

First, the Physics of running a race in Track and Field. A runner must get up to speed by accelerating, and every runner has a finite energy supply. Therefore the relevant equations that must be used are

$$F = m\,a \qquad \text{and} \qquad \frac{dE}{dt} = \sigma - f \cdot v$$

Second we will investigate the Physics of Binary Stars. Here it is gravity that controls the orbits of the stars, and here it is necessary to keep track of the positions and velocities of each star as they orbit one another. It is a bit of a challenge to map out the orbital dynamics of the stars, and to animate the paths of the two stars, but the Physics is that of gravitation. We start simply with two stars of equal mass, with Star 1 at (x, y) and Star 2 at $(-x, -y)$. The forces acting are then

$$\mathrm{M}x'' = -\frac{\mathrm{MGM}}{(2\,x)^2 + (2\,y)^2}\,\frac{2\,x}{\sqrt{(2\,x)^2 + (2\,y)^2}} \qquad \mathrm{M}y'' = -\frac{\mathrm{MGM}}{(2\,x)^2 + (2\,y)^2}\,\frac{2\,y}{\sqrt{(2\,x)^2 + (2\,y)^2}}$$

Third, we complicate the Binary Star problem. Now we take two stars of *unequal* mass, and after a certain time, we allow the star of greater mass to explode. In this way, we can use the computer to track the orbits of the two stars before the explosion, and then determine whether the stars remain gravitationally bound after the explosion. Again, it is gravity that controls the dynamics of the system, but the interesting aspect of the problem of the Exploding Star is that the mass of one of the components changes discontinuously at a certain time.

Fourth is an engineering system, that of Heat Exchangers, where two fluids -- water and oil -- exchange heat without the two fluids mixing. Here it is heat transfer, *as the water* heats up by absorbing heat $dQ = \dot{m}\, c\, dT$ and the heat transferred in a distance dx is $dQ = h\,\pi\,D\,(T^* - T)\,dx$.

Again, once the system is successfully modeled, it is possible to change the system conditions, and in the problems section for this chapter the heat-transfer characteristics of a *counterflow* heat-exchanger are examined.

One may ask why differential equations are almost exclusively used in this text, and it is because a differential equation allows a system variable to be tracked in space and time. Using *Mathematica*, you may use **DSolve** to obtain a mathematical function, or **NDSolve** to find a numerical solution that by means of an *interpolation function* describes the system behavior. In either case, the criterion for good modeling of a dynamical system is whether the Physics and Mathematics are reasonably close to the behavior of the real physical system.

3.2 Track and Field

Following, is a list of world records for men in running, from 100-meters to 10,000 meters. The question before us is whether a Physics model of running can account for all the data.

Distance (m)	Time (s)	Avg Velocity (m/s)	Record Holder
100	9.77	10.23	Asafa Powell (JAM) 2005
200	19.32	10.35	Michael Johnson (USA) 1996
400	43.18	9.26	Michael Johnson (USA) 1999
800	101.1	7.91	Wilson Kipketer (DEN) 1997
1000	132.0	7.57	Noah Ngeny (KEN) 1999
1500	206.0	7.28	Hicham ElGuerrouji (MOR) 1999
1 mile	223.1	7.21	Hicham ElGuerrouji (MOR) 1999
2000	284.8	7.02	Hicham ElGuerrouji (MOR) 1999
3000	440.7	6.81	Daniel Komen (KEN) 1996
5000	757.4	6.60	Keninisa Bekelese (ETH) 2004
10000	1580.3	6.33	Keninisa Bekelese (ETH) 2004

Let us first of all put this data into a form where we can see the average velocity to run a race of a given distance. *Mathematica* can manipulate and plot data in an easy to read form.

```
vel = {{100, 10.2}, {200, 10.35}, {400, 9.26}, {800, 7.9}, {1000, 7.57},
    {1500, 7.28}, {1609, 7.21}, {2000, 7.02}, {3000, 6.81}, {5000, 6.6}}
velc = ListPlot[vel, PlotRange → {{-100, 5100}, {5, 11}}];
```

average velocity (m / s)

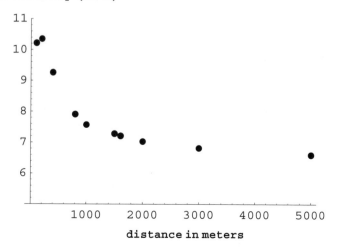

distance in meters

Let us first of all take an unbiased look at the data. For the short distance runs, 100-m and 200-m, the average velocities are very fast at over 10 m/s, diminishing almost exponentially to about 6 m/s as the run distance increases.

What this means is that the runner must go flat-out in a short distance run (the dash), and for a middle-distance or long-distance run, the runner must go at a slower pace so as not to burn up all the available energy.

Building the Differential Equations

Based on the above observations, let us model the acceleration phase of any race utilizing Newton's Law:

$$\frac{dv}{dt} = f - k\,v$$

f is the force per unit mass, and k is the resistance we encounter when we try and go faster.

We must also keep account of our energy

$$\frac{dE}{dt} = \sigma - f\,v$$

The energy in the runner's system per unit mass diminishes from an initial value E_o due to the power exerted in running $f \cdot v$ with an energy resupply rate of σ as the body begins to burn glucose, carbohydrates, and fat. These are converted into ATP which is the chemical source of muscular energy.

A Theory of Competitive Running

In September 1973, Joseph Keller published "A Theory of Competitive Running" in <u>Physics Today</u> (vol 26, pp 43-47). Keller utilized the above two equations with best-fit values of F, k, σ, and E_o and obtained a reasonable match with the world records of the time.

As it turns out, all the world records in Men's running have been broken over the last 30 years, so the purpose of this section is to revisit Keller's theory with a slightly-changed set of running parameters to see how closely the theory matches the world of 2008.

We choose the following physiological constants as reasonable for a male world-class runner

$F = 1.25 * 9.8$ m/s^2 (maximum acceleration of runner is 1.25 g)

$k = 1.12$ s^{-1} (internal and external resistance increases with velocity)

$\sigma = 44.5$ W/kg (rate of ATP resupply to muscles) [7% greater than Keller]

$E_o = 2400$ J/kg (initial energy and oxygen supply in muscles)

(These are the same constants chosen by J. Keller, except for σ, which is 7% larger than the 1973 value.)

We will now solve the first differential equation to find out how fast a world-class athlete could run the 100- and 200-meter dash. Then we will find out how far you can run at maximum speed before all the available energy is used up.

The Dash

For a short-distance run, the runner is accelerating all the way and maximum force F is exerted during the entire race. Then $\frac{dv}{dt} = F - k\,v$

Utilizing *Mathematica*,

```
In[67]:=  F = 1.25 * 9.8; k = 1.12; s = 44.5; E0 = 2400;
          sol =
          DSolve[{v'[t] == F - k*v[t], v[0] == 0}, v[t], t] // FullSimplify
          sol1 = DSolve[{x"[t] == F - k*x'[t], x[0] == x'[0] == 0}, x[t], t] //
            FullSimplify
          Plot[Evaluate[v[t] /. sol, {t, 0, 10}],
            PlotRange → {{0, 10.1}, {0, 11}},
            AxesLabel → {"Time (s)", "Velocity(m/s)"}];
```

$Out[68]= \{\{v[t] \to 10.9375 - 10.9375\,e^{-1.12\,t}\}\}$

$Out[69]= \{\{x[t] \to -9.76563 + 9.76563\,e^{-1.12\,t} + 10.9375\,t\}\}$

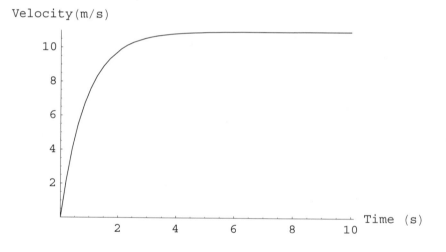

How long does it take us (according to the physics model) to do the 100- and 200-meter dash? We have the velocity with time, and assuming we accelerate the whole distance,

$$v[t] = 10.9375*(1 - Exp[-1.12t])$$

$$x[t] = 10.9375 \cdot t - 9.7656(1 - Exp[-1.12t])$$

Mathematica carries the solution **x[t]** in memory, so let us find the time at which our hypothetical runner crosses the 100-m and the 200-m line.

```
In[71]:=   sol100 = FindRoot[100 == x[t] /. sol1, {t, 10}]
           sol200 = FindRoot[200 == x[t] /. sol1, {t, 20}]
```

Out[71]= $\{t \to 10.0357\}$

Out[72]= $\{t \to 19.1786\}$

This compares favorably with the world record time 9.77 s in the 100-meter, and 19.32 s in the 200-meter dash.

How far can the runner keep this up, running as fast as possible, before running out of energy ?

Let's use the second equation

$$\frac{dE}{dt} = \sigma - F \cdot v \quad \text{and integrate to find} \quad E1 = F \cdot x1 - \sigma \cdot t$$

```
In[29]:=   Clear[x]; F = 1.25 * 9.8; k = 1.12; s = 44.5; E0 = 2400;
           x1[t_] = (F / k) * t - (F / k^2) * (1 - Exp[-k * t]) ;
           t1 = FindRoot[2400 == F * x1[t] - s * t, {t, 10}]
           Print["xmax = ", x1[t /. t1]]
```

Out[31]= $\{t \to 28.1572\}$

```
           xmax = 298.204
```

This is the furthest distance that a runner with the above physical constants can run, going flat-out. So the dash merely depends on the runners ability to accelerate and maintain maximum force throughout the run. *However,* if we want to run farther than 300 meters, then we will have to run *slower* (otherwise all the available energy is expended and we can go no further).

So here is the race strategy as given by Joseph Keller: accelerate in the first stage (time t) to velocity v1 and maintain that constant velocity throughout the remainder (time t2) of the race. Of course this is an oversimplification, but let us see if it produces reasonable results (one can always complicate the model later).

The Run

After the initial acceleration phase, the equations of motion are

$$\frac{dv}{dt} = f - k\,v = 0, \quad \text{or,} \quad f = k\,v$$

and, at the constant running speed, $\quad \frac{dE}{dt} = \sigma - f\,v = \sigma - k\,v^2$

Notice that f is now much less than F, and with a reduced running speed, the rate of energy loss in the run-phase is diminished. So, for a run of distance z > 300 m, we accelerate for time t, reaching a velocity v1, and travel a distance x1, while burning energy E1. Then we maintain the constant velocity v1 for time t2.

We will solve for the time t2 at constant velocity, subject to the condition that all energy is used-up by the end of the race. Then, the total distance travelled is z = x1 + x2 = x1 + v1·t2 and the total time is t + t2.

The average velocity for running the race is then $w = \frac{z}{(t+t2)}$.

Also, for races less than 300 m, there is only the acceleration phase, and the distance travelled is x1 in time t. Writing the equations in general terms (so we can vary F, k, σ, and E_o if we wish),

```
F = 1.25 * 9.8; k = 1.12; s = 44.5; E0 = 2400;
x1[t_] = (F / k) * t - (F / k^2) * (1 - Exp[-k * t]) ;
v1[t_] = (F / k) * (1 - Exp[-k * t]);
E1[t_] = F * x1[t] - s * t ; E2[t_] = E0 - E1[t] ;

(* and since   E2 = (kv² - s) t2   *)
           E0 - E1[t]
t2[t_] = ─────────────── ;
         (k * v1[t]^2 - s)
z[t_] = v1[t] * t2[t] + x1[t] ;
t3[t_] = t2[t] + t ;
w[t_] = z[t] / t3[t] ;
p = ParametricPlot[{z[t], w[t]},
    {t, 0.77, 4.2}, PlotRange → {{0, 5500}, {5, 11}}];
r = ParametricPlot[{x1[t], x1[t] / t}, {t, 3, 26.8},
    PlotRange → {{0, 5500}, {5, 11}}];
```

We will now use the above equations to find the times to run specific distances.

```
In[56]:=  (* -- Predicted times at various race distances-- *)
          sol400 = FindRoot[400 == v1[t] * t2[t] + x1[t], {t, 1}];
          t400 = (t2[t] + t) /. sol400;
          sol800 = FindRoot[800 == v1[t] * t2[t] + x1[t], {t, 1}] ;
          t800 = (t2[t] + t) /. sol800;
          sol1000 = FindRoot[1000 == v1[t] * t2[t] + x1[t], {t, 1}];
          t1000 = (t2[t] + t) /. sol1000;
          sol1500 = FindRoot[1500 == v1[t] * t2[t] + x1[t], {t, 1}] ;
          t1500 = (t2[t] + t) /. sol1500 ;
          sol1609 = FindRoot[1609 == v1[t] * t2[t] + x1[t], {t, 1}] ;
          t1609 = (t2[t] + t) /. sol1609;
          sol2000 = FindRoot[2000 == v1[t] * t2[t] + x1[t], {t, 1}] ;
          t2000 = (t2[t] + t) /. sol2000;
          sol3000 = FindRoot[3000 == v1[t] * t2[t] + x1[t], {t, 1}];
          t3000 = (t2[t] + t) /. sol3000;
          sol5000 = FindRoot[5000 == v1[t] * t2[t] + x1[t], {t, 1}] ;
          t5000 = (t2[t] + t) /. sol5000 ;
          sol10000 = FindRoot[10000 == v1[t] * t2[t] + x1[t], {t, 1}];
          t10000 = (t2[t] + t) /. sol10000;
          Print[" t100 = 10.035"]; Print [" t200 = 19.178"];
          Print[" t400 = ", t400]; Print[" t800 = ", t800];
          Print[" t1000 = ", t1000];
          Print[" t1500 = ", t1500]; Print[" t1609 = ", t1609];
          Print[" t2000 = ", t2000]; Print[" t3000 = ", t3000];
          Print[" t5000 = ", t5000]; Print["t10000 = ", t10000]
```

	Theory
Race Dist	time(s)
t100 =	10.035
t200 =	19.178
t400 =	42.5421
t800 =	103.215
t1000 =	134.363
t1500 =	212.905
t1609 =	230.091
t2000 =	291.835
t3000 =	450.087
t5000 =	767.063
t10000 =	1560.06

Let's plot the theory line against the world record data points and see how we do.

```
Show[{velc, p, r}, AxesLabel → {"Distance(m)", "Velocity(m/s)"}]
```

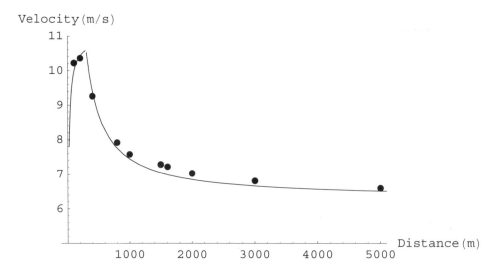

This appears to be a rather good fit, with the dots for the middle-distance runs actually exceeding the theory line. Only the 400 meter run is marginally below the theory line, which is already 3 % faster than the original theory of Joseph Keller. (Runners have certainly improved their racing performance dramatically over the last 30 years!) The last question is how well do the theoretical and actual record numbers compare?

Track and Field World Records -- Men 2007

Distance (m)	Theory Time	Record Time (s)	% Difference	
100	10.03	9.77	− 2.6	
200	19.18	19.32	0.7	
400	42.54	43.18	1.5	(record likely to fall)
800	103.2	101.1	− 2.1	
1000	134.4	132.0	− 1.9	
1500	212.9	206.0	− 3.4	
1 mile	230.1	223.1	− 3.2	
2000	291.8	284.8	− 2.6	
3000	450.1	440.7	− 2.2	
5000	767.1	757.4	− 1.4	
10000	1560.1	1580.3	1.2	

All the track records for men may be accounted for with the Physics model of running up to 10,000 meters. Beyond this range, when running fatigue sets in -- as the runner's system tries to clear the lactic acid build-up, acidosis of the muscles begins to occur -- this is often referred to as THE WALL. Many long-distance runners have learned to cope with the fatigue by training and diet, and also by running at a somewhat reduced pace. This can be accounted for in the Physics model by allowing k (the internal resistance) to increase for runs beyond 10,000 meters. For track and field records for women, see Challenge problem 3-6.

Technically, the supply of ATP to the runner's muscles is much more complicated than just a constant number. The biochemistry of running is carefully discussed in Henry Bent's article "Energy and Exercise" in the *Journal of Chemical Education*, vol 55, p 797 (1978), and in even greater detail by Jack Wilmore and David Costill (2004) *Physiology of Sport and Exercise* [Human Kinetics publications].

A capsule summary is as follows: In the first 3-5 seconds, almost all the ATP is "burned-up" ATP→ ADP + Energy. Then, over the next 10-20 seconds, the "burned" ATP is reconstituted by creatine phosphate   ADP + CP → ATP . This is all the "ready-reserve" in the runner's system, and after this, the runner burns glycogen, first anaerobically and then aerobically to supply ATP to the muscles on a continuous basis.

A more complete (and complicated) theory of competitive running would factor in the non- constant rate of energy resupply to the runner.

3.3 The Binary Pulsar 1913 + 16

Some 20,000 lightyears from the Earth is an amazing object: two neutron stars (or *pulsars*) locked in an orbit so tight that the two stars complete an orbit in just 7.75 hours. The mass of each star is very nearly equal to 1.4 M_{sun}, and astronomers have determined the eccentricity of the orbit is $\epsilon = 0.617$.

What we would like to do with *Mathematica* is the following: First, find the "semi-major" axis, the distance between the centers of the two stars' orbits. Then do a basic polar plot of the orbit.

Second, solve the equations of motion, where Star1 is at (x,y) and Star 2 is at (-x,-y). Then we will Table the results, and finally Animate the plot.

The Physics of Binary Stars

When two stars are locked in a binary system, the parameters that describe the system are

$\underline{a}$ the semi-major axis, which is the <u>average</u> of the periastron (closest) and apastron (furthest) separations of the two stars. $a = \frac{1}{2}(d_{peri} + d_{Ap})$

ϵ the orbital eccentricity, which is the <u>elongation</u> of the orbit of either star. The closer ϵ is to one, the further the orbit is away from circular.

v_p the periastron velocity $v_p = \frac{2\pi r}{T}\sqrt{\frac{1+\epsilon}{1-\epsilon}}$

v_A the apastron velocity $v_A = \frac{2\pi r}{T}\sqrt{\frac{1-\epsilon}{1+\epsilon}}$

where r is the *average radius* of the star's orbit measured from the CM

d_p the periastron separation of the stars $d_p = a(1-\epsilon)$

d_A the apastron separation of the stars $d_A = a(1+\epsilon)$

We have almost everything we need, for all these quantities,

except $\underline{a}$ which we will now find by using the Newton-Kepler Law

$$T^2 = \frac{4\pi^2 a^3}{G(M1+M2)}$$

```
In[25]:=  Clear[x, v, a, t];  T = 7.75 * 3600;  M1 = M2 = 1.4 * 2 * 10^30;
          G = 6.67 * 10^-20;  (* distances in km, time in seconds*)
          FindRoot[T^2 == 4 π^2 * a^3 / (G * 2 M1), {a, 10^6}]
```

```
Out[28]=  {a → 1.9456 × 10^6}
```

```
In[15]:=  dp = 1.945 * 10^6 * (1 - .617)          da = 1.945 * 10^6 * (1 + .617)
```

```
Out[15]=      744935.                                   3.14507 × 10^6
```

$In[17]:=$ $a = 1.9456 * 10^6$; $\epsilon = .617$; $T = 7.75 * 3600$; $r = a/2$;

$$vp = \frac{2\pi r}{T} \sqrt{\frac{(1+\epsilon)}{(1-\epsilon)}} \qquad\qquad va = \frac{2\pi r}{T} \sqrt{\frac{(1-\epsilon)}{(1+\epsilon)}}$$

$Out[18]=$ 450.148   106.621

We now know everything about the orbits of the two stars. We will now polar plot the orbits of the two stars. For each star, $r = \dfrac{a\,(1 - \epsilon^2)}{1 + \epsilon \cos\theta}$

$In[1]:=$ `Needs["Graphics`Graphics`"];`
`a = (1.945 * 10^6) / 2; ` ϵ ` = .617;`
`stars1 = PolarPlot[a * (1 - ` ϵ `^2) / (1 + ` ϵ ` * Cos[` θ `]),`
`   {` θ `, 0, 2 ` π `}, AspectRatio ` $\rightarrow$ ` 0.5];`
`stars2 = PolarPlot[a * (1 - ` ϵ `^2) / (1 - ` ϵ ` * Cos[` θ `]),`
`   {` θ `, 0, 2 ` π `}, AspectRatio ` $\rightarrow$ ` 0.5];`
`Show[{stars1, stars2}, AspectRatio ` $\rightarrow$ ` 0.3]`

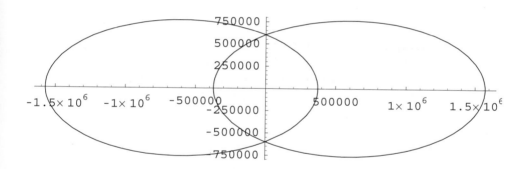

The two stars approach within 744,000 km of each other at periastron (the point of closest approach.) At this time, each star is moving at 450 km/s. The two stars move past each other at periastron like two bullet trains moving in opposite directions at one million miles per hour. The bending of space is so great at this distance (which is only half the diameter of our sun) that the binary system emits *gravitational waves,* and *with this loss of energy, the orbit of the two stars is slowly collapsing.* (See problem 1 in Chapter 3 for the length of time for this binary system to collapse into a black hole.)

Due to the *symmetry* of this problem, with the masses of the two stars equal, we may write the differential equation of motion of Star 1 to apply at (x,y) with the understanding that Star 2 will be at (-x, -y). Therefore we only have to solve one set of Differential Equations. (The case of stars of unequal masses will be treated in Section 3.4).

Building the Differential Equation

The gravitational forces acting on Star 1 at (x, y) from Star 2 at $(-x, -y)$ are

$$Mx'' = -\frac{MGM}{(2x)^2 + (2y)^2} \frac{2x}{\sqrt{(2x)^2 + (2y)^2}} \qquad x_o = 1.572 \times 10^6 \text{ km} \qquad \dot{x}_o = 0$$

$$My'' = -\frac{MGM}{(2x)^2 + (2y)^2} \frac{2y}{\sqrt{(2x)^2 + (2y)^2}} \qquad y_o = 0 \qquad \dot{y}_o = 106.6 \text{ km/s}$$

Notice that the gravitational attraction between the two stars is MGM/r^2 where the distance r between the two stars is $\sqrt{(2x)^2 + (2y)^2}$ and the Cosine of the angle from the CM to Star 1 is $2x/r$. We will start the two stars at their Apastron distance, each 1.57×10^6 km from the CM.

Programming in *Mathematica*, with distance in km and time in seconds,

```
Clear[x, y, vel, t] ;
sol =
 NDSolve[{x''[t] == -60^2 * 2.8 * 10^30 * 6.67 * 10^-20 * (2 x[t]) /
     (4 x[t]^2 + 4 y[t]^2)^1.5, y''[t] == 60^2 * 2.8 * 10^30 *
     6.67 * 10^-20 * (-2 y[t]) / (4 x[t]^2 + 4 y[t]^2)^1.5,
   x[0] == 1.5725 * 10^6, y[0] == 0, x'[0] == 0, y'[0] == 106.6 * 60},
  {x, y}, {t, 0, 500}]
InterpFunc1 = x /. sol[[1]]; InterpFunc2 = y /. sol[[1]];
InterpFunc3 = x' /. sol[[1]]; InterpFunc4 = y' /. sol[[1]];
sep[t_] = 2 * √InterpFunc1[t]^2 + InterpFunc2[t]^2 ;
vel[t_] = √InterpFunc3[t]^2 + InterpFunc4[t]^2 ;
Print["max velocity=", vel[232.5] / 60,
 " km/s   separation =", sep[232.5], " km"]

T2 = Table[{t, InterpFunc1[t], Chop[InterpFunc2[t]],
    Chop[InterpFunc3[t] / 60], InterpFunc4[t] / 60, vel[t] / 60},
   {t, 0, 465, 15}] // TableForm
```

From In[344]:= max velocity = 450.464 km/s   separation = 744250. km

Table of values for the binary pulsar system. The orbital period is 465 min. The maximum velocity of Star 1 and Star 2 is 450 km/s (1 million mph) at a distance of only 744,000 km.

t (min)	x (km)	y (km)	x' (km/s)	y' (km/s)	v (km/s)
0	1.572×10^6	0	0	106.6	106.6
15	1.564×10^6	95784.	-17.	106.0	107.4
30	1.541×10^6	190625.	-34.	104.4	109.9
45	1.503×10^6	283535.	-51.	101.7	114.1
60	1.448×10^6	373438.	-69.	97.78	119.9
75	1.377×10^6	459104.	-88.	92.31	127.5
90	1.290×10^6	539076.	-107	85.06	137.
105	1.184×10^6	611545.	-127	75.55	148.4
120	1.059×10^6	674181.	-149	63.06	162.2
135	914677.	723841.	-172	46.48	178.9
150	747933.	756077.	-198	23.95	199.4
165	557697.	764229.	-224	-7.74	225.1
180	342719.	737653.	-252	-54.5	258.4
195	104400.	658041.	-275	-128.	303.5
210	-144018.	492294.	-267	-250.	366.1
225	-340834.	192498.	-136	-414.	436.5
240	-337889.	-201197	142.	-411.	435.2
255	-138375.	-497527	268.	-246.	364.4
270	110193.	-660721	274.	-126.	302.3
285	348035.	-738789	251.	-53.2	257.5
300	562431.	-764383	224.	-6.85	224.4
315	752099.	-755566	197.	24.56	198.9
330	918313.	-722857	172.	46.92	178.5
345	1.062×10^6	-672848	148.	63.4	161.9
360	1.186×10^6	-609950	127.	75.80	148.1
375	1.292×10^6	-537281	106.	85.25	136.7
390	1.379×10^6	-457157	87.6	92.46	127.3
405	1.450×10^6	-371376	69.0	97.89	119.8
420	1.504×10^6	-281390	51.2	101.8	114.0
435	1.542×10^6	-188422	33.7	104.5	109.8
450	1.565×10^6	-93548.	16.6	106.1	107.3
465	1.572×10^6	2246.4	-0.3	106.6	106.6

To better represent the orbits of the pulsars as they move about the CM, an Animated series of plots is given by the following **Do Loop.** A dozen stills from this Animation are shown. To see the entire Animation, access the program *BinaryStarsOrbit* on the CD, and run it in *Mathematica.*

```
In[27]:= g1 = ParametricPlot[{Table[{InterpFunc1[t], InterpFunc2[t]}],
           Table[{-InterpFunc1[t], -InterpFunc2[t]}]}, {t, 0, 465},
           AspectRatio → Automatic, PlotStyle → GrayLevel[0.5],
           PlotRange → {{-2*10^6, 2*10^6}, {-1*10^6, 1*10^6}},
           AxesStyle → GrayLevel[0.65], Ticks → None];
```

```
In[28]:= Do[(g2 = ListPlot[{{InterpFunc1[i], InterpFunc2[i]},
           {-InterpFunc1[i], -InterpFunc2[i]}},
           PlotRange → {{-2*10^6, 2*10^6}, {-1*10^6, 1*10^6}},
           DisplayFunction → Identity, PlotStyle → PointSize[0.03]];
         Show[g1, g2];), {i, 0, 437, 23}]
```

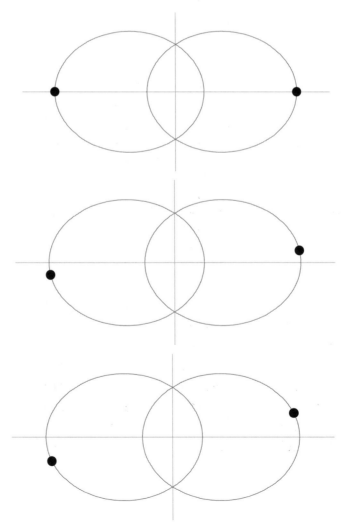

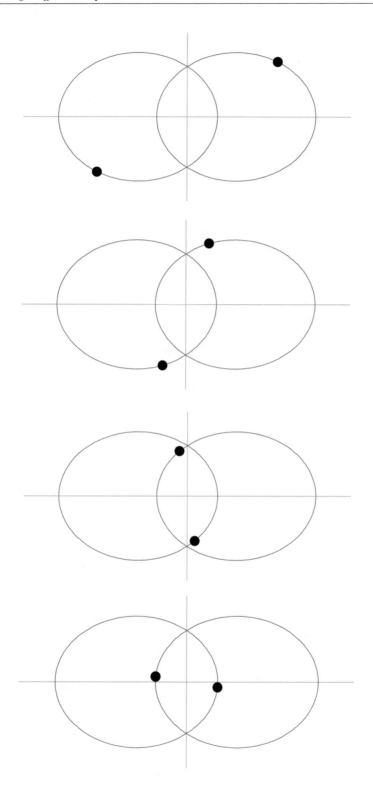

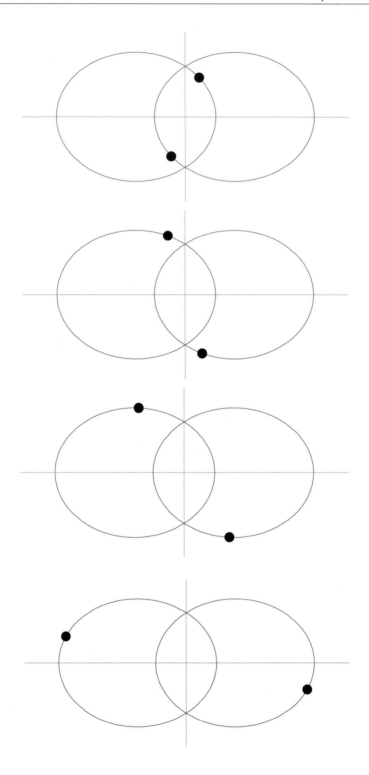

3.4 Exploding Star in a Binary System

A 15 M_{sun} star is in a binary system with an 8 M_{sun} star. The stars are 1.5×10^8 km apart, in circular orbits about their Center-of-Mass. Suddenly, the 15 M_{sun} star goes supernova in a spherically symmetric explosion, reducing its mass to 3 M_{sun}. What are the velocities of the stars after the explosion ? Does the system become unbound ?

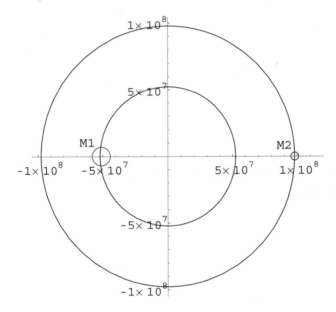

Before the explosion of M1, the two stars are in circular orbits about the CM. Let's first of all find the orbit radius a of star M1 and orbit radius b of M2. Then use the Newton-Kepler Law to find the orbit time T of each star, then find V1 and V2.

```
In[55]:=  M1 = 15; M2 = 8; r = 1.5 * 10^8;
          sol2 = Solve[{M1 * a == M2 * b, r == a + b}, {a, b}]

Out[56]=  {{a → 5.21739 × 10^7, b → 9.78261 × 10^7}}

In[57]:=  G = 6.67 * 10^-20; (* x and y in km, T in seconds *)
          FindRoot[T^2 == 4 π^2 * r^3 / (G * (M1 + M2) * 2 * 10^30), {T, 10^6}]

Out[58]=  {T → 6.58984 × 10^6}

          T = 6.5898 * 10^6; a = 5.22 * 10^7; b = 9.78 * 10^7;
          v1 = 2 π * a / T; v2 = 2 π * b / T; Print["v1→", v1, "   v2→", v2 ]

          v1→49.7712   v2→93.2495 km/s
```

Building the Differential Equation

We are now ready to write the differential equations of the two stars. Since both stars orbit the center-of-mass at $(0,0)$ we will call the coordinates of star M1 (x,y) and the coordinates of star M2 (w,z). Then the accelerations of the two stars are

$$x'' = -M2\ G\ \frac{(x-w)}{\left((x-w)^2+(y-z)^2\right)^{3/2}} \qquad x_o = -5.22\times10^7 \qquad x_o' = 0$$

$$y'' = -M2\ G\ \frac{(y-z)}{\left((x-w)^2+(y-z)^2\right)^{3/2}} \qquad y_o = 0, \qquad y_o' = -49.8$$

$$w'' = -M1^*\ G\ \frac{(w-x)}{\left((x-w)^2+(y-z)^2\right)^{3/2}} \qquad w_o = +9.78\times10^7 \qquad w_o' = 0$$

$$z'' = -M1^*\ G\ \frac{(z-y)}{\left((x-w)^2+(y-z)^2\right)^{3/2}} \qquad z_o = 0, \qquad z_o' = +93.3$$

* Where M1 is 15 M_{Sun} before $t = 0$, and 3 M_{Sun} after $t = 0$.

The positions of the two stars are shown from 960 hours before the explosion up to the moment of DETONATION $(t = 0)$. It is at this time that Star M1 explodes.

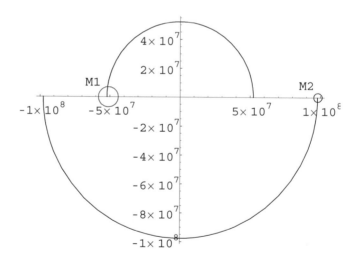

The subsequent paths of the two stars is determined by their gravitational interaction. The mass of M1 after the explosion is 3 M_{sun}. M2 remains at 8 M_{sun}. The programming in *Mathematica* proceeds as follows:

```
In[109]:=
    (* distance in km, time in hr *)
    Clear[x, y, v, t]
    sol = NDSolve[{x''[t] == -3600^2 * 16 * 6.67 * 10^10 *
        (x[t] - w[t]) / ((x[t] - w[t])^2 + (y[t] - z[t])^2)^1.5,
        y''[t] == -3600^2 * 16 * 6.67 * 10^10 *
        (y[t] - z[t]) / ((x[t] - w[t])^2 + (y[t] - z[t])^2)^1.5,
        w''[t] == -3600^2 * 30 * If[t < 0, 1, .2] * 6.67 * 10^10 *
        (w[t] - x[t]) / ((x[t] - w[t])^2 + (y[t] - z[t])^2)^1.5,
        z''[t] == -3600^2 * 30 * If[t < 0, 1, .2] * 6.67 * 10^10 *
        (z[t] - y[t]) / ((x[t] - w[t])^2 + (y[t] - z[t])^2)^1.5,
        x[0] == -5.22 * 10^7, y[0] == 0, w[0] == 9.78 * 10^7, z[0] == 0,
        x'[0] == 0, y'[0] == -49.8 * 3600, w'[0] == 0, z'[0] == 93.3 * 3600},
       {x, y, w, z}, {t, -960, 1960}]
    InterpFunc1 = x /. sol[[1]]; InterpFunc2 = y /. sol[[1]];
    InterpFunc3 = x' /. sol[[1]]; InterpFunc4 = y' /. sol[[1]];
    InterpFunc5 = w /. sol[[1]]; InterpFunc6 = z /. sol[[1]];
    InterpFunc7 = w' /. sol[[1]]; InterpFunc8 = z' /. sol[[1]];
    rad[t_] = √((InterpFunc1[t] - InterpFunc5[t])^2 +
        (InterpFunc2[t] - InterpFunc6[t])^2);
    vel1[t_] = √(InterpFunc3[t]^2 + InterpFunc4[t]^2);
    vel2[t_] = √(InterpFunc7[t]^2 + InterpFunc8[t]^2);
    xcm[t_] =
       (3 * InterpFunc1[t] + 8 * InterpFunc5[t]) / 11 * If[t < 0, 0, 1];
    ycm[t_] = (3 * InterpFunc2[t] + 8 * InterpFunc6[t]) / 11 *
        If[t < 0, 0, 1];
    ParametricPlot[{{x[t], y[t]} /. sol, {w[t], z[t]} /. sol},
       {t, -912, 1500}, AspectRatio -> Automatic,
       Prolog -> {{Text["M1", {-6.2 * 10^7, 10^7}],
          Hue[.91], Circle[{-5.12 * 10^7, 10}, 7000000]},
          {Text["M2", {11 * 10^7, .8 * 10^7}], Hue[0.67],
          Circle[{9.8 * 10^7, 100}, 3000000]}}];
    T2 = Table[{t, rad[t] / 10^6, vel1[t] / 3600,
        vel2[t] / 3600, xcm[t] / 10^6, ycm[t] / 10^6},
       {t, -168, 600, 24}] // TableForm
```

The **IF** command allows us to plot the evolution of the orbit from 7 days before the explosion (where the orbits are circular) to 25 days after the explosion (as the orbits diverge). That's it. *Mathematica* easily handles 4 simultaneous differential equations.

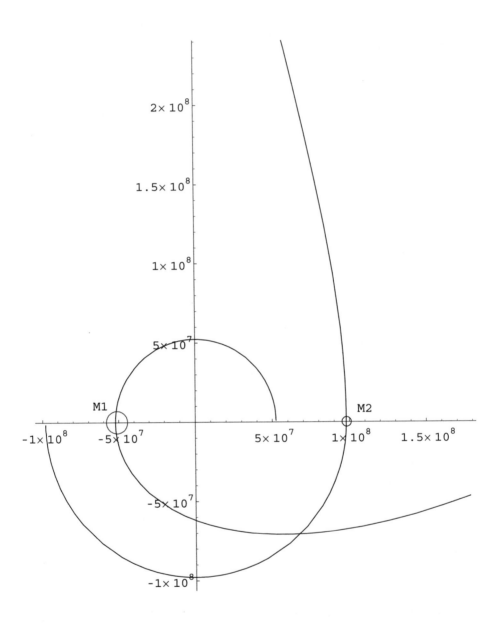

M2 tears off into space at high velocity. The binary system is now unbound.

Time and Distance between stars in 10^6 kilometers, Velocities, Location of CM in 10^6 km

t (hr)	r (10^6 km)	v1 (km/s)	v2 (km/s)	CM - X	CM - Y
-168	150.027	49.7868	93.2873	0	0
-144	150.02	49.7902	93.2906	0	0
-120	150.014	49.7932	93.2934	0	0
-96	150.009	49.7956	93.2958	0	0
-72	150.005	49.7975	93.2976	0	0
-48	150.002	49.7989	93.2989	0	0
-24	150.001	49.7997	93.2997	0	0
0	150.	49.8	93.3	56.8909	0.
24	150.266	49.7997	93.2495	56.8909	4.68916
48	151.06	49.7989	93.0994	56.8909	9.37833
72	152.371	49.7976	92.8545	56.8909	14.0675
96	154.182	49.7958	92.522	56.8909	18.7567
120	156.47	49.7935	92.1113	56.8909	23.4458
144	159.208	49.7909	91.633	56.8909	28.135
168	162.367	49.7881	91.0984	56.8909	32.8241
192	165.913	49.785	90.5189	56.8909	37.5133
216	169.815	49.7817	89.9052	56.8909	42.2025
240	174.039	49.7784	89.2674	56.8909	46.8916
264	178.554	49.775	88.6144	56.8909	51.5808
288	183.329	49.7715	87.9538	56.8909	56.27
312	188.338	49.7681	87.292	56.8909	60.9591
336	193.552	49.7648	86.6344	56.8909	65.6483
360	198.949	49.7615	85.9853	56.8909	70.3375
384	204.506	49.7583	85.348	56.8909	75.0266
408	210.203	49.7552	84.7249	56.8909	79.7158
432	216.023	49.7522	84.118	56.8909	84.4049
456	221.949	49.7493	83.5286	56.8909	89.0941
480	227.968	49.7465	82.9573	56.8909	93.7833
504	234.066	49.7438	82.4047	56.8909	98.4724
528	240.233	49.7412	81.871	56.8909	103.162
552	246.458	49.7387	81.356	56.8909	107.851
576	252.732	49.7364	80.8595	56.8909	112.54
600	259.049	49.7341	80.3811	56.8909	117.229

$In[128]:=$ **vy_cm = 4.689 * 10^6 / 24 / 3600**

  54.2708

An extremely interesting result from the above table is that the Center-of-Mass of the binary system shifts from (0,0) to $(56.8 \times 10^6, 0)$ right after detonation. Then the CM moves vertically upward at 54 km/s as the stars separate.

For more on exploding stars in binary systems, see Romas Mitalas *Am J Physics* (Mar 1980) "Supernovae in Binary Systems" pp 226-231.

3.5   Heat Exchanger

Before we get to the very interesting subject of the heat-exchanger, let us take a somewhat simpler case of extracting heat from a hot environment. Say we start with water flowing through a tube of diameter D and total length L being used to extract heat from an operating automobile engine.

If water flowing at 0.8 m/s enters a 4-cm diameter tube at 25 °C and the walls of the tube inside the engine are maintained at 100 °C, how much is the water heated in passing 4-meters total distance through the engine? And how much heat is extracted per second?

To solve this problem, we need to set-up a basic differential equation, where the heat extracted from the engine block as the fluid flows a distance dx is $dQ = h \, \pi D \, dx \, (T^* - T)$ and the water absorbs dQ and heats up by an amount dT $dQ = \dot{m} \, c \, dT$

Setting these two equations equal, we have our basic differential equation

$$\frac{dT}{dx} = \frac{h \, \pi \, D \, (T^* - T)}{\dot{m} \, c} \quad \text{(Eq 1)}$$

where $T^* = 100 \,°C$, $D = 0.04$ m, $c = 4180$ J/kg °C, $\dot{m} = 1$ kg/s, L = 4 m, and $T_0 = 25 \,°C$.

Before solving, we first have to find h, the convective heat transfer coefficient. This depends on the heat transfer characteristics of water, and is given in the engineering literature by the *Petukhov equation*

$$h = \frac{k}{D} \, \frac{(f/8) \, (RN - 1000) \, PR}{1 + 12.7 \, (f/8)^{.50} \, (PR^{2/3} - 1)} \quad \text{for } 3000 < RN < 5 \times 10^6 \text{ and } PR > 0.5$$

where RN is Reynolds number, PR is the Prandtl number and

the friction factor $f = (0.79 \, \text{Ln} \, RN - 1.64)^{-2}$

For water at 25 °C, $\nu = 0.9 \times 10^{-6}$ m²/ s, PR = 6.0, and k = 0.612 W/m °C

Thus $RN = \frac{.8 * .04}{.9 \times 10^{-6}} = 35{,}550$ (turbulent flow) and   f = 0.0227

Now, solving the above Eq 1 for T(x)

```
DSolve[{T'[x] == h * π D / mc * (100 - T[x]), T[0] == 25}, T[x], x] //
FullSimplify
```

Out [23]= $\left\{ \left\{ T[x] \rightarrow 100 - 75 \, e^{-\frac{D \, h \, \pi \, x}{m \, c}} \right\} \right\}$

We now have the form of T(x), but we still need to find the heat transfer coefficient h

```
In[10]:=  k = .612; D1 = .04; PR = 6.0; RN = 35550; f = .0227;

          k    (f / 8) * (RN - 1000) * PR
     h = ─── ────────────────────────────────
          D1   1 + 12.7 √f / 8  (PR^{2/3} - 1)

Out[11]=  3519.76
```

```
In[12]:=  Clear[T, t]; mc = 1 * 4180; h = 3519.76; D1 = .04;
          sol =
          DSolve[{T'[x] == h * π * D1 / mc * (100 - T[x]), T[0] == 25}, T[x], x]
          Plot[Evaluate[T[x] /. sol], {x, 0, 10}, PlotRange → {0, 100}]
          T[x_] = T[x] /. sol;
          T[4]
```

$$Out[13]= \{\{T[x] \rightarrow 100. - 75. \, e^{-0.105815 \, x}\}\}$$

```
Out[16]=  {50.8819}
```

The heat transferred per second is $\dot{Q} = \dot{m} \, c \, \Delta T = 1 * 4180 * 25.88 = 108$ kW

A Parallel-flow Heat Exchanger

A heat exchanger is a device for transferring heat from one fluid to another without mixing. In a *parallel-flow* heat exchanger, the cold fluid enters from the left and exits heated on the right, and the hot fluid also enters from the left and exits after cooling on the right. If hot oil at 200 °C with a mass flow rate of 3 kg/s and a specific heat of 1900 J/kg ·°C enters the outer shell from the left, and cool water at 20 °C with a mass flow rate of 0.9 kg/s and specific heat 4180 J/kg ·°C enters the inner tube from the left, what is the rate of heat exchange in this heat exchanger? We will assume a heat-transfer coefficient U = 460 W/m²·°C, and the heat-exchange surface area is A = π D L. Where D = 10 cm, and length L = 40 m.

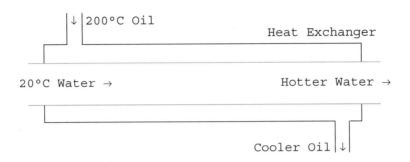

A parallel-flow Heat Exchanger

Building the Differential Equations

For a heat exchanger, the cold fluid is heated dT by absorbing heat energy dQ through an area $\pi D\, dx$.
In the inner tube,

$$dQ = (\dot{m}\, c)_c\, dT = U\, \pi\, D\, (T^* - T)\, dx \qquad \text{Cold side}$$
$$\text{Heat Absorbed} = \text{Heat Transferred}$$

In the outer shell, the hot fluid (which is moving along $+ dx$) *loses* heat energy dQ through the same area.
In the outer shell,

$$dQ = (\dot{m}\, c)_h\, dT^* = -\, U\, \pi\, D\, (T^* - T)\, (dx) \qquad \text{Hot side}$$
$$\text{Heat Lost} = \text{Heat Transferred}$$

When we write the differential equations, with T for the cold fluid and T^* for the hot fluid,

$$(\dot{m}\, c)_c \frac{dT}{dx} = U\, \pi\, D\, (T^* - T) \qquad \text{and}$$
$$(\dot{m}\, c)_h \frac{dT^*}{dx} = U\, \pi\, D\, (T^* - T)$$

Now, programming in *Mathematica*,

```
In[155]:=
      mh = 3.0 * 1900; mc = 0.9 * 4180; D0 = .10; L = 40; U = 460; A = π * D0 * L;
      sol = NDSolve[{mc * T'[x] == U * π * D0 * (S[x] - T[x]),
          mh * S'[x] == - U * π * D0 * (S[x] - T[x]),
          T[0] == 20, S[0] == 200}, {S, T}, {x, 0, 80.0}]
      InterpFunc1 = S /. sol[[1]]; InterpFunc2 = T /. sol[[1]];
      Table[{x, Chop[InterpFunc1[x]], Chop[InterpFunc2[x]]},
          {x, 0, 40, 5}] // TableForm
      Plot[{S[x] /. sol, T[x] /. sol}, {x, 0, 40},
          AxesLabel → {"Dist(m)", "Temp(°C)"},
          Epilog → {Text["Oil", {17, 168}], Text["Water", {27, 85}]}];
```

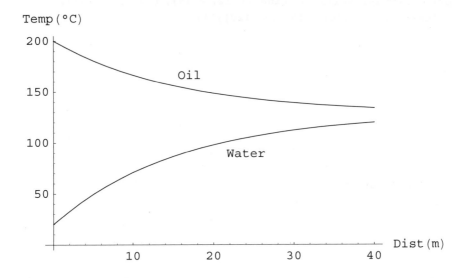

The Parallel-flow heat exchanger

x (m)	S (hot)	T (cold)
0	200.	20.
5	180.462	49.603
10	166.258	71.1242
15	155.932	86.77
20	148.425	98.1444
25	142.967	106.414
30	138.999	112.425
35	136.115	116.796
40	134.018	119.973

```
Plot[4, {x, -12, 10.5}, PlotRange → {-0.91, .6}, Axes → None,
  Prolog → {Text["134°C Oil ↓", {5.65, -.67}],
    Text["↓ 200°C Oil ", {-6.42, 0.15}],
    Text["Heat Exchanger", {6.8, .07}], Text["120°C Water →",
      {7.78, -.25}], Text["20°C Water →", {-9.7, -.25}],
    Line[{{-11.0, -.5}, {7.59, -.5}}], Line[{{8.58, -.5}, {9.3, -.5}}],
    Line[{{7.6, -.5}, {7.6, -.7}}], Line[{{8.5, -.5}, {8.5, -.7}}],
    Line[{{-9.5, .2}, {-9.5, 0}}], Line[{{-8.5, .2}, {-8.5, 0}}],
    Line[{{9.3, -.12}, {9.3, 0}}], Line[{{9.3, -.5}, {9.3, -.39}}],
    Line[{{-11, -.12}, {-11, 0}}], Line[{{-11, -.5}, {-11, -.39}}],
    Line[{{-11.0, 0}, {-9.5, 0}}], Line[{{-8.51, 0}, {9.3, 0}}]},
  Epilog → {GrayLevel[0.6], Line[{{-12, -.39}, {10.5, -.39}}],
    Line[{{-12, -.120}, {10.5, -.120}}]}]
```

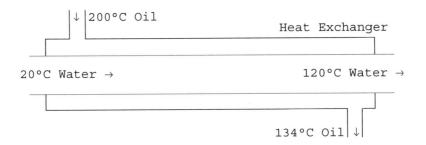

A Parallel-flow Heat Exchanger
The hot oil is cooled from 200 °C to 134 °C
while the water is heated from 20 °C to 120 °C.
Only heat is exchanged. The fluids do not mix.

3.6 Challenge Problems

1. Gravity Waves and Pulsar 1913+16

In 1964, ten years before the discovery of the binary pulsar, Philip C. Peters published a paper "Gravitational Radiation from Two Point Masses" (*Physical Review* 136, B1224-1232).

In this paper, the author derives a series of results from general relativity theory. Let us solve each of the following for the binary pulsar (M1 = M2 = 2.8×10^{30} kg, $\epsilon = 0.617$, $P_o = 27900$ s, $a = 1.95 \times 10^9$ m, $c = 3 \times 10^8$ m/s, $G = 6.67 \times 10^{-11}$ m^3/kg·s^2)

A. Rate of Energy Loss due to Gravitational Radiation

$$\frac{dE}{dt} = -\frac{32}{5} \frac{G^4 \, M1^2 \cdot M2^2 \, (M1 + M2)}{c^5 \cdot a^5 \, (1-\epsilon^2)^{7/2}} (1 + \frac{73}{24} \epsilon^2 + \frac{37}{96} \epsilon^4)$$

B. Rate of Decrease of "Semi-major" axis of Binary System

$$\frac{da}{dt} = -\frac{64}{5} \frac{G^3 \, M1 \cdot M2 \, (M1 + M2)}{c^5 \cdot a^3 \, (1-\epsilon^2)^{7/2}} (1 + \frac{73}{24} \epsilon^2 + \frac{37}{96} \epsilon^4)$$

C. Rate at which the Orbital Period decreases

$$\frac{dP}{dt} = -\frac{96}{5} \frac{G^3 \, M1 \cdot M2 \, (M1 + M2)}{c^5 \cdot P^{5/3} \, (1-\epsilon^2)^{7/2}} (\frac{4\pi^2}{G \, (M1+M2)})^{4/3} (1 + \frac{73}{24} \epsilon^2 + \frac{37}{96} \epsilon^4)$$

D. Time (in s) for the two stars to spiral in (and become a black hole)

$$T_{collapse} = \frac{12}{19} \frac{(1.55 \times 10^9)^4}{6.86 \times 10^{19}} \int_0^{.617} \epsilon^{29/19} \frac{(1+\frac{121}{304} \epsilon^2)^{.5137}}{(1-\epsilon^2)^{1.5}} \, d\epsilon$$

Astronomers have observed that the orbital period of the binary pulsar decreases at 76.5 μ sec/ year. Using the Newton–Kepler Law, this results in a decrease in the semi-major axis of 3.5 m/year.

Solve the above equations of General Relativity to find out how much of the decrease in the orbital period and in the semi-major axis is due to the emission of gravity waves.

2. Halley's Comet Equations

Use the equations of Section 3.3 to find all the orbital parameters of Halley's comet, given only its velocity of 54.45 km/s at closest approach to the Sun, which was 0.59 AU in 1986. The mass of the Sun is 2×10^{30} kg.

Find orbital time T, eccentricity ϵ, semi-major axis a, and maximum and minimum velocity and distance of the comet from the Sun. Then polar plot Halley's orbit to scale.

3. The Largest Stars

Plaskett's binary star system consists of two stars that revolve in a circular orbit about a center of gravity midway between them. This means the masses of the two stars are equal.

If the orbital velocity of each star is 220 km/s and the orbital period of each is 14.4 days, find the mass M of each star.

4. The Alaska Pipeline

The Alaska pipeline is 800 miles long, stretching from Prudhoe Bay in the North to the port of Valdez in the South. About 1 million barrels of oil per day is pumped through the 1.2-meter diameter pipeline (accounting for 5% of the United States daily consumption).

The oil in the pipeline is hot, at 50°C, and to prevent heat loss, the pipeline is wrapped with 10-cm of fiberglass insulation of conductivity k = 0.035 W/m·°C.

For the above-ground pipeline, find the heat-loss per meter and the temperature of the surface of the fiberglass-wrapped pipeline if, on a cold winter's night, the air temperature is − 40 °C, and the convective heat transfer coefficient is h = 12 W/ m²·°C.

5. Counterflow Heat Exchanger

In a *counterflow* heat exchanger, the cold fluid enters from the left and exits heated on the right, and the hot fluid enters from the right and exits after cooling on the left. If hot oil at 200 °C with a mass flow rate of 3 kg/s and a specific heat of 1900 J/kg ·°C enters the outer shell from the right, and cool water at 20 °C with a mass flow rate of 0.9 kg/s and specific heat 4180 J/kg ·°C enters the inner tube from the left, what is the rate of heat exchange in this heat exchanger? We will keep the same heat-exchange parameters as in Section 3-5, the heat-transfer coefficient U = 460 W/m²·°C, and the heat-exchange surface area is A = π D L. Where D =10 cm, and the length is L = 40 m.

Be sure to compare the effectiveness of this counter-flow heat exchanger with the effectiveness of the parallel-flow heat exchanger of Section 3-5. Which system, *parallel* or *counterflow*, does the better job of extracting heat from the hot fluid ?

6. Women's Records in Track and Field

Following, is a list of world records for women in running, from 100-meters to 10,000 meters.

Distance (m)	Time (s)	Avg Velocity (m/s)	Record Holder
100	10.49	9.53	Florence Joyner (USA) 1988
200	21.34	9.37	Florence Joyner (USA) 1988
400	47.60	8.40	Maria Koch (GER) 1985
800	113.28	7.06	Jarmila Kratochvilova (CZ) 1983
1000	149.34	6.69	Maria Mutola (MOZ) 1995
1500	230.46	6.51	Qu Yunxia (CHINA) 1993
1 mile	252.56	6.37	Svetlana Mastercova (RUS) 1996
2000	325.36	6.15	Sonia O'Sullivan (IRE) 1994
3000	486.11	6.17	Wang Junxia (CHINA) 1993
5000	864.68	5.78	Elvan Abeylegesse (TUR) 2004
10000	1771.78	5.64	Wang Junxia (CHINA) 1993

See if the Physics model of running will account for all the above world records. Try the following physiological constants

$$F = 1.2 * 9.8 \text{ m/s}^2, \ k = 1.14 \text{ s}^{-1}, \ \sigma = 36.5 \text{ W/kg}, \ E_o = 2300 \text{ J/kg}$$

Based on the data fit, which of the above world records is most likely to be broken ?

Chapter 4
Partial Differential Equations and Quantum Mechanics

4.1 Introduction to Partial Differential Equations

Some of the most important Physics equations are Partial Differential Equations, or PDEs. Let's examine one of the most famous, the Partial Differential Equation for Heat Diffusion, to see how an analytic solution may be obtained with standard mathematical techniques, and then obtain a computer solution of the same equation using *Mathematica.*

Partial Differential Equation #1

The partial differential equation for heat diffusion

$$\frac{\partial u}{\partial t} = \alpha \frac{\partial^2 u}{\partial x^2} \qquad \alpha = .96;$$

Boundary condition $u(x, t) = 0$ at $x = 0$ and $x = 20$
Initial condition $u(x, 0) = 70$ for $0 < x < 20$

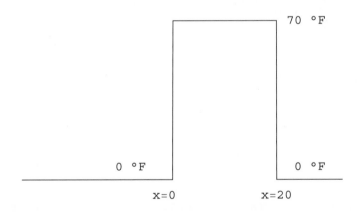

Temperature distribution at $t = 0$

The Physics of this heat problem is very straightforward. Over time, the heat constrained between $x = 0$ and $x = 20$ will diffuse outward and be lost to the system, so the final steady-state distribution will be a flat zero across the board. The question is, how long will it take for the heat to leave the system, and what does the temperature distribution look like as a function of time?

Classical Solution by separation of variables

$$\frac{\partial u}{\partial t} = \alpha \frac{\partial^2 u}{\partial x^2}$$

Boundary condition $u\,(x,\, t)\; =\; 0$ at $x = 0$ and $x = 20$

Initial condition $u\,(x,\, 0)\; =\; f[x]\; =\; 70$ for $0 < x < 20$

Let $u\,(x,\, t)\; =\; X\,(x)\;\cdot\; T\,(t)$

Then $\dfrac{\partial u}{\partial t}\; =\; X\,T'\; =\; \alpha\,\dfrac{\partial^2 u}{\partial x^2}\; =\; \alpha\,X''\,T$

$$\frac{T'}{\alpha\,T}\; =\; \frac{X''}{X}\; =\; -\lambda^2$$

Thus $T\,(t)\; =\; A\,e^{-\alpha\lambda^2 t}$ and $X\,(x)\; =\; B\,\mathrm{Sin}\,\lambda\,x$

Now match the boundary conditions at $x = 0$ and $x = L$,
where $u\,(0,\,t)\; =\; u\,(L,\,t)\; =\; 0$,

$$X\; =\; B_N\;\mathrm{Sin}\,\frac{N\,\pi\,x}{L} \qquad N = 1,\, 2,\, 3\; \ldots$$

and $X\,(x)\; =\; \displaystyle\sum_{N=1}^{\infty} B_N\;\mathrm{Sin}\,\frac{N\,\pi\,x}{L}$ note $\lambda = \dfrac{N\,\pi}{L}$

Then the solution is given by

$$u\,(x,\, t)\; =\; \sum_{N=1}^{\infty} A\,e^{-\alpha\lambda^2 t}\cdot B_N\;\mathrm{Sin}\,\frac{N\,\pi\,x}{L}\; =\; \sum_{N=1}^{\infty} C_N\cdot e^{-\alpha\,\frac{N^2\,\pi^2}{L^2}\,t}\cdot\mathrm{Sin}\,\frac{N\,\pi\,x}{L}$$

We find the C_N by utilizing the Euler formula,

$$C_N\; =\; \frac{2}{L}\int_0^L f\,(x)\cdot\mathrm{Sin}\,\frac{N\,\pi\,x}{L}\,dx \qquad \text{where } u\,(x,\, 0)\; =\; f\,(x)\; =\; 70$$

$$C_N\; =\; \frac{2}{L}\int_0^L 70\cdot\mathrm{Sin}\,\frac{N\,\pi\,x}{L}\,dx\; =\; \frac{2*70}{N\,\pi}\,(1 - \mathrm{Cos}\,N\,\pi)\; =\; \frac{4*70}{N\,\pi} \quad (N\ \text{odd})$$

$$C_N\; =\; \frac{4*70}{(2\,k+1)\,\pi} \qquad k = 0,\, 1,\, 2,\, 3.\,..$$

For a grand solution of

$$u\,(x,\, t)\; =\; \frac{4*70}{\pi}\sum_{k=0}^{\infty}\frac{1}{(2\,k+1)}\,e^{-\alpha\,\frac{(2\,k+1)^2\,\pi^2}{L^2}\,t}\cdot\mathrm{Sin}\,\frac{(2\,k+1)\,\pi\,x}{L}$$

Now, let us compare the analytic solution to the *Mathematica* solution. Using **NDSolve** we solve for **u[x,t]** and then 3-D plot the result

```
In[11]:=
  L = 20; α = .96;
  eq4 = {D[u[x, t], t] - .96 * D[u[x, t], x, x] == 0,
    u[x, 0] == Which[x ≤ 0, 0, 0 < x < 20, 70, x ≥ 20, 0],
    u[0, t] == 0, u[20, t] == 0};
  sol4 = NDSolve[eq4, u[x, t], {x, 0, 20}, {t, 0, 40.0}];
  Plot3D[Evaluate[u[x, t] /. sol4[[1]]], {t, 0, 40.0}, {x, 0, 20}],
    PlotRange → All, AxesLabel → {"t", "x", " "}];
```

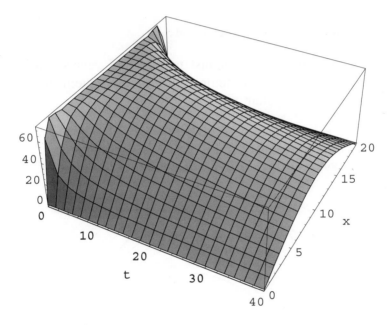

3-D representation of temperature in the slab versus time.

The *Mathematica* solution is retained in memory, so let us compare the *exact* solution to the numerical solution, by comparing the temperature at $x = 10$ (down the center-line) as a function of time.

$$\texttt{Table}\Big[$$

$$\texttt{Evaluate}\Big[\{t, u[x, t], \frac{4 * 70}{\pi} \sum_{k=0}^{10} \frac{(-1)^k}{(2k+1)} * \texttt{Exp}\Big[-\frac{\alpha (2k+1)^2}{(L/\pi)^2} t\Big]\}\Big] /.$$

$$\texttt{sol4}[[1]] /. x \to 10, \{t, 2.0, 40.0, 2.0\}\Big]\Big] // \texttt{TableForm}$$

t (da)	*Mathematica* T (x = 10)	Exact T (x = 10)
4.	69.9683	69.9569
8.	68.4981	68.4986
12.	64.7889	64.7891
16.	60.0307	60.0324
20.	55.0783	55.0785
24.	50.3008	50.3015
28.	45.8394	45.8404
32.	41.7318	41.7333
36.	37.9752	37.9763
40.	34.549	34.55

The agreement is very very good.

Note that we may also plot just temperature versus time as a 2-D graphic, and, as expected, the temperature of the system decreases over time as heat leaves the system.

```
In[47]:=  f[x_ , t_] = u[x, t] /. sol4[[1]];
          Pictures = Table[f[x, t], {t, 0.5, 30.5, 3}]
          Plot[Evaluate[Pictures], {x, 0, 20}]
```

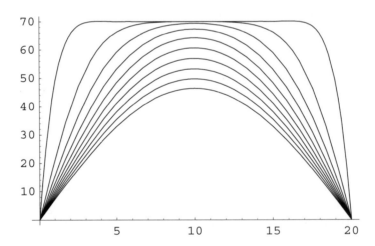

The temperature distribution with time. Notice how the originally rectangular temperature distribution becomes more and more like a sine function as time goes on.

Note that it is not whether the *Mathematica* or the exact solution is better, it is that the *Mathematica* computer solution is easier to use.

Partial Differential Equation #2

$$\frac{\partial u}{\partial t} = \alpha \frac{\partial^2 u}{\partial x^2} \qquad \text{Let } \alpha = 1, \quad L = 1$$

Boundary condition u (x, t) = 0 at x = 0 and x = 1
Initial condition   u (x, 0) = f (x) = 100 * Sin (π x) for 0 < x < 1

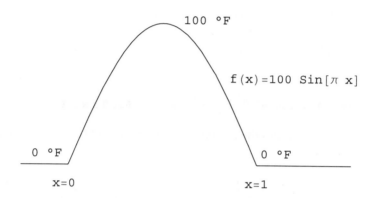

100 °F

f (x) = 100 Sin [π x]

0 °F

0 °F

x = 0

x = 1

Temperature distribution at t = 0

Classical Solution by separation of variables

Let   u (x, t) = X (x) · T (t)

Then $\dfrac{\partial u}{\partial t}$ = X T′ = $\alpha \dfrac{\partial^2 u}{\partial x^2}$ = α X″ T

$$\frac{T'}{\alpha T} = \frac{X''}{X} = -\lambda^2$$

Thus   T (t) = A e$^{-\alpha \lambda^2 t}$ and   X (x) = B Sin λ x

Now match the boundary conditions at x = 0 and x = L,
where u (0, t) = u (L, t) = 0,

$$X = B_N \; \text{Sin} \; \frac{N \pi x}{L} \qquad N = 1, 2, 3 \ldots$$

and   X (x) = $\displaystyle\sum_{N=1}^{\infty} B_N \; \text{Sin} \; \frac{N \pi x}{L}$ note $\lambda = \dfrac{N \pi}{L}$

Then the solution is given by

$$u (x, t) = \sum_{N=1}^{\infty} A e^{-\alpha \lambda^2 t} \cdot B_N \; \text{Sin} \; \frac{N \pi x}{L} = \sum_{N=1}^{\infty} C_N \cdot e^{-\alpha \frac{N^2 \pi^2}{L^2} t} \cdot \text{Sin} \; \frac{N \pi x}{L}$$

We find the C_N by utilizing the Euler formula,

$$C_N = \frac{2}{L} \int_0^L f(x) \cdot \text{Sin} \, \frac{N \pi x}{L} \, dx \qquad \text{where } u(x, 0) = f(x) = 100 \, \text{Sin}[\pi x]$$

$$C_N = \frac{2}{1} \int_0^1 100 \, \text{Sin}[\pi x] \cdot \text{Sin}[N \pi x] \, dx$$

By *orthogonality* of the sine functions
$$C_N = 100 \quad \text{if } N = 1$$
$$C_N = 0 \quad \text{if } N = 2, 3, 4 \ldots$$

The exact solution is
$$u(x, t) = 100 \, e^{-\pi^2 t} \cdot \text{Sin} \, \pi x \qquad \text{for } 0 \le x \le 1$$

Now, let us compare the analytic solution to the *Mathematica* solution, using **NDSolve**

```
In[4]:=  eq2 = {D[u[x, t], t] - D[u[x, t], x, x] == 0,
            u[x, 0] == 100 Sin[π * x], u[0, t] == 0, u[1, t] == 0};
         sol2 = NDSolve[eq2, u[x, t], {x, 0, 1}, {t, 0, 0.5}];
         Plot3D[Evaluate[u[x, t] /. sol2[[1]], {t, 0, 0.5}, {x, 0, 1}],
            PlotRange → All, AxesLabel → {"t", "x", " "}];
         Table[Evaluate[{t, u[x, t], 100 * E^(-π^2 * t)} /. sol2[[1]] /.
            x → 0.5, {t, 0, .5, .1}]] // TableForm
```

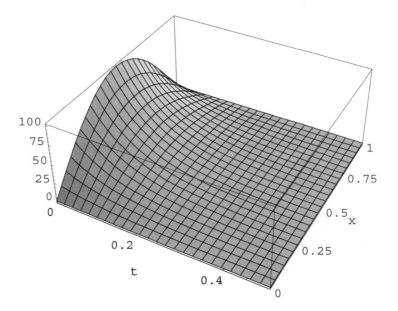

3-D representation of decrease of sine-temperature function with time

Comparing the *Mathematica* **NDSolve** and the exact solutions at x = 0.5,

t	*Mathematica* T (x = 0.5)	**Exact Solution** T (x = 0.5)
0	100.	100
0.1	37.2729	37.2708
0.2	13.8913	13.8911
0.3	5.17645	5.17733
0.4	1.9287	1.92963
0.5	0.718935	0.719188

Again, the numbers are very very close.

We will also show the temperature change in the system with time, as a 2-D representation

```
In[86]:=  f[x_, t_] = u[x, t] /. sol2[[1]];
          Pictures = Table[f[x, t], {t, 0, 0.3, .05}]
          Plot[Evaluate[Pictures], {x, 0, 1},
            Epilog → {Text["t=0", {0.72, 88}], Text["t=.05", {.67, 61}],
              Text["t=.10", {.62, 41}], Text["t=.15", {.58, 26}]}]
```

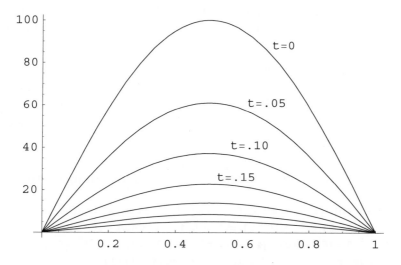

Temperature decrease with time, as heat diffuses out of the system.

With the numbers between *Mathematica* and the exact solutions so close, we will now proceed to analyze even more complicated heat problems using **NDSolve**. We will compare the *exact* solution, whenever available, with the *Mathematica* solution of the Partial Differential Equation.

4.2 PDE – Copper Rod

One end of a 40-cm long copper rod of cross-section 10 cm² is buried in ice at 0 °C, and the free end is heated with a propane torch to 400 °C. How long before the temperature gradient is *linear* (within 1%) from 400 °C to 0 °C along the length of the rod? Assuming heat-transfer by conduction only (if we insulate the rod to reduce convection and radiation losses), how much heat energy per second is transmitted by the copper rod from the flame to the ice?

The thermal conductivity of copper is 390 W/m/°C. The thermal diffusivity of copper is $\alpha = 0.0061\,\text{m}^2$/min. The PDE we are solving is

$$\frac{\partial u}{\partial t} = \alpha \frac{\partial^2 u}{\partial x^2} \qquad \alpha = .0061\,\text{m}^2\,/\,\text{min}$$

Boundary condition u (x, t) = 0 at x = 0 and u (x, t) = 400 at x = 0.4 m
Initial condition   u (x, 0) = 0 for 0 ≤ x < 0.4 m

Note that we have what are apparently two contradictory conditions: we want the initial temperature of the rod to be 0 °C all along its length at t = 0, *and* we want the free end to be heated to 400 °C. How do we reconcile these two conditions?

The answer is that we allow the free end to heat up rapidly with a temperature function **u[.4,t]= 400-400*Exp[-1000t]**. Now both the boundary conditions and the initial condition are satisfied, and we also have the physical effect of rapidly heating the end of the rod with the propane torch.

The *Mathematica* programming to find **u[x,t]** is straightforward

```
eq4 = {D[u[x, t], t] - .0061 * D[u[x, t], x, x] == 0,
    u[x, 0] == 0, u[0, t] == 0, u[.4, t] == 400 - 400 * Exp[-1000 t]};
sol400 = NDSolve[eq4, u[x, t], {x, 0, .4}, {t, 0, 20}];
Clear@f;  f[x_, t_] = u[x, t] /. sol400[[1]]
```

However the graphing of **u[x,t]** is more complex

```
Show[
  Plot3D[Evaluate[u[x, t] /. sol400[[1]], {t, 0, 10}, {x, 0, .4}],
    PlotRange → All, AxesLabel → {"t", "x", " "},
    Ticks →
      {Table[{2 * k, 2 * (5 - k)}, {k, 0, 5, 1}], Automatic, Automatic},
    DisplayFunction → Identity] /. SurfaceGraphics[arr_, opts___] :→
    SurfaceGraphics[Reverse /@ arr, opts],
  DisplayFunction → $DisplayFunction,
  ViewPoint -> {2.428, -2.280, 1.010}]
Pictures = Table[f[x, t], {t, 0, 11}]
Plot[Evaluate[Pictures], {x, 0, 0.4}, Frame → True,
  Prolog → {Text["t=0 min", {0.35, 10}], Text["1 min" , {.30, 85}],
    Text["2", {.259, 123}], Text["3", {.241, 146}]}]
```

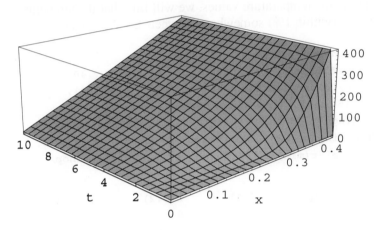

3-D representation of temperature in the copper rod

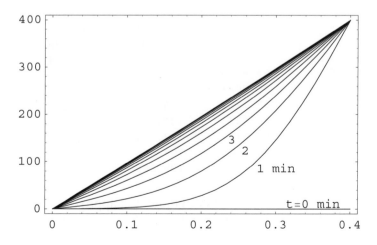

2-D plot of temperature vs distance along the rod, from t = 0 to t = 11 min

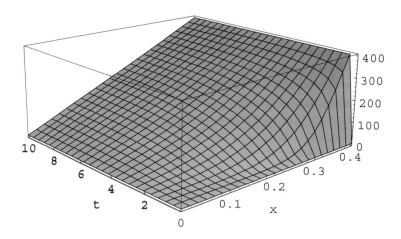

If we **Table** the temperature values, we will find that the temperature gradient becomes linear (within 1%) somewhere between $t = 10$ min and $t = 14$ min.

```
TableForm[Transpose@Prepend[
    Transpose@Table[f[x, t], {t, 0, 15, 1}, {x, 0, .40, 0.10}],
    Table[StyleForm[ToString@t, FontWeight -> "Bold"],
    {t, 0, 15, 1}]],
  TableHeadings -> {None, Prepend[Table[StyleForm["x(" <>
        ToString@x <> ")", FontWeight -> "Bold"], {x, 0, .40, 0.10}],
    StyleForm["t(min)", FontWeight -> "Bold"]]}] // Chop
```

t(min)	x(0)	x(0.1)	x(0.2)	x(0.3)	x(0.4)
0	0	0	0	0	0
1	0	2.62697	28.0163	146.007	399.968
2	0	21.3455	80.0686	208.763	399.968
3	0	43.1353	117.609	240.328	399.968
4	0	60.3177	143.438	259.682	399.968
5	0	72.6146	161.169	272.462	399.968
6	0	81.165	173.335	281.115	399.968
7	0	87.0546	181.68	287.028	399.968
8	0	91.103	187.411	291.084	399.968
9	0	93.8899	191.354	293.874	399.968
10	0	95.8083	194.068	295.793	399.968
11	0	97.1227	195.926	297.107	399.968
12	0	98.023	197.199	298.007	399.968
13	0	98.6414	198.074	298.625	399.968
14	0	99.0665	198.675	299.051	399.968
15	0	99.3581	199.088	299.342	399.968

Note that when the temperature gradient is approximately linear that the straight-line equation of heat conduction applies:

$$\dot{Q} = \frac{k\,A\,\Delta T}{\Delta x} = \frac{390 * .001 * 400}{0.4} = 390 \text{ Watts}$$

4.3 Heat Conduction in the Earth

A Temperature Wave into the Earth

A classic problem in mechanical engineering (Incropera and DeWitt, *Heat Transfer*) is to determine how deep water pipes should be buried such that they will never freeze.

If we assume that in most northern latitudes of the United States (Alaska excepted) that the temperature at the Earth's surface varies from + 35 °C in summer to −15 °C in winter then T (surface) = 10 + 25 Sin [2π t] where t is measured in years

Let us assume that the Partial Differential Equation for heat transfer in soil is

$$\frac{\partial u}{\partial t} = 2.4 \, \frac{\partial^2 u}{\partial x^2} \qquad \text{where} \quad \alpha = 2.4 \, m^2 / yr$$

with initial conditions $u(x, 0) = 10$ [the ground temperature **u** = 10 at all depths at t = 0],
the surface temperature at any time t is $u(0, t) = 10 + 25 \, Sin[2\pi \, t]$, and
an additional boundary condition (D[u[x, t], x] /. x → 4) = 0 must also be imposed

Solving with *Mathematica* we find a highly attenuated temperature wave propagating into the Earth.

```
In[12]:=  eq3 = {D[u[x, t], t] - 2.40 * D[u[x, t], x, x] == 0, u[x, 0] == 10,
             u[0, t] == (10 + 25 * Sin[2 π * t]), (D[u[x, t], x] /. x → 4) == 0};
          sol1 = NDSolve[eq3, u[x, t], {x, 0, 4}, {t, 0, 3}];
          Plot3D[Evaluate[u[x, t] /. sol1[[1]], {x, 0, 2}, {t, 0, 3}]];
```

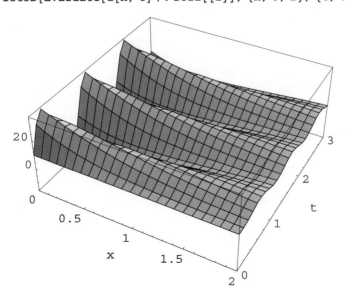

Shown above is the temperature variation at the Earth's surface on the left and how it propagates into the earth with x in meters. The temperature variation is phase-shifted as one goes deeper into the Earth, because it takes a certain length of time before the heat diffuses to a given depth.

How deep, then, should we bury the water pipes? A table of values will determine the answer. At whatever depth the temperature never goes negative is the depth at which the water pipes wont freeze. From the table, any depth greater than x = 0.8 meters is safe from a −15 °C surface freeze in winter.

```
Clear@f; f[x_, t_] = u[x, t] /. sol1[[1]]; TableForm[Transpose@
  Prepend[Transpose@Table[f[x, t], {t, 0, 2.0, .1}, {x, 0, 1.0, 0.2}],
    Table[StyleForm[ToString@t, FontWeight → "Bold"], {t, 0, 2.0, .1}]],
TableHeadings -> {None, Prepend[
    Table[StyleForm["x(" <> ToString@x <> ")", FontWeight → "Bold"],
      {x, 0, 1.0, 0.2}], StyleForm["t(yr)", FontWeight → "Bold"]]}]
```

t(yr)	x(0)	x(0.2)	x(0.4)	x(0.6)	x(0.8)
0	10.	10.	10.	10.	10.
0.1	24.694	19.188	15.460	13.070	11.631
0.2	33.776	27.877	22.987	19.122	16.201
0.3	33.777	30.392	26.782	23.332	20.265
0.4	24.696	25.462	24.828	23.373	21.518
0.5	10.002	14.846	17.637	18.911	19.124
0.6	−4.692	2.5371	7.8350	11.487	13.798
0.7	−13.77	−6.801	−0.905	3.8337	7.4515
0.8	−13.77	−9.627	−5.294	−1.192	2.4257
0.9	−4.693	−4.878	−3.687	−1.719	0.5822
1.	10.000	5.6206	3.2783	2.4217	2.5836
1.1	24.695	17.849	12.924	9.6239	7.6337
1.2	33.778	27.132	21.554	17.118	13.781
1.3	33.779	29.917	25.862	22.027	18.658
1.4	24.704	25.140	24.195	22.465	20.389
1.5	10.010	14.617	17.184	18.257	18.302
1.6	−4.684	2.3697	7.5001	11.001	13.184
1.7	−13.76	−6.927	−1.159	3.4640	6.9814
1.8	−13.76	−9.723	−5.490	−1.480	2.0578
1.9	−4.685	−4.952	−3.841	−1.947	**0.2880**
2.	10.009	5.5615	3.1540	2.2361	2.3429

It is worthwhile to check our answer against the **exact** analytic solution which is

$$T(x,t) = 10 + 25 \exp(- x \sqrt{\pi\, f / \alpha})* \mathrm{Sin}[2\pi\, f\, t - x \sqrt{\pi\, f / \alpha}]$$

Then, the maximum depth at which freezing will occur is found by solving

```
In[2]:=  Solve[0 == 10 + 25 * Exp[-x √(π * 1 / 2.4)] * (-1), x]

Out[2]=  {{x → 0.800874}}
```

4.4   Quantum Physics

One of the most puzzling and also the most interesting aspects of twenti-eth-century Physics is the advent of Quantum mechanics. How is it possible that an electron or an atom or a nucleus may be <u>both</u> a particle and a wave? The answer is *that is just how Nature is*. The value of Quantum Physics is that it accurately describes Nature at the atomic level. Given a particular physical situation, Quantum theory will give the correct energy *eigenstates* of the system and the *probability* of a process occuring.

The fundamental equation of Quantum Physics is Schrodinger's equation

$$\frac{-\hbar^2}{2\,m}\,\psi'' + V\,\psi = E\,\psi$$

We shall only use the one-dimensional case in this and the following section, because the mathematical difficulties rapidly become formidable. For a very instructive discussion on Schrodinger's equation, see *"Feynman's Derivation of Schrodinger's Equation"* by David Derbes in the July 1996 American Journal of Physics.

The Schrodinger equation may also be written as

$$\psi'' = -\frac{2\,m}{\hbar^2}\,(E - V)\,\psi$$

or more succinctly as

$$\psi'' = -k^2\psi \qquad \text{where} \qquad k = \frac{\sqrt{2\,m\,(E - V)}}{\hbar}$$

where the wavefunction solutions are of the form

$$\psi = A\,e^{i\,k\,x} \qquad \text{where A, k, and } \psi \text{ may be } \underline{\text{complex}}$$

In the following sections, we show how *Mathematica* may be used in quantum computations to match wavefunctions and their derivatives -- and find the probability of a process occuring. We will consider three examples: the Quantum Step, the Quantum Barrier, and the Quantum Well. Then we will utilize the *Mathematica* program with the *Shooting Method* to find the energy eigenstates and the wavefunctions for the Quantum Oscillator and the Hydrogen Atom.

4.4 A Quantum Step

A 7-eV electron encounters a 5-eV potential step. What is the probability that the electron is transmitted? What is the probability that it is reflected?

```
In[9]:= Plot[{7, 5 * UnitStep[x - 1]}, {x, -5, 5}, PlotRange → {0, 10.1},
           Prolog → {Text["7 eV", {4.2, 7.5}],
           Text["5 eV", {4.6, 4.2}], Text["Region 1", {-3.4, 1.5}],
           Text["Region 2", {2.8, 1.5}]}]
```

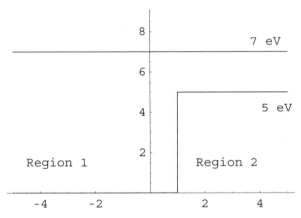

Potential Step

Classically, we would expect every electron encountering the potential step to be transmitted, albeit at a lower kinetic energy. However, Quantum Physics says that there is a probability that the electron will be reflected.

We are looking for solutions of the form

$$\psi_1 = e^{ikx} + r e^{-ikx} \qquad \text{where} \quad k = \frac{\sqrt{2\,m\,E}}{\hbar}$$

$$\psi_2 = t\, e^{i\alpha x} \qquad \text{where} \quad \alpha = \frac{\sqrt{2\,m\,(E-V)}}{\hbar}$$

$$k = \sqrt{2 * 9.1 * 10^{-31} * 7 * 1.6 * 10^{-19}} \Big/ (6.626 * 10^{-34} / (2\,\pi)) \Big/ 10^{10}$$

$$\alpha = \sqrt{2 * 9.1 * 10^{-31} * 2 * 1.6 * 10^{-19}} \Big/ (6.626 * 10^{-34} / (2\,\pi)) \Big/ 10^{10}$$

```
Out[10]= 1.35386
```

```
Out[11]= 0.723668
```

```
In[10]:=  Clear[r, t]; k = 1.3538; α = .7236;
          sol = NSolve[{1 + r == t,  k * (1 - r) == α * t}, {r, t}]
          R = Abs[r]^2 /. sol
          T = (α / k) * Abs[t]^2 /. sol
          psi = Re[Exp[I * k * x] + r * Exp[-I * k * x]] /. sol;
          bar = Re[t * Exp[I * α * x]] /. sol;
          Plot[psi * If[x < 0, 1, 0] + bar * (UnitStep[x]), {x, -3.5, 13}]
```

Out[12]= {{r → 0.30336, t → 1.30336}}

Out[13]= {R=0.0920273}

Out[14]= {T=0.907973}

We find, using *Mathematica*, that 9.2% of the incident electron beam is reflected, and 90.8% is transmitted.

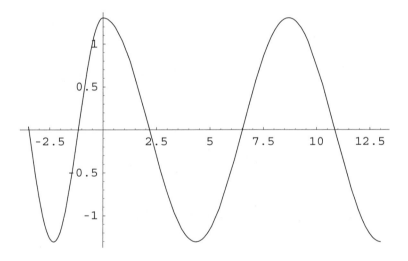

The real part of the wavefunction

The wavefunction of the transmitted wave is shown proceeding to the right with longer wavelength (less kinetic energy). The wavefunction to the left consists of both the incident and reflected wave.

4.4 B   Quantum Barrier

A 3-eV electron encounters a 5-eV potential barrier. The barrier is 2 Angstroms wide. What is the probability the electron tunnels through?

```
In[219]:=  Plot[{3, 5 * UnitStep[x - 1.1] - 5 * UnitStep[x - 3.1]},
           {x, -3, 4}, PlotRange → {0, 8.1},
           Prolog → {Text["3 eV", {3.8, 3.4}], Text["5 eV", {3.6, 5}]}]
```

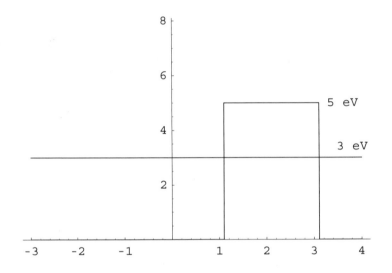

The Quantum wavefunctions in the three regions are :

$$\psi_1 = e^{ikx} + r e^{-ikx} \qquad \text{where } k = \frac{\sqrt{2 m E}}{\hbar}$$

$$\psi_2 = a e^{\alpha x} + b e^{-\alpha x} \qquad \text{where } \alpha = \frac{\sqrt{2 m (V - E)}}{\hbar}$$

$$\psi_3 = t e^{ikx}$$

$$k = \sqrt{2 * 9.1 * 10^{-31} * 3 * 1.6 * 10^{-19}} \Big/ (6.626 * 10^{-34} / (2 \pi)) \Big/ 10^{10}$$

$$\alpha = \sqrt{2 * 9.1 * 10^{-31} * 2 * 1.6 * 10^{-19}} \Big/ (6.626 * 10^{-34} / (2 \pi)) \Big/ 10^{10}$$

```
Out[63]= 0.886308
```

```
Out[64]= 0.723668
```

```
In[204]:=  Clear[a, b, r, t];
           k = .886; α = .724; L = 2;
           sol = NSolve[
             {1 + r == a + b,  a * Exp[α * L] + b * Exp[-α * L] == t * Exp[I * k * L],
              I * k - I * k * r == α * a - α * b,  α * a * Exp[α * L] - α * b * Exp[-α * L] ==
              I * k * t * Exp[I * k * L]}, {a, b, r, t}]
           R = Abs[r]^2 /. sol
           T = Abs[t]^2 /. sol
           psi = Re[Exp[I * k * x] + r * Exp[-I * k * x]] /. sol;
           bar = Re[a * Exp[α * x] + b * Exp[-α * x]] /. sol;
           tran = Re[t * Exp[I * k * x]] /. sol;
           Plot[
            psi * If[x < 0, 1, 0] + bar * (UnitStep[x] - UnitStep[x - 2]) +
            tran * If[x > 2, 1, 0], {x, -10, 10}]

Out[206]=  {{a → 0.039386 + 0.0712345 i,  b → 1.12157 - 0.955549 i,
              r → 0.160958 - 0.884315 i,  t → -0.00927323 - 0.438172 i}}

Out[207]=  {R=0.80792}

Out[208]=  {T=0.19208}
```

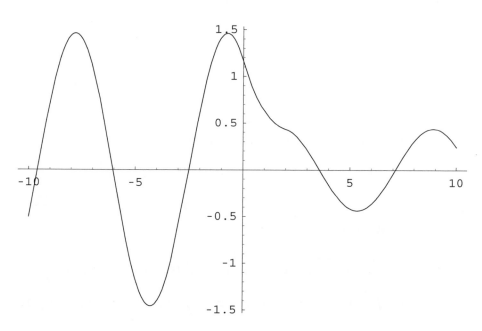

The real part of the wavefunction. 19% of the incident electron beam is transmitted. Classically, we would expect <u>none</u> of the electron beam to be transmitted.

4.4 C Quantum Well

A 3-eV electron encounters an 8-eV potential well. The well is 3 Angstroms wide. What is the probability of transmission? What is the probability the electron is reflected?

```
Plot[{3, -8*UnitStep[x - 1.1] + 8*UnitStep[x - 3.1]},
   {x, -3, 4}, PlotRange → {-8.3, 4.1}, Axes → None,
   Prolog → {Text["3 eV", {3.8, 3.4}], Text["-8 eV", {3.6, -7.5}]}]
```

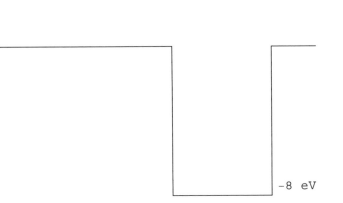

The Quantum wavefunctions in the three regions are :

$$\psi_1 = e^{ikx} + r\,e^{-ikx} \qquad \text{where } k = \frac{\sqrt{2\,m\,E}}{\hbar}$$

$$\psi_2 = a\,e^{i\alpha x} + b\,e^{-i\alpha x} \qquad \text{where } \alpha = \frac{\sqrt{2\,m\,(E + |V|)}}{\hbar}$$

$$\psi_3 = t\,e^{ikx}$$

$In[16] :=$ $k = \sqrt{2*9.1*10^{-31}*3*1.6*10^{-19}} \bigg/ (6.626*10^{-34}/(2\pi)) \bigg/ 10^{10}$

$\alpha = \sqrt{2*9.1*10^{-31}*11*1.6*10^{-19}} \bigg/ (6.626*10^{-34}/(2\pi)) \bigg/ 10^{10}$

$Out[16] =$ 0.886308

$Out[17] =$ 1.69715

We may use *Mathematica* to match the wavefunctions and their derivatives at $x = 0$ and $x = L$

```
Clear[a, b, r, t];
 k = .886; α = 1.697; L = 3;
sol = NSolve[
   {1 + r == a + b, a * Exp[I * α * L] + b * Exp[-I * α * L] == t * Exp[I * k * L],
    k - k * r == α * a - α * b, α * a * Exp[I * α * L] - α * b * Exp[-I * α * L] ==
     k * t * Exp[I * k * L]}, {a, b, r, t}]
R = Abs[r]^2 /. sol
T = Abs[t]^2 /. sol
psi = Re[Exp[I * k * x] + r * Exp[-I * k * x]] /. sol;
bar = Re[a * Exp[I * α * x] + b * Exp[-I * α * x]] /. sol;
tran = Re[t * Exp[I * k * x]] /. sol;
Plot[psi * If[x < 0, 1, 0] + bar * (UnitStep[x] - UnitStep[x - 3]) +
   tran * If[x > 3, 1, 0], {x, -10, 10}]
```

Out [192]=
$$\{\{a \to 0.637615 - 0.0402901\,i,\ b \to -0.154181 - 0.128322\,i,$$
$$r \to -0.516566 - 0.168612\,i,\ t \to -0.60168 + 0.585415\,i\}\}$$

Out [193]= $\{R = 0.295271\}$

Out [194]= $\{T = 0.704729\}$

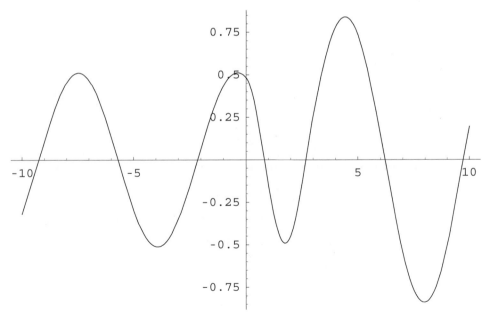

The real part of the wavefunction. 70% of the incident electron beam is transmitted. Classically, we would expect <u>all</u> of the electron beam to be transmitted.

4.4 D Quantum Oscillator

A particle of mass m is inside a potential well of depth $V = \frac{1}{2}kx^2$. How do we find the allowed quantum energy levels ?

```
Plot [ x² / 2, {x, -5, 5}, Prolog → {Text["V= 1/2 kx²", {3.4, 11.4}],
    {GrayLevel[0.5], Line[{{-2.8, 1}, {2.7, 1}}],
    Line[{{-3.68, 3}, {3.56, 3}}], Line[{{-4.62, 5}, {4.4, 5}}]}}]
```

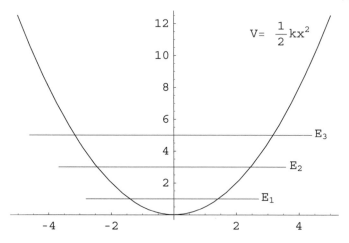

Discrete energy levels in a harmonic oscillator potential

Let's apply the one-dimensional Schrodinger equation to find the allowed energy levels, $E_1, E_2, E_3, ...$

$$\frac{-\hbar^2}{2m}\psi'' + V\psi = E\psi$$

for a particle of mass m in a harmonic-oscillator potential $V(x) = \frac{1}{2}kx^2$

$$\psi'' = -\frac{2m}{\hbar^2}(E - \frac{1}{2}kx^2)\psi$$

$$\psi'' = -\frac{2m}{\hbar^2}(E - m\omega^2\frac{x^2}{2})\psi \qquad \text{setting } \omega = \sqrt{k/m}$$

$$\psi'' = -(\frac{2m}{\hbar^2}E - \frac{m^2\omega^2}{\hbar^2}x^2)\psi$$

$$\text{Let } \beta = \frac{2m}{\hbar^2}E \quad \text{and} \quad \alpha = \frac{m\omega}{\hbar}$$

$$\psi'' = -(\beta - \alpha^2 x^2)\psi$$

$$\text{Then if } \epsilon = \frac{\beta}{\alpha} \text{ and } z = \sqrt{\alpha}\, x,$$

$$\frac{d^2\psi}{dz^2} = -(\epsilon - z^2)\psi$$

This is the equation we wish to solve using *Mathematica*.

Notice that we have chosen $\epsilon = \dfrac{\beta}{\alpha} = \dfrac{2\,E}{\hbar\omega} = \dfrac{2\,E}{h\,\nu}$ where ν is the frequency of oscillation. So once we determine the *eigenvalues* for ϵ, then we have the allowable energies for the particle in the well. That is, there are only certain values of ϵ that will allow ψ to go to zero as $z \to \pm\infty$.

Note also that in the harmonic oscillator equation, $\dfrac{d^2 \psi}{dz^2} = -\,(\epsilon - z^2)\,\psi$ that as $z \to 0$ then $\psi'' = -\epsilon\,\psi$ with solutions of cosines and sines. Therefore, for starting values of ψ choose $\psi(0) = 1$ and $\psi'(0) = 0$ <u>or</u> $\psi(0) = 0$ and $\psi'(0) = 1$. (We will normalize ψ later).

<u>The Shooting Method for solving Schrodinger's Equation</u>
The strategy we shall follow in finding the energy levels and acceptable wavefunctions is as follows:
We shall start with some initial value of ϵ with $\psi(0) = 1$ and $\psi'(0) = 0$ and then observe whether the wavefunction goes to zero as $z \to \infty$. This is the criterion for a particle to be bound in a quantum well.
As an example, let's start with $\epsilon = 1$

```
In[11]:=  Clear[y, z];  ε = 1;
          sol1 =
          DSolve[{y"[z] == -(ε - z^2) y[z], y[0] == 1, y'[0] == 0}, y[z], z]
          Plot[Evaluate[y[z] /. sol1, {z, -5, 5}]];
```

$$Out[12] = \left\{\left\{ y[z] \to e^{-\frac{z^2}{2}} \right\}\right\}$$

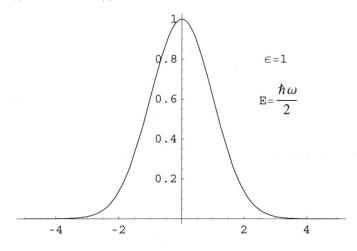

$\epsilon = 1$

$E = \dfrac{\hbar\omega}{2}$

An acceptable wavefunction is found for $\epsilon = 1$
We now normalize the wavefunction using the substitution $z = \sqrt{\alpha}\,x$

$In[7]:=$ **Assuming$\left[\text{Re}[\alpha] > 0, \text{Solve}\left[A^2 \int_{-\infty}^{\infty} \text{Exp}[-\alpha * x^2] \, dx \, == 1, A\right]\right]$**

$Out[7]=$ $\left\{\left\{A \rightarrow -\dfrac{\alpha^{1/4}}{\pi^{1/4}}\right\}, \left\{A \rightarrow \dfrac{\alpha^{1/4}}{\pi^{1/4}}\right\}\right\}$

Therefore, $\psi_1(x) = \dfrac{\alpha^{1/4}}{\pi^{1/4}} e^{-\alpha x^2/2}$ and $E_1 = 1\,\dfrac{\hbar\omega}{2}$

The next eigenvalue occurs at $\epsilon = 3$,

$In[1]:=$ **Clear[y, z]; ϵ = 3;**
        sol2 =
        DSolve[{y″[z] == -(3 - z^2) y[z], y[0] == 0, y′[0] == 1}, y[z], z]
        Plot$\left[\text{Evaluate}[y[z] \, /. \, \text{sol2}, \{z, -5, 5\}]\right.$,

        Prolog → $\left\{\text{Text}["\epsilon=3", \{3, .58\}], \text{Text}\left["E=\dfrac{3\hbar\omega}{2}\,", \{3.1, .4\}\right]\right\}$];

$Out[2]=$ $\left\{\left\{y[z] \rightarrow e^{-\frac{z^2}{2}} z\right\}\right\}$

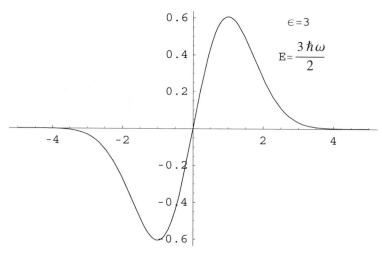

The wavefunction for $\epsilon = 3$.

$In[9]:=$ **Assuming$\left[\text{Re}[\alpha] > 0, \text{Solve}\left[A^2 \int_{-\infty}^{\infty} \alpha * x^2 * \text{Exp}[-\alpha * x^2] \, dx \, == 1, A\right]\right]$**

$Out[9]=$ $\left\{\left\{A \rightarrow -\dfrac{\sqrt{2}\,\alpha^{1/4}}{\pi^{1/4}}\right\}, \left\{A \rightarrow \dfrac{\sqrt{2}\,\alpha^{1/4}}{\pi^{1/4}}\right\}\right\}$

Therefore, $\psi_2(x) = \sqrt{2}\,\dfrac{\alpha^{1/4}}{\pi^{1/4}}\,\sqrt{\alpha}\,x\,e^{-\alpha x^2/2}$ $E_2 = 3\,\dfrac{\hbar\omega}{2}$

The next eigenvalue occurs at $\epsilon = 5$,

```
ε = 5;
sol3 = DSolve[{y″[z] == -(5 - z^2) y[z], y[0] == 1, y′[0] == 0}, y[z], z]
Plot[Evaluate[y[z] /. sol3, {z, -5, 5}]];
```

$Out[5]= \left\{\left\{y[z] \rightarrow -e^{-\frac{z^2}{2}}\left(-1 + 2\, z^2\right)\right\}\right\}$

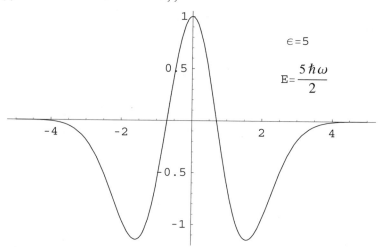

The wavefunction for $\epsilon = 5$.

$In[8]:=$ $\mathbf{Assuming}\left[\mathbf{Re}[\alpha] > 0,\right.$

$\qquad \mathbf{Solve}\left[\,A^2 \displaystyle\int_{-\infty}^{\infty} \mathbf{Exp}[-\alpha * x^2] * (1 - 2\,\alpha * x^2)^2\ dx\ ==\ 1,\ A\right]\Big]$

$Out[8]= \left\{\left\{A \rightarrow -\dfrac{\alpha^{1/4}}{\sqrt{2}\ \pi^{1/4}}\right\},\ \left\{A \rightarrow \dfrac{\alpha^{1/4}}{\sqrt{2}\ \pi^{1/4}}\right\}\right\}$

$$\psi_3(x)\ =\ \frac{\alpha^{1/4}}{\sqrt{2}\ \pi^{1/4}}\ (1 - 2\,\alpha x^2)\ e^{-\alpha x^2/2} \qquad\qquad E_3\ =\ 5\ \frac{\hbar\omega}{2}$$

Therefore the wavefunctions for the first three energy levels are

$$\psi_1(x) = \frac{\alpha^{1/4}}{\pi^{1/4}}\, e^{-\alpha x^2/2} \qquad\qquad E_1 = 1\ \frac{\hbar\omega}{2}$$

$$\psi_2(x) = \sqrt{2}\ \frac{\alpha^{1/4}}{\pi^{1/4}}\,\sqrt{\alpha}\ x\, e^{-\alpha x^2/2} \qquad\qquad E_2 = 3\ \frac{\hbar\omega}{2}$$

$$\psi_3(x) = \frac{\alpha^{1/4}}{(4\,\pi)^{1/4}}\,(1 - 2\alpha x^2)\, e^{-\alpha x^2/2} \qquad E_3 = 5\ \frac{\hbar\omega}{2}$$

Notice that once the energy eigenvalue ϵ is found that the wavefunctions are easy to obtain. However, how does one find the *eigenvalues* ?

The easiest way is just to examine a graph, as one gradually increases the value of ϵ. Following are the wavefunctions for $\epsilon = 0.9$ and $\epsilon = 1.1$. They both cascade off to infinity. But there is a change in direction in going from 0.9 to 1.1. Therefore ϵ is between 0.9 and 1.1

```
ε = 0.9; DSolve[{y″[z] == -(ε - z^2) y[z], y[0] == 1, y′[0] == 0}, y[z], z]
Plot[Evaluate[y[z] /. %, {z, -5, 5}]];
```

$Out[24]= \left\{\left\{ y[z] \rightarrow e^{-0.5 z^2} \left(\text{Hypergeometric1F1}\left[0.025, \frac{1}{2}, 1. z^2 \right] \right) \right\}\right\}$

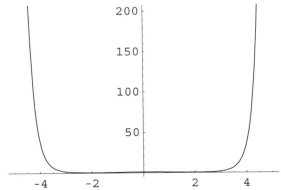

unacceptable wavefunction ψ goes to $+\infty$ for z > 4 ($\epsilon = 0.9$ is too small)

```
ε = 1.1; DSolve[{y″[z] == -(ε - z^2) y[z], y[0] == 1, y′[0] == 0}, y[z], z]
Plot[Evaluate[y[z] /. %, {z, -5, 5}]];
```

$Out[36]= \left\{\left\{ y[z] \rightarrow e^{-0.5 z^2} \left(\text{Hypergeometric1F1}\left[-0.025, \frac{1}{2}, 1. z^2 \right] \right) \right\}\right\}$

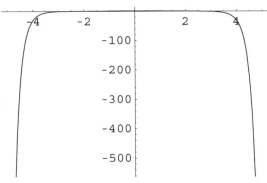

unacceptable wavefunction ψ goes to $-\infty$ for z > 4 ($\epsilon = 1.1$ is too large)

Therefore the "shooting method" allows one to rapidly home-in on the eigenvalues. In the next section, we will show how the shooting method may be used with *Mathematica* to find acceptable wavefunctions and energy levels for the hydrogen atom.

4.5 Hydrogen Atom

The Bohr Atom

In 1913, Niels Bohr formulated the quantum theory of the hydrogen atom. Bohr envisioned an electron circling the central nucleus like a planet about the sun. In a quantum formulation, only orbits with angular momentum $L = n\hbar$ would be allowed, thus only *certain energies* could be absorbed or emitted by the electron when changing orbits.

Let us now use the Bohr postulate $L = mvR = n\hbar$ to find the size of the Hydrogen atom and its energy levels. For a single electron held in orbit by the Coulomb force,

$$\frac{mv^2}{R} = \frac{Ze^2}{4\pi\epsilon_0 R^2} \quad \text{and} \quad v = \frac{Ze^2}{4\pi\epsilon_0 \hbar n}$$

Since $L = mvR = n\hbar$,

$$R = \frac{n\hbar}{mv} = \frac{n^2\hbar^2 4\pi\epsilon_0}{Zme^2} = n^2 \frac{a_0}{Z}$$

where

$$a_0 = \frac{\hbar^2 4\pi\epsilon_0}{me^2} = .529177 \text{ Å} \quad \text{(the Bohr radius)}$$

Thus, for Hydrogen with $Z = 1$,

$$E_n = \frac{1}{2}mv^2 - \frac{Ze^2}{4\pi\epsilon_0 R^2} = -\frac{Z^2 e^2}{8\pi\epsilon_0 a_0 n^2} = \frac{-13.6057 \text{ eV}}{n^2}$$

The Schrodinger Equation for the Hydrogen Atom

In 1926, Erwin Schrodinger solved the following equation for the *radial wavefunction* $y[r]$

$$\frac{-\hbar^2}{2m}\left[\frac{2}{r}y'[r] + y''[r]\right] + V[r]y[r] = E\ y[r]$$

where the potential is $V[r] = \frac{-Ze^2}{4\pi\epsilon_0 r} + \frac{\hbar^2}{2m}\frac{\ell(\ell+1)}{r^2}$

to start, we shall choose $\ell = 0$ and find the wavefunctions, energy states, and the *average* radius, for $n = 1, 2, 3$ and compare the results to Bohr.

We use the following values for the fundamental constants

$\hbar = 1.054572 \times 10^{-34}$ J· s $m_e = 9.109389 \times 10^{-31}$ kg $\frac{\hbar^2}{2m} = 3.80998$ eV ·A^2

$\epsilon_0 = 8.854188 \times 10^{-12}$ C^2/ N· m^2 $e = 1.602177 \times 10^{-19}$ C $\frac{e^2}{4\pi\epsilon_0} = 14.3996$ eV · A

We write Schrodinger's equation as

$$\frac{2}{r}y'[r] + y''[r] = \frac{-2m}{\hbar^2}[E - V[r]]\,y[r]$$

$$\frac{2}{r}y'[r] + y''[r] = \frac{-2m}{\hbar^2}[E + \frac{Ze^2}{4\pi\epsilon_0 r}]\,y[r]$$

or, with energy E in eV and distance r in Angstroms

$$\frac{2}{r}y'[r] + y''[r] = -.262468[E + \frac{14.3996}{r}]\,y[r]$$

This is the equation we will solve with *Mathematica*.

First, however, we need to define the *boundary conditions*. In Schrodinger's theory for a zero angular momentum state $\ell = 0$, the wavefunction will have some value as $r \to 0$, and to be a bound state, the wavefunction will go to zero as $r \to \infty$. We will arbitrarily pick y[.0001] = 1 and y'[30] = 0. In this way, we avoid the infinity at $r = 0$, and we also establish an *effective infinity* at large r. We will now choose some starting value of E (say, –20 eV) and increase the value of E until the wavefunction approaches 1 as $r \to 0$.

Now, with the wavefunction well-behaved, we may relax the inner boundary condition and choose y[0] = 1 and y'[30] = 0 at that energy *eigenvalue,* to get an ultra-precise value of the wavefunction.

Let's see how this plays out in practice: (refer to the graphics on next page) Say we choose E = –13.60 eV, the wavefunction heads to some extreme negative value as $r \to 0$
Say we choose E = –13.61 eV, the wavefunction goes to some extreme positive value as $r \to 0$
(Therefore the true energy eigenvalue will be somewhere midway between these two values.)

```
sol1 =
 DSolve[{2*y'[r] + r*y''[r] == -.262468 * (-13.60 r + 14.3996) *y[r],
    y[0.0001] == 1, y'[30] == 0}, y[r], r]
Plot[Evaluate[y[r] /. sol1, {r, .000000001, .000001}]];
```

Out[81]= $\{\{y[r] \rightarrow e^{-1.88933\,r}$
$\qquad (2.19781 \text{ HypergeometricU}[-0.000205318, 2, 3.77866\,r] -$
$\qquad 8.23596 \times 10^{-42} \text{ LaguerreL}[0.000205318, 1, 3.77866\,r])\}\}$

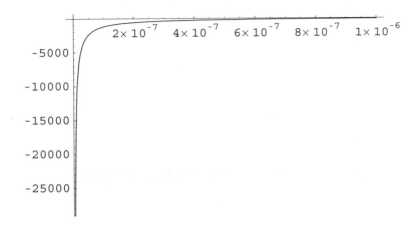

unacceptable behavior of the wavefunction $\psi \rightarrow -\infty$ **as** $r \rightarrow 0$ **E = -13.60 eV**

```
sol1 =
 DSolve[{2 * y'[r] + r * y''[r] == -.262468 * (-13.61 r + 14.3996) * y[r],
   y[0.0001] == 1, y'[30] == 0}, y[r], r]
Plot[Evaluate[y[r] /. sol1, {r, .000000001, .000001}]];
```

Out[116]=
$\{\{y[r] \rightarrow$
$\qquad e^{-1.89002\,r} (0.699228 \text{ HypergeometricU}[0.000162202, 2, 3.78005\,r] +$
$\qquad 3.17319 \times 10^{-42} \text{ LaguerreL}[-0.000162202, 1, 3.78005\,r])\}\}$

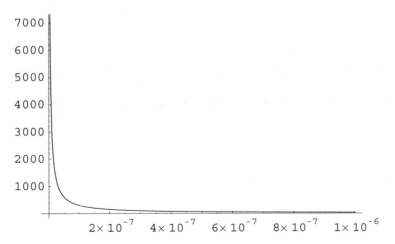

unacceptable behavior of the wavefunction $\psi \rightarrow +\infty$ **as** $r \rightarrow 0$ **E = -13.61 eV**

The Shooting method as $r \to 0$

The above two graphs tell us that the proper eigenvalue is between −13.60 and −13.61 eV. We may proceed by the *bisection method* to get an ever more accurate value. When we get close to an energy *eigenvalue*, then the graph of the wavefunction is well-behaved and y goes to a value near 1 as $r \to 0$.

On magnification, the slope is a straight-line.

```
sol1 =
 DSolve[{2 * y'[r] + r * y"[r] == -.262468 * (-13.605585 r + 14.3996) * y[r],
   y[0.0001] == 1, y'[30] == 0}, y[r], r]
Plot[Evaluate[y[r] /. sol1, {r, .001, 3}]];
Plot[Evaluate[y[r] /. sol1, {r, .000000001, .000001}]];
```

Out [18]= $\{\{y[r] \to e^{-1.88972\,r} (1.00019 -$
9.17501×10^{-38} LaguerreL$[8.18262 \times 10^{-9}, 1, 3.77943\,r])\}\}$

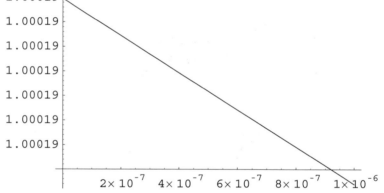

Acceptable behavior of the wavefunction as r → 0

After several trials, we find the energy eigenvalue E1 = − 13.605585 eV.

We will now run the *Mathematica* program one more time *this time with the proper initial conditions* y[0] = 1 and y′[30] = 0.

```
sol1 = DSolve[
    {2 * y'[r] + r * y"[r] == -.262468 * (-13.605585 r + 14.3996) * y[r],
     y[0] == 1, y'[30] == 0}, y[r], r] // Chop
Plot[Evaluate[y[r] /. sol1, {r, .001, 3}],
    Prolog → {Text["n=1, ℓ=0", {1.8, .56}]}];
Plot[Evaluate[r² * y[r]^2 /. sol1, {r, .001, 3}],
    Prolog → {Text["n=1, ℓ=0", {1.8, .026}]}];
```

Out[76]= $\{\{y[r] \rightarrow 1.\, e^{-1.88972\, r}\}\}$

We may now graph the wavefunction for $n = 1, \ell = 0$

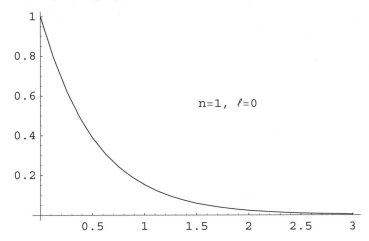

We may also evaluate $r^2 y^2[\, r\,]$ to find out where the probability is greatest of finding the electron

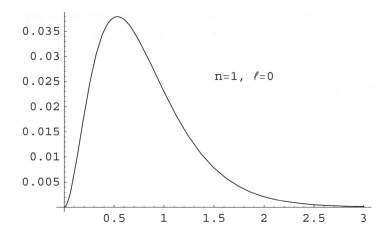

```
Table[{r, r² * y[r]^2 /. sol1}, {r, .51, .54, .001}] // TableForm
```

r (A)	$r^2 y^2 [r]$
0.51	0.0378471
0.511	0.0378523
0.512	0.0378573
0.513	0.037862
0.514	0.0378663
0.515	0.0378704
0.516	0.0378742
0.517	0.0378777
0.518	0.0378809
0.519	0.0378839
0.52	0.0378866
0.521	0.037889
0.522	0.0378911
0.523	0.0378929
0.524	0.0378944
0.525	0.0378957
0.526	0.0378967
0.527	0.0378975
0.528	0.0378979
0.529	0.0378981
0.53	0.037898
0.531	0.0378977
0.532	0.037897
0.533	0.0378961
0.534	0.037895
0.535	0.0378936
0.536	0.0378919
0.537	0.0378899
0.538	0.0378877
0.539	0.0378852
0.54	0.0378825

From the above table of values, we see that the maximum value of the *Probability function* $r^2 y^2$ occurs at r = 0.529 A or, more precisely, we maximize $r^2 (e^{-1.88972\, r})^2$

In[83]:= **Maximize[{r² Exp[-1.88972 r]^2, 0 < r < 10}, r]**

Out[83]= {0.037898, {r → 0.529179}}

which is the Bohr radius.

We will now find wavefunctions for n=2, ℓ=0 and n=3, ℓ=0 and n=2, ℓ=1.
Using the Shooting Method as before,

```
sol1 = DSolve[
    {2/r *y'[r] +y"[r] == -.262468* (-3.4013963256 + 14.3996/r) *y[r],
    y[0.0001] == 1, y'[30] == 0}, y[r], r] // Chop
Plot[Evaluate[y[r] /. sol1, {r, .001, 7}], PlotRange → {-.15, 1.00},
    Prolog → {Text["n=2, ℓ=0", {2.3, .5}]}];
Plot[Evaluate[r² *y[r]^2 /. sol1, {r, .001, 7}],
    Prolog → {Text["n=2, ℓ=0", {5.4, .054}]}];
```

Out[179]= $\{\{y[r] \rightarrow e^{-0.944859\,r}\ (1.00019 - 0.945037\,r)\}\}$

We may now graph the wavefunction for n=2, ℓ=0

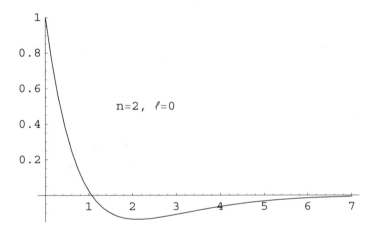

We evaluate $r^2y^2[\,r\,]$ to find where the probability is greatest of finding
the electron

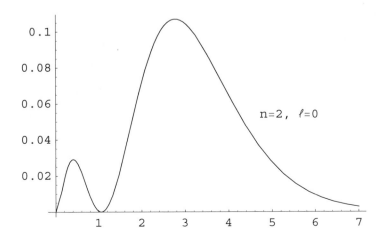

```
Clear[r, y];
sol1 = DSolve[
    {2 * y'[r] + r * y"[r] == -.262468 * (-1.51173167895 r + 14.3996) * y[r],
    y[0] == 1, y'[50] == 0}, y[r], r] // Chop
Plot[Evaluate[y[r] /. sol1, {r, .001, 12}], PlotRange → {-.15, 1.00},
    Prolog → {Text["n=3, ℓ=0", {3.3, .5}]}];
Plot[Evaluate[r² * y[r]^2 /. sol1, {r, .001, 12}],
    Prolog → {Text["n=3, ℓ=0", {3.2, .11}]}];
```

Out[201]= $\{\{y[r] \to e^{-0.629906\,r}\,(1. - 1.25981\,r + 0.264521\,r^2)\}\}$

The wavefunction for n=3, ℓ=0

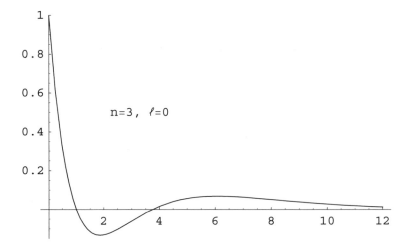

n=3, ℓ=0

The probability function $r^2 y^2[\,r\,]$ for n=3, ℓ=0

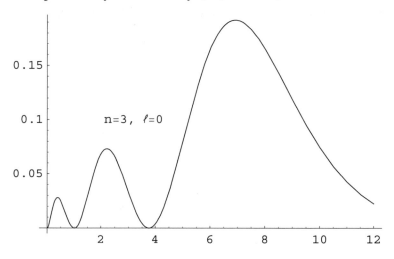

n=3, ℓ=0

For n = 2, ℓ = 1, we need to revisit the original Schrodinger equation for the hydrogen atom.

$$\frac{2}{r}y'[r] + y''[r] = -.262468\,[\,E + \frac{14.3996}{r}\,]\,y[\,r\,] + \frac{\ell\,(\ell+1)}{r^2}y[\,r\,]$$

We will solve this equation using *Mathematica* with the proviso that for any $\ell > 0$ angular momentum state that $y[\,r\,] \to 0$ as $r \to 0$. It is amazing, but true, that the energy eigenvalue for the n =2, ℓ=1 state is the same as the n =2, ℓ=0 state. We show this in the expanded graphic where the wavefunction y approaches zero as $r \to 0$.

```
DSolve[{ 2/r * y'[r] + y''[r] == -.262468 * (-3.4013963256 + 14.3996/r) * y[r]
       + 1 (1 + 1)/r² * y[r], y[0.0001] == 0.00001, y'[50] == 0}, y[r], r] // Chop
Plot[Evaluate[y[r] /. %, {r, .000000001, .000001}]];
```

Out[7]= $\{\{y[r] \to 0.100009\,e^{-0.944859\,r}\,r^{1.}\}\}$

We may now evaluate the wavefunction and the probability function for n =2, ℓ=1

```
sol1 = DSolve[{ 2/r * y'[r] + y''[r] ==
       -.262468 * (-3.4013963256 + 14.3996/r) * y[r] + 1 (1 + 1)/r² * y[r],
       y[0.000001] == 0.0000001, y'[50] == 0}, y[r], r] // Chop
Plot[Evaluate[y[r] /. sol1, {r, .001, 7}],
   Prolog → {Text["n=2, ℓ=1", {4.8, .022}]}];
Plot[Evaluate[r² * y[r] ^2 /. sol1, {r, .001, 7}],
   Prolog → {Text["n=2, ℓ=1", {5, .0017}]}];
```

Out[33]= $\{\{y[r] \to 0.1\, e^{-0.944859\, r}\, r^{1.}\}\}$

The wavefunction for n=2, ℓ=1

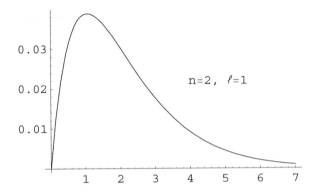

n=2, ℓ=1

The probability function $r^2 y^2[\,r\,]$ for n=2, ℓ=1

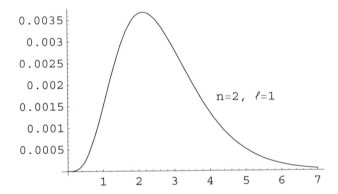

n=2, ℓ=1

It is now time to normalize all the wavefunctions

For n=1, ℓ=0, y[r] = A $e^{-1.88972\, r}$
For n=2, ℓ=0, y[r] = A $e^{-0.944859\, r}$ (1. $-$ 0.944859 r)
For n=2, ℓ=1 y[r] = A $e^{-0.944859\, r}$ r
For n=3, ℓ=0 y[r] = A $e^{-0.629906\, r}$ (1. $-$ 1.25981 r + 0.264521 r^2) }}

(*n=1,1=0*) Solve$\left[A^2 \int_0^\infty r^2 * \text{Exp}[-1.88972*2*r]\ dr == 1, A\right]$

(*n=2,1=0*) Solve$\left[A^2 \int_0^\infty r^2 * (1 - .944859\, r)^2\, \text{Exp}[-.944859*2*r]\ dr == 1, A\right]$

(*n=2,1=1*) Solve$\left[A^2 \int_0^\infty r^2 * r^2\, \text{Exp}[-.944859*2*r]\ dr == 1, A\right]$

(*n=3,1=0*)
Solve$\left[A^2 \int_0^\infty r^2 * (1 - 1.25981\, r + .264521\, r^2)^2\, \text{Exp}[-.629906*2*r]\ dr == 1, A\right]$

Out[1]= {{A $\to$ 5.19549}}   {A $\to$ 1.83688}   {A $\to$ 1.00204}   {A $\to$ 0.9998689}}

To summarize, the first four radial wavefunctions for the hydrogen atom, and their energy levels in terms of the Bohr radius $a_0 = 0.529179$ Å are

$n=1, \ell=0$ $y_{10}[r] = 5.19549$ e^{-r/a_0} $E_1 = -13.6056$ eV

$n=2, \ell=0$ $y_{20}[r] = 1.83688$ $(1 - \frac{r}{2a_0})$ $e^{-r/2a_0}$ $E_2 = -3.4014$ eV

$n=2, \ell=1$ $y_{21}[r] = 1.00204$ r $e^{-r/2a_0}$ $E_2 = -3.4014$ eV

$n=3, \ell=0$ $y_{30}[r] = 0.99987$ $(1 - \frac{2r}{3a_0} + \frac{2r^2}{27a_0^2})$ $e^{-r/3a_0}$ $E_3 = -1.51173$ eV

It is worth noting that the Niels Bohr theory, when corrected for the finite mass of the nucleus, and the relativistic mass increase of the electron ($v = .0073$ c in the ground state), agrees with spectroscopic data to 3 parts in 100,000. The Bohr theory was ground-breaking for its time because it showed that atoms could only be described in quantum terms.

However, <u>the Bohr theory only works for hydrogen and singly-ionized helium atoms</u>. Schrodinger's theory predicts exactly the same results as Bohr for hydrogen, and when the Dirac relativistic theory of the electron is applied, corresponds exactly with the hydrogen spectrum.

The value of the Schrodinger theory is that it may be applied to all quantum-mechanical systems.

A question that we should answer here is how well the above "Shooting Method" compares with the exact quantum mechanical solution for the Hydrogen atom.

As given in Leonard Schiff's <u>Quantum Mechanics</u> (1968) or in Harald Enge's <u>Introduction to Atomic Physics</u> (1972) the exact quantum mechanical wavefunctions for hydrogen, which we evaluate with $Z = 1$ and $a_0 = 0.529179$ Å are

$n=1, \ell=0$ $y_{10}[r] = \left(\frac{Z}{2a_0}\right)^{3/2} e^{-Zr/a_0}$ $= 5.19549 \, e^{-r/a_0}$

$n=2, \ell=0$ $y_{20}[r] = \left(\frac{Z}{2a_0}\right)^{3/2} 2\,(1 - \frac{Zr}{2a_0}) \, e^{-Zr/2a_0} = 1.83688 \,(1 - \frac{r}{2a_0}) \, e^{-r/2a_0}$

$n=2, \ell=1$ $y_{21}[r] = \left(\frac{Z}{2a_0}\right)^{3/2} \frac{Zr}{\sqrt{3}\,a_0} \, e^{-Zr/2a_0}$ $= 1.00205 \, r \, e^{-r/2a_0}$

$n=3, \ell=0$ $y_{30}[r] = \left(\frac{Z}{3a_0}\right)^{3/2} 2\,(1 - \frac{2Zr}{3a_0} + \frac{2Z^2 r^2}{27a_0^2}) \, e^{-Zr/3a_0}$

$\qquad\qquad\qquad = 0.99987 \,(1 - \frac{2r}{3a_0} + \frac{2r^2}{27a_0^2}) e^{-r/3a_0}$

The exact wavefunctions and those derived via the Shooting Method correspond to 6-decimal places. Any very small inaccuracy is due to our choice of an "effective infinity" for the wavefunction to go to zero at large r. Thus the choice of the Shooting Method will allow evaluation of the energy eigenvalues and the wavefunctions for potentials where Schrodinger's equation cannot be solved analytically.

4.6   Deuterium

Deuterium is the simplest nuclear system.   Here we have one proton bound to one neutron to form the nucleus of heavy hydrogen. Let us assume the attractive potential between the two particles is 40 MeV acting to some distance a. Let us find the wavefunction for Deuterium, given that the energy needed to break apart (photodissociate) this nucleus is only 2.225 MeV.

```
Plot[{.566738 * Sin[.953 r] * If[r < 1.898, 1, 0] +
   .85384 * Exp[-.232 r] * UnitStep[r - 1.9],
  -.4 * UnitStep[r] + .4 * UnitStep[r - 1.9]}, {r, 0, 4},
  PlotRange → {-0.42, 0.58}, PlotStyle → {GrayLevel[0], GrayLevel[0.45]}]
```

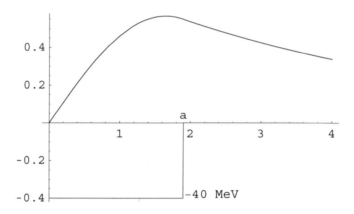

The wavefunction for Deuterium is barely anchored in the 40-MeV well.

Schrodinger's wave equation in center-of-mass coordinates is

$$\frac{-\hbar^2}{2\,m}\,\psi'' + V(r)\psi = E\psi$$

with solution $\psi = A\,\text{Sin}\,kr$ for $0 < r < a$

$\psi = B\,e^{-\alpha r}$ for $r \geq a$

Using the reduced mass for *m* and the binding energy –2.225 MeV for E,

$$k = \sqrt{1.673 * 10^{-27} * (40 - 2.225) * 1.6 * 10^{-13}} \Big/ (6.626 * 10^{-34} / (2\pi)) \Big/ 10^{15}$$

$$\alpha = \sqrt{1.673 * 10^{-27} * 2.225 * 1.6 * 10^{-13}} \Big/ (6.626 * 10^{-34} / (2\pi)) \Big/ 10^{15}$$

```
     k=0.953539   α=0.23142   in units of Fm⁻¹
```

We may now find the range of the 40 MeV attractive potential that produces –2.225 MeV binding energy. Match the wavefunction components and their derivatives at $r = a$.

$$A \sin ka = B \, e^{-\alpha a} \qquad \text{and}$$

$$kA \cos ka = -\alpha B \, e^{-\alpha a} \qquad \text{at } r = a$$

Dividing the first equation by the second gives us an equation in **a**

```
In[7]:=  FindRoot[Tan[k * a] == - k / α, {a, 2, 1, 3}]

Out[7]= {a → 1.89873}
```

We may now find **A** and **B** from the normalization condition and the continuity of the wavefunction at **a**

```
In[6]:=  a = 1.898; k = .953; α = .231;

        FindRoot[{1 == A² ∫₀ᵃ (Sin[k * r])² dr + B² ∫ₐ^∞ Exp[-2 α * r] dr,

        A * Sin[k * a] == B * Exp[-α * a]}, {{A, .5}, {B, .5}}]

Out[8]= {A → 0.566738, B → 0.853839}
```

Let's find the wavefunction probability ψ^2 of finding the particle within the radius **a** of the attractive potential

```
In[9]:=  A = 0.566738; B = 0.853839;

        Pr1 = A² ∫₀ᵃ (Sin[k * r])² dr     Pr2 = B² ∫ₐ^∞ Exp[-2 α * r] dr

Out[10]=    Pr1 = .34342              Pr2 =  0.65658
```

This says that the Deuteron only spends 34% of the time entirely within the range of the attractive nuclear potential. The Deuteron is more of a wave than a particle.

Let's now find the *effective radius* of the Deuteron.

```
In[12]:=  rad = A² ∫₀ᵃ r * (Sin[k * r])² dr + B² ∫ₐ^∞ r * Exp[-2 α * r] dr

         Print[rad, " Fermis"]

            3.1134 Fermis
```

Which considerably exceeds the range of the attractive nuclear potential. The wavefunction for the Deuteron is barely anchored in the 40-MeV well.

4.7 Challenge Problems

1. Ramsauer Effect

An interesting quantum-mechanical effect is observed if low-velocity electrons are sent into a gas of Xenon, Argon, or Krypton. Depending on the energy of the electrons, they will not be scattered at certain energies, rather they will transmit through the gas. As a simple model, assume that surrounding each gas nucleus there is an attractive square-well potential of − 8.4 eV depth and width 2 Angstroms. What is the transmission T for 1 eV electrons?

2. Intermediate Quantum Step

We can get 100 % transmission across a potential step if we put in an intermediate quantum step of the appropriate height V and width L. For a 7-eV electron beam impinging on a 5-eV potential step, how high and how wide does the intermediate step need to be? Express the answer in eV and Angstroms.

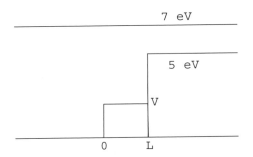

Intermediate Quantum Step

3. Energy Eigenvalues

An electron is trapped in a quantum well. If this is a square well of depth 5 eV and width 2 Angstroms, what is the ground state energy of the electron above the bottom of the well? The ground-state wavefunction for the electron will be a Cosine wave, because this allows a maximum probability at the center of the well. Notice that it is necessary to match the Cosine wave in the well with an exponentially decreasing wavefunction outside the well.

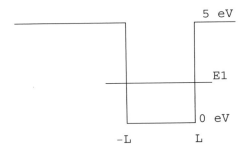

Electron in a Quantum Well

4. Cooling Concrete in Dams

The dams on the Columbia River are Grand Coulee, McNary, John Day, and Bonneville. Each one may appear to be a quiescent object holding back millions of tons of water. It is inside the dam that all the activity is: the flow of water through Grand Coulee powers generators that provide over 6000 MW of electrical power, and even inside the walls there is heat retained from the formation of the dam.

To show how long it takes the concrete to equilibrate to the outside temperature, let us consider a 20-foot-wide slab of concrete. When freshly poured, as chemical reactions proceed toward solidification, the concrete may have a temperature of 130 °F. Let us assume the temperature of the slab is constant throughout at t = 0 and we further assume that the outside temperature is constant at 60 °F on both walls of the slab. In this way, we may model the cooling of the concrete as a one-dimensional heat-flow problem

$$\frac{\partial T}{\partial t} = \alpha \frac{\partial^2 T}{\partial x^2} \qquad \text{where} \qquad \alpha = \frac{k}{\rho c}$$

Given the following thermal properties of concrete,

thermal conductivity k = 1.5 BTU/hr/ft²·°F, density ρ = 150 lb/ ft³,

and specific heat C = 0.25 BTU/ °F·lb,

What is the temperature in the center of the slab after 40 days?

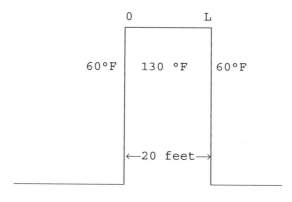

concrete slab at t = 0

5. Heat Transfer in Steel

A thick steel slab at 550 °F has its surface suddenly cooled to 100 °F. How long before the temperature at 1-inch depth reaches 200 °F ?

The Partial Differential Equation for heat transfer in steel is given by

$$\frac{\partial T}{\partial t} = \alpha \frac{\partial^2 T}{\partial x^2} \qquad \text{where} \qquad \alpha = 0.45 \text{ ft}^2 / \text{hr}$$

Chapter 5 Applications

5.1   Laser Pulse Dynamics

In their 1988 book, *Lasers*, Peter Milonni and Joseph Eberly model the generation of a laser pulse as a collapse of a population inversion of electrons in a higher energy state. If the laser is "pumped-up" so that twice as many electrons are in an excited state, relative to the lower energy state, then as a few of the electrons de-excite, more and more photons are generated *in phase and with the same energy*. So, no matter how few photons were originally in the laser, the output signal is increased exponentially until the upper energy state is unloaded. This process will occur over several nanoseconds, and may be modeled by the following equations: (We set x equal to the intensity of the beam, and y as the population inversion)

$$x' = (y - 1) * x \quad \text{and} \quad y' = -x * y \quad \text{where} \quad x_o = .0001 \text{ and } y_o = 2$$

Notice that no laser amplification is achieved if $x_o = 0$. If the starting value of x is zero, then the first equation says that x' can never increase.

These equations are easy to program in *Mathematica*.

```
sol = NDSolve[{x'[t] == (y[t] - 1) * x[t],
   y'[t] == -x[t] * y[t], x[0] == .0001, y[0] == 2}, {x, y}, {t, 0, 20}]
InterpFunc1 = x /. sol[[1]]; InterpFunc2 = y /. sol[[1]];
Plot[Evaluate[{10 * x[t], y[t]} /. sol, {t, 0, 20}],
  Epilog → {Text["Laser" , {13, 2.5}], Text["Intensity", {13.5, 2.25}],
    Text["y, 10x", {1.8, 3}], Text["Inversion", {17.5, .6}]}];
```

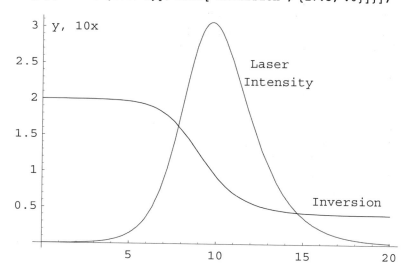

Laser pulse height and population inversion vs time in nanoseconds

An interesting question is "How much power is contained in this laser pulse?" Following Milloni and Eberly, (*Lasers*, Chapter 12) we use these values for x and y

$$x = \frac{I}{ch\nu\, N_t} \quad \text{and} \quad y = \frac{N_2}{N_1} \quad \text{where } N_t \text{ for ruby is } 3.7 \times 10^{17} \text{ atoms/cm}^3 \quad \lambda = 6943 \text{ A.}$$

Then, the peak intensity is given by $I = (c\, h\, \nu)\, N_t\, (0.3) \approx 10^9$ Watts/cm² in the laser beam. This very large number is due to the fact that we have overloaded the upper population level with approximately 10^{18} excited electrons. This can be accomplished in most lasers by delaying the onset of the laser "download" by making the cavity of the laser *non-resonant*.

This allows the excited-electron population to build to an extremely high level, such as $y = 2$. The laser cavity is then turned *resonant* at $t = 0$, and the laser pulse builds in intensity and then escapes. This process of allowing the laser to build in power is called Q-switching.

```
tbl = Table[{t, InterpFunc1[t], InterpFunc2[t]}, {t, 5, 15, .5}] //
    TableForm
```

t (ns)	Intensity	Inversion
5	0.0144311	1.97113
5.5	0.0233538	1.95293
6.	0.0373577	1.924
6.5	0.0586797	1.87903
7.	0.0896394	1.81143
7.5	0.131455	1.71479
8.	0.182211	1.58588
8.5	0.235043	1.42861
9.	0.278974	1.25556
9.5	0.303583	1.08436
10.	0.304422	0.930539
10.5	0.284446	0.802567
11.	0.25102	0.701738
11.5	0.21194	0.624984
12.	0.173081	0.567687
12.5	0.137892	0.525328
13.	0.107872	0.494135
13.5	0.0832608	0.471183
14.	0.0636328	0.454281
14.5	0.048279	0.441819
15.	0.0364335	0.432617

5.2 The B-Z Chemical Reaction

<u>The Belousov-Zhabotinsky chemical reaction</u>

A series of chemical reactions using bromates to oxidize Cerium or Manganese will under proper conditions lead to an oscillating chemical reaction. The color of a solution will change from green to blue to violet and red then repeat the sequence of colors as many as 20 times. This curious reaction was discovered by Boris Belousov in 1951.

The reaction was shown to occur in several other chemical systems by Anatol Zhabotinsky in 1958. The sequence of colors is due to a complicated set of color-absorbing ions whose concentrations change in a periodic manner as the reaction proceeds.

The differential equations of the reaction are (using x, y, and z as the different interacting ions) as follows:

$$x' = 15 \, (.0005y - xy + x - x^2)$$
$$y' = 2500 \, (-.0005\,y - xy + z)$$
$$z' = \ x - z$$

with initial conditions $x(0) = 1, \ y(0) = 0, \ z(0) = 0$

Because of the vastly different concentrations of x, y, and z, and their remarkably fast rates of change, this is a STIFF DIFFERENTIAL EQUATION. It is to *Mathematica*'s credit that the adaptive algorithm used in **NDSOLVE** is able to test for stiffness, and then solve the reaction equations with mathematical accuracy.

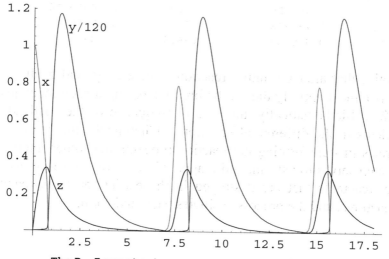

The B – Z reactants : x = green, y = red, z = blue

```
In[1]:=  sol =
            NDSolve[{x'[t] == 15 (.0005 y[t] - (x[t] * y[t]) + x[t] - x[t]^2),
             y'[t] == 2500 (-.0005 y[t] - x[t] * y[t] + z[t]),
             z'[t] == x[t] - z[t], x[0] == 1, y[0] == 0,
             z[0] == 0}, {x, y, z}, {t, 0, 18}]

In[2]:=  InterpFunc1 = x /. sol[[1]]; InterpFunc2 = y /. sol[[1]];
         InterpFunc3 = z /. sol[[1]];
         Chop[Table[{t, InterpFunc1[t], InterpFunc2[t], InterpFunc3[t]},
            {t, 0, 8, .4}] // TableForm]
         Plot[Evaluate[{x[t], y[t] / 150, z[t]} /. sol],
           {t, 0, 18}, Prolog → {Text["x", {.6, .8}],
            Text["y/120", {2.8, 1.1}], Text["z", {1.44, .24}]}]
```

t (s)	x	y	z
0	1.	0	0
0.4	0.685862	0.409051	0.281224
0.8	0.076063	3.69684	0.328543
1.2	0.000503115	161.497	0.221106
1.6	0.000502946	170.611	0.148378
2.	0.000503665	137.303	0.0996265
2.4	0.000505021	100.475	0.0669479
2.8	0.000507189	70.4543	0.0450434
3.2	0.000510536	48.3675	0.0303612
3.6	0.000515663	32.8393	0.0205209
4.	0.000523536	22.1636	0.0139269
4.4	0.00053575	14.9075	0.00951006
4.8	0.000555046	10.0043	0.00655455
5.2	0.000586499	6.69879	0.00458173
5.6	0.000640594	4.4684	0.00327323
6.	0.000743342	2.95276	0.00242138
6.4	0.000984731	1.89301	0.00190396
6.8	0.00202584	1.03856	0.00173055
7.2	0.0500508	0.106964	0.00519576
7.6	0.786846	0.195808	0.154469
8.	0.510139	0.636694	0.325821

Notice that after the initial reactions where x, y, and z all peak, the reactants almost entirely die out before undergoing a resurgence just after t = 7.2 s. Mathematically, this is a consequence of the vastly different coefficients in the differential equations. Chemically, this is due to very rapid rates in transforming the reaction products into other intermediates which themselves transform, and reappear in the original reactions. So the original reactants that are destroyed in the earlier reactions reappear in later reactions, and the process oscillates back-and-forth.

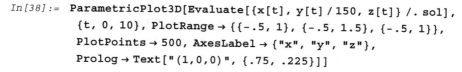

```
In[38]:= ParametricPlot3D[Evaluate[{x[t], y[t] / 150, z[t]} /. sol],
         {t, 0, 10}, PlotRange → {{-.5, 1}, {-.5, 1.5}, {-.5, 1}},
         PlotPoints → 500, AxesLabel → {"x", "y", "z"},
         Prolog → Text["(1,0,0)", {.75, .225}]]
```

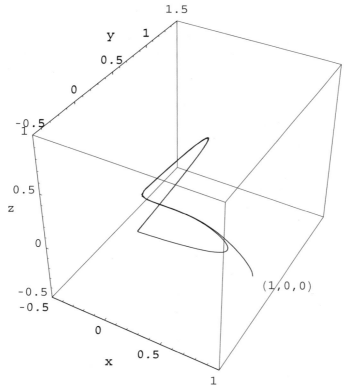

The time track of the concentrations x, y, z in the B-Z reaction.
After $t = 2$ s, the figure-8 curve traces itself over and over.

Starting at $x = 1$, $y = 0$, $z = 0$ at $t = 0$, the curve decreases in x and increases along z until it makes a sharp turn to a very-high value of y after $t = 0.8$ s. Then the curve tracks down in z and decreases in y as it approaches 0,0,0 near $t = 6.8$ s. Then the curve begins to increase again in x, and traces the distorted figure-8 track over and over as the cycle repeats. Notice that the y-axis is compressed by a factor of 150.

In[3]:=

```
sol = NDSolve[{x'[t] == 15 (.0005 y[t] - (x[t] * y[t]) + x[t] - x[t]^2),
    y'[t] == 2500 (-.0005 y[t] - x[t] * y[t] + z[t]), z'[t] == x[t] - z[t],
    x[0] == 1, y[0] == 0, z[0] == 0}, {x, y, z}, {t, 0, 12}]

Module[{gridSpec},
 gridSpec = Table[{k, GrayLevel@0.65}, {k, -0.5, 1, 0.5}];
 ParametricPlot3D[Evaluate[{x[t], y[t] / 150, z[t]} /. sol],
   {t, 0, 10}, PlotRange → {{-.5, 1}, {-.5, 1.5}, {-.5, 1}},
   PlotPoints → 500, AxesLabel → {"x", "y", "z"},
   FaceGrids → {{{0, 0, -1}, {gridSpec, gridSpec}},
     {{0, 1, 0}, {gridSpec, gridSpec}}},
   ViewPoint -> {1.446, -2.952, 0.803}];]
```

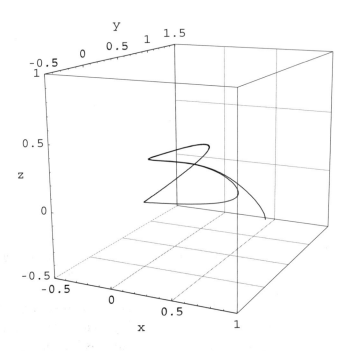

With grid-lines and a different perspective.

After $t = 2$ s, the figure-8 curve traces itself over and over.

5.3   Foxes and Rabbits

Vito Volterra in the 1930s wrote down the equations for the interaction of two species, one a predator (foxes) and their prey (rabbits). Since the foxes depend upon the rabbits as food, the two populations change in time – if there is a sufficient number of rabbits, the foxes will increase in time, however if there are too few rabbits, then the fox population will decrease in time.

Let us take a somewhat simplified situation, where we let loose some rabbits into a very large grassland. We let their numbers increase to 1000, and then we introduce 8 foxes.

We will plot the number of rabbits x and foxes y versus time, and then show that the two populations are <u>periodic</u> in time, and find the time for the cycle to repeat.

With time measured in days, the Predator-Prey equations are

for rabbits $\dot{x} = \alpha\, x - \beta\, xy$ $x_o = 1000$

for foxes $\dot{y} = -\epsilon\, y + f\, xy$ $y_o = 8$

where $\beta = 0.001$ (each fox averages one rabbit per day when $x = 1000$)

   $\alpha = 0.01$ (the rabbit population initially increases at 10 rabbits per day)

   $\epsilon = 0.015$ (the fox mortality rate if rabbits arent caught)

   $f = \beta/40$ (the conversion efficiency of rabbits into new foxes)

Mathematica easily solves the Predator-Prey equations

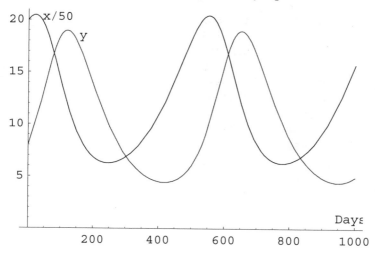

Number of rabbits x / 50 and number of foxes y versus time in days

```
In[17]:=
Clear[α, β, ε, f, x, y, t]; α = .01; β = .001; ε = .015; f = β / 40;
sol = NDSolve[{x'[t] == α * x[t] - β * x[t] * y[t],
    y'[t] == -ε * y[t] + f * x[t] * y[t],
    x[0] == 1000, y[0] == 8}, {x, y}, {t, 0, 1000}]
InterpFunc1 = x /. sol[[1]]; InterpFunc2 = y /. sol[[1]];
InterpFunc3 = x' /. sol[[1]]; InterpFunc4 = y' /. sol[[1]];
Plot[Evaluate[{x[t] / 50, y[t]} /. sol, {t, 0, 1000}], PlotRange → {0, 22},
    Prolog → {Text["x/50", {90, 20.3}], Text["y", {168, 18.5}]},
    PlotStyle → {{RGBColor[0, 0, 1]}, {RGBColor[1, 0, 0]}}];
Table[{t, InterpFunc1[t], InterpFunc2[t]}, {t, 0, 540, 30}] //
    TableForm
```

If we **Table** the results, we will find the <u>maximum</u> number of foxes and the <u>minimum</u> number of rabbits. We see that the number of foxes increases to about 19 and the rabbit population falls to about 310. (The rabbits dont do too well on fox-rabbit encounters). However, as the number of rabbits decreases, the fox population must also fall (food supply almost gone) so then as the number of predators decreases, the rabbits begin to increase, and the cycle begins again. The **Table** or the graph shows that the ecological game cycle repeats about every 530 days.

t (days)	x (rabbits)	y (foxes)
0	1000.	8.
30	1018.76	10.9358
60	938.01	14.6178
90	776.128	17.7836
120	600.05	18.9738
150	463.246	17.9678
180	376.912	15.6537
210	331.091	12.9886
240	314.341	10.5336
270	319.266	8.50771
300	342.19	6.94452
330	381.889	5.80337
360	438.566	5.02802
390	512.977	4.57554
420	605.404	4.43271
450	713.992	4.63164
480	831.82	5.27189
510	942.229	6.54639
540	1014.11	8.72517

Another way of looking at this ecological game cycle is to plot the number of rabbits x versus the number of foxes y. In a **Parametric Plot**, we can see how the two populations move along a closed path.

```
In[24]:= ParametricPlot[{x[t], y[t]} /. sol,
            {t, 0, 550}, Prolog → Text["Δ", {1000, 8}],
            PlotRange → {{300, 1040}, {4, 19}},
            AxesLabel → {"x rabbits", "y foxes"}];
```

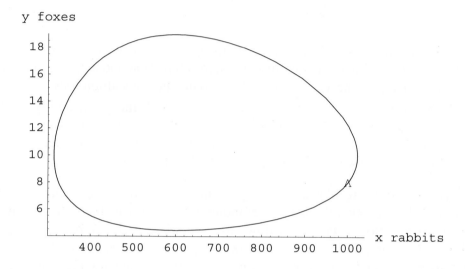

The number of rabbits and the number of foxes as a function of time in days. The cycle starts at Δ (1000, 8) and moves counter-clockwise.

5.4 PDE – Nerve Conduction

Nerve Conduction

In 1952, after more than 20 years of work, Alan Hodgkin and Andrew Huxley published a theory of nerve conduction. (Journal of Physiology, <u>117</u>, 500-544).

The first thing that Hodgkin and Huxley strove to understand was how a nerve fiber with an electrical resistance of 100 million ohms could conduct electrical signals. What they found, by using radioactive tracer elements was that the inside of a nerve cable (or nerve *axon*) has an abundance of potassium ions and outside the nerve axon is a super-abundance of sodium ions.

When the nerve receives a stimulus at the end, the <u>permeability</u> of the membrane that surrounds the nerve changes, allowing sodium (Na^+) in and potassium (K^+) to surge out. Hodgkin-Huxley found that the applied voltage changed the <u>rate</u> at which the conductivity of the membrane changed.

Thus there is a <u>feedback mechanism</u> whereby the initiating voltage causes a current flow through the nerve membrane which changes the voltage, which allows current flow ever further down the line. This is the *action potential*, a 100-mV voltage pulse that moves down the nerve axon at a rate determined by the capacitance of the nerve membrane, and the rate at which the conductivity changes.

By a series of very careful measurements on nerve fiber in squid, Hodgkin and Huxley were able to determine how the conductivity of the nerve membrane changed as a function of voltage.

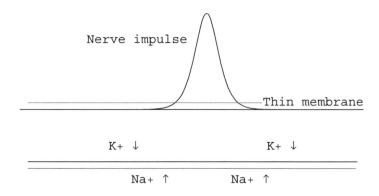

**The nerve axon with potassium flow outward
and sodium flow inward produces the nerve impulse**

The Voltage-Clamp experiments

To determine how the conductivity of the nerve changes with voltage, Hodgkin and Huxley placed a fine wire down the center of the axon, and then applied a current impulse or voltage step. In this way, the potassium and sodium conductivity could be determined as a function of voltage and the nerve impulse would occur at the same time all along the axon. The expressions they found for the conductivity are as follows:

The potassium conductivity is $g_K = .036 * n^4$

The sodium conductivity is $g_{NA} = .120 * m^3 h$

where n, m, and h are found to change with time as

$$n' = \alpha (1 - n) - \beta n \qquad m' = \gamma (1 - m) - \delta m \qquad h' = \epsilon (1 - h) - \eta h$$

and α, β, γ, δ, ϵ, η are functions of voltage and at 6.3 °C are given by

$$\alpha = \frac{.01 * (10 - V)}{Exp[(10 - V)/10] - 1} \qquad \beta = 0.125 * Exp[-\frac{V}{80}]$$

$$\gamma = \frac{10 * (25 - V)}{Exp[(25 - V)/10] - 1} \qquad \delta = 4.0 * Exp[-\frac{V}{18}]$$

$$\epsilon = .07 * Exp[-\frac{V}{20}] \qquad \eta = \frac{1}{Exp[(30-V)/10] + 1}$$

These correspond to the opening and closing of channels in the protein structure of the nerve membrane as the voltage changes.

We are now ready to put together the *conductivity equations* that show how the current flows through the nerve membrane. The current due to the potassium and sodium ion flow is

$$i_{ion} = g_K (V + 12) + g_{NA} (V - 115) + g_L (V - 10.6) \quad \text{or}$$

$$i_{ion} = .036 * n^4 (V + 12) + .120 * m^3 h (V - 115) + .0003 (V - 10.6)$$

where g_L is a leakage term for all other ions, like Calcium, that keep the ion current at zero when the nerve is at rest. The current is measured in mA, the voltage is in mV. We now add in the capacitance of the membrane and the initial current to obtain the H–H *voltage-clamp equations*

$$i_{init} = C_m \frac{dV}{dt} + .036 * n^4 (V + 12) + .120 * m^3 h (V - 115) + .0003 (V - 10.6)$$

We will apply a 0.1 mA initial current pulse to the nerve and we will see how the voltage changes in time as sodium surges in and potassium surges out.

Again, V is measured from its rest potential of zero, and is in millivolts. $C_m = 0.001$ mF is the membrane capacitance, and time is measured in milliseconds. Using *Mathematica*,

```
Clear[V, z, m, h, n];
vsol1 = NDSolve[{.001 * V'[t] + .036 * n[t]^4 * (V[t] + 12) +
    .120 * m[t]^3 * h[t] * (V[t] - 115) + .0003 * (V[t] - 10.6) ==
  0.1 * Which[t < .3, 0, .3 ≤ t ≤ .4, 1, t > .4, 0], n'[t] ==
    .01 * (10 - V[t])
  ─────────────────── * (1 - n[t]) - .125 * Exp[- V[t] ] * n[t],
  Exp[(10 - V[t]) / 10] - 1                          80

  m'[t] ==    .1 * (25 - V[t])
           ─────────────────── * (1 - m[t]) - 4.0 * Exp[- V[t] ] * m[t],
           Exp[(25 - V[t]) / 10] - 1                        18

  h'[t] == .07 * Exp[- V[t] ] * (1 - h[t]) -        1        * h[t],
                        20                  Exp[(30 - V[t]) / 10] + 1

  n[0] == .3177, m[0] == .0529, h[0] == .5961, V[0] == 0},
  {V, n, m, h}, {t, 0, 4.2}]
Plot[Evaluate[V[t] /. vsol1], {t, 0, 3.9},
 PlotRange → {{0, 4.2}, {0, 105}}, Prolog →
 {Text["100-mV", {3.1, 82}], Text["Action Potential", {3.4, 72}],
  Text["mV", {0.18, 101}], Text["t(ms)", {4.0, 6.7}]}]
InterpFunc1 = V /. vsol1[[1]]; InterpFunc2 = n /. vsol1[[1]];
InterpFunc3 = m /. vsol1[[1]]; InterpFunc4 = h /. vsol1[[1]];
V1[t_] = InterpFunc1[t]; n1[t_] = InterpFunc2[t];
m2[t_] = InterpFunc3[t]; h3[t_] = InterpFunc4[t];
 Plot[{36 * n1[t]^4, 120 * m2[t]^3 * h3[t]}, {t, 0, 4.0}]
 Plot[0.1 (UnitStep[t - .3] - UnitStep[t - 0.4]), {t, 0, 2.0}]
```

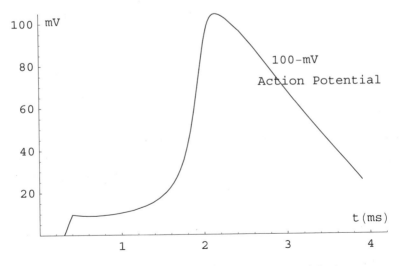

In the voltage-clamp experiments, the action potential arises at all points along the axon.

It is interesting to compare the 100-mV *action potential* to the sodium and potassium conductance and also to the rather small current of 0.1 mA for 0.1 ms that triggered the 100-mV voltage response all along the voltage-clamped nerve axon. It is the rapid increase in sodium conductance that starts the *action potential* and the steady increase of the potassium conductance that brings the action potential back to zero.

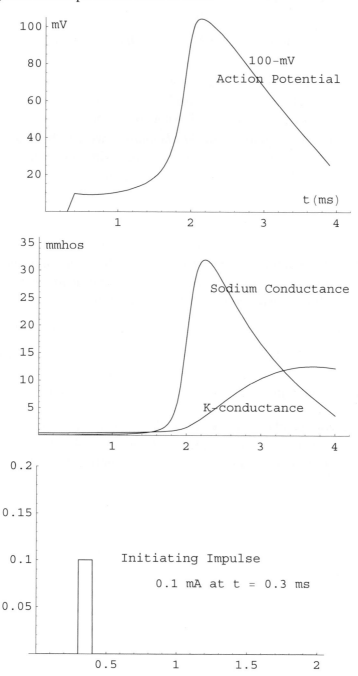

The Full Hodgkin–Huxley Equations of Nerve Conduction

The next question is whether the H–H model of nerve conduction can account for the actual speed of propagation of the *action potential* down the nerve axon. For this, Hodgkin and Huxley used the Partial Differential Equation of cable conduction with additional terms for sodium and potassium conductance. The PDE is

$$\frac{a}{2\rho}\frac{\partial^2 V}{\partial x^2} = C_m \frac{\partial V}{\partial t} + .036\, n^4\,(V+12) + .12\, m^3\, h\,(V-115) + .0003\,(V-10.6)$$

where a = 0.0238 cm is the radius of the axon, and ρ = 35.4 Ω-cm is the resistivity of the membrane. Hodgkin-Huxley actually solved this equation using only a desk calculator! But we have a computer.

Programming this equation into *Mathematica*, we will be able to find the voltage along the nerve fiber as it starts as a half-sine pulse of 0.5 ms duration with a peak of 60 mV. There is now no wire in the axon, and we will see the voltage response of the nerve in time and space as the signal moves down the axon.

A note of caution is in order. We must be careful to set the initial conditions such that the voltage at t = 0 is zero everywhere, and that the initiating voltage ramps up from there. We must also be careful to set the boundary conditions such that the *maximum* x in the PDE solution term

`(D[u[x,t],x]/.x→ 4) == 0` is greater than the region we are examining.

We will solve the H–H Partial Differential Equation and plot the voltage response at x=0, x=0.5, x=1.0, x=1.5, x=2.0, and x=2.5 cm and from the observed maximum points, we will find the speed of the action potential.

Using the voltage-clamp values of α, β, γ, δ, ϵ, η at T = 6.3 °C, Hodgkin and Huxley obtained a nerve-conduction velocity of 12.4 m/s. The *Mathematica* results on the next page show the nerve conduction velocity at 6.3 °C to be 12.5 m/s.

```
Clear[h];  (*Action potential calculation at T=6.3 °C*)
c = .001; a = .0238; ρ = 35.4;
```

$$eq3 = \left\{ \frac{a}{2*\rho} \ D[u[x, t], x, x] == \right.$$

$$c * D[u[x, t], t] + .036 * n[x, t]^4 * (u[x, t] + 12) +$$
$$.120 * m[x, t]^3 * h[x, t] * (u[x, t] - 115) + .0003 * (u[x, t] - 10.6),$$
$$u[x, 0] == 0, u[0, t] == (60 * Sin[2\pi*t] * If[0 < t < 0.5, 1, 0]),$$
$$\left.(D[u[x, t], x] /. x \to 4) == 0\right\};$$

$$eq4 = \left\{ D[n[x, t], t] == \frac{.01 * (10 - u[x, t])}{Exp[(10 - u[x, t]) / 10] - 1} * (1 - n[x, t]) - \right.$$
$$\left..125 * Exp[-u[x, t] / 80] * n[x, t], n[x, 0] == .3177\right\};$$

$$eq5 = \left\{ D[m[x, t], t] == \frac{.1 * (25 - u[x, t])}{Exp[(25 - u[x, t]) / 10] - 1} * (1 - m[x, t]) - \right.$$
$$\left.4.0 * Exp[-u[x, t] / 18] * m[x, t], m[x, 0] == .0529\right\};$$

$$eq6 = \left\{ D[h[x, t], t] == .07 * Exp[-u[x, t] / 20] * (1 - h[x, t]) - \right.$$
$$\left.\frac{1}{Exp[(30 - u[x, t]) / 10] + 1} * h[x, t], h[x, 0] == .5961\right\};$$

```
soll = NDSolve[{eq3, eq4, eq5, eq6},
    {u[x, t], n[x, t], m[x, t], h[x, t]}, {x, 0, 4}, {t, 0, 3.5}];
Plot3D[Evaluate[u[x, t] /. soll[[1]]], {x, 0, 1.5}, {t, 0, 3.5}],
DisplayFunction → $DisplayFunction, ViewPoint → {2.5, -1.5, .95}];
Clear@f; f[x_, t_] = u[x, t] /. soll[[1]]
Pictures = Table[f[x, t], {x, 0, 3.0, .5}]
Plot[Evaluate[Pictures], {t, 0, 3.5}, Frame → True, PlotRange → All]
```

March of the action potentials

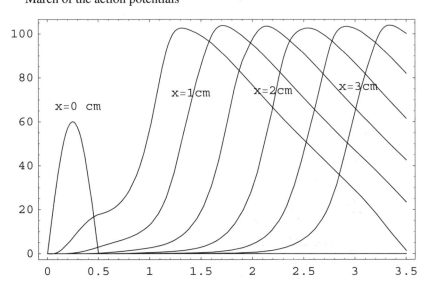

If we evaluate the time at which the maximum is reached at x =1 cm, it is at t =1.7 ms, the time of the maximum voltage at x = 3 cm is 3.3 ms. Therefore, at T = 6.3 °C, the speed of the action potential is $v = \frac{\Delta x}{\Delta t} = \frac{2 \text{ cm}}{1.6 \text{ ms}} = 12.5$ m/s.

If we examine a 3-D plot of voltage in time and space, we see the initiating half-sine voltage input on the lower left, followed within a millisecond by the rising hill of the action potential. The action potential then heads off upwards and to the right as it moves down the nerve axon.

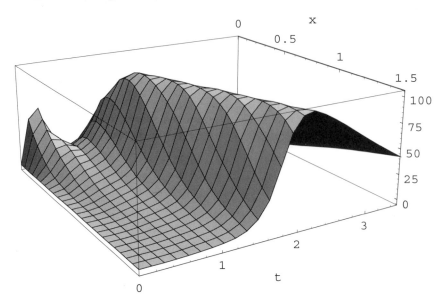

3-D plot of the action potential

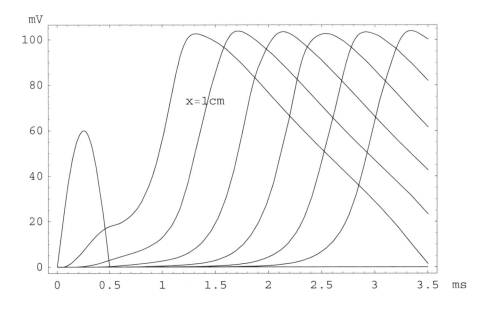

If we take a cross-section of the above 3-D curve, and station ourselves at, say x = 1 cm, then we will see the action potential rise from zero, peak at 100-mV, and then descend almost linearly to zero.

Finally, we may check the conduction velocity v by creating a Table of values of voltage versus time and distance. The maximum voltages (102-103 mV) occur at (x = 1.0 cm, t = 1.7 ms), (x = 2.0 cm, t = 2.5 ms), and (x = 3.0 cm, t = 3.3 ms).

Therefore the conduction velocity is $v = \frac{\Delta x}{\Delta t} = \frac{1\ cm}{0.8\ ms} = 12.5$ m/s.

```
TableForm[Transpose@Prepend[
    Transpose@Table[f[x, t], {t, 0, 3.5, .1}, {x, 0, 3.0, 0.5}],
    Table[StyleForm[ToString@t, FontWeight → "Bold"],
      {t, 0, 3.5, .1}]] // Chop,
  TableHeadings -> {None, Prepend[
    Table[StyleForm["x(" <> ToString@x <> ")", FontWeight → "Bold"],
      {x, 0, 3.0, 0.5}], StyleForm["t(ms)", FontWeight → "Bold"]]}]
```

t(ms)	x(0)	x(0.5)	x(1.)	x(1.5)	x(2.)	x(2.5)
0	0	0	0	0	0	0
0.1	35.27	0.689	0.010	-0.00	-0.00	-0.00
0.2	57.03	4.562	0.075	-0.00	-0.00	-0.00
0.3	57.03	10.05	0.553	0.002	-0.00	-0.00
0.4	35.26	14.87	1.575	0.063	-0.00	-0.00
0.5	-0.01	17.80	2.956	0.232	0.004	-0.00
0.6	-0.02	19.27	4.417	0.528	0.029	-0.00
0.7	-0.00	22.48	5.781	0.926	0.083	0.002
0.8	-0.00	28.50	7.332	1.394	0.171	0.011
0.9	-0.00	38.98	9.507	1.941	0.292	0.028
1.	-0.00	56.15	13.00	2.634	0.448	0.055
1.1	-0.00	79.16	19.26	3.617	0.647	0.094
1.2	-0.00	97.41	30.73	5.178	0.913	0.147
1.3	-0.00	**102.5**	48.83	7.804	1.302	0.218
1.4	-0.00	101.5	69.20	12.18	1.916	0.318
1.5	-0.00	99.19	85.72	19.27	2.932	0.465
1.6	-0.00	95.90	98.97	30.68	4.623	0.697
1.7	-0.00	91.65	**103.7**	47.98	7.414	1.071
1.8	-0.00	86.82	102.3	70.25	11.93	1.685
1.9	-0.00	81.76	99.09	87.61	19.10	2.689
2.	-0.00	76.62	95.19	97.66	30.11	4.326
2.1	-0.00	71.47	90.76	**103.0**	45.94	6.958
2.2	-0.00	66.38	85.96	102.6	66.81	11.20
2.3	-0.00	61.40	80.96	99.47	88.52	18.31
2.4	-0.00	56.53	75.88	95.48	99.28	29.91
2.5	-0.00	51.78	70.81	91.04	**102.3**	46.46
2.6	-0.00	47.14	65.81	86.25	102.1	64.84
2.7	-0.00	42.59	60.90	81.26	99.54	84.33
2.8	-0.00	38.06	56.11	76.19	95.81	99.26
2.9	-0.00	33.50	51.44	71.12	91.43	**103.3**
3.	-0.00	28.79	46.87	66.11	86.64	102.4

The Hodgkin–Huxley Equations of Nerve Conduction

After 50 years, the Hodgkin and Huxley equations are still the best descriptor of nerve conduction. This is because the equations were developed using an *empirical model* where the conductivity of the nerve was determined from experimental measurements, without regard to the precise structure of the nerve membrane. One of the first new applications of the H-H theory was to *myelinated nerve,* which is the coated nerve – with gaps – found in the higher animals (humans included !). One of the first computations of nerve velocity in myelinated nerve was done by Richard Fitzhugh in 1962.

If you would like to try and model myelinated nerve conduction, the details are given in R. Fitzhugh *Biophysical Journal* <u>2</u> , 11-21 "Computation of Impulse Initiation and Conduction in Myelinated Nerve".

For more information on nerve conduction, see Problem #5-4 in the Challenge problems for this chapter, where temperature effects on nerve conduction and impulse initiation are explored.

For the very latest on the molecular structure of nerve membrane and potassium-sodium flow, see recent issues of the *Biophysical Journal.*

5.5 Fusion Reactor

Thermonuclear Fusion

Suppose we are able to magnetically confine a deuterium plasma at a particle density $N = 10^{16}$ deuterons/cm^3 and at temperature $T = 10^8$ Kelvin degrees in a Maxwell-Boltzmann probability distribution. Given that the reaction cross-section for deuterons of kinetic energy E coming from an accelerator and striking stationary deuterons is [§]

$$\sigma(E) = \frac{288}{E} \exp(-45.8/\sqrt{E}) \text{ barns (with E in keV)}$$

We shall first use this result to compute the rate of energy generation per cubic meter from deuterons alone, given that each D + D fusion liberates 3.65 MeV of energy.

Secondly, let us take into account that for every <u>two</u> D + D reactions we obtain <u>one</u> tritium atom which reacts T + D $\rightarrow \alpha$ + n with an energy release of 17.6 MeV. If all tritium reacts as soon as it is generated, what is the total rate of fusion energy (D + D and T + D) per cubic meter of plasma ?

Let's begin by writing all variables in terms of center-of-mass coordinates.

$$m_* = \frac{m_1\,m_2}{m_1 + m_2} = \frac{m_D}{2} = \text{reduced mass}$$

$$E_* = \tfrac{1}{2}m_*v^2 = \frac{E}{2} = \text{relative K.E. of colliding nuclei}$$

The Maxwell-Boltzmann probability distribution is given by

$$f(v)\,dv = \left(\frac{m_*}{2\pi kT}\right)^{3/2} \exp\left(-\frac{m_*\,v^2}{2\,kT}\right) 4\pi\,v^2\,dv$$

$$\sigma(E_*) = \frac{288}{2\,E_*} \exp(-45.8/\sqrt{2\,E_*}) \text{ reaction cross-section in barns } (E_* \text{ in keV})$$

Now consider a collision of two deuterium nuclei at velocity v. An incoming particle (1) moving at v relative to particle (2) encounters $\overline{v\,\sigma N_2}$ nuclei in one second. The bar represents an average over the Maxwell-Boltzmann distribution.

The reaction rate is $\quad R = N_1 N_2\,(\overline{v\,\sigma})_{DD} = N_1\,N_2 \int_0^\infty (v\,\sigma)f(v)\,dv$

[§] Arnold, Phillips, Sawyer, Stovall, and Tuck (1954) Physical Review <u>93</u>, 483

Since the reacting particles are identical, we must be careful not to count each collision twice

$$R = \frac{N_D^2}{2} \int_0^\infty (v\,\sigma)\left(\frac{m_*}{2\pi kT}\right)^{3/2} \exp\left(-\frac{m_* v^2}{2\,kT}\right) 4\pi\,v^2\,dv$$

Now, expressing everything in terms of $E_* = \frac{1}{2} m_* v^2$,

$$R = \frac{N_D^2}{2}\left(\frac{2}{m_*\,\pi}\right)^{1/2} \frac{a}{(kT)^{3/2}} \int_0^\infty e^{-b\,E_*^{-1/2}}\, e^{-E_*/kT}\, dE_*$$

$a = 288$ keV·barns, $b = \dfrac{45.8}{\sqrt{2}}$ keV$^{1/2}$, $kT = 8.62$ keV, $m_* = \dfrac{m_D}{2} = 931.5$ MeV/c^2

The above integral cannot be solved analytically. Therefore we will use *Mathematica* to numerically evaluate the integral from $E_* = 1$ to $E_* = 101$ keV

$$\int_1^{101} e^{-b\,E_*^{-1/2}}\, e^{-E_*/kT}\, dE_*$$

```
In[1]:=   α = 8.62; b = 45.8/√2;      f[x_] = Exp[-x/α] * Exp[-b/√x];

          NIntegrate[f[x], {x, 1, 101}]

Out[2]=  0.00278303
```

Then the reaction rate of $D + D$ per cubic cm per second is

$$R_{DD} = \frac{10^{32}}{2}\left(\frac{2*9\times10^{20}}{\pi*931500}\right)^{1/2} \frac{288\times10^{-24}}{(8.62)^{3/2}}\,(.002783)$$

```
In[3]:=   n = 10^16; (*R = # reactions/cm³/s*)

          R = n²/2 (2*9*10^20 / π*931500)^(1/2) * (288*10^-24 / (8.62)^(3/2)) * .002783

Out[3]=  R = 3.92722 × 10^13
```

At 3.65 MeV per $D + D$ reaction, $P_{DD} = 23$ MW/ m^3. If all tritium reacts as soon as it is produced, with $U_{DT} = 17.6$ MeV and $U_{DD} = 3.65$ MeV, then the total reactor power from both D+D and D+T is

$$P = R_{DD}\left[\, U_{DD} + \frac{U_{DT}}{2}\,\right] = 78\,\text{MW}/\text{m}^3.$$

Maxwell–Boltzmann and the Gamow Peak

The Maxwell–Boltzmann distribution tells us that the most probable energy of the colliding deuterium particles is

$$\overline{E} = kT = (1.38 \times 10^{-23} \text{ J/K} * 10^{8}\text{K}) / (1.6 \times 10^{-16}\text{J/keV}) = 8.62 \text{ keV}.$$

However, this is not the energy at which most D + D reactions take place. To find the most probable energy of reaction, we need to find the maximum of the function inside the reaction integral

$$R_{DD} = \frac{N_D^2}{2} \left(\frac{2}{m_* \pi} \right)^{1/2} \frac{a}{(kT)^{3/2}} \int_0^\infty e^{-bE_*^{-1/2}} \, e^{-E_*/kT} \, dE_*$$

We want to maximize $g(E_*) = e^{-bE_*^{-1/2}} \, e^{-E_*/kT}$

$In[15] := \ \alpha = 8.62; \ \beta = \dfrac{45.8}{\sqrt{2}}; \ g[x\_] = Exp[-x / \alpha] * Exp[-\beta x^{-1/2}];$

$\quad Plot[\{Exp[-x / \alpha] * Exp[-\beta x^{-1/2}]\}, \{x, 1, 100\}]$

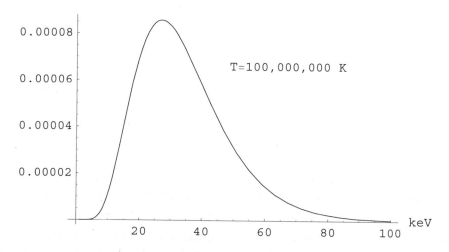

Reaction probability as a function of energy

$In[17] := \ \textbf{Maximize}[\{g[x], \ x > 0\}, \ x]$

$Out[17] = \ \{0.0000856866, \{x \to 26.9082\}\}$

This is the so-called "Gamow peak". So even though the average energy of the particles in the deuterium plasma is 8.62 keV, most reactions occur for only the most energetic deuterons, those with three times the average energy. The most probable energy of reaction occurs at 26.9 keV.

This gives us some idea of the difficulty in achieving nuclear fusion. Let us plot the Gamow function at T = 50,000,000 K (which is 3 times the core temperature of the sun) and also at T = 100,000,000 K, to see what effect doubling the temperature has on the reaction probability.

```
In[77]:=  α = 8.62; β = 45.8/√2; δ = .5 α;
          Plot[{Exp[-x/α] * Exp[-β x^-1/2], Exp[-x/δ] * Exp[-β x^-1/2]},
          {x, 1, 100}, PlotRange → {0, .0001}, AxesLabel → {"keV", " "},
          Prolog → Text["100,000,000 K", {53, .000066}]]
```

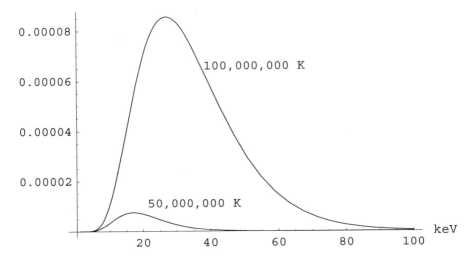

Difficulty of starting a "nuclear fire." The power output at each temperature is proportional to the area under each curve

The power output of our theoretical fusion reactor is only 3 MW/ m³ at 50 million degrees, however it increases to 78 MW/ m³ at 100 million degrees K. But achieving actual fusion power is far more difficult. The first problem is *Bremsstrahlung* (German for braking-radiation) and this is the effect of electrons being accelerated in the electric field of the deuteron ions. As they are accelerated, the electrons emit radiation (X-rays !) and remove heat from the plasma. A formula for loss of energy due to Bremsstrahlung radiation is

$$P_{Brem} = 4.8 \times 10^{-37} \, N_{ion} N_e \sqrt{T(keV)} \quad W/m^3$$
(For $N_{ion} = N_e = 10^{22} / m^3$, $T = 8.62$ keV, $P_{Brem} = 140$ MW/m³).

This means that we will have to heat a deuterium plasma to a temperature greater than 10^8 K or we will have to start with a deuterium-tritium mix. Otherwise the Bremsstrahlung losses will cool the plasma and the nuclear fire will go out.

Another problem is the reaction products of the deuterium-tritium reaction $D + T \rightarrow \alpha + n$. The α-particle carries about 4 MeV of kinetic energy and is easily captured in the plasma, but the neutron with no electric charge and 14 MeV of energy will escape the reaction chamber. So the problem is not only achieving temperature, but maintaining it, and also shielding against a flood of X-rays and high-energy neutrons from the reactions themselves.

The figure-of-merit for achieving and maintaining fusion reactions is the Triple-Product

$\qquad$ N T t = (Particle density # ions/ m^3) × (Temperature keV) × (Confinement time sec)

Some of the best results to date in world fusion research are with Tokamak magnetic confinement machines where a huge electric current of approx 7 million amperes is started in deuterium or $D + T$ gas, heating it into the keV range and simultaneously creating a magnetic field which confines the plasma. Best efforts in temperature are in the Japanese Tokamak, JT-60, where T= 45 keV and confinement times of t = 28 seconds have been reached however the particle density was only 10^{13} deuterons/cm^3.

The next generation of fusion machines are presently being constructed. The ITER (International Tokamak Experimental Reactor) is presently being built in the South of France. It's completion date is in 2015. An inertial confinement device, the NIF (National Ignition Facility) which will use lasers to ignite a plasma will begin operations at Lawrence Livermore in California in 2010.

For more on thermonuclear fusion the reader is referred to the very excellent article by J Rand McNally (Jan 1982) "*Physics of Fusion Fuel Cycles*" Nuclear Technology/Fusion $\underline{2}$, pp 1-28.

5.6   Space Shuttle Launch

The workhorse of the United States space program from 1980 to 2008 is the Space Shuttle. The Shuttle consists basically of a 40-meter long airplane strapped to a main fuel tank and two solid-fuel booster rockets.

In MKS units, the Equations of Motion of the Space Shuttle are

$$my'' = (T - D)\cos(\epsilon t) - mg_o \left(\frac{R}{R+y}\right)^2 \qquad y_o = 0 \qquad y'_o = 0$$

$$mx'' = (T - D)\sin(\epsilon t) \qquad\qquad\qquad\qquad x_o = 0 \qquad x'_o = 0$$

This will take us from launch to booster-rocket separation. Let's look carefully at each term in these equations. First, the mass of the Shuttle, its rockets, and its fuel $m = m_o - \alpha t$. The mass declines as the fuel is burnt, at a rate $\alpha = 9800$ kg/s during the first 120 seconds. The Thrust of the Shuttle, with its solid boosters and main engine is $T = 28.6 \times 10^6$ N. The Drag the shuttle experiences during its lift-off phase is $D = .5\rho A C_D v^2$, where the air density declines with height as $\rho = \rho_o e^{-y/8000}$ with y (the height) measured in meters. The angle the Shuttle makes in-flight with respect to its launch point is $\theta = \tan^{-1}(x/y)$ and ϵ is the angle the thrust makes with the launch angle, so that the Shuttle will move out along x and allow the booster rockets to fall into the Atlantic Ocean. Last is the gravity term $mg_o \cdot r^{-2}$ where Earth's gravity becomes weaker at great height, however this will only amount to a 2% reduction in g until the Shuttle reaches a height y >50 km.

We can now fill in the details for the first 120 seconds of the Space Shuttle launch. The total mass at lift-off is $m_o = 2.04 \times 10^6$ kg. The initial angle $\theta_o = 0°$ (vertical), and the Shuttle is steered to an angle of approx 50° from the vertical over 120 seconds by allowing the angle from the vertical to increase by $\epsilon = 0.007$ radians every second. We will take the air density $\rho_o = 1.2$ kg/m^3 at the surface, and A and C_D in the drag term to be 100 m^2 and 0.3 respectively.

We are now ready to program these equations into *Mathematica*, and see how the computerized version of the Space Shuttle performs

```
m0 = 2.04 * 10^6; Cd = 0.3; T = 28.6 * 10^6; α = 9800; ε = .007; g = 9.8;
sol = NDSolve[{ (m0 - α * t) * y''[t] ==
    (T - 18 * (x'[t]^2 + y'[t]^2) * Exp[-y[t] / 8000]) * Cos[ε * t] -
    (m0 - α * t) * g * (6400 / (6400 + y[t] / 1000))^2, (m0 - α * t) * x''[t] ==
    (T - 18 (x'[t]^2 + y'[t]^2) * Exp[-y[t] / 8000]) * Sin[ε * t],
    x[0] == y[0] == 0, x'[0] == 0, y'[0] == 0}, {x, y}, {t, 0, 120}]
```

In[26]:=
```
InterpFunc1 = x /. sol[[1]]; InterpFunc2 = y /. sol[[1]];
InterpFunc3 = x' /. sol[[1]]; InterpFunc4 = y' /. sol[[1]];
tbl = Table[{InterpFunc1[t], InterpFunc2[t]}, {t, 0, 120, 2}];
ListPlot[tbl,
    Prolog → AbsolutePointSize[4], AspectRatio → 0.65];
```

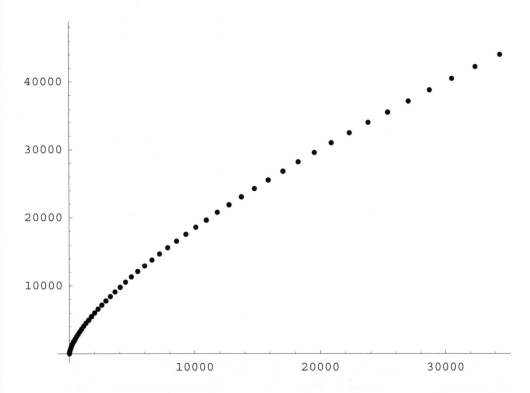

The trajectory of the Space Shuttle from launch to $t = 120$ seconds, just before booster rocket separation. If we **Table** the results, we will find the height achieved is $y = 47$ km, and the distance downrange is $x = 39$ km.

In[7]:=
```
Table[{t, Chop[InterpFunc2[t]], Chop[InterpFunc1[t]],
    Chop[√(InterpFunc3[t]^2 + InterpFunc4[t]^2)],
    Chop[InterpFunc4[t]], Chop[InterpFunc3[t]]},
    {t, 0, 120, 4}] // TableForm
```

The (x,y) coordinates and velocity of the Space Shuttle during its 120-second first-stage flight

t (s)	y (m)	x (m)	v (m / s)	y' (m / s)	x' (m / s)
0	0	0	0	0	0
4	34.4714	1.05687	17.4315	17.4134	0.795176
8	140.711	8.53576	36.0239	35.8796	3.22112
12	322.873	29.0812	55.8538	55.3696	7.33858
16	584.989	69.5807	76.9921	75.8506	13.2083
20	930.953	137.164	99.5051	97.2874	20.8907
24	1364.52	239.205	123.458	119.644	30.4468
28	1889.29	383.318	148.915	142.887	41.9378
32	2508.76	577.371	175.946	166.987	55.427
36	3226.3	829.486	204.627	191.922	70.9805
40	4045.23	1148.06	235.044	217.678	88.6692
44	4968.81	1541.79	267.295	244.252	108.571
48	6000.35	2019.69	301.492	271.655	130.772
52	7143.18	2591.16	337.763	299.908	155.369
56	8400.79	3265.98	376.251	329.043	182.471
60	9776.76	4054.43	417.115	359.103	212.202
64	11274.9	4967.29	460.528	390.139	244.7
68	12899.2	6015.92	506.678	422.204	280.118
72	14654.	7212.35	555.762	455.355	318.627
76	16543.6	8569.3	607.989	489.648	360.41
80	18572.8	10100.3	663.576	525.135	405.668
84	20746.3	11819.6	722.746	561.859	454.617
88	23069.3	13742.4	785.731	599.86	507.485
92	25547.	15885.	852.769	639.165	564.519
96	28184.4	18264.5	924.107	679.796	625.98
100	30987.1	20899.1	1000.01	721.766	692.146
104	33960.4	23808.3	1080.74	765.082	763.318
108	37109.6	27012.8	1166.62	809.75	839.824
112	40440.2	30534.5	1257.96	855.772	922.022
116	43957.6	34397.1	1355.14	903.156	1010.31
120	47667.2	38625.7	1458.58	951.913	1105.13

We can visualize the flight of the Shuttle a little better if we Animate the plot, and this is on the Compact Disc as program *SpaceShuttleLaunch*.

For more on the Space Shuttle dynamics, see *The Space Shuttle Operator's Manual* by K.M. Joels, G.P. Kennedy and D. Larkin [Ballantine, 1988]

5.7 Challenge Problems

1. Space Shuttle Phase 2

After solid-booster separation, the mass of the Shuttle and main fuel tank is 690,000 kg. The thrust of the main engine is $T = 5 \times 10^6$ N. If the Space Shuttle main engine burns for 380 seconds with a fuel consumption of 1400 kg/s, find the velocity and height of the Shuttle when the second-phase burn is complete. During the second phase of flight, the Shuttle continues to lean over towards the horizontal, but only at .001 radians per second.

From Section 5.6, the height at booster rocket separation is 47,660 m, and it is 38,620 meters downrange. At the start of the second phase the upward y-velocity is 951 m/s and the x-velocity is 1105 m/s. The heading of the Shuttle is 0.84 radians from vertical at the start of second-phase burn.

During Phase 2 the Shuttle begins to move outward, over the horizon. So be sure when calculating the height to take into account the curvature of the Earth.

2. B–Z Chemical Reaction

The ingredients for a B-Z reaction are the following: 1.415 grams of Sodium bromate, 150 ml 1M H_2SO_4 , 0.175 grams of Cerium-Ammonium-Nitrate, 4.29 grams of malonic acid, all carefully dissolved in distilled water. [See Shakhashiri in the References for the equipment needed and how to safely prepare this solution]. With Ferroin added to the solution, the Cerium ion Ce^{+4} is blue and the Ce^{+3} ion is red. As the reaction proceeds, the solution color changes from green to blue to violet to red, and repeats over and over again as the concentration of the reactants change.

Let $x = [HBrO_2]$, $y = [Br^-]$, and $z = [Ce^{+4}]$ The equations of this B-Z reaction are

$$\epsilon x' = x + y - q x^2 - xy$$
$$7y' = -y + 2hz - xy$$
$$p z' = x - z$$

where $\epsilon = 0.21$, $p = 14$, $q = 0.006$, $h = 0.75$, $x(0) = 100$, $y(0) = 1$, $z(0) = 10$

Show how the concentration of z varies in time (and therefore, the color change blue-red-blue) and, for good measure, also plot x and y as functions of time. What is the period of oscillation of this reaction ? Also try varying the initial concentrations: take $x(0) = 17$, $y(0) = 3$, $z(0) = 5$.

3. Equations of Winemaking

When making wine, the first step is to put the juice from the crushed grapes or grape-juice concentrate into pure water. The amount of natural sugar in the solution may be gauged with a hydrometer. If, for example, our hydrometer reads 1.095 then there are 1140 grams of sugar in a gallon of the grape-juice + pure water mixture.

We now add yeast to the mix. To be scientific we will only add 100 yeast to one gallon. Then place a fermentation lock on the container to allow CO_2 to escape and no foreign material to fall in. Now, here is the winemaking process as seen by the yeast:

1] The yeast break up the sugar into ethyl alcohol + carbon dioxide

$$C_6H_{12}O_6 \longrightarrow 2\ C_2H_5OH + 2\ CO_2 \uparrow$$

2] The yeast extract energy from this process and their numbers grow exponentially at a rate $\quad x' = a\,x \quad$ where $\quad a = 0.125/\,hour$

(the yeast number x increases by a factor of e in 8 hours).

3] Because their numbers increase so rapidly, the yeast soon get in each others way.

Therefore the rate of growth of the yeast population really goes as

$$x' = (a - bx)\,x \quad \text{where } b = 10^{-7}a$$

4] All this time, the yeast have been faithfully producing ethanol. The alcohol percentage is increasingly lethal to the yeast at concentrations beyond 4 %, so the full differential equations are

$$x' = (a - bx - cy_*)\,x \qquad x = \#\ \text{yeast}$$
$$y' = k\,x \qquad\qquad\quad y = \text{alcohol per cent}$$
$$z' = -\gamma\,x \qquad\qquad\quad z = \text{remaining sugar (grams)}$$

initial conditions are x(0) =100, y(0) =0, z(0) =1140 for a one-gallon batch

$$a = 0.125, \quad b = a/10^7, \quad c = a/10, \quad y_* = y - 4 \ \ (\text{before } y = 4, c = 0)$$

$\gamma = 10^{11}$ sugar molecules/ second/ yeast $= \dfrac{180*10^{11}}{6*10^{23}}$ grams/second/yeast.

$k = 3.4 \times 10^{-13}$ percent/ second/ yeast

How long before the yeast reach their maximum population, x_{max} ?

Solve the above equations of winemaking by tracking x , y , and z for 70 days. What is the final alcohol percentage in the wine at this time ?

4. H–H Nerve Conduction

Alan Hodgkin and Andrew Huxley found that the conductivity of the nerve membrane varies with temperature as $3^{(T-6.3)/10}$. That is, for a nerve fiber at a temperature different than 6.3 °C all the conductivity coefficients $\alpha, \beta, \gamma, \delta, \epsilon, \eta$ are multiplied by the above factor.

Find the conduction velocity of a nerve impulse at T = 18.5 °C by using the H–H nerve conduction model of Section 5.4 with $\alpha, \beta, \gamma, \delta, \epsilon, \eta$ all multiplied by 3.82.

Then use the H–H model to show that when a nerve fiber is at 18.5 °C that a 29-mV, 0.5 ms half-sine pulse will elicit an action potential whereas a 28-mV, 0.5 ms half-sine pulse will <u>not</u> produce an action potential. This is called the "all or nothing response" of a nerve fiber.

5. Flight to the Stars

As a final problem, let's evaluate the possibility of achieving interstellar flight utilizing the most powerful physical processes known.

Assume that matter and antimatter can somehow be stored in a rocket and brought together, producing gamma-ray photons. If the radiation is reflected straight back and the process is 100% efficient, then consider an exploratory trip to Alpha Centauri 4.4 lightyears distant. Accelerate 1/2 way at 1 g, decelerate 1/2 way at 1 g, stay 1 year then accelerate and decelerate at 1 g home.

(A) How long does the trip take as seen by the astronauts on the spaceship, and by the people back home on Earth ?

(B) If the total mass of the returning spaceship is 5000 kg (same as Saturn V re-entry vehicle) find the original mass of the starship when it was assembled in space in orbit above the Earth.

Utilize the Impulse–Momentum theorem and the Equations of Special Relativity,

$$m\frac{dv}{d\tau} = -\frac{dm}{d\tau}c = mg \qquad \text{and} \qquad dt = \gamma\, d\tau$$

where Earth-time is t and ship-time is τ and $\gamma = \left(1 - v^2/c^2\right)^{-1/2}$

Appendices

Appendix A
Mathematica, Maple, and MatLab

The Mathematical Softwares   *Mathematica*   Maple   MatLab

In comparing the different mathematical softwares, it is instructive to see how each solves the non-linear differential equation

$$\frac{dy}{dt} = y^2 + t^2 \quad \text{with} \quad y(0) = 0, \text{from } t = 0 \text{ to } t = 1.$$

Mathematica

```
sol = NDSolve[{y'[t] == y[t]^2 + t^2, y[0] == 0}, y, {t, 0, 1}];
Plot[y[t] /. %, {t, 0, 1}]
```

Maple

```
> dsolve ({diff (y (t), t) = y (t)^2 + t^2, y (0) = 0}, y (t), numeric);
> plot (y (t), (t = 0..1));
```

MatLab (with Maple Kernel)

```
>> y = dsolve ('Dy = y (t)^2 + t^2', 'y (0) = 0')
>> ezplot (y, [0, 1])
```

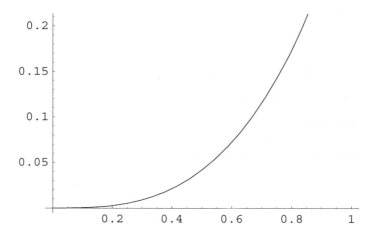

Graphical solution of $y' = y^2 + t^2$, $y_0 = 0$

Mathematical Softwares   *Mathematica*   Maple   MatLab

As a second example, see how each solves the van der Pol non-linear differential equation

$$y'' + \mu \, (y^2 - 1) \, y' + y \; = 0 \quad \text{with} \quad \mu = 2, \; y(0) = 1, \text{and } y' = 0,$$

from $t = 0$ to $t = 15$.

Mathematica

```
NDSolve[{y''[t] + 2 * (y[t]^2 - 1) * y'[t] + y[t] == 0,
   y'[0] == 0, y[0] == 1}, y, {t, 0, 15}];
Plot[y[t] /. %, {t, 0, 15}]
```

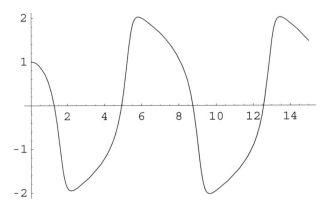

Maple

```
> numsol := dsolve
     ({diff (y (t), t$2) + 2 * (y (t)^2 - 1) * diff (y (t), t) + y (t) = 0,
        D (y) (0) = 0, y (0) = 1}, y (t), numeric);
> with (plots) : odeplot (numsol, [t, (y (t)], 0. .15);
```

MatLab (with standard m-file)

First, set $\frac{dy1}{dt} = y2$, then $\frac{dy2}{dt} = 2 * (1 - y1^2) * y2 - y1$

then write the m-file

```
vanderpol.m
   function yp = vanderpol (t, y)
   yp = [y (2); 2 * (1 - y (1)^2) * y (2) - y (1)];
```
then call this subroutine with the program
```
>> tspan = [0, 15];
>> y0 = [1, 0];
>> [t, y] = ode45 ('vanderpol', tspan, yo);
>> plot (t, y (:, 1))
```

As a third example, consider the predator-prey equations

$$x' = .01x - .001\, xy \quad \text{and} \quad y' = -.015\, y + (.001/40)\, xy \qquad x_0 = 1000, \quad y_0 = 8$$

Mathematica

```
sol = NDSolve[{x'[t] == .01 x[t] - .001 x[t] * y[t],
    y'[t] == -.015 * y[t] + (.001 / 40) * x[t] * y[t],
    x[0] == 1000, y[0] == 8}, {x, y}, {t, 0, 1000}];
p = Plot[Evaluate[{x[t] / 50, y[t]} /. sol,
    {t, 0, 1000}, PlotRange → {0, 22}]];
q = ParametricPlot[{x[t], y[t]} /. sol, {t, 0, 550},
    PlotRange → {{300, 1040}, {4, 19}}]
Show[GraphicsArray[{p, q}]]
```

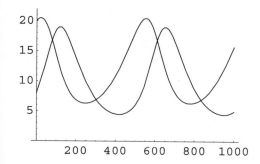 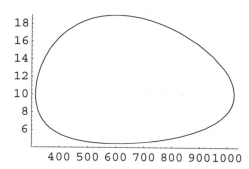

Maple

```
> eq1 := diff (x (t), t) = .01 * x (t) - .001 x (t) * y (t)
> eq2 := diff (y (t), t) = -.015 * x (t) - (.001 / 40) x (t) * y (t)
> sol1 :=
dsolve ({eq1, eq2, x (0) = 1000, y (0) = 8}, {x (t), y (t)}, numeric);
> with (plots) :
> odeplot
    (sol1, [[t, x (t) / 50], [t, y (t)]], 0. .1000, color = [RED, BLUE]);
> odeplot (sol1, [x (t), y (t)], 0. .550, color = [BLACK])
```

MatLab (with standard m-file)

Set $y(1) = x$, and $y(2) = y$

$$\frac{dx}{dt} = .01\, x - .001\, xy, \quad \frac{dy}{dt} = -.015\, y + (.001 / 40)\, xy$$

predprey.m

```
function yp = predprey (t, y)
yp = [.01 * y (1) - .001 * y (1) * y (2);
    -.015 * y (2) + (.001 / 40) * y (1) * y (2)];
```

then call this subroutine with the program

```
>> tspan = [0, 1000];    >> y0 = [1000, 8];
>> [t, y] = ode23 ('predprey', tspan, y0);
>> plot (t, y)       >> plot (y (:, 1), y (:, 2))
```

As a final example, consider **Animations** in *Mathematica,* Maple, and MatLab

In[6]:= *Mathematica*

```
Do[Plot[(Sin[π * x] * Cos[π * t]), {x, 0, 1},
   PlotRange → {-1, 1}], {t, 0, 2, .1}]; (* jump rope *)
```

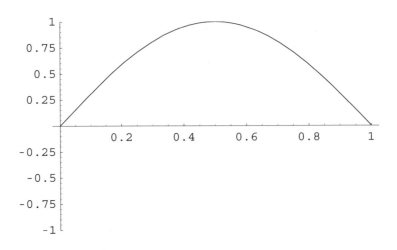

Maple Animation

```
> u := sin (π * x) * cos (π * t)
> with (plots) :
> animate (u (x, t), x = 0. .1, t = 0. .2, frames = 20,
   view = [0. .1, -1..+1], color = BLACK);
```

MatLab Movie (with standard m-file)

```
to run the program
```

>> movie (M, 4, 10)   [run 4 times with 10 frames per second]

```
subroutine (M - File)
x = 0 : .1 : 1;
moviein (20);
for j = 1 : 20
  t = (j - 1) * .1
  u = cos (pi * t) * sin (pi * x)
  plot (x, u)
  M = ( : , j) = getframe;
end
```

Cost and Capability comparison for the Mathematical Software (June 2007)

	Universities and Commercial	other full-capability software	Student Edition
Mathematica 5/6	$999	*Mathematica* for classroom $260	$130
Maple 11	$899	Maple for Faculty $180	$125
MatLab 7	$700	(includes PDE toolbox)	$ 99

	Linear and NL ODEs	Stiff DEs	Partial DEs	Animation	Full Graphing Capability
Mathematica 5/6	✓	✓	✓	✓	✓
Maple 11	✓	✓	✓	✓	✓
MatLab 7	✓	✓	✓	✓	✓

Note 1. Wolfram Inc also offers Calculation Center 3 for $250. This software may be suitable at the high-school level, however it does not have the full capability of *Mathematica* and is not recommended at the college level.

Note 2: All Student editions are full-capability. However print-outs may include the words "Student Edition"

Note 3: Many of the mathematical software packages are available for less at www.Academicsuperstore.com, www.CampusTech.com, or www.Ebay.com.

Note 4: *Mathematica* solves PDEs numerically with NDSolve.
  Maple solves PDEs with pdsolve.
  MatLab utilizes the program pdepe, in the PDE toolbox.

Note 5: Maple may need certain processing limits reset to solve stiff DEs.
  MatLab uses the special program ode15s or ode23s to solve stiff DEs
  Mathematica's NDSolve automatically tests for stiffness and applies proper techniques

Note 6: Full graphics capability implies 2-D and 3-D graphs and fitting functions to data.

Note 7: Trial versions of all three mathematical softwares are available
  Mathematica www.wolfram.com (15 days)
  Maple (thru sales rep) www.maplesoft.com
  MatLab www.mathworks.com (15 days)

Sources for more information on *Mathematica*, Maple, and Matlab

Websites for *Mathematica,* Maple, and MatLab

Mathematica www.wolfram.com   (see also library.wolfram.com)
Maple   www.maplesoft.com (see also www.mapleapps.com)
MatLab   www.mathworks.com

Books on *Mathematica,* Maple, and MatLab

Mathematica
Heikki Ruskeepaa (2004) *Mathematica* Navigator [Elsevier, 2 ed]
Stephen Wolfram (2003) *Mathematica* Book 5 [Wolfram Publishing]
Martha Abell, James Braselton (2004) Differential Equations with
Mathematica [Academic 3 ed]
Ferdinand Cap (2003) Mathematical Methods in Physics & Engineering
[CRC Press]

Maple
Andre Heck (2003) Introduction to Maple [Springer 3 ed]
Martha Abell, James Braselton (2000) Differential Equations with Maple V
[Academic 3 ed]
Jon Davis (2001) Differential Equations with Maple [Birkhauser]
John Putz (2003) Maple Animation [CRC Press]

MatLab
Duane Hanselman, Bruce Littlefield (2003) Mastering MatLab 7 [PrenticeHall]
Stephen Chapra, Raymond Canale (2002) Numerical Methods for Engineers
[McGraw-Hill 4 ed]
William Palm (2004) Introduction to MatLab 7 for Engineers [Prentice-Hall]
Sergey Lyshevski (2003) Engineering and Scientific Computations using
MatLab [Wiley]

Appendix B The Compact Disc

The *Mathematica* Compact Disc

Contained on the Compact Disc are all the *Mathematica* program notebooks and Animations in the book. To run any of these programs, it is recommended that they be downloaded to your computer and then run in *Mathematica*.

This way, you may change any file as you wish. However, to re-run any changed file, be sure to press Shift-Enter to re-do all the calculations, otherwise, the old values may still be there.

To run any of the Animations, just double-click on the first of the string of pictures, and you may adjust the run-speed by clicking on the double arrows at the bottom of the page.

Mathematica Animations

`Animation-1.nb`	moving waves and pulses	(Appendix C)
`BaseballAnimation.nb`	flight of a baseball	(Section 1-4)
`BinaryStarsOrbit.nb`	elliptical orbits of stars	(Section 3-3)
`BouncingBall.nb`	follow the bouncing ball	(Section 1-5)
`HalleysComet.nb`	elliptical orbit of comet	(Section 1-2)
`JupitersMoon.nb`	Ganymede orbiting Jupiter	(Problem 1-7)
`LunarLander.nb`	descent to the moon	(Problem 1-4)
`NerveConduction.nb`	action potential pulse	(Section 5-4)
`SpaceShuttleLaunch.nb`	trajectory of shuttle	(Section 5-5)
`VoyageratJupiter.nb`	gravity-assist from Jupiter	(Section 1-3)

If you would like a larger Animation, the operational command in *Mathematica* is ImageSize. The largest available is **ImageSize → 6*72.** See *VoyageratJupiter.nb* or *HalleysComet.nb* for details.

Mathematica Files Chapter 1

`1-1 Motion-of-the-Planets.nb`	`Problem 1-1 Satellite`
`1-2 HalleysNeartheSun.nb`	`Problem 1-2 DragRacer`
`1-3 VoyageratJupiter.nb`	`Problem 1-3 SolarSailing`
`1-4 BaseballwithAirFriction.nb`	`Problem 1-4 LunarLander`
`1-5 Bounce.nb`	`Problem 1-5 MickeyMantle`
`1-6 SkyDivingwithParachute.nb`	`Problem 1-6 SatelliteOrbits`
	`Problem 1-7 JupitersMoon`

Mathematica Files Chapter 2

2-1 Vibration.nb	Problem 2-1 NYSubway
2-2 ShockAbsorbers.nb	Problem 2-2 LunarSubway
2-3 StepResponse.nb	Problem 2-3 SpeedBump
2-4 RCcircuit.nb	Problem 2-4 RLC-circuit
2-5 VariableStars.nb	Problem 2-5 ImpulseResponse

Mathematica Files Chapter 3

3-1 ModelingwithDiffEqs.nb	Problem 3-1 Pulsar-GenRel
3-2 TrackandField.nb	Problem 3-2 HalleysCometEq
3-3 Pulsar1913+16.nb	Problem 3-3 TheLargestStars
3-4 ExplodingStar.nb	Problem 3-4 AlaskaPipeline
3-5 HeatExchanger.nb	Problem 3-5 HeatExchanger
	Problem 3-6 WomensTrack

Mathematica Files Chapter 4

4-1 PartialDiffEqs.nb	Problem 4-1 RamsauerEffect
4-2 PDE-CopperRod.nb	Problem 4-2 IntermediateStep
4-3 PDE-Heatwave.nb	Problem 4-3 EnergyEigenvalues
4-4 QuantumStep.nb	Problem 4-4 CoolingConcrete
4-4 QuantumBarrier.nb	Problem 4-5 PDE-Steel
4-4 QuantumWell.nb	
4-4 QuantumOscillator.nb	
4-5 HydrogenAtom.nb	
4-6 Deuterium.nb	

Mathematica Files Chapter 5

5-1 LaserPulseDynamics.nb	Problem 5-1 SpaceShuttlePhase2
5-2 B-ZChemicalReaction.nb	Problem 5-2 B-ZReactions
5-3 FoxesandRabbits.nb	Problem 5-3 Winemaking
5-4 PDE-NerveConduction.nb	Problem 5-4 H-HNerveConduction
5-5 FusionReactor.nb	Problem 5-5 FlighttotheStars
5-6 SpaceShuttleLaunch1.nb	

More *Mathematica* Files

* z–1 PDE-Wave.nb	* z–3 AIDSEpidemicUnitedStates.nb
* z–2 PDE-Incropera.nb	* z–4 BinaryStarswithVaporTrails.nb

Appendix C *Mathematica* Commands

<u>The Mathematica Commands</u>. To evaluate any set of *Mathematica* commands one needs to press the keys SHIFT-ENTER on the keyboard.

Basic Functions (note capitalization of functions -- and **square** brackets)

$$\sin(2x) = \mathbf{Sin[2x]} \qquad\qquad \cos(x) = \mathbf{Cos[x]}$$

$$e^{-x} = \mathbf{Exp[-x]} \qquad\qquad \sin^{-1}(x) = \mathbf{ArcSin[x]}$$

$$\ln(x) = \mathbf{Log[x]} \qquad\qquad \log_{10}(x) = \mathbf{Log[10,x]}$$

$$i = \mathbf{I} \qquad\qquad \tanh(x) = \mathbf{Tanh[x]}$$

Summations (utilize Basic Input Palette)

$$In[9] := \sum_{N=1}^{\infty} \frac{1}{N^2}$$

$$Out[9] = \frac{\pi^2}{6}$$

Complex Numbers

```
In[63] := z = 7 + 8 I ;
          Re[z]
          Im[z]
          Abs[z]
```

$$Out[64] = 7$$

$$Out[65] = 8$$

$$Out[66] = \sqrt{113}$$

Solving Algebraic Equations (note double equals in equations)

$$In[59] := \mathbf{Solve[x^2 + 14\,x + 9 == 0,\ x]}$$

$$Out[59] = \left\{ \left\{ x \to -7 - 2\sqrt{10} \right\},\ \left\{ x \to -7 + 2\sqrt{10} \right\} \right\}$$

$$In[58] := \mathbf{NSolve[x^3 + 2\,x + 3 == 0,\ x]}$$

$$Out[58] = \{\{x \to -1.\},\ \{x \to 0.5 - 1.65831\,i\},\ \{x \to 0.5 + 1.65831\,i\}\}$$

Solving Transcendental Equations (note starting value of 1000)

In[60]:= FindRoot[40 * Log[10, z] + .001 z == 150, {z, 1000}]

Out[60]= {z → 4372.16}

Solving simultaneous equations

In[61]:= Solve[{x + y == 5, x - y == 1}, {x, y}]

Out[61]= {{x → 3, y → 2}}

CALCULUS

Differentiation

In[51]:=   D[x², x]

Out[52]= 2 x

Integration

$$In[1]:= \int_0^1 x^2 \, dx$$

Out[1]= $\dfrac{1}{3}$

Numerical Integration

In[5]:= α = 8.62; β = 45.8/√2;
f[x_] = Exp[-x / α] * Exp[-β x⁻¹ᐟ²];
NIntegrate[f[x], {x, 1, 101}]

Out[7]= 0.00278303

Maximize

In[10]:= α = 8.62; β = 45.8/√2;
Maximize[{Exp[-x / α] * Exp[-β x⁻¹ᐟ²], x > 0}, x]

Out[11]= {0.0000856866, {x → 26.9082}}

Minimize

$$In[12]:= \text{FindMinimum}\left[400 * \left(\frac{1}{x^{12}} - \frac{1}{x^6}\right), \{x, 1\}\right]$$

Out[13]= {-100., {x → 1.12246}}

Differential Equations

In [15] := DSolve[{y''[x] + 3 y'[x] - 4 y[x] == 0, y[0] == 1, y'[0] == 0},
y[x], x] // Simplify
Plot[Evaluate[y[x] /. %], {x, 0, 5}]

Out [15] = $\left\{\left\{y[x] \rightarrow \dfrac{e^{-4x}}{5} + \dfrac{4 e^{x}}{5}\right\}\right\}$

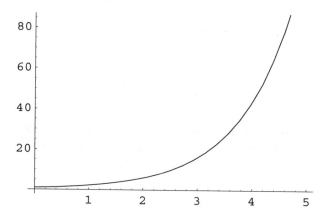

Numerical Solution of any differential equation

In [1] := sol = NDSolve[{y''[t] + 2 * (y[t]^2 - 1) * y'[t] + y[t] == 0,
y'[0] == .001, y[0] == 1}, y, {t, 0, 15}];
Plot[y[t] /. sol, {t, 0, 15}]
Table[{t, y[t] /. sol, y'[t] /. sol}, {t, 0, 1, .2}] // TableForm

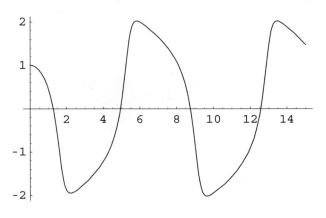

t	y	y '
0	1.	0.001
0.2	0.980234	-0.198464
0.4	0.920465	-0.400748
0.6	0.818515	-0.624963
0.8	0.666473	-0.910267
1.	0.445819	-1.3259

Partial Differential Equations

(*The partial differential equation for heat diffusion*)

$$\frac{\partial u}{\partial t} = \alpha \frac{\partial^2 u}{\partial x^2} \qquad \alpha = .96;$$

Boundary condition $u(x, t) = 0$ at $x = 0$ and $x = 20$

Initial condition $u(x, 0) = 100$ for $0 < x < 20$

This PDE is written in *Mathematica* as

```
In[79]:=  α = .96;
      eq4 = {D[u[x, t], t] - .96 * D[u[x, t], x, x] == 0,
        u[x, 0] == Which[x ≤ 0, 0, 0 < x < 20, 100, x ≥ 20, 0],
        u[0, t] == 0, u[20, t] == 0};
      sol4 = NDSolve[eq4, u[x, t], {x, 0, 20}, {t, 0, 40.0}];
      Plot3D[Evaluate[u[x, t] /. sol4[[1]]], {t, 0, 40.0}, {x, 0, 20}],
        PlotRange → All, AxesLabel → {"t", "x", " "}];
```

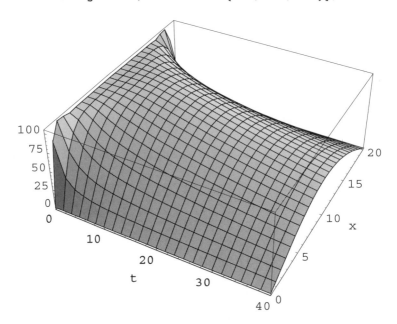

3-D Plot of temperature function $u[x, t]$

Animation (requires use of a Do-Loop)

In[83]:= **Do[Plot[(Sech[(x - t)]) ^2, {x, -5, 5}, PlotRange → {0, 1}],**
{t, -.5, 8, .5}];

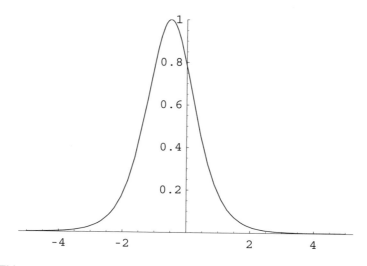

This creates a sequence of pictures, that when scrolled through rapidly,
give the illusion of motion ...   Double-click on the plot to see the animation ...

Notes:

1. If you would like to suppress the In/Out labels on calculations, go to the
<u>Kernel</u> and turn the **Show In/Out Names** option <u>off</u>

2. There are many more *Mathematica* functions than shown in this brief
introduction. See Chapters 1, 3, and 5 for more animations. See Chapters 4
and 5 for more Partial Differential equations. See Chapters 1 thru 5 for many
more solutions to linear and non-linear differential equations.

3. For a direct comparison of *Mathematica* commands with Maple and,
MatLab, turn to Appendix A.

4. In *Mathematica* version 6, the word **Do** in a Do-Loop is replaced with
the word **Animate**.

5. Some changes have been made in *Mathematica* 6 to the "front-end" or
user-interface. Thus programs written in *Mathematica* 6 will not run in
Mathematica 5, however *Mathematica* 5 programs (with some revision) will
run in *Mathematica* 6. To avoid confusion, please download any programs
you wish to run from the Compact Disc, using *Mathematica*-5 or *Mathematica*-6 files as appropriate for your computer.

Problem Solutions

Problem1-1 Changing Satellite Speeds

We will start the satellite at the same location as in Section 1.3, except this time, we will give the satellite an initial velocity $v_y = 20$ km/s and $v_x = 0$.

```
M = 1.9 * 10^27; G = 60^2 * 6.67 * 10^-29; (*dist in 10^3 km,time in min*)
sol = NDSolve[{
    x''[t] == M * G * (.013 * 60 t - x[t]) /
        ((x[t] - .013 * 60 t)^2 + y[t]^2)^1.5,
    y''[t] == M * G * (-y[t]) / ((x[t] - .013 * 60 t)^2 + y[t]^2)^1.5,
    x[0] == 10^3 * Cos[40 π / 180], y[0] == -10^3 * Sin[40 π / 180],
    x'[0] == 0, y'[0] == .020 * 60}, {x, y}, {t, 0, 2400}]

InterpFunc1 = x /. sol[[1]]; InterpFunc2 = y /. sol[[1]];
InterpFunc3 = x' /. sol[[1]]; InterpFunc4 = y' /. sol[[1]];
vel[t_] = √(InterpFunc3[t]^2 + InterpFunc4[t]^2) * 1000 / 60;
dist[t_] = √((InterpFunc1[t] - .78 t)^2 + InterpFunc2[t]^2);
ParametricPlot[{x[t], y[t]} /. sol, {t, 0, 2400},
    Prolog → Circle[{490, 1}, 70], AspectRatio → Automatic];
```

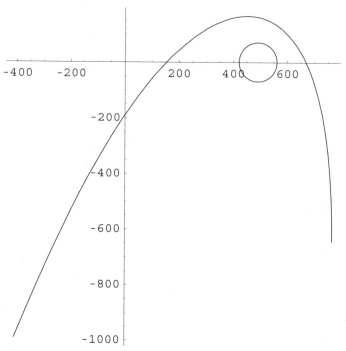

Path of satellite in-front-of Jupiter. All distances in thousands of kilometers. Satellite is slowed from 20 km/s going in to 11 km/s going out.

```
Table[{t, InterpFunc1[t], InterpFunc2[t], vel[t],
      (.78 t), dist[t]}, {t, 0, 2400, 60}] // TableForm
```

t(min)	x(1000 km)	y(1000 km)	v(km/s)	x(Jup)	distance to Jup(1000km)
0	766.0	-642.7	20.	0	1000.
60	765.3	-570.2	20.31	46.8	917.342
120	763.1	-496.4	20.69	93.6	833.544
180	759.0	-421.2	21.15	140.	748.468
240	752.4	-344.4	21.71	187.	661.959
300	742.6	-265.6	22.43	234.	573.874
360	728.3	-184.5	23.37	280.	484.149
420	707.4	-100.8	24.67	327.	393.034
480	675.9	-14.46	26.54	374.	301.933
540	625.8	71.959	29.22	421.	216.925
600	543.7	143.72	31.36	468.	162.461
660	438.4	166.07	28.27	514.	182.776
720	348.3	144.53	23.68	561.	257.624
780	278.3	107.30	20.65	608.	347.046
840	221.6	65.350	18.70	655.	438.477
900	173.4	21.973	17.37	702.	528.972
960	131.2	-21.69	16.40	748.	617.883
1020	93.45	-65.23	15.67	795.	705.169
1080	58.92	-108.4	15.09	842.	790.946
1140	27.01	-151.3	14.62	889.	875.368
1200	-2.77	-193.8	14.23	936.	958.582
1260	-30.8	-236.0	13.90	982.	1040.72
1320	-57.3	-277.8	13.62	1029	1121.9
1380	-82.6	-319.3	13.37	1076	1202.22
1440	-106.	-360.5	13.16	1123	1281.76
1500	-130.	-401.4	12.97	1170	1360.6
1560	-152.	-442.0	12.80	1216	1438.8
1620	-174.	-482.4	12.65	1263	1516.42
1680	-194.	-522.6	12.51	1310	1593.51
1740	-215.	-562.6	12.39	1357	1670.1
1800	-235.	-602.3	12.27	1404	1746.24
1860	-254.	-641.9	12.17	1450	1821.96
1920	-273.	-681.3	12.07	1497	1897.29
1980	-291.	-720.5	11.99	1544	1972.25
2040	-309.	-759.6	11.91	1591	2046.87
2100	-327.	-798.5	11.83	1638	2121.17
2160	-344.	-837.3	11.76	1684	2195.18
2220	-361.	-876.0	11.69	1731	2268.9
2280	-378.	-914.6	11.63	1778	2342.36
2340	-394.	-953.0	11.57	1825	2415.56
2400	-410.	-991.3	11.52	1872	2488.53

Problem 1-2   Drag Racing

Equations of motion for the drag racer

$$v' = \mu g - kv^2 \quad \text{and} \quad x'' = \mu g - kx'^2$$

Terminal velocity v_t is reached when $kv^2 = \mu g$ therefore $k = \dfrac{\mu g}{v_t^2}$

To see the form of the velocity,
take a trial case with $\mu = 2$, $v_{term} = 100$ m/s, $g = 9.8$ m/s^2

```
In[29]:=  Clear[x, v, t];
          sol =
            DSolve[{v'[t] == 2 * 9.8 - .00196 * v[t]^2, v[0] == 0}, v[t], t] //
            FullSimplify
          Plot[Evaluate[v[t] /. sol, {t, 0, 10}]];
          v[t] /. sol[[1]] // Chop
```

Out[6]= $\{\{v[t] \rightarrow 100. \; \text{Tanh}[0.196 \; t]\}\}$

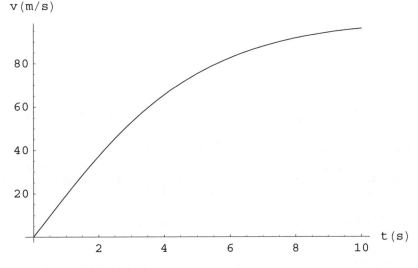

Out[8]= $100. \; \text{Tanh}[0.196 \; t]$

In terms of μ, g, v_t the velocity is $\quad v[t] = v_t \, \text{Tanh}\left[\dfrac{\mu \, g}{v_t} \, t\right]$

To see the form of the distance travelled, solve the equation $x'' = \mu g - kx'^2$
with the same values $\mu = 2$, $v_{term} = 100\,m/s$, $g = 9.8\,m/s^2$, $k = .00196$

$In[13]:=$ **soll = DSolve[{x"[t] == 2 * 9.8 - .00196 x'[t]^2, x[0] == x'[0] == 0},**
             x[t], t] // FullSimplify
             Plot[Evaluate[x[t] /. soll, {t, 0, 10}]];

$Out[13]=$ $\{x[t] \to 510.204\ Log[Cosh[0.196\ t]]\}\}$

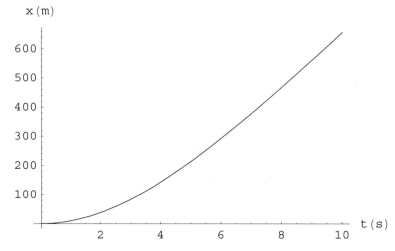

In terms of μ, g, v_t

the distance travelled is $\quad x[t] = \dfrac{v_t{}^2}{\mu g}\ Log\Big[Cosh\Big[\dfrac{\mu g}{v_t}\,t\Big]\Big]$

Now we wish to solve the following two equations for μ and v_t

$$v[t] = v_t\ Tanh\Big[\frac{\mu g}{v_t}\,t\Big] \quad \text{and} \quad x[t] = \frac{v_t{}^2}{\mu g} * Log\Big[Cosh\Big[\frac{\mu g}{v_t}\,t\Big]\Big]$$

Since we know $x = 400\,m$, $v = 143\,m/s$ when $t = 4.5\,s$,

$$143 = v_t\ Tanh\Big[\frac{9.8\,\mu}{v_t} * 4.5\Big] \quad \text{and} \quad 400 = \frac{v_t{}^2}{9.8\,\mu}\ Log\Big[Cosh\Big[\frac{9.8\,\mu}{v_t} * 4.5\Big]\Big]$$

we use *Mathematica* to solve these two equations

$In[68]:=$ **FindRoot$\Big[\Big\{$143 == v * Tanh$\Big[\dfrac{44.1\,u}{v}\Big]$,**

            400 == $\dfrac{v^2}{9.8\,u}$ * Log$\Big[$Cosh$\Big[\dfrac{44.1\,u}{v}\Big]\Big]\Big\}$, {{u, 2}, {v, 100}}$\Big]$

$Out[68]=$ $\{u \to 5.20113,\ v \to 160.362\}$

The answer to the required coefficient of friction is a rather surprising $\mu = 5.2$ and the terminal velocity at constant μg acceleration is $v \to 160.36$ m/s .

It would be most interesting to have actual accelerometer data from a high–speed drag race. My guess is the coefficient of friction would exceed 6 at the beginning of the race, and tail off during the race because the tires would not have the same grip on the road.

It is also possible to solve this problem symbolically. Using *Mathematica*, we solve $\qquad v' = \mu g - kv^2 \quad$ and $\quad x'' = \mu g - kx'^2$

```
In[21]:=  Assuming[{Re[µg] > 0, Re[k] > 0},
            DSolve[{v'[t] == µg - k*v[t]^2, v[0] == 0}, v[t], t]]
          Assuming[{Re[µg] > 0, Re[k] > 0},
            DSolve[{x''[t] == µg - k*x'[t]^2, x[0] == x'[0] == 0}, x[t], t]]
```

Out[21]=

$$\left\{\left\{v[t] \to \frac{\sqrt{\mu g}\ \text{Tanh}\left[\sqrt{k}\ t\ \sqrt{\mu g}\ \right]}{\sqrt{k}}\right\}\right\}$$

Out[22]=

$$\left\{\left\{x[t] \to \frac{\text{Log}\left[\text{Cosh}\left[\sqrt{k}\ t\ \sqrt{\mu g}\ \right]\right]}{k}\right\}\right\}$$

Then, as before, with $v = 143$ m/s and $x = 400$ m, when $t = 4.5$ s,

```
In[24]:=  FindRoot[{143 ==
```
$$\frac{\sqrt{\mu g}\ \text{Tanh}\left[\sqrt{k}\ *4.5*\ \sqrt{\mu g}\ \right]}{\sqrt{k}},$$
```
          400 ==
```
$$\frac{\text{Log}\left[\text{Cosh}\left[\sqrt{k}\ *4.5*\ \sqrt{\mu g}\ \right]\right]}{k}\}, \{\{µg, 2\}, \{k, 1\}\}]$$

Out[24]= $\{µg \to 50.9711,\ k \to 0.00198208\}$

Then, $\mu = \dfrac{50.9711}{9.8} = 5.2$ and $v_t = \dfrac{\sqrt{\mu g}}{\sqrt{k}} = 160.36$ m/s

Problem1-3 Solar Sailing

In[5]:= `r = r₀ = 1.5 * 10¹¹ ; (*distance from sun, 1 AU*)`
          `a = .00592 ;   (*solar gravity at 1 AU*)`
          `g = .00018 ;   (*solar sail acceleration at 1 AU*)`

At Earth's orbit, r = r_0 and we solve for u(x) as a function of sail angle x

In[6]:= $u[x\_] = \dfrac{2\, r_0 * (g * Sin[x * \pi / 180] * Cos[x * \pi / 180]\,\hat{}\,2)}{\cdot r^{.5}\ (a - g * Cos[x * \pi / 180]\,\hat{}\,3)\,\hat{}\,.5}$;   (*EQ 1*)

          `Plot[u[x], {x, 0, 90}];`
          `Maximize[{u[x], 0 < x < 90}, x]`

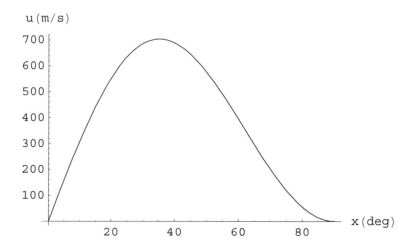

Out[8]= `{703.349, {x → 35.0941}}`

outward velocity (m / s) vs. sail angle in degrees

Maximum velocity near the Earth's orbit is 703 m/s achieved with a sail angle of 35 degrees. To compute the time to go from Earth's orbit to Mars, notice that u diminishes as $r^{-1/2}$

          `b = 703.35 ;  r = 1.5 * 10¹¹ ;`

          $T = \displaystyle\int_{r}^{1.5\,r} \dfrac{z^{.5}}{r^{.5} * b}\, dz$

          `Print["T = ", T / (3.15 * 10^7), " years"]`

          `T = 3.77837 years`

Problem1-4 Lunar Lander

In solving this problem of how to get down to the lunar surface at zero downward velocity, starting from an initial height 10,000 m and an initial downward velocity of 100 m/s, we will adopt the strategy of not turning on the jets for t seconds, and then fire the rockets for T seconds.

Then, $v = v_o + gt$ and $h = 10000 - v_o\, t - .5\, gt^2$

before firing the rockets

and $2.3\, T = v_o + gt$ and $\frac{1}{2} 2.3\, T^2 = h_o - v_o\, t - .5\, gt^2$

after firing the rockets

By substituting for T in the $\frac{1}{2} 2.3\, T^2$ expression, we solve for t. Using *Mathematica*,

```
v0 = 100; g = 1.7;

NSolve[ .5 * 2.3 * (v0 + g * t)^2 / 2.3^2  == 10000 - v0 * t - .5 * g * t^2, t]

{{t → -152.388}, {t → 34.741}}

T =  v0 + g * 34.74
     ---------------
          2.3

T = 69.1557
```

We have now solved for t, the free-fall time, and T the rocket burn time. How much fuel have we expended? F = 2 * 69.15 = 138.3 kg. We are ok. The question now is, do we touch down at zero downward velocity at the surface? Let's plot the entire descent and then Animate the motion. (assume a sideways velocity of $x' = 2$ m/s for visualization)

```
sol = NDSolve[{y'[t] == - (100 + 1.7 t -
        4 * If[t < 34.74, 0, 1] * (t - If[t < 34.74, 0, 34.74])),
    x'[t] == 2, y[0] == 10^4, x[0] == 0}, {x, y}, {t, 0, 104}];
InterpFunc1 = x /. sol[[1]]; InterpFunc2 = y /. sol[[1]];
InterpFunc3 = y' /. sol[[1]];
```

```
ParametricPlot[{x[t], y[t]} /. sol, {t, 0, 103}]
```

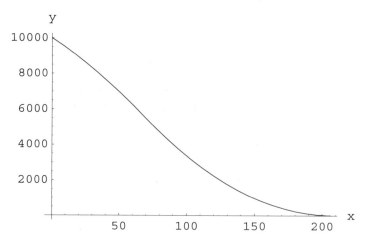

```
Table[{t, InterpFunc1[t], InterpFunc2[t], InterpFunc3[t]},
    {t, 0, 102, 6}] // TableForm
```

t (s)	x (m)	y (m)	↓ v (m/s)
0	0.	10000.	-100.
6	12.	9369.4	-110.2
12	24.	8677.6	-120.4
18	36.	7924.6	-130.6
24	48.	7110.4	-140.8
30	60.	6235.	-151.
36	72.	5301.58	-156.16
42	84.	4406.02	-142.36
48	96.	3593.26	-128.56
54	108.	2863.3	-114.76
60	120.	2216.14	-100.96
66	132.	1651.78	-87.16
72	144.	1170.22	-73.36
78	156.	771.455	-59.56
84	168.	455.495	-45.76
90	180.	222.335	-31.96
96	192.	71.9753	-18.16
102	204.	4.41531	-4.36

We may now Animate the plot

```
Needs["Graphics`MultipleListPlot`"]
Do[MultipleListPlot[
  Table[{InterpFunc1[t], InterpFunc2[t]}, {t, 0, i, 4}],
  SymbolShape → {PlotSymbol[Triangle, 5]}, SymbolStyle → {Hue[2 / 3]},
  PlotRange → {{0, 202}, {-10, 10000}}], {i, 0, 104, 4}]
```

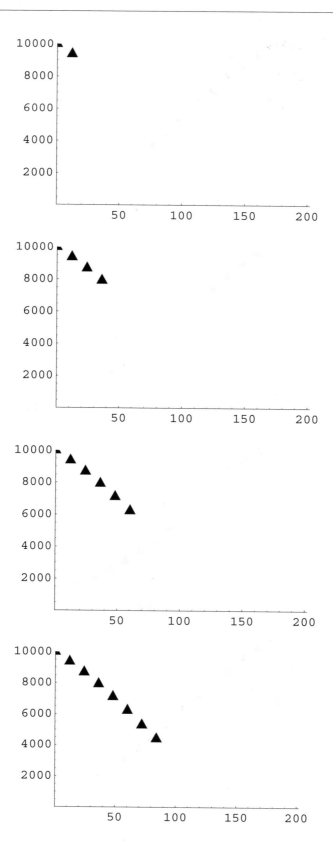

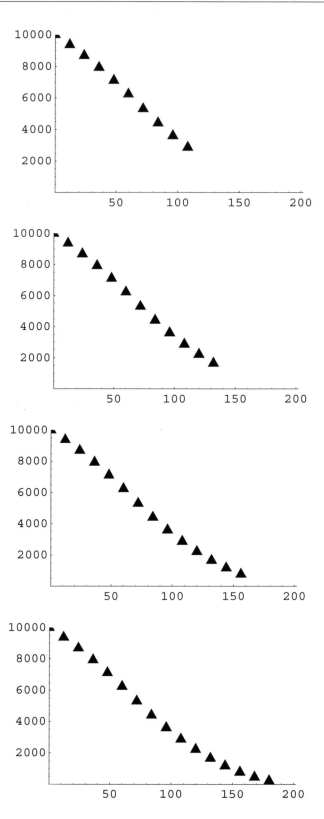

Problem 1-5 Mickey Mantle Home Run

In 1963, Mickey Mantle hit a home run into the third deck at Yankee Stadium. The baseball struck the facade over the third deck some 106 feet above the playing field, and 365 feet from home plate. If the (x,y) coordinates of the ball were (365, 107) then this would have been the first home run ever hit out of Yankee Stadium.

We will assume the same parameters of Section 1-4: a constant drag coeff $C_D = 0.4$, air density = 0.077 lb/ft^3, diameter of baseball = 0.238 ft, and weight of the baseball mg = 0.3125 lb. The drag force is $D = \frac{1}{2}\rho A C_D\, v^2$

To get the required height of the baseball, we will try an initial angle $\theta = 33°$

$$x'' = -.00219\, x' \sqrt{x'^2 + y'^2} \qquad\qquad x_o = 0 \qquad x'_o = 220 * \cos 33°$$

$$y'' = -32 - .00219\, y' \sqrt{x'^2 + y'^2} \qquad y_o = 0 \qquad y'_o = 220 * \sin 33°$$

Assuming, as in Section 1-4, that the velocity of the baseball off the bat is 220 ft/s, then the initial x- and y-velocity of the baseball are

```
vx = 220. * Cos[33 * π / 180]        vy = 220. * Sin[33 * π / 180]

    vx = 184.508                        vy = 119.821
```

```
sol = NDSolve[{y''[t] == -32 - .00219 y'[t] * √x'[t]^2 + y'[t]^2 ,
    x''[t] == -.00219 x'[t] * √x'[t]^2 + y'[t]^2 , x[0] == y[0] == 0,
    x'[0] == 184.5, y'[0] == 120}, {x, y}, {t, 0, 5.4}]

InterpFunc1 = x /. sol[[1]]; InterpFunc2 = y /. sol[[1]];
InterpFunc3 = x' /. sol[[1]]; InterpFunc4 = y' /. sol[[1]];
tbl = Table[{InterpFunc1[t], InterpFunc2[t]}, {t, 0, 5.4, .2}];
ListPlot[tbl, Prolog → AbsolutePointSize[4], AspectRatio → 0.3];
```

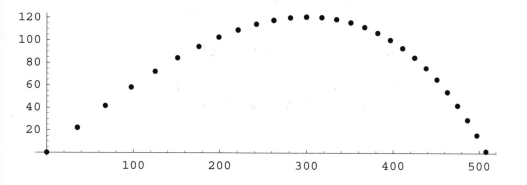

If we **Table** the results, we will find the distance the baseball travelled and the time. From the Table, the baseball passed above the point (365, 107) at $t = 3.2$ seconds.

$$\texttt{Table}\big[\{\texttt{t, Chop[InterpFunc1[t]], Chop[InterpFunc2[t]],}$$
$$\sqrt{\texttt{InterpFunc3[t]}\,\char`\^\,\texttt{2 + InterpFunc4[t]}\,\char`\^\,\texttt{2}}\,\},$$
$$\{\texttt{t, 0, 5.4, .2}\}\big]\;\texttt{// TableForm}$$

The (x,y) coordinates and velocity of the baseball during its 5.4-second flight

t (s)	x (ft)	y (ft)	v (ft / s)
0	0	0	220.091
0.2	35.2363	22.297	197.612
0.4	67.5446	41.5133	178.854
0.6	97.4139	58.0463	162.984
0.8	125.222	72.2017	149.413
1.	151.266	84.219	137.718
1.2	175.783	94.2888	127.591
1.4	198.966	102.565	118.802
1.6	220.973	109.173	111.179
1.8	241.932	114.217	104.592
2.	261.952	117.783	98.9408
2.2	281.122	119.944	94.148
2.4	299.517	120.764	90.1501
2.6	317.2	120.297	86.8926
2.8	334.221	118.591	84.3258
3.	350.625	115.69	82.4007
3.2	366.447	111.634	81.0673
3.4	381.717	106.463	80.2728
3.6	396.459	100.212	79.9619
3.8	410.693	92.9191	80.0774
4.	424.436	84.6197	80.5613
4.2	437.703	75.3504	81.3567
4.4	450.506	65.1483	82.4089
4.6	462.854	54.0509	83.6668
4.8	474.758	42.0962	85.0836
5.	486.227	29.3225	86.6174
5.2	497.268	15.7685	88.2312
5.4	507.891	1.47279	89.8934

Mantle's prodigious drive would have travelled somewhere around 508 feet.

For more on baseball home runs, see Dan Valenti *Clout! The Top Home Runs in Baseball History* [Stephen Greene Publisher, 1989]

Problem 1-6 Satellite Orbits

Let us first find the velocity of the satellite in its original circular orbit

```
r = 6.6 * 10^6; M = 6 * 10^24;  G = 6.67 * 10^-11;
v0 = NSolve[v^2 / r == M * G1 / r^2, v]

{{v → -7786.94}, {v → 7786.94}}
```

Then we set up the Newtonian equations for the new orbit

$$\ddot{x} = -\frac{GM\,x}{(x^2 + y^2)^{3/2}} \qquad x_0 = 6600 \text{ km} \qquad \dot{x}_0 = 0$$

$$\ddot{y} = -\frac{GM\,y}{(x^2 + y^2)^{3/2}} \qquad y_0 = 0 \qquad \dot{y}_0 = 1.08 * 7.787 \text{ km/s}$$

```
M = 6 * 10^24; G = (60^2) * 6.67 * 10^-20; (*dist in km, time in min*)
sol = NDSolve[{x''[t] == M * G * (-x[t]) / (x[t]^2 + y[t]^2)^1.5,
    y''[t] == M * G * (-y[t]) / (x[t]^2 + y[t]^2)^1.5,
    x[0] == 6.6 * 10^3, y[0] == 0,
    x'[0] == 0, y'[0] == 1.08 * 7.787 * 60}, {x, y}, {t, 0, 1000}]
ParametricPlot[{x[t], y[t]} /. sol, {t, 0, 800}];
```

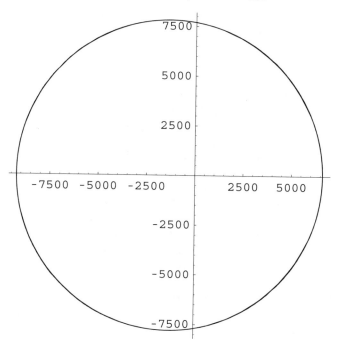

New orbit of the satellite, after rocket burn. Perigee remains at 6600 km from Earth center.

New apogee is at 9230 km from Earth center. (See **Table** for values of distance and velocity)

```
InterpFunc1 = x /. sol[[1]]; InterpFunc2 = y /. sol[[1]];
InterpFunc3 = x' /. sol[[1]]; InterpFunc4 = y' /. sol[[1]];
Rad[z_] = √(InterpFunc1[z]^2 + InterpFunc2[z]^2) ;
Dist[z_] = √(InterpFunc1[z]^2 + InterpFunc2[z]^2) - 6400;
Vel[z_] := √(InterpFunc3[z]^2 + InterpFunc4[z]^2) / 60;
Table[{t, Rad[t], Dist[t], Vel[t]}, {t, 0, 150, 5}] // TableForm
```

t (min)	R (km)	D (km ↑ E)	v (km/s)
0	6600.	200.	**8.40996**
5	6667.74	267.744	8.33638
10	6859.09	459.086	8.13306
15	7143.51	743.507	7.84223
20	7482.99	1082.99	7.51115
25	7841.59	1441.59	7.17815
30	8190.14	1790.14	6.86891
35	8506.99	2106.99	6.59864
40	8776.94	2376.94	6.37559
45	8989.74	2589.74	6.204
50	9138.72	2738.72	6.08589
55	9219.9	2819.9	6.0222
60	9231.29	2831.29	6.0133
65	9172.64	2772.64	6.05923
70	9045.33	2645.33	6.15973
75	8852.67	2452.67	6.31412
80	8600.35	2200.35	6.52079
85	8297.33	1897.33	6.77639
90	7957.04	1557.04	7.07424
95	7598.87	1198.87	7.40177
100	7249.52	849.522	7.73705
105	6943.01	543.009	8.04588
110	6717.51	317.506	8.28288
115	6607.24	207.24	**8.40206**
120	6631.16	231.163	8.37601
125	6785.03	385.03	8.21099
130	7043.27	643.275	7.94324
135	7369.07	969.065	7.62043
140	7725.03	1325.03	7.28464
145	8079.57	1679.57	6.96559
150	8408.66	2008.66	6.68149

Notice that the period of the new orbit (when v = 8.4 km/s again) is 115 minutes.

Problem 1-7 Jupiter's Moon

Let us write the equations of motion for Ganymede orbiting Jupiter.

First, find the orbital velocity of Ganymede, in a circular orbit,

$$\frac{mv^2}{r} = \frac{mGM}{r^2} \qquad v = 11.26 \text{ km/s}$$

Then program the Equations of motion for Ganymede orbiting Jupiter

$$\ddot{x} = -\frac{GM\,(x - 13\,t)}{\left((x - 13\,t)^2 + y^2\right)^{3/2}} \qquad x_0 = 0 \qquad \dot{x}_0 = 24.26 \text{ km/s}$$

$$\ddot{y} = -\frac{GM\,y}{\left((x - 13\,t)^2 + y^2\right)^{3/2}} \qquad y_0 = -10^6 \text{ km} \qquad \dot{y}_0 = 0$$

This takes into account the motion of Jupiter, which is moving along $+x$ at 13 km/s.

```
In[39]:=  G = 6.673 * 10^-11; M = 1.9 * 10^27; r = 10^9;
          Solve[v^2 / r == G M / r^2, v]

Out[41]=  {{v → -11260.}, {v → 11260.}}


M = 1.9 * 10^27; G = (60^2) * 6.673 * 10^-29; (*dist in 10^3 km, time in min*)
sol = NDSolve[{
   x''[t] == G * M * (.013 * 60 t - x[t]) / ((x[t] - .013 * 60 t)^2 + y[t]^2)^1.5,
   y''[t] == G * M * (-y[t]) / ((x[t] - .013 * 60 t)^2 + y[t]^2)^1.5,
   x[0] == 0, y[0] == -10^3, x'[0] == .02426 * 60, y'[0] == 0},
   {x, y}, {t, 0, 24000}]

InterpFunc1 = x /. sol[[1]]; InterpFunc2 = y /. sol[[1]];
InterpFunc3 = x' /. sol[[1]]; InterpFunc4 = y' /. sol[[1]];
Table[{t, Chop[InterpFunc1[t]], InterpFunc2[t], (.013 * 60 * t),
   √(InterpFunc3[t] - .78)^2 + InterpFunc4[t]^2 * 1000 / 60,
   √(InterpFunc1[t] - .78 t)^2 + InterpFunc2[t]^2},
   {t, 0, 10000, 400}] // TableForm
ParametricPlot[{x[t], y[t]} /. sol, {t, 0, 23000}];
```

Table of values with Time in minutes, Distance in thousands of kilometers

t(min)	x	y	v(km/s)	x(Jup)	dist from Jup (1000 km)
0	0	-1000.	11.26	0	1000.
400	578.96	-963.7	11.26	312.	1000.
800	1138.5	-857.4	11.26	624.	1000.
1200	1660.7	-688.9	11.26	936.	1000.
1600	2130.4	-470.4	11.26	1248.	1000.
2000	2535.9	-217.8	11.26	1560.	1000.
2400	2870.7	50.616	11.26	1872.	1000.
2800	3132.9	315.39	11.26	2184.	1000.
3200	3326.3	557.28	11.26	2496.	1000.
3600	3459.4	758.72	11.26	2808.	1000.
4000	3545.2	905.09	11.26	3120.	1000.
4400	3600.1	985.76	11.26	3432.	1000.
4800	3642.9	994.88	11.26	3744.	1000.
5200	3692.9	931.79	11.26	4056.	1000.
5600	3769.4	801.07	11.26	4368.	1000.
6000	3889.2	612.20	11.26	4680.	1000.
6400	4066.5	378.89	11.26	4992.	1000.
6800	4310.9	118.08	11.26	5304.	1000.
7200	4627.5	-151.2	11.26	5616.	1000.
7600	5015.7	-409.6	11.26	5928.	1000.
8000	5470.2	-638.3	11.26	6240.	1000.
8400	5980.6	-820.6	11.26	6552.	1000.
8800	6532.4	-943.4	11.26	6864.	1000.
9200	7108.3	-997.7	11.26	7176.	1000.
9600	7689.1	-979.5	11.26	7488.	1000.

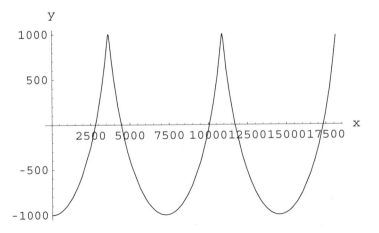

The x-y coordinates of Jupiter's moon. Believe it or not, this is the exact path of the moon as Jupiter moves along + x at 13 km/s. To see how this results in a circular orbit, **Animate** the plot. To see the animation, access *Jupiter'sMoon* on the CD.

Do Loop

```
Needs["Graphics`MultipleListPlot`"]
Do[
 MultipleListPlot[Table[{InterpFunc1[t], InterpFunc2[t]}, {t, i, i}],
  {{0 + .78 * t, 10}} /. {t → i}, PlotRange → {{0, 7600}, {-1100, 1100}},
  SymbolShape → {MakeSymbol[Circle[{0, 0}, 50]],
    MakeSymbol[Circle[{0, 0}, 140]]},
  AspectRatio → Automatic, Ticks → None], {i, 0, 9600, 640}]
```

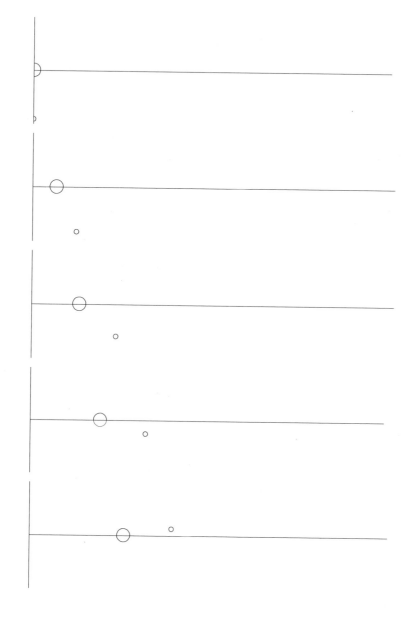

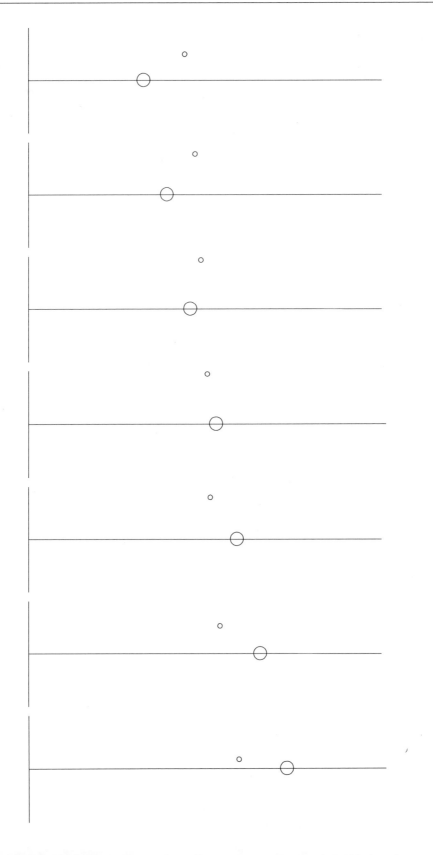

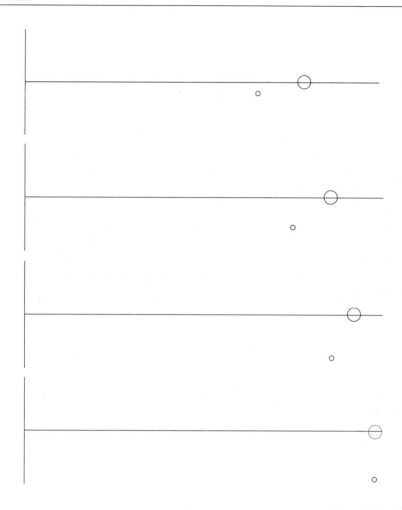

To slow down the animation, just click on the downward-facing chevron. Or, you can change the Cell option "AnimationDisplayTime." Select the cell bracket that is grouping all 6 cells that contains the graphs, go to Format > Options Inspector, then search for "AnimationDisplayTime."

Problem 2-1 New York to Leningrad

A famous physics problem is to consider travel by gravity by tunneling from one city to another over great distances through the Earth. One interesting possibility would be to tunnel from New York to Leningrad, a distance of some 12,000 km over the Earth's surface or 108° of a great circle around the Earth.

Such a tunnel would reach a depth of 2640 km and would have to be impervious to the several thousand degree temperatures inside the Earth. However, if such a tunnel could be built, what would be the transit time from New York to Leningrad by a train traveling in an evacuated tunnel? Use the very interesting fact that gravity inside the Earth is nearly constant at 10 m/s^2 to a depth of 3000 km.

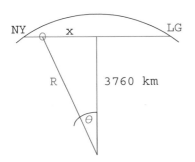

The train leaves the NewYork station at x= −5177 km

and arrives in Leningrad at x= +5177 km some time later.

At any time, the gravitational acceleration of the train along x is

$$\ddot{x} = -g\, \text{Sin}\, \theta = -g\, \frac{x}{R} = -\frac{.010\, x}{\sqrt{x^2 + 3760^2}}$$

$$\ddot{x} = -\frac{.010\, x}{(x^2 + 3760^2)^{1/2}} \qquad x_0 = -5177\ \text{km} \qquad \dot{x}_0 = 0$$

The travel time is found by using *Mathematica* to solve the above non-linear Differential Equation for x, and plotting the result versus time.

```
In[1]:=
   sol1 = NDSolve[{x"[t] == -.01*x[t] / (x[t]^2 + 3760^2)^0.5,
       x[0] == -5177, x'[0] == 0}, x, {t, 0, 6000}]
   Plot[Evaluate[-x[t] /. sol1, {t, 0, 1200}],
       AxesLabel → {"Time(s)", "Dist(km)"}];
   InterpFunc1 = x /. sol1[[1]]; InterpFunc2 = x' /. sol1[[1]];
   Plot[Evaluate[x'[t] /. sol1, {t, 0, 1200}],
       AxesLabel → {"Time(s)", "Vel(km/s)"}];
```

If we plot the distance travelled versus time, we find the train crosses the midpoint ($x = 0$) at $t \simeq 1200$ seconds.

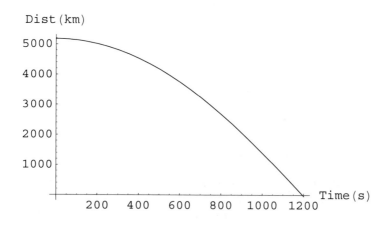

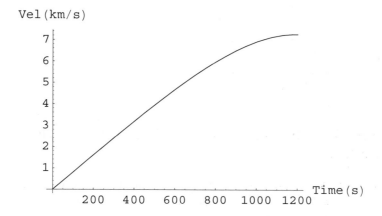

Therefore the travel time is twice 19.9 minutes, the time to reach $x = 0$, and is 39.8 minutes from NewYork to Leningrad. The velocity at center-point is over 7 kilometers/second, or about 16,000 miles per hour.

```
In[80]:= Table[ {t / 60, InterpFunc1[t], Chop[InterpFunc2[t]]},
          {t, 0, 2400, 120}] // TableForm
```

t(min)	x(km)	v(km/s)
0	-5177.	0
2	-5118.78	0.969671
4	-4944.59	1.93156
6	-4655.87	2.87685
8	-4255.25	3.7945
10	-3746.89	4.66942
12	-3137.15	5.47994
14	-2435.53	6.19429
16	-1656.16	6.76723
18	-819.341	7.14159
20	47.6908	7.26368
22	912.926	7.11212
24	1744.66	6.71313
26	2516.4	6.12208
28	3208.55	5.39526
30	3807.59	4.57636
32	4304.42	3.69585
34	4692.97	2.77454
36	4969.27	1.82694
38	5130.82	0.863801
40	5176.3	-0.106241

As a check, we may query the Interpolation Function values to find x and v at any time.

```
In[5]:= Print[ "At t=39.8 min x=", InterpFunc1[39.8 * 60], "km",
        "      At t=19.9 min, vmax=", InterpFunc2[19.9 * 60] /
        1.609 * 3600, " mph"]

        At t=39.8 min x=5177.km      At t=19.9 min, vmax=16252.8 mph
```

The total travel time is 39.8 minutes from New York to Leningrad.

[World travelers will note that the name of the Russian city has changed from Leningrad to St. Petersburg.]

Problem 2-2 Lunar Subway

Assume the moon is of constant density. Find the maximum velocity and transit time from the North pole to the South pole of the moon. The radius of the moon R_{moon}= 1700 km. The surface gravity of the moon g_{moon} = 1.7 m/s^2.

The moon appears to be made of the same material as the crust of the Earth. Thus the assumption of a constant-density moon is not a bad one. Then the mass inside a distance x from the center of the moon is $m = \frac{4}{3}\pi\rho x^3$, and the gravitational acceleration of an object inside the moon is

$$\ddot{x} = -\frac{mG}{x^2} = -\frac{4}{3}\pi\rho x G = -g\frac{x}{R_{moon}}$$

and the equation we will solve with *Mathematica* is

$$\ddot{x} = -.0017\frac{x}{R_{moon}} \qquad x_o = -1700 \text{ km} \qquad \dot{x}_o = 0$$

```
In[1]:=  sol = DSolve[{x"[t] == -.0017/1700*x[t], x[0] == -1700, x'[0] == 0},
            {x[t], x'[t]}, t] // Chop
         Plot[Evaluate[-x[t] /. sol, {t, 0, 1560},
            AxesLabel → {"Time(s)", "Dist(km)"}]];
         x[t_] = x[t] /. sol[[1]]; Plot[Evaluate[x'[t] /. sol,
            {t, 0, 1560}], AxesLabel → {"Time(s)", "Vel(km/s)"}];
         v[t_] = x'[t] /. sol[[1]]; Print[" v(t)= ", v[t]]

Out[1]=  {{x[t] → -1700. Cos[0.001 t]}}
```

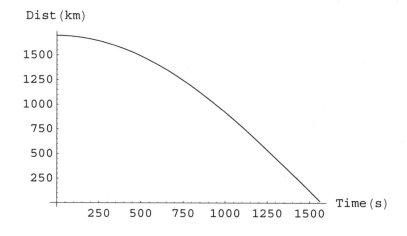

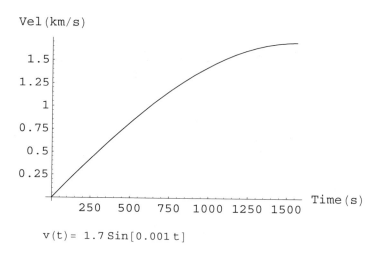

$$v(t) = 1.7\,\mathrm{Sin}[0.001\,t]$$

If we Table the results, we find that the transit time from North-to-South pole of the moon is 52 minutes.

In[31]:= **Table[{t / 60, x[t], v[t]}, {t, 0, 3120, 120}] // TableForm**

t (min)	x (km)	v (km / s)
0	-1700.	0
2	-1687.77	0.203511
4	-1651.27	0.404094
6	-1591.02	0.598866
8	-1507.89	0.785025
10	-1403.07	0.959892
12	-1278.07	1.12095
14	-1134.69	1.26589
16	-974.984	1.39263
18	-801.258	1.49933
20	-616.008	1.58447
22	-421.898	1.64682
24	-221.72	1.68548
26	-18.3534	1.6999
28	185.277	1.68987
30	386.244	1.65554
32	581.654	1.5974
34	768.699	1.51628
36	944.689	1.41335
38	1107.09	1.2901
40	1253.57	1.14829
42	1382.02	0.989962
44	1490.59	0.817398
46	1577.72	0.633078
48	1642.16	0.439653
50	1682.99	0.239904
52	1699.6	0.0367047

The maximum velocity at $t = 26$ minutes is $v = 1.7$ km/s.

Problem 2-3 Speed Bump

Let's say we're cruising along the highway in our favorite automobile, when up ahead is a *speed bump*. Should we speed-up or slow down? You had best slow-down, if you wish to preserve the integrity of your car, and the following problem will show the reason.

Assume the highway engineers have placed a 0.8-meter long, 0.1 meter-high sinusoidal *speed bump* in your lane, and you approach the bump at 2 m/s, what is the suspension-system response?

Now take a computer-simulated run at the speed-bump at 4 m/s and at 8 m/s. What is the suspension-system response at each of these velocities? Assume m = 1000 kg, c = 3400 N-s/m, k = 80,000 N/m.

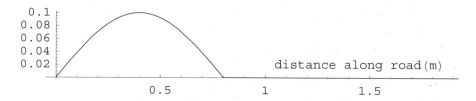

Here is how we see the speed-bump.

```
In[14]:=  L = 0.8; A = .10; v = 2; m = 1000; k = 80000; c = 3400;
          ω0 = √(k / m) ; ω = π * v / L;
          bump = Plot[.10 * Sin[π * x / L] * If[x ≤ 0.8, 1, 0],
            {x, 0, 2}, AspectRatio → 0.15,
            Prolog → {Text["distance along road(m)", {1.50, .022}]}]
          road = Plot[.10 * Sin[(π * v / L) * t] * If[t ≤ 0.4, 1, 0], {t, 0, 1},
            Prolog → {Text["time along road(s)", {.75, .028}]}]
```

```
0.1
0.08
0.06
0.04
0.02                              time along road(s)

        0.2       0.4       0.6       0.8
```

-- Road surface vs time in seconds --

This is how the suspension system sees the road surface at a speed of 2 m/s.

Let's now look at the suspension system's response on encountering the speed-bump at v = 2 m/s.

We want to solve the system

$$m\,y'' + c\,y' + k\,y \;=\; k\,A\,\mathrm{Sin}[\tfrac{\pi\,v}{L}t] \;+\; c\,\omega\,A\,\mathrm{Cos}[\tfrac{\pi\,v}{L}t] \quad \text{where } t \le 0.4 \text{ s}$$

$$m\,y'' + c\,y' + k\,y \;=\; 0 \quad \text{for } t \ge 0.4 \text{ s}$$

```
In[5]:=

L = 0.8; A = .10; v = 2; m = 1000; k = 80000; c = 3400; ω0 = √(k/m);
ω = π * v / L; sol1 = NDSolve[{m*y''[t] + c*y'[t] + k*y[t] ==
    k*A*Sin[(π*v/L)*t] * If[t ≤ 0.4, 1, 0] +
      c*ω*A*Cos[(π*v/L)*t] * If[t ≤ 0.4, 1, 0], y[0] == 0, y'[0] == 0},
  y, {t, 0, 1.0}]; InterpFunc1 = y /. sol1[[1]];
InterpFunc2 = y' /. sol1[[1]];
Acc[t_] = (k*A*Sin[ω*t] * If[t ≤ 0.4, 1, 0] + c*ω*A*Cos[ω*t] *
      If[t ≤ 0.4, 1, 0] - c*InterpFunc2[t] - k*InterpFunc1[t]) / m;
plot2 = Plot[Evaluate[y[t] /. sol1], {t, 0, 1},
    PlotStyle → {{RGBColor[1, 0, 0]}}, AspectRatio → 0.2,
    AxesLabel → {"time(s)", "y(m)"}];
Show[{road, plot2}, PlotRange → All, AspectRatio → 0.2,
  AxesLabel → {"t(s)", "y(m)"}, Prolog →
    {{Text["Road", {0.05, 0.03}]}, {Text["Response", {.35, .12}]}}]
```

Road profile and response at velocity 2 m/s. Notice how the suspension system "carries" the front-end of the automobile up-and-over the bump, letting the front-end down gently as the rear wheels make contact with the "bump". Thus the "underbelly" of the auto will not make contact with the "bump".

```
In[9]:=  Table[{t, Chop[InterpFunc1[t]], Chop[InterpFunc2[t]],
           Acc[t], Chop[.10 * Sin[(π * v / L) * t] * If[t ≤ 0.4, 1, 0]]},
         {t, 0, 0.9, .05}] // TableForm
```

t (s)	y (m)	Y' (m / s)	Acc	Road
0	0	0	2.67035	0
0.05	0.00429707	0.187594	4.54697	0.0382
0.1	0.0195637	0.423024	4.5417	0.0707
0.15	0.0457914	0.610445	2.67411	0.0923
0.2	0.0784193	0.667736	-0.54384	0.1
0.25	0.109583	0.547693	-4.25963	0.0923
0.3	0.130198	0.249815	-7.49657	0.0707
0.35	0.132356	-0.179307	-9.38444	0.0382
0.4	0.111457	-0.656432	-9.35502	0
0.45	0.0719014	-0.892601	-2.71727	0
0.5	0.0255245	-0.930394	1.12138	0
0.55	-0.018219	-0.793988	4.15708	0
0.6	-0.051828	-0.535292	5.96626	0
0.65	-0.070808	-0.22021	6.41342	0
0.7	-0.074013	0.0854375	5.63062	0
0.75	-0.063330	0.327851	3.95173	0
0.8	-0.042861	0.473066	1.82051	0
0.85	-0.017840	0.510009	-0.306808	0
0.9	0.00649635	0.448927	-2.04606	0

Let's now look at the suspension system response and the accelerations experienced in contacting the speed bump at 10 m/s.

```
Clear[x, t, L, A];
L = 0.8; A = .10; v2 = 10; m = 1000; k = 80000; c = 3400; ω = π * v2 / L; bump =
  Plot[.10 * Sin[π * x / L] * If[x ≤ 0.8, 1, 0], {x, 0, 2}, AspectRatio → 0.15,
    Prolog → {Text["distance along road(m)", {1.45, .03}]}]
road = Plot[.10 * Sin[(π * v2 / L) * t] * If[t ≤ 0.08, 1, 0], {t, 0, .2},
    AspectRatio → 0.15,
    Prolog → {Text["time along road(s)", {.145, .03}]}]
sol2 = NDSolve[{m * y''[t] + c * y'[t] + k * y[t] == k * A * Sin[(π * v2 / L) * t] *
        If[t < 0.08, 1, 0] + c * ω * A * Cos[(π * v2 / L) * t] * If[t < 0.08, 1, 0],
    y[0] == 0, y'[0] == 0}, y, {t, 0, 1.0}];
InterpFunc1 = y /. sol2[[1]]; InterpFunc2 = y' /. sol2[[1]];
InterpFunc3 = y'' /. sol2[[1]];
plot3 = Plot[Evaluate[y[t] /. sol2],
    {t, 0, 0.8}, PlotStyle → {{RGBColor[1, 0, 0]}},
    AspectRatio → 0.2, AxesLabel → {"time(sec)", "y(m)"}];
Show[{road, plot3}, PlotRange → All,
  Prolog → {{Text["Road", {0.05, 0.03}]},
    {Text["Response", {.25, .06}]}, {Text["Time(s)", {0.6, .02}]}}]
```

Road profile and response at velocity 10 m/s. At the higher velocity, the suspension-system sees the speed bump as a speed "spike". Let's Table the data and see how much of a jolt the driver and passengers take on hitting the speed bump at 10 m/s.

In[43]:= `Table[{t, InterpFunc1[t], Chop[InterpFunc2[t]], InterpFunc3[t],`
`          Chop[.10 * Sin[(π * v2 / L) * t] * If[t ≤ 0.08, 1, 0]]},`
`          {t, 0, .32, .02}] // TableForm`

t (s)	y (m)	y' (m/s)	Acc (m/s^2)	road
0	0	0	13.3518	0
0.02	0.002870	0.288809	13.8865	0.0707
0.04	0.010985	0.49432	5.44059	0.1
0.06	0.021136	0.478534	-7.1021	0.0707
0.08	0.028557	0.232136	-14.295	0
0.1	0.032576	0.169397	-3.1820	0
0.12	0.035326	0.105556	-3.1849	0
0.14	0.036804	0.042655	-3.0894	0
0.16	0.037051	-0.01742	-2.9048	0
0.18	0.036137	-0.07302	-2.6427	0
0.2	0.034169	-0.12270	-2.3163	0
0.22	0.031276	-0.16534	-1.9399	0
0.24	0.027608	-0.20007	-1.5284	0
0.26	0.023330	-0.22634	-1.0968	0
0.28	0.018613	-0.24390	-0.6597	0
0.3	0.013631	-0.25279	-0.2310	0
0.32	0.008557	-0.25329	0.17661	0

At the higher speed (10 m/s), the chassis doesnt have time to "clear" the speed bump... At this speed we are attempting to drive "right through" the speed bump. Not a good idea.

If there is any downward movement of the chassis, we will scrape the bottom of our auto. What will also happen at 10 m/s is that the driver will feel a "shock" of 1.4 gees as the front tires run up and over the "speed bump". Running over a speed bump at speeds greater than 20 mph would definitely jar your teeth and put terrific stress on your suspension system.

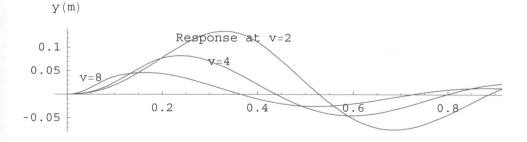

Because the suspension system takes a certain amount of time to respond, the suspension-system response to the speed bump is good at velocity 2 m/s, fair at 4 m/s, and terrible at 8 m/s.

Problem 2-4   RLC circuit

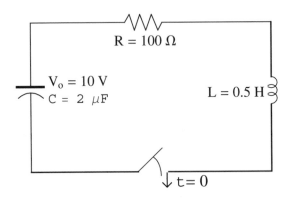

The differential equation for the RLC-series circuit is

$$\frac{d^2 V}{dt^2} + RC\,\frac{dV}{dt} + V = 0 \quad \text{where} \ \ V_o = 10 \text{ volts and } \ I_o = 0.$$

It is easy enough to plot the voltage response using **NDSolve**, however since this is a *linear equation*, we can utilize **DSolve** to obtain the complete equation of the voltage response versus time. Then we can just read the time-constant and frequency of oscillation of the circuit.

```
In[28]:=  Clear[v, t, R, L, c]; L = 0.5; R = 100; c = 2 * 10^-6;
          sol = DSolve[{.000001 v''[t] + .0002 v'[t] + v[t] == 0,
               v[0] == 10, v'[0] == 0}, v[t], t] // FullSimplify
          Plot[v[t] /. sol, {t, 0, .030}, PlotRange → {-8, 10}]
```

Out[29]= $\{\{v[t] \rightarrow e^{-100.\, t}\,(10.\,\text{Cos}[994.987\,t] + 1.00504\,\text{Sin}[994.987\,t])\}\}$

The natural response of an RLC series circuit

Given that the voltage across the capacitor with time is given by

$$\{\{v[t] \rightarrow e^{-100. \, t} \, (10. \, \text{Cos}[994.987 \, t] + 1.00504 \, \text{Sin}[994.987 \, t])\}\}$$

We find the time-constant from the exponential term: $e^{-t/\tau}$, $\tau = .01\text{s} = 10$ ms. The oscillation frequency of this *underdamped* circuit is $f = \frac{\omega}{2\pi} = \frac{994.987}{2\pi}$ $= 158.4$ Hz.

Given the time-constant, we may plot the curve with an exponential envelope

```
Plot[{v[t] /. sol, 10 Exp[-100 t], -10 Exp[-100 t]}, {t, 0, .030},
   PlotRange → {-8, 10}, AxesLabel → {"T(s)", "Voltage"}]
```

Out[43]= $\{\{v[t] \rightarrow e^{-100. \, t} \, (10. \, \text{Cos}[994.987 \, t] + 1.00504 \, \text{Sin}[994.987 \, t])\}\}$

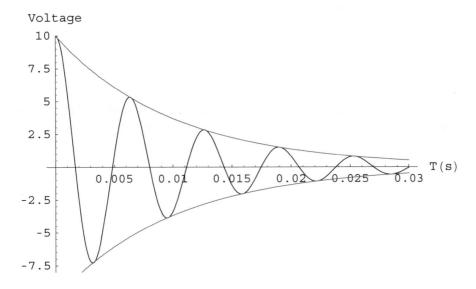

The response of the RLC-circuit with an exponential envelope.

As in Section 2-1, we may also determine the amount of damping per cycle.

Given that $V[t] = e^{-100. \, t} \, (10. \, \text{Cos}[994.987 \, t] + 1.005 \, \text{Sin}[994.987 \, t])$

After one cycle, $994.987 \, t = 2\pi$

Thus the time of one cycle is $t = .00631$ sec.

At that time, $e^{-100 \, t} = e^{-.631} = .532$

Thus the voltage amplitude decreases by a factor of $\text{Exp}[-.631] = .532$ every cycle.

Problem 2-5 Impulse Response

The differential equation of a mechanical system is given by

$$y'' + 3\,y' + 2\,y\ = F(t) \qquad\qquad y_o = 0, \quad y_o' = 0.$$

This corresponds to a system with mass m = 1, spring constant k =2, and damping coefficient c = 3, at rest at t = 0.

Let's find the system response to each FΔt

(A) a unit impulse of (F=1 acting for Δt=1 second)

(B) a unit impulse of (F=2 acting for Δt=0.5 second).

(C) a unit impulse of (F=10 acting for Δt=0.1 sec.)

(D) a unit impulse from the Dirac Delta function

(A) The unit impulse of FΔt=1, F=1 acting for 1 second

```
Clear[y];
q1 = Plot[{1 * UnitStep[t] - 1 * UnitStep[t - 1]},
   {t, -1.1, 2.75}, PlotRange → {0, 1.5}, AxesLabel → {None, "Force"}]
sol1 = NDSolve[{y''[t] + 3 y'[t] + 2 y[t] == UnitStep[t] - UnitStep[t - 1],
   y[0] == y'[0] == 0}, y, {t, -1.8, 4}];
r1 = Plot[y[t] /. sol1, {t, -1., 3.0}, PlotRange → {0, 0.28},
   AxesLabel → {"t(s)", "y"}]
```

In[24]:= Show[GraphicsArray[{q1, r1}]]

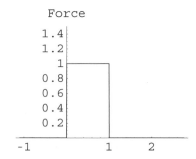
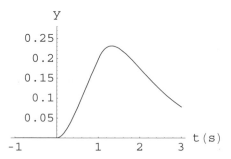

The Impulse   The Response

(B) A force F=2 acting for 0.5 seconds

```
p2 = Plot[{2 * UnitStep[t] - 2 * UnitStep[t - .5]},
   {t, -1., 2.}, PlotRange → {0, 2.5}]
sol2 = NDSolve[
    {y''[t] + 3 y'[t] + 2 y[t] == 2 * UnitStep[t] - 2 * UnitStep[t - .5],
    y[0] == y'[0] == 0}, y, {t, -1.8, 4}];
r2 = Plot[y[t] /. sol2, {t, -1., 3.0}, PlotRange → {0, 0.28},
   AxesLabel → {"t(s)", "y"}]
```

In[57]:= `Show[GraphicsArray[{p2, r2}]]`

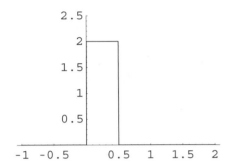
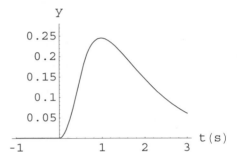

The Impulse   The Response

(C) The force is 10, the time is 0.1, the impulse is still one

```
Clear[y]; p10 = Plot[{10 * UnitStep[t] - 10 * UnitStep[t - .1]},
   {t, -1.0, 1}, PlotRange → {0, 10.2}]
sol10 = NDSolve[
   {y''[t] + 3 y'[t] + 2 y[t] == 10 * UnitStep[t] - 10 * UnitStep[t - .1],
   y[0] == y'[0] == 0}, y, {t, -1.8, 4}]
r10 = Plot[y[t] /. sol10, {t, -1., 3.0},
   PlotRange → {0, 0.28}, AxesLabel → {"t(s)", "y"}]
```

In[52]:= `Show[GraphicsArray[{p10, r10}]]`

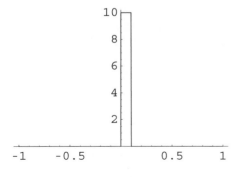
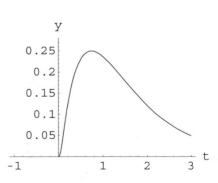

The Impulse   The Response

(D) The unit impulse from the Dirac Delta function: F→∞ as Δt → 0, but FΔt =1. Note that with the Dirac function applied at t = 0, FΔt = mΔv =1 and since m=1, the system must leave t =0 with a velocity y′ = 1. *We must modify the initial conditions*

```
In[7]:=  DSolve[{y''[t] + 3 y'[t] + 2 y[t] == DiracDelta[t],
            y[0] == 0, y'[0] == 1}, y[t], t]
         Plot[y[t] /. %, {t, -1.8, 3}, PlotRange → {0, .25}]

Out[8]=  {{y[t] → e^{-2 t} (-1 + e^t) UnitStep[t]}}
```

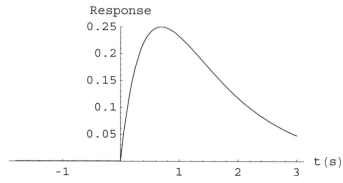

Response of system to Dirac-Delta function at t = 0. (Almost identical to Force of 10 for 0.1 s)

(D) The unit impulse from the Dirac Delta function F→∞ as Δt → 0, but FΔt =1. Note that with the Dirac function applied at t =.000001s, we dont need to worry about the initial conditions. The system is at rest at t =0, and the Dirac function chimes in after that time. At t = 0, the system has position y = 0 and velocity y′ = 0. *We dont need to modify the initial conditions.*

```
In[1]:=  DSolve[{y''[t] + 3 y'[t] + 2 y[t] == DiracDelta[t - .000001],
            y[0] == 0, y'[0] == 0}, y[t], t]
         Plot[y[t] /. %, {t, -1.8, 3}, PlotRange → {0, .25}]

Out[2]=  y[t] → e^{-2 t} (-1 + e^t) UnitStep[-1 × 10^{-6} + t]
```

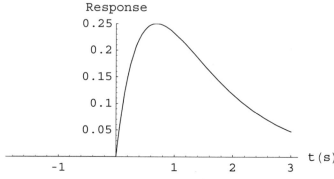

Response of system to Dirac-Delta function at t =10^{-6} s. (Indistinguishable from above curve)

Problem 3-1 Gravity Waves and Pulsar 1913+16

The equations of General Relativity may be used to find the power radiated by gravitational waves, the rate of decrease of the semi-major axis of the binary Pulsar, the rate of decrease of the orbital period, and the total time for system collapse.

$G = 6.67 * 10^{-11}; \ c = 3 * 10^8; \ M1 = M2 = 1.4 * 2 * 10^{30};$

$M = M1 + M2; \ a = 1.95 * 10^9; \ P0 = 27900; \ \epsilon = .617;$

(* Note: for $\epsilon = .617$, we need η *)

$$\eta = 1 / (1 - \epsilon^2)^{7/2} * \left(1 + \frac{73}{24} \epsilon^2 + \frac{37}{96} \epsilon^4 \right)$$

11.8428

A. Power emitted in gravitational waves (in Watts)

$$P = 6.4 * (G^4 * (M1^2 * M2^2) * (M1 + M2)) / (a^5 * c^5) * 11.84$$

7.53487×10^{24}

B. Rate of decrease of semi-major axis of system (in meters per year)

$$\text{adot} = \frac{64 * G^3 * (M1 * M2) * M}{5 * a^3 * c^5} * 11.84 * 3.156 * 10^7$$

3.45836

C. Rate of decrease of period of binary pulsar system (seconds per year)

$$\text{Pdot} = \frac{96 * G^3 * (M1 * M2) * M}{5 * c^5} * \left(\frac{4 \pi^2}{G * M} \right)^{4/3} * \frac{11.84}{(27900)^{5/3}} * 3.156 * 10^7$$

0.000074895

D. Time for system to in-spiral and collapse (in years)

$$T = \frac{12 * (1.55 * 10^9)^4}{3.156 * 10^7 * 19 * 6.86 * 10^{19}} \int_0^{.617} x^{29/19} \frac{(1 + (121. / 304) x^2)^{.5137}}{(1 - x^2)^{1.5}} \, dx$$

3.04533×10^8

Note that the results from General Relativity theory account for 98 per cent of the observed value of 76.5 μs orbital period decrease per year.

For more details on gravitational radiation, see
Charles Misner, Kip Thorne, and John A. Wheeler (1973) *Gravitation*
Bradley Carroll, Dale Ostlie (1996) *Introduction to Modern Astrophysics*

Certain constants β, η, **k** for the PSR 1913+16 system are evaluated as follows from the 1964 paper by Philip C. Peters, "Gravitational Radiation from Two Point Masses" *Physical Review* <u>136</u>, B1224-1232.

$$G = 6.67 * 10^{-11}; \quad c = 3 * 10^8; \quad M1 = M2 = 2.8 * 10^{30}; \quad M = M1 + M2;$$

$$\beta = \frac{64}{5} \frac{G^3 * M1 * M2 * (M1 + M2)}{c^5}$$

$$6.86255 \times 10^{19}$$

$In[76] :=$ $\epsilon = .617;$

$$\eta = \frac{1}{(1 - \epsilon^2)^{7/2}} * \left(1 + \frac{73}{24} \epsilon^2 + \frac{37}{96} \epsilon^4\right)$$

$Out[77] = $ 11.8428

For this particular binary pulsar system, it is necessary to find the constant k in the equation

$$a(\epsilon) = k * \frac{\epsilon^{12/19}}{(1 - \epsilon^2)} * \left(1 + \frac{121}{304} \epsilon^2\right)^{870/2299}$$

At the present time, $a = 1.95 \times 10^9$ meters and $\epsilon = .617$, and to evaluate **k**, we have

$$NSolve\left[1.95 * 10^9 == k * \frac{\epsilon^{12/19}}{(1 - \epsilon^2)} * \left(1 + \frac{121}{304} \epsilon^2\right)^{870/2299}, k\right]$$

$$\{\{k \rightarrow 1.55248 \times 10^9\}\}$$

Problem 3-2 Halley's Comet Equations

Use the equations of Section 3.3 to find all the orbital parameters of Halley's comet, given only the observed velocity of 54.45 km/s at closest approach to the Sun, which was 0.59 AU on its last approach in 1986. The mass of the Sun is 2×10^{30} kg.

Let us use the following equations from Section 3.3

(1) perihelion velocity $v_p = \frac{2 \pi a}{T} \sqrt{\frac{1 + \epsilon}{1 - \epsilon}}$

(2) the Newton-Kepler Law $T^2 = \frac{4 \pi^2 a^3}{G (M1 + M2)}$

(3) perihelion distance $d_p = a (1 - \epsilon)$

Notice first, *and this is very important,* that the first of these equations contains *two* singular points -- at $\epsilon = 1$, and $T = 0$. Now you and I would never choose these values for eccentricity and time period, but the computer is looking at all possible values for a solution. It is best, therefore, to put Equations (2) and (3) into Equation (1).

Thus, $v_p = \sqrt{GM} \sqrt{\frac{1 + \epsilon}{d_p}}$ and we now use *Mathematica* to solve for ϵ

with *perihelion* velocity $v_P = 54.45$ km/s and distance $d_P = 88.5 \times 10^6$ km.

```
In[1]:= Clear[ε, a, T, d];
        vp = 54.45; d = 88.5 * 10^6; G = 6.67 * 10^-20; M = 2 * 10^30;
        (* note dist in km, time in seconds*)
        sol = NSolve[vp == √GM * √(1 + ε) / d, ε]

Out[4]= {{ε → 0.966904}}
```

We now solve for the orbital time T and the semi-major axis a

```
In[22]:= ε = .967;
         NSolve[{d == a * (1 - ε), T == √4 π² * a^3 / (G * M) }, {a, T}]

Out[23]= {{a → 2.68182 × 10^9, T → 2.38916 × 10^9}}
```

In terms of AU and years,

```
In[28]:= Print["Au= ", 2.68182 * 10^9 / (1.5 * 10^8),
         " Tyears= ", 2.38916 * 10^9 / (3.156 * 10^7)]

         Au= 17.8788  Tyears=75.7022
```

Since we now have the main parameters ϵ, T, and a, it is relatively easy to find the comet's distance and velocity at Aphelion

In[31]:= **a = 17.88; ϵ = .967;**

$$\text{NSolve}\left[\left\{\text{da} == a * (1 + \epsilon), \text{va} == \text{vp} * \frac{(1 - \epsilon)}{(1 + \epsilon)}\right\}, \{\text{da, va}\}\right]$$

Out[32]= {{da → 35.17, va → 0.913498}}

Table of values for Halley's comet

Semi-major axis **a**	17.88 AU	Orbital Time **T**	75.7 years
Perihelion distance	0.59 AU	Perihelion velocity	54.45 km/s
Aphelion distance	35.17 AU	Aphelion velocity	0.91 km/s
Orbital eccentricity	$\epsilon = .967$		

```
Needs["Graphics`Graphics`"];
a = 17.88; ε = .967;
comet = PolarPlot[a * (1 - ε²) / (1 - ε * Cos[θ]), {θ, 0, 2 π}]
```

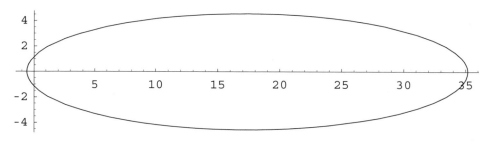

The orbit of Halley's comet. $r = \dfrac{a\,(1-\epsilon^2)}{1 + \epsilon\,\text{Cos}\,\theta}$ $a = 17.88$ AU, $\epsilon = .967$

In 1986, the spacecraft *Giotto* approached to within 600 km of the nucleus of Halley's comet. From the deviation in its trajectory, the mass of comet Halley was estimated at about 10^{15} kg. Halley's comet has about the same mass as Mt Everest.

Problem 3-3 The Largest Stars

Plaskett's binary star system consists of two stars that revolve in a circular orbit about a center of gravity midway between them. This means the masses of the two stars are equal.

If the orbital velocity of each star is 220 km/s and the orbital period of each is 14.4 days, find the mass M of each star.

For a circular orbit, $v = \frac{2\pi r}{T}$

the radius of the orbit $r = \frac{vT}{2\pi} = \frac{(220)\,(14.4)\,(86400)}{2\pi} = 43.56 \times 10^6$ km

The interstellar separation $a = r + r = 87.12 \times 10^6$ km

```
r = 43.56 * 10^6; Needs["Graphics`Graphics`"];
stars = PolarPlot[r, {t, 0, 2 π}, PlotStyle → Dashing[{.03}],
   AspectRatio → 1, AxesStyle → GrayLevel[0.7],
   PlotRange → {{-5 * 10^7, 5 * 10^7}, {-5 * 10^7, 5 * 10^7}},
   Prolog → {{Hue[.67], Circle[{-4.35 * 10^7, 10}, 6000000]},
    {Hue[0.67], Circle[{4.35 * 10^7, 100}, 6000000]}}]
```

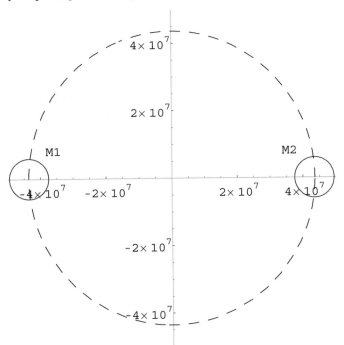

The orbits of the two stars comprising Plaskett's binary system.
The radius r = 43.56 × 10⁶ km, and the separation of the two stars is
a = 87.12 × 10⁶ km.

We may now use the Newton-Kepler Law to find the masses of the stars.

$$T^2 = \frac{4\,\pi^2\,a^3}{G\,(M1 + M2)}$$

In[50]:=   (* dist in km, time in s, mass in kg *)
            a = 87.12 * 10^6; T = 14.4 * 86400; G = 6.67 * 10^{-20};
            sol = Solve[T^2 == 4 * π^2 * a^3 / (G * 2 M), M]

Out[51]=   { {M → 1.26417 × 10^{32}} }

In[57]:=   Print["The mass of each star, in solar masses is M →",
            (M / (2 * 10^{30})) /. sol]

            The mass of each star, in solar masses is M →{63.2084}

Therefore the total mass of Plaskett's binary system is approx 126 solar masses. The two blue-giant stars comprising Plaskett's system are among the most massive known.

Problem 3-4 The Alaska Pipeline

For the above-ground portion of the Alaska pipeline, the 1.2 meter-diameter pipe is wrapped with 10-cm of fiberglass insulation of thermal conductivity $k = 0.035$ W/m·°C.

We will find the heat loss per meter of pipeline when the oil inside the pipe is at 50°C, and the air temperature outside is – 40 °C with a convective heat transfer coefficient of $h = 12$ W/ m²·°C.

Solution to Alaska Pipeline problem:

The rate at which heat flows through the insulation is equal to the rate at which heat is abstracted away by the air currents at the surface of the pipeline. Therefore

$$\dot{Q} = \frac{2\,\pi\,k\,L\,(T_i - T_2)}{\ln\,(r_o / r_i)} = 2\pi r_o h\,L\,(T_2 - T_{air})$$

We are solving for T_2 the temperature at the surface of the pipeline

```
In[3]:=  k = .035; Ri = 0.60; Ro = 0.70; L = 1; h = 12; Ti = 50; TA = -40;
         sol = Solve[
             k * 2 Pi * 1 * (Ti - T2) / Log[Ro / Ri] == h * 2 Pi * Ro * 1 * (T2 - TA) , T2]

Out[4]=  {{T2 → -37.6313}}
```

The temperature at the surface of the pipeline (with insulation) is just 2.4 degrees above the outside air temperature ! With or without gloves on, we wouldn't be able to detect the difference.

Now for the amount of heat lost per meter of pipeline

$$\dot{Q} = 2\pi r_o h\,L\,(T_2 - T_{air}) = 2\pi\,(0.7\text{ m})(12\text{ W/ m}^2\text{·°C})(1\text{ m})(2.4\text{ °C}) = 126\text{ W}$$

```
In[14]:=  Qdot = 2 π * .7 * 12 * 1 * (-37.6 - TA)

Out[14]=  126.669
```

This is not really significant. If we take the specific heat of the oil as $C_V = 2000$ J/kg·°C then the average temperature drop ΔT in the oil (which is moving at 3 m/s) in one meter due to heat loss is

$$\rho\,A L\,C_V\,\Delta T = (126\text{ W})\,(1/3\text{ s}) = 42\text{ J} \quad (\rho = 870\text{ kg/m}^3,\ A = \pi\,r_i^2,\ L = 1\text{ m})$$

$$\Delta T = \frac{42 \text{ J}}{\rho \text{ AL } C_V} = 21 \times 10^{-6} \text{ °C/m} \text{ or } 2.1°C \text{ in } 100 \text{ km !!}$$

In[15]:= $\Delta T = 42 / (870 * \pi * .6^2 * 2000)$

Out[15]= 0.0000213426

And this is on the coldest of cold nights. It is interesting to see what would happen if the pipe-line wasn't insulated. Then the heat loss would be due to both convection and radiation. And for 0.5" stainless steel with k = 15 W/m·°C and emissivity 0.4, we can solve the following very difficult equation

$$\dot{Q} = \frac{2 \pi \text{ k L } (T_i - T_2)}{\ln (r_o / r_i)} = 2\pi r_o h \text{ L } (T_2 - T_{air}) + \epsilon \sigma 2\pi r_o L (T_2^4 - T_{air}^4)$$

Using *Mathematica* we find T_2= 321.994 K (or 48.99 °C). (Good thing there's insulation!)

In[13]:= k2 = 15; σ = 5.67 * 10^-8; Ti = 323; TA = 233; Ri = 0.6; Ro = 0.612;
           sol = Solve[2 π * k2 * 1 * (Ti - T2) / Log[Ro / Ri] ==
              h * 2 π * Ro * 1 * (T2 - TA) + .4 * σ * 2 π * Ro * 1 * (T2^4 - TA^4), T2]

Out[14]= {{T2 → -3907.14}, {T2 → 321.994},
           {T2 → 1792.57 - 3301.22 i}, {T2 → 1792.57 + 3301.22 i}}

Then, $\dot{Q} / L = 2\pi r_o h (321.994 - 233) + \epsilon \sigma 2\pi r_o (321.994^4 - 233^4)$

$\dot{Q} / L = 4107 \text{ W } + 680 \text{ W } = 4787$ Watts per meter

The heat loss per meter due to both convection and radiation for uninsulated pipe is

In[18]:= Qdot = h * 2 π * Ro * 1 * (321.994 - 233) +
            .4 * σ * 2 π * Ro * 1 * (321.994^4 - 233^4)

Out[18]= 4786.96

Problem 3-5 Counterflow Heat Exchanger

In a *counterflow* heat exchanger, the cold fluid enters from the left and exits heated on the right, and the hot fluid enters from the right and exits after cooling on the left. If hot oil at 200 °C with a mass flow rate of 3 kg/s and a specific heat of 1900 J/kg ·°C enters the outer shell from the right, and cool water at 20 °C with a mass flow rate of 0.9 kg/s and specific heat 4180 J/kg ·°C enters the inner tube from the left, what is the rate of heat exchange in this heat exchanger? We will keep the same heat-exchange parameters as in Section 3-5, the heat-transfer coefficient U = 460 W/m²·°C, and the heat-exchange surface area is A = π D L. Where D = 10 cm, and the length is L = 40 meters.

Be sure to compare the effectiveness of this counter-flow heat exchanger with the effectiveness of the parallel-flow heat exchanger of Section 3-5. Which system does the better job of extracting heat from the hot fluid ?

Building the Differential Equations

For a heat exchanger, the cold fluid is heated dT by absorbing heat energy dQ through an area π D dx.

In the inner tube,

$$dQ = (\dot{m}\, c)_c\, dT = U\, \pi\, D\, (T^* - T)\, dx \qquad \text{Cold side}$$

Heat Absorbed = Heat Transferred

In the outer shell, the hot fluid (which is moving along – dx) *loses* heat energy dQ through the same area.

In the outer shell,

$$(\dot{m}\, c)_h\, dT^* = -\, U\, \pi\, D\, (T^* - T)\, (-\, dx) \qquad \text{Hot side}$$

Heat Lost = Heat Transferred

When we write the differential equations, with T for the cold fluid and T* for the hot fluid,

$$(\dot{m}\, c)_c\, \frac{dT}{dx} = U\, \pi\, D\, (T^* - T) \qquad \text{and}$$

$$(\dot{m}\, c)_h\, \frac{dT^*}{dx} = U\, \pi\, D\, (T^* - T)$$

The *Mathematica* program for the counterflow heat exchanger is

```
(* T[x]= water Temp,  S[x] = oil temp *)
mh = 3.0 * 1900; mc = 0.9 * 4180; A = 12.56; D0 = .10; L = 40; U = 460;
sol = NDSolve[{mc * T'[x] == U * π * D0 * (S[x] - T[x]),
    mh * S'[x] == U * π * D0 * (S[x] - T[x]),
    T[0] == 20, S[40] == 200}, {S, T}, {x, 0, 40.0}]
InterpFunc1 = S /. sol[[1]]; InterpFunc2 = T /. sol[[1]];
Table[{x, Chop[InterpFunc1[x]], Chop[InterpFunc2[x]]},
    {x, 0, 40, 5}] // TableForm
Plot[{S[x] /. sol, T[x] /. sol}, {x, 0, 40}, Prolog →
    {Text["Cold Fluid Out 140°C", {31, 124}], Text["→", {15, 60}],
    Text["←", {25, 165}], Text["Hot Fluid In 200°C", {32, 184}],
    Text["Cold Fluid In 20°C", {8.5, 34}],
    Text["Hot Fluid Out 120°C", {8.5, 110}]},
    Frame → True, GridLines → Automatic];
```

The *Mathematica* program for the Parallel-flow heat exchanger is

```
(* T[x]= water Temp,  S[x] = oil temp *)
mh = 3. * 1900; mc = .9 * 4180; A = 12.56; D0 = .10; L = 40; U = 460;
sol = NDSolve[{mc * T'[x] == U * π * D0 * (S[x] - T[x]),
    mh * S'[x] == - U * π * D0 * (S[x] - T[x]),
    T[0] == 20, S[0] == 200}, {S, T}, {x, 0, 80.0}]
InterpFunc1 = S /. sol[[1]]; InterpFunc2 = T /. sol[[1]];
Table[{x, Chop[InterpFunc1[x]], Chop[InterpFunc2[x]]},
    {x, 0, 40, 5}] // TableForm
Plot[{S[x] /. sol, T[x] /. sol}, {x, 0, 40},
    Prolog → {Text["Oil", {15, 164}], Text["Water", {23, 90}]},
    AxesLabel → {"Dist(m)", "Temp(°C)"},
    Frame → True, GridLines → Automatic];
```

Note that the only difference between the two programs is the direction of flow of the oil. We now examine the performance of the two heat exchangers.

The Counterflow Heat Exchanger

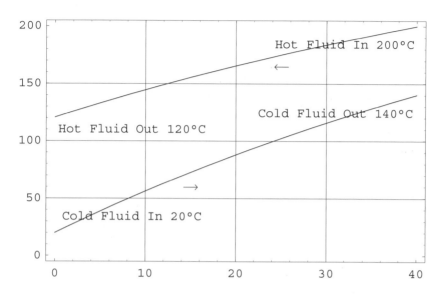

x(m)	S(hot)	T(cold)
0	120.564	20.
5	132.905	38.6981
10	144.465	56.2142
15	155.295	72.6229
20	165.44	87.9943
25	174.944	102.394
30	183.847	115.883
35	192.187	128.52
40	200.	140.358

Hot-side temperature declines from 200 °C to 120 °C while cold-side temperature increases from 20 °C to 140 °C. The counterflow set-up for the heat exchanger allows the cold fluid (water) to heat up to a temperature *greater than the temperature of the outgoing hot fluid* (oil). Let us see how much heat is extracted from the system in the counterflow mode:

$$Q = (\dot{m}\,c)_c\ \Delta T_c\ =\ (3762)\,(120)\ W =\ 451.4\ kW$$

The *effectiveness* ϵ in removing heat from the system is

$$\epsilon\ =\ \frac{\Delta T_c}{(T_h - T_c)_{max}}\ =\ \frac{120}{180} =\ 66.7\ \%$$

Let us compare this commendable performance with the parallel-flow system of Section 3.5. We keep everything the same, except the direction of flow, so that the hot fluid and the cold fluid both enter on the left-hand-side and both exit on the right-hand-side.

The Parallel-flow Heat Exchanger

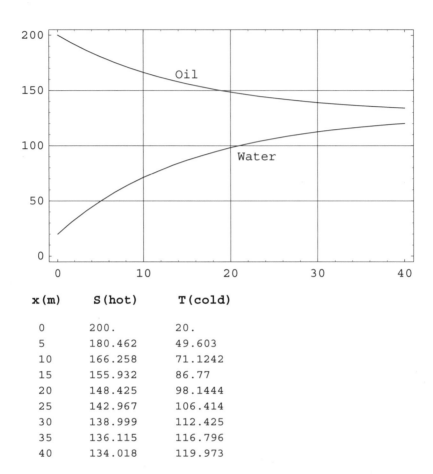

x(m)	S(hot)	T(cold)
0	200.	20.
5	180.462	49.603
10	166.258	71.1242
15	155.932	86.77
20	148.425	98.1444
25	142.967	106.414
30	138.999	112.425
35	136.115	116.796
40	134.018	119.973

Hot-side temperature declines from 200 °C to 133 °C while cold-side temperature increases from 20 °C to 121 °C. The parallel-flow set-up for the heat exchanger allows the cold fluid (water) to heat up to a temperature *less than* the temperature of the outgoing hot fluid (oil). Let us see how much heat is extracted from the system in the parallel-flow mode:

$$Q = (\dot{m}\,c)_c\ \Delta T_c\ =\ (3762)\,(100)\ W = 376.2\ kW$$

The *effectiveness* ϵ in removing heat from the system is

$$\epsilon\ =\ \frac{\Delta T_c}{(T_h - T_c)_{max}}\ =\ \frac{100}{180} = 55.6\ \%$$

The counterflow arrangement for a heat exchanger is more effective in removing heat from the system by approximately 11 %.

An interesting visual comparison between counterflow and parallel-flow heat exchangers is that if we reverse the direction of flow of the oil, then we change the performance by 11 percent. The counterflow heat exchanger does better in removing heat from the oil, because of its more uniform temperature gradient.

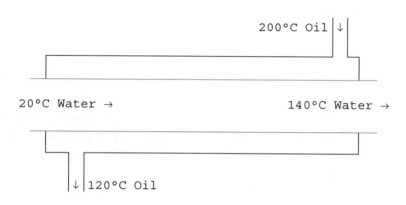

A counterflow heat exchanger. The water is heated from 20 °C to 140 °C, while the oil is cooled from 200 °C to 120 °C.

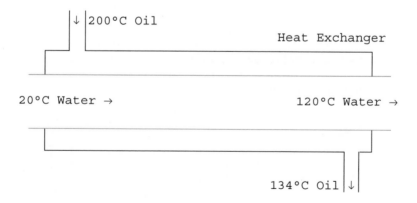

A parallel-flow heat exchanger. The water is heated from 20 °C to 120 °C, while the oil is cooled from 200 °C to 134 °C.

Problem 3-6 Women's Track and Field

Following, is a list of world records for women in running, from 100-meters to 10,000 meters.

Distance (m)	Time (s)	Avg Velocity (m/s)	Record Holder
100	10.49	9.53	Florence Joyner (USA) 1988
200	21.34	9.37	Florence Joyner (USA) 1988
400	47.60	8.40	Maria Koch (GER) 1985
800	113.28	7.06	Jarmila Kratochvilova (CZ) 1983
1000	149.34	6.69	Maria Mutola (MOZ) 1995
1500	230.46	6.51	Qu Yunxia (CHINA) 1993
1 mile	252.56	6.37	Svetlana Mastercova (RUS) 1996
2000	325.36	6.15	Sonia O'Sullivan (IRE) 1994
3000	486.11	6.17	Wang Junxia (CHINA) 1993
5000	864.68	5.78	Elvan Abeylegesse (TUR) 2004
10000	1771.78	5.64	Wang Junxia (CHINA) 1993

Let us first of all put this data into a form where we can see the average velocity to run a race of a given distance. *Mathematica* can manipulate and plot data in an easy to read form.

```
In[157]:= velocity = {{100, 9.53}, {200, 9.37}, {400, 8.40}, {800, 7.06},
            {1000, 6.70}, {1500, 6.51}, {1609, 6.37}, {2000, 6.15},
            {3000, 6.17}, {5000, 5.78}, {10000, 5.64}, {20000, 5.09}};
         velc = ListPlot[velocity, PlotRange → {{-100, 5100}, {5, 10}}]
```

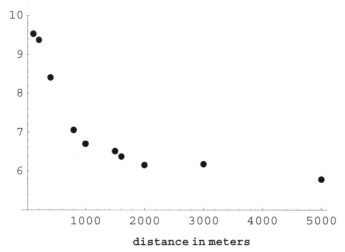

Let us apply the Joseph Keller theory to see how well we can match the track records for women. We choose the following physiological constants as representative for a female world-class runner

$F = 1.20 * 9.8$ m/s^2 (maximum acceleration of runner is 1.2 g)

$k = 1.14$ s^{-1} (internal and external resistance increases with velocity)

$\sigma = 36.5$ W/kg (rate of oxygen resupply to muscles)

$E_o = 2300$ J/kg (initial energy and oxygen supply in muscles)

We will apply these constants to the differential equations we solved before

$$\frac{dv}{dt} = f - k\,v \quad \text{and} \quad \frac{dE}{dt} = \sigma - f\,v$$

with solutions

for x < 300 m, $\quad v1 = \frac{F}{k}(1 - e^{-kt})$ and $\quad x1 = \frac{F}{k}\,t - \frac{F}{k^2}(1 - e^{-kt})$ all acceleration

for x > 300 m, $\quad f = k\,v$ and $\quad \frac{dE}{dt} = \sigma - k\,v^2$ constant velocity

$$z = x1 + v \cdot t2 \quad \text{and} \quad E2 = (k\,v^2 - \sigma) \cdot t2$$

```
In[185]:=
    Clear[x]; F = 1.2*9.8; k = 1.14; s = 36.5; E0 = 2300;
    sol1 = DSolve[{x''[t] == F - k*x'[t], x[0] == x'[0] == 0}, x[t], t] //
        FullSimplify
    sol2 = DSolve[{v'[t] == F - k*v[t], v[0] == 0}, v[t], t] //
        FullSimplify
```

Out[186]= $\{\{x[t] \to -9.04894 + 9.04894\ e^{-1.14\,t} + 10.3158\ t\}\}$

$\{\{v[t] \to 10.3158 - 10.3158\ e^{-1.14\,t}\}\}$

Again, t is the acceleration time, and t2 is the time at constant velocity. E1 is the energy expended in the acceleration phase, and E2 is the energy expended in the run-phase.

We will solve for the time t2 at constant velocity, subject to the condition that all energy is used-up by the end of the race. Then, the total distance travelled is $z = x1 + x2 = x1 + v1 \cdot t2$ and the total time is $t + t2$. The average velocity for running the race is then $w = \frac{z}{(t+t2)}$.

Also, for races less than 300 m, there is only the acceleration phase, and the distance travelled is x1 in time t.

```
In[188]:=   F = 1.2 * 9.8; k = 1.14; s = 36.5; E0 = 2300;
            x1[t_] = (F / k) * t - (F / k^2) * (1 - Exp[-k * t])
            v1[t_] = (F / k) * (1 - Exp[-k * t])
            E1[t_] = F * x1[t] - s * t ;
            E2[t_] = E0 - E1[t] ;
            (* and since  E2 = (kv²- s) t2  *)
                       E0 - E1[t]
            t2[t_] = ─────────────── ;
                     (k * v1[t]^2 - s)
            z[t_] = v1[t] * t2[t] + x1[t] ;
            t3[t_] = t2[t] + t ;
            w[t_] = z[t] / t3[t] ;
            (*Table[{z[t],v1[t],t,t3[t]},{t,.737,1.637,.03}]//
              TableForm*)
            p = ParametricPlot[{z[t], w[t]}, {t, 0.735, 4.2},
                PlotRange → {{0, 5500}, {5, 10}}];
            r = ParametricPlot[{x1[t], x1[t] / t}, {t, 3, 26.8},
                PlotRange → {{0, 5500}, {5, 10}}];
```

The distance run and the velocity gained in the acceleration time t are

Out[188]= $x[t] \rightarrow -9.04894 \ (1 - e^{-1.14\,t}) + 10.3158\,t$

               $v[t] \rightarrow 10.3158 \ (1 - e^{-1.14\,t})$

Let's find the total time (t + t2) required to run each race using the Keller theory

```
(*Predicted times at various race distances -- *)
sol100 = FindRoot[100 == x1[t], {t, 10}];
sol200 = FindRoot[200 == x1[t], {t, 20}];
sol400 = FindRoot[400 == v1[t] * t2[t] + x1[t], {t, 1}];
t400 = (t2[t] + t) /. sol400;
sol800 = FindRoot[800 == v1[t] * t2[t] + x1[t], {t, 1}] ;
t800 = (t2[t] + t) /. sol800;
sol1000 = FindRoot[1000 == v1[t] * t2[t] + x1[t], {t, 1}];
t1000 = (t2[t] + t) /. sol1000;
sol1500 = FindRoot[1500 == v1[t] * t2[t] + x1[t], {t, 1}] ;
t1500 = (t2[t] + t) /. sol1500 ;
sol1609 = FindRoot[1609 == v1[t] * t2[t] + x1[t], {t, 1}] ;
t1609 = (t2[t] + t) /. sol1609;
sol2000 = FindRoot[2000 == v1[t] * t2[t] + x1[t], {t, 1}] ;
t2000 = (t2[t] + t) /. sol2000;
sol3000 = FindRoot[3000 == v1[t] * t2[t] + x1[t], {t, 1}];
t3000 = (t2[t] + t) /. sol3000;
sol5000 = FindRoot[5000 == v1[t] * t2[t] + x1[t], {t, 1}] ;
t5000 = (t2[t] + t) /. sol5000 ;
sol10000 = FindRoot[10000 == v1[t] * t2[t] + x1[t], {t, 1}];
t10000 = (t2[t] + t) /. sol10000;
Print["  t100 = ", t /. sol100];
Print ["  t200 = ", t /. sol200]; Print["  t400 = ", t400];
Print["  t800 = ", t800]; Print[" t1000 = ", t1000];
Print[" t1500 = ", t1500]; Print[" t1609 = ", t1609];
Print[" t2000 = ", t2000]; Print[" t3000 = ", t3000];
Print[" t5000 = ", t5000]; Print["t10000 = ", t10000]
```

```
          t100 = 10.5711

          t200 = 20.2649

          t400 = 46.431

          t800 = 113.762

         t1000 = 148.404

         t1500 = 235.819

         t1609 = 254.953

         t2000 = 323.705

         t3000 = 499.953

         t5000 = 853.025

        t10000 = 1736.38
```

In[207]:= Show[{velc, p, r}, AxesLabel → {"Dist(m)", "Velocity(m/s)"}]

Let's run the *Mathematica* program and see how well we do

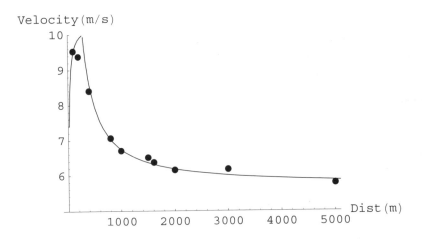

Velocity (m/s)

This appears to be a rather good fit, with only two dots away from the theory line (at 200 meters and 3000 meters). Assuming the theory to hold, we may expect the 200-m record to be broken, and the 3000-m record to stand for a number of years. The last question is how well do the theoretical and actual record numbers compare?

Track and Field World Records for Women 2007

Dist (m)	Theory Time	Record Time (s)	% Difference	
100	10.57	10.49	− 0.7	
200	20.26	21.34	4.8	(world record most likely to fall)
400	46.4	47.60	2.5	
800	113.7	113.28	− 0.4	
1000	148.4	149.34	0.6	
1500	235.8	230.46	− 2.3	
1 mile	254.9	252.56	− 0.6	
2000	323.7	325.36	0.9	
3000	499.9	486.11	− 2.8	(world record least likely to fall)
5000	853.0	864.68	1.3	
10000	1736.4	1771.78	2.0	

All the track records for women may be accounted for with the Physics model of running up to 10,000 meters. Beyond this range, when running fatigue sets in -- as the runner's system tries to clear the lactic acid build-up, acidosis of the muscles begins to occur -- this is often referred to as THE WALL. Many long-distance runners have learned to cope with the fatigue by training and diet, and also by running at a somewhat reduced pace. This can be accounted for in the Physics model by allowing k (the internal resistance) to increase for runs beyond 10,000 meters.

Problem 4-1 Ramsauer Effect

A 1-eV electron encounters an 8.4 eV potential well. The well is 2 Angstroms wide. What is the probability of transmission? What is the probability the electron is reflected?

```
Plot[{1, -8.4 * UnitStep[x - 1.1] + 8.4 * UnitStep[x - 3.1]},
  {x, -3, 4.1}, PlotRange → {-8.5, 4.1}, Axes → None,
  Prolog → {Text["1 eV", {3.78, 2.45}], Text["-8.4 eV", {3.7, -7.5}]}]
```

1 eV

−8.4 eV

The potential well near a noble-gas atom for 1-eV electron beam

The Quantum wavefunctions in the three regions are :

$$\psi_1 = e^{ikx} + r e^{-ikx} \qquad \text{where } k = \frac{\sqrt{2 \, m \, E}}{\hbar}$$

$$\psi_2 = a \, e^{i\alpha x} + b \, e^{-i\alpha x} \qquad \text{where } \alpha = \frac{\sqrt{2 \, m \, (E + |V|)}}{\hbar}$$

$$\psi_3 = t \, e^{ikx}$$

In[122]:=

$$k = \sqrt{2 * 9.1 * 10^{-31} * 1 * 1.6 * 10^{-19}} \Big/ (6.626 * 10^{-34} / (2 \pi)) \Big/ 10^{10}$$

$$\alpha = \sqrt{2 * 9.1 * 10^{-31} * 9.4 * 1.6 * 10^{-19}} \Big/ (6.626 * 10^{-34} / (2 \pi)) \Big/ 10^{10}$$

Out[122]= $k = 0.51171$

Out[123]= $\alpha = 1.56887$

```
Clear[a, b, r, t]
k = .51171; α = 1.56887; L = 2;
sol = NSolve[
   {1 + r == a + b, a * Exp[I * α * L] + b * Exp[-I * α * L] == t * Exp[I * k * L],
    k - k * r == α * a - α * b, α * a * Exp[I * α * L] - α * b * Exp[-I * α * L] ==
     k * t * Exp[I * k * L]}, {a, b, r, t}]
R = Abs[r]^2 /. sol
T = Abs[t]^2 /. sol
psi = Re[Exp[I * k * x] + r * Exp[-I * k * x]] /. sol;
bar = Re[a * Exp[I * α * x] + b * Exp[-I * α * x]] /. sol;
tran = Re[t * Exp[I * k * x]] /. sol;
Plot[psi * If[x < 0, 1, 0] + bar * (UnitStep[x] - UnitStep[x - 2]) +
   tran * If[x > 2, 1, 0], {x, -10, 10}]
```

Out[114]=
$\{\{a \to 0.663071 - 0.00177808\,i,\ b \to 0.336895 - 0.00349941\,i,$
$\quad r \to -0.0000344849 - 0.0052775\,i,\ t \to -0.514851 + 0.857263\,i\}\}$

Reflection

Out[115]= {0.0000278531}

Transmission

Out[116]= {0.999972}

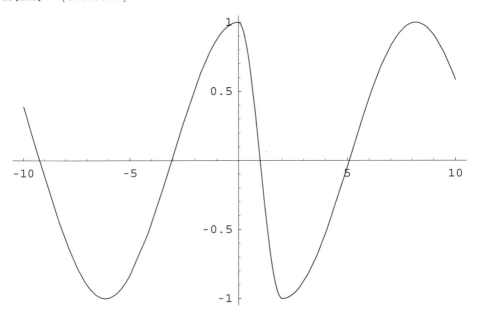

The real part of the wavefunction. Virtually 100% of the incident electron beam is transmitted.

Problem 4-2 Intermediate Quantum Step

We can get 100 % transmission across a potential step if we put in an intermediate quantum step of the appropriate height V and width L. For a 7-eV electron beam impinging on a 5-eV potential step, how high and how wide does the intermediate step need to be? Express the answer in eV and Angstroms.

```
Plot[{7, 2 * (UnitStep[x - 1] - UnitStep[x - 3]), 5 * UnitStep[x - 3]},
   {x, -3, 7}, PlotRange → {-1, 10.1}, Axes → None]
```

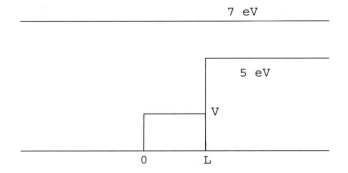

In the three regions,
$$\psi_1 = e^{ikx} \qquad\qquad k = \sqrt{2\,m\,(E-0)}\,\big/\,\hbar$$
$$\psi_2 = ae^{i\beta x} + be^{-i\beta x} \qquad \beta = \sqrt{2\,m\,(E-V)}\,\big/\,\hbar$$
$$\psi_3 = t\,e^{i\gamma x} \qquad\qquad \gamma = \sqrt{2\,m\,(E-5)}\,\big/\,\hbar$$

Note that the only places where two different wavefunctions and their derivatives will match is at a maximum or minimum, where $\psi'(x) = 0$, or where $\psi(x) = 0$. The distance in phase between these matching points is a multiple of $\pi/2$, or, the minimum width of the step potential is found from $\beta L = \pi/2$. Then, matching wavefunctions and their derivatives at $x = 0$, and $x = L$, we will have 5 equations in 5 unknowns. We will solve for $\{a, b, \beta, L, \text{and } t\}$ and then check our answer by finding the transmitted quantum flux, $T = (\gamma/k)t^2$. If this is 1, then the values of β and L are correct.

$In[1]:=$ $k = \sqrt{2 * 9.1 * 10^{-31} * 7 * 1.6 * 10^{-19}}\,\Big/\,(6.626 * 10^{-34} / (2\,\pi))\,\Big/\,10^{10}$

$\gamma = \sqrt{2 * 9.1 * 10^{-31} * (7 - 5) * 1.6 * 10^{-19}}\,\Big/\,(6.626 * 10^{-34} / (2\,\pi))\,\Big/\,10^{10}$

$Out[1]=$ $k = 1.35386$ $\gamma = 0.723668$

```
In[195]:=  Clear[a, b, β, L, t]; k = 1.35386; γ = .723668;
           (*Note 5 equations in 5 unknowns*)
           sol = FindRoot[{1 == a + b,
               a * Exp[I * β * L] + b * Exp[-I * β * L] == t * Exp[I * γ * L],
               k == β * a - β * b, β * a * Exp[I * β * L] - β * b * Exp[-I * β * L] ==
               γ * t * Exp[I * γ * L], β * L == π / 2},
              {β, 1}, {L, 1}, {a, 1}, {b, 1}, {t, 1}] // Chop


           T = (γ / k) * Abs[t]^2 /. sol
           psi = Re[Exp[I * k * x]] /. sol;
           bar = Re[a * Exp[I * β * x] + b * Exp[-I * β * x]] /. sol;
           tran = Re[t * Exp[I * γ * x]] /. sol;
           Plot[psi * If[x < 0, 1, 0] + bar * (UnitStep[x] - UnitStep[x - 2]) +
              tran * If[x > 2, 1, 0], {x, -10, 10}]
```

Out[198]= {β → 0.989821, L → 1.58695, a → 1.18389,
             b → -0.183891, t → 1.24758 + 0.560688 i}

Out[199]=

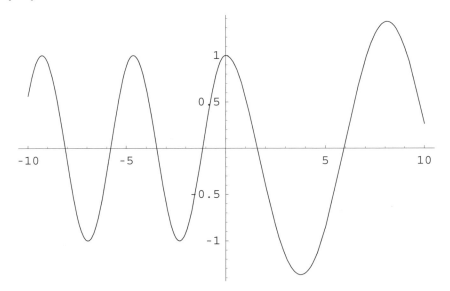

The real part of the wavefunction. 100% of the incident electron beam is transmitted.

```
           Clear[V]; L = L /. sol; β = β /. sol; E1 = 7;
           sol1 = Solve[β ==
```

$$\sqrt{2 * 9.1 * 10^{-31} * (7 - V) * 1.6 * 10^{-19}} \Big/ (6.626 * 10^{-34} / (2\,\pi)) \Big/ 10^{10}, V]$$

           {{V → 3.25834}}

```
In[3]:=  Print["L=", L, "A   β=", β, "   V=", V /. sol1, "eV"]
```

           L =1.58695 A   β =0.989821   V =3.25834 eV

Problem 4-3 Energy Eigenvalues

An electron is trapped in a quantum well. If this is a square well of depth 5 eV and width 2 Angstroms, what is the ground state energy of the electron above the bottom of the well?

The ground-state wavefunction for the electron will be a Cosine wave, because this allows a maximum probability at the center of the well. Notice that it is necessary to match the Cosine wave in the well with an exponentially decreasing wavefunction outside the well.

```
In[2]:=  Plot[{-5 * UnitStep[x - 1.1] + 5 * UnitStep[x - 3.1]},
         {x, -3, 4}, PlotRange → {-6.3, 3.1}, Axes → None]
```

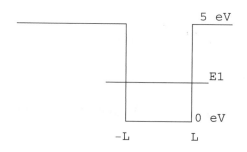

5 eV

E1   Electron in a Quantum Well

0 eV

−L   L

For the ground-state wavefunction, ψ_2 = A Cos β x   in the well,

and ψ_3 = C e^{-kx} outside the well.

If we set $\psi_2 = \psi_3$ and their derivatives equal at x = L

$$A \cos \beta L = C e^{-kL}$$
$$-\beta A \sin \beta L = -k C e^{-kL}$$

Then, $\beta \tan \beta L = k$

Where $\beta = \dfrac{\sqrt{2 m E}}{\hbar}$ and $k = \dfrac{\sqrt{2 m (V - E)}}{\hbar}$

The above equation may easily be solved in *Mathematica*, however let us first make sure E and V are in eV, and L is in Angstrom units.

$$\beta \, \text{Tan} \, \beta L = k$$

$$\text{Tan} \, \beta L = \frac{k}{\beta}$$

$$\text{Tan}\left(\frac{\sqrt{2\,m\,E}\ L}{\hbar}\right) = \frac{\sqrt{V - E}}{\sqrt{E}}$$

or $\qquad \text{Tan}\left(\gamma \ \sqrt{E}\right) = \dfrac{\sqrt{V - E}}{\sqrt{E}}$

where $\quad \gamma = \dfrac{\sqrt{2*9.1 \times 10^{-31} * 1.6 \times 10^{-19}}}{(6.626 \times 10^{-34}/2\pi)} * 1 \times 10^{-10} = .5117\,\text{eV}^{-1/2}$

In[19]:= $\qquad \gamma = \dfrac{\sqrt{2*9.1*10^{-31}*1.6*10^{-19}}}{(6.626*10^{-34}/(2\pi))} * (1 \times 10^{-10})$

Out[19]= 0.51171

In[14]:= **Clear[V]; V = 5.0; γ = .51171;**

sol = FindRoot$\left[\text{Tan}\left[\gamma * \sqrt{\text{E1}}\ \right] == \dfrac{\sqrt{(V - E1)}}{\sqrt{\text{E1}}}, \{\text{E1, 1}\}\right]$

Out[16]= $\{\text{E1} \to 2.43473\}$

The ground state energy of the electron is 2.434 eV above the bottom of the well.

A question that should be asked is, whether there are any higher energy eigenstates in the 5-eV square well. The next-higher wavefunction that can fit into the well is $\psi = \text{Sin} \, \beta x$. For this wavefunction to match the decaying exponential at the well wall, $\beta x > \pi/2$. (The wavefunction in the well must have a downward slope when it meets the wall).

So for there to be a higher energy eigenstate, $\beta L > \pi/2$, or, for L = 1 A,

$.5117 \sqrt{E} > \pi/2$,   and   E2 > 9.42 eV.

In[20]:= **Solve$\left[.5117 \sqrt{\text{E2}} == \pi/2, \ \text{E2}\right]$**

Out[20]= $\{\{\text{E2} \to 9.42343\}\}$

There are no higher energy eigenstates in this 5-eV deep, 2A-wide well.

Problem 4-4 Cooling Concrete in Dams

Given a concrete slab with the following thermal properties,
thermal conductivity k = 1.5 BTU/hr/ft²·°F, density ρ = 150 lb/ ft,
specific heat C = 0.25 BTU/ °F·lb

What is the temperature in the center of the slab after 40 days?

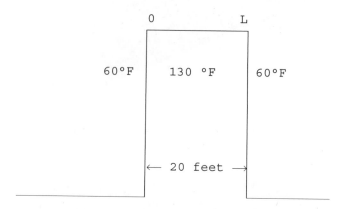

We will use *Mathematica* to numerically solve the PDE

$$\frac{\partial T}{\partial t} = \alpha \frac{\partial^2 T}{\partial x^2}$$
with (T at x = 0 and x = L) = 0   and T (x, 0) = 70

Where T is the temperature in the slab above the constant outside
temperature. The plan is to solve this equation numerically with **NDSolve**
and then compare the answer with the exact solution (which for the center
of the slab is

$$T\left(\frac{L}{2}\right) = \frac{4*70}{\pi} \sum_{k=0}^{10} \frac{(-1)^k}{(2k+1)} * \mathrm{Exp}\left[-\frac{\alpha\,(2k+1)^2}{(L/\pi)^2} t\right]$$

The diffusivity constant $\alpha = \frac{k}{\rho C} = 0.04$ ft²/ hr = 0.96 ft²/ day with L = 20 ft.

Now, programming *Mathematica* to solve the PDE with the appropriate initial
and boundary conditions,

```
In[1]:=  L = 20; α = .96; eq4 = {D[u[x, t], t] - .96*D[u[x, t], x, x] == 0,
         u[x, 0] == Which[x ≤ 0, 0, 0 < x < 20, 70, x ≥ 20, 0],
         u[0, t] == 0, u[20, t] == 0};
       sol4 = NDSolve[eq4, u[x, t], {x, 0, 20}, {t, 0, 40.0}];
       Plot3D[Evaluate[u[x, t] /. sol4[[1]]], {t, 0, 40.0}, {x, 0, 20}]];
```

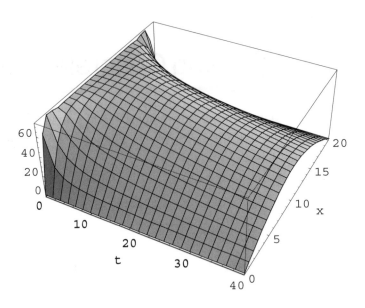

The gradual cooling of the concrete slab.

```
Table[Evaluate[
```

$$\left\{t, \; u[x, t], \; \frac{4 * 70}{\pi} \sum_{k=0}^{10} \frac{(-1)^k}{(2k+1)} * \text{Exp}\left[-\frac{\alpha \, (2k+1)^2}{(L / \pi)^2} \, t\right]\right\} \; /. \; \text{sol4}[[1]] \; /.$$

```
x → 10, {t, 2.0, 40.0, 2.0}]] // TableForm
```

t (da)	T (x = 10)	T (Exact)
2.	70.0021	70.
4.	69.9683	69.9569
6.	69.5539	69.5497
8.	68.4981	68.4986
10.	66.851	66.853
12.	64.7889	64.7891
14.	62.4749	62.4741
16.	60.0307	60.0324
18.	57.5485	57.5489
20.	55.0783	55.0785
22.	52.6557	52.6558
24.	50.3008	50.3015
26.	48.0267	48.0277
28.	45.8394	45.8404
30.	43.7413	43.7423
32.	41.7318	41.7333
34.	39.8108	39.812
36.	37.9752	37.9763
38.	36.2222	36.2233
40.	34.549	34.55

The agreement between the numerical solution of *Mathematica* and the exact solution is remarkable. If one desired greater accuracy in the numerical solution, then the command **MaxSteps** → **20000** could be added at the end of the **NDSolve** command line.

The final result, with the outside temperature maintained at 60 °F for 40 days and 40 nights, is the center temperature in the slab is a toasty 94.5 °F. Such a temperature differential would cause thermal stresses in the dam, therefore cooling pipes are installed within the concrete to allow the slabs to cool at a uniform rate from center to surface.

For a most interesting description of the methods employed, see Clarence Rawhouser's article "Cooling the Concrete in Grand-Coulee Dam" in <u>Mechanical Engineering</u> (1940) vol. 62, pages 715-718.

For readers who are mathematicians, note that *Mathematica* program only summed the exact solution to a maximum k = 10. This is more than is necessary. A maximum value of k = 5 is entirely satisfactory due to the rapid convergence of the exponential term.

As an additional example of the capabilities of *Mathematica*, notice that the program can evaluate an infinite series

$$In[77]:= \frac{4.0 * 70}{\pi} \sum_{k=0}^{\infty} \frac{(-1)^k}{(2k+1)}$$

$$Out[77]= 70.$$

Problem 4-5 Heat Transfer in Steel

<u>Temperature input into a semi-infinite block of steel</u>
    A thick steel slab at 550 °F has its surface suddenly cooled to 100 °F. How long before the temperature at 1-inch depth reaches 200 °F ?

    The Partial Differential Equation for heat transfer in steel is

$$\frac{\partial u}{\partial t} = \alpha \, \frac{\partial^2 u}{\partial x^2} \quad \text{where} \quad \alpha = 0.45 \, \frac{ft^2}{hr} = 0.0075 \, \frac{ft^2}{min}$$

Initial conditions
    `u[x,0]==550`   the block is originally at 550 °F throughout

    `u[0,t] == (100+ 450*Exp[-1000*t])`

The latter condition is chosen so that the surface is originally at 550 °F, and is then rapidly cooled to 100 °F. In this way, the initial conditions at x=0 are not inconsistent.

    Boundary conditions are   `D[u[x,t],x]/.x→1) == 0,` so that the spatial derivative vanishes at large x. Since we are looking at a depth of 1-inch, x→1 foot will suffice. The *Mathematica* solution follows:

```
In[47]:=  eq3 = {D[u[x, t], t] - (0.45 / 60) * D[u[x, t], x, x] == 0,
           u[x, 0] == 550, u[0, t] == (100 + 450 * Exp[-1000 * t]),
           (D[u[x, t], x] /. x → 1) == 0};
          sol1 = NDSolve[eq3, u[x, t], {x, 0, 1}, {t, 0, 6}];
          Plot3D[Evaluate[u[x, t] /. sol1[[1]], {t, 0, 6}, {x, 0, 1}]]
```

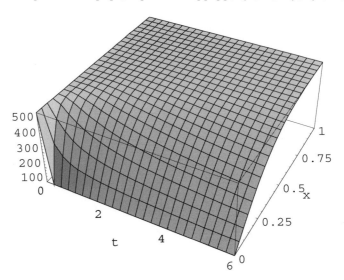

Shown above is the temperature variation in the steel block with u(0 , t) = 100 for t > 0. At a depth of 0.083 ft (1 inch) it should take 5.765 minutes for the temperature to fall from 550 °F to 200 °F. (See Table)

```
Clear@f;  f[x_, t_] = u[x, t] /. sol1[[1]];
TableForm[Transpose@Prepend[
    Transpose@Table[f[x, t], {t, 0, 6, .25}, {x, 0, 0.415, 0.083}],
    Table[StyleForm[ToString@t, FontWeight → "Bold"], {t, 0, 6, .25}]],
  TableHeadings -> {None, Prepend[
    Table[StyleForm["x(" <> ToString@x <> ")", FontWeight → "Bold"],
      {x, 0, 0.415, 0.083}], StyleForm["t(min)", FontWeight → "Bold"]]}]
```

		0 "	1 "	2"	3 "
t(min)		x(0)	x(0.083)	x(0.166)	x(0.249)
0		550.	550.	550.	550.
0.25		100.076	467.188	545.933	550.133
0.5		100.076	402.375	524.115	548.097
0.75		100.076	358.574	497.414	541.203
1.		100.076	328.188	471.965	530.931
1.25		100.076	306.187	449.425	518.995
1.5		100.076	289.526	429.886	506.552
1.75		100.076	276.412	412.995	494.261
2.		100.076	265.745	398.316	482.464
2.25		100.076	256.845	385.446	471.311
2.5		100.076	249.262	374.072	460.841
2.75		100.076	242.694	363.935	451.049
3.		100.076	236.935	354.828	441.904
3.25		100.076	231.824	346.592	433.358
3.5		100.076	227.246	339.101	425.362
3.75		100.076	223.114	332.244	417.87
4.		100.076	219.358	325.939	410.836
4.25		100.076	215.925	320.118	404.222
4.5		100.076	212.781	314.728	398.001
4.75		100.076	209.879	309.714	392.13
5.		100.076	207.189	305.033	386.579
5.25		100.076	204.687	300.652	381.323
5.5		100.076	202.355	296.541	376.338
5.75		100.076	**200.171**	292.669	371.602
6.		100.076	198.123	289.015	367.095

An EXACT solution follows: In either the numerical or Exact case, it takes 5.76 minutes for the temperature at $x = 0.083$ ft to fall from 550 °F to 200 °F.

```
In[13]:=  α = 0.45 / 60 ; x1 = .083 ;  (*time T1 in minutes *)
```

$$\text{NSolve}\left[\left\{\frac{200 - (100)}{550 - (100)} == \text{Erf}[z], \; z == \frac{x1}{2 \sqrt{\alpha * t1}}\right\}, \{z, t1\}\right]$$

```
Out[14]=  {{t1 → 5.76635, z → 0.199557}}
```

Problem 5-1 Space Shuttle Phase 2

The second phase of the Space Shuttle launch carries the Shuttle up to the point of orbit insertion. The main engine delivers 5×10^6 N of thrust by burning 1400 kg of liquid H_2 and O_2 per second.

If we take into account the *curvature of the Earth,* we must rewrite the Equations of Motion of the Space Shuttle such that gravitational force decreases with the actual height h of the Shuttle, and is measured toward the center of the Earth. We still measure x and y from the launch point and the height of the Shuttle above the surface (see diagram) is $h = \sqrt{(R + y)^2 + x^2} - R$.

$$my'' = (T - D)\,Cos\,(.84 + \epsilon t) - mg_o \left(\frac{R}{R + h}\right)^2 Cos\,\lambda \quad y_o = 47660 \quad y'_o = 951 \quad \epsilon = .001$$

$$mx'' = (T - D)\,Sin\,(.84 + \epsilon t) - mg_o \left(\frac{R}{R + h}\right)^2 Sin\,\lambda \quad x_o = 38620 \quad x'_o = 1105$$

We may now program these equations into *Mathematica,*

```
m0 = 690000; T = 5 * 10^6; α = 1400; ε = .001; g0 = 9.8; R = 6.4 * 10^6;
m[t] = m0 - α * t; r[t] = √(R + y[t])^2 + x[t]^2; g[t] = g0 * R^2 / r[t]^2;
sol = NDSolve[{
   m[t] * y''[t] == T * Cos[.84 + ε * t] - m[t] * g[t] * (R + y[t]) / r[t],
   m[t] * x''[t] == T * Sin[.84 + ε * t] - m[t] * g[t] * (x[t]) / r[t],
   y[0] == 47660, x[0] == 38620, y'[0] == 951,
   x'[0] == 1105}, {x, y}, {t, 0, 380}]
```

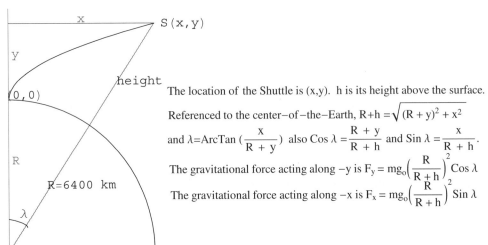

The location of the Shuttle is (x,y). h is its height above the surface.

Referenced to the center−of−the−Earth, $R+h = \sqrt{(R + y)^2 + x^2}$

and $\lambda = ArcTan\left(\dfrac{x}{R + y}\right)$ also $Cos\,\lambda = \dfrac{R + y}{R + h}$ and $Sin\,\lambda = \dfrac{x}{R + h}$.

The gravitational force acting along −y is $F_y = mg_o\left(\dfrac{R}{R + h}\right)^2 Cos\,\lambda$

The gravitational force acting along −x is $F_x = mg_o\left(\dfrac{R}{R + h}\right)^2 Sin\,\lambda$

Location of the Shuttle with respect to the center-of-the-Earth

```
In[33]:=
    InterpFunc1 = x /. sol[[1]]; InterpFunc2 = y /. sol[[1]];
    InterpFunc3 = x' /. sol[[1]]; InterpFunc4 = y' /. sol[[1]];
    height[t_] = √((R + InterpFunc2[t])^2 + InterpFunc1[t]^2) - R;
    tbl = Table[{InterpFunc1[t] / 1000, height[t] / 1000}, {t, 0, 380, 10}];
    ListPlot[tbl, Prolog → AbsolutePointSize[4], AspectRatio → 0.65,
       Epilog → {Text["Distance Downrange(km)", {875, 60}],
         Text["Height of Shuttle(km)", {158, 226}]}];
```

The trajectory of the Space Shuttle from $t = 120$ seconds to $t = 500$ s, just before main fuel-tank separation. Notice that as the Shuttle approaches the end of the second-stage that it begins to gain height. This is because the velocity of the Shuttle is basically along x and the curve of the Earth is falling away beneath the Shuttle.

If we **Table** the results, we will find the height achieved is h = approx 230 km, and the velocity after the second-phase burn is approx 5500 m/s.

```
In[49]:= Table[
          {t, Chop[InterpFunc2[t] / 1000], Chop[InterpFunc1[t] / 1000],
           height[t] / 1000,
           Chop[√(InterpFunc3[t]^2 + InterpFunc4[t]^2)]},
          {t, 0, 380, 10}] // TableForm
```

The (x,y) coordinates and velocity for Space Shuttle second-phase flight

t (s)	y (km)	x (km)	height (km)	v (m / s)
0	47.66	38.62	47.775	1457.88
10	56.930	49.939	57.123	1469.44
20	65.726	61.806	66.021	1486.04
30	74.054	74.237	74.480	1507.66
40	81.923	87.247	82.510	1534.25
50	89.338	100.85	90.122	1565.73
60	96.309	115.06	97.328	1602.01
70	102.84	129.91	104.14	1642.97
80	108.94	145.40	110.56	1688.5
90	114.62	161.57	116.62	1738.48
100	119.88	178.41	122.33	1792.8
110	124.74	195.97	127.69	1851.38
120	129.20	214.26	132.72	1914.11
130	133.27	233.30	137.43	1980.93
140	136.96	253.12	141.85	2051.81
150	140.27	273.74	145.99	2126.7
160	143.22	295.20	149.87	2205.62
170	145.81	317.52	153.51	2288.58
180	148.06	340.73	156.92	2375.64
190	149.97	364.86	160.12	2466.88
200	151.56	389.95	163.15	2562.42
210	152.84	416.04	166.03	2662.4
220	153.81	443.17	168.78	2767.01
230	154.50	471.37	171.43	2876.47
240	154.91	500.70	174.01	2991.06
250	155.06	531.21	176.55	3111.08
260	154.97	562.94	179.10	3236.93
270	154.65	595.96	181.69	3369.04
280	154.12	630.34	184.36	3507.93
290	153.4	666.14	187.16	3654.22
300	152.59	703.43	190.15	3808.61
310	151.43	742.31	193.39	3971.96
320	150.38	782.87	196.93	4145.27
330	149.02	825.22	200.85	4329.72
340	147.78	869.47	205.24	4526.77
350	146.42	915.77	210.18	4738.16
360	145.16	964.26	215.78	4966.01
370	143.88	1015.1	222.16	5213.
380	142.72	1068.5	229.46	5482.48

After the main fuel tank is jettisoned, the 110,000 kg Shuttle is moving at 5500 m/s on a downrange trajectory. Its height is approx 230 km. At this point, the orbit maneuvering system takes over with two 27,000 N thrust engines and increases the speed to approx 7700 m/s which is appropriate for a 400-km orbit above the Earth.

Problem 5-2 B-Z Chemical Reaction

I. Solve the B-Z equations for $x = [HBrO_2]$, $y = [Br^-]$, and $z = [Ce^{+4}]$

$$\epsilon\, x' = x + y - q\, x^2 - xy$$
$$7y' = -y + 2hz - xy$$
$$p\, z' = x - z$$

where $\epsilon = 0.21$, $p = 14$, $q = 0.006$, $h = 0.75$, $x(0) = 100$, $y(0) = 1$, $z(0) = 10$

```
ε = .21; p = 14; q = .006; h = .75;
sol = NDSolve[{ε * x'[t] == x[t] + y[t] - q * x[t] ^2 - x[t] * y[t],
    7 y'[t] == -y[t] - x[t] * y[t] + 2 h * z[t], p * z'[t] == x[t] - z[t],
    x[0] == 100, y[0] == 1, z[0] == 10}, {x, y, z}, {t, 0, 100}]
InterpFunc1 = x /. sol[[1]]; InterpFunc2 = y /. sol[[1]];
InterpFunc3 = z /. sol[[1]];
Chop[Table[{t, InterpFunc1[t], InterpFunc2[t], InterpFunc3[t]},
    {t, 0, 80, 2}] // TableForm]
Plot[Evaluate[{x[t] / 4, y[t], z[t]} /. sol],
  {t, 0, 100}, PlotRange → {0, 45},
  PlotStyle → {{RGBColor[0, 1, 0]},
    {RGBColor[1, 0, 0]}, {RGBColor[0, 0, 1]}}]
```

Again, as in Section 5-2, *Mathematica* handles stiff differential equations with no problem.

Let's Table the values to get an accurate read on the time of oscillation. Scanning the Table, we find maximum x at 2 and 56 s, maximum y at 12 and 66 s, and maximum z at 4 and 58 s. The time of oscillation of x, y, and z is therefore 54 seconds.

t(s)	x	y	z
0	100.	1.	10.
2	**122.521**	0.299541	25.2651
4	71.9039	0.687823	**35.1345**
6	1.16892	7.06071	33.7346
8	1.07679	13.9677	29.3908
10	1.06385	16.5633	25.6205
12	1.06249	**16.8911**	22.3513
14	1.06579	16.0882	19.5174
16	1.07202	14.7776	17.0615
18	1.08069	13.294	14.9335
20	1.09171	11.8114	13.0901
22	1.10527	10.4148	11.4938
24	1.12168	9.1409	10.1119
26	1.1414	8.00059	8.91642
28	1.16509	6.99132	7.88296
30	1.1936	6.10419	6.99055
32	1.22806	5.32765	6.22112
34	1.27003	4.64947	5.5592
36	1.32164	4.05772	4.99161
38	1.386	3.54121	4.50728
40	1.46773	3.08959	4.09711
42	1.57423	2.69316	3.75401
44	1.71838	2.34255	3.47313
46	1.9255	2.02771	3.25273
48	2.25613	1.73545	3.0966
50	2.91417	1.44012	3.02339
52	5.37809	1.04398	3.12871
54	105.093	0.0963774	6.47907
56	**127.011**	0.27174	23.7781
58	79.1861	0.620734	**34.5446**
60	1.21035	5.98529	34.2246
62	1.07966	13.5059	29.817
64	1.06441	16.431	25.9901
66	1.06235	**16.928**	22.6717
68	1.06528	16.2059	19.7952
70	1.07125	14.9282	17.3022
72	1.07967	13.4519	15.1421
74	1.09044	11.9638	13.2708
76	1.10372	10.5557	11.6502
78	1.11981	9.26806	10.2473
80	1.13917	8.11368	9.03348

II. Solve the B-Z equations

$$\epsilon\, x' = x + y - q\,x^2 - xy$$
$$7\,y' = -y + 2hz - xy$$
$$p\,z' = x - z$$

where $\epsilon = 0.21$, $p=14$, $q= 0.006$, $h =0.75$, $x(0) = 17$, $y(0) = 3$, $z(0) = 5$

```
ε = .21; p = 14; q = .006; h = .75;
sol = NDSolve[{ε * x'[t] == x[t] + y[t] - q * x[t]^2 - x[t] * y[t],
    7 y'[t] == -y[t] - x[t] * y[t] + 2 h * z[t], p * z'[t] == x[t] - z[t],
    x[0] == 17, y[0] == 3, z[0] == 5}, {x, y, z}, {t, 0, 100}]
InterpFunc1 = x /. sol[[1]]; InterpFunc2 = y /. sol[[1]];
InterpFunc3 = z /. sol[[1]];
Chop[Table[{t, InterpFunc1[t], InterpFunc2[t], InterpFunc3[t]},
    {t, 0, 90, 2}] // TableForm]
Plot[Evaluate[{x[t] / 4, y[t], z[t]} /. sol],
 {t, 0, 100}, PlotRange → {0, 45},
 Prolog → {Text["x/4", {15, 30}], Text["y", {29, 14}],
   Text["t(s)", {96.5, 2.3}], Text["Z", {31.2, 26}]},
 PlotStyle →
  {{RGBColor[0, 1, 0]}, {RGBColor[1, 0, 0]}, {RGBColor[0, 0, 1]}}]
```

Again, as in Section 5.2, *Mathematica* handles the stiff-differential equations with no problem. The effect of varying the reactants is to delay the onset of the oscillations. x, y, and z have to build up to a certain level before the color-changes begin.

Let's Table the values to get an accurate read on the time of oscillation.

t(s)	x	y	z
0	17.	3.	5.
2	1.65736	2.50447	4.67582
4	1.64866	2.51232	4.27255
6	1.70065	2.39096	3.92624
8	1.80322	2.20369	3.63636
10	1.96856	1.98605	3.40276
12	2.23804	1.75373	3.22862
14	2.74323	1.50346	3.12704
16	4.18261	1.18636	3.15243
18	28.8258	0.375095	4.02654
20	**138.115**	0.197779	18.876
22	96.9563	0.478375	32.1197
24	7.25794	2.49496	**35.6603**
26	1.09234	11.8139	31.1106
28	1.06688	15.8651	27.1124
30	1.06229	**16.9474**	23.6447
32	1.06401	16.511	20.6386
34	1.06913	15.3556	18.0331
36	1.07682	13.9146	15.7753
38	1.08686	12.417	13.8193
40	1.09935	10.9781	12.1251
42	1.11453	9.65103	10.6584
44	1.13282	8.45526	9.38907
46	1.15478	7.39267	8.29143
48	1.18118	6.4564	7.34313
50	1.21302	5.63566	6.52494
52	1.25167	4.91834	5.82036
54	1.29899	4.2923	5.21528
56	1.35763	3.74603	4.69779
58	1.43149	3.26882	4.25797
60	1.5266	2.85075	3.88791
62	1.6531	2.48237	3.58176
64	1.82977	2.15408	3.33633
66	2.0976	1.85452	3.15263
68	2.57319	1.56537	3.04098
70	3.81791	1.23611	3.04638
72	18.9211	0.518451	3.64735
74	**141.401**	0.174242	17.0551
76	102.118	0.440834	31.1047
78	18.9609	1.71081	**36.0066**
80	1.09873	11.128	31.5625
82	1.06805	15.6137	27.5044
84	1.06238	**16.9237**	23.9846
86	1.06364	16.6004	20.9332
88	1.06847	15.4952	18.2884

Judging from peaks in x, y, and z the period of oscillation is about 54 seconds.

Problem 5-3 Equations of Winemaking

We will solve the equations of winemaking

$$x' = (a - bx - cy_*) x \qquad x = \text{\# yeast}$$
$$y' = k\,x \qquad\qquad\qquad y = \text{alcohol per cent}$$
$$z' = -\gamma\,x \qquad\qquad\quad z = \text{remaining sugar (grams)}$$

with initial conditions $x(0) = 100,\ y(0) = 0,\ z(0) = 1140$

$a = 0.125,\ b = a/10^7,\ c = a/10,\ y_* = y - 4$ (before $y = 4, c = 0$)

$\gamma = 10^{11}$ sugar molecules/second/yeast $= \dfrac{180 * 10^{11}}{6 * 10^{23}}$ grams/second/yeast.

$k = 3.4 \times 10^{-13}$ percent/ second/ yeast

We will measure time in hours. Then, programming the three equations into *Mathematica*, we will first plot out the yeast population x for 10 days, and then **Table** the values of x, y, and z for 70 days.

```
a = 1 / 8.;  (*e-folding time of 8 hr*) k = 3600 * 3.4 * (10^-13);
b = a / (10^7);  (*max yeast population of 10^7 *)
c = a / 10; d = 3600 * 10^11 * 180 / (6 * 10^23);
sol1 =
 NDSolve[{x'[t] == (a - b * x[t] - c * If[y[t] < 4, 0, 1] * (y[t] - 4)) * x[t],
   y'[t] == k * x[t], z'[t] == -d * x[t], x[0] == 100,
   y[0] == 0, z[0] == 1140}, {x, y, z}, {t, 0, 1680}]
InterpFunc1 = x /. sol1[[1]]; InterpFunc2 = y /. sol1[[1]];
InterpFunc3 = z /. sol1[[1]];
ListPlot[Table[{t / 24, InterpFunc1[t]}, {t, 0, 240, 2}]]
```

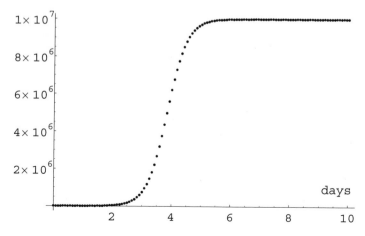

The yeast population reaches its maximum in about 6 days.

```
Plot[Evaluate[{x[t] / 10^6, y[t]} /. sol1, {t, 0, 1200}],
 Prolog → {Text["yeast population", {310, 9.0}], Text["(millions)",
   {330, 8.3}], Text["Alcohol percent", {925, 8.0}],
  Text["Hours", {1145, 0.8}]}]
```

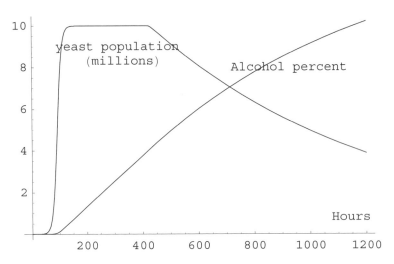

The progress of winemaking. Yeast population and
alcohol percentage for the first 50 days.

Let us now **Table** the functions x, y, and z
x = # of yeast, y = Alcohol percent, z = Sugar supply in grams

```
tbl = Table[{t / 24, InterpFunc1[t] / 10^6, Chop[InterpFunc2[t]],
    InterpFunc3[t]}, {t, 0, 1680, 48}] // TableForm
```

Time(da)	x Millions of yeast	y Alcohol %	z Sugar(gm)
0	0.0001	0	1140.
2	0.0401811	0.000393267	1139.97
4	6.19419	0.0945953	1131.65
6	9.98479	0.635363	1083.94
8	9.99996	1.22273	1032.11
10	10.	1.81025	980.272
12	10.	2.39777	928.432
14	10.	2.98529	876.592
16	10.	3.57281	824.752
18	9.91893	4.15985	772.955
20	9.37113	4.72673	722.936
22	8.83648	5.26144	675.755
24	8.33227	5.76565	631.267
26	7.85684	6.24108	589.316
28	7.40853	6.68939	549.76
30	6.98581	7.11211	512.461
32	6.5872	7.51072	477.289
34	6.21134	7.88658	444.125
36	5.85692	8.241	412.853
38	5.52273	8.57519	383.366
40	5.20761	8.89031	355.561
42	4.91046	9.18746	329.342
44	4.63028	9.46764	304.62
46	4.36607	9.73185	281.308
48	4.11695	9.98097	259.326
50	3.88204	10.2159	238.599
52	3.66053	10.4374	219.054
54	3.45166	10.6463	200.624
56	3.25471	10.8432	183.247
58	3.069	11.0289	166.86
60	2.89389	11.204	151.409
62	2.72876	11.3692	136.839
64	2.57306	11.5249	123.101
66	2.42624	11.6717	110.146
68	2.2878	11.8101	97.9309
70	2.15726	11.9407	86.4126

See Cyril J Berry's 1994 book <u>First Steps in Winemaking</u> for all the steps necessary to make a good wine. As noted in Berry's book, different strains of yeast will have different tolerances to the alcohol percentage. Also, the dryness or sweetness of the wine is determined by how long the fermentation process takes. The alcohol percentage in the wine may be gauged at any time by utilizing the hydrometer. Winemaking is an art as well as a science.

Problem 5-4 H-H Nerve Conduction

Let us use the equations of Section 5-4 to find the form of the action potential at 18.5 °C. With $\alpha, \beta, \gamma, \delta, \epsilon$, and η all multiplied by the temperature factor $3^{(T-6.3)/10} = 3.82$, we will find the conduction velocity.

We will then use the H–H model to show that when a nerve fiber is at 18.5 °C that a 29-mV, 0.5 ms half-sine pulse will elicit an action potential whereas a 28-mV, 0.5 ms half-sine pulse will <u>not</u> produce an action potential.

```
c = .001; a = .0238; ρ = 35.4;
eq3 = {  a
        ————  D[u[x, t], x, x] ==
        2 * ρ
      c * D[u[x, t], t] + .036 * n[x, t]^4 * (u[x, t] + 12) +
      .120 * m[x, t]^3 * h[x, t] * (u[x, t] - 115) + .0003 * (u[x, t] - 10.6),
    u[x, 0] == 0, u[0, t] == (60 * Sin[2 π * t] * If[0 < t < 0.5, 1, 0]),
    (D[u[x, t], x] /. x → 3.2) == 0};
eq4 = {D[n[x, t], t] ==  .0382 * (10 - u[x, t])
                        ————————————————————————— * (1 - n[x, t]) -
                        Exp[(10 - u[x, t]) / 10] - 1
    .4775 * Exp[- u[x, t]
                 ————————] * n[x, t], n[x, 0] == .3177};
                 80
eq5 = {D[m[x, t], t] ==  .382 * (25 - u[x, t])
                        ————————————————————————— * (1 - m[x, t]) -
                        Exp[(25 - u[x, t]) / 10] - 1
    15.28 * Exp[- u[x, t]
                 ————————] * m[x, t], m[x, 0] == .0529};
                 18
eq6 = {D[h[x, t], t] == .2674 * Exp[- u[x, t]
                                    ————————] * (1 - h[x, t]) -
                                    20
         3.82
    ————————————————————————— * h[x, t], h[x, 0] == .5961};
    Exp[(30 - u[x, t]) / 10] + 1
sol1 = NDSolve[{eq3, eq4, eq5, eq6},
    {u[x, t], n[x, t], m[x, t], h[x, t]}, {x, 0, 3.2}, {t, 0, 3.6}];
Plot3D[Evaluate[u[x, t] /. sol1[[1]], {x, 0, 2.0}, {t, 0, 2}],
    PlotRange → {0, 100}, PlotPoints → 30, AxesLabel → {"x", "t", " "},
    DisplayFunction → $DisplayFunction, ViewPoint → {2.5, -1.5, .95}];
Clear@f; f[x_, t_] = u[x, t] /. sol1[[1]]
Pictures = Table[f[x, t], {x, 0, 2.5, .5}]
Plot[Evaluate[Pictures], {t, 0, 3.0}, Frame → True,
    PlotRange → All, Prolog → { Text["x=0", {.25, 64}],
        Text["1cm", {0.88, 74}], Text["x=2.5cm", {2., 74}]}]
```

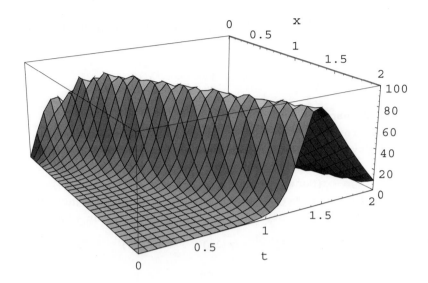

Response of nerve axon to 60-mV input

If we plot the nerve impulse as it is seen at various points along the axon

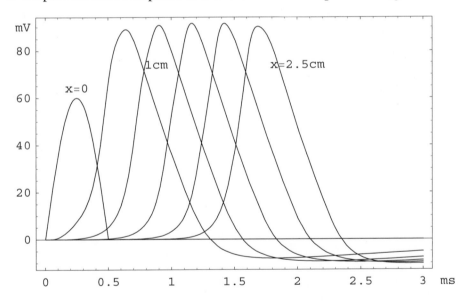

Note that the pulse is "sharper" at 18.5°C -- the conduction velocity is ~ 18.8 m/s

The H–H model shows that the action potential at 18.5 °C is much "sharper" than the impulse at 6.3 °C, and travels much faster. The peak of the action potential passes $x = 1$ cm at 0.9 ms, and passes the 2.5-cm mark at 1.7 ms, then the conduction velocity at 18.5 °C is $v = \frac{1.5 \text{ cm}}{0.8 \text{ ms}} = 18.8$ m/s. This is the same value found by Hodgkin and Huxley.

   We may also check the conduction velocity by creating a table of values of voltage versus time and distance. The maximum voltages (89-90 mV) occur at (x = 1.0 cm, t = 0.9 ms) and (x = 2.5 cm, t = 1.7 ms). Therefore the conduction velocity is $v = \frac{\Delta x}{\Delta t} = \frac{1.5\ cm}{0.8\ ms} = 18.8$ m/s

```
TableForm[Transpose@
    Prepend[Transpose@Table[f[x, t], {t, 0, 3, .1}, {x, 0, 2.5, 0.5}],
      Table[StyleForm[ToString@t, FontWeight → "Bold"],
        {t, 0, 3, .1}]] // Chop,
    TableHeadings -> {None, Prepend[
      Table[StyleForm["x(" <> ToString@x <> ")", FontWeight → "Bold"],
        {x, 0, 2.5, 0.5}], StyleForm["t(ms)", FontWeight → "Bold"]]}]
```

t(ms)	x(0)	x(0.5)	x(1.)	x(1.5)	x(2.)	x(2.5)
0	0	0	0	0	0	0
0.1	35.26	0.62	-0.00	-0.00	-0.00	-0.00
0.2	57.06	4.43	0.071	-0.00	-0.00	-0.00
0.3	57.06	11.5	0.572	0.007	-0.00	-0.00
0.4	35.26	28.1	1.889	0.071	0.000	-0.00
0.5	0.00	66.5	5.176	0.277	0.007	-0.00
0.6	0.00	**87.1**	14.09	0.846	0.037	0.000
0.7	0.00	84.8	40.02	2.446	0.126	0.004
0.8	0.00	68.9	75.69	6.835	0.388	0.017
0.9	0.00	51.9	**90.81**	19.97	1.155	0.057
1.	0.00	36.2	80.18	52.71	3.343	0.178
1.1	0.00	21.7	63.51	82.96	9.499	0.545
1.2	0.00	8.87	46.82	**89.37**	28.45	1.634
1.3	0.00	0.39	31.46	75.14	61.18	4.778
1.4	0.00	-4.2	17.02	58.18	**90.70**	13.82
1.5	0.00	-6.4	5.236	41.92	85.12	36.48
1.6	0.00	-7.4	-2.14	26.79	69.97	74.77
1.7	0.00	-7.9	-5.79	12.95	52.98	**90.09**
1.8	0.00	-8.0	-7.64	2.359	37.10	81.84
1.9	0.00	-8.0	-8.61	-3.63	22.34	65.04
2.	0.00	-7.8	-9.09	-6.57	9.231	47.68
2.1	0.00	-7.6	-9.30	-8.12	-0.03	31.88
2.2	0.00	-7.4	-9.34	-8.94	-4.77	17.16
2.3	0.00	-7.1	-9.27	-9.36	-7.23	4.395
2.4	0.00	-6.8	-9.13	-9.55	-8.54	-3.45
2.5	0.00	-6.5	-8.94	-9.59	-9.26	-7.39
2.6	0.00	-6.2	-8.71	-9.55	-9.65	-9.18
2.7	0.00	-5.9	-8.45	-9.44	-9.82	-9.93
2.8	0.00	-5.6	-8.18	-9.29	-9.84	-10.2
2.9	0.00	-5.3	-7.90	-9.10	-9.78	-10.2
3.	0.00	-5.0	-7.60	-8.88	-9.64	-10.1

Now consider the response of the nerve axon to a 29-mV pulse.

```
c = .001; a = .0238; ρ = 35.4;
eq3 = { ─────── D[u[x, t], x, x] ==
         2 * ρ
     c * D[u[x, t], t] + .036 * n[x, t]^4 * (u[x, t] + 12) +
         .120 * m[x, t]^3 * h[x, t] * (u[x, t] - 115) + .0003 * (u[x, t] - 10.6),
     u[x, 0] == 0, u[0, t] == (29 * Sin[2 π * t] * If[0 < t < 0.5, 1, 0]),
     (D[u[x, t], x] /. x → 3.2) == 0};
                          .0382 * (10 - u[x, t])
eq4 = {D[n[x, t], t] == ───────────────────────── * (1 - n[x, t]) -
                          Exp[(10 - u[x, t]) / 10] - 1
     .4775 * Exp[-u[x, t] / 80] * n[x, t], n[x, 0] == .3177};
                          .382 * (25 - u[x, t])
eq5 = {D[m[x, t], t] == ───────────────────────── * (1 - m[x, t]) -
                          Exp[(25 - u[x, t]) / 10] - 1
     15.28 * Exp[-u[x, t] / 18] * m[x, t], m[x, 0] == .0529};
eq6 = {D[h[x, t], t] == .2674 * Exp[-u[x, t] / 20] * (1 - h[x, t]) -
              3.82
     ───────────────────────── * h[x, t], h[x, 0] == .5961};
     Exp[(30 - u[x, t]) / 10] + 1
sol1 = NDSolve[{eq3, eq4, eq5, eq6}, {u[x, t], n[x, t], m[x, t], h[x, t]},
     {x, 0, 3.2}, {t, 0, 3.6}];
Plot3D[Evaluate[u[x, t] /. sol1[[1]], {x, 0, 1.5}, {t, 0, 2.5}],
     PlotRange → {0, 100}, PlotPoints → 30, AxesLabel → {"x", "t", " "}];
Clear@f; f[x_, t_] = u[x, t] /. sol1[[1]]
Pictures = Table[f[x, t], {x, 0, 2.5, .5}]
Plot[Evaluate[Pictures], {t, 0, 3.0}, Frame → True, PlotRange → All]
```

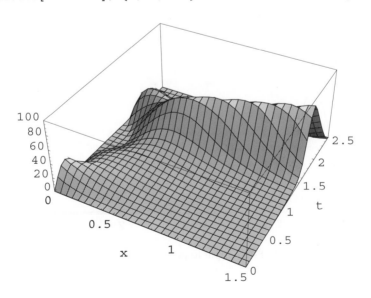

The action potential of the 29-mV input

If we run the above program with a 28-mV input, we get this response

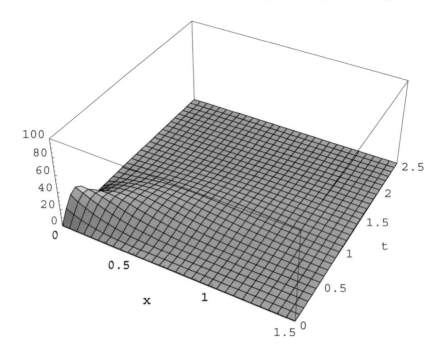

The 28-mV impulse falls away to nothing

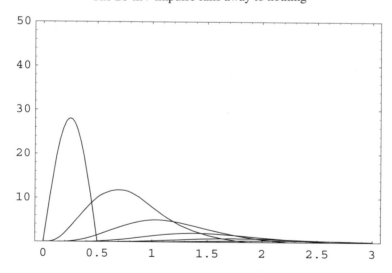

Failure of 28-mV input to produce an action potential

The "all-or-nothing" response of a nerve fiber is essential to keep noise off the line. If the threshold of minimum input is not met, then the nerve does not conduct.

Problem 5-5 Flight to the Stars

In this final problem, we will find the time required to get to the nearest star at 1 g acceleration.

Let us first put all the distances in lightyears, and measure time in years. Thus $c = 1$ ly/y, and $g = 9.81 \text{m/s}^2 = 1.032$ ly/y^2. We shall continue to measure mass in kilograms.

Now, we will carefully find the velocity v of the starship as a function of Earth-time t.

$$\frac{dv}{d\tau} = \frac{d}{dt}\{\gamma\, v\} = g$$

we will use *Mathematica* to find $\mathbf{v[t]}$

In[18]:= $\gamma[t\_] = (1 - v[t]^2 / c^2)^{-1/2}; c = 1;$
     $\texttt{DSolve[\{D[\gamma[t] * v[t], t] == g, v[0] == 0\}, v[t], t];}$
     $\texttt{Simplify[\%, \{t > 0, g > 0\}]}$

Out[20]= $\left\{\left\{v[t] \rightarrow -\dfrac{g\,t}{\sqrt{1 + g^2\,t^2}}\right\}, \left\{v[t] \rightarrow \dfrac{g\,t}{\sqrt{1 + g^2\,t^2}}\right\}\right\}$

Now that we have $\mathbf{v}$ as a function of acceleration, g, and Earth-time t, we may find the length of time necessary to travel half-way to Alpha-Centauri.

In[32]:= $\texttt{g = 1.032; c = 1.0; } v[t\_] = \dfrac{g * t}{\sqrt{1 + g^2\,t^2 / c^2}}\,;$

     (* Given v[t], we integrate to time t to find x[t]*)

     $x[t\_] = \displaystyle\int_0^t v[t]\, dt$

Out[34]= $\left\{\left\{x[t] \rightarrow -0.968992 + 0.968992\,\sqrt{1. + 1.06502\,t^2}\,\right\}\right\}$

Even more conveniently we may solve the following for the time t at which $x = 2.2$ lightyears

$$\texttt{Solve}\left[2.2 == \int_0^t v[t]\, dt,\ t\right]$$

$\{\{t \rightarrow -3.01721\}, \{t \rightarrow 3.01721\}\}$

Therefore, it takes 3 years (Earth-time) for the starship to reach midpoint.

At that time, the velocity of the starship is .9515 c.

In[69]:= v[3]

Out[69]= 0.951593

```
Plot[{v[t], .95}, {t, 0, 3}, PlotRange → {{-0.3, 3.06}, {0, .99}},
  AxesLabel → {"EarthTime(y)", "Velocity (v/c)"}]
```

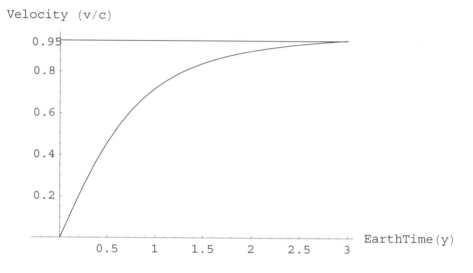

The approach to lightspeed

We have the velocity and distance traveled for the starship in Earthyears. Let us now find the distance traveled and the time for the space-travelers. Let z[t] be the distance the astronauts see as being traveled, and τ[t] be the time elapsed.

In[35]:= $z[t\_] = \int_0^t \frac{v[t]\, dt}{\gamma[t]}$ // FullSimplify; Simplify[%, t > 0]

Out[36]= $z[t] \rightarrow 0.030522 + 0.484496\, Log[0.938946 + 1.\, t^2]$

In[215]:= $\tau[t\_] = \int_0^t \frac{dt}{\gamma[t]}$ // FullSimplify

Out[216]= $\tau[t] \rightarrow 0.968992\, ArcSinh[1.032\, t]$

Note that z[3] = 1.14 lightyears and τ[3]= 1.79 years as seen on the ship.

Therefore the astronauts see the trip to Alpha-Centauri as taking 3.58 years, and a distance of only 2.28 lightyears.

We may **Table** our results and check on the astronauts every 3 months from our Earth-bound observatory.

```
Table[{t, τ[t], x[t], z[t], v[t], γ[t]}, {t, 0, 3, .25}] // TableForm
```

Earth-time t (yr)	Ship-time τ (yr)	Distance traveled (our frame) x (ly)	Distance traveled (ship-frame) z (ly)	v/c	γ
0	0	0.	0	0	1
0.25	0.247306	0.031735	0.031222	0.249819	1.0327
0.5	0.480113	0.121396	0.114372	0.458552	1.1252
0.75	0.690152	0.256343	0.227435	0.612077	1.2645
1.	0.875796	0.423468	0.351328	0.718153	1.4370
1.25	1.03909	0.612604	0.474741	0.790341	1.6322
1.5	1.18323	0.81677	0.592387	0.839978	1.8429
1.75	1.31139	1.03137	0.702351	0.874842	2.0643
2.	1.42626	1.25338	0.804336	0.899939	2.2934
2.25	1.53005	1.48079	0.898739	0.918448	2.5281
2.5	1.62455	1.71223	0.986212	0.932411	2.7670
2.75	1.71118	1.94673	1.06746	0.943162	3.0090
3.	1.79106	2.18362	1.14315	0.951593	3.2534

We now have sufficient information to answer the questions posed in part (A).
The trip as seen by the astronauts will take $4\tau + 1$ years = 8.2 years
The trip as seen from Earth will take $4t + 1$ years = 13 years.

At this time it is worthwhile checking to be sure we have done everything right by computing the acceleration of the spacecraft, say from t=1.000 y to 1.001 yr and seeing if this matches the correct relativistic formula for measurements made in the *accelerated* frame: $a' = a_x \gamma^3$ (see John David Jackson *Classical Electrodynamics*, or Paul Lorrain and Dale Corson *Electromagnetic Fields and Waves*, for derivation of relativistic formulas).

$$In[39]:= \frac{v[1.001] - v[1.000]}{1.001 - 1.000} * \gamma[1.0005]^3$$

$$Out[39]= 1.032$$

This is g in ly/y^2 so the astronauts see the engines providing the necessary acceleration whereas we on Earth would see the spacecraft as accelerating only at 1/3 g.

At $t = 3$ y Earth-time, or $\tau = 1.8$ y Ship-time the spacecraft is up-to-maximum speed. It is now time to turn the craft through 180° and begin the slow-down to Alpha-Centauri. However, how much matter and antimatter have we burned? This may be computed with the thrust equation, noting that time is in τ

$$-\frac{dm}{d\tau} c = mg \qquad \text{where the starting mass of the ship is M}$$

In[43]:= **Clear[c, g, v];**
DSolve[{m'[τ] * c == - m[τ] * g, m[0] == M}, m[τ], τ]

Out[44]= $\left\{\left\{m[\tau] \rightarrow e^{-\frac{g\,\tau}{c}}\,M\right\}\right\}$

Thus at the end of the first-stage burn, the mass of the starship is reduced by a factor of **Exp[g*τ/c]**. For g = 1.032 ly/y^2, τ = 1.8 y, c = 1 ly/y,

In[42]:= **Exp[1.032 * 1.8]**

Out[42]= 6.40834

For each stage (and there are four of them) we reduce the mass of the previous stage by a factor of 6.4. Thus if the *returning* spacecraft has a mass of 5000 kg, then the original mass of the starship is

$$M = 6.4^4 * 5000 \text{ kg} = 8.4 \text{ million kg} = 8400 \text{ metric tons}$$

This actually compares favorably with the at-launch mass of the 3000 metric ton Saturn-V rocket, until one realizes that the α-Centauri rocket requires

4200 metric tons of matter <u>and</u>
4200 metric tons of antimatter !

Notes on the Text

1.1 Motion of the Planets about the Sun

A question that naturally arises when solving Newton's equations with numerical methods is "How difficult is it to determine the track of an asteroid?" Clearly one can find the orbital path of an asteroid if you are given a portion of its orbit that is well-defined. Then the *initial conditions* are known. You may then develop a 3-body problem, using the coordinates of the asteroid, the sun, and the moving Earth. This will also be a 3-dimensional problem, because the Earth and the asteroid, in general, will not be in the same plane. But it will be solvable, by computer, <u>if</u> the starting distances and original velocity vector of the asteroid are well-determined.

Problem 1-3 Solar Sailing

A new material, Kapton, has been developed which is essentially a thin film of micron thickness which can be aluminized with a coating approximately 0.1 microns in thickness. With a weight density of 1 gram per square meter for the sail, solar sailing could become a reality, if a reliable means for packaging and unfurling the sail can be found.

For more on solar sailing, see *Space Sailing* by Jerome Wright (1992), *Solar Sailing* by Colin McInnes (1999), *or* the article by T C Tsu (1959) *"Interplanetary Travel by Solar Sail."* American Rocket Society 422-427.

2.1 Damped and Driven Oscillations

In vibration analysis, engineers utilize the *damping ratio* $z = \dfrac{c}{2\sqrt{mk}}$

For $z > 1$, the motion is overdamped,

for $z = 1$, the motion is critically damped,

and for $z < 1$, the motion is underdamped.

Using the *damping ratio,* the frequency of underdamped motion is

$$\omega_d = \omega_o\sqrt{1 - z^2}$$

where $\omega_o = \sqrt{\dfrac{k}{m}}$ and the logarithmic decrement is $\delta = \dfrac{c\pi}{m\,\omega_d} = \dfrac{2\pi z}{\sqrt{1-z^2}}$

2.2 Shock Absorbers

$e^{-\delta}$ is the amount of damping in one cycle.

For new shock absorbers $\delta \simeq 4$, for an older automobile $\delta \simeq 2$.

2.5 Variable Stars

The mathematics of the (Eddington) variable star equations are very interesting. For $x^3 \ddot{x} = -Ax + B$ this is an oscillating system as long as $B > Ax_o$. Then at $t = 0$, the stellar envelope will be accelerated towards $+x$. (Pressure exceeds gravity).

When x gets large enough, then $Ax > B$ and $\ddot{x}$ is negative. (Gravity exceeds pressure). Notice also that when $Ax = B$, the system is passing through its equilibrium point. At that time, $x = B/A$ and putting in the numbers from Section 2.5, $\bar{x} = 1.50 \times 10^7$ km.

Problem 2-5 Impulse Response

Theoretically, we could model a piano sonata as the response of the piano strings to the impulse functions of the hammers striking the strings.

3.3 The Binary Pulsar

Astronomers find the *eccentricity* of a binary system by measuring the Doppler shift due to the velocity of the stars. The maximum and minimum velocities of the stars are in the ratio $\quad \frac{v_P}{v_A} = \sqrt{\frac{1-\epsilon}{1+\epsilon}}$

3.5 Heat Exchanger

In the engineering literature, the usual method for finding the effectiveness of a heat-exchanger is to evaluate

parallel-flow

$$\epsilon = \frac{1 - \text{Exp}\left[-NTU*(1+C^*)\right]}{(1+C^*)} \quad \text{where } NTU = \frac{UA}{(\dot{m}c)_{min}} \quad \text{and} \quad c^* = \frac{(\dot{m}c)_{min}}{(\dot{m}c)_{max}}$$

counter-flow

$$\epsilon = \frac{1 - \text{Exp}\left[-NTU*(1-C^*)\right]}{1 - C^* \text{Exp}\left[-NTU*(1-C^*)\right]} \quad \text{where } NTU = \frac{UA}{(\dot{m}c)_{min}} \quad \text{and} \quad c^* = \frac{(\dot{m}c)_{min}}{(\dot{m}c)_{max}}$$

Problem 3-5 Heat Exchangers

It is amazing, but true, that if we reverse the direction of flow of oil in the parallel-flow heat exchanger, that we will now have a counterflow heat exchanger that will extract 11% more heat energy. This is because of the more uniform thermal gradient in the counterflow system.

4.1 Partial Differential Equations

The PDE for waves on a string is

$$\frac{\partial^2 u}{\partial t^2} = c^2 \frac{\partial^2 u}{\partial x^2}$$

Boundary condition $u(x, t) = 0$ at $x = 0$ and $x = L$

Initial condition $u(x, 0) = f[x] = 0$ for $0 < x < L$

and $D[u(x, 0), t] = g[x]$

As an example, take a piano string under tension T and with mass per unit length of ρ. Then with the hammer striking the string at $t = 0$, with an impulse function of fairly small width, $g(x)$ is the initial velocity of the string. See the *Mathematica* file **PDE-Wave** on the CD for the necessary code to solve the Wave Equation.

4.3 PDE – Heat Conduction in the Earth

In their book on Heat Transfer, Incropera and DeWitt treat the case of 20 °C soil subjected to a – 15 °C surface freeze for 60 days. For $\alpha = 0.138 \times 10^{-6}$ m^2/s, *Mathematica* shows that the ground will be frozen (T = 0 °C) to a depth of 0.68 meters. (See **PDE-Incropera** on the CD)

5.2 The B–Z Chemical Reaction

According to John Tyson (1976, see References), the B-Z reaction proceeds along two pathways:

One set of reactions, Process 1, consumes Br^-, and when $[Br^-]$ falls below a critical level, then

Process 2 takes over. But Process 2, during which $Ce^{+3} \to Ce^{+4}$, produces more Br^-. Then the Ce^{+4} oxidizes malonic acid to produce the reactants of Process 1. Thus the reactions oscillate back and forth (with color changes) until all the malonic acid is consumed.

5.4 Nerve Conduction

A 0.1 mA current acting for 0.1 ms delivers a charge of 10^{-8} Coul. If this charge is delivered to a capacitor $C = 10^{-6}$ Farads, then the capacitor has a voltage of $V = Q/C = 10$ mV. If we measure current in mA, time in ms, and capacitance in mF, then

$$V = \frac{(0.1\,\text{mA})\ (0.1\,\text{ms})}{.001\,\text{mF}} = 10\,\text{mV}.$$

Richard Feynman in his <u>Lectures on Physics</u> Volume 1, Chapter 3 compares nerve conduction to the knocking-over of a line of dominos. An initial tap converts the stored energy into a travelling wave.

5.6 Space Shuttle Launch

The Thrust T of the Shuttle engines is actually throttled back toward the end of the first-and-second stage burns, to keep the acceleration within strict (structural) limits. This can be modeled in the Diff Eqs by making T(t) a user-defined function of time.

Problem 5-5 Flight to the Stars

The origins of this problem, accelerated relativistic flight, are somewhat uncertain. Sebastian von Hoerner in his 1962 paper in <u>Science</u> **137** , 18-23 "The General Limits of Space Travel" derives the relevant equations in analytic form. However there is a very interesting book <u>The Realities of SpaceTravel</u> (selected papers of the British Interplanetary Society, McGraw-Hill, 1957) that would indicate that J. Ackeret and L. R. Shepherd had worked out the solutions to this problem as early as 1952.

A Final Note

This is the end of the text, however I hope you will find, as I have, that the computer with modern mathematical software is a wonderful tool for investigating Physics, and physics-related problems. If you have found new applications for mathematics applied to the physical sciences, please send them along.

Ideas for new problems and their solutions are always welcome.

stevevw@u.washington.edu

svanwyk@oc.ctc.edu

References

Chapter 1

1-1 David and Judith Goodstein, Ed. (1996) Feynman's Lost Lecture [Norton]

1-2 Jerry Marion (1970) Classical Dynamics Chapter 8 [Academic Press]

1-3 James van Allen (2003) *"Gravitational Assist"* Amer. J. Phys. $\underline{71}$, 448

1-4 Robert Adair (2002) The Physics of Baseball [Harper-Collins, 3ed]

1-5 PSSC Physics (any edition) the Physical Science Study Committee

1-6 Dan Poynter, Mike Turoff (2003) Parachuting, the Skydivers Handbook
     [Para Publishing, 9ed]

Chapter 2

2-1 Richard Feynman (1964) The Feynman Lectures on Physics Volume 1,
     Chapter 23   [Addison–Wesley]

2-2 Robert Vierck (1979) Vibration Analysis [International Textbooks, 2ed]

2-3 Ray Wylie, Louis Barrett (1995) Advanced Engineering Mathematics
     Chapter 8 [McGraw–Hill, 6ed]

2-4 Brice Carnahan, Herbert Luther, James Wilkes (1969) Applied Numerical
     Methods [Wiley]

2-5 Sir Arthur Eddington (1926) The Internal Constitution of the Stars
     [Cambridge University Press]

2-5 Robert Christy (1964) *"Calculation of Stellar Pulsation"* Reviews of
     Modern Physics $\underline{36}$, 555

2-5 Bradley Carroll, Dale Ostlie (1996) Introduction to Modern Astrophysics
     Chapter 14 [Addison–Wesley]

Chapter 3

3-2 JackWilmore, David Costill (2004) Physiology of Sport and Exercise
     [Human Kinetics Publications]

3-2 Henry Bent (1978) *"Energy and Exercise"* J. Chemical Education $\underline{55}$, 797

3-2 William Pritchard (1993) *"Mathematical Models of Running"* SIAM
     Review $\underline{35}$, 359

3-3 http://www.JohnstonsArchive.net/astro/index.html
     -- For data and references on the relativistic binary pulsar PSR 1913+16

3-4 Romas Mitalas(1980) *"Supernova in Binary Systems"* Am. J Phys $\underline{48}$, 226

3-5 Jack Holman (1997) Heat Transfer Chapter 10 [McGraw–Hill, 8ed]

Chapter 4

4-1 Horatio Carslaw, John Jaeger(1959) <u>Heat Conduction in Solids</u> [Oxford]

4-3 Frank Incropera, David DeWitt (2002) <u>Introduction to Heat Transfer</u> [Wiley]

4-4 David Derbes (1996) "*Feynman's Derivation of Schrodinger's Equation*" American Journal of Physics <u>64</u> , 881

4-4 Stephen Gasiorowicz (1996) <u>Quantum Physics</u> Chapter 5 [Wiley, 2ed]

4-5 Leonard Schiff (1968) <u>Quantum Mechanics</u> Chapter 13 [McGraw-Hill]

4-5 Kenneth Krane (1988) <u>Introductory Nuclear Physics</u> [Wiley]

4-5 Richard Robinett (1997) <u>Quantum Mechanics with Visualized Examples</u> [Oxford]

Chapter 5

5-1 Peter Milonni, Joseph Eberly (1988) <u>Lasers</u> [Wiley]

5-1 WilliamWagner,Bela Lengyel (1963)"*Evolution of Giant Pulse in a Laser*" Journal of Applied Physics <u>34</u>, 2040

5-2 Bassam Shakhashiri (1985) <u>Chemical Demonstrations</u> Volume 2, Chapter 7 [U. Wisconsin Press]

5-2 John Tyson (1976) "*The Belousov-Zhabotinski Reaction*" Chapter 2 in <u>Lecture Notes on Biomathematics</u> Volume 10 [Springer]

5-3 George Gause (1934) <u>The Struggle for Existence</u> [Dover reprint 2003]

5-3 Narendra Goel, Samaresh Maitra, Elliot Montroll (1971) "*On the Volterra Model of Interacting Populations*" Reviews of Modern Physics <u>43</u> , 231

5-4 Alan Hodgkin, Andrew Huxley (1952) "*A Quantitative Description of Membrane Current and Excitation in Nerve*" Journal of Physiology <u>117</u> , 500

5-4 Russell Hobbie (1997) <u>Intermediate Physics for Medicine and Biology</u> [AIP Press]

5-5 J Rand McNally (1982) "*The Physics of Fusion Fuel Cycles*" Nuclear Technology/Fusion <u>2</u>, 1

5-6 Kerry Joels, David Larkin, Greg Kennedy (1988) <u>Space Shuttle Operator's Manual</u> [Ballantine]

Constants

M_{sun} = 1.989×10^{30} kg R_{sun} = 7×10^5 km

M_{Earth} = 5.98×10^{24} kg R_{Earth} = 6380 km

M_{Moon} = 7.35×10^{22} kg R_{Moon} = 1740 km

M_{Jupiter} = 1.90×10^{27} kg R_{Jupiter} = 71000 km

G = Gravitational constant = $6.673 \times 10^{-11} \dfrac{\text{N} \cdot \text{m}^2}{\text{kg}^2}$ = $6.673 \times 10^{-11} \dfrac{\text{m}^3}{\text{kg} \cdot \text{s}^2}$

with distance in km, G = $6.673 \times 10^{-20} \dfrac{\text{km}^3}{\text{kg} \cdot \text{s}^2}$

h = 6.626075×10^{-34} J-s

$\hbar$ = 1.054572×10^{-34} J-s 1 eV = 1.602177×10^{-19} J

m_p = 1.672×10^{-27} kg

m_n = 1.675×10^{-27} kg q_e = 1.602177×10^{-19} C

m_e = 9.109389×10^{-31} kg

c = 3.00×10^8 m/s σ = 5.67×10^{-8} W/m^2-K^4

k = 1.3807×10^{-23} J/ K

1 AU = 149.6×10^6 km = 149.6×10^9 m

1 yr = 3.1557×10^7 s ϵ_o = 8.854188×10^{-12} C^2/N·m^2

1 g = 9.81 m/s^2 = 32.16 ft/s^2

Index

Numbers after each entry correspond to location in the text by Section, Problem number, Appendix, or CD.

Marketing Essentials in Hospitality and Tourism

Foundations and Practices

Marketing Essentials in Hospitality and Tourism

Foundations and Practices

Stowe Shoemaker

Donald Hubbs Distinguished Professor
Associate Dean of Research
Conrad N. Hilton College of Hotel and Restaurant Management
University of Houston

Margaret Shaw

Professor of Marketing
School of Hospitality and Tourism Management
University of Guelph
Ontario, Canada

PEARSON
Prentice
Hall

Upper Saddle River, New Jersey
Columbus, Ohio

Library of Congress Cataloging-in-Publication Data

Shoemaker, Stowe.

Marketing essentials in hospitality and tourism : foundations and practices / Stowe Shoemaker, Margaret Shaw.

  p. cm.

ISBN 0-13-170827-9

1. Hospitality industry—Marketing. 2. Tourism—Marketing. I. Shaw, Margaret II. Title.

TX911.3.M3S553 2008

647.94068—dc22   2007015278

Editor-in-Chief: Vernon R. Anthony
Senior Editor: William Lawrensen
Editorial Assistant: Lara Dimmick
Production Editors: Jane Bonnell and Alexandrina Benedicto Wolf
Production Coordination: Linda Zuk, WordCraft, LLC
Senior Design Coordinator: Miguel Ortiz
Interior Design: Janice Bielawa
Cover Designer: Anthony Gemmellaro
Cover Image Credits:
  Businesswoman in Lobby: Getty/Iconica/ Photographer: Caroline von Tuempling
  Mandalay Bay Hotel (Las Vegas): Getty/Robert Harding World Imagery; Photographer: Neil Emmerson
  Corporate Jet: Getty/National Geographic; Photographer: Michael Melford
  World Map (Bkgd.): Getty/Photodisc Green; Photographer: Siede Preis
  Cruise Ship at Twilight: VEER/Royalty-Free (Bon Voyage CD)
  Great Wall of China: VEER/Royalty-Free (World Landmarks and Travel CD)
  Couple Drinking/Nighttime Skyline: Food Pix; Photographer: Ray Kachatorian
Production Manager: Deidra Schwartz
Director of Marketing: David Gesell
Marketing Manager: Leigh Ann Sims
Marketing Coordinator: Alicia Dysert

This book was set in Minion by Carlisle Publishing Services. It was printed and bound by Quebecor. The cover was printed by Phoenix Color Corp.

Image credits appear on page 611.

Pearson Education Ltd.   Pearson Education Australia Pty. Limited
Pearson Education Singapore, Pte. Ltd.   Pearson Education North Asia Ltd.
Pearson Education Canada, Ltd.   Pearson Educación de Mexico, S.A. de C.V.
Pearson Education—Japan   Pearson Education Malaysia, Pte. Ltd.

10 9 8 7 6 5 4 3 2 1
ISBN-13: 978-0-13-170827-3
ISBN-10:   0-13-170827-9

This book is dedicated to
Antonia Levak (age 7) and Elisha Levak (age 5).
You were a valuable source of inspiration
in the development and completion of this text.
Thank you.

—*Margaret Shaw*

This book is dedicated to four important people in my life:
my twin sister, Jane Martin,
my two mentors, Robert C. Lewis and Mark Renaghan,
and my wife, Martha McArdell Shoemaker, the love of my life.
Without their help, support, and deep friendship,
I would never have been able to write this book.
My deepest thanks, respect, and admiration go to each of them.

—*Stowe Shoemaker*

Brief Contents

Contents

Preface

Marketing begins and ends with the customer. Marketing is not just a series of tactical actions; rather, it is a way of thinking about how to incorporate the customers' views into all organizational decisions. This book illustrates this role of marketing by:

- Providing a clear definition of marketing
- Demonstrating through examples, interviews with industry professionals, and web exercises how hospitality marketing has moved beyond both the 4 Ps of product marketing and the 7 Ps of services marketing to the 13 Cs of loyalty marketing
- Introducing the importance of strategy as part of the marketing function
- Showing how to understand and evaluate the competition
- Providing a clear understanding of the different types of hospitality customers (individuals, groups, tourism markets)
- Detailing the various tactics that hospitality businesses can use to build a competitive advantage, for example: positioning, segmentation, sales, e-distribution
- Defining how to develop a marketing plan

We do all of the above and more by not only examining the strategies and tactics of large international firms, but also examining the independent hotels, restaurants, and tourism entities that make up much of the hospitality industry.

MARKETING ESSENTIALS

This book is geared toward a first course in hospitality marketing as a basic alternative to the more detailed fourth edition of *Marketing Leadership in Hospitality and Tourism*. The goal of this text is to provide the first-year student with an introduction to marketing and many of its components. It is written in a way that is easy to understand and filled with activities to engage and excite the reader about careers in marketing. Like the more detailed fourth edition of *Marketing Leadership in Hospitality and Tourism*, this book is filled with international examples.

Executive interviews include conversations with not only senior executives but also with those who have recently started their careers. All of the interviews are related to material discussed in the chapter at hand and all end with advice to the student on paths to take for a career in hospitality marketing.

Key words are highlighted in the text and defined in the margins to reinforce learning.

WHY A SEPARATE BOOK ON HOSPITALITY AND TOURISM MARKETING?

With all of the generic marketing texts available, the reader may wonder, why a separate book on hospitality and tourism marketing? Aren't the principles of marketing the same regardless of the industry? While we certainly believe that there are similarities across industries, we believe that the differences need to be highlighted and explored. One of the major differences pertains to the nature of services: their intangibility, their heterogeneity, and the simultaneous production and consumption of them. These characteristics present unique challenges for our industry. The subject of *Marketing Essentials in Hospitality and Tourism* is how to meet these challenges. Typically, generic marketing books do not cover such material.

Another major difference is the role that the guest experience plays in the hospitality purchase. Customers are buying not only rooms or meals, but memories. The role of marketing is to help define and create these memories. The ways in which marketing executives accomplish this goal are covered in this book. Again, generic marketing books do not cover such material.

We believe that textbooks should not merely educate students about the strategies and tactics used in marketing; they should also educate students about the industry they plan to enter. *Marketing Essentials in Hospitality and Tourism* is filled with examples that both educate students and excite them about the opportunities the hospitality industry presents. The book also is filled with numerous examples that illustrate how tourism organizations use marketing to gain competitive advantage. These *Tourism Marketing Applications* are highlighted throughout the chapters.

OUR APPROACH TO MARKETING

This first edition of *Marketing Essentials in Hospitality and Tourism* brings together what we, the industry, and academics have learned in the past few years—and a lot of what we are still learning. The foundation of marketing starts at the highest level by deeds and actions, not by words alone. These deeds and actions must permeate down to the lowest level of the organization. At the highest level, marketing shapes the corporate effort; at the lowest level, successful marketing means the porter doesn't mop where the customer is walking.

Taking a long-range perspective rather than an operational how-to approach—because marketing is long-range for any organization that seeks survival and growth—we explore the latest trends in marketing, distribution, and communication, including web blogs and websites.

THE LATEST TRENDS IN CUSTOMER LOYALTY

This first edition examines the latest trends in marketing theory and practice, especially the emphasis on *CRM—customer relationship management.* Its focus on the 13 Cs (introduced in Chapter 3 of the text) of customer relationship marketing helps to reinforce how various marketing actions can be used to *customize* the service or product for the guest, and also demonstrate how firms can create value for and from customers by using *customer insight* and *channel management*.

Our thesis is that the sole purpose of marketing is to create value for and from the customer. There is logic to marketing and this logic can be learned. There are ways to understand customer behavior. There are ways to understand marketing problems and opportunities. There are underlying principles that appear time and time again.

THE LATEST IN PRICING AND REVENUE MANAGEMENT

Pricing and revenue management are two areas critical to the hospitality industry. Chapter 15, the comprehensive pricing chapter, introduces the reader to the various types of costs that are calculated and explains why knowledge of these costs is important to those in marketing. The chapter also talks about the components of value and shows how value pricing, compared to cost-based pricing, is the future. Finally, the chapter explains why revenue management works and how to think strategically about pricing.

REACHING AN INTERNATIONAL MARKET

Throughout the text we include perspectives and examples from companies headquartered around the globe. Included in this text are:

- An interview with the Director of Worldwide Communications for InterContinental Hotels Group PLC
- An interview with the director of e-commerce and distribution for Rocco Forte Hotels
- An interview with the revenue manager of the Westin Guangzhou, China
- Tourism marketing applications highlighting KwaZulu-Natal in South Africa, Footprint Vietnam Travel, and other international locations
- E-commerce challenges in China

DISTRIBUTION CHANNELS

There are two areas of concern for hospitality marketing students learning about distribution channels. One is how to get the hospitality product to where the customer is; and the second is how to get the customer to the hospitality product. We provide students with an understanding of how to reach these two goals. In terms of *getting the product to where the customer is*, we discuss:

- branded hospitality companies
- franchising
- management contracts

In terms of *getting the customers to where the product is,* we discuss:

- reservation services
- representation firms
- consortia
- incentive travel organizations
- corporate travel departments
- travel management companies
- global distribution systems (GDS)
- traditional offline travel agents
- central reservation systems
- Internet channels

We have worked hard to make this first edition of *Marketing Essentials in Hospitality and Tourism* user-friendly and engaging. It is our goal to create excitement and enthusiasm about marketing. We hope we have succeeded.

FEATURES OF *MARKETING ESSENTIALS IN HOSPITALITY AND TOURISM*

Tourism Marketing Applications

These tourism sidebars provide numerous examples of how marketing is used to develop tourism. The tourism industry and the hospitality industry are interdependent. Both offer experiences to customers, both can be considered intangible, and simultaneous production and consumption occur in both fields. Our approach is to present the concepts of hospitality marketing and demonstrate the application of these concepts using hotels, restaurants, and tourism destinations.

Tourism Marketing Application

The province of KwaZulu-Natal is located in South Africa. The goal of the Tourism Commission for this region is for the region to be recognized as Africa's premier tourism destination. To reach this goal, the commission created a website, www.zululand.kzn.org.za (accessed March 25, 2006) that provides information about the various destinations in the region. Each destination is positioned on different features, such as cultural attractions or natural wildlife reserves.

Interviews with Executives in the Industry

Each chapter begins with an interview with either a senior executive in the industry or a recent college graduate. These interviews provide insight into how the theory presented in each chapter is translated into real life. They also provide insight to various career paths. The executives work in restaurants, hotels, casinos, and tourism destinations.

Marketing in Action

Carrie Ballew
Corporate Sales Manager, Four Seasons Hotel Silicon Valley

Advertisements and Illustrations

The examples that appear in this book come from many sources. Concepts presented throughout the text are based on accepted principles and solid research. The examples and the advertisements used to illustrate the marketing concepts covered are largely those of well-known international companies. Our worldwide readers will be familiar with their names and will be able to identify with these companies.

Exhibit 13-4 The Waterlot Inn also tangibilizes its service by evoking tradition.

Source: Fairmont Hotels Bermuda. Used by permission.

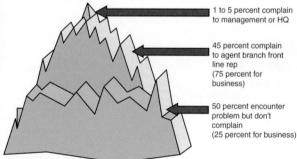

Exhibit 4-11 The tip of the iceberg phenomenon

Source: Goodman, J. (1999). *Basic facts on customer complaint behavior and the impact of service on the bottom line (p. 5)*. White paper published by TARP. Retrieved from www.e-satisfy.com/research2.asp. Used by permission

Case Studies

Case studies, one at the end of each chapter, are illustrative of the chapter's context and give life to it in real-world situations. All the cases are based on actual events, although in some instances names, places, and figures may be disguised. Each case is preceded by case study discussion questions and the Instructor's Manual provides discussion of the cases for optimal learning.

.... case study

Castle Spa

Discussion Questions

1. In your own words describe the vision and mission of Castle Spa.
2. What is the definition of a spa?
3. Who are the primary, secondary, and tertiary markets for Castle Spa? Do you feel these markets as described in the case are well defined?
4. Briefly describe the Castle Spa pricing strategy? In your opinion is this a good pricing strategy? Is there any for improvement?
5. How would you assess the competitive environment in which Castle Spa operates?
6. Should Sonja consider personal selling (which is somewhat expensive) as part of her communications mix strategy to reach out to the incoming residents of the new multimillion-dollar luxury condominium? What do you think of her idea to focus on sales promotions to boost nonpeak period sales in this new potential market?
7. How can Sonja go about building her loyal customer base? Is there evidence in the case that Sonja truly understands who her customer base really is?

Sonja Drury, CEO and owner of Castle Spa Inc., sat in Goody's restaurant in late September sipping a ginger tea and staring out the expansive floor-to-ceiling windows onto the historic Downwind market of downtown Vancouver. Directly across the street, the framework of a new multimillion-dollar luxury condominium complex was beginning to take shape.

In six months' time, the condominium complex was due to be completed, and Sonja wanted to make sure that Castle Spa captured this market before its competition, the Devon Spa, which was located a mere three blocks away.

In anticipation of an influx of new clients that match Castle Spa's target market, the spa management team was busily brainstorming ways to make sure the experience that these potential new clients had on their first visit would ensure that they would become loyal clients in the future. But how to ensure that their needs were best met? How might their booking demand differ from Castle's current client base, given their proximity to the Devon Spa and unusually high income bracket?

Sonja was concerned the condominium clients were likely to increase sales in periods that were already booked, such as high-demand Saturdays and evenings in the latter part of the week. If this were the case, all spa clients were likely to become frustrated with overcrowded lounges, a decrease in service quality due to service providers having to rush through treatments to accommodate back-to-back bookings, and an overall decline in the ambiance of relaxation and personalized service that Castle strived to uphold. How could Castle Spa develop a marketing strategy to maneuver the condo business into nonpeak periods, to ensure

This case was contributed by/Ana Yuristy and Cheryl Pylypiuk, and developed under the supervision of Margaret Shaw, PhD, School of Hospitality and Tourism Management, University of Guelph, Ontario, Canada. Names and places have been disguised. All monetary figures are in Canadian dollars. For conversion, use 1$C/0.85$U.S. All rights reserved. Used by permission.

Web Browsing Exercise

Go to www.learningandteaching.info/learning/dissonance.htm to learn more about cognitive dissonance. Who first investigated this theory, and under what applications? How can cognitive dissonance apply to learning in school? How can we overcome cognitive dissonance?

Web Browsing Exercises

These web exercises throughout the chapters encourage readers to seek current information on the topic under discussion. Information in hospitality marketing is constantly changing, and what is critical now may not be critical tomorrow. The web exercises also keep the reader abreast of the latest information while providing web addresses that the reader can refer to in the future. We have been very careful to use websites for firms that are long-term players in the industry.

Discussion Questions

Each chapter ends with a series of discussion questions that help students synthesize the material covered in the chapter.

Discussion Questions

1. Taking time to review the marketing communications mix, briefly describe different ways to reach out to current and potential new customers for hospitality and tourism product offerings.

2. Why is personal selling a good communications tool for the more complex purchases such as meetings, conventions, banquets, and group tour sales?

3. What is the personal selling sales process all about? What are the steps involved in this process?

4. Describe in your own words what is meant by sales management. What is being "managed"?

5. Why are sales promotions an important part of the marketing communications mix?

6. What makes a good sales promotion?

SUPPLEMENTS FOR THE INSTRUCTOR

This textbook offers the instructor a multitude of options for covering the various dimensions of marketing. Online Instructor Materials are available to qualified instructors for downloading. To access supplementary materials online, instructors need to request an instructor access code. Go to **www. prenhall. com**, click the **Instructor Resource Center** link, and then click **Register Today** for an instructor access code. Within 48 hours after registering, you will receive a confirming e-mail including an instructor access code. Once you have received your code, go to the site and log on for full instructions on downloading the materials you wish to use.

Instructor's Manual with Test Item File

The instructor's manual, available in print and downloadable formats, includes examples of syllabi that can be used for teaching various aspects of marketing. For example, we provide a plan for those wishing to concentrate on marketing strategy—for both the quarter and semester system. We also include discussion outlines for each class period, learning objectives, teaching tips, answers to-end-of-chapter questions, and discussion of the end-of-chapter case studies. The Test Item File includes multiple-choice, true/false, and short-answer questions.

TestGen Test Bank

The full Test Item File is available in this test generation software program.

PowerPoint Slides

A complete set of PowerPoint slides is provided for each chapter. The slides are comprehensive and include copies of some exhibits. They have been designed so instructors can either use the slides as is or can incorporate their own material into the slides provided.

FOR THE STUDENT

Companion Website

Discussions with our students led us to develop a study guide that students can access via a dedicated website (**www.prenhall.com/shoemaker**). This site contains review questions (multiple choice and true/false) with immediate feedback to test the students' understanding of the concepts in the text. Essay questions provide the opportunity to apply knowledge. PowerPoint slides, key-term searches, and chapter objectives provide the main points in each chapter.

Acknowledgments

Many people—friends, former students, colleagues, industry people, and customers—both advertently and inadvertently, have contributed to this book. Many will never realize how helpful they have been. We can mention only a few. We are grateful to these individuals, and to many others unmentioned, especially some great graduate and undergraduate students from around the world and executives with whom we have been privileged to work.

Those who have contributed directly to a case are noted on the first page of that case. Many of these contributors are graduate and undergraduate students at the University of Massachusetts/Amherst and the University of Guelph, Ontario, Canada. Still others are faculty and/or graduate students at other universities, as noted on the case.

We would especially like to thank Ashley Trevitz, a student at the University of Houston Honors College. Ashley served as the overall project director and provided many ideas for the book. She also obtained all of the permissions, a notable effort in itself. Jennifer Aiyer deserves our appreciation for her work on reviewing the manuscript, writing the instructor's manual and all the supplementary material, and being the teaching assistant for Stowe Shoemaker during the writing of this book. Mary Michele White, an honors student at the University of Houston who is now with Expedia.com, deserves a huge thank-you for her help and good cheer. This book could not have been finished without the help of these three remarkable women. They made coming to work each day a pleasure.

Still others contributed to this book in their own way through their support, experience, and knowledge in various discussions about the industry and its customers. These helpful individuals include Frank and Jane Emanuel, former owners of the Middlebury Inn; Jill Reith and Jeff Jordan, both of International Game Technology;

Ran Kivetz of Columbia University; John Shields of Hyatt Hotels; Nan Moss, formerly of Hyatt Hotels and Resorts; Amy Weyman of Hyatt Hotels and Resorts; Judd Goldfeder of the Customer Connection; Jennifer Ploszaj of InterContinental Hotels and Resorts; Valerie Cotter of University of Pennsylvania; Adam Burke of Hilton Hotels; Rick Mansur and Doug Leiber of JC Resorts; Dave Hanlon of Empire Resorts; Jim Eyster of Cornell University; Emanuel Berger of the Victoria-Jungfrau; Bill Carroll of Cornell University; John Deighton of Harvard Business School; Tom Kline and his team in the Office of Executive Education in the School of Hotel Administration at Cornell University; Pennie Beach, owner of Basin Harbor Club; the students at Ecole Hoteliere de Lausanne, whose questions guided the structure of this book; Neil Young of MEI, and Meg Galliano of Las Vegas, Nevada.

Each of the industry executives profiled in the book deserves special thanks for taking the time to contribute their wisdom. Thanks also to the firms that graciously supplied advertisements and other information for the book.

Linda Zuk, of WordCraft, was a wonderful editor and a pleasure to work with as we put together the final version. Every author should be as fortunate as we were to work with someone of Linda's qualifications. Her constant "good cheer" made last-minute changes less painful for us. More importantly, her professionalism, knowledge, and insightful comments improved the quality of this book. We are grateful to Judy Casillo, Vern Anthony, and Bill Lawrensen for their help and championing of this book at Prentice Hall.

And, of course, we thank our reviewers, whose helpful ideas and suggestions were well considered and often utilized: Haze Dennis, West Valley-Mission Community College District; Jamal Feerasta, University of Akron; Stephen Fries, Gateway Community College; Elinor Garely, Borough of Manhattan Community College/City University of New York; David Grulich; Brevard Community College; Terry Jones, Community College of Southern Nevada; Ian W. McVitty, Algonquin College; and Ezat Moradi, Houston Community College.

Finally, to all of our readers:

> A truly good book teaches me better than to read it. I must soon lay it down, and commence living on its hint. What I began by reading, I must finish by acting.

> —Henry David Thoreau

About the Authors

Stowe Shoemaker is the Donald Hubbs Distinguished Professor and the Associate Dean of Research at the University of Houston's Conrad N. Hilton College of Hotel and Restaurant Management. He holds a PhD from Cornell University School of Hotel Administration, an MS from the University of Massachusetts, and a BS from the University of Vermont. In addition to his role at the University of Houston, Dr. Shoemaker is on the executive education faculty at Cornell University in the School of Hotel Administration.

Dr. Shoemaker's research has appeared in numerous hospitality journals, and his research has been honored as best yearly piece of research four times. Dr. Shoemaker is co-author of a Harvard Business School Case Study on Hilton HHonors. Prior to moving to the University of Houston, Dr. Shoemaker taught at the University of Nevada, Las Vegas. Before earning his PhD, Dr. Shoemaker spent 15 years working in the hotel industry and in consulting.

Margaret Shaw is Professor of Marketing in the School of Hospitality and Tourism Management at the University of Guelph, Ontario, Canada. She holds a PhD from Cornell University School of Hotel Administration, where she also received her BS degree. Dr. Shaw has an MBA from the Johnson Graduate School of Management, Cornell University.

Prior to joining the University of Guelph, Dr. Shaw served seven years on the faculty at the University of Massachusetts/Amherst. She also spent several years in the hospitality industry in various sales and marketing positions with both the Sheraton and Hyatt hotel companies.

Marketing Essentials in Hospitality and Tourism

Foundations and Practices

part I

Introduction to Hospitality Marketing

overview

As an industry we have lost our connection to the consumer. Early on, hotels had a relationship with the customer. The customer would call the hotel directly, make a reservation (sometimes with the general manager), check into the hotel, have very little staff interaction, and return if the services or experiences met or exceeded their expectations. Today, customers make reservations through the Internet or with a central reservation service that handles multiple hotels. Examples include well-known firms such as Hotels.com, Expedia, and Orbitz and lesser-known firms such as Supranational and

continued on pg 4

The Concept of Marketing

learning objectives

After reading this chapter, you should be able to:

1. Explain why marketing is much more than advertising and salespeople trying to sell you something you do not want.

2. Explain the purpose of a business: to create and keep customers.

3. Identify several types of management orientations in hospitality and tourism and explain why the firms that have a marketing orientation will be successful over the long run.

4. Understand that a firm that exhibits a marketing orientation practices marketing leadership.

5. Understand how to become a marketing leader.

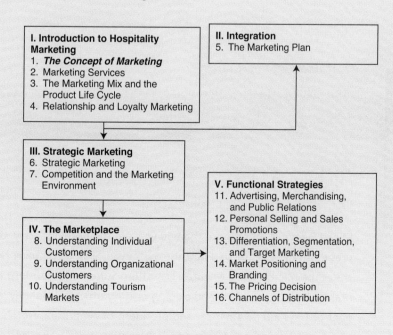

I. Introduction to Hospitality Marketing
1. *The Concept of Marketing*
2. Marketing Services
3. The Marketing Mix and the Product Life Cycle
4. Relationship and Loyalty Marketing

II. Integration
5. The Marketing Plan

III. Strategic Marketing
6. Strategic Marketing
7. Competition and the Marketing Environment

IV. The Marketplace
8. Understanding Individual Customers
9. Understanding Organizational Customers
10. Understanding Tourism Markets

V. Functional Strategies
11. Advertising, Merchandising, and Public Relations
12. Personal Selling and Sales Promotions
13. Differentiation, Segmentation, and Target Marketing
14. Market Positioning and Branding
15. The Pricing Decision
16. Channels of Distribution

Preferred Hotels. The relationship now is with the online reservations service, not necessarily with people at the hotel or individual hotels. The direct relationship with the hotel now does not begin until the consumer arrives at the front desk. In some cases (for example, the Hyatt Regency in Chicago), the guest does not even have to go to the front desk; he can check in and check out using the self-service kiosk. When things go right, he may rebook, or he may go to another hotel because of a deal. When things go wrong, he screams at the hotel staff and tells his friends about his terrible experience.

This disconnect between the consumer and the hotel is troubling and a result of poor marketing leadership in the industry. This book is about repairing, regaining, and keeping the relationship with the customer. This book is about marketing leadership in the hospitality industry. More specifically, it is about the causes and effects of marketing leadership. This book is about why people such as Ray Kroc and companies such as McDonald's initially succeeded, and why others do not succeed or succeed less well. Or, why Bill Marriott and Marriott Hotels succeeded and Howard Johnson ran into trouble. Marketing is about what it takes to succeed in the hospitality business (or any business).

After reading this chapter, you will understand why marketing begins and ends with the customer and why everyone in the organization should be focused on the customer. In short, you will understand the concept of marketing, the marketing philosophy, and the elements of marketing leadership. You will also understand why these elements are critical not only to the success of the hospitality industry, but to your own success as well. In the chapters that follow, you will learn how marketing leadership works.

Marketing Executive Profile

Michael A. Leven
Vice Chairman, Marcus Foundation

Mike Leven has spent more than 50 years in the hotel industry. Until recently, he was president and CEO of US Franchise Systems (USFS), a company he formed, that franchises the Microtel Inns & Suites, Hawthorn Suites, and Best Inns and Best Suites hotel brands. He was president and chief operating officer of Holiday Inn Worldwide, president of Days Inn of America, and president of Americana Hotels. Currently he sits on two hotel boards: Las Vegas Sands (Venetian Hotel) and Hersha Hospitality. Leven cofounded the Asian American Hotel Owners Association (AAHOA). He is a former international president of the Hotel Sales & Marketing Association International (HSMAI) and a member of HSMAI's Hall of Fame. Leven holds a bachelor of arts from Tufts University and a master of science from Boston University. He also holds honorary doctorate degrees from Johnson & Wales and Niagara University.

 Marketing in Action

Michael A. Leven
Vice Chairman, Marcus Foundation

How do you define marketing?

For many years, the topic of conversation was, What is the difference between sales and marketing? because originally in the hotel business, there was no marketing—it really evolved through the selling function. I picked up a pamphlet once a number of years ago, and it very adequately defined the difference between sales and marketing, which is that sales is getting rid of something and marketing is knowing what to get rid of. From that, I developed my own definition of marketing as knowing what to get rid of. When you talk about establishing a product brand or line that you're going to deliver to customers, either from a service standpoint or a product standpoint, what marketing does is define what the customers' requirements are so that you can design the product or service to fill that requirement and then sales will sell it.

What does the marketing concept mean to you and your organization?

I think the marketing concept is essentially one that integrates the customer require-ments with the requirements of the business to create more and more customers. Marketing to me is an all-encompassing function that each department in the corpo-ration, all the way to the accounting department, participates in when the objectives that come from marketing are defined. The marketing concept is essentially a total business concept, without which you can't be successful in creating customers or repeating customers. The marketing culture is created by the leadership of the orga-nization. It's not created by the chief marketing officer. It's created by the top senior person who puts the customer requirement on the front end of the burner and acts that way personally through every relationship that individual has with his staff in every department. Even the way they collect an account or the way an accounts receivable is passed through is part of a genuine marketing effort to develop the customer response mechanism. When you, as a leader, integrate your customer philosophy through each department as it works for you, that is how you generate the marketing culture.

Can you provide an example of a hospitality company that is truly marketing oriented?

There are some examples of specific marketing activities that go on in various corporations that one might identify as appropriate activities, but there are no companies that represent true marketing philosophies. They all have some level of marketing conditions in them, but I do not feel that any of them really fit the profile as having the total marketing concept environment. The hotel business has been, for all practical purposes, an operation business and not a marketing business. You see examples of this from time to time, but it's generally not what would fall into the real definition of a marketing company. There are some companies that tend to move forward in this area, but since most hotel companies are run by finance people or development people, the marketing culture has had to be built from underneath them, and it is extremely difficult to do that. At Coca-Cola, they just eliminated their chief marketing officer position. That speaks to what the leadership thinks is really important.

One of the great marketers in the hotel business is a guy named Harris Rosen in Florida, and he's not even recognized by major companies. They look at him as a salesman of hotel rooms; I look at him as a great marketer. He builds a culture within his organization and his company, all the way down to the shoes that employees wear, their health care program, their uniform, and their benefits, which is part of the marketing effort—it's not a human resource decision. This is a decision as to how I am going to deliver services that I need to deliver through my distribution point or my people. It's a marketing decision and how he prices and what he does—I would say that he's the best I've seen, and he's not in a major corporation.

Starbucks is another company that is absolutely a prototypical marketing example. Most of the great marketing decisions are intuitive, and Howard Schultz's decision to build an environment—creating an environment where people wanted to be there and coffee became incidental—that's one of the most esoteric but most magnificent marketing decisions of this generation. Every time I go by, I am stunned by the number of people of all ages and demographics who would be there just to be there. He hit it right on the button—he created an environment. With respect to hotels, W has done some of that for a particular market segment. You don't see that in your traditional companies—there is more of a "copycat" type of marketing.

What are traps organizations fall into that take them away from a customer orientation?

The answer to that is that nobody has enough drive to change it. The biggest trap is fear; the biggest trap is having to report to somebody. I was 24 years old starting as a sales representative at the Hotel Roosevelt in New York City, and I also had a secondary job there—my job was internal promotions. I was taking care of the signs in the elevators and the lobby—the various promotional cards and whatnot. I didn't really know what I was doing, but we had a bar there called the Club Car Bar that you had to walk by to go to the corridor to get to Grand Central Station. So there's a tremendous amount of traffic at 5.00 P.M. when everyone walks by the windows in the bar. I was there about a month or two and I was wandering around developing notes for all of these signs and things, and I saw that they had two rather large display cases at the front entrance of the bar (which was made to look like the club car of a train) and in the display cases were two vases of flowers. I looked at it and said to myself, "Why in the world would you have flowers in the window of a club car?"

The next day I looked in the yellow pages and I looked up the Lionel trains office and I stopped by and said that I would like to see their sales promotion manager. I asked if they could give me some trains so that I could replace the ridiculous flowers and put some Lionel trains in the window. It would be good promotion for Lionel and would be more applicable to the bar. The guy said, "Sure," and the next day I got a couple of boxes of trains and the key to the window and I took out the flowers and put in the trains. About six hours later, I got a call from the controller. He got all aggravated because he was the one who put the flowers in, and he started yelling and screaming at me about putting the trains in the window. That was my first of many disputes with controllers over the years.

From a marketing perspective, people are going to run into individuals who do not quite understand what it's all about. I think this is a part of studying hospitality marketing. I think the focus has to be on how people have changed the parameters—how and why do these amenities and changes get in—so that they can have a better understanding of the process and the ability to get things through and to understand the results of the Howard Schultz kinds of decisions. You can't do without it, but you have to be prepared to fight for it.

Used by permission from Mike Leven.

WHAT IS MARKETING?

For many, the term *marketing* conjures up images of selling and advertising with the hope of capturing as many new customers as possible. Selling and advertising represent only two of the many areas that make up the marketing process. Just because a restaurant advertises does not mean that it is marketing. Similarly, even though a hotel has four salespeople selling, it does not mean that the hotel is marketing. Again, selling and advertising are only two parts of the total practice of marketing. True marketing, which encompasses all parts of the organization, determines the nature of consumer demand and then develops, promotes, and delivers the products and services that will satisfy that demand. The word *delivers* is important because this is how front line staff become involved in marketing.

All phases of marketing focus on the customer. Sales- and advertising-oriented managers think in terms of the virtues of their product and how they can persuade the customer to buy. They often think in terms of what they have to offer the customer, usually in terms of the physical end of the business. In contrast, marketing-oriented managers think in terms of customers' whole buying process: from understanding the customer's needs when designing the service or product, to delivering the service and supporting the product after the sale is over. They think in terms of what the customer wants.

marketing The process of identifying evolving consumer preferences; then capitalizing on them through the creation, promotion, and delivery of products and services that satisfy the corresponding demand. This is done by solving customers' problems and giving them what they want or need at the time and place of their choosing and at the price they are willing and able to pay.

customer value The customer perception of a fair return in goods, services, or money for something exchanged. Marketing creates value for customers by understanding and delivering on their needs and wants. Value is a judgment assigned by consumers to the expected or completed consumption of goods and services.

We define **marketing** *as identifying evolving consumer preferences; then capitalizing on them through the creation, promotion, and delivery of products and services that satisfy the corresponding demand. This is done by solving the right customers' problems and giving them what they want or need at the time and place of their choosing and at the price they are willing and able to pay.* Notice that the definition includes needs as well as wants. This is because customers often know their needs (e.g., I need a quick, easy, and inexpensive way to communicate with friends and colleagues around the globe) before their wants (e.g., I want an e-mail account).

Notice also that the definition of marketing also includes the words *right customer*. Successful organizations focus on those customers that are profitable. To illustrate this point, consider two customers—Tom and William—and one hotel room. Tom comes to this five-star hotel frequently and purchases additional items from the hotel, whereas William comes to this hotel because of an Internet special. William is a one-time customer with a low chance of a repeat visit because he normally stays in three-star properties. Naturally, Tom should get the one available room because William is the wrong customer for this hotel given his normal lodging stay pattern. The practice today, however, is to give the room to the first person to book.

The term *price they are willing to pay* relates to perceived value. After all, customers are willing to pay more if the value is there. We pay more for Coca-Cola than we do for generic soda when, in reality, both are sugared water. We define **customer value** as the attitudes and beliefs about the specific goods and services customers receive from the firm and what they pay for these goods and services.

Marketing is involved in creating value. It brings customers into the hospitality organization and keeps them coming back. Returning customers provide value to the organization by giving the organization revenue, positive word of mouth, constructive criticism when things go wrong, and the like.

The organization that focuses on the customer and the changing nature of customers' wants and needs exhibits marketing leadership. Great marketing leaders in the hospitality industry recognize opportunities in the marketplace well before their competitors do. Marketing leadership does not just occur in the executive suite, but flows through every phase of the operation of the hospitality enterprise. Marketing is a way of thinking: If you are not thinking about the customer, you are not thinking about marketing.

Marketing and the Hospitality Industry

McDonald's is a household word. Why is this? Some will say it is the products, the Big Mac or the Chicken McNuggets or the french fries or milkshakes. Others might repeat the McDonald's slogan, QSC—quality, service, and cleanliness. Both would be wrong. McDonald's had no monopoly on any of these features, yet for a long time its competition could not catch up. The reason *McDonald's* was a household word around the world was because for a long time its lifeblood was filled with the concept of marketing, and it practiced it in nearly everything it did.

Some fast-food chain manuals instruct management that the front windows must be cleaned every six hours, but the McDonald's manual states, "the front windows will never be dirty." In fact, some believe that the most innovative thing that Ray Kroc, founder of McDonald's, did was put large windows in front of every store. It was not enough that a McDonald's was clean inside; it was that people could look in and see that it was clean. Before McDonald's, you took your chances when you walked into a low-priced restaurant. Kroc went one step further. He also insisted that not only the outside of McDonald's be litter-free, but also the area right next to the restaurant, even if McDonald's employees had to do the cleaning.

What do clean windows and clean sidewalks have to do with marketing? In hospitality, everything that management does affects the customer's image of the product, and everything that affects the customer is marketing, good or bad. We cannot repeat that enough, because it may be the most important thing that you will ever learn about marketing. This is the **concept of marketing.**

Practicing the concept of marketing means that organizations recognize that marketing and management in a service business, such as a hotel or restaurant, are one and the same. Practicing the concept of marketing means practicing marketing leadership, which recognizes that it is marketing that shapes the total organization. It means putting oneself in the customer's shoes. It means identifying, profiling, and selecting market segments that can be served profitably. This translates into profitable products and services. Practicing the concept of marketing means making the business do what suits the customer's interests.

Does this mean that marketing has replaced operations and accounting? Of course not. The person focused on a career in operations will probably take fewer marketing courses, but she must learn to apply the principles of marketing in her operations courses. When a menu is designed, the first question to be asked is, How will the customer react to it? When a hotel room is configured, the first question to be asked is, How will the customer use it? When prices are established, the first question should be, How will the consumer perceive the price–value relationship? When engineering is taught and electric consumption is measured, the first question is, Is the lighting appropriate for the customer? When food, liquor, or labor cost controls are taught, we must ask ourselves how these potential cost reductions may affect the customer. We should always ask, What is the value of the product or service to the customer? The foundation is the concept of marketing; the application is its practice. Understanding this perceived customer value is the driver of marketing.

The marketing concept is based on the premise that the customer is king; the customer has a choice; the customer does not have to buy your product or service. Thus, the best way to earn a profit is to serve the customer better. The marketing concept stresses the importance of customers and emphasizes that marketing activities begin and end with them.

In attempting to satisfy customers, businesses must consider not only short-run, immediate needs but also broad, long-term desires. Trying to satisfy customers' current needs by sacrificing their long-term desires will only create future dissatisfaction. To meet these short- and long-run needs and desires, a firm must coordinate all of its activities. Production, finance, accounting, personnel, and marketing departments must work together.

concept of marketing
The art of creating customer value and helping customers to be better off by fulfilling their expectations and solving their problems.

The marketing concept is not a second definition of marketing. It is a way of thinking or a culture—a management philosophy guiding an organization's overall activities. This philosophy affects all efforts of the organization, not just marketing activities.[1] McDonald's CEO Jim Cantalupo gave an example of implementing the marketing concept when he stated, "McDonald's is changing its philosophy from building more stores to getting more customers in our existing stores."

McDonald's planned to do this by listening to the customer's want and needs. Specifically, it chose to add healthier items such as premium salads and fruit to its menu. In addition, in early 2003 McDonald's created a Global Advisory Council on Healthy Lifestyles, whose goal is to help guide the company toward activities that address the need for balanced, healthy lifestyles, something that consumers are striving for. McDonald's also plans to retrain employees and managers "in an effort to provide service that is more speedy, accurate, and polite."[2] All this exemplifies marketing.

Marketing and the Firm

No company can operate without a profit. But, let's put first things first. No company can begin to operate without customers. Today we are in the customer business. Without customers, we cannot succeed. And the way to have customers is to offer them solutions to their problems. To keep them coming back, you must continue to solve their problems, satisfy their needs and wants, and add value to the product or service. Although we often think of price when we hear the word *value,* in Chapter 3 we show that value is much more than price. For example, when organizations save the customer time by having a speedy check-in, this can be thought of as a temporal value. Resorts often provide activities in which guests can interact with other guests. These gatherings provide social value for the guests. Other types of value are emotional, which is exhibited when guests are called by their first names and made to feel important.

The customer determines what a business is. What the business thinks it produces is not of first importance. It is what the customer is buying—or what the customer values—that is of first importance. This determines what a business is, what it produces, and whether it will prosper. Las Vegas has seen tremendous growth because the city realized that it was selling not just gambling, but an overall experience. Las Vegas has focused on the customer.

Unfortunately, some businesses, including hospitality firms, have still not discovered the customer or have lost a previous focus on the customer. You will read many current examples of this throughout this book.

Twofold Purpose of Marketing

The twofold purpose of marketing is to create and keep a customer. Creating a customer is sometimes relatively easy. This usually occurs through the use of

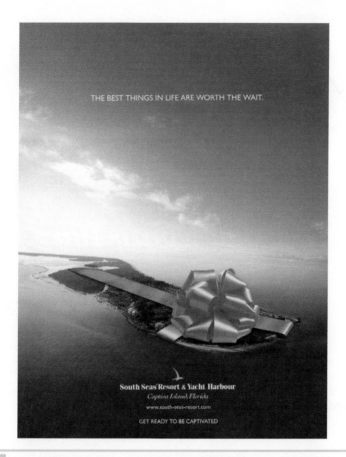

THE BEST THINGS IN LIFE ARE WORTH THE WAIT.

South Seas Resort & Yacht Harbour
Captiva Island, Florida
www.south-seas-resort.com

GET READY TO BE CAPTIVATED

Exhibit 1-1   This advertisement induces the customer to visit the South Seas Resort
Source: South Seas Resort & Yacht Harbour. Used by permission.

promotions and discounts, as exemplified by advertisements that get the customer to buy the product at an introductory price. Promotions may get the customer to purchase the product once, but they won't, by themselves, get the customer either to buy additional products or services on this first visit or to return. That will depend on how well the organization handles the customer after she gets in the door. Thus, *creating* and *keeping* are two words that define marketing. Compare, for example, the two advertisements shown in Exhibit 1-1 and Exhibit 1-2. The first is an advertisement that induces or encourages a first-time trial of a product. The second encourages a repeat purchase.

The marketing challenge is creating new customers and turning these new customers into repeat customers. Firms earn repeat customers by establishing relationships with them. These relationships can range from financial (e.g., the frequent guest programs that reward guests with points that can then be used for free stays), to social (e.g., the vice president of marketing playing golf with a top travel manager), to structural (the organization keeps extensive records on the guest's preferences so that the guest's needs will be taken care of no matter who happens to be working that day.) We discuss more about these and other types of relationships in Chapter 4.

Exhibit 1-2 Middlebury Inn employees encourage their guests to "stay another night" with these buttons
Source: The Middlebury Inn. Used by permission.

Solving Customers' Problems

There is a basic idea of marketing that we must understand. We mentioned it earlier, but it is worthwhile repeating. Simply put, consumers do not buy something unless they have a problem to solve and believe that a purchase will provide the solution to the problem. This problem may be an actual problem (e.g., I buy gasoline because my car is out of gas) or a perceived problem (e.g., I buy an iPod because all my friends have iPods and it is a problem that I do not own one). To further illustrate this point, consider what Charles Revlon, founder of Revlon cosmetics, said: "In the factory we make cosmetics; in the store we sell hope."

It should be noted, however, that customers also buy products and services because they are attracted to specific features. For instance, one of the authors owns an Apple iPod, even though he already owns another MP3 player. He did not need the iPod, but was attracted to the amount of memory it stored, so he purchased it. Also, all of his students had one so he had to have one!

Consumers also buy products and services that they hope will help them achieve certain images and dreams. The clothes we buy announce who we are, as do the hotels and restaurants we choose to visit. If we can think of goods and services that we want to sell as solutions to problems, we are a long way down the road to successful marketing. Thinking this way forces us to stand in the customer's shoes, to think like the customer thinks, and to understand what problems the customer is trying to solve. Consumers may not be able to articulate what they really want or need, but they do know what their problems are. For instance, the meeting planner may not know exactly

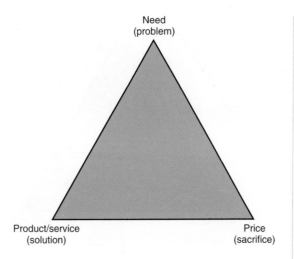

Exhibit 1-3 The trade-off of problem solutions

what a proper coffee service should look like, but he will know what one does not look like—cold coffee, hot water in a coffee pot so the tea tastes like coffee, and access to the coffee from only one side of the table. These are the kinds of problems that hoteliers can solve while at the same time demonstrating that they understand the problem, too.

Often, however, the customer does know the solution to a problem. This can be illustrated as follows. Perhaps you are driving down a highway and you need a place to sleep. This is a need, and a basic one at that. Needs create problems—namely, how to satisfy them—so what you do next is seek a solution. You know that solution will have a cost. You have to give up something or make a sacrifice to get the solution. What emerges is a trade-off situation (portrayed in Exhibit 1-3) that we refer to as the **consumer trade-off model.**

This is the trade-off thought process a consumer faces when contemplating a purchase. In general, the decision-making process gets more complicated as the cost of the item increases—both in terms of dollar amount and the associated risk of making a wrong decision. A consumer may spend months selecting a honeymoon destination, and seconds selecting a can of soda or tube of toothpaste. Nevertheless, the process takes place and the depth of deliberation depends on numerous factors that will be discussed in more depth in later chapters.

For the moment, however, let's continue the illustration and assume that a solution presents itself: A sign on the highway announces a motor inn ahead with guest rooms at $59.50. Guest rooms provide a solution for the need to sleep, and $59.50 is an investment you are willing to make. You decide to head for the motor inn rather than continue driving.

Now the situation becomes a little more complicated. You expect that the solution is at hand; that is, you expect that you can get a good night's sleep at this motor inn. You expect, of course, that there will be a bed in the room, a bathroom, and other appointments. You also expect that the bed will be comfortable and that the room will be quiet so that you will sleep well. You may not verbalize these expectations, but subconsciously they exist. You also have, consciously or unconsciously, made another decision: You have decided that spending $59.50 is worth the risk that your expectations will be met, that the solution will solve your problem, and that the value you will receive will be

consumer trade-off model The idea that if a solution to a customer's problems, needs, or wants meets the customer's expectation, and the value of that product or service justifies the sacrifice or risk, that sacrifice or risk is justifiable, and a high level of satisfaction is likely.

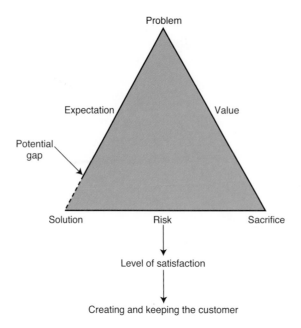

Exhibit 1-4 Expansion of the trade-off model

worth the sacrifice. The expanded consumer trade-off model looks like that in Exhibit 1-4. This figure is similar to Exhibit 1-3, except we have added these expectations to the triangle. The expectations part of the triangle makes marketing in hospitality or other service industries unique.

Unlike buying an automobile, consumers cannot take a hotel guest room for a "test drive" before they buy it. Instead, they purchase the guest room solution at the same time they consume it. Therefore, the consumer has the expectation that the product or service will solve the problem "on the spot." Once the decision is made to purchase, there is no backing out. (We discuss this concept in more detail in Chapter 2.) It then follows that the greater the sacrifice, the greater the risk, the greater the expectation, and the more demanding the customer is of the solution. To put it another way, if the solution meets the expectation and the value justifies the sacrifice, the risk becomes reasonable, and a higher level of satisfaction becomes more likely. The result is a higher likelihood that we have created a customer. To put this in realistic terms, consider each element of Exhibit 1-3 in terms of going to eat at McDonald's versus a fancy, expensive restaurant in New York City where luncheon checks can exceed $150 per person.

Notice now what happens when the solution does not meet the expectation. This is the "potential gap" on the lefthand side of Exhibit 1-4 as indicated by the dotted line. We have made the dotted line, let's say, "Pizza Hut length." We might make it a hair shorter for "McDonald's length," but for the fancy restaurant it might run almost to the peak of the triangle. The point is that the greater the expectation, the greater the risk and the greater the potential that it will not be fulfilled. We will explore this point in more detail in later chapters. For now, you can see why marketing and management are so intertwined. If management cannot fulfill customers' expectations, it won't create and keep customers.

Notice in Exhibit 1-4 where satisfaction occurs—under the solution, not between sacrifice and solution. This location indicates that satisfaction is a result of marketing, not part of the process in creating and keeping the customer. If the gap is too great between the expectation and fulfillment of the solution, there will be a low level of satisfaction. This exhibit should help you understand why marketing must address the issue of meeting or exceeding expectations. If it does this, satisfaction will be high and the guest will return. The reverse is not necessarily true. For instance, consider bathroom amenities. These amenities may create high levels of satisfaction in themselves, but if they don't help solve the customer's problem(s) (i.e., the reason the sacrifice was made), the customer will not repurchase. For example, a hotel might have had a long check-in, delayed room service, but great bathroom amenities. The amenities alone are not enough to make the guest return to a poorly run hotel.

In essence, Exhibit 1-4 represents the concept and definition of marketing—solving customers' problems, giving them what they want or need at the time and place of their choosing and at the price they are willing to pay (or a sacrifice they are willing to make). Each step in Exhibit 1-4 represents the process of marketing.

Naturally, the solution to any problem rarely exists in a vacuum. That's why marketing becomes more complex than the example presented. If a solution to the problem of needing a night's sleep were only a room and a bed at the right price in the right place, then there would be little need for marketing. Solutions aren't that simple; they involve many needs other than a simple bed in a simple room. The instant one motor inn provides something different from another motor inn, competition is created. Marketing is then tested. Instead of "here's a bed" (solution to problem), marketing creates "here's a heavenly bed," the *only* solution to your problem. Exhibit 1-5 illustrates this concept. This marketing effort turned out to be fairly successful for Westin Hotels, one of the well-known brands of Starwood Hotels.

The goal for marketers is *to present the best solution to the problem at the lowest risk.* Marketing, however, does not stop there, especially in the hospitality industry. The creation of expectations might be classified as traditional marketing (selling and advertising). However, some believe that marketing does end there and that it is now operations management's job to ensure that those expectations are fulfilled.

Because the fulfillment of expectations is primarily operations management's responsibility, operations management is totally involved in the marketing effort. Traditional marketing only brings the customer to the door; it is up to nontraditional marketing to *create* and *keep* the customer. For example, a well-known upscale hotel in a major metropolitan area recently authorized the guest service staff to "make it right for the customer." What did this mean? That the guest service staff was authorized to refund up to $500 per guest without management's approval for any problem encountered during their stay. If a customer had a problem with the telephone, room service, wake-up call, or anything else, it was handled immediately. This is nontraditional marketing.

The trade-off model is critical to the understanding of marketing. The concept will be developed further in the chapters to follow, but first we need to see how marketing influences the total picture.

Exhibit 1-5 Westin's marketing of its "heavenly bed" created a need that only Westin could fulfill
Source: Westin Hotels & Resorts. Used by permission.

FIRMS' ORIENTATIONS

All companies, firms, organizations, and other business entities operate under a basic belief, or philosophy. The orientation may be spoken, written or just simply implied. An organization's philosophy is a part of its corporate culture—it emphasizes that "this is the way we do business around here." This is what drives the firm, what makes it work.

The hospitality industry encompasses many orientations at various times and places, including operations, product/service, selling, and bottom-line orientations, or some combination of these four. Then there is, of course, a marketing orientation.

Operations Orientation

An **operations orientation** is categorized by its emphasis on a "smooth operation," as exemplified by the statement, This is a great business to be in, if only the customers didn't get in the way. Operations manuals provide prescriptions for direction and behavior for almost every conceivable occurrence—until the customer decides to do something differently.

Operations-oriented hotels and restaurants sometimes forget the customer in the interest of a smooth operation. Although these facilities run well, customers are fickle and procedures cannot be written for every kind of demand or problem. This does not mean that manuals are not desirable for operations purposes. In fact, in today's large chains it would be impossible to obtain consistency in service delivery without some of them. Problems occur, however, when the manual becomes the "be-all and end-all" and there is no room for deviation on the customer's behalf. Or, what may be even worse, sometimes the manual is written only from an operations efficiency or cost perspective and without consideration for the customer. Consider a true story that occurred in a deluxe hotel in Paris. A former student of one of the authors was working in this hotel as a floor supervisor in housekeeping. She was told that at 14:00 every day (2:00 P.M.) the guest's room should be refreshed. When the student asked what to do if a "do not disturb" sign was on the door, she was told "Ignore the sign. Guests sometimes leave the sign on the door by mistake. Plus, this is the way we do things here and rooms need to be refreshed." Naturally, guests often complained to the former student, but her supervisors always said, "Policies are policies."

Operations philosophies, like all philosophies, come down from top management. When the company or the company's executives are very bottom line or profit driven, they tend to follow procedures based mainly on cost considerations, overlooking their impact on customers. Consider, for example, the restaurant that has a slow night. Typically, management will send wait personnel home (or in restaurant language, "cut the floor") and close part of the dining room. The section that is often closed is the quieter section, away from the main dining room. This section may also be the most desirable from the customer's perspective. The part that remains open, of course, is closest to the kitchen because it is most convenient to serve. Similarly, you may have had the experience of saying to a dining room hostess, "Can we have that table over there?" (instead of the one you were led to) and received the response, "I'm sorry, but it's not that waiter's turn." These practices, of course, can lead to even emptier dining rooms.

One of the authors experienced a similar experience during the off-season on Cape Cod, Massachusetts, with very different results. Upon entering the restaurant in midwinter, the author and her husband were approached by the owner, who said, "I had two people call in sick this evening. If you need to be seated right away and have dinner served quickly, let me call the restaurant down the road and they will take care of you. If you would like to be my guest for a glass of wine while we get your table ready and can be a bit patient with the service, I would be very grateful if you stayed." This is marketing at its best. The author and his wife had a very relaxed evening and recommended the restaurant many times over.

In another example, consider a large hotel of a major chain that claimed to be customer oriented but rewarded all of its people based on bottom-line results. This hotel was losing occupancy in the face of increasing competition and was responding by cutting costs and raising prices. Management spent $17 million on a new entryway that had little if any effect on business. The problems were inside where staff had been cut to save payroll. Normal check-in time took 10 minutes, and that's when there wasn't a line. Checkout was about as bad. The widely advertised indoor pool didn't open until 9:00 A.M. (to save payroll), long after the business clientele had left for the day or were in meetings. The brightest light in the room was 67 watts (to save energy costs), difficult to use for paperwork or reading. The widest writing space in most rooms, including suites, was 16 inches.

These types of procedures are established in the name of operational or cost efficiency. Perhaps a better phrase might be "customer blindness efficiency." Hotels and restaurants that operate by these kinds of procedures pride themselves on their operational efficiencies, rather than on their solutions to customers' problems. Obviously, such efficiencies may well cause problems instead of solving them for guests seeking a hassle-free experience.

Product/Service Orientation

product/service orientation A style of hospitality management with a primary emphasis on creating great products or services to attract customers.

Hospitality properties that operate under a **product/service orientation** place their emphasis on the product or service. These properties market according to the concept of "build it and they will come." They trumpet that their property has the best food, the finest chefs, the designer-decorated lobbies, or even the best location. In terms of service they argue that they have the finest service available and that they know what service is best for the customer. The problem with the product/service orientation approach is that the firm may offer products and services that the customer does not want. Marketing is about finding out firsthand from customers what their needs and wants are. Exhibit 1-6, "The Coffee Break Dilemma," demonstrates how hotel managers learned that their perception of a good coffee break differed from that of the meeting planner customer.

The example in Exhibit 1-6 shows how far management can be from the real needs and problems of customers.

Exhibit 1-7 lists examples of how this emphasis on the product/service orientation without regard to the customers' wants, needs, and problems can creep into

Tourism Marketing Application

Disneyland Paris is an example of a tourist destination that initially opened with a product orientation. The belief was, "If we build it, people will come." The destination was an exact copy of the destination built in the United States. No consideration was given to the consumption habits of Europeans; for example, restaurants did not offer wine or beer. In addition, tour operators have a tremendous influence on the travel behavior of European travelers. Yet, because their needs were not met, sales suffered. Finally, no consumer research was undertaken to gain an understanding of Europeans' wants and needs. U.S. culture is very different from European Culture.

Exhibit 1-6   The Coffee Break Dilemma

The executive committee of hotel managers diligently set out to design the "mother of all coffee breaks," including trilevel presentations, mirrors, ice carvings, lighting, flavored coffees, and so on. When costs are broken out, the break had to be priced at $32 per person to make a profit. The meeting planners then turned in their "perfect" coffee break design: a simple coffee break with the table for the coffee positioned about 10 feet from the back wall so people could access it from both sides. The purpose, they explained, was to alleviate congestion. Their meetings could resume more quickly if the placement of the table was simply moved a few feet!

Exhibit 1-7   Examples of Product/Service Orientation

A joint promotional brochure for some of Atlanta's hotels directed at meeting planners contained pictures of the hotels along with the following copy:

- At the Westin, after the hotel had spent $31 million, you could have a "more than successful" meeting because of "sumptuous new fabrics, elegant furnishings, Italian marble and breathtaking views" and, for "meeting inspiration" there were "twinkling arches and hidden courtyards."
- At the InterContinental, you could have "a chauffeur-driven Rolls Royce meet your private airplane," and you could dine in "surroundings of hand-polished wood, antique crystal and gleaming silver."
- At the Hyatt Regency, you would get to enjoy the hotel that, "20 years later continues to fulfill John Portman's vision."*
- At the Marriott Marquis, as a small group, you could "practically have the place to yourself . . . with its awesome atrium, 10 restaurants, and lounges."

- At the Ritz-Carlton, on the other hand, they "concentrate on the fine points of innkeeping like luxurious rooms, afternoon tea served in English bone china and richly paneled walls graced by 18th and 19th century oils."

In the same brochure, there was the Colony Square Hotel, a Preferred Hotel, which "assigns individuals who work behind the scenes to oversee every detail of your meeting—from planning to follow-up." The ad also contained a picture of and a quote from the manager of the Coca-Cola U.S.A. Training Center, who said, "The thing I appreciate most is the flexibility. They respond quickly to our last minute requests."

Tens of thousands of dollars were spent on these ads. Which of all of the attributes listed by the various properties do you suppose meeting planners most preferred?

*In case you didn't know (and few today would), John Portman is the architect who created this hotel with the first-of-its-kind atrium lobby in the 1960s.

---

advertising. Most of the commentary in the brochure in Exhibit 1-7 has little to do with a successful meeting, which is the "problem" the meeting planner wants to "solve." Italian marble? Hand-polished wood? Who is John Portman? (He is the architect who created the atrium lobby concept in the late 1960s.) Again, the product that meeting planners are really looking for is a successful meeting. Items such as soundproof meeting rooms; Internet access; and coffee breaks served simply, efficiently, and on schedule are the types of products and services that facilitate a successful meeting. "Twinkling arches and hidden courtyards" do not create value for the meeting planner customer.

Selling Orientation

A **selling orientation** in hotel and restaurant companies is one in which the effort to obtain customers emphasizes finding someone who will come through the doors, as opposed to marketing a solution to a designated market's needs. Hospitality companies with this kind of orientation often have large sales forces and/or large advertising budgets. Hotels are very conscious of their open periods, and they push their salespeople to "go out and fill the rooms" and to meet their sales quotas. In the case of restaurants, they may run frequent promotions and special offers. Everything is based on the sell, sell, sell edict rather than on identifying customers' needs and wants.

selling orientation A style of hospitality management with a primary emphasis on salespeople and promotions that communicate a message to customers in order to sell products.

Bottom-Line Orientation

Whereas the orientations mentioned previously may all have their place in different companies, the bottom-line orientation without regard to its impact on customers does not. Certainly, it is important to yield a satisfactory return on invested capital; as such, one needs to be concerned with profitability. However, to do this without thinking of the impact on the customer is a dangerous orientation that has destroyed a number of companies in many industries. Twenty years ago this was very prevalent in hospitality companies, many of which did not survive. It is not as prevalent today. We would like to think the bottom-line mentality is gone completely, but unfortunately it still prevails in some places—from small owner-operated restaurants to some large hospitality organizations.

The bottom line, of course, is profit. But a bottom-line orientation means that most things, if not everything, are done in the name of profit. In these cases, profit (or its reciprocal, cost) is the basis for decisions, not the test of their validity as we noted earlier. This manifests itself most often in terms of cost—food cost, labor cost, beverage cost, marketing cost, refurbishment cost, and so forth. We discuss the fallacy of this more in the pricing chapter, but suffice it to say for now that when the emphasis is on cost, the customer is soon forgotten.

Liquor bars in airports run a 14 percent liquor cost, but they have a captive audience. Some exclusive resorts get away with the same, but they have other attractions. For many businesses the mandate is cost-driven prices, not price-driven costs, which should be the case.

Web Browsing Exercise

Search the websites of both Boston Market (www.bostonmarket.com/index.jsp) and Sheraton Hotels (www.starwoodhotels.com/sheraton/index.html). How have these companies incorporated their new customer orientation into their websites?

There is obviously nothing wrong with managing a good operation, having a good product or service, or employing an effective sales force (these should be advanced as targeted goals!). Well-run and successful companies accomplish all of these things well. A truly marketing-oriented company, however, views these achievements as subsets of marketing; that is, they are accomplished with the customer as the focal point. The operations manager says, "I run a tight ship," but only after making sure that the customers' needs and wants have been considered. The service manager considers first what the service will do for that customer. And the sales manager sells benefits that will solve customers' problems and make their experiences hassle-free.

Marketing Orientation

As defined by Kohli and Jaworski, **marketing orientation** is "simply the implementation of the marketing concept."[3] The marketing concept is based on the premise that the customer is king: He or she has choices, and the best way to earn a profit is to serve the customer. A marketing orientation reflects a willingness to recognize and understand the consumer's needs and wants, and a willingness to adjust any of the marketing mix elements, including product, to satisfy those needs and wants. In other words, communicating effectively to customers is part of the concept.

Like all the orientations we have discussed, a marketing orientation is reflected in the culture of the organization. A yes answer to the following questions indicates a firm with a marketing orientation:

- Does the firm focus on the customer?
- Do the various departments within the firm work together to respond effectively to customers' wants and needs?
- Is the market intelligence on customers, competitors, and market conditions shared throughout the organization?
- Does the firm's reward system support the emphasis on the customer?

For a marketing orientation to work, you have to "put your money where your mouth is." Having a marketing orientation without practicing the marketing concept is a good start, but it will not succeed in the long run. When the company dies, many will say, "They were such nice people, I wonder why they didn't make it." On the other hand, practicing the marketing concept without a marketing orientation is like giving lip service to marketing; marketing becomes company policy without permeating the firm. How many times have we seen the poster "The customer is number one" in the hotel employee cafeteria, only to have it ignored by both management and staff? A marketing philosophy and the marketing concept must coexist before we can define the firm as a true marketing company. Marketing-oriented companies and marketing-oriented people are the ones who are truly successful in the highly competitive hospitality marketplace.

MARKETING LEADERSHIP

The guiding philosophy in any firm is established by the top management, which provides the leadership and direction for the organization. These leaders must believe in the marketing philosophy and the marketing concept and ensure that they pervade all levels of the organization. **Marketing leadership** accepts change as a constant. It recognizes not only the needs and wants of the customer, but also the fact that the customer changes; the customer is not in a static state and any successful company must change with, if not before, the customer. Business obituaries are replete with companies that failed to recognize changes in the marketplace.

An excellent example is that of the "old" Howard Johnson. In 1965 Howard Johnson's annual sales were greater than the combined sales of McDonald's, Burger King, and

marketing leadership
A characteristic of a hospitality enterprise that integrates marketing into every phase of its operation through opportunity, planning, and control. Marketing leadership combines a vision for the future with systematic planning for solving customers' problems.

Tourism Marketing Application

Golf Tourism and Planning for a Growth Market

Golf tourism is a growing industry worldwide. One company that designs golf courses around the world is Golfplan. Rather than just designing the course (product orientation), the company practices a marketing orientation by helping developers with customer research, competitive analyses, ecological studies, and land use planning, among other activities. Golfplan also works to effectively reduce the annual expenses of managing and operating the course.

KFC. By 1970 its sales were about the same as McDonald's. In 1984, when the company was broken up, Howard Johnson's sales were less than $750 million; McDonald's had grown to almost $3.5 billion.

When the customer changed, Howard Johnson did not; slowly but surely its customer base eroded. As things got worse, Howard Johnson concentrated on cutting costs rather than recognizing its customers' problems and finding out what the customer wanted. Howard B. Johnson, son of the founder, said they ran a very tight operation. They were on top of the numbers daily. Others said that if he'd eaten in his own restaurants more instead of lunching at Club 21 in New York City, he might have learned something.

In the following sections we discuss briefly the three components of marketing leadership: opportunity, planning, and control.

Opportunity

Great success stories in business almost always include tales of visionary leaders who saw and grasped opportunity, people like Kemmons Wilson, founder of Holiday Inns; Bill Marriott of Marriott International; and Ray Kroc, founder of McDonald's. These men were visionaries. They didn't create the need, wants, or problems, but recognized them as opportunities.

Very few opportunities are as grand as these. To find the smaller ones, marketing concept managers don't look for opportunity first; they look for consumer problems because they are easier to identify. What is perceived as a consumer problem may well be the symptom of an operations problem. This, then, becomes an opportunity. For example 60 watt lightbulbs are real consumer problems to this day, caused by the management problem of wanting to cut costs.

Opportunity continues to be the lifeblood of successful marketing. It doesn't start with fancy draperies or upholstered walls; it starts with consumers' problems. Look for a problem, the real problem, and you will find an opportunity. No industry, no business, and no product enjoys an automatically ensured growth. Only seeking, finding, and successfully exploiting opportunities can ensure growth.

Planning

Another element of practicing the marketing concept is planning. Planning is defining what has to be done and allocating the resources to do it. It means proacting rather than reacting. It means shaping your own destiny.

Although one would expect planning to be a given in most companies, it is not difficult to find companies that do not plan, plan haphazardly, or plan only as an exercise. Good planning flows from good leadership. Growth must be carefully planned. Opportunities must be sought and planned for in a systematic manner. This means planning with the customer in mind. Companies such as Golfplan, highlighted in the Tourism Marketing Application, assist in the planning and development for the golf course industry.

Many hotels develop annual marketing plans (though restaurants rarely do), but they sometimes have very little to do with planning. These marketing plans often turn into promotional objectives, advertising and sales allocations, budgets, and day-by-day occupancy forecasts. Rarely do they address the creation of customers or changes in current operating procedures to keep these customers coming back. Although financial planning is often routine, true marketing planning has yet to achieve that status. This is odd because without customers, there are no finances to manage.

Control

Control is the third element of leadership, but it is also the glue that holds the others together and makes them work. When control declines, leadership and planning flounder. Control in the marketing sense means control of your destiny through leadership, planning, and opportunity. This requires a solid understanding of the customer, the market, and the product.

Control is the feedback loop of the system that tells whether the system is working and provides information to management on who the market is and what the customer's problems, expectations, perceptions, and experiences are. Control is knowing whether perceptions equal reality, why the customers come or don't come, how they use the product, how their complaints are handled, and whether they return. In short, control is knowing and serving the customer. Control is also knowing your employees because, as we will see later, every employee is an integral part of the marketing effort.

Control in marketing means a good management information system, which we will address later in greater detail.

MARKETING IS EVERYTHING

Technology is transforming choice, and choice is transforming the marketplace. Almost unlimited customer choice accompanied by new competitors is seen as a threat by many marketers. But the threat of new competitors is balanced by the opportunity of new customers. Nicely articulated in the *Harvard Business Review* in the early 1990s are thoughts from Regis McKenna to help close this first chapter of *Marketing Essentials in Hospitality and Tourism:*

> These new customers don't know about the old rules, the old understandings or the old way of doing business—and they don't care. What they do care about is a company that is willing to adapt its products or services to fit their strategies. This represents the evolution of marketing to the market-driven company.[4]

We have written, and will write more, about creating and keeping a customer and about customer loyalty. But today, with so much choice, it is extra hard to maintain loyalty. The only way to keep a customer is to integrate him or her into the company and create and sustain a relationship between the customer and the company. This is marketing's job.

> The old notion of marketing was based on certain assumptions and attitudes but marketing today is not a function. It is a way of doing business. Marketing has to be all-pervasive, part of everyone's job description. Its job is to integrate the customer

into the design of the product and to design a systematic process for interaction that will create substance in the relationship.[5]

. . . Technology permits information to flow in both directions between the customer and the company. It creates the feedback loop that integrates the customer into the company, allows the company to own a market, permits customization, creates a dialogue and turns a product into a service and a service into a product.[6]

. . . The critical dimension of the company—including all the attributes that together define how the company does business—are ultimately the functions of marketing. That is why marketing is everyone's job, why marketing is everything and everything is marketing.[7]

Better yet, we might add, the customer is everything. The rest of this book is about customers.

Chapter Summary

This chapter introduced marketing as a philosophy and a way of life of the hospitality firm. We defined marketing in terms of the customer, and we demonstrated how a marketing orientation or concept, or the lack of it, affects the entire organization.

Marketing is far more than selling and advertising, the traditional concepts of the field. In fact, it has been shown that advertising and selling, equated by some with the term *marketing,* are only subsets of marketing. The philosophy of marketing is needed before any communications vehicles are employed. In some cases these activities may not even be necessary to marketing, as demonstrated by a number of successful establishments that rarely advertise or practice direct selling.

The other side of this coin should also be apparent. You don't have to be a marketing professional to engage in marketing. Marketing is an integral part of management and the day-to-day business of running an operation.

Those readers for whom this chapter is their first real introduction to marketing may be a little bewildered with this concept of marketing. Not to worry. In services industries, of which the hospitality industry is certainly a part, more than 80 percent of marketing may be nontraditional marketing. In Chapter 2 we will explain why.

Key Terms

bottom-line orientation, p. 20
concept of marketing, p. 9
consumer trade-off model, p. 13
customer value, p. 8
marketing, p. 8
marketing leadership, p. 21
marketing orientation, p. 20
operations orientation, p. 17
product/service orientation, p. 18
selling orientation, p. 19

Discussion Questions

1. What is the true definition of marketing? In other words, what is marketing all about?
2. Why is the purpose of business to create and keep customers?
3. What is meant by customer value?
4. Give an example of the consumer trade-off model from your own experience.
5. Briefly describe each of the management orientations described in this chapter. Have you experienced any of these as an employee at a hospitality or tourism establishment?

Endnotes

1. Grainger, D. (2003, April 14). Can McDonald's cook again? The great American icon ain't what it used to be. *Fortune*, 120–127.
2. Day, S., & Elliott, S. (2003, April 8). At McDonald's, an effort to restore lost luster. *The New York Times*, C1.
3. Kohli, A. K., & Jaworski, B. J. (1990, April). Market orientation: The construct, research propositions, and managerial implications, *Journal of Marketing*, 54, 1–18.
4. McKenna, R. (1991, January–February). Marketing is everything. *Harvard Business Review*, 65.
5. Ibid., 69
6. Ibid., 78.
7. Ibid., 79.

.... case study

Sojourn in Jamaica

Discussion Questions

1. Apply the consumer trade-off model to the Saltzers' experience at the Elegant Hotel.
2. How would you describe the management orientation at the Elegant Hotel?
3. To what degree does management understand and practice the concept of marketing at the Elegant Hotel?

It was a balmy evening in March when the Air Jamaica 727 touched down at the Montego Bay airport in Jamaica. As Mr. and Mrs. Saltzer stepped off the flight, they marveled at the 75 degree weather, the short-sleeve shirts everyone was wearing, and the bright moon in the sky. They had boarded the plane at John F. Kennedy Airport in New York where the temperature was 10 degrees above zero, the ground was covered with snow, and the wind was howling. They were happy to leave all this behind and planned to totally forget it for the next two weeks as they enjoyed a sojourn in Jamaica.

It was a short taxi ride to the Elegant Hotel just outside of Montego Bay. They had chosen the hotel from an ad in a popular magazine that read, "Where Adults Go When They Run Away from Home. If you're looking for Jamaica's most spectacular vacation setting, no other place comes close." They were soon there, checked in, and unpacked. Having had dinner on the plane, they went downstairs to explore the hotel, after which they stopped for a drink at the circular bar off the lobby. "Wow," they thought, "this is going to be neat. We really made the right decision." They then retired to bed so as to get an early start in the morning.

THE NEXT MORNING

The Saltzers were up bright and early and looked out over the grounds from the balcony of their room before going downstairs for breakfast. The hotel was shaped like a U facing the ocean, and from their room in the center they had a broad view of the ocean with extensions of the U projecting out on either side of them. Immediately below them, the Saltzers could see a large, inviting pool with a pool bar with the bar stools in the water. To the right of the pool was a large dining area with about 30 round tables with umbrellas, about 10 of which were occupied. In this section, and closer to the hotel, was an extensive breakfast buffet setup. Farther out, beyond this area and the pool, was a beach house where sports facilities were offered, and beyond this the sandy beach with sailfish, other boats, and water sports equipment.

The Saltzers headed for the outside dining area as soon as they got downstairs. Although they wanted only juice and coffee and not the buffet (they were on European plan), they wanted to sit out in the sun, watch the ocean, and forget the New York weather. The maitre d' quickly seated them as far out toward the ocean as possible. He then suggested that they partake of the buffet at their leisure. "We only want juice and coffee," they responded. "Oh," he said, "in that case you can't sit here." "What do you mean we can't sit here?" said Mrs. Saltzer. "This is only for people having the buffet," replied the maitre d'. "Then where can we sit?" said Mr. Saltzer. "You have to go inside to the coffee shop," was the reply.

Reluctantly, the Saltzers got up and went inside to the coffee shop. They found a booth on the outside of the room by a window. There was no sun shining on them, but they could peer out and see the ocean in the distance. "Oh, well," they said, "we'll be out there soon enough." A waitress came and handed them two menus. "Just orange juice and coffee," they said. "Oh," she replied, "if you just want orange juice and coffee, you can't sit here."

"Why not?" they said. "This room is only for people ordering from the menu," was the reply. "Then where can we sit?" "You can sit in the other room at the counter."

The Saltzers went into the other room. There was no sun, no ocean, no windows, and no view. "How do we know

we're not still in New York?" they said. They decided to skip breakfast and head for the waterfront.

At lunchtime the Saltzers ran into the same problem. There again was a buffet outdoors that 30 or 40 people were partaking of, and they were told they had to go into the bar restaurant if they didn't want the buffet. They found this restaurant to be empty except for one other couple. They sat down and waited about 20 minutes for the waitress. When she came, they ordered two club sandwiches. These came about 45 minutes later; the delay almost caused them to miss their planned trip into the interior of Jamaica on a train where they visited the Appleton rum distillery.

THAT EVENING

The train was not air conditioned, and although the Saltzers enjoyed the trip, they returned to the hotel at 5:00 P.M. very hot and perspiring from the 90 degree weather. They went straight to their room, put on their bathing suits, and headed for the pool. On one side of the pool was the pool bar; on the other side were lounge chairs. The Saltzers put their robes, towels, reading materials, and so on, on two lounge chairs and jumped into the pool. After swimming for a few minutes, they swam up to the pool bar and sat on two of the stools; the others were unoccupied. They ordered two of the hotel's signature rum drinks.

"Sorry, the bar is closed," said the bartender, pointing to a sign above the back bar which stated that the bar closed at 6:00 P.M. "But it's only 5:35," they protested, pointing to the clock above the back bar. "I know, but it takes us time to check out and close up." "Come on," they pleaded, "we just got here. Give us a break." "Well, okay, let me see your chits." "What chits?" "The small cardboard piece you got when you checked in that show that you are a registered guest and can charge to your room." "Oh, yeah," they said. "they're in our room. We didn't think of them. Besides, if we had them in our bathing suits they'd be soaking wet by now." "Well, I have to have your chits," said the bartender. "Do you think we just parachuted into the pool?" said Mrs. Saltzer.

The bartender finally agreed to call the front desk and verify that the Saltzers were guests of the hotel. After he did this, he served them their drinks and proceeded to close up the bar. At about 6:00 the Saltzers, still hot, went back into the pool. At the end of the pool was a very large sign that someone else in the pool called to their attention. The sign read, "POOL CLOSES AT 6:00." "That's ridiculous," thought Mr. Saltzer as he kept on swimming, "no one closes a pool at 6:00 in Jamaica when the temperature is 90 degrees, especially at an expensive resort like this."

At about 6:05 a man, obviously a hotel employee, walked along the end of the pool where Mr. Saltzer was swimming and where the sign was and, catching Mr. Saltzer's attention, said to him, "Do you see that sign?" "No, I can't read," said Mr. Saltzer. The man went away. About 10 minutes later he returned carrying two large buckets in each hand. He dumped them into the pool about five feet from Mr. Saltzer, who quickly realized that the pool was being chlorinated. He got out quickly.

The Saltzers had dinner in the hotel dining room that night. They weren't sure whether it was a good dinner or not. By this time they were so upset that they just picked at their food. After a brief discussion, they decided to check out the next morning, although they didn't know where they would go.

The next morning they packed their bags and carried them downstairs to the lobby where they found about 40 people waiting in line to check out. There was one desk clerk behind the desk handling these checkouts. Also behind the desk, however, were three other employees engaged in conversation but not helping with the checkouts.

After about an hour the Saltzers reached the head of the line. "Why are you checking out today when your reservation is for two weeks?" asked the desk clerk. "This place stinks," said Mr. Saltzer. "Gee, that's what a lot of people are saying," said the desk clerk, "Why do *you* say that?" "Just give us our bill," said Mr. Saltzer.

As the Saltzers rode in the taxi toward Ocho Rios, about 50 miles down the northern coastline, looking for a place to finish their vacation, they discussed the comments they had heard from others while standing in line to check out. "I guess we were lucky," they said, "even if it did cost us $500 for two nights."

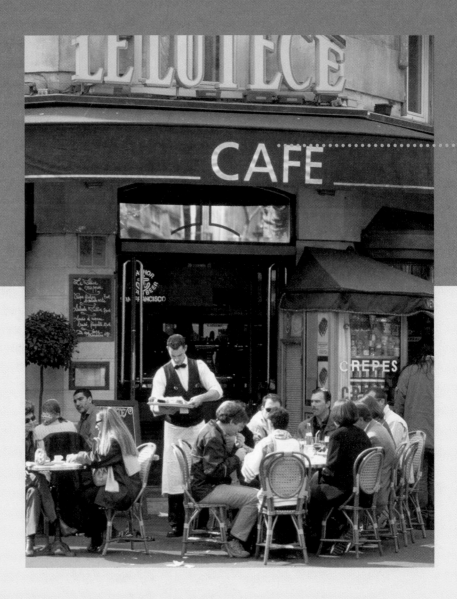

overview

Chapter 2 focuses on services, the cornerstone of hospitality in both marketing and operations. Hospitality is first and always a service industry. The differences between goods and services and the impact these differences have on both customers and marketers are fully covered in this chapter. The chapter also examines the four components of a service and explains the characteristics of services that make each of the components unique. A model of service "quality gaps" is introduced. Finally, we discuss how marketers can use this information to increase revenues and create happy customers.

chapter 2

Marketing Services

learning objectives

After reading this chapter, you should be able to:

1. Explain the basic differences between goods and services.

2. Explain why hospitality and tourism are considered part of the service industry.

3. Show how the marketing of services is different from the marketing of goods and what this means for those in marketing.

4. Shape the different components of the hospitality product to ensure that the guest has a great experience.

5. Use the knowledge of the gap model of service quality to ensure that customers' expectations are met, and when they are not met, to know why.

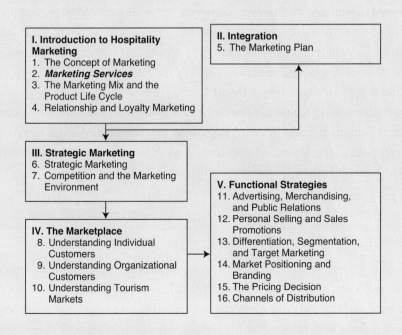

I. Introduction to Hospitality Marketing
1. The Concept of Marketing
2. *Marketing Services*
3. The Marketing Mix and the Product Life Cycle
4. Relationship and Loyalty Marketing

II. Integration
5. The Marketing Plan

III. Strategic Marketing
6. Strategic Marketing
7. Competition and the Marketing Environment

IV. The Marketplace
8. Understanding Individual Customers
9. Understanding Organizational Customers
10. Understanding Tourism Markets

V. Functional Strategies
11. Advertising, Merchandising, and Public Relations
12. Personal Selling and Sales Promotions
13. Differentiation, Segmentation, and Target Marketing
14. Market Positioning and Branding
15. The Pricing Decision
16. Channels of Distribution

Ivan Artolli

General Manager, Hotel Amigo, Brussels, Belgium

Ivan Artolli began his hospitality career as a bellman in the Adriatico Palace Hotel. Over the years he has held many positions at several hotels all over Europe, such as assistant front office manager of the Hotel de Paris, public relations manager at Countess I. Batthyany, and director of sales and marketing at two hotels. Ivan is currently the general manager of Hotel Amigo in Belgium, where he has worked for almost six years.

 ## Marketing in Action

Ivan Artolli

General Manager, Hotel Amigo, Brussels, Belgium

What does it take to deliver a consistent level of service at a five-star hotel?

This is really a double question. What is top service at a five-star hotel, and how do you make it consistent? There are several keys elements that contribute to a top performance.

The distinguished friendliness of your team is one of the key elements—your team's ability to communicate openness, welcome your guests cheerfully with a sincere smile, and adopt the proper body language. Your team's relevance to your guest is another key element. The interests of every one of your front line team members need to be tuned to your guest's interests, wants, and needs. Be knowledgeable and prepared for what you are going to be asked. This produces reassurance and positive attitudes in staff members' minds and contributes to their capability to charm your clients. Guests can tell by your employees' enthusiasm if they have experienced the various recommendations they are suggesting to them.

Genuineness is yet another key element. If your team adopts your service standards and honestly believes in their values, your perceived realness will elevate you in the eyes of your guests. If you and your team share these values, your service will be perceived to be at the highest level and you will be consistent in its delivery.

How do you create a service-oriented culture among your employees?

You need to be an inspirational leader and turn your department or hotel into an inspiring place to work. It has been proven that employees feel that they work for people before working for a company. You need to share a positive attitude, courage, enthusiasm, and passion with your staff. After a while this situation generates positive energy and motivates all employees to excel in their jobs. Success generates enthusiasm, which becomes highly contagious and trickles down to all levels of your team.

Services are intangible. How do you make the services tangible at your property?

When it comes to five-star service, the client's expectations and perceptions are the only reality that matters. A competitive edge is created when a hotel takes steps to assess how clients perceive the quality of its service standards and acts to improve upon those perceptions with the objective to exceed clients' expectations and create "customer delight." Measuring the perception of service quality is a relatively easy undertaking thanks to companies specialized in measuring service standards using mystery visits and first-hand experience. This is standard procedure for a company such as Leading Hotels of the World or groups such as Four Seasons or Rocco Forte Collection.

Guest comments and guest questionnaires are also very effective tools to determine in which areas you need to focus and what additional training your team might require.

Delivering good service to your clients is also based on the type of services you deliver to your team members. Measuring your staff satisfaction is also a powerful management tool, that you should not underestimate when you are looking at delivering an outstanding and consistent level of five-star service.

What recommendations can you make for students who want to be general managers some day?

I have already come across a few stars in recent years. It is relatively simple to recognize the potential of young people when they join a well-established hotel for a training course. You will notice that they perform the simplest tasks with a smile and lots of care and attention. They are sensitive to the outside world, they have a logical approach, they are eager to please, and nothing is too much to be asking for. They treat everyone with utmost respect, clients or colleagues. Last but not least, they are passionate and ready to take as much guidance as you can give.

What kind of recommendations can you make for students who want to get into hotel sales? That is, what do you look for when you hire salespeople?

There are very few born salespeople. But a great number of people are very good at interacting with other people. They are good listeners and encourage others to tell them more. A natural smile, interest, enthusiasm, and curiosity also can be found in many students. A student with a sales attitude will prepare for a meeting by preparing a number of questions he or she will want to ask you. To be asked about your impression of the meeting is also a significant sign and a positive indication. Trying to close the sale is a challenging level for a student to reach, but it is as simple as asking for that job. *Used by permission from Ivan Artolli.*

Today, people working in either the service industry or knowledge work (e.g., accountants, lawyers, and teachers) account for more than three fourths (75 percent) of the workforce in all developed countries—and their share is increasing. In 1955 these people represented less than one third of the work force. Today, in the United States the service sector generates over three fourths of the **gross domestic product (GDP).**

gross domestic product (GDP) The market value of all final goods and services produced within a country in a given period of time. The most common approach to measuring and understanding GDP is the expenditure method:
GDP = consumption + investment + government spending + (exports − imports)

The hospitality industry is part of the service industry, and therefore, the marketing of services must be compared and contrasted with the marketing of goods or products. The basic concept of marketing will not change, however; it remains the fulfillment of consumers' needs and wants, regardless of the industry. A number of aspects of the service component of the hospitality product affect the marketing effort, however. One of these is that buyers may not be aware of their needs; thus, they do not notice their need until it is missing. This factor places a burden on the marketer to anticipate buyers' needs. There is really only one way to do this, and that is by putting yourself in the customer's shoes and thinking like a customer. Exhibit 2-1, "An Orange Juice, Coffee and Danish Cart" demonstrates this point.

It was argued in Chapter 1 that every act of management in the hospitality industry is also an act of marketing. When this notion is contrasted with the management of a manufacturing plant and the marketing of the goods produced by that plant, it can be seen that many differences exist between the two types of industries. These differences, and the differences between services marketing and goods marketing, are discussed next.

services The nonphysical and intangible aspects of a product that management does, or should, control.

goods The physical factors of a product over which management has direct, or almost direct, control.

intangibility The attributes of services that the customer cannot grasp with any of the five senses; that is, customers cannot taste, feel, see, smell, or hear a service until they have consumed it. The intangible aspects of a service product are difficult to grasp conceptually prior to purchase.

SERVICES VERSUS GOODS

The important differences between **services** and manufactured **goods** are **intangibility, perishability, heterogeneity,** and **inseparability of production and consumption.** It is important to understand these differences because they have major implications for marketing practice. We discuss each in detail next.

Exhibit 2-1   An Orange Juice, Coffee, and Danish Cart

Mike Leven is vice chairman of the Marcus Foundation. He is former chairman and CEO of US Franchise Systems (Microtel Inns & Suites and Hawthorn Suites). He tells a story demonstrating the need to put oneself in the customer's shoes by recounting an incident that took place when he was president of Americana Hotels.

I was in the lobby of our Jamaica property with some other of our executives about 6:00 A.M. one morning, waiting for the limo to take us to the airport. Also waiting for the same limo to catch the same flight were half a dozen guests who had just checked out. Suddenly, down the hallway came the F&B manager and a waiter wheeling a cart with fresh orange juice, coffee, and

Danish. I thought to myself, "Now here's a management that knows how to take care of its customers," until I watched them wheel the cart right past the guests and in front of us. I immediately grabbed the cart, turned it around, and wheeled it back in front of the guests. I couldn't help but think, "If only we treated our customers as well as we treat ourselves."

The absence of the coffee cart might have had absolutely no effect on the guests' perception of the hotel's service quality. After all, that is not what they came to Jamaica to buy. On the other hand, the presence of the coffee cart could have considerable impact, because the customers would leave with a warm feeling about the service quality delivered by the hotel.

Used by permission from Mike Leven.

Intangibility

Intangibility refers to the fact that there are parts of the service experience that the customer cannot grasp with any of the five senses (tasting, feeling, seeing, smelling, and hearing) without first purchasing the product. Services are experienced, rather than possessed. Unlike a car, there is no passing of the title of ownership when a service is purchased. Buyers leave the transaction empty-handed. They do not, however, go away *empty-headed*. They have memories of the experience that they can recall and share with associates and friends.

The fact that services are essentially experiences has major implications for consumers. Experiences cannot be easily replicated, as you probably know by your own life. Think of the most wonderful evening at a restaurant or movie with a significant other. Now, answer the following questions: Did you try to repeat that experience? Was it exactly the same as the first experience?' The second experience may have been better or worse, but it was not the same. Now think of visiting a Best Buy. Envision looking at two IPods that are the same model and same color. Does one feel different from the other? The answer is probably no.

The inability to touch, see, feel, or hear the service without purchasing it suggests that customers are buying based on what they think they might receive; in other words, they are buying the service based on their expectations, which are formed in part by word of mouth from their friends and advertisements from the firm. Next we discuss how expectations are formed.

Customers' expectations are formed in a variety of ways. When buyers have not purchased the service previously, they may rely on similar experiences to create their expectations. For a customer who always stays at a Ritz-Carlton, the expectations of a first visit to a Four Seasons will be based on what he has received at the Ritz-Carlton.

A second way expectations are formed is through traditional marketing communication methods—advertising, public relations, and promotional events. Contemporary methods include all of these plus online (Internet) marketing. Hospitality companies use these methods in many cases, but there are inherent problems—it is not easy to advertise or sell an intangible service. You can use words, but often these are as abstract as the service itself and may serve only to compound the intangibility (e.g., the finest, the ultimate), or you can use tangible clues, sometimes called "tangibilizing the intangible." One example of tangible clues is shown in Exhibit 2-2. Here, the picture of the airplane in the ballroom helps show how big the ballroom is. Without the airplane, it would be very hard to determine the size of the ballroom by looking at a picture of it.

A third way expectations are set is to ask friends and relatives about their experiences. This is why word of mouth is so important. Research has shown that, excluding firsthand experience, buyers of services assign greater credibility to word of mouth than any other source of information. To create positive word of mouth, a hospitality organization must create positive experiences for customers. Thus, it is clear that one of the most important elements of marketing a service is the consistent delivery against customers' expectations. Customers who experience poor room service may not complain only about room service; they are as likely to complain about the total service in that hotel. The same analogy can be made for restaurants. Research also shows that people who

perishability The life cycle of the hospitality service. A guest room has a life cycle of 24 hours. A luncheon meal at a restaurant has a life cycle of about one to two hours midday.

heterogeneity The inconsistent delivery of service levels provided by different employees and affected by different types of customers.

inseparability of production and consumption Consumption of the service while it is being produced. This concept reflects the notion that the consumer is part of the "assembly line." In other words, an empty guest room produces nothing.

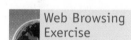

Web Browsing Exercise

Browse "Hotel and Conference Center" on a search engine. What are some of the similarities offered by these properties? What are some of the differences? Are there differences between branded and unbranded properties? What are the different types of management structures you see? What features seem to be promoted? Are they tangible or intangible features? How do they convey the intangible side of the business?

Exhibit 2-2 This advertisement of an airplane in the Mirage tangibilizes the intangible

Source: MGM Mirage. Used by permission.

are happy with an experience tell about 14 people. However, people who have had a bad experiences tell many more people. Word of mouth is therefore a double-edged sword.

Finally, customer expectations may be set by previous purchases of the service. If we have a wonderful time on the first visit to a hotel, we expect that the next experience will be equally wonderful. The first service experience creates expectations for future experiences. All of these factors increase the risk for customers.

The intangibility of services also creates challenges for marketers. It is not easy to display and communicate intangible services. Marketers must convince prospective buyers that they offer the right solution to buyers' needs or problems, while at the same time not overpromising. Practitioners need to be careful about promising what they cannot deliver. A Radisson hotel may advertise itself as a "hotel and conference center." Similarly, a dedicated conference center such as Arrowwood in Westchester County, New York, also advertises itself as a conference center. Yet, a conference center that is dedicated solely to conferences is a very different structure from one that is part of a hotel. A customer who has just held a meeting at the Arrowwood Conference Center would be very disappointed at the Radisson version of a conference center. Conversely, the Radisson customer would presumably be overwhelmed by the cost of holding a similar meeting at the Arrowwood Conference Center. Buyers who have only the advertising or promise of the seller on which to rely may be buying something completely different from their expectations. The visit to the Radisson and the Arrowwood conference centers will be very different intangible experiences, just as guests will have very different experiences at any hotel or restaurant. Thus, hotels often advertise the intangible as shown in Exhibit 2-3. This advertisement, while stating the specifics of the convention center, is selling the intangibles of the area.

Business centers represent another service offered to hotel customers with varying degrees of comparability. Hotel A may have a state-of-the-art business center with

With a setting this spectacular,
who needs a motivational speaker?

From enchanting Gulf Coast views and cascading fountains to a beautiful European spa
and a peaceful garden solarium, sources of inspiration abound at Beau Rivage.

Plan your meeting online at beaurivage.com
or call 1-800-239-2771.

A FEELING LIKE NO OTHER
Beau Rivage
RESORT & CASINO • BILOXI

AN MGM⊙MIRAGE RESORT

Exhibit 2-3   This advertisement for Beau Rivage, while stating the specifics of the convention center, is selling the intangibles of the area

Source: MGM Mirage. Used by permission.

secretarial help, fax machines, computers, copiers, and so on. Hotel B's business center may be the administrative assistant to the general manager. Both advertise "business centers," or tangible goods, as an amenity to their guests, but they offer very different levels of service and intangible experiences.

Terms such as *conference center, fitness center,* and *business center* create perceptions that may be far from reality. Good marketing seeks to equate perception with reality because, as we will see later, perception *is* reality for the consumer.

Tourism Marketing Application

Expedition Tourism Travel to Unusual Places

Lindblad Expeditions is a tour company that specializes in expedition travel. Expedition travel focuses on visiting out-of-the-way places not visited by many tourists (for example, the North Pole). Lindblad's expedition leaders include naturalists, historians, and undersea specialists.

In addition, they have partnered with National Geographic's ship the *Endeavor*. Here, the physical product is the ship on which they are sailing, the food they eat, the maps and guide books they receive, and the like. The service product and the service delivery are the expedition leaders that supply the knowledge to educate guests on what they are seeing. The service environment includes the classrooms where the lectures take place, as well as the unusual destinations they visit.

Perishability

The second primary characteristic of services is their perishability. Consider an airline seat or a hotel room; if a seat is not sold on a particular flight or a room is not sold on a particular night, the opportunity to sell it is gone forever. The seat or hotel room cannot be stored and sold the next day. It is therefore the task of marketing management to create demand for these rooms or seats so that there is always 100 percent occupancy.

The issue of perishability is compounded by the fact that most services have fixed capacity. There are only a certain number of seats on a plane and only a certain number of rooms in a hotel. Similarly, those selling professional services (e.g., attorneys and doctors) have only a limited amount of time in which to perform their services.

One implication of perishability and fixed capacity is that it can affect the quality of service the consumer receives. A front desk clerk may be swamped during a busy check-in period and therefore be incapable of providing good service to all customers. An hour later, however, he may be standing around with little to do. When a guest checks in at this time, she may be treated to exceptional service, or she may be treated with indifference because the front desk clerk is trying to catch up on paperwork. These fluctuations in demand can cause uneven service quality. The same is true, of course, in restaurants where the kitchen may be so backed up that service is very slow.

Perishability and fixed capacity also affect product availability. Hotels, airlines, and restaurants are often unable to increase capacity in the short run. Not having enough rooms at certain times presents problems, and so does having too many rooms vacant. The challenge, then, is to "manage" both demand and capacity. "Seat sales" for airlines, and weekend package and off-season rates for hotels, are methods used to do this. This means that those customers who might not normally be able to afford a given service may be in the position to purchase the service at a lower promotional rate if they are willing to change the time they visit. While marketers often discount the price during slow times, they also increase prices in times of heavy demand. In other words, the challenge for the marketer is to balance the demand and the capacity as much as possible.

Bars and restaurants also experience peak demand. A software company has created a system that enables bars to change the prices of beer depending on how many people are standing at the bar. When lots of people are standing at the bar, the price of beer goes up;

Tourism Marketing Application

An article in *USA Today* described how it is easy to manipulate pictures. For example, it explains how a wide-angle lens can be used to make rooms look larger and describes airbrushing in an ocean view where none exists.

Source: www. usatoday.com/travel/hotels/hotsheet/2005-05-24-column_x. htm (accessed March 12, 2006).

when not many people are at the bar, the price goes down (see Exhibit 2-4). The early bird specials offered by restaurants are an attempt to move demand away from the peak demand of dinner times and increase demand during traditionally slow times.[1]

Heterogeneity

The third primary characteristic of services is their heterogeneity. Whereas goods can be manufactured so that each product is the same (that is, there is homogeneity), services are not always similar. Here we are not referring to a lack of service caused by insufficient staff; instead, we are referring to changes in service that are a result of differences among employees and among customers themselves, as well as customers' perceptions of these differences.

On any given day, a hotel guest may come in contact with as many as a dozen or more employees. The switchboard operator, guest service agent, restaurant manager, and housekeeper may all have some contact with the guest. To have all these contacts take place without incident is a challenge. Jan Carlzon, formerly CEO of the Scandinavian airline SAS, called each of these employees and customer contacts a "**moment of truth.**" Carlzon estimated that each of SAS's 10 million customers came in contact with approximately five SAS employees for an average of 15 seconds on each trip. This

moment of truth
Anytime an employee has contact with a customer; that is, when the service product meets the service delivery.

Exhibit 2-4   The Beer Stock Exchange

WHAT IS A BEER STOCK EXCHANGE?

Beer Stock Exchange is the name of a cool new German party idea. Based on the system of a stock exchange, Beer Stock Exchange has been developed to add interactive entertainment in small bistros, public houses, clubs, and discotheques of all sizes. The prices of drinks and other consumables float up and down automatically based on actual sales and reflecting customer demand. Customers are informed of current prices through multi-media display—TV monitors or projector screens.

HOW DOES A BEER STOCK EXCHANGE EVENT WORK

There are no given prices for drinks in a Beer Stock Exchange party location—instead prices vary according to visitors' demand for drinks. Increasing demand means rising prices; decreasing demand means falling prices.

WHY SHOULD I SET UP A BEER STOCK EXCHANGE?

Beer Stock Exchange is an excellent marketing and promotional tool as well as an enhancement to the atmosphere in any bar, pub, club, or disco. The system may be used in a number of ways: as a regular feature to attract more customers on otherwise quiet nights, for short periods to add real 'buzz' to a busy evening, or even as a main theme for locations designed around the Stock Exchange concept.

WHY DO CUSTOMERS LOVE A BEER STOCK EXCHANGE?

Surely most of your customers have never been before to a venue having set up something similar to our 'Beer Stock Exchange.' This new kind of party idea creates—in contrast to normal parties—a very special atmosphere. Customers will be curiously watching the changing prices, discussing with their friends about the best moment to buy, and will be really satisfied when they achieved to buy their drinks for a great price. It's just a lot of fun!

THE STOCK MARKET CRASH

Something very special about our 'Beer Stock Exchange' software is the possibility to trigger a Stock Market Crash. This means that all prices will be reset to their minimum for a predetermined period of time. Of course your guests will love it—everyone will be trying to get some drinks for little money. . . .

WHAT DO I NEED TO SET UP A BEER STOCK EXCHANGE?

You just need a Pentium 200 class PC with a TV-out graphic-card, a videorecorder, and some TV screens and/or projection screen and of course our 'Beer Stock Exchange' software.

WHY DO YOU CALL IT BEER STOCK EXCHANGE?

In the beginning our software was developed in German language only and was called 'Bierbörse.' We decided to make it international, translated 'Bierbörse' into English, and got 'Beer Stock Exchange.' That's it. . . .

More questions? E-mail: info@partyboerse.net

Source: The Beer Stock Exchange. Retrieved February 22, 2004, from http://english.partyboerse.net/ bierboerse_allgemeines.html. Used by permission.

yielded, he estimated, a total of 50 million contacts a year. "These 50 million 'moments of truth' are the moments that ultimately determine whether SAS will succeed or fail as a company."[2]

Consistency of service is very difficult because of the interactive nature of providing a service. Manuals may describe exactly what every employee in a large restaurant is supposed to do in any given situation, but they can never predict what various individuals with various backgrounds, various orientations, and various personalities will actually do in a given situation.

Not only are employees different, but customers may also be different. Consider, for example, an elderly gentleman at the front desk of a hotel who needs considerable help in understanding where things are in the hotel, how to work the electronic key in his door, and how to get assistance when he needs it. The service-oriented guest service agent patiently and graciously explains these things to the man, who will depart the hotel to tell all his friends how nice the employees are. Unfortunately, the woman waiting in line behind this man may be a businessperson anxious to get to a meeting. As a frequent traveler, this woman knows her way around and only wants to register, get her key, and be on her way. She may depart from the hotel and tell her friends about the wait in line and the slow service.

The knowledge and experience of the customer will also affect how the customer feels about the purchase. One customer says, "Look at the full glass of wine they give you." Another says, "Don't they know that a glass of wine should never be more than half full?" One customer says, "That bellman was very helpful when he took me to my room and explained all the hotel's services." Another says, "The last thing I need is someone to carry my bag to my room the last 50 yards of my 2,000-mile journey." Obviously, there is a wide variation in consumers' opinion of quality, and what satisfies one may very well not satisfy another.

In a later chapter we discuss market segmentation, which is used to group these different customers into smaller groups that are more similar. The dissimilarity of the service experience means that one cannot be 100 percent confident that the experience the firm promotes is what will actually be produced.

Many organizations have attempted to overcome the differences in both employees and customers through the use of self-service technologies. Gas stations used to be known as service stations. Today, most of us have forgotten the time when customers never pumped their own gas, but instead a service employee pumped it for them. (New Jersey state law requires, however, that employees pump gas, not customers!)

Exhibit 2-5 shows an example of a self-service kiosk by Technology Portals, Inc., that was developed for the state of Pennsylvania to use at welcome centers for tourists visiting the Philadelphia area. This new system allows the experienced traveler to avoid the line at the help desk altogether and at the same time frees up space at the help desk for those who really want to talk to someone.

The question for marketers is, When is less service more, and vice versa? Or, to put it another way, When should service be more personal, and when should it be less personal? Following the lead of the banking industry, airlines began to install kiosks in terminals for customers to get their boarding passes, get upgrades, change seats, and do everything a live person would do for them at the ticket counter. These machines are

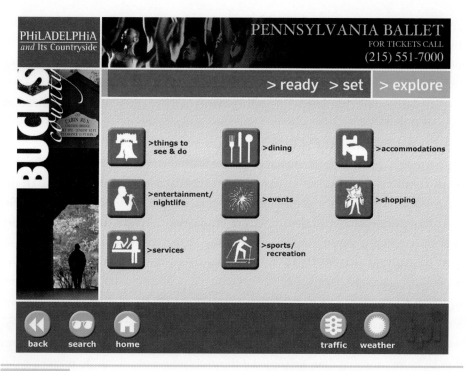

Exhibit 2-5   This advertisement shows an example of a self-service kiosk by Technology Portals, Inc.

Source: Technology Portals. Used by permission.

faster than, and as reliable as, their human counterparts. In this case, less personal service is better!

It is not hard to see that marketers face many problems when trying to cope with the differences (heterogeneity) of services, or even when trying to provide solutions to consumers' problems. The answer, at least partially, lies in knowing your market and your customers well so you can customize the desired services in a manner that is consistent with their expectations. This means that the emphasis should be on understanding the customer, not on the service.[3] We will address this subject in more detail in later chapters.

Inseparability of Production and Consumption

The fourth primary characteristic of services is the fact that services are consumed at the same time they are purchased. Unlike goods, services are consumed in the same place and/or time as they are produced. When the customer orders a beer, the bartender immediately produces it from the tap. In contrast, a car may be manufactured in Ohio and sold in Phoenix. With services, the customer is also part of the production process. If the customer does not order the beer, it is not produced. If the customer is mean, the employee may take a little longer to "produce" the beer from the tap. With goods, the customer is not involved in the production.

The fact that production and consumption occur at the same time emphasizes the idea that *management is marketing in the hospitality industry*. Because the buyer must be present to experience (consume) the service provided (produced) by the seller, the "moment of truth" between the buyer and seller may in fact be bigger than the service itself. Therefore, it is critical that service personnel be trained to produce the service that meets or exceeds the customer's expectations, and of course, solves his problem(s). Traditionally this training would be considered the role of human resources.

In the hospitality industry, the entire product (service and goods) is consumed on premise with the seller (producer) on hand. One individual can represent the service of a particular establishment and cause a customer not to return. Each employee is literally part of the product because each employee is producing the service while the customer consumes. The efficient service of a very friendly desk clerk can be ruined if the room service waiter is late with the order or the housekeeper in the hallway is unfriendly.

Another aspect of the inseparability of production and consumption is that every time we purchase a service, regardless of how many times we have made the purchase, there is always the possibility of a completely new experience—despite any attempt to make each experience the same.

For the "moments of truth" to result in a satisfied experience for guests, employees must embrace a marketing philosophy. Employees may be given "smile training," but still be restricted in their ability to solve consumers' problems. Smiles in these cases may be more annoying than welcomed if the problem is not solved.[*] To overcome this problem, some hotels empower their employees to take appropriate action on the spot and advertise this as a point of differentiation. Exhibit 2-6 illustrates this point.

There are numerous examples of firms that do not instill the marketing philosophy in their staff. These firms are easy to spot, as many rules for using the service are operations driven, not customer driven. Examples include, "Sorry, that section is closed," or "It's not that waiter's turn," or "The pool doesn't open until 10:00," or, on a menu, "No substitutes allowed." Rules such as these become part of the service because they restrict the consumer's consumption.

The Hospitality Product

hospitality product
The combination of goods, services, environment, and experience that the hospitality customer buys.

Tying together the preceding information on service, we have come up with the following description of the **hospitality product:**

> The hospitality product is something that prospective buyers, for the most part, cannot experience before buying. Buyers do not know whether they will get what they think they are buying. After they buy, they must wait until the seller produces it before they can consume it, sometimes after having paid for it. It may be available today at $50 but not be available tomorrow at $150. The seller is not totally sure that he or she can produce it, and the buyer must consume it contemporaneously according to the seller's rules. What's more, if buyers don't like what they get, they can't take it back, exchange it, or in most cases, get their money back.

---

[*]Consider an actual experience: A room service waiter in a top hotel arrived an hour and 15 minutes late with breakfast, after three follow-up phone calls by the guest who was about to miss an appointment, with a large smile and "How are you today?" but no explanation or regrets.

> # "I would
> ## have missed
> ### the big match, if
> ### the quick-thinking
> Marriott receptionist hadn't
> picked up the ball and run with it.
> It was the day of the crucial Australia-New Zealand
> clash, and I was most disappointed to discover that it
> clashed with my business meeting. I happened to mention this in passing to the
> receptionist. On my return, I was amazed to find that she had gone to the
> trouble of ringing her brother, getting him to tape the match and bring it
> to the hotel for me. Marriott calls this Empowerment. It means they encourage
> all their staff to take this sort of initiative on behalf of their guests.
> As far as I am concerned, it is certainly a winning approach.
>
> *ALWAYS IN THE RIGHT PLACE AT THE RIGHT TIME.* **Marriott.**
> HOTELS · RESORTS · SUITES
> Over 275 locations worldwide, for reservations call free on 0800-22 12 22 or contact your travel agent.

Exhibit 2-6 This advertisement shows Marriott empowering employees

Source: Marriott International, Inc. Used by permission.

Exhibit 2-7 Differences between Services and Manufactured Goods

Functional Characteristics	Manufactured Goods	Services
Unit definition	Precise	General
Ability to measure	Objective	Subjective
Creation	Manufactured	Delivered
Distribution	Separated from production	Same as production
Communication	Tangible	Intangible
Pricing	Cost basis	Limited cost basis
Flexibility of producer	Limited	Broad
Time interval	Months to years	Simultaneous or shortly after
Delivery	Consistent	Variable
Shelf life	Days to years	Zero
Customer perception	Standardized—what you see	Have to consume to evaluate
Marketing	Traditional, external	Nontraditional, largely internal

Although this characterization may appear to be a little extreme, perhaps we can better understand the challenge of hospitality marketing if we consider the product in this light.

Exhibit 2-7 summarizes the differences between services and manufactured goods.

COMPONENTS OF THE HOSPITALITY PRODUCT

There are four major components that customers receive when purchasing and using a hospitality product—the physical product, the service environment, the service product, and the service delivery.[4] These combine to form the hospitality experience. Although consumers of the hospitality product may not always seek an experience, per se, that is what they come away with and remember most. Consider the value of experience from the viewpoint of a coffee bean. It begins life as a *commodity*. At about three dollars a pound, it translates to 3 or 4 cents a cup. Someone grinds and packages it, turning it into a *good,* costing 5 to 25 cents a cup. Brewed and served in a diner, it becomes a *service* at perhaps 1 dollar per cup. Then we add the *environment* of, say, an upscale restaurant or an espresso bar, and consumers gladly pay 2 to 5 dollars a cup.

So coffee is a commodity and a good. It is purchased as a service in an environment. But how can one place charge more than another, up to $20? Because each adds a distinctive *experience*. Experiences are a distinct economic offering, as different from services as services are from goods.[5] Examples of the four components in four different industries—a restaurant, a casino and a hotel—are shown in Exhibit 2-8.

How are experiences different from goods and services? Experiences are *memorable,* experiences *unfold over a period of time*, and experiences are *naturally personal.* Thus, experiences can create new and greater economic value. They are not merely entertainment, such as the experiences one finds at theme restaurants like Planet Hollywood, where food is just a prop for what has become known as an "eatertainment" experience. Rather, they *engage* the customer, connecting with him in a personal, memorable way. Effective service providers use experiences to increase the attractiveness of their offering—to bring customers back to the same service. As services increasingly become copied, successful hospitality operators create memorable experiences to acquire and keep customers. Components of the hospitality product are shown in Exhibit 2-9.

Physical Product

The physical product is the tangible component of the service. It is mostly the physical product over which management has direct, or almost direct, control. It is management's decisions or practices that directly affect the physical product. In some cases management expertise determines the quality level of the product, as in the case of a

Exhibit 2-8	Four Components of a Service			
Industry	Physical Product	Service Product	Service Environment	Service Delivery
Full-service five-star restaurant	Food served	Plan for how order is to be taken by wait staff	Use of pressed and starched tablecloths and fine china and silverware	How the waiter actually takes the order
Casino	Game of roulette	Procedures for dealing the game	Atmospherics of the casino	Friendliness and competency of the dealer
Hotel	Firmness of the mattress	Procedures for turn-down service	Colors and decor of the room	Attitude of service personnel

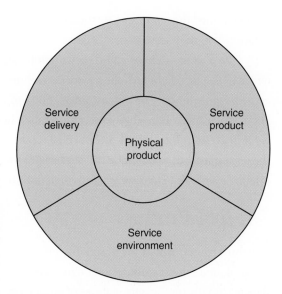

Exhibit 2-9   Components of the hospitality product

Source: Rust, R. T., & Oliver, R. L. (1993). Service quality: Insights and managerial implication from the frontier. *Service Quality: New Directions in Theory and Practice.* Copyright 1993 by Sage Publications. Adapted by permission of Sage Publications, Inc.

chef. Alternatively, the quality of the product may depend on management's willingness to spend or not spend money in pursuit of the target market it wishes to serve. In this category we place beds, food, room size, furnishings, bathroom amenities, elevator service, heating and air-conditioning, TVs, things that don't work, and so forth. We also define price as tangible, although it is a cost of service as well as a good because it tangibilizes the intangible. To the consumer, price is very tangible in any purchase decision. In hospitality the physical components, generally speaking, satisfy or don't satisfy the basic needs of customers, as illustrated in Chapter 1. They provide solutions to basic problems.

Service Environment

The service environment, also known as the "servicescape," is defined as the "physical environment in which the service is delivered."[6] The servicescape at check-in may be the new business desks with one clerk at each. Because services are intangible, customers often look to the servicescape for clues on how to behave.

In the category of service environment, we place environmental items that may or may not be tangible, although they are something the customer feels. That feeling is central to the hospitality product that is being marketed. For example, putting electronic locks on bedroom doors is something very physical and tangible, but we do not sell the electronic lock to the customer. What we sell, instead, is the benefit of the feature—a feeling of security, a very important factor for many hotel customers (i.e., the customer needs a room with a lock, but wants a room with an electronic lock for an enhanced sense of security).

The three distinct elements that make up the service environment are the surrounding area, the layout, and signs and symbols. Such things as lighting, background music, decor, architecture, and so forth, create the surrounding area. The spatial layout can also be used to create a certain atmosphere. For instance, one casino in Las Vegas purposely made the walkways in the casino narrow so customers could rub shoulders with each other. Such closeness is believed to make people feel that they are part of something special. David Kranes, who has written extensively about the spatial layouts of casinos, stated that as consumers we are "seeking to inhabit space which empowers us, which feels most rewarding, most secure, most natural, and most intimate." He further argued that emotions customers in a casino might feel, such as chaos or disorientation, will only serve to drive people away.[7]

The feeling of chaos and disorientation can occur because of a lack of signage, which is the third element of the service environment with which marketers must be concerned. Signs not only direct customers, but also inform customers about procedures. For example, one of the authors recalls standing in line for about 15 minutes waiting to get into a restaurant only to learn when he reached the hostess that he was not in line for the restaurant, but to purchase the restaurant's merchandise. There were no lines to get into the restaurant. There were also no signs pointing the customer in the right direction.

Service Product

The third component of the hospitality product is the service product. It includes non-physical, intangible attributes that management should control. Items in this category depend heavily on the personalities of employees, such as friendliness, speed, attitude, professionalism, responsiveness, and so on. This is the core performance or service purchased by the patron. The service product defines how the service works in theory, which is why we often say "plan your work." The idea is that management develops and plans all the actions employees will undertake when hosting guests. For instance, consider the following scripts employees may be told to use when talking with guests:

■ Hostess:
   "We will be asking you throughout your visit how we can do things better. Please be aware that our goal is to provide a wonderful dining experience; if we fall short of that goal, please do not hesitate to tell us."
■ Wait person:
   "We have great desserts here. A woman named Cynthia makes them locally. Cynthia has lived in the area for ages and follows a family recipe."

   About Our Fish: *As you may know, one should not eat oysters in months that have an R. Therefore, we will not be serving oysters tonight as we only serve the freshest fish here."*

Service Delivery

The final component of the hospitality product is the service delivery. This refers to what happens when the customer actually consumes the service. We refer to this as "working your plan"; that is, performing all the steps that were developed before the guest arrived. The moment of truth is when the service product meets the service delivery.

Interrelationships of the Components of the Hospitality Product

Now that all the components have been discussed, we can use room service to demonstrate the complexity of the interrelationships among the four components of the hospitality product. Management must first decide to offer room service. Obviously, this decision depends on many things including the particular property and the target market. The first question to be answered, of course, is whether offering room service will be desired by customers at this property. If management decides that the answer is yes, it will then analyze demand, cost, resources, and facilities. If customers expect room service and it is not offered, there will be dissatisfaction. The decision to offer room service is not the end. There are still many opportunities to fulfill or not fulfill expectations. First, there is the service product element.

- How many times should the phone ring before the room service department answers it?
- What questions are asked by the room service attendant?
- When the meal is finished and the tray is put out in the hallway, how long does it stay there before someone takes it away?

Next is the service delivery part of room service.

- What is the attitude of the person who answers the room service call?
- Is it delivered when promised?
- What is the attitude of the room service waiter?
- Did he remember the rolls, the sugar, and enough cream for the coffee?

Now, let's look at the goods element.

- Is the orange juice fresh and the coffee hot?
- Is the silverware clean? Is the bacon burned or the toast soggy?
- Is the price fair?

Finally, there is the service environment.

- Is there a table on which to place the food without rearranging the bedroom, perhaps even a balcony where one can enjoy the view?
- Are there chairs to sit on that enable one to reach the table?
- Is the tray well presented? Is the table set with a tablecloth and fresh flowers?

If all four components of the service are executed well (and remember, it is the target market that determines what "well" is), does the customer say, "Boy, this is a well-managed hotel"? Probably not. But if one thing is not executed well, she may well say the opposite. Why should this be? Because it is *expected* to be executed well. That is the solution to the customer's problem. That is how the customer measures the price–value relationship. This is why the "risk" was taken. *All* these lead to the level of guest satisfaction. There is no opportunity to return room service for another room service in the same way a good can be exchanged. You can see now that room service, like every other aspect of hospitality operations, is *marketing*—it solves or causes problems, and it can keep or lose a customer.

SERVICE QUALITY AND SERVICE GAPS

Service gaps can occur when expectations do not meet reality for the customer. Exhibit 2-10 is a model that illustrates potential gaps in service delivery and where they may occur. This model is based on the premise that customers' evaluation of a service purchase (e.g., their satisfaction) is determined by how well the actual purchase experience compares to what they expected. The service experience often does not exactly meet expectations; it may exceed or fall below expectations. The range in which the service experience falls below expectations but customers are still satisfied is known as the **zone of tolerance.** We discuss what leads to the zone of tolerance after we examine each of the potential service gaps.

There are five gaps identified in the gap model of service quality. Gaps are sometimes measured using the **SERVQUAL model.** The role of management is to understand how the five gaps are created and then develop strategies to close them. Following are descriptions of each of these gaps.

Gap 1: Gap between services expected by the customer and management's perceptions of customers' expectations.

This gap refers to the difference between what the company thinks customers' wants and needs are and the actual wants and needs of the customers. This gap occurs because there is no communication between management and customers and between contact employees and managers. This gap is usually the result of not enough customer or employee research. Many companies do not ask their customers what their problems, wants, needs, or expectations are. If customers do communicate their wants and needs to employees in general conversation and there is no mechanism for these comments to get passed up "the chain of command," or if management never asks employees what they are hearing, comments will remain with the employee and be forgotten. This often happens with complaints. Marketing needs to think of complaints as "the customer telling us what he wants." When the customer speaks, management needs to want to hear what employees are being told.

Gap 2: Gap between management's perceptions of customer expectations and service quality specifications.

This gap occurs when the service provider fails to design service procedures to meet the expectations of guests. An example is the promise of early check-in. One of the authors of this text was arriving in Chicago at 6:18 A. M. after an overnight flight from the West Coast. When booking the room, he asked the hotel if he would be able to check in early and was told no that to do so he should have booked a room for the night before. When he asked why he was not allowed to check in early, he was told that the housekeeping staff does not start work in time to allow early check-ins. Clearly, many rooms were checked out early for other customers to make their flights in the morning, but the housekeeping staff arrived at 8 A.M. sharp. This is an example of an "operationally driven mentality," which is knowing what guests want, but not providing it because operational concerns take priority over guests' concerns.

Gap 2 also occurs because the reward system for employees is designed in conflict with guests' needs. For instance, managers who are rewarded based on meeting their budget will be less likely to go over budget to satisfy a guest than will managers who are rewarded based on overall guest satisfaction. A final reason for Gap 2 is a poor service design. When one of the authors asked a waiter why it took so long to get served, the waiter replied, "We closed the kitchen next to this dining area to save costs, and now I have to walk all the way to the main kitchen, which is on the other side of the building."

Gap 3: Gap between service quality specifications and service delivery.

A gap here suggests that what was planned by management was not carried out by the staff. One of the main reasons for this gap is a poor

---

service gaps Areas in service delivery that are inadequate, missing, or poorly executed and result in customer dissatisfaction.

zone of tolerance The area between desired service and adequate service that the customer will tolerate, even if not totally satisfied.

SERVQUAL model A model developed to help identify potential gaps in service delivery that could or should be corrected by management.

 Web Browsing Exercise

Look up SERVQUAL on a search engine. Develop a brief analysis of how it is used in different industries. What are the similarities in SERVQUAL use between the hospitality industry and other industries? What are the differences?

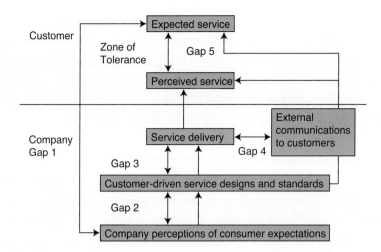

Customer

Zone of Tolerance

Gap 5

Expected service

Perceived service

Company
Gap 1

Service delivery

External communications to customers

Gap 4

Gap 3

Customer-driven service designs and standards

Gap 2

Company perceptions of consumer expectations

Exhibit 2-10 The SERVQUAL model illustrates potential gaps that may occur during service delivery

Source: Adapted form Zeithaml, V. A., & Bitner, M. J. (2003). *Services marketing* (3rd ed., international). Boston: McGraw-Hill Education.

human resources policy. A poor policy leads to the hiring of "warm bodies," as opposed to the right person for the job. It also leads to an unwillingness to properly train employees and poor scheduling of employees. If too many employees are on duty, service can suffer because every employee thinks it is someone else's job to wait on the customer. If too few employees are working relative to the demand, then service can also suffer.

Gap 4: Gap between the service delivered and the service promised.

This gap occurs because the communications mix presented to customers exceeds what the provider can actually deliver. Advertising is frequently to blame. To make the service appealing, hotel companies often include the picture of their largest and nicest room, even though many of the rooms in the hotel look nothing like the room in the brochure. Resorts also tend to show very attractive men and women sitting by the pool, even though the actual guests may look quite different! Finally, hotel salespeople are also known for "saying anything" to get the business, only to have to work out the impact of the business on the hotel at a later date.

Tourism Marketing Application

Smartertravel.com offers advice on top destinations that provide great value during the off-season. In 2004–2005, these destinations included Bermuda, Napa Valley, Santa Fe, Tahiti, and Sweden. This website is an example of a firm that spreads by word of mouth and helps reduce the risk of the consumer purchase.

Source: Retrieved December 1, 2006, from www.smartertravel.com/advice/advfeatures/advice.php?id=10946.

Gap 5: Gap between the perceived service and actual service. Expected service is affected by consumers' past experiences, the unspoken promise (e.g., the promise made because it is a hotel with a specific brand name), exposure to communications (e.g., advertising and promotion), and word-of-mouth recommendations. Thus, consumers have a certain level of expectations even before they arrive at a hotel or restaurant. We are most concerned with Gap 5 because this gap is the result of all the others. In other words, any gaps in the other four measures will likely cause a Gap 5. This gap is hard to manage because customers may not be aware of the fact that they are getting excellent service. This is because much of the service experience is intangible. Certainly customers can determine quality from the tangible parts of the service (e.g., the thread count of the sheets on the bed), but determining quality from the service experience—the interaction with an employee—is much more difficult. Luckily, an understanding of the dimensions of service quality enables marketers to constantly remind customers of the excellent service they are receiving. The dimensions of service quality should be incorporated into each of the four components of the service product. This is shown in Exhibit 2-8.

Exhibit 2-11 summarizes the gaps and suggests ways to close them, and Exhibit 2-12 introduces the RATER system in four components.

Dimensions of Service Quality

The dimensions of service quality are reliability, assurance, tangibility, empathy, and responsiveness. These dimensions lead to the acronym RATER. All dimensions together result in the total experience the customer takes away. Clearly, a hospitality mission statement that contains these elements and has them fulfilled by all employees would be a long way on the road to keeping customers.

Although the definitions of these words are self-explanatory, the key is to incorporate each of the words in the service experience to remind the customer that he is receiving great service. The way to do this is as follows:

1. Develop a list of each interaction the guest has with a hotel representative for each "service component" being investigated. The representative does not have to be an employee. In the case of in-room dining, the first contact is most likely an object—the menu. The last interaction is mostly likely an employee—the in-room dining staff collecting the used dishes. Two lists of encounters must be developed: one from the employee point of view and a second from the customer point of view.
2. Combine the preceding lists as if each list were one side of a zipper. Find out where both lists overlap and then put them together. The result should be a complete list or script of the total service encounter from both the guest's and hotel representative's perspectives. This list should be in chronological order.
3. Examine each action and determine what, if any, RATER component can be applied.
4. Apply the RATER component and develop procedures to ensure that the addition of a RATER component is not a one-time activity, but something that occurs on a regular basis.

Exhibit 2-13 provides an example of using the RATER components for in-room dining.

Exhibit 2-11   The Gap Model of Service Quality

Gap	Why Gap Exists	How to Close Gap
Gap 1: *Gap between services expected by the customer and management's perceptions of customers' expectations*	1. Inadequate marketing research 2. Lack of upward communication 3. Inadequate service recovery processes	1. Undertake formal marketing research. 2. Undertake informal research (managing by walkng around). 3. Train staff to understand that "complaints tell us what the customers' needs are": create a culture where complaints are welcome.
Gap 2: *Gap between management's perceptions of customer expectations and service quality specifications*	1. Poor service design 2. Absence of customer-defined standards	1. Incorporate customer satisfaction as part of employee rewards. 2. Develop service blueprints.*
Gap 3: *Gap between service quality specifications and service delivery*	1. Deficiencies in human resources policies 2. Failure to match supply and demand 3. Problems with service intermediaries	1. Develop a strategic approach to hiring. 2. Develop people: provide needed equipment so employees can do their jobs. 3. Manage intermediaries.
Gap 4: *Gap between the service delivered and the service promised*	1. Lack of integrated marketing communication 2. Overpromising 3. Inadequate horizontal communications	1. Integrate marketing communication. 2. Undersand exactly what can and cannot be done by the organization. 3. Organize meetings between departments.
Gap 5: *Gap between the perceived service and actual service*	1. Not closing any of Gaps 1–4 2. Not incorporating the RATER system when interacting with guests	1. Incorporate RATER in all interactions.

* A service blueprint is defined as a picture or a map that visually displays the service by simultaneously depicting the process of the service delivery. It does this by breaking down the service into logical components and steps from both the employees' and the customers' perspectives.

Source: Adapted from Zeithaml, V., & Bither, M. J. (2003). *Services Marketing* (3rd ed., International). Boston: McGraw-Hill Education.

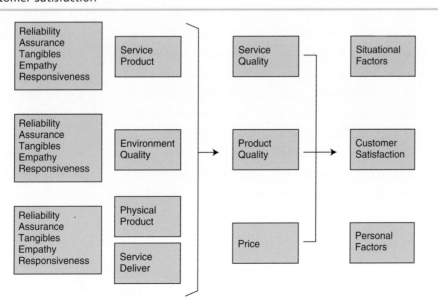

| Exhibit 2-13 | Incorporating the Dimensions of Service Quality into In-Room Dining |

RELIABILITY

To demonstrate reliability, a firm must deliver on its promise. The promise must be made explicit to the guest, as the guest may not know the promise without being told. For in-room dining the promise is delivering the order in a stated amount of time. When the guest places the order, the employee receiving the order should state the current time of the order and then tell the guest the exact time the order will be delivered. Clearly, it is important to then deliver the items at the stated time. Delivering the items early is not always recommended unless the guest gives permission to do so, as the guest may be inconvenienced by the early, arrival of the order.

ASSURANCE

The goal of assurance is to remove the anxiety associated with the fact that both consumption and production occur simultaneously. If the in-room dining waiter brings the wrong meal, the only choice for the guest is to eat the wrong meal or wait for a new order to arrive. Employees can provide assurance to customers by repeating the order at the time the order is taken and then again when the order is delivered. Reviewing the order at the time of delivery removes the guest's worry that something will be wrong only to be discovered once the waiter has left the room.

TANGIBILITY

Because services are intangible, firms must attempt to make them tangible. This is done by the use of tangible cues. For instance, consider the menu for in-room dining. This first exposure to in-room dining provides the first tangible cue of what the service may be like. Consider the impact on customers' expectations of their in-room dining experience of a marked up and poorly designed dirty menu versus that of a clean and well-designed menu. In seminars with industry executives, many admitted that it had been a long time since they had examined the in-room dining menus in their properties. If you are currently working in hotel, ask to examine the in-room dining menu in the guest rooms.

EMPATHY

Empathy refers to the ability to understand the thoughts, feelings, and experiences of another person and to be able to communicate these thoughts and feelings to that person in such a way that a sense of trust and comfort is created. To be empathetic, one needs to listen. When ordering, in-room dining customers very often mention foods to which they are allergic or foods they particularly like. The employee taking the order needs to remember these comments and warn customers of potentially problematic foods or inform customers of foods they may particularly enjoy. The distinction between an "order taker" and "salesperson" illustrates this point in a simple but compelling way: One who lacks empathy is an order taker; one who has empathy is a salesperson.

RESPONSIVENESS

The final component of service quality is responsiveness. This refers to how well the organization is able to take care of customers' needs. The organization's policies and procedures should be designed to improve the customer's visit. Often, policies and procedures are designed for the organization's benefit. One of the authors has stayed in hotels in which an item that is sold in the dining room is not available on the in-room dining menu because of operational issues. Such hotels are not being very responsive to customers needs.

Source: Shoemaker, S. (2003, September). Loyalty 360: Adding value through quality: Improving the value component by reinforcing the quality image. *EHLITE magazine 1* (4). Ecole Hôtelière de Lausanne Institute of Technology & Entrepreneurship.

Zone of Tolerance

The zone of tolerance fills the gap between expected and actual service. The smaller the gap, the closer is the expected to the perceived. Note, however, that there is a "desired" service level, which may be less than expected, and an "adequate" service level, which may be just acceptable. Between the two is the zone in which customers may tolerate the level of service but not be particularly thrilled by it. The service cannot be returned, so although the customer may be somewhat satisfied, he will not be totally satisfied. Thus, he may not complain, but he also may never return.

Overall, these gap models are valuable tools in tracing customer satisfaction and dissatisfaction. Gap models can be helpful in identifying where problems exist. Management can come up with ideas to "fix the gap" in their service delivery.

Chapter Summary

The world's economy can be broadly divided into manufacturing and service sectors. The hospitality industry, the service sector, has grown to become three times larger than its manufacturing counterpart in many countries. In this chapter we argued that the fundamental differences between services and manufactured goods necessitate different approaches to their marketing. Services are essentially intangible, heterogeneous, and perishable. In addition, services are produced and consumed at the same time. Most important, consumption of the hospitality product results in a total experience, the only thing the customer takes away.

The components of services, and the fact that management and marketing are inseparable in hospitality, create unique challenges when it comes to acquiring and keeping customers. This is because most of what hospitality products do for the customer takes place at the property. The successful management of those challenges requires a further extension of marketing, which is often referred to as relationship marketing, the subject of Chapter 4.

Key Terms

goods, p. 32

gross domestic product (GDP), p. 32

heterogeneity, p. 33

hospitality product, p. 40

inseparability of production and consumption, p. 33

intangibility, p. 32

moment of truth, p. 37

perishability, p. 33

service gaps, p. 46

services, p. 32

SERVQUAL model, p. 46

zone of tolerance, p. 46

Discussion Questions

1. How are services different from goods? Briefly describe each of the four ways services differ from goods.

2. Why are hospitality and tourism considered part of the service industry? (Isn't a hamburger purchased at a quick-service restaurant simply a hamburger?)

3. Briefly describe the four basic components of the hospitality product.

4. What is meant by service gaps?

5. Explain the meaning of the gap model of service quality, which is sometimes referred to as the SERVQUAL model. How does this model help the marketer in hospitality and tourism?

Endnotes

1. See Kimes, S. E., Chase, R. B., Choi, S., Lee, P. Y., & Ngonzi, E. N. (1992). Restaurant revenue management: Applying yield management to the restaurant industry. *Cornell Hotel and Restaurant Administration Quarterly, 39* (3), 32–39.

2. Carlzon, J. (1987). *Moments of truth.* Cambridge, MA: Ballinger.

3. See Lewis, R. C., & Nightingale, M. (1991). Targeting service to your customer. *Cornell Hotel and Restaurant Administration Quarterly, 32* (2), 12–27.

4. Shoemaker, S. (1996), Managing service quality. In V. Eade (Ed.), *Casino management* (pp. 73–104). Las Vegas: UNLV International Gaming Institute.

5. Pine, J. B., II, & Gilmore, J. H. (1997, August 4). How to profit from experience. *The Wall Street Journal*, p. A24.

6. Zeithaml, V., & Bitner, M. J. (1996). *Services marketing.* New York: McGraw-Hill, 518.

7. Kranes, D. (1995). Play grounds. *Journal of Gambling Studies. 11* (1), 91–102.

.... case study

Little Things Mean a Lot

Discussion Questions

1. What service gaps can you identify in this case study?
2. What are your thoughts about the management?
3. Do you think Joan and her husband Jim will return to the Meadow Lodge Resort based on the general manager's response to their concerns about the hotel?
4. If you were Joan and Jim, would you return?

"Eureka!" cried Jim Jackson to his wife Joan. "We won a free weekend from the drawing at that party we went to last week! It's a weekend at the Meadow Lodge Resort (a property of a major upscale hotel chain), with deluxe accommodations in a Signature Service room. We also get a free Sunday brunch. All we have to do is call the concierge. Let's go next week; I need a break."

The Meadow Lodge Resort was not far from where Jim and Joan Jackson lived. Next to it was a modern athletic center with, among other things, four tennis courts. Tennis was the Jackson's favorite sport. The resort also featured indoor and outdoor swimming pools and other health center amenities. The Jacksons had often talked about going there for a weekend just to get away from the "rat race."

FRIDAY NIGHT

The Jacksons arrived at Meadow Lodge about 5:00 P.M. on a Friday afternoon in early June. They planned to play some tennis, do a little swimming, take a whirlpool and sauna, drink wine at the outside pool patio cafe, catch up on some reading, watch a recent movie, and, in general, just relax. They arrived in their room full of anticipation.

"Uh, oh," said Jim as they entered the room. "There's only one comfortable chair in the room for reading." He called housekeeping and another one was delivered promptly. There was still, however, only one reading lamp. He tried to position the desk lamp for the other chair, but it didn't provide enough light. Also, there wasn't any table for either chair on which to put a drink or a coffee cup. "So much for reading," he said as he positioned the desk chair so at least one of them had a "table."

The Jacksons proceeded to unpack and put their clothes in the two dresser drawers. It wasn't that easy because the drawers had no runners and kept going askew and falling out. "I wonder where they got this furniture," said Joan. Jim picked up the TV program to see if he could catch the French Open tennis tournament. "This TV program starts tomorrow," he said. "They've already replaced this week's." He called housekeeping again. The same program was delivered. "I think I know where there's this week's," said the houseman, who quickly went and got one. "Let's play tennis," said Jim. After tennis, they decided to have a drink on the outdoor pool patio, but unexpectedly found it closed.

When they returned to their room, the phone rang. It was a friend of the Jacksons who lived nearby. They had agreed to meet him for a drink at 8:00. "Meet you in the piano lounge," said Jim. "The room directory says it's open until 10:00; it should be nice and quiet." They went downstairs, met their friend, and went into the piano lounge, which was part of the lobby but sectioned off with comfortable chairs and couches. There was no one else there. As they sat down, the piano player said, "The bartender just went home. If you go to the bar and ask, they'll bring you a drink."

The three got up and went to the bar. "We can't stay here," said Jim. There was a TV blaring some nonsense in one corner and a large-screen TV doing the same in the center, although there was only one couple in the room, who was not watching it. They tried to find a corner to get away from the TV, but there was no escaping. They asked the bartender to serve them in the piano lounge. He said he couldn't do that but offered to give them drinks to take out themselves. Jim signed the check, and as they walked out, they picked up some munchies on a table by the bar. "Do you want that charged to your room?" said the bartender. "We thought they were free snacks," they said. "That's okay," said the bartender, "Go ahead and take them and I won't charge you." They went back to the lounge to have their drinks and enjoy the music, but the piano player left at 9:00, about the time they got there.

At 9:45 the Jacksons' friend left and they headed for the casual restaurant, which closed at 10:00. The hostess greeted them with, "We close in 15 minutes." "Does that mean you'll throw us out at 10:00?" Joan asked. "No, but we close in 15 minutes," was the response as she led them to a table, repeating the admonishment one more time. There was only a party of two in the restaurant that looked as though it could seat 150. The hostess seated them in a dark corner close to the kitchen entrance. "Do you have another table?" they asked and were led to a table near the entrance. They ordered a light meal and finished it promptly.

The Jacksons went back to their room and turned on a pay movie. It was too fuzzy to watch so they turned it off and went to sleep. The next morning the desk clerk removed the charge from their bill and told them if it happened again to call maintenance.

SATURDAY

The Jacksons got up at 7:00 A.M. and made some coffee in the in-room coffee maker. They found orange juice and cream in the minibar but it was so warm they were hesitant to use it. "Let's go for a swim," said Joan. "Check first to see if it's open," said Jim. Joan looked in the room directory and saw that the pool opened at 7:00. "But it's closed from 8:00 to

9:00," she said. "What's the point in going, it's already after 8:00. I wonder why they would do that with all these people in the hotel on weekend packages?" They decided to spend the day driving around the countryside instead.

Back in their room that evening, the Jacksons had a glass of wine and tried to decide where to go for dinner. Not wanting to get back in the car and drive again, they decided to go back to the same restaurant. This time they went earlier, about 8:00. There was a party of six and a deuce in the dining room. They were seated in the same dark corner next to the party of six and they once again asked to move. They were moved again near the door and right in front of where the buffet was being set up for Sunday brunch. "What's the use?" they said. The hostess, a different one, overheard them and came back to ask if they would like to move. "Yes," they said. "We'd like to sit near a window." The hostess went and talked to the waitress, who, shortly after, came to take their order. "The hostess said we could move." "Well," was the reply, "she talked to me, but I'm the only one on tonight and that section is closed." The pained expression on the Jacksons' faces must have moved her because she finally said, "Well, all right," and took them to a window table. She wasn't too pleasant, however, during the rest of the meal. The Jacksons finished dinner and went to their room, where they finally found a movie to watch that was clear.

SUNDAY

On Sunday the Jacksons slept late. They packed their bags before going downstairs to have their free Sunday brunch in the same dining room where they had been twice for dinner. This time there were about 20 people in the room. They were led to the same dark corner table. As they muttered, the waitress overheard them. "Would you like to move?" she asked. "We'd really like to sit by the window on such a beautiful day," they replied. "Fine," she said, "but I'll have to set up the table because that section is closed." "Never mind," replied the Jacksons, "we'll stay here." The brunch was delightful.

After brunch the Jacksons checked out. "Well, that's the last time for that place," said Jim as they pulled out of the driveway, "even if it was free." "Not really," said Joan, as she looked at the bill. "We spent over $200. Besides, it wasn't so bad. It's a nice hotel, and the people were certainly nice and responsive when you asked." "I know," said Jim, "but it's the little things that make the difference."

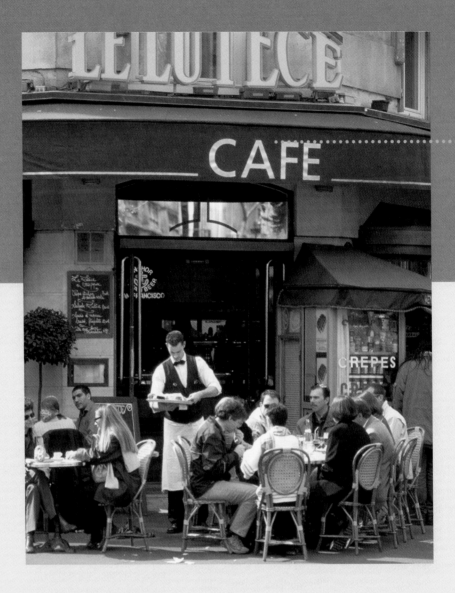

overview

In Chapter 2 we discussed the differences between goods marketing and services marketing. In this chapter we introduce the concept of the hospitality marketing mix. We first discuss the original marketing mix and then introduce the components that have been added to account for the nature of the service business. We end the chapter by examining how products and services reinvent themselves over time.

chapter 3

The Marketing Mix and the Product Life Cycle

learning objectives

After reading this chapter, you should be able to:

1. Discuss the basics of the marketing mix and explain how to apply this mix to parts of the hospitality business.

2. Explain why the traditional marketing mix does not completely apply to the hospitality business.

3. List the various stages of the product and service life cycles. Consumers are always changing, and the products they use also change.

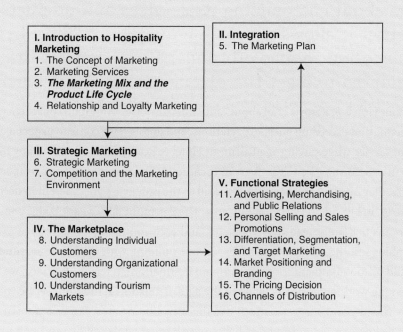

I. Introduction to Hospitality Marketing
1. The Concept of Marketing
2. Marketing Services
3. ***The Marketing Mix and the Product Life Cycle***
4. Relationship and Loyalty Marketing

II. Integration
5. The Marketing Plan

III. Strategic Marketing
6. Strategic Marketing
7. Competition and the Marketing Environment

IV. The Marketplace
8. Understanding Individual Customers
9. Understanding Organizational Customers
10. Understanding Tourism Markets

V. Functional Strategies
11. Advertising, Merchandising, and Public Relations
12. Personal Selling and Sales Promotions
13. Differentiation, Segmentation, and Target Marketing
14. Market Positioning and Branding
15. The Pricing Decision
16. Channels of Distribution

Nicole Tabori
Food and Beverage Manager, The Ritz-Carlton, St. Louis

Currently, Nicole works as the F&B manager for The Ritz-Carlton in St. Louis. Prior to that, she was the restaurants manager for the Fairmont Jasper Park Lodge in Jasper, Alberta, Canada. Nicole has also worked as an assistant food service manager for Sodexho at John Abbott College in Montreal, where she was responsible for overall operations management. In addition, she has worked at some of the world's finest hotels and restaurants, including the Ritz-Carlton Montreal and the Beau-Rivage Palace and the Bleu Lézard Restaurant, both located in Lausanne, Switzerland. Nicole graduated from the L'Ecole Hôtelière de Lausanne in 2003 and holds a Bachelor of Science Degree from the International Hospitality Management Program. She is proficient in six languages (English, French, Hebrew, Spanish, Italian, and German) and is a seasoned world traveler.

 ## Marketing in Action
....

Nicole Tabori
Restaurant Manager, The Ritz-Carlton, St. Louis
....

Could you talk about how your hotel/restaurant has gone about customizing its services and products or whether it allows for some regional flexibility in its product and service offerings? How has your firm attempted to meet the diverse needs of customers?

Fairmont Hotels & Resorts is all about "authentically local." Although consistency is key in meeting and exceeding guest expectations, and guests have come to expect a certain level of service while staying at a Fairmont Hotel & Resort, it is still a company that believes in promoting its environment in each hotel. For example, at their properties in Alberta, they offer AAA Alberta Beef, and all of the properties are different in decor and infrastructure to suit the surrounding area. The Fairmont Jasper Park Lodge consists of 446 rooms that are spread over 900 acres, and each room is housed in its own cabin. Log cabins suit the environment of Jasper; one building with all the rooms situated in it would be too dominant for this mountain destination. In terms of service, we have a very young workforce, mostly between the ages of 18 and 24; thus, our service offering is very friendly and energetic, much like the surrounding area. Jasper is a town where everyone knows everyone and says hello to each other while walking through town.

In terms of meeting the diverse needs of our customers, we begin with the basics in our operations; all guests want to be treated with respect and integrity. They want bread offered with their meals and cutlery to eat with. We then build from this into a more personalized approach. Fairmont Hotels & Resorts created the Fairmont President's Club for this exact reason so that we have profiles of our guests with their likes and dislikes clearly noted. With this system any hotel can add preferences to a guest's profile, and any hotel can access the information for guests staying in the hotel. Guests can also access their profiles and add or remove information. This is how we know that Mr. X, for example prefers his martinis made with a certain type of vodka or that he has two children five and seven years old and that their names are Susie and Jon. The company also puts importance on using the guest's names instead of just Sir or Madam, which goes a long way in making them feel that they are special and appreciated.

How do restaurants match their services to their products?

I believe that each restaurant has its own technique for matching its service to its product, but that they are closely tied if you want your restaurant to be successful. One needs to understand that what you sell in a restaurant is not the food or the service; it is the overall experience. A guest needs to leave your restaurant feeling excellent about the experience. I believe that we sell experiences, not food, but that it is the quality of the food and service that make up that experience. If your restaurant has a more casual ambiance, then the food and service need to match; food needs to be more of the "meat and potatoes," and service needs to be friendly and warm. If your restaurant is fine dining, then this needs to be reflected in the food and service as well; service needs to be more formal and the food more upscale. For example, John Smith taking his family out on a Wednesday night for pizza wants an amazing pizza and efficient, friendly service. However, if John takes his wife to a fine dining restaurant for their anniversary dinner on Saturday night, he may want to experience foie gras and have privacy.

For students interested in entering the restaurant industry, what sort of educational and work background do you think would be most useful?

I think that an education in hospitality management or restaurant management is key, but that you need a balance between education and work experience. To be successful, one needs to understand the operational aspects of running a restaurant, but also be able to understand financials and how one relates to the other. I have seen many people come up the ranks from busperson to server to manager and be very successful, and others who have not. I have also seen people with great educational backgrounds fail because they did not understand what running a restaurant is all about. It is important to understand a dining experience from a guest's perspective and know what your guest's expectations are. Passion is key as the hours are long. While most people are vacationing and relaxing, a hotelier is working. People with passion, ambition, and a willingness to learn can go far.

Used by permission from Nicole Tabouri.

ORIGINAL MARKETING MIX: THE FOUR Ps

four Ps A term applied to a common marketing mix for goods: product, price, place, and promotion.

The marketing mix was originally developed by Professor Neil Borden of Harvard in what has come to be known as the **four Ps.**[1] Borden originally developed six elements— product planning, pricing, distribution, promotion, servicing, and marketing research; McCarthy later reduced these to four elements—product, price, place (distribution), and promotion.[2] Many generations of students learned the four Ps as price, product, promotion, and placement.

The four Ps were mainly developed for the marketing of goods. Consistent with our previous arguments in Chapter 2, the marketing of hospitality services is different from the marketing of goods and therefore requires a different approach to the marketing mix. We will add to these four to account for the nature of hospitality; that is, its intangibility, inseparability of production and consumption, and the difference that can occur in the service delivery due to the nature of the person providing the service. To show the difference between goods and services, we use the term *hospitality marketing mix*, which was first developed by Renaghan.[3]

hospitality marketing mix The mix of marketing activities that are directed toward an identified target market. The elements of the marketing mix include the product/service mix, the presentation mix, the pricing mix, the communications mix, the distribution mix, the people mix, and the process mix. The hospitality marketing mix is sometimes referred to as the seven Ps: product, price, place, promotion, process, people, and physical attributes.

product/service mix The combination of products and services, whether for free or for sale, that are aimed at the needs of the target market.

HOSPITALITY MARKETING MIX: THE SEVEN Ps

The **hospitality marketing mix** includes all the marketing activities undertaken by the firm to understand customers' wants, needs, and problems. The firm then develops products and services to fill these wants and needs. It also includes all other marketing activities directed to the target audience. Current generations of students are learning this new seven P concept—product, physical attributes, price, promotion, placement, people, and process.

The first element in the hospitality marketing mix is the **product/service mix**, which is aimed at satisfying the needs of the target market. For example, the product/service mix of a Four Season's hotel is luxurious accommodations and a high staff-to-guest ratio that ensures top-level service. In contrast, the product/service mix for Formule1 hotels (with rooms designed to sell for $35) is a very small staff-to-guest ratio to keep costs low. The product/service mix is discussed in detail later in this chapter.

presentation mix All of the elements used by the firm to increase the tangibility of the product/service mix in the perception of the target market.

The second element is the **presentation mix**. This represents all of the elements used by the hospitality firm to increase the tangibility of the product/service mix. The Heavenly Bed by Westin is part of the presentation mix, as are the flowers that are often found in the lobby of many hotels. This mix is discussed in Chapters 11, 12, and 13.

pricing mix The combination of prices that consumers are offered to purchase a product or service.

The third element is the **pricing mix.** This element is the combination of prices that customers pay for products or services. We say "combination of prices" because a firm never offers just one price. Prices change by type of room, date of arrival, length of stay, whether the customer is a loyal customer, or whether the customer is part of a group. This mix is the subject of Chapter 15.

communications (promotion) mix All communications between the firm and the target market that increase the tangibility of the product/service mix, that establish or monitor consumer expectations, or that persuade or induce consumers to purchase a product or spread information on it by word of mouth.

The fourth element in the marketing mix is the **communications (promotion) mix.** The communications mix is all communications between the firm and the target market that increase the tangibility of the product or service. It also establishes or monitors consumer expectations, builds relationships, or persuades consumers to purchase. This area is similar to that of traditional marketing, although with some new twists because of the intangibility of the product. Except for these new twists, this part of the marketing mix is no different from the promotion element of the four Ps. The word *communication*, however, covers a larger area than the word *promotion*. In fact, we will show that promotion is only one division of communications. The communications mix is the subject of Chapters 11 and 12. It is often referred to as *promotion*, however, to keep with the notion of seven Ps.

The fifth element is the placement (distribution) mix. The **distribution mix** is all channels available between the firm and the target market that increase the probability of getting the customer to the product. This is how the customer can buy and use the services offered.

The general concept of services, as opposed to goods, is that rather than the good being taken to the customer (e.g., through retail outlets), the customer must come to the service. Thus, a hotel or restaurant chain that has 500 locations nationwide is "distributing" the product so that the customer can come to it.

The sixth element is the **people mix.** The people mix refers to the employees that work in the hospitality organization. Recall from Chapter 2 our discussion of the gaps that may occur during the service delivery. Employees are very involved in Gap 3. This gap occurs when the firm knows what customers want and has designed policies and procedures to deliver the expected service, but then, because of human resources policies, the firm does not have the people to deliver the service. We do not discuss this mix in this book. This mix is the subject of classes in human resources and organizational management.

The final element is the **process mix.** The process mix refers to the activities designed to deliver the desired services to the guest. In Chapter 2 we discussed the service product and the service delivery—two of the components of the hospitality product. The process determines how these two components affect the guest experience. For instance, one of the authors recently was on a cruise when a guest wanted to take his latte away from the coffee bar. He was told he was not allowed to do so because each of the cups cost $25 and management was afraid that the cups would never return. There were no paper cups at the coffee bar because they had not been ordered. This subject is addressed in classes on services management and design and layout of hospitality organizations. We do not go into detail about the process mix in this book.

All of the elements of the marketing mix directly affect the customer. The firm can decide what kind of products to offer (e.g., the kinds of hotels and restaurants it wants to build), the method of service delivery to offer (e.g., changes in the service delivery and service product), the magazines or radio stations to use to carry its message and of course, the prices to be charged. Finally, the firm also selects the distribution channels.

It is important to realize that all of the marketing mix activities will take place only after we have studied the external environment and learned the needs and wants of our customers. These two topics are discussed in Chapters 7, 8, 9, and 10.

THE 13 Cs

The trouble with the four Ps and seven Ps is that an organization that truly practices the marketing concept is involved in many activities beyond those presented. To account

distribution mix (placement) All channels available between the firm and the target market that increase the probability of getting the customer to the product.

people mix The people who work in an organization and how their attitudes, work ethic, and disposition affect the service delivery.

process mix The activities designed to deliver the desired services to the guest.

Exhibit 3-1	The 13 Cs of Marketing

- **Customer:** Everything the organization does should be designed for and with the customer.
- **Categories of Offerings:** Hotels offer many types of rooms; in addition, they can offer customers many ways to spend their money, whether it be at a spa, restaurants, or a lounge.
- **Capabilities of the Firm:** A company should focus on attracting those customers it is best able to serve.
- **Cost, Profitability, and Value:** It is important to examine the cost to serve every guest and focus on those that are the most profitable; for instance, if a hotel has one room available and two guests want the room, it should go to the guest who has spent and will continue to spend the most money with the hotel.
- **Control of Process:** The goal is to make the service process as homogeneous as possible and remove any variability that could interfere with guests' expectations.
- **Collaboration within the Firm:** Departments within an organization should work together to best take care of the customer. An example is housekeeping and front desk; another example is the chef and the wait staff. Traditionally, these different departments have had adversarial relationships because each has different goals (get people to their room as quickly as possible versus make sure all rooms are thoroughly cleaned and checked).
- **Customization:** Each product and service should be customized for the guest; in the simplest example, this involves ensuring a specific bed type and filling the minibar with products that the guest requests.
- **Communications:** This is more than advertising and promotion. It involves creating a two-way dialogue with the customer, which means that the timing of the communication, the information presented, and the form of the communication (e.g., e-mail, telephone, or mail) is determined by the customer.
- **Customer Measurement:** All customers are not equal; some are more valuable than others. In order to determine the value of the customer, it is necessary to collect data on the customer.
- **Customer Care:** Once the organization is designed around the customer's wants, needs, and problems, it is necessary to ensure that all effort is focused on taking care of the customer.
- **Chain of Relationships:** Organizations need to take a holistic view of the customer and at times may need to partner with another organization to best care for the customer. An example is hotels that partner with cab companies to take guests to and from the airport.
- **Capacity Control:** If a room is not sold on a given day, the revenue is lost forever. This is unlike products that if not sold today can be sold tomorrow. The challenge becomes managing capacity to maximize revenue.
- **Competition:** Because firms do not operate in a vacuum.

for this, one author suggested looking at the 11 Cs.[4] Sean Darlington of British Airways added a 12th C to this list—capacity control. We add another C—competition. The 13 Cs of marketing are shown in Exhibit 3-1. Notice that the 13 Cs define and articulate the marketing concept much better than the four Ps and seven Ps do. In fact, one of the Cs is customer. There is no direct mention of the customer in any of the Ps. This book is all about the customer. The typology of both the four Ps and seven Ps focuses on creating a product or service *for* the customer, whereas the 13 Cs suggest that the focus is on creating a product or service *with* the customer. Many of the 13 Cs are discussed in more detail later in this text.

HOSPITALITY PRODUCT/SERVICE MIX

The product/service mix is the combination of products and services that are offered to satisfy the needs of the target market. The product/service mix is what customers see, get, and experience when they go to a hotel, restaurant, or other hospitality entity. From here on we may use the word *product* as a generic term to describe product/service or the offering of a hospitality entity. We have covered this ground before, but repeat: "People don't buy quarter-inch drills; they buy quarter-inch holes." It is what the drills do, not what they are, that makes people buy them. Recall that we are in the business of solving consumers' problems. The products customers buy need to solve their problems or they will not buy the products again.

Designing the Hospitality Product

If a product is defined in terms of what it does for the customer, then it becomes immediately obvious that the design of a product begins with what the customer wants done. In the case of goods, that is often easy to determine. People who buy tires want safety and endurance. People who buy music want good sound reproduction. People who buy a Mercedes want status, comfort, and safety. But what do people want when they buy hotel rooms and restaurant meals? A comfortable bed and a good meal? Of course this is what they want, but we know that it goes far beyond those basic minimums. They also are looking for an experience.

As discussed, the product comprises the four elements discussed in Chapter 2—the physical product, the service product, service delivery, and the service environment. Consumers do not purchase individual elements (e.g., just the bed); rather, they purchase a bundle or unified whole. When buying the bed, they are also buying all the other features of the room as well as the check-in procedure, the environment of the surroundings, and the service staff. It is clear that a delicate balance exists in the mix and that management must be aware of how the various elements of the product or service interact. Every element is an important part of the product, and a change in one element can affect the perception of the entire product. If the Internet is not working in the guest room, then the guest is likely to be unhappy for the whole stay. Similarly, a very good meal can be ruined with a nasty waitperson. Thus, it is useful to break the bundle down into its component parts that management can help influence. This is referred to as the **bundle purchase concept.**

In Chapter 2, when we introduced the four components of the hospitality product, we treated the product or service as though it were just one product. The bundle purchase concept suggests that instead of having four components of one product (physical product, service product, service delivery, and service environment), we would have eight components for two products (e.g., the hotel room and the restaurant), 12 for three products (hotel room, restaurant, and spa), and 16 for four products (hotel room, restaurant, spa, and room service.) This helps explain why hospitality marketing is much more exciting than traditional goods marketing.

Bundle Purchase Concept

In the bundle purchase concept, the consumer is buying not one product, but many. As discussed, if the guest is buying a hotel that offers a restaurant, there are two products, each with its own set of physical product, service product, service delivery, and service environment. We can break the bundle purchase into different types of products: the formal product, the core product, and the augmented product. We discuss each of these types next.

Formal Product

The **formal product** can be defined as what customers think they are buying. This may be as simple as a bed or a meal, or it may be as hard to define as quality or elegance; it may be as intangible as environment or as specific as location. The formal product, in fact, might be defined as what the customer cannot easily articulate. Because of this, it is easy to be misled by what the customer says he wants.

bundle purchase concept When consumers buy a hospitality product, they are purchasing a bundle or a unified whole and not an individual element. A hotel guest room includes a bed, a bathroom, security, the check-in procedure, housekeeping services, and so on.

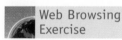
Web Browsing Exercise

Visit Hampton Inn's website (http://hamptoninn.hilton.com/). What type of bundle do they offer to their guests? What are its component parts?

formal product What the customer thinks she is buying

Hotel and restaurant customers frequently name location and good food as their primary reasons for choosing a particular hotel or restaurant. In many cases, however, this is only because location and good food are elements that easily come to mind. In fact, if these were really major reasons for choosing a hotel or restaurant, we could forget the bundle concept and concentrate only on location and good food. Concentrating only on these two items would be a serious mistake, however. For instance, the customer who goes to the latest "hip" nightclub is buying, in part, the belief that she may meet someone "hip." In another example, people buy Nike golf balls in the belief that they will be able to drive the ball like Tiger Woods—a Nike spokesperson.

Core Product

core product What the customer is really buying.

The **core product** is what the customer is really buying. It is the quarter-inch hole. For meeting planners, it is the fact that participants are able to bond and build trust with each other. It could also be the fact that a successful meeting will make them look good. For the restaurant patron it could be a quiet place to sign an important business deal. Understanding customers' problems is where product design should begin. Too often it begins, instead, with management's problems.

Consider, for example, the case of large banquet rooms that can be divided into smaller meeting rooms through the use of folding accordion doors. The invention of doors of this kind came about some years ago because of a critical management problem: how to accommodate both large and small groups in the same space. The solution solved the problem for management, but caused one for the customer. In many hotels, even today, one can sit in a small meeting room and listen not only to what is occurring in that room, but also to what is occurring in the rooms on either side, not to mention the banquet kitchen. This is a bothersome and ongoing problem for meeting groups. Today, better folding or collapsing doors almost eliminate this problem, but they can be found largely only in the newer or more recently renovated hotels.

The formal product—what customers think they are buying—is the meeting room and the seating capacity and a quiet, controlled, hassle-free, successful meeting. The formal product is the meeting space, and unless it meets a minimum standard, it will be unacceptable. The core product is how the entire team of the hotel deals with the meeting planner's problems and works in concert with the planner to come up with solutions to make that successful meeting happen.

Augmented Product

augmented product The total of all the benefits received or experienced by the customer.

The **augmented product** is the combination of all the benefits received or experienced by the customer. It is how the service delivery, the service product, and the service environment come together. It is also the way the customer uses the product. The augmented product may include both tangible and intangible attributes. These attributes range from the way things are done (service delivery), the assurance that they will be done (the service product), the timeliness, the personal treatment, and the no-hassle experience, to the size of the bath towels (the physical product), the cleanliness of the restrooms, the decor (the service environment), and the honored reservation.

The augmented product even includes the weather. As any resort manager can testify, there is nothing worse than three or four rainy days with all your guests locked inside on their vacation, or a ski area with unsuitable ski conditions. The frequent effect is that customers go away angry over something management can do nothing about. Or can it? For a marketing-oriented management the answer is yes. This is a customer problem that management should anticipate and for which it can prepare by developing alternative activities and "solutions." For example, when one of the authors was working at a summer resort in Vermont, there were three very rainy days. To entertain the guests on one of the days, a sail boat was put in the dining room and a shipwreck party was held. Guests enjoyed the party and forgot about the rain. On another day an outing to a museum was arranged, with guests of the resort getting a private tour.

The augmented product should solve all the customers' problems, and even some they haven't thought of yet. Hotels that supply different types of cell phone chargers are anticipating a customer problem before it may actually occur. In designing the product, it is critical to understand the augmented concept and how the different parts of the product help solve consumer problems. This is different from simply adding to the product for the sake of adding to the product. Mints on pillows don't make up for poor lighting. Elaborate bathroom amenities don't make up for a businessperson's having no place to write or work on her laptop, or for a couple not having two comfortable chairs in which to sit and relax.

The success of the all-suite hotel concept is based on the augmented product. This concept provides guests with a total living experience rather than simply meeting their basic needs. The success of McDonald's was based on the augmented product, which included, among other things, cleanliness and fast service. In fact, the success of any hospitality enterprise begins with an understanding of the core product and its augmentation to solve consumers' problems.

Take a look at the advertisement in Exhibit 3-2. What do you consider to be the formal, core, and augmented products?

COMPLEXITY OF THE PRODUCT/SERVICE MIX

Now that we understand what a product is and what it does, it should be easy enough to go out and design a hotel or restaurant that will solve consumers' problems. Things are not that simple. Obviously, we have many consumers with many different problems, and we can never hope to satisfy all the consumers or solve all the problems. We narrow the problem down, of course, by looking for specific types of customers (target) and grouping customers with similar wants, needs, and problems (segmentation), which is why these strategies are so critical to effective marketing. Even within these submarkets, however, we can never hope to be all things to all people. We address segmentation and target marketing in greater depth in Chapter 13.

It is clear, however, that we need to go beyond the basic and formal product when designing the hospitality product or service. Certain characteristics of products and services are taken for granted by consumers, especially those concerned with basic functional performance. If these are missing in a product, the user may well be upset.

Exhibit 3-2 What do you consider to be the formal, the core, and the augmented products for this advertisement of The Canyons?

Source: The Canyons. Used by permission.

But if they are present, the seller gets no special credit because, quite logically, every other seller is assumed to be offering the equivalent. TVs in hotel guest rooms are a good example. Wireless Internet access, not just cable hookup, is becoming this type of product. This is why we discussed the use of the RATER system in Chapter 2. Firms can use the RATER system to get credit for what they do normally; specifically, the RATER system reminds the guest of the quality of service being given by the service provider.

One thing is for sure—the product or service shouldn't *cause* problems. Consider, for example, the experiences of Stanley Turkel, a hotel consultant, described in Exhibit 3-3. Too often, designers, architects, or whomever, don't understand how people use a hotel guest room.

Today's hospitality customers are much more well traveled and sophisticated than those of previous generations. Thus, the basic functions served by a hotel are taken for granted. Customers expect a good location, clean room and bath, comfortable beds, and pleasant service from all hotels. However, customers look for other benefits that may be unique to a particular hotel, benefits that separate one hotel from others. For example, business travelers will look for a hotel that offers services that will increase their productivity such as wireless Internet access and a desk that is large enough to work on. A family on vacation will look for services that allow them to be together as a family but also perhaps to separate to enjoy their own activities, as advertised in Exhibit 3-4.

Web Browsing Exercise

Using your favorite web browser, type in "resorts with children's programs." What do you learn about the different resorts that offer kids' programs? What is the core product? The augmented product? Is their offering unique? Why or why not? Is it sustainable?

Standardization of Products

When we discuss hospitality products (formal, core, or augmented), we can think of them as being standard products, standard products with modification, or customized products. Standard products are important in services because production happens at the same time as consumption. Recall that in Chapter 2 we discussed how the firm may know what the customer wants, but may not be able to deliver on this because employees are different and as such operate in different ways. For example, different cooks prepare the same item differently. In fact, the same cook can prepare the item differently depending on the day. This is an obvious problem for the customer. One way to overcome the differences in preparation and help guarantee that the customer gets the same product across different purchase situations is to standardize the way an employee

---

Exhibit 3-3   User-Friendly Guest Rooms

Staying recently in a five-star, five-diamond hotel, the guest rooms were oversized and expensively decorated by a well-known hotel designer who, apparently, never stayed in a hotel room:

- The blackout draperies did not protect against the morning sunrise and allowed a "halo of light" to penetrate around the periphery of the draperies.
- The lamp on the light table did not encourage reading in bed. Its 65 watt bulb was inadequate even when the lamp shade was tilted.

- The guest room doors had no electronic door locks, but the old fashioned metal key and lock cylinder.
- A low-quality clock radio-alarm perched on the night table, but the variety and quality of music was nil.
- The television did not provide recent release movies.
- The desk chair and upholstered chair were too low to allow comfortable dining from the room service table.
- There was no telephone on the desk. One had to sit, unsupported, on the edge of the bed to make calls.

Source: Adapted with permission from Stanley Turkel, MHS, ISHC, a New York-based hotel consultant who can be reached at 917-628-8549. (stanturkel@aol.com).

Exhibit 3-4 Smugglers' Notch has been named America's number one family destination
Source: www.smuggs.com. Used by permission.

prepares the item. For instance, McDonald's standardizes its product and the method of producing the product so its hamburgers are all the same no matter where you purchase them.

standard products
Products that appear similar and standard to the customer, albeit in different locations, especially for a branded product.

Standard products have the advantage of closing Gap 4—delivering what is promised. Holiday Inns' original motels are an example of a successful standard product. No matter where in the country customers stayed at a Holiday Inn, they could just about find their way blindfolded to the front desk, their room, the lounge, or the dining room. Today, Microtel and the Accor properties Formule1 and Etap are prime examples of hotels that use standardization to protect quality. In addition, the fast-food chains such as McDonald's and Burger King are also prime examples.

A problem with standard products, however, and one that befell Holiday Inns for a while, is the emphasis placed on cost savings, which means that sometimes more expensive variations in the product in certain markets are ignored. Eventually this results in a loss of customers who either want something different or want a more modified or customized product. Even McDonald's, which has been successful with a highly standard product, allows its **franchisees** to make variations on the theme. The effect has been a major contribution to their success. In Canada, for example, McDonald's allows franchisees to incorporate a maple leaf (the national symbol) into the traditional golden arches. Canadian consumers feel that the restaurant has addressed their nationalist needs, while serving the same fare as the next town across the border in the United States.

franchisee An organization or person that purchases a brand name to distribute the product or service.

In the same vein, McDonald's in Japan added promotional products such as its Gratin Croquette Burger that appeal to Japanese tastes. In France, McDonald's added a hot ham-and-cheese sandwich dubbed the Croque McD, a fast-food version of its native version of the ham and cheese sandwich, called the Croque Monsieur.

Standard Products with Modifications

The **standard product with modifications** is a compromise between the standard product and the customized product. An example is the concierge floor of a hotel. In such cases, the scale economies of building and furnishing a standard room remain unchanged; the modifications are easily added to only those rooms requiring them, and an additional charge is often extracted for them.

> **standard product with modifications** A core product that has been embellished with new elements.

This strategy has one considerable advantage: The modifications, or added amenities, are sometimes easily added, removed, or changed as the market changes. Thus, the property maintains a flexibility that in itself may be perceived as a desirable attribute because the property can more easily meet customer requirements and encourage new uses of the product. Another advantage that accrues is differentiation within the product class, while maintaining the same strategic position.

In restaurants, one example of a standard product with modifications is different-size portions of menu items. Such a policy has a high level of flexibility as well as the ability to cater directly to changing market needs. (The popularity of "doggie bags" indicates that many people don't eat all they order or that the portion size was too big.) Burger King's "We do it your way" campaign had considerable impact on McDonald's method of doing it only their way.

Starbucks coffee shops are a very successful standard product with modifications. To see the different modifications, all one has to do is stand in line at a Starbucks and listen to how customers modify what they order. For example, a single-shot latte is usually made in a tall cup and comes with caffeinated coffee. However, many have been known to modify this to a Grande size cup with extra foam and decaffeinated coffee. Exhibit 3-5 describes how Starbucks added additional products and mechanisms to increase the sales of its product.

Customized Products

Customized products are based on the premise of designing the product to fit the specific needs of a particular target market, or even one individual's needs. Price may not be a large consideration for buyers of customized products, because they expect to

> **customized products** Products designed to fit the specific needs of a particular target market.

Exhibit 3-5	Starbucks: New Products

As competition increased, so too did the demand for new twists to classic products. Starbucks responded by introducing the iced Frappuccino blended beverages. These new products, some of which did not contain coffee, were a way to attract and lure in the anticoffee customer—the person who otherwise would not have come to Starbucks. By mixing up their product presentation, they began attracting a whole new kind of customer.

In an effort to extend its long-term growth, Starbucks did such things as form alliances with Breyer's Ice Cream and Coca-Cola to promote their product in new, innovative ways. They now have the best selling coffee ice cream in the United States. Starbucks also worked with the Jim Beam company to create a high-quality brand of Starbucks coffee liqueur.

pay a premium to have it exactly the way they want it. On the other hand, Marriott demonstrated with Courtyard a customized product for the price-conscious business traveler. Business traveler rooms with all the business amenities are examples of customized products that are now branded with names such as Smart Room (Sheraton), The Room that Works (Marriott), The Guest Office (Westin), Wyndham ByRequest (Wyndham), The Business Plan (Hyatt), The Business Class Room (Loews), and The Quiet Zone (Crowne Plaza).

Hotel bathrooms may be the latest version of customized products in the hotel industry. In the world's deluxe hotels, the bathroom frills once considered lavish (telephones, TV, marble) have now become commonplace. Designers are looking for new attractions. Drew Limsky wrote in *USA Today* that what brings consumers back to hotels are the bathrooms. He provided an example of the Maison Orleans located in New Orleans' French Quarter. The hotel has a three-foot-wide Jacuzzi bathtub that according to one guest "is so serene, so beautiful. You could party in there. There is nothing like it in the city." Limsky additionally quoted the editor of a business travel magazine who stated that hotel bathrooms "become a sanctuary, because most people don't have full marble baths or Bulgari products at home. It's very pampering."[5]

INTERNATIONAL PRODUCT OR SERVICE

The question of standardization versus customization is an even more complex one when dealing with international markets. "Think global, act local," say some. "Think local, act global," say others. Theodore Levitt, the noted retired Harvard Business School professor, says that high-touch products are as global as high-tech ones and that global organization should seek global standardization. The answer to the question of standardization versus customization, like all product decisions, should lie in the needs and wants of the marketplace and the amount of difference among the markets being served.

Web Browsing
Exercise

Compare and contrast the websites of Jollibee (www. jollibee.com.ph/default. htm) and McDonald's (www.mcdonalds.com/). How are they similar? How are they different? Does their commitment to standardization come through in their websites? Explain.

Jollibee, a burger chain in the Philippines, however, has taken advantage of McDonald's pattern of standardization. First, it borrowed every trick in McDonald's book from child-friendly spokes-characters to prime locations. Instead of selling a generic burger, however, it caters to a local penchant for sweet and spicy flavors, which it also puts in its fried chicken and spaghetti. It is the dominant chain in the Philippines with over 900 outlets. In 2004 Tony Tan Caktiong, the chairman of the organization, was named the Ernst and Young's 2004 World Entrepreneur of the Year.

Each international expansion requires designing the product line for the location and the markets to be served. Products such as hotels and restaurants may need to be adapted to different countries and cultures. The product goals for each country and market must be clearly outlined and related to the local situation as well as to the overall corporate goals. On the other hand, TGI Friday's adapted to French customs when it opened in Paris with a dismal performance. Research finally revealed, "If we're going to go to an American concept, we want it to be American."

> [The international] product decision must be made on the basis of careful analysis
> and review. The nature, depth, and breadth of the product line; the possibilities of

new product development and product innovation . . . the adaptation and customization of products to suit local conditions vis-à-vis standardization, . . . and a planned screening and elimination of unsuccessful products bear heavily on success in foreign markets.[6]

MAKING THE PRODUCT DECISION

Standardizing, modifying, and customizing are important marketing decisions in designing the hospitality product or service. Although the examples used here have been on a fairly large scale, these decisions also apply to all parts of the product. To illustrate this point, let's examine a relatively minor product decision in the light of the criteria that have been proposed.

If a restaurant has rack of lamb on its menu, or Dover sole, or Caesar salad, does it carve, fillet, or mix these in the kitchen or at tableside? To do it in the kitchen is to standardize it. This provides cost efficiencies, and presumably the finished product offered to the customer is identical to the one offered when the work is done at tableside.

The decision, however, is a marketing one, not a cost one. To perform the work at tableside has elements of both the core and augmented product in it. First, we would have to identify the target market. Does this market expect, want, and appreciate the additional effort and cost to customize the product at tableside? Is it willing to pay an additional price for it? What does the modified or customized product do for customers? Does tableside service make customers feel better and more prestigious? Does it impress their guests, add perceived quality, or add romance or mysticism to the product? Or, does it simply delay the service delivery?

What business are we in? Are we in the business of serving quality food at a fair price, or providing a dining experience? Are we providing elegance, flair, or entertainment? Finally, do we have the capabilities? Is the staff properly trained, or can they be trained? If trained to do the carving, filleting, or mixing properly, can they do it with flair and finesse? If not, we may defeat the entire purpose.

The hospitality product or service includes everything we have to offer the guest whether "free" or for sale. It contains the basic elements of what guests think they are buying, what they really hope to get, and the total augmentation of the product that makes up the entire experience in purchasing it. From the budget motel in North Overshoe to the Bristol Hotel in Paris, from the hot dog vendor at Fenway Park to the Restaurant Zum Zee in Switzerland, the hospitality product or service determination is a marketing decision based on the target market. The problem for the marketer is to determine the effective demand for the various product features and the total benefit bundle.

There is one more thing to be said about designing the hospitality product, which has been said before but bears repeating: No matter how successful your product is now, never forget that the customer changes. The hospitality product requires constant evaluation and reevaluation. We will discuss this in more detail in the next section on the product life cycle.

A Hassle-Free Hotel in Bermuda

Discussion Questions

1. How would you describe the formal, core, and augmented product for the Hamilton Princess Hotel?
2. Discuss the various elements of the presentation mix experienced by Margaret and Laura at the Hamilton Princess Hotel.
3. What stage in the product life cycle do you think the Hamilton Princess Hotel is in? Why?
4. Do you think Margaret and Laura will return to the hotel? To Bermuda?

Margaret Manley and Laura Ferncroft, two widowed sisters in their 60s, decided to have a last "fling" in Bermuda, where Laura had previously lived for 20 years with her late husband. They chose the upscale Hamilton Princess Hotel on the waterfront in the capital city of Hamilton. This was February, off-season in Bermuda, commonly called "Rendezvous Season" by the Bermuda Tourist Board in its advertising, so they knew it wouldn't be too busy. The Princess Hotel had been there for decades and was considered a "classic."

Laura arranged the trip with her travel agent and obtained a one-week package rate that included breakfast and a superior bedroom. Arriving at the hotel, they found the central hotel doors blocked by ladders and the doorman busy chatting to the workmen. Their taxi driver took them through the automatic left-hand door. Registration, even with no one ahead of them, took a good 20 minutes, as the clerk left the desk twice for lengthy periods, with no explanation. A welcoming attitude was definitely not in evidence. "Well, we're off to a good start," said Margaret, sarcastically, as they headed for their room.

It was an attractively decorated room in the older part of the hotel and looked out on a spacious lawn and garden area. On the right one had a view across the harbor to the north shore of Warwick parish. After unpacking, Margaret, an inveterate reader, sat down with a book she had started on the plane. Unfortunately, she couldn't see the printing in the evening dusk. After checking all the lightbulbs in the room, she found none greater than 60 watts. Housekeeping was contacted for some 100 watt bulbs, which took an hour to arrive.

This case was contributed by Barbara Brooks and developed with Robert C. Lewis. All rights reserved. Used by permission.

Before dinner, Laura decided she would enjoy a relaxing hot bath, but when she went to push the lever for the drain stopper, it wouldn't budge. After some substantial effort, she called and carefully explained the problem to housekeeping, who said maintenance would be contacted. Forty minutes later, the maintenance man arrived, and Laura noted that he carried no tools. After fiddling around the tub for a few minutes, he decided to get his took kit—another 15-minute delay. About 20 minutes after restarting work, he announced, "All set now." Laura, ever cautious, said, "Let me try it." She did and was still unable to move the lever. It turned out the maintenance man had worked on clearing the drain and not on the lever at all. Laura had taken it for granted that, since he asked no questions, he had been told what the problem was. "Oops," he said, "I'll have to replace that fixture," and left again. In about 15 minutes he was back and 30 minutes later the lever was fixed. Of course there was no time for the predinner bath.

DINNER

Margaret and Laura then went downstairs for dinner. There were three choices listed in the room directory—a quite pricy fine dining room; a theme bar and grill; and a bright casual dining room with a reasonable menu, high ceilings, and windows that looked out on the harbor. Not caring for the first two, they chose the latter, but when they arrived at the door, it was locked and the room was dark. Looking around in amazement, they finally noticed a small sign on the door, "Closed for Renovations." They decided to go out for dinner.

The last straw for that evening was Laura's discovery, when attempting to take a pre-bed bath, that there was no hot water! "Oh, just run the bath for 15 minutes and it will get hot," they were informed. The cause was cited as a computer breakdown. After 20 minutes of wasted water, the temperature hadn't changed from barely lukewarm, so Laura gave up and endured a tepid bath. In the midst of it the bathroom fluorescent light fixture started to flicker badly, one tube conking our completely, resulting in a half-lighted bathroom. "What a marvelous way to end one's first day in this supposedly classy hotel," she cracked to her sister.

BREAKFAST

The next morning Margaret and Laura had breakfast in the dining room. When a check was given them after finishing, Laura explained to the waiter that breakfast was included in their package. "I don't know about that," he said. "I'll have to check it out." They waited about 10 minutes until he returned and told them they were listed as on the European plan. "This is too much," said Laura, "particularly since our plan is prepaid." Margaret and Laura went to the front desk where it took about 20 minutes for a rather indifferent clerk, then the front office manager, then an assistant manger, to check the situation. This consisted of all three staring at a computer screen, indifferent to the guests who, by that time, were sorely in need of some friendly reassurance while waiting.

It was finally discovered that the Air Canada package did indeed include breakfast. Nonetheless, at the end of the week, when Margaret and Laura were checking out, they found seven breakfasts on their bill. They repeated the previous process until the clerk agreed to remove the charges. It took Laura only a couple of minutes to figure out the amount on a piece of paper, but it took the clerk a good five minutes to make the adjustments on the computer. "So much for that rendezvous," they said to each other as they took a cab to the airport. "At least we'll be able to 'dine out'* on this for ages."

* A British expression for a good story to tell at dinnertime with others.

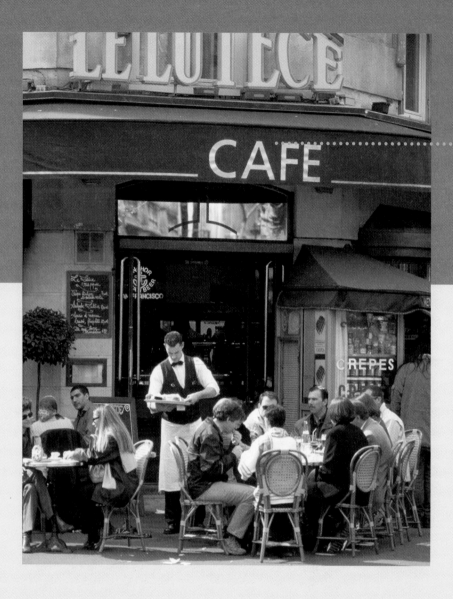

overview

The concept of relationship marketing (also known as loyalty marketing) is introduced in this chapter. In this concept, customers are considered valuable assets. Firms should develop an ongoing relationship with customers in order to develop repeat purchase, positive word of mouth, emotional bonding, and partnership activities (for example, wanting the customer to let the firm know when a service failure has occurred.) Relationship/loyalty marketing is also applied to employees to develop the same level of commitment from employees that the firm hopes to receive from its guests. The chapter deals with the "whys" and "hows" of relationship/loyalty marketing and how this will shape the future of marketing. We also discuss how all this relates to CRM—customer relationship management.

chapter 4

Relationship and Loyalty Marketing

learning objectives

After reading this chapter, you should be able to:

1. Articulate how relationship/loyalty marketing is different from traditional marketing and why hospitality firms need to be concerned with loyalty.

2. Determine the lifetime value of the customer and how to classify customers based on this figure.

3. Track both customer complaints and the effectiveness of service recovery strategies.

4. Design loyalty programs for both guests and employees.

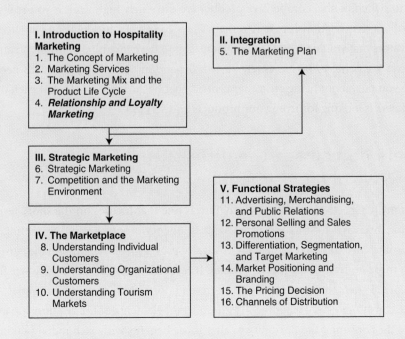

I. Introduction to Hospitality Marketing
1. The Concept of Marketing
2. Marketing Services
3. The Marketing Mix and the Product Life Cycle
4. ***Relationship and Loyalty Marketing***

II. Integration
5. The Marketing Plan

III. Strategic Marketing
6. Strategic Marketing
7. Competition and the Marketing Environment

IV. The Marketplace
8. Understanding Individual Customers
9. Understanding Organizational Customers
10. Understanding Tourism Markets

V. Functional Strategies
11. Advertising, Merchandising, and Public Relations
12. Personal Selling and Sales Promotions
13. Differentiation, Segmentation, and Target Marketing
14. Market Positioning and Branding
15. The Pricing Decision
16. Channels of Distribution

Adam Burke

Senior Vice President and Managing Director, Hilton HHonors Worldwide

Adam Burke was named senior vice president and managing director of Hilton HHonors Worldwide (HHW) in July 2004. In this capacity, he is responsible for the worldwide marketing and operations of the Hilton HHonors guest loyalty program, currently available at more than 2,700 Hilton Family hotels worldwide. HHW is jointly owned by Hilton Group, PLC, and Hilton Hotels Corporation. Previously, Burke served as vice president of marketing for Hilton HHonors Worldwide, a position he held since June 2001.

 Marketing in Action

Adam Burke

Senior Vice President and Managing Director, Hilton HHonors Worldwide

How would you define customer loyalty?

It's very much customer defined. A lot of people talk about customer centricity, but I think it takes a real discipline to truly "walk the talk." The way I would define customer loyalty is: The relationship you have with your customer is so differentiated that the barriers to switching to a competitive product are extremely high. Unless you really drop the ball on that relationship, they remain invested in your products and services. The motivators that would give an individual reason to have that type of loyalty to an organization are going to be extremely varied, depending on each individual's needs. So when you talk about being customer centric, that has to stem from what each of your customers is looking for from your products and services.

Where do you see the difference between frequency and loyalty?

From my perspective, a frequency program is one that focuses on the short-term win of using incentives to "buy the business." A frequency program is simply getting a particular incentive for a particular stay. A loyalty program, on the other hand, involves a much more holistic view of the customer. It recognizes that it's not a "one size fits all" approach. I guess that's one of the key differentiators—that a loyalty program looks to let the consumer define what the most important elements of the relationship are and allows the consumer the flexibility to alter those benefits with each individual interaction. We'd like to know as much about our customers as they're willing to tell us.

And with that, there is an incumbent responsibility to not only use that information responsibly, but also use it to further develop the relationship. When you start dating someone, they're not going to blame you if you don't know their birthday, their parents' names, their favorite color, etc., but if you're still forgetting birthdays three years in, the likelihood is you may not have a relationship anymore.

Why is loyalty good for the hotel company versus being good for the customer?

Starting with the franchiser or the developer, it's important to recognize that we do live in an era of consolidation. Fewer and fewer companies control more and more of the hotels and the hotel brands, and as such, network size is critically important. It has to do with shelf space, because if a guest wants to consolidate their stays with a particular chain, you have to have not only broad distribution, but the price points and product types that will meet all of the guest's needs as well. For example, if you talk about flying a flag from the Hilton Family of Hotels, you know that customers will always find the right products for their needs because we run the gamut from midprice with limited food and beverage in Hampton Inn and Hampton Inn & Suites, all the way up to luxury in Conrad. As a developer, you know that customers have a reason to consolidate their purchases within the Hilton Family because it represents one of the largest networks in the world. That's hugely important.

From a customer perspective, the big win is that we are increasingly focused on not having just one point of distinction. What we try to focus on is that any single point of differentiation is ultimately replicable, and so the way we're positioning the program for the future is all around elements of choice. Making sure our customers know that however they wish to utilize the program, we want to give them the flexibility to customize it on a per-stay basis. So, for example, you can still earn both points and miles for the same stay (unique to Hilton HHonors among the major programs), but you can now decide to earn points and miles one of three different ways, and you can change it with each stay. Similarly, when you talk about the benefits that you receive on property, rather than saying that Gold and Diamond VIPs get one set standard, we are creating greater elements of choice so that our top members can choose what types of benefits they want to get with each stay. One time you may want one item because you're traveling by yourself on business; the next time you may want something completely different because you're traveling for a week with your family. That approach also carries through the communications channels we have, what type of special offers you receive, how you book with us, etc. We're essentially trying to create a program to give every individual the flexibility to custom tailor it to their needs, so I think that's the win from the consumer perspective.

What advice can you give to students who wish to pursue a career in marketing?

The first thing I would say is, if you want to pursue a career in marketing, particularly as it relates to the hospitality industry, you've got to start with a broad understanding of the operation overall. I think the pitfall of a lot of marketers is that perhaps they look too much into the theoretical world of how to run a good marketing program and not enough into the nuts and bolts of their industry and what the implications are for how you can integrate that into the overall operation. So if I am talking about the hotel space, it's critical that you have an understanding of the front office operation. It's critical that you have an understanding of the franchise environment. It's critical that you have an understanding of the financial part of the operation and how to structure things in a way that makes sense within those parameters. It's also equally critical, as a marketer, that you are completely versed in the customer service and reservations operations of the company. My first piece of advice is, if you really want to pursue a marketing career in the hospitality industry, don't forget that, first and foremost, you are in the hospitality industry.

The second thing I would say is that you have to have a commitment to viewing business from the customer's perspective. There's a lot written out there about CRM [customer relations management]—about what it is and what it should be. But ultimately, if you really want to plan on long-term success, then I believe that you've really got to be willing to make the commitment to put the customer at the center of your decision-making process. That means that you stop looking at things from a short-term focus and start talking about how you cultivate true loyalty over a long-term relationship with your customers. This requires taking a very different look at things like financial investment, cost containment, and return on investment (ROI), and it means that you have to be willing to take some risks to really cultivate that long-term relationship.

A huge opportunity in working with a loyalty program is recognizing your role as one of the strongest customer advocates within your organization. Let's face it—across a variety of industries, there are a lot of more conservative ways of looking at the business. Those perspectives aren't necessarily wrong, but they don't always take the customer perspective into account. So if you're sitting in a meeting where some of those more traditional approaches come up, I think you have to be willing to view yourself as the voice of the customer in those settings because that's who you're representing. That's not to say that you shouldn't be financially responsible—far from it. It's just your opportunity to ensure that the voice of the customer is an integral part of your organization's business decisions. *Used by permission from Hilton Hotels Corporation.*

WHAT IS LOYALTY AND WHY IS IT IMPORTANT?

The success of hospitality firms in the future will be based to a large extent on their ability to create loyalty—both with its external shareholders (i.e., customers) and with its internal stakeholders (i.e., employees). When relationship/loyalty marketing is successful, loyalty is created. What do we mean by loyalty? Loyalty occurs when:

> The customer feels so strongly that you can best meet his or her relevant needs that your competition is virtually excluded from the consideration set and the customer buys almost exclusively from you—referring to you as their restaurant or their hotel. The customer focuses on your brand, offers, and messages to the exclusion of others. The price of the product or service is not a dominant consideration in the purchase decision, but only one component in the larger value proposition.[1]

A second definition of loyalty:

> A loyal customer is one who values the relationship with the company enough to make the company a preferred supplier. Loyal customers don't switch for small variations in price or service, they provide honest and constructive feedback, they consolidate the bulk of their category purchasers with the company, they never abuse company personnel, and they provide enthusiastic referrals.[2]

Loyalty is important because it provides critical inoculation across multiple areas. For instance, loyal customers are less likely to ask about price when making a reservation. They are also less likely to shop around; hence, competitive offers face a higher hurdle. The customer becomes more forgiving when you make a mistake because there is goodwill equity. In fact, loyal customers are more likely to report service failures. *Loyalty begets loyalty.* Further, marketing and sales costs are lower, as are transaction costs. As we will discuss, research has shown that increasing customer retention by 2 percent is the equivalent of cutting operating costs by 10 percent.[3]

A focus on loyalty means that the firm treats the customers as assets. Customers are the most important assets a company can have. Firms spend large amounts of money on such things as insurance and elaborate alarm systems to protect their physical property such as buildings, furniture, and warehouses. In the same way, firms need to spend money to protect their customers. This new subfield of marketing is called relationship/loyalty marketing. **Relationship marketing** sees the customer as an asset. Its function is to attract, maintain, and enhance customer relationships. To put this in perspective, consider the hypothetical abbreviated balance sheets in Exhibit 4-1.

Exhibit 4-1 shows two balance sheets: a traditional balance sheet and a customer-driven balance sheet. As discussed in finance, total assets always equal total liabilities plus shareholders' equity. To improve shareholders' equity, you must either increase assets, decrease liabilities, or do both. Looking at the second balance sheet, you can see that if management can increase both the number of customers that are likely to come and the number of those that definitely come, while at the same time decreasing the number of customers who will either defect because of not being satisfied or not come at all because they have heard negative comments, then shareholder equity will increase and the sheet will be balanced.

The customer balance sheet in Exhibit 4-1 demonstrates that the goal of an organization should be threefold: (1) to get more customers, (2) to keep more customers coming back (that is, reduce the number of customers defecting), and (3) to lose fewer potential customers because unhappy current customers say unflattering things about the property or organization. This is essentially what marketing is all about. But relationship/loyalty marketing adds a new dimension; specifically, the goal is not only to encourage guests to return, but to get them to tell their friends how wonderful the property or organization is. Exhibit 4-2 describes one way to measure this word of mouth.

It should be clear that relationship/loyalty marketing is about creating long-term customer value and long-term relationships. Relationship/loyalty marketing is not database marketing, nor is it frequent guest programs, although both database marketing and frequent guest programs certainly play a part in creating long-term customer value and long-term relationships. It means thinking in terms of the customers we have, rather than just in terms of the ones we hope to acquire. This is crucial in the hospitality

relationship marketing An ongoing process of identifying and creating new value for individual customers for mutual value benefits and then sharing the benefits from this over a lifetime of association, also known as **loyalty marketing**.

Exhibit 4-1	Customers as Assets on the Balance Sheet

Hypothetical Standard Balance Sheet for a Hotel

CURRENT ASSETS:

Cash and investments	$120,000
Receivables	30,000
Inventory	50,000
Total current assets	200,000

FIXED ASSETS:

Properties and land	25,000,000
Investments	2,500,000
Total assets	27,700,000

LIABILITIES:

Current debt	100,000
Accounts payable	50,000
Long-term debt	20,000,000
Shareholder's equity: capital stock	7,550,000
Total liabilities and equity	27,700,000

Hypothetical Customer Balance Sheet for the Same Hotel

CURRENT ASSETS:

Probable customers: 2,000 @ $2,000 a year	$4,000,000
Receivables: 1,000 customer prospects @ $2,000 a year	2,000,000
Inventory (guaranteed customer deposits for future year): 1,400 @ $500	700,000
Total current assets	6,700,000

FIXED ASSETS:

Investments: 10,500 loyal customers @ $2,000	21,000,000
Total assets (6,700,000 + 21,000,000)	27,700,000

LIABILITIES:

Current debt: loss of upset customers 500 @ $2,000	1,000,000
Accounts payable: pay back irate customers	100,000
Long-term debt: loss of customers by word of mouth	19,050,000
Shareholders' equity: capital stock	7,550,000
Total liabilities and equity	27,700,000

industry. Competition is standing by all too ready and willing to take the customers you can't keep. Levitt compared the relationship to something like a marriage:

> The sale merely consummates the courtship. Then the marriage begins. How good the marriage is depends on how well the relationship is managed by the seller. That determines whether there will be continued or expanded business or troubles and divorce, and whether costs or profits increase. It is not just that once you get a customer you want to keep him. It is more a matter of what the buyer wants. He wants a vendor who will keep his promises, who'll keep supplying and stand behind what he promised. The age of the blind date or the one-night stand is gone. Marriage is both more convenient and more necessary .... In these conditions success in marketing, like success in marriage, is transformed into the inescapability of a relationship.[4]

Like a good marriage or a good relationship, both parties have to "get something." The firm hopes to get repeat patronage, positive word of mouth, and the assurance that guests will tell management when things go wrong. For the guest, the expectation is that the firm will do everything in its power to ensure an error-free purchase experience and a customized experience, and will look after the guest's best interest.

Relationship/loyalty marketing, then, can be defined as *an ongoing process of identifying and creating new value for individual customers for mutual value benefits and then*

| Exhibit 4-2 | Measuring Word-of-Mouth Influence |

One metric used to determine the word-of-mouth influence is to ask the following question: "How likely is it that you would recommend _____ to a friend or colleague?" This is done on a scale of 0 to 10. The percentage of people who respond with a 9 or 10 are called promoters, whereas those who score from 0 to 6 are called detractors. Those who score 7 or 8 are indifferent. The net promoter score is the difference between the percentage that are promoters and the percentage that are detractors.

Source: Adapted from letter to the editor by K. Reichneld, *Harvard Business Review,* 80(11), p.126.

sharing the benefits from this over a lifetime of association. In this sense, although it is part of traditional marketing, it differs from it in the following ways:

- It seeks to create *new* value for customers and *share* the value so created.
- It recognizes the key role of *individual* customers in defining the value they want; that is, value is created *with* customers, not *for* customers.
- It requires that a company define its organization to support the value that individual customers want.
- It is a continuously cooperative effort between buyer and seller.
- It recognizes the value of customers over their purchasing *lifetimes.*
- It seeks to build a chain of relationships between the organization and its main stakeholders to create the value those customers want.
- It focuses on the processes and whatever else is needed to advance the customer relationship. [5]

Relationship/loyalty marketing is most applicable under the following conditions:

- There is an ongoing and periodic desire for service by the customer.
- The service customer controls the selection of the service supplier.
- There are alternative supplier choices.
- Customer loyalty is weak and switching is common and easy.
- Word of mouth is an especially potent form of communication about a product.

These conditions are obviously quite prevalent in the hospitality industry. We don't sell one-time services, and the consumer has many choices. In an era of heavy hotel building and restaurant openings, or excess capacity, any hotel or restaurant is especially vulnerable to competition. Just about everyone likes to try a new place. The question is, Will they come back? Do we offer a competitive product on dimensions that are meaningful to customers and solve customer problems? Is our product difficult for competitors to duplicate? Do we have a meaningful relationship? This is what relationship/loyalty marketing is all about, and when the preceding conditions pertain, the opportunities to practice it are abundant.

Exhibit 4-3 summarizes how relationship/loyalty marketing is different from traditional marketing. Later in this book we will talk about segmentation and target marketing. We take relationship/loyalty marketing one step further to develop customer-specific segments that are unique to each customer.

The goals of relationship/loyalty marketing also include having customers spend more money while on the property and tell management when things go wrong, instead of just walking away and never coming back. Of course, when guests do tell management

---

Exhibit 4-3   Comparison between Relationship/Loyalty and Traditional Marketing

RELATIONSHIP/LOYALTY MARKETING

- Orientation to customer retention
- Continuous customer contact
- Focus on customer value
- Long time scale
- High customer service emphasis
- High commitment to meeting customer expectations
- Quality is concern of all staff

TRADITIONAL MARKETING (A.K.A. TRANSACTIONAL MARKETING)

- Orientation to single sales
- Discontinuous customer contact
- Focus on product features
- Short time scale
- Little emphasis on customer service
- Limited commitment to meeting customer expectations
- Quality is the concern of the production staff

Source: Adapted from the notes of Stowe Shoemaker.

Exhibit 4-4   Undoing a Relationship

A couple had gone each year since their marriage to a certain hotel within 25 miles of their home to celebrate their anniversary. They spent about $600 each visit. The time was early December, when the hotel traditionally ran 20 percent occupancy. They also went there other times or ate there, sometimes with friends.

On the fifth year, they arrived about 7:00 P.M. and were welcomed by the desk clerk, whom they had come to know. They went to their room and ordered champagne and dinner. They also had their remaining meals in the hotel and spent time in the hotel lounge. When they checked out three days later, they asked for the 10 percent room discount they were entitled to as members of an organization with which the well-known hotel chain had an agreement. The same clerk's response was that he could not grant the discount because they did not ask for it when they checked in, according to the standard operating procedure. When they got

home, they wrote to the general manager, asking for the discount. In a two-page letter, the discount was refused in no uncertain terms. They then wrote to the vice president of marketing of the company, including a copy of the manager's letter.

Shortly afterward, they received a polite letter from the general manager starting, "We are always glad to know of our customers' complaints because it helps us to improve our operations." It included an invitation to spend a weekend at the hotel as his guest. No mention was made of the 10 percent discount—the main purpose of the letter. Nor did management mention the changes that had been made to ensure that the problem did not happen again. The couple never went back and never will, nor will many of their friends.

Consider the cost of granting the discount, about $25, against the business lost.

---

when things go wrong, management must not only solve the problem, but also put in place systems to ensure that the same mistake does not happen again. Consider the situation in Exhibit 4-4, which examines the poor response of management to a guest's complaint. It was a poor response because management did not address the guest's complaint, nor did management mention anything about new systems to ensure that the problem would not happen again. Rather, management just tried to buy loyalty by giving away a free room. We address the issue of complaint management in more detail later in this chapter.

MAINTAINING A GOOD RELATIONSHIP WITH THE CUSTOMER

Good relationships involve commitment and trust and, as in our personal lives, there are numerous antecedents that bring this about. There are also numerous consequences when they do not exist. Exhibit 4-5 shows a model of service relationships developed by Bowen and Shoemaker based on their research in the luxury hotel industry. This model illustrates the antecedents and consequences of commitment and trust

Exhibit 4-5	Model of Service Relationships

Antecedents that affect trust and commitment are as follows:

1. A perception of fair costs. This means that we cannot arbitrarily practice yield management on our best customers. If we do, we violate their trust.
2. A reduction of switching costs.
3. Customer beliefs that the provider's word can be relied on.
4. Understood values and goals, such as the extent to which two parties have beliefs in common. For example, Scandia is known as an environmentally friendly hotel and I support the environment, so I will stay there.

Consequences that result from trust and commitment, are as follows:

1. Increased product usage (e.g., using the hotel's restaurant versus going elsewhere).

2. Voluntary activities such as spreading strong word of mouth and making business referrals.

Note that the preceding are all positive outcomes. Following are negative outcomes that result from a lack of trust and commitment:

1. Uncertainty (e.g., a lack of trust in the service delivery).
2. Natural and reactive opportunistic behavior, in which one party takes advantage of the other, such as a hotel changing rates according to demand (natural) or a customer getting irate when the room desired is not available (reactive).

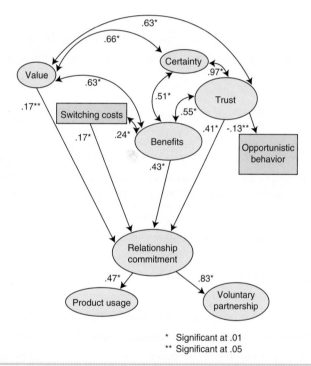

* Significant at .01
** Significant at .05

Source: Adapted from Bowen, J. T., & Shoemaker, S. (1998, February). Loyalty: A strategic commitment. *Cornell Hotel and Restaurant Administration Quarterly,* 14-17. Adapted from SAGE Publications. Used by permission.

in service relationships, as indicated by the plus and minus signs. We explain the model after first defining what we mean by trust and commitment.[6]

Trust is the belief that an individual or exchange partner can be relied on to keep his or her word and promise. Trust is an antecedent of loyalty because the customer trusts the organization to do the things that it is supposed to do, implicitly or explicitly. Any actions taken to increase feelings of trust will lead to commitment. Conversely, any action taken to decrease trust will lead to a lack of commitment.

Commitment is the belief that an ongoing relationship is so important that the partners are willing to work at maintaining it and are willing to make short-term sacrifices to realize long-term benefits.

Clearly this focus on loyalty is different from the old-fashioned view of marketing that focused on "getting customers in the door."

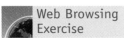
lifetime value of a customer The total value of a customer based on repeat purchases anticipated and word of mouth to others, minus the costs associated with serving that customer.

CUSTOMER RELATIONSHIP MANAGEMENT (CRM)

Customer relationship management (CRM) is the term used to describe the components of relationship marketing. Payne and Frow defined CRM as follows:

> CRM is a strategic approach that is concerned with creating improved shareholder value through the development of appropriate relationships with key customers and customer segments. CRM unites the potential of relationship marketing strategies and IT to create profitable, long-term relationships with customers and other key stakeholders. CRM provides enhanced opportunites to use data and information to both understand customers and cocreate value with them. This requires a cross-functional integration of processes, people, operations, and marketing capabilities that is enabled through information, technology, and applications.[7]

The fundamental reason relationship/loyalty marketing and CRM are important is because of the lifetime value of the customer, which we discuss next.

LIFETIME VALUE OF A CUSTOMER

The **lifetime value of a customer**, in short, is the net profit received from doing business with a given customer during the time that the customer continues to buy from you. According to members of the Harvard Business School faculty, "The lifetime value of a loyal customer can be astronomical especially when referrals are considered." These researchers calculated that the lifetime revenue from a loyal pizza customer can be greater than $8,000.[8]

Exhibit 4-6 gives another view of profits from loyal customers. This figure shows that customers' profitability increases over their lifetimes because they tend to spend while staying on the property, they bring other guests via positive word of mouth, and they cost less to serve than do customers who do not favor a particular hotel.

Exhibit 4-7 shows the questions that form the bases of understanding customer profitability. Exhibits 4-6 and 4-7 illustrate that we have switched our thinking from product and service profitability to customer profitability.

With the answers to the questions in Exhibit 4-7, we can now begin to develop a customer analysis, in a sense, for each customer. Which customers are profitable, now

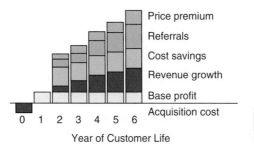

Exhibit 4-6 Profits throughout the hotel guest's life

Exhibit 4-7   What One Needs to Know to Understand Customer's Profitability

1. How much does it cost to get a new customer?
2. How much does it cost to keep that customer?
3. What is the revenue from that customer—each visit, annually, lifetime?
4. What is the cost of serving that customer—each visit, annually, lifetime?
5. What are the retention rates of customers?
6. If we are losing loyal customers for reasons beyond our control (e.g., death, moving away), are we replacing them, and with whom? (We will address this later, but keep in mind that grandiose statements such as, "We have 80 percent repeat customers" can be dangerous if you are not also acquiring new ones. See next question.)
7. What is a repeat customer? Is it someone who comes only on Saturday night when the restaurant is always full? One that stays every time he is in town but that's only once a year? One that wants us to stay as we are when we know we have to change? One that stays in this hotel but in other cities goes to a different chain? One that is loyal as long as we give him an upgrade? You can imagine that this list could go on and on. You can also turn the questions around to make them positive.
8. What is the revenue and profitability from repeat customers versus those we might replace them with? (For example, in hotels, do they buy the cheapest rooms and always eat out—that is, what is the total revenue and profitability?)
9. What other opportunities are there for revenue from our customers for things that we don't now provide?
10. Do our customers really want all the things in our package bundles?
11. If our customers weren't here, where would they be or where might they go?
12. What value do our customers get from us?
13. Do we have a relationship with our customers, or are they just customers?
14. How frequently do we communicate (Internet, telephone, mail, and mass media) with our customers? Is this favorable to them and to us?

---

and in the future? Some customers add value, whereas others have negative value; that is, they require too much time and effort to ever make them profitable. With this kind of analysis, marketing management can better discern who are the true loyal customers and, perhaps more important, why.

Exhibit 4-8 shows what has come to be known as the customer **loyalty ladder**. The obvious effort is to move the customer from the state of awareness (suspect/prospect) to being a brand advocate. The strategies one would select for each rung of the ladder are discussed next.[9]

On the first two rungs of the ladder are suspects and prospects. A suspect is anyone who might buy your product or service. They are called suspects because you suspect they might buy your service, but you don't know enough about them to be sure. A prospect is someone who has a need for your product or service as well as the ability to buy. The prospect has heard of you, but has not yet purchased. The strategy for this group is to overcome apprehension with empathy or encouragement, client "success stories," site visits, or product/service guarantees. As told in the Tourism Marketing Application, Hampton Inns was one of the first hospitality organizations to offer a service guarantee and saw tremendous growth of its first-time users.

In the search for prospects, it is possible to come across disqualified prospects. These are customers that you have learned enough about to know that they do not need or do not have the ability to buy your product or service. This group may also represent customers who are unprofitable and likely to remain so.

On the next rung of the customer loyalty ladder is the first-time customers. These may have come from the competition or may be new to the product or service category. The strategy for this group is to meet or exceed their expectations, as discussed in Chapter 3. At this stage, management should try to build a promise for return visits; that is, thank them for their business and invite them to return.

loyalty ladder Moving a customer from a state of awareness of your product or service to being an advocate who is sincerely loyal and promotes your product or service to others.

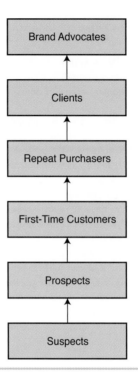

| **Exhibit 4-8** | Adding the customer loyalty ladder to the consumer buying process |

Tourism Marketing Application

Hampton Inns Builds First-Time Users with Service Guarantee

Hampton Inns gave an unconditional guarantee starting in the early 1990s. In the first year they sold about 157,000 rooms to first-time users because of the guarantee. This grossed an additional $7 million. Of those who invoked the guarantee, 3,300 returned the same year and 61 percent of these said the reason was the guarantee. The payout on the guarantee was $350,000. The total additional revenue was $8 million.

Repeat customers are on the next rung of the ladder. These are customers who have purchased two or more times. The strategy for this group is to provide value added benefits with each repeat purchase and to seek regular feedback.

Repeat customers eventually become clients. These are customers who buy regularly all or most of the firm's products or services. The firm's strong ongoing relationship makes them almost immune to the competition. The strategy for this group is to tailor the service to their needs. You want to customize the care and service provided as much as possible. It is also important that your employees not take this business for granted. As with members of the other groups, it is important to continually seek input and feedback.

On the final rung of the ladder are the brand advocates. These customers are like clients, but they also encourage others to buy from you. They help do your marketing for you and bring additional customers. The strategy for this advocacy group is to encourage them to sell for you. This can be done through letters of endorsement from the advocates and referral acknowledgments. Communicate with them regularly and frequently. They have become part of the sales team. Don't lose them!

BUILDING LOYALTY

Evolution of Customer Loyalty

Exhibit 4-9 shows the evolution of building loyalty. Initially, the focus of marketing was purely on sales, not loyalty. The goal was to get as many new customers as possible. In this phase, there was little targeting, little measurement, and lots of discounts. As marketing developed, marketing managers began to focus on specific market segments, but the mentality was still on pushing traffic to the property.

Frequency programs were the next element of the evolution of loyalty (we discuss frequency programs in more detail later in the chapter). These programs initially were used to reward frequent purchasers. S&H Green Stamps started this movement in the 1950s by giving consumers stamps every time they purchased certain brands. These stamps were then pasted into books that were exchanged for free merchandise. Today, companies award points or airline miles that are exchanged for free travel, as well as other items. The points are also used to keep track of customers' purchase behaviors across a multiple of retail outlets. Consider, for example, the number of locations where it is possible to purchase a room at a Marriott Hotel. With so many locations, it would be almost impossible to keep track of individual purchase behavior by name alone. The membership number provides an easy way for firms to track behavior. The points then determine a person's status, which means that more frequent purchasers get certain

Web Browsing Exercise

Go to the TARP website (www.tarp.org). What types of studies does the firm undertake? What do they say about the latest trends in complaint management? What types of firms are represented in their research?

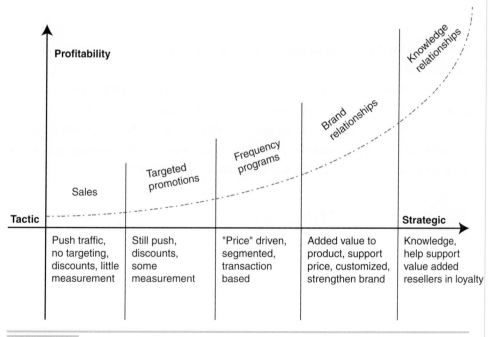

Exhibit 4-9   Evolution of building loyalty

benefits not available to less frequent purchasers. For example, airlines often hold the most desirable seats (e.g., those up front or in the bulkhead) for frequent travelers, whereas less frequent travelers get the less desirable seats.

The advent of a mechanism to track multiple stays enables marketing executives to provide customized services, such as preferred seating on an airplane. In addition, management can customize the service. There is tremendous opportunity to learn about particular customers and their specific problems and to tailor the service to solve those problems. For example, at Caneel Bay Resort on St. John in the Caribbean, guests are allowed to store anything that they do not want to carry back home, including such things as suntan lotion, swim suits, and snorkel gear. For their next visit, these belongings are placed in their rooms before arrival. Other methods of customization include asking hotel guests if they want help to get to their rooms or are happy to find their own way and carry their own bags. Video checkout and bed turn-down *on request* are still others.

In the final phase of the evolution of relationship marketing the firm fully understands the customer.[10] Although customization also occurs in this phase, the focus is on knowing the customer and using that knowledge to augment the product to better serve the customer. Service augmentation means building extras into the service that help improve the customer's stay. We call this final phase *knowledge relationships*. This is the one-to-one marketing popularized by Peppers and Rodgers.[11] An example of one hospitality firm that is beginning to move in this direction is Wyndham Hotels and their Wyndham ByRequest. This program is not a point-or mileage-based program found in many hotels. Rather, guests customize their hotel stay by providing Wyndham with information on their room preference (e.g., smoking versus nonsmoking, bed type, and the like) and the types of snacks they would like to see in the honor bar.[12] The information provided by the customer is then stored in a central database so that the room will be customized in any Wyndham the customer chooses.

The Loyalty Circle

You should now be convinced that customer loyalty is important. A natural question then is, How do we create loyalty? The loyalty circle, as shown in Exhibit 4-10, shows the three main areas in which firms must focus their efforts to create loyalty. The three main functions on the circle are process, value, and communication. Notice that the customer might exit the circle and hence the relationship at different points along the circle. The goal of hoteliers is to keep the customer in the circle by executing equally well the three functions of the circle. Equality is the key to the loyalty circle. If hoteliers are

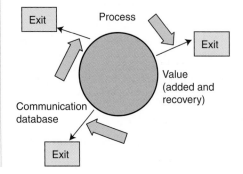

Exhibit 4-10   The loyalty circle

great on creating value, for example, but do not effectively communicate with the customer, then that customer may leave the relationship.

On one side of the loyalty circle is the process, which is "how the service works." It involves all activities from both the guest's perspective and the hotelier's perspective. Ideally, there should be no gaps in this process. For the guest, the process includes everything that happens from the time she begins buying the service (e.g., making a reservation) to the time she leaves the property (e.g., picking up her car from a valet.) All interactions with employees are part of this process.

For the hotel, the process includes all interactions between the employees and the guests; the design of the service operations; the hiring and training of service personnel; and the collection of information to understand customers' needs, wants, and expectations. Many of the specifics of the process were discussed in Chapter 3 when we introduced the gap model of service quality. One way to monitor the process is through the use of consumer research and intelligence techniques. One such technique is mystery shopping. With this technique people who are hired by the service organization act as customers and report everything that occurred during the purchase of the service. Another way is to conduct focus groups with customers. Focus groups consist of 10 to 12 customers (or noncustomers) who are asked to focus on one or two topics of interest to the service organization. A moderator leads the discussion and keeps participants focused on the topics of interest to management. A third way is to undertake large-scale survey research with current as well as past customers.

A second component of the loyalty circle is value creation. Value creation is subdivided into two parts: value added and value recovery. Valued added strategies increase loyalty by providing guests with more than just the core product; that is, for hotels, offering more than just a place to sleep. Valued added strategies increase the long-term value of the relationship with the service firm by offering greater benefits to customers than can be found at competing firms that charge a comparable price. Features that pertain to value added are of seven types: *financial* (e.g., saving money); *temporal* (e.g., saving time); *functional* (e.g., the product does what it was designed to do); *experiential* (e.g., enhancing the experience such as by getting an upgrade); *emotional* (e.g., more recognition or a more pleasurable service experience); *social* (e.g., an interpersonal link with a service provider), and *customer concern*. Temporal value is important because business travelers have stated that they value their time at $100 per hour and anything that saves them time saves them money.

Consider the check-in process of a hotel. Research reveals that many frequent business travelers want to go immediately to their rooms and do not want to wait in line to check in. If they have to wait in line for 15 minutes, they mentally figure they have spent $25 to check in. Waiting in line is especially annoying if the guest is a member of the hotel's frequent guest program and all the guest's information is already stored on file. Certain technologies allow guests to check in, receive their room numbers, unlock their rooms, and have charges automatically billed to their credit cards without having to check in with the front desk. Moving these guests to this form of check-in would have the benefit of shortening the line for those guests who want to speak with a front desk clerk. This new check-in procedure speeds up and improves the process (*functional value*) and adds value because it saves the guests' time (*temporal value*).

The importance of value added strategies in creating customer loyalty is illustrated in a study conducted by Shoemaker and Bowen of business travelers who spend more than $120 per night for a hotel room and take six or more business trips per year. The study revealed that 28 percent of the 344 who spend more than 75 nights per year in hotels (38 percent of the total sample) claimed that the feature "is a good value for the price paid" is important in the decision to stay in the same hotel chain when traveling on business. A similar percentage rated the features "collects your preferences and uses that information to customize your current and future stays" and "accommodates early morning check-in and late afternoon checkout" important in the decision to stay with the same chain. Both of these tactics are examples of features that add value to the core product offering.

Value recovery strategies are designed to rectify a lapse in service delivery. The goal is to ensure that the guest's needs are taken care of without further inconveniences. Empowering employees to solve problems and offering a 100 percent guarantee are examples of value recovery strategies. The key to value recovery strategies is that the complaints be taken seriously by the hotel and that processes be put in place so that the same mistakes do not happen again. We discuss complaints later in this chapter.

The final component of the loyalty circle is communication. This side of the circle incorporates database marketing, newsletters, and general advertising. It involves all areas of how the hotel communicates with its customers. When communicating with guests, firms must be sure that external communications do not overpromise what the service can deliver; this would create Gap 4, as discussed in Chapter 2. It is also critical that the communiqué reflect the needs of the customer and that he does not receive offers in which he has no interest.

If marketers can focus the organization on these components, they will create loyal customers who will return over and over again. If they do not focus on these components, they will be forced to focus on getting more and more customers to replace those who have left the circle.

Frequent Guest Programs

frequency program
Any program that rewards guests with points, miles, stamps, and so forth, that they can redeem for free or discounted products or service.

loyalty program A strategy undertaken by a firm to encourage an emotional bond with the customer so that she gives the firm a majority of her business, provides positive word of mouth, acts in partnership with the firm, and spends more time at the firm's establishment than nonloyal guests would.

Two questions are often asked: Are frequency programs the same as loyalty programs? and, Do frequent guest programs build loyalty? The answer to the first question is no, **frequency programs** are not the same as **loyalty programs.** We state this even though many firms call their frequent guest programs loyalty programs. We define a loyalty program as *a strategy undertaken by a firm to manage the three components of the loyalty circle in order to create an emotional bond with the customer so that she gives the firm's establishment a majority of her business, provides positive word of mouth, acts in partnership with the firm, and spends more with the firm than a nonloyal guest would.*

In contrast, we define a frequency program as *any program that rewards guests with points, miles, stamps, or "punches," that they can redeem for free or discounted merchandise.* The potential trap is to confuse purchase frequency with customer loyalty; that is, to confuse the ends with the means. Frequency in itself does not build loyalty, as we define loyalty; rather, loyalty builds frequency. Frequency can create loyalty if the firm uses the information gathered on frequent visits to focus on the components of the loyalty circle; however, if the firm ignores this opportunity, then it ignores the

Tourism Marketing Application

Frequency programs are also used to market resort destinations. Consider, for example, Vail, Colorado. The Vail Resort Company operates the lifts and many of the resorts and restaurants in Vail, Beaver Creek, Breckenridge, Keystone (all in Colorado), and Heavenly (Lake Tahoe, California). The Peaks program enables members to earn rewards on their purchases at multiple restaurants, hotels, mountain activities, and other related activities. The "soft" rewards enable members to purchase ski tickets in advance, charge all their purchases on a credit card so they do not have to carry cash, and receive exclusive offers and e-mail letters. They can redeem points for activities such as lift tickets and ski lessons.

Source: (www.snow.com/info/peaks.asp (accessed March 18, 2006).

"Today's lecture is on loyalty."

emotional and psychological factors that build real commitment. Without that commitment, focus on the "dcustomers eal," not the brand or product relevance. An example is the coffee shops that provide a free coffee after so many purchases. For example, after 12 purchased coffees, patrons of It's a Grind receive a free cup of coffee. Because there is no mechanism in place to track specific consumers' requests or identify the consumer, the consumer focuses only on the deal.

With frequency only programs, sales may increase, as they would with price discounts. Repeat purchase may also increase, but the focus is on the rewards, not on product superiority or brand relevance. This behavior focus makes bribing the customer the line of reasoning. Over time, the economics of bribery begin to collapse necessitating greater and greater bribes and eventually eroding the brand image and diminishing product or service differentiation.

What Loyalty Programs Don't Do

Sometimes the wrong things are expected from loyalty programs. These programs are not "quick fixes." They will not fix an essential problem in the operation that may be costing customers. They won't show a profit in the short run. These are dedicated long-term efforts. They are not a promotion that is temporary or, worse, becomes part of the product and only adds to the cost. And they won't bring in new customers. The brand has to overcome the barriers to first trial before the loyalty program can kick in.

What Makes Loyalty Programs Work

According to Richard Dunn at Loyaltyworks, the following elements are essential to a successful frequency program:

- A vital database—the relationship foundation
- Targeted communications—the relationship dialogue
- Meaningful rewards—relationship recognition
- Simplicity—easy to participate and understand
- Attainability—motivational rewards must be attainable (e.g., upgrades)
- Sustainability—don't let it lapse; keep it active
- Measurability—make sure it is working in the right ways
- Management—full commitment and behind it all the way
- Manageability—don't let it get out of hand
- Profitability—is it really working in the long term?

Further caveats from Dunn are these:

- Don't treat the program like a promotion.
- Don't focus excessively on rewards, but on the relationship.
- Don't short-change the communications component.
- Don't underestimate the importance of internal support.
- Don't pretend to care more than you really do.
- Tailor the value of benefits to specific customers based on their achieved or expected value.

Tourism Marketing Application

Frequency programs are not just for hotels, restaurants, or cruises; they are used to market baseball as well. In the spring of 2006, the Los Angeles Dodgers created the Think Blue rewards program, which gives fans the opportunity to earn such experiential rewards as sitting in the dugout, meeting players, and getting access to tickets. Points are earned at concessions within the ballpark and at partner merchants in the Los Angeles area. Additionally, fans can earn points online at the Think Blue mall, which features more than 250 e-retailers such as Lands' End, Overstock.com, and Teleflora.

Source: www.colloquy.com/cont_breaking_news.asp?industry=Entertainment&region=All (accessed March 18, 2006).

CUSTOMER COMPLAINTS AND SERVICE RECOVERY

When we discussed the loyalty circle, we mentioned that a component of value is service recovery; that is, how the firm manages and addresses service failures. Firms that are committed to relationship/loyalty marketing are proactive in terms of service recovery strategies. By this we mean that they encourage guests to speak out when things go wrong; after all, there can be no service recovery if the firm does not know when something goes wrong. Proactive firms believe in the adage, "A complaint is a gift."[13]

Customer complaints deserve special treatment in this chapter because they are one of the most misunderstood and mishandled areas of customer relations in the hospitality industry. Let us look first at what customer complaints are:

- **Inevitable.** Nothing is perfect. The diversity of hospitality customers and the heterogeneity of the hospitality product absolutely ensure that there will be complaints. This will be true even when everything goes according to plan. Of course, when everything doesn't go according to plan (and it almost never does), there will be additional problems and there will be complaints.

- **Healthy.** An old army expression is, "If the troops aren't griping look out for trouble." An absence of complaints may be the best indication (along with declining occupancies or covers) that something is wrong. Hospitality customers are never totally satisfied, especially over a period of time. Probably, instead, they are simply not talking to you or you are not talking to them. The communication process is not working; the relationship is deteriorating. By the time it explodes, it will be too late. Some say, If it isn't broken, don't fix it. First, you have to know if it's broken; the ones who know first are your customers. And, incidentally, the ones they tell first are your employees, which means that you should listen to your employees as well.

- **Opportunities.** Customer complaints are opportunities to learn of customers' problems, whether they are idiosyncratic or caused by the operation itself. If it's broken, you have an opportunity to fix it. If it's not broken, you have an opportunity to make it better, to be creative, to develop a new product, to learn new needs, and to keep old customers. A study by TARP found that customers who complain and are satisfied are up to 8 percent more loyal than are those who had no problem at all.[14]

- **Marketing Tools.** If marketing is to give customers what they want, then marketing must know what they want. All the customer surveys in the world won't tell you as much as customer complaints will tell you.

- **Advertising.** Yes, advertising. The advertising is negative if you don't resolve the problems, and there is nothing more devastating in the hospitality business than negative word of mouth. It is positive if you fix the problem. Research has shown that one of the best and most loyal customers are the ones who had a complaint that was satisfactorily resolved. And these customers love to tell others about it.

The initial research on complaint handling was undertaken by TARP, a research firm located outside Washington, D.C. Tom Peters, author of *On Achieving Excellence*, stated, "TARP is perhaps (America's) premier customer service research firm."[15] The initial research undertaken by TARP occurred in the late 1970s and has been replicated in multiple industries and 20 countries. The replicated studies revealed that the initial findings still hold today. Some of the major findings are as follows:[16]

- On average, across all industries, 50 percent of consumers will complain about a problem to a front line person.
- Only 1 to 5 percent of customers will escalate their complaints to a local manager or corporate headquarters. For packaged goods and other small ticket items, TARP found that 96 percent of consumers either do not complain or complain to a retailer where they bought it.
- Complaint rates vary by the type of problem. Problems that result in out-of-pocket monetary loss have high complaint rates (e.g., 50 to 75 percent), whereas mistreatment, quality, and incompetence problems evoke only 5 to 30 percent complaint rates to the front line.
- On average, twice as many people are told about a bad experience than they are about a good experience.

TARP also developed what they called the tip of the iceberg phenomenon. The basic idea is that top management, which is ultimately responsible for the organization, rarely hears directly what the customer is asking for. Unless there is a mechanism to get these comments throughout the organization, management will never hear them. This is shown in Exhibit 4-11.

Complainants want to feel that management is sincere and will make a concerted effort to correct the situation. If this belief is supported, they will probably choose the same hotel again. The tendency of complainants, however, is not to believe. In a study one of the authors found that 29 percent of the still unsatisfied complainants indicated they would have been satisfied simply with a proper response from management rather than what they believed were token gestures.

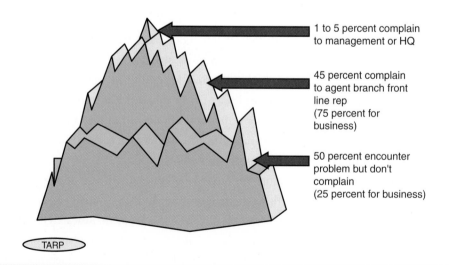

Exhibit 4-11 The tip of the iceberg phenomenon

Source: Goodman, J. (1999). *Basic facts on customer complaint behavior and the impact of service on the bottom line (p. 5).* White paper published by TARP. Retrieved from www.e-satisfy.com/research2.asp. Used by permission

What to Do about It

Practicing relationship marketing through consumer complaint handling is not the easiest task in the world. Many discontented customers will not take the trouble to complain. Actually encouraging complaints becomes the necessary objective.

Research has shown that people do not complain for three primary reasons: It is not worth the time and effort, they don't know where or how to complain, and they believe that nothing will be done even if they do complain. Marketing's task is to overcome these obstacles by making it easy to complain, making it known where and how to complain, and truly doing something about the complaint if it is reasonable (and over three fourths of all complaints appear to fall into that category). This means setting up specific procedures.[17] Such an action will also constitute internal marketing; when employees see management taking complaints seriously, they will feel more inclined to do likewise.

Categorically, there are four ways firms usually handle complaints, as shown in Exhibit 4-12. It should be clear that the cell in the lower right corner is where the service organization should be. Firms that take the active/corrective approach indicated in this cell essentially follow the eight steps to complaint handling listed below.

There are many reasons for encouraging customers to complain. The one we would like to point out here is the opportunity that this unhappy customer is giving to the company to solve the problem and transform her into a satisfied customer with repurchase intentions. According to a study conducted on the hotel industry, the way complaints are handled is the major factor determining whether someone will return for another night's stay.[18]

The following steps for effective complaint handling were extracted from the book *A Complaint Is a Gift:*[19]

1. **Say thank you.** Before you apologize about the incident thank the customer for sharing this valuable information with you.
2. **Explain why you appreciate the complaint.** Tell the customer how grateful you are to learn about this problem and to be given another chance to resolve it.
3. **Apologize for the mistake.** Apologize to the customer about the incident and any inconvenience it may have caused. Tell him that you sincerely hope you will be given a chance to serve him again in the future.

Exhibit 4-12	Complaint Matrix: How Companies Respond to Complaints

Company's Response

	Defensive	Corrective
Passive	Management apologizes to unhappy customer and does nothing.	Management solves complaint but does nothing to prevent it from happening again.
Active	Management encourages customers to complain but does nothing about the complaints.	Management solves complaint, finds out why it happened, and fixes cause. It then follows up with the customer and encourages him or her to return.

Source: Reprinted with permission from "Marketing Complaints: A Management Tool," published by the American Management Association, by Roland T. Rust and Bala Subramanian, 1992, *Marketing Management* 1(3), pp. 40–45.

4. **Promise to do something about the problem immediately.** Once you have apologized, tell the customer that you two should try to work things out immediately.
5. **Ask for necessary information.** Ask the customer all the essential information about the incident so you can more effectively try to resolve it.
6. **Correct the mistake, promptly.** This would be the ideal situation, so that the customer could walk out satisfied and remember what an effective recovery system your company has.
7. **Check customer satisfaction.** Before the customer leaves, check once more whether she is entirely satisfied with the outcome and see whether or how you can be of help. Thank her again for the opportunity she gave you.
8. **Prevent future mistakes.** Make sure you take the appropriate actions so this incident does not happen again.

Once complaint-handling procedures have been put in place, it is important to measure how well the procedures are working. Specifically, it is important to know the following:

- What percentage of guests has a problem during their stay?
- Of those that have a problem, what percentage is reporting the problem to management?
- Of those that report a problem, what percentage is claiming their problem was solved to their satisfaction?
- Of those that report a problem, what percentage is claiming their problem was not solved to their satisfaction?
- What is the impact of complaint procedures on customers' willingness to return, overall satisfaction, and net promoter score?

Each company must devise its own system for soliciting and handling complaints. Handling satisfied customers is easy; handling dissatisfied customers is the acid test of marketing and management.

EMPLOYEE RELATIONSHIP MARKETING, OR INTERNAL MARKETING

employee relationship marketing Applying marketing principles to those who serve the customer (directly or indirectly) and building a mutual bond of trust with them. This is sometimes referred to as **internal marketing**.

Employee relationship marketing, often referred to as **internal marketing,** means applying marketing principles and the components of the loyalty circle to the people who serve the customers. The emphasis of internal marketing is on the employee as the internal customer who also has needs, wants, and problems. What these customers are buying is their jobs. Thus, the job is the product that satisfies the needs and wants of these internal customers so that they, in turn, will better satisfy the needs and wants of the external customers.

To create customer value, the company must create employee value. Employees, after all, manage the process, provide the imagination, implement the policies, and derive the insight to help deepen customer bonding. This is especially true when we are dealing with so many intangibles that make for customer value and can never be explicit in the policies ordained. Without the commitment of employees, relationship marketing is doomed to fail. This may mean looking at employees in a new light, like customers. Thus, the employer must also create value for the employee.

Taco Bell found that 20 percent of the stores with the lowest employee turnover rates had double the sales and 55 percent higher profits than the 20 percent of the stores with the highest employee turnover rates.[20]

At first glance, this may appear to be a strange way to look at marketing; it is certainly not the way we look at the marketing of goods. However, a closer look makes the case obvious: One of the first tasks of marketing and management is to have the employees believe in their jobs, which is the product that they represent to the customer. The successful hospitality firm must first sell its jobs to employees before it can sell its services to customers. If it does not do this, it ends up with dissatisfied customers (employees) who will, one way or another, express their dissatisfaction to the paying customers. Paying customers, in turn, find that their problems are not adequately solved, so they go elsewhere. Clearly, this is not the way to keep customers. Thus, what is practiced in the creation and keeping of customers needs to also be practiced in the creation and keeping of employees. Relationships with customers often depend on employees to go beyond standard policies and procedures to make a big difference in problem resolution and the feeling a customer has for the company.

Noncontact Employees

We have discussed internal marketing from the point of view of the customer-contact employee. This is certainly the most obvious way to look at it. But it doesn't stop there. The engineers, housekeepers, night porters, cooks, dishwashers, storeroom people, and even accountants are part of the internal marketing effort in hospitality. Management's task is to get employees to realize and feel that they are part of the effort.

A dishwasher, for example, can stop a chipped glass or stained plate from going back into the dining room. The storeroom person can make certain that the right degree of freshness is received. The housekeeper can make sure that all the lightbulbs in a room are working. The engineer—well, you get the point. *All* employees in hospitality are part of the marketing effort. If you said this to back of the house employees, most of them would probably say, "Huh?" What influences customers must be *marketed* to employees. Consider, for example, housekeepers talking (or even shouting) in a hallway at 7:00 on a Saturday morning. One could tell them not to do this; better, one could remind or train the staff about the effect this has on sleeping customers in their guest rooms.

All employees need to fully understand how what they do affects the customers and people with whom they interact—fellow employees as well as customers. They should also be aware of new products, new developments, and new customer promises even before the customer. If a customer wishes to claim an advertised benefit, it is self-evident that the employee to whom the claim is made should know exactly what the customer is talking about and be enthused about promoting it. Too many times have we called the 800 reservation number of a national chain to request a rate advertised in a full-page newspaper ad to find that the reservation agent knew nothing about the promotion. "It's not on my screen," is a frequent response.

Here's a story that makes the point:

Mike Leven was once honored by Washington State University as Hotel Marketer of the Year. Two students from the hotel school picked him up at the Seattle airport. "Tell us about marketing," they said. "Just watch," said Mike. When they got to the Holiday Inn in Pullman (the other end of the state), there on the hotel sign in large neon lights were the words, "Welcome Mike Leven, Hotel Marketer of the Year." "Neat," thought Mike. When he went to check in, however, the clerk proclaimed that he had no reservation. More than a little tired, and now a little irate, Mike asked how that could be when his name was in neon lights at the front of the hotel. "How would I know that?" responded the clerk, "I came in the back door."

Clearly, figuratively speaking, employees need to come in the front door to understand the customer. Okay, but was this the clerk's fault or management's?

Keeping employees aware of what is going on is not just so they can respond to customers. Unawareness leads to embarrassment, disappointment, reduced motivation, and lack of support for marketing efforts. Employees are, in fact, as much vital recipients of marketing campaigns as are consumers.

The success of the internal marketing concept ultimately lies with management. Lower-level employees cannot be expected to be customer conscious if the management above them does not display the same focus. Operations-oriented managers who concern themselves primarily with policies and procedures, often instituted without regard to the customer, undermine the firm's internal marketing effort, reducing employees' jobs to mechanical functions that offer little in the way of challenge, self-esteem, or personal gratification. Moreover, by requiring employees to adhere rigidly to specific procedures, the operations-oriented manager ties their hands and restricts their ability to satisfy the customer. All this means that the organization that practices relationship marketing has to change from the traditional style.

Chapter Summary

A technique of nontraditional marketing is relationship marketing. Relationship marketing creates customer bonding and understanding that is an integral part of any company's sustenance and growth. Relationships must be developed and sustained so they build loyalty and increase the lifetime value of a customer. In a services business, this is hardly possible without similar employee relations. The principle of internal marketing is to market hospitality jobs to employees just as we market hotels or restaurants to customers.

Establishing good customer relations also involves creating an atmosphere in which customers' complaints are sincerely addressed. We call this service recovery. One way to do this is to talk to the customer and to make it easy for the customer to talk back. Complaints are healthy, customer problems are opportunities, and marketers must be opportunists. Innovative relationship marketing will give a sustainable competitive advantage to tomorrow's leaders and their companies.

Web Browsing Exercise

Go to the website for ICLP (international customer loyalty programs, www.iclployalty.com). What kind of firm is this? Who are their clients? What do they believe are their strengths and weaknesses?

Key Terms

employee relationship marketing, p. 106

frequency program, p. 100

internal marketing, See *employee relationship marketing.* p. 106

lifetime value of a customer, p. 94

loyalty ladder, p. 95

loyalty marketing: See *relationship marketing.* p. 89

loyalty program, p. 100

relationship marketing, p. 89

Discussion Questions

1. How would you describe relationship/loyalty marketing?

2. What is the loyalty ladder? Are you an advocate of any particular product or brand? Why?

3. What are loyalty programs? How are they different from frequent guest programs?

4. Can you explain why customer complaints can be good? Why can the lack of customer complaints be problematic?

5. Describe in your own words employee relationship (internal) marketing. Have you encountered good or bad internal marketing in your own hospitality or tourism work experience?

Endnotes

1. Shoemaker, S., & Lewis, R. (1999). Customer loyalty: The future of hospitality marketing. *Hospitality Management, 18,* 349.

2. Reichheld, K. (2002). [Letter to the editor]. *Harvard Business Review, 80*(11), 126.

3. Shoemaker & Lewis.

4. Levitt, T. (1981). Marketing intangible products and product intangibles. *Harvard Business Review,* May–June, 94–102. Copyright 1981 by the President and Fellows of Harvard College; all rights reserved.

5. Adapted from Gordon, I. (1988). *Relationship marketing.* Toronto, Canada: John Wiley & Sons, 9.

6. Adapted from Bowen, J. T., & Shoemaker, S. (1998, February). Loyalty: A strategic commitment. *Cornell Hotel and Restaurant Administration Quarterly,* 14–17.

7. Payne, A., & Frow, P. (2005). A strategic framework for customer relationship management. *Journal of Marketing, 69,* 157–176.

8. Heskett, J. L., Jones, T. O., Loveman, G. W., Sasser, W. E. Jr., & Schlesinger, L. A. (1994, March–April). Putting the service-profit chain to work. *Harvard Business Review,* 164.

9. Shoemaker & Lewis.

10. Much of the information found in this phase of the evolution of loyalty comes from Shoemaker, S., & Bowen J. (2003). Antecedents and consequences of

customer loyalty: An update. *Cornell Hotel and Restaurant Administration Quarterly, 6* (4), 31–52.

11. Peppers, D., & Rogers, M. (1996). *The one to one future.* New York: Doubleday.

12. Piccoli, G., & Applegate, L. M. (2003). Wyndham International: Fostering high-touch with high tech [Case study]. Case no. 9-803-092, pp. 1–42. Boston: Harvard Business School.

13. Barlow, J., & Møller, C. (1996). *A complaint is a gift.* San Francisco: Berrett-Koehler.

14. Goodman, J. (1999). *Basic facts on customer complaint behavior and the impact of service on the bottom line.* White paper published by TARP. Retrieved from www.e-satisfy.com/research2.asp

15. TARP. (2002). [Electronic brochure]. Retrieved July 22, 2004, from www.tarp.com/clients.html.

16. Goodman.

17. For more discussion of this subject, see Lewis, R. C., & Morris, S. V. (1987, February). The positive side of guest complaints. *Cornell Hotel and Restaurant Administration Quarterly,* 13–15.

18. Gilly, M. C. (1987). Post-complaint processes: From organizational response to repurchase behavior. *The Journal of Consumer Affairs, 21,* n-2. Extracted from Barlow, & Møller.

19. Barlow, J., & Møller, C. (1996). *A complaint is a gift.* San Francisco: Berrett-Koehler Publishers.

20. Heskett, et al., 19.

.... case study

Harrah's CRM Strategy

Discussion Questions

1. Why does Harrah's consider itself to be in the entertainment business? What is Harrah's major source of revenue?

2. Briefly describe Harrah's CRM (customer relationship management) Total Rewards program. How does it work?

3. What are some of the objectives of Harrah's loyalty program?

4. Apply the customer loyalty ladder concept described in this chapter to Harrah's Total Rewards program. Does it fit? How well does it fit?

5. What kind of information did Harrah's collect about its customers? How did management define its "ideal player"? Why?

_____

They (Harrah's) understand precisely what is going on with customers, how to motivate them and how to sell them more.[1]

—*Sergio Zyman, former Coca-Cola branding Guru*

CRM is the vertebrae of everything we do. I don't think most retailers take full advantage of what they already know about their customers or use that knowledge to modify future behaviour.[2]

—Gary Loveman, president and CEO, Harrah's Entertainment

The late 1990s witnessed a rapid expansion of casinos, especially in Las Vegas and Atlantic City, New Jersey, due to relaxations in the U.S. state and federal gaming laws. New laws had legalised gaming on riverboats and Indian reservations,[3] and this led to intense rivalry between casino operators who started spending millions of dollars in opening new extravagant properties that featured shopping malls and hotels to attract customers. Harrah's Entertainment Incorporated (Harrah's),[4] an entertainment company which had business interests in casinos, food & beverages and hotel rooms, decided to follow a different approach as it realized that about 90 percent of its revenues came from its casino business and not from associated ventures.

Instead of opening lavish properties, Harrah's initiated a customer relationship management (CRM) program that aimed at developing long-term relationships with its customers that would enable the company to capture a bigger market share in the gaming business. In 1997, Harrah's discovered that its customers spent only 36% of their annual gaming budget with the company. It realized that increasing customer spending would translate into significant increase in revenues.

At the heart of Harrah's CRM program was a loyalty program called Total Rewards (derived from an earlier Total Gold program) that rewarded customers with comps of food and lodging to stimulate loyalty. The Total Rewards program was aimed at gathering information about customers and using it to customize the company's marketing programs for each customer. Richard Mirman, Harrah's Senior Vice President of Marketing, explained, "We have an advantage in that we know who the customers are, what they're worth and we can touch them in ways that our competitors can't even think about."

Customers who enrolled in the Total Rewards program were given an electronic card which they inserted into the machines that they played. This card allowed Harrah's to keep track of its customers' preferences and collect data about them. Harrah's made extensive use of IT to aid the effective use of collected customer data. A massive data warehouse and business intelligence (BI) initiative was undertaken that enabled Harrah's to collect and consolidate customer data. Harrah's then used decision science tools from business analytical software on the collected data to better understand its customers and gain insight into their gaming preferences. "Harrah's is an outstanding example of a company that is aggressively and smartly using technology to understand the behavior of its customers" according to a press release following a national award on the use of it.

THE CRM PROGRAM

Harrah's competitors, such as MGM Mirage, Mandalay, and Caesars, were opening lavish properties that had spas, shopping malls, and extravagant hotel rooms, apart from gambling, in order to lure customers. Harrah's, however, decided to pursue a different strategy. The company knew that the majority of its revenues came from its casinos and not from hotels, stores and spas, as shown in Exhibit 1.

Harrah's decided to make information the key aspect of the company's marketing efforts. Instead of investing in extravagant properties, Harrah's began investing in IT that allowed the company to track and analyse every individual customer's transactions. The database generated out of this effort was then used to design customer programs that would

Exhibit 1	Revenue from Business Segments (In U.S. dollars)		
Business Segment	Revenues (2004)	Revenues (2003)	Revenues (2002)
Casino	4,077,694	3,458,396	3,285,877
Food and beverage	665,515	596,772	572,775
Rooms	390,077	339,037	317,914
Management fees	60,651	72,149	66,888
Other	217,195	190,092	148,635
Less: Casino promotional allowances	(862,806)	(707,581)	(644,223)
Total Revenue	**4,548,326**	**3,948,865**	**3,747,866**

Source: Harrah's Annual Report, 2004.

help deliver better customer service, which in turn would aid in keeping customers loyal to Harrah's casinos.

Enhancing customer loyalty and capturing a larger share of customers' gambling budgets was the main aim of Harrah's CRM strategy. The CRM program helped Harrah's understand the potential value of each customer and identify a suitable contact strategy that would bring him/her back to Harrah's casino more frequently. Harrah's CRM strategy focused on four key elements. The first and the most important was the Total Gold/Total Rewards customer loyalty program. The second element was decision science based tools. The Hotel Revenue Management initiative constituted the third element; the final element was Personal Contact Management.

CUSTOMER LOYALTY PROGRAMS

The Total Gold program was based on a premise that the best way to improve business performance was not to attract new customers but instead to get the existing customers to spend more. Harrah's realized that capturing a higher percentage of its existing customers' gaming budget would significantly improve the company's profits. This led Harrah's to shift its emphasis toward building long-lasting relationships with its customers.

Customers who agreed to join the Total Gold program had to fill out a membership form from which Harrah's obtained data such as their name, address and telephone number. The customers were then given an electronic card, which was inserted into the machines that they played on. The card had a magnetic strip that identified each customer. Each time customers used their cards, Harrah's got to know many details, such as when the customer had visited, which property of Harrah's they had used, what games they had played, and the result of those games. This information was then stored in a centralized repository and used to recognize potential customers with whom long-lasting relations could be established. The information was also used to reward customers with "goodies" such as casino compliments, discounts on meals, hotel accommodations and invitations to special events. Commenting on the benefits of the program, Philip Satre, CEO, explained, "The Total Gold player reward and recognition program and our industry-leading customer databases give Harrah's important tools for enhancing relationships with our target customers and making them loyal to Harrah's wherever they choose to play." They were able to do this because

Harrah's had obtained a patent for "real time" data on customer casino play technology.

In 1998, Gary Loveman was hired as chief operating officer of Harrah's. He was an MIT trained economist who had spent four years as Professor at Harvard Business School. Loveman had spent a great deal of time at Harvard studying how businesses can identify their most profitable customers and get them to shop more often. After joining Harrah's, Loveman immediately identified weaknesses in the Total Gold program. First, there was nothing different about Harrah's rewards as compared to its competitors. Second, there was no consistency in the rewards earned by the customers. They earned different rewards at different properties. Finally, customers were not given adequate incentives to consolidate their play at Harrah's and aspire for higher levels of benefits and services. Loveman wanted consolidation of play to enable a customer to combine the reward points earned at various properties.

Research conducted by Loveman revealed that the Total Gold customers spent only 36 percent of their total annual gaming budgets at Harrah's. Loveman immediately saw an opportunity and decided upon a strategy that would focus on encouraging customers to spend more of their gaming budget at Harrah's. He decided to customise the relationship Harrah's had with each of its customers so that the experience a customer had at Harrah's was of a higher quality in comparison to its competitors. Harrah's began treating millions of customers differently depending upon their value to the company. This was made the key aspect of the Total Rewards program, which was essentially a successor of the Total Gold program. The focus was now on a customer's worth over a period of time instead of his/her worth during one visit.

The Total Rewards program was a tiered approach to customer loyalty that promoted another important facet of Loveman's strategy—aspiration. Based on their value, customers were divided into three tiers: Gold, Platinum and Diamond. The customers earned points, called Reward Credits, every time they played and were allotted different tiers depending upon the number of credits they earned. The levels of service offered to customers progressively improved from Gold card holders to Diamond card holders, who were offered the greatest levels of service. Deliberate efforts were made to highlight the difference in service levels to customers holding Gold cards so that they could aspire to achieve higher levels of service. Commenting on the aspiration program, Mirman, then Chief Marketing Officer at Harrah's, said, "It's in customers' best interests to continue to play with us, not go to our competitor, and achieve that status.

Because once you get that status, the quality of your experience increases dramatically."

Another important objective of the Total Rewards program was consolidation of play. Harrah's found that customers considered luck a major factor in deciding where to play. They decided to go to another casino when they were unlucky and started losing. Harrah's extended the rewards program to all games so that customers could just move to another game whenever they found they were not lucky in a particular game. This helped to stop customers from moving to a competitor's casino.

Cross-market playing at various properties of Harrah's was another objective of the Total Rewards program. Cross-market play happened when customers played at one of Harrah's properties other than their "home" casino. Customers could visit any of Harrah's properties in the U.S. and use the same Total Rewards card. This gave Harrah's customers enough incentive to stick to the Harrah's brand wherever they went in the country. Harrah's strategy of encouraging cross-market playing was supported by the company's branding strategy that focused on building the Harrah's brand and not promoting any particular property of Harrah's. This enabled a consistent brand experience across all touch points.

DECISION SCIENCE TOOLS

The Total Rewards program enabled Harrah's to develop a huge database containing a record of customers' personal information and their transactions. Loveman felt that the use of decision-science-based analytical tools on this database would enable him to gain valuable insights into customers' spending habits and preferences. Customised offers to each customer were then designed based on this information. Loveman explained, "We're going to keep offering you things and asking you and talking with you . . . [until] we get to the point where we know what you want."

Estimating a customer's worth over a period of time was the primary purpose of incorporating decision science tools into Harrah's CRM strategy. The tools allowed Harrah's marketing analysts to query the customer database and determine each customer's preferences and predict the services and rewards each customer would prefer. Decision science tools made use of data such as age, distance from casino, types of games played, number of bets placed, average bet and total amount of money deposited in order to come out with a detailed gambling profile which would then be used to personalize a marketing program that would entice the customer to revisit Harrah's. This data was used to estimate how much money the company could earn from an individual customer over a period of time. These tools also helped Harrah's acquire important information like: 80 percent of its casino revenues came from customers who spent only between $100 and $500 per visit and that an "ideal player" would be a woman, 62 years of age, who lived within 30 minutes of a casino. Such customers would have spare time, disposable cash and easy access to a casino.

Decision science tools also helped Harrah's in segmenting the entire customer base. Customers broadly fell under four categories: Prospect, Non-Loyal, Loyal and Attritor. The Prospect segment mainly consisted of nongamers or gamers who were loyal to another casino. Harrah's marketing treatment to this segment lowered their switching cost, thereby inciting them to come to Harrah's. The marketing treatment for the Non-Loyal segment encouraged customers to consolidate their play at Harrah's. The Loyal segment was encouraged to maintain loyalty and the Attritor (a non-word derived from "attrition") segment was given a marketing treatment that would reinstill loyalty. The differential marketing treatment meted out to each segment was aimed at strengthening customer loyalty.

Decision science tools that acted upon the data warehouse were used by analysts in Harrah's marketing department to customize offerings to its customers. Harrah's analysts used the Cognos Impromptu[5] to access the data warehouse and run predefined reports and execute queries. For instance, Harrah's could use the Cognos tool to obtain a list of customers who reside close to one of Harrah's properties. It was the Cognos tool that made Harrah's realize that customers liked to move from property to property instead of being loyal to a single property. This led Harrah's to bring in the element of encouraging cross-market play in the Total Rewards program. Using the tool, Harrah's was also able to link staff reports to sales receipts, which helped in identifying its best staff. The company was also able to anticipate seasonal customer behaviour. Commenting on the benefits of Cognos, Tracy Austin, Vice President of IT Development at Harrah's, said, "Using Cognos in our Total Rewards card program, we can track each player's gaming patterns and preferences across our 21 locations in the U.S. By uncovering trends in the data, Cognos BI allows us to determine the destination, dining, and gaming choices of our customers. Using this insight, we are able to build complimentary and special offer programs that are customized to the individual customers."

SAS business analytic software was instrumental in carrying out market segmentation, profiling and predictive modelling. Harrah's used SAS on its data warehouse in order to segment its entire customer base in 90 segments based on demographics. The tool also helped Harrah's to develop a detailed profile for each customer, based on the gaming habits, which would allow it to study how profitable the customer was to the company. Predictive analysis was then carried out by SAS on these profiles in order to identify those customers who were most likely to respond to Harrah's campaigns. Harrah's then targeted these customers with specific campaigns, tailored to each individual customer, in order to steal him/her away from competitors and increase loyalty towards Harrah's. SAS allowed Harrah's to be more precise and effective in its communication with customers, which in turn increased the profitability of its campaigns. Commenting on the benefits of SAS, David Norton, Senior Vice President, Relationship Marketing, said, "Our profitability around marketing interventions is much higher because of the precision of understanding what SAS provides." By using SAS, Harrah's was able to group data into clusters and seek trends.

HOTEL REVENUE MANAGEMENT

Harrah's Hotel Revenue Management System (HRMS) was aimed at offering customers rooms at attractive tariffs. Apart from achieving higher room occupancy rates, the revenue management initiative allowed Harrah's to maximize profit on each room. In order to achieve this, trade-offs were made between gaming revenues and room tariffs. The gaming revenue from a customer was tracked from the Loyalty program. For example, a customer who had a history of wagering $4,000 a night at Harrah's casinos, might be offered a complimentary room at one of the Harrah's properties, but another customer unknown to Harrah's may be turned down, even if he/she was willing to spend $400 a night for that room. The aim was to optimise the sum of gaming and room revenues. Norton explained, "It's not about filling each room. It's about maximizing out the profit from each room."

The Hotel Revenue Management System (HRMS) determined the availability and the room rates across Harrah's properties. Exhibit 2 shows a diagrammatic representation of HRMS. The system made use of current and historical demand in order to come out with a demand forecast for a particular time of the year. Customer profitability and competitive conditions were then taken into consideration by the system in order to work out the room rates.

The HRMS allowed Harrah's to ensure that rooms were made available to highly profitable customers at the right price. This was in tune with Harrah's CRM strategy that aimed at capturing a bigger piece of its profitable customers' gaming budget. Filling Harrah's hotel properties with profitable customers also helped in boosting hospitality revenues. Tim Stanley, Chief Information Officer at Harrah's, commented, "Our goal is to never turn down a good customer but not leave the hotel empty."

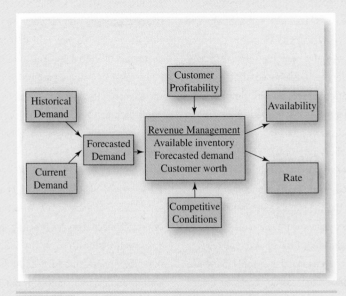

Exhibit 2   Hotel Revenue Management System
Source: www.bcdma.org.

PERSONAL CONTACT MANAGEMENT

After the analysis of its customer database, Harrah's arrived at a conclusion that over 25 percent of its revenues came from just 5 percent of its customer base. The Personal Contact Management (PCM) initiative allowed Harrah's to target its marketing efforts at these "high value" customers. Harrah's called this segment "VIP customers" and devised a strategy that ensured additional visits from these customers, which, in turn, delivered additional revenues. This initiative was an important element of Harrah's CRM strategy as it focused on drawing more revenues from existing customers.

An important component of PCM initiative was a dedicated sales force, called Host, who served the VIP customers. These customers were assigned personal hosts who were responsible for delivering a highly personalized experience to them. Hosts were expected to know a great deal about their guests and create a personal bond with them. A host, for example, was required to know which room at a particular property was a VIP guest's favorite. The host would book the room and stock it with items that suited the guest's tastes and preferences.

A host's ability to deliver highly personalized service to a VIP guest depended on the in-depth information he/she had about the guest. A Player Contact System (PCS) was implemented to provide information such as guest preferences, visit histories, worth, relationships and contact information. Apart from providing hosts with a centralized source of customer information, the system also allowed Harrah's to prioritise hosts' tasks and track their performance.

In addition to the above, several other measures helped enhance the overall experience that a customer went through at Harrah's. Friendly and quick services were identified as major factors that customers expected from their hosts. Loveman decided to associate employee remuneration with these two factors. Employees were required to take courses that would enable them to deliver higher levels of service. Customer satisfaction scores were instituted for each property and these scores were independent of its financial performance.

USING INFORMATION TECHNOLOGY FOR CRM

The extensive use of information technology (IT) was central to the success of Harrah's CRM initiatives. The company's ability to develop and nurture relationships with its customers was a result of its ability to capture information and use it to its advantage. With millions of customers visiting its casinos annually, a significant amount of information was generated. Harrah's ability to manage this information and act on it with the help of IT gave the company a competitive advantage. John Boushy, the CIO of Harrah's, explained, "Our objective is to use information as a strategic offensive weapon and that starts with a clear understanding of who the customer is and what they want."

A massive data warehouse and BI initiative formed the backbone of Harrah's CRM strategy. In 1994 Harrah's started developing a system called Winner Information Network (WINet). WINet was used to consolidate customer data and facilitate information sharing across Harrah's properties Exhibit 3 shows the WINet architecture. WINet went on to drive the loyalty programs at Harrah's and helped in ensuring consistency in customer experience across the company's properties. Commenting on its benefits, Boushy said, "WINet is the engine of Total Rewards. It is the database that allows us to deliver the individually tailored benefits that keep customers coming back to our properties."

Harrah's gathered data about its customers using a magnetic card, which had a unique player identification code on it. Code readers were present on all gaming tables, which when cards were inserted into them, identified customers. Gaming tables were, in turn, connected to AS 400 transactional systems on each property. A UNIX gateway was then used to consolidate all customer data into a single repository, known as the Patron Database (PDB). This database was then used as a source for the data warehouse, called Marketing Workbench (MWB), where the analytical work was carried out. Decision science tools were used on this data warehouse to study customer preferences and predict the services and rewards each customer would respond to.

The PDB was essentially a national customer database that recorded customer activity across Harrah's various points of sale at its casinos, restaurants and hotels. It also acted as an operational data store that catered to queries, which were generated during day-to-day operations such as hotel check-ins. All customer information gathered at various properties across the country were tested for validity and loaded into the PDB. The PDB allowed Harrah's to have a single view of each customer and get a more comprehensive picture. This view comprised gaming habits at casinos and lifestyle preferences at hotels and restaurants. Employees across various properties could now get a consistent picture of each customer

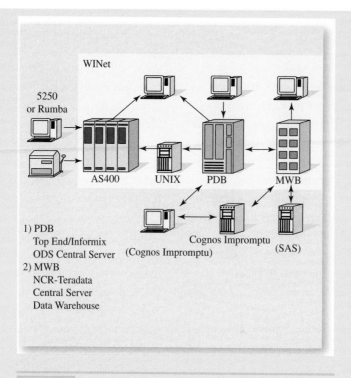

Exhibit 3   WINet Architecture

Source: www.teradata-j.com.

and ensure that the customer gets the same service wherever he/she went.

On the other hand, MWB served as the data warehouse where analytical work such as customer segmentation, profiling and predictive analysis were carried out. The data warehouse stored daily data for three months, monthly information for two years and yearly information since 1994. Harrah's used decision science tools including Cognos and SAS on this data warehouse inorder to create customized marketing campaigns.

In order to further personalise its relationships with customers, Harrah's used Internet technologies. In June 2001, Harrah's launched eTotal Rewards program that allowed Harrah's customers to view benefits and complimentary offers on the Internet. Harrah's had to integrate its back-end CRM infrastructure with Internet technologies in order to extract information from the database and make them available on its website for the customers to see. Customers could view offers personalized to them online. They could even check for availability and make online reservations at Harrah's hotels and casinos. Commenting on the eTotal Rewards program, Norton said, "We're seeing great return from eTotal Rewards. In addition to increased revenue and cost savings from customer

self-service, the biggest difference is how the channel is working to strengthen our customer relationships. We can now provide customers a personalized experience on the Web, with deals targeted and tailored to their level of loyalty instead of generic offerings. By offering our customers a fully integrated Total Rewards program across multiple channels, we have taken customer service and our CRM capabilities to a new level."

THE BENEFITS

Harrah's CRM initiatives were the first of its kind in the U.S. gaming industry, and the benefits reaped by the company were impressive. When the Total Rewards program began in the year 2000, Harrah's was able to capture only 36% of its customers' gaming budget. By 2002, this figure rose to 43%. The company witnessed a 13% jump in profits in the very first year of the Total Rewards initiative even after spending $251 million as rewards. The loyalty card programs had 12 million enrollments in 1997. By 2003, the enrollments went up to 26 million. Harrah's generated 75 percent of its revenues from its loyalty card enrollments in 2003, up from 50% when the loyalty program started. This data revealed the popularity of the loyalty programs.

Exhibit 4   Harrah's Five-Year Financial Summary (In $ millions except common stock data and financial percentages and ratios)

Operating Data	2004	2003	2002	2001	2000
Revenues	4,548.3	3,948.9	3,747.9	3,317.4	2,977.8
Income from operations	791.1	678.8	708.7	521.8	188.2
Income/loss from continuing operations	329.5	261.1	282.2	173.8	(46.4)
Net income/loss	367.7	292.6	235.0	209.0	(12.1)
Common Stock Data					
Earnings/loss per share—diluted					
Income from continuing operations	2.92	2.36	2.48	1.50	(0.40)
Net income/loss	3.26	2.65	2.07	1.81	(0.10)
Cash dividends declared per share	1.26	0.60	-	-	-
Financial Position					
Total assets	8,585.6	6,578.8	6,350.0	6,128.6	5,166.1
Long-term debt	5,151.1	3,671.9	3,763.1	3,719.4	2,835.8
Stockholders' equity	2,035.2	1,738.4	1,471.0	1,374.1	1,269.7
Financial Percentages/Ratios					
Return on revenues—continuing	7.2%	6.6%	7.5%	5.2%	(1.6)%
Return on average invested capital					
Continuing operations	8.2%	8.0%	8.9%	7.5%	2.4%
Net income/loss	8.0%	7.6%	6.9%	7.3%	2.9%
Return on average equity					
Continuing operations	17.5%	16.0%	19.3%	12.9%	(3.2)%
Net income/loss	19.5%	18.0%	16.1%	15.5%	(0.8)%
Ratio of earnings to fixed charges	2.7	2.6	2.7	2.0	2.0

Source: Harrah's Annual Report, 2004.

The CRM initiative also helped Harrah's to increase its same-store sales, that is, sales from stores that have been functioning for more than one year. Same store sales, an indication of customer loyalty, went up by 14% in 1999, an increase of $242 million over 1998. Due to its CRM initiative, Harrah's was able to increase its same-store sales between 1999 and 2002 another 65%.

Another important benefit of the CRM program was the increase in revenues through cross-market play. Revenues from cross-market play went up from 13% in 1997 to 22% in 1999. In 2002, Harrah's generated more than $1 billion in revenues from cross-market plays, which was up by 20% as compared to the 2001 figures. By 2004, the revenues rose to $1.44 billion. Harrah's two properties in Las Vegas, considered a very competitive market, posted a combined EBITDA of 37%, because of increased cross-market play.

The CRM initiative's impact on the overall earnings of Harrah's was significant. In 2002, Harrah's 26 casinos across 13 states generated revenues of $4.14 billion, which was

12% higher than the previous year. This was in the aftermath of September 11 terrorist attacks when the U.S. entertainment industry was in the doldrums. Harrah's total revenues grew from $2,977.8 million reported in fiscal 2000 to $4,548.3 million in fiscal 2004. (Exhibit 4 shows Harrah's five-year financial summary).

Endnotes

1. Daniel McGinn and Steve Friess, "From Harvard to Las Vegas," *Newsweek*, April 18, 2005.
2. Susan Reda, "Harrah's Hits the Jackpot with CRM," *Stores*, June 2003.
3. Indian reservations are lands reserved for American Indian tribes in the U.S. and are governed by tribal governments. There are more than 300 Indian reservations in the U.S.
4. Las Vegas, Nevada–headquartered Harrah's Entertainment Inc. generated revenues of $4.54 billion in 2004. In July 2004, Harrah's reached an agreement to acquire Caesars Entertainment Inc. for $9.4 billion. The completion of the deal made Harrah's the world's largest gaming company.
5. Developd by Cognos Inc., the Impromptu is a database query and reporting tool that distributes reports over the web.

Integration

overview

Now that we have defined hospitality marketing and explained how hospitality marketing is different from traditional goods marketing, it is important that we introduce the concept of the marketing plan. The marketing plan is the management tool that sets forth the specific action steps that the marketing department, and in fact the whole organization, must take in the forthcoming year in order to be financially successful. The marketing plan is a working document used to guide the organization in its strategic direction toward the achievement of detailed goals and objectives. We also discuss

continued on pg 122

chapter 5

The Marketing Plan

learning objectives

After reading this chapter, you should be able to:

1. Create a marketing plan and understand all the components necessary to make a successful marketing plan.

2. Develop marketing action plans that are critical to the marketing plan.

3. Develop a marketing forecast, a marketing budget, and marketing controls and explain why they are all critical components of the marketing plan.

4. Develop a mission statement that ties into the marketing plan.

5. Identify the type of data needed to develop a marketing plan.

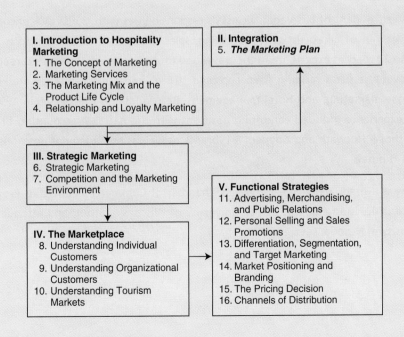

I. Introduction to Hospitality Marketing
1. The Concept of Marketing
2. Marketing Services
3. The Marketing Mix and the Product Life Cycle
4. Relationship and Loyalty Marketing

II. Integration
5. *The Marketing Plan*

III. Strategic Marketing
6. Strategic Marketing
7. Competition and the Marketing Environment

IV. The Marketplace
8. Understanding Individual Customers
9. Understanding Organizational Customers
10. Understanding Tourism Markets

V. Functional Strategies
11. Advertising, Merchandising, and Public Relations
12. Personal Selling and Sales Promotions
13. Differentiation, Segmentation, and Target Marketing
14. Market Positioning and Branding
15. The Pricing Decision
16. Channels of Distribution

data collection in this chapter. Workable, effective, and realistic marketing plans are developed only through the gathering of complete and adequate information followed by its thorough and objective analysis.

Finally, we discuss the mission statement. After all of the data collection and analysis has been completed, the mission statement, including broad objectives and positioning strategy for the future is created for the business unit. This mission statement follows the conclusions of the research, not the other way around.

Marketing Executive Profile

Terry Jicinsky
Senior Vice President of Marketing, Las Vegas Convention and Visitors Authority

Terry Jicinsky is senior vice president of marketing for the Las Vegas Convention and Visitors Authority (LVCVA). The LVCVA, with a current annual budget of just over $190 million, is notably the largest convention and visitors bureau in the world. The organization is responsible for marketing the brand of Las Vegas and Southern Nevada as one of the nation's premier vacation, gaming, and convention destinations.

A 17-year resident of Las Vegas, Jicinsky's responsibilities with the LVCVA encompass the oversight of all advertising, marketing, sales, and public relations efforts. With close to 22 years of experience in the travel and tourism industry, Jicinsky's career path has covered aspects ranging from consumer travel research to Internet marketing, database marketing, and hotel management. Before joining the LVCVA in 1992, his work experience included consulting positions with the national accounting firms of Laventhol & Horwath and Coopers & Lybrand, as well as management positions with Marriott hotels.

Jicinsky is a graduate of the Las Vegas Chamber of Commerce Leadership Program, class of 2002; is an alumnus of the University of Wisconsin, Stout; and attended the University of Nevada, Las Vegas.

Marketing in Action

Terry Jicinsky

Senior Vice President of Marketing, Las Vegas Convention and Visitors Authority

How is your marketing plan for the Las Vegas Convention and Visitors Authority developed? What is the process you go through?

In the example of Las Vegas, it's clearly a collaborative effort in that all of our stakeholders and partners play a role in developing that **marketing plan.** For example, our sales force is integral to helping us understand the market segments we're trying to reach with the marketing plan. Our advertising agency plays a key role in helping us understand the branding message we're sending out. Our board of directors, representing the community as a whole, plays a role in helping us understand how we tie in with the hotel product and the attractions in Las Vegas.

marketing plan The working document a firm uses to implement its marketing strategy. This is typically done on an annual basis.

How many market segments are there that you go after in Las Vegas?

We go after almost every market segment there is, but at the broadest level, we focus on two very broad-based market segments, those being the leisure/pleasure traveler and the convention/trade show traveler. As Las Vegas evolves from a gaming destination to a full-service resort destination, the gaming message over the years has become a smaller and smaller component of the overall branding campaign. So now, it's part of the mix, but in each of those market segments, it's becoming less and less of the primary component. As a proportion of total expenditure, the gaming budget continues to decrease, but nongaming becomes a larger portion of the total travel wallet.

When you develop the marketing plan, what are the critical issues you consider?

First and foremost, we consider what the hotel community and the attractions are offering our consumer. As a destination marketing organization, we are in that position where we actually market something that we don't control; rather, the hotels and attractions control the actual experience. The first element of building our branding campaign and our marketing campaign is understanding how the product is evolving and making sure our message is in tune with what the actual product offering is. It's a branding campaign about the destination as a whole that is reflective of every stakeholder we have.

How often is your marketing plan developed, and what does that entail?

We produce a five-year marketing plan that is updated every 18 months. It is actually a full rewrite of the campaign. We will add another 18 months to the back of the campaign so that we always have a working five-year document, or a five-year road map. In reality, it's really in the current 18 months that we have tangible action plans in place, and that's why we update them every 18 months. The integral part of the marketing plan itself is a very robust research program that focuses on tracking our successes on a month-to-month basis. For example, we do a U.S. population–based advertising awareness survey, where we measure the awareness of the campaign itself across the United States with the traveling public, and we create an index of how we compare from month to month in the awareness factor. A sister program to that is what we call a perception study, done every six months, which is a much more in-depth consumer research program, where we measure the people's perception of Las Vegas as a vacation destination, and then we tie that back to goals and objectives of the advertising campaign.

Getting the All Star game for the NBA must be a huge feather in your cap.

Attracting the 2007 NBA All Star game was really a milestone for our destination. First and foremost, it is the first time the NBA has agreed to play an All Star game in a city that doesn't have an NBA franchise. We were very excited to be the first community to do that. Second, it really raises the profile of Las Vegas in the sporting world. Throughout our history, we've always found a high correlation between sporting enthusiasts and people who like to gamble. There's that degree of risk taking, that degree of involvement in your leisure activity whether you're actually playing a sport or sitting in an arena as a fan. There's that same connection to the casino experience, whether you're actually participating in the activity. Anytime we can tie our brand to a sporting event proves to be very successful for us.

So these major sporting events really fit into the whole brand awareness mentality?

The National Finals Rodeo is another example of that. Ten days out of the month of December every year, Las Vegas turns into a Country and Western community, and we have the premier rodeo event of the world headquartered in Las Vegas. The NASCAR race we have every spring is another example of tapping into that NASCAR world and cobranding Las Vegas and NASCAR racing.

Does Las Vegas market to diversity submarkets?

Right now we have programs specific to four diversity submarkets: the Hispanic market, the African American market, the Asian market, and the gay and lesbian market. We have had most of our success with the Hispanic market to be centered around the Spanish holidays, as well as generational family travel, where the grandparents, parents, and adult children travel together. So, many of our programs in this market segment are specific to these criteria. For the African American segment, most of the programs are focused on special events, or special entertainers, so we have jazz concerts that might have specific entertainers that are heavily followed by the African American community. This fall, we have a syndicated radio personality, Tom Joyner, who will be doing live broadcasts from Las Vegas for a week, and then we have specific minority marketing programs targeting the African American market in partnership with his airing of the show in Las Vegas.

The gay and lesbian market is our newest market, and we're still in the process of developing programs. We are looking for opportunities that will capitalize on special events that may be of interest to this market, as well as some customer service issues that are letting them know that this marketing segment is welcome in the community and will be treated appropriately.

What insights could you give to students that might help them with marketing plans?

Specific to marketing and destination, I think the biggest thing that Las Vegas has learned over the years is that, while understanding the product and understanding the individual stakeholders' needs, the biggest thing to take into account when developing a marketing plan is understanding individual consumer experience. So, in addition to just focusing on your product offering and the tangible things, the most successful marketing plans really understand the experience of visiting the destination. There is a difference between simply marketing your destination based on the "what" and the "who." Really, marketing is based on the "why," and how it transcends the visitor experience.

Used by permission from Terry Jicinsky.

The marketing plan of a business unit (e.g., Marriott Hobby Airport) is derived from its strategy and mission statement, which, in turn, derives from the corporate (Marriott International) strategy and corporate mission statement. In many hospitality firms, the corporate level and the business unit level are one and the same, so the strategic plan and the marketing plan will be at one level only. Marketing plans are quite common in hotel chains and large restaurant companies, but not so common in smaller businesses, especially restaurants. These smaller firms typically do not do strategic planning, nor do they develop annual marketing plans or even have mission statements. This can be a mistake.

We focus our discussion of the marketing plan on the business unit (e.g., individual property), with the hope that it can be used as a guide for both those properties that are not currently developing a marketing plan and those that want to be sure they are developing their plan properly.

REQUIREMENTS FOR A MARKETING PLAN

There are three key elements to a successful marketing plan:

1. It is workable.
2. It is realistic and flexible.
3. It has measurable, achievable goals.

Too many plans fail in one or more of these respects. The marketing plan has to remain simple and easy to execute. Two-hundred-page marketing plans with a list of 100 action steps may be impressive, but they are not workable. Too many businesses confuse activity with productivity. The result is poor performance and frustration. The marketing plan that is the simplest, listing just the key items to be completed, will be the most focused and successful.

The marketing plan must also be realistic and flexible. Although an analogy to a road map is somewhat old, it is still useful here. A road map is useful if one is lost in a highway system, but not in a swamp whose topography is constantly changing. A simple compass that indicates the general direction and allows you to use your own ingenuity in overcoming difficulties is far more valuable in a swamp.[1]

The topography of the hospitality industry changes rapidly these days, as the wants and needs of customers change and as new firms enter and exit the market. Marketing plans need to always be flexible to account for these changes. Thus, marketing plans must constantly be reviewed and reevaluated. This is not to say that they should be changed at the slightest sign of a market change (e.g., the announcement that a property is changing its brand affiliation). It simply means that you must not be locked into your developed plan. If the situation changes and there is evidence that the plans you initially made are no longer the most effective, you must be ready to change. In the aftermath of the terrorist attacks in 2001 many people stopped flying. This meant that the geographic radius of the targeted customer suddenly became smaller. The fly-in market was replaced with the drive-in market.

The marketing plan must also be realistic in terms of the time and resources available for execution. This would seem to be a fairly obvious statement, but it is often violated. Owners' demands, corporate demands, management's demands, and those of others lead to many marketing plans that simply have little or no chance of success because the firm lacks the resources to carry out the objectives. Wild-eyed dreams and wishful thinking will not overcome the realities of the marketplace just because someone higher up says, "Raise the numbers" (a hotel industry expression meaning increase occupancy and average rate).

Finally, the marketing plan should have measurable goals. The marketing plan needs to assign specific responsibilities, with times and dates for accomplishment of measurable and achievable goals, to individuals and departments within the organization. Examples might be: "Raise occupancy four percentage points" or "Raise REVPAR four dollars." Continuous follow-up ensures that these responsibilities will be met or changed, as need be. This provision requires that the plan be thoroughly understood by everyone in the organization. A good plan indicates how marketing activities are integrated with all of the other activities of the operation. What this means is that responsibility for implementing the marketing plan does not stop at the door of the marketing office. Although the details of the entire plan will not go to every person in the workforce, the essence of the plan should do exactly that. Too often, plans are made yet key individuals know nothing about them.

A Bangkok hotel, for example, planned to attract a market segment of German families with children on vacation at a package rate. The promotion was a success and the families came, but no one had made adequate plans, as promised in the promotion, for children's activities, baby-sitters, or even extra beds to be placed in the rooms. In another example, a Valentine's Day promotion at a large New York City hotel was part of a well-written marketing plan. The hotel sold out on the promotion, making it a financial success. Unfortunately, the front office manager forgot to tell the garage that the promotion included free parking. The result was a one and a half hour wait to park and another hour to retrieve a car the next day. A full hotel with angry customers is not an example of a well-executed marketing plan.

Any plan that succeeds in attracting the market but fails to fulfill its promises (recall that this was Gap 4 that was discussed in Chapter 2) will be self-defeating in the long term. Personnel cannot deliver marketing promises if they don't know what those promises are or don't have the tools to deliver them.

A good marketing plan provides direction for an operation. It states where you are going and what you are going to have to do to get there. It builds employee and management confidence through shared effort and teamwork toward common goals. It recognizes weaknesses, emphasizes strengths, and deals with reality. It seeks and exploits opportunities. And last but certainly not least, a good marketing plan gets everyone into the act. Like everything else we have said in this book, the test of the marketing plan is embodied in the question, How will the customer be served?

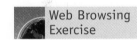

Web Browsing Exercise

Use your favorite search engine and type in the words "marketing plan." Examine in detail one or two of the sites you find. Be prepared to discuss in detail in class what you found and what you learned.

DEVELOPMENT OF THE MARKETING PLAN

As in strategic planning, the marketing plan begins with a situational analysis. Here, however, we are dealing with greater specifics. Our goal is to decide how our marketing resources will be used to best attract and serve the specific markets.

Therefore, it is best to begin with a short, simple version of the mission statement of the individual property, which sets forth its broad mission, keeping in mind the corporate philosophy and the master strategy. For example, consider the hotel that picks up on the corporate mission and adds a short mission for each of its functional departments, as shown in Exhibit 5-1.

To explain how the marketing plan is developed, we have created a fictional five-star hotel (four-star deluxe by French nomenclature) that we have located in Paris. We will use this hotel as an example throughout this chapter. We start with words from the mission statement for this property:

- Grand expectations
- Pleasant surprises
- We will consistently serve our guests, employees, and owners by exceeding expectations and continually enhancing our standards of service excellence.

Exhibit 5-1   Sample Mission Statement

CORPORATE

The corporation is a socially sensitive organization committed to its team members and customers, and to the communities in which we do business. We are committed to the recognition and satisfaction of human needs through integrity, quality, and communications. As an industry leader, we create a quality product and a quality work environment, which will result in superior financial performance and long-term asset appreciation.

FUNCTIONAL AREAS

- **Human Resources:** We are the human tools that ensure the success of our team members by providing a healthy environment through competitive recruitment, development recognition, and compensation.
- **Housekeeping:** We will do our best to provide you with clean and comfortable accommodations each and every day.

- **Engineering:** We are committed to doing it right the first time.
- **Food & Beverage:** We provide each and every guest with an exceptional dining experience, using the freshest ingredients available and presenting our products in an interesting, efficient, friendly, and professional manner.
- **Front Office:** We are empowered and committed to providing quality service and a friendly atmosphere—at your home away from home.
- **Reservations:** There are no hesitations when you make reservations—no matter which channel you use.
- **Sales:** We will demonstrate and encourage a level of service that inspires our fellow team members. We will respect each team member as we work together to achieve excellence and have fun in the process.

- ▓ Listening, not hearing
- ▓ Doing, not just acting
- ▓ Anticipating, not just serving
- ▓ Caring, genuinely

This will establish the context within which the marketing plan will be developed, which will be restated later as mission objectives.

At this point the marketing plan should include a statement about what has worked well in the past and broad objectives for the coming year, such as to increase actual market share from 8 to 10 percent* or, qualitatively, to reach or expand into new markets or to project a better image.

The next step is to complete the first major portion of the plan, data collection.

DATA COLLECTION

Data collection can be divided into two parts: external and internal. Following are explanations of what this means in terms of developing the marketing plan. We discuss these topics only briefly here; Chapters 6 and 10 address them in more detail.

External Environment

External data deal with the environment, including domestic and international trends. There are also numerous industry trends to be considered, such as the growth or decline of various market segments, building trends affecting future supply, room occupancy, eating-out trends, and new concept trends. We discuss much of this later in the

---

*An interesting side note: Jack Welch, CEO of General Electric for many years and considered by many to have been one of the top CEOs in industry, is noted for saying, "I don't want my fair share; I want my UNfair share."

text. Let it suffice here to say that the marketing plan should include data on any competitor in the forthcoming year from which we can reasonably expect to take customers or to which we could potentially lose customers.

Then there are external impacts such as state, regional, or national tourism promotions; major new tourist attractions; new industries in the area; new office buildings being built; airline routes added or removed; plant closings; companies merged and moved; new origin markets of visitors; and new convention centers being built. Every factor does not affect every operation; the key is to recognize those that may affect yours. The marketing plan has to deal with these factors, prepare for them, and whenever possible, capitalize on them or counteract them.

For example, we know of one restaurateur who operated a very successful restaurant for a number of years until business began declining quickly. Because this operation was in a rural area and some distance off a main highway, the operator concluded that people were simply not traveling as often or as far because of the cost of gasoline. Closer analysis revealed, however, that his competition was doing better than ever. In fact, the tastes of the market had changed, and new markets had emerged. Instead of adapting to the market by changing his menu, which had remained pretty much the same for 30 years, he watched his business gradually disappear.

Competitive Environment

The second area of external data collection deals with competitive data. It is important that the local marketing team collect data on all feasible competitors within logical boundaries. Understand that "logical boundaries" may mean the hotel or restaurant across the street or one that is 3,000 miles away. The competition for the convention market for the Hotel del Coronado in San Diego includes the Homestead Resort in Virginia; the Greenbrier Resort in West Virginia; the Cloister in Sea Island, Georgia; the Breakers Resort in Florida; and the Hyatt Regency in Maui, not to mention many others.

A motel in North Overshoe, Maine, competes with a motel in South Skislope, Maine, even though they are 30 miles apart. The Club Med in Eleuthera, Bahamas, competes with the all-inclusive resorts in Jamaica. A restaurant in the city competes with the one in the suburbs. And McDonald's competes with the convenience store, but, by the same token, neither one competes with the French restaurant located between them. **Competition,** as defined by the marketing plan, is any business competing for the same customer with the same or a similar product or a reasonable alternative that the customer has an opportunity to purchase at the same time and in the same context. The best way to understand who is the competition is to ask your current customers where they would have stayed had they not stayed with you. Another way is to call a hotel you know is full on a given night and ask where they recommend you stay instead.

The marketing team must take an objective stance when it comes to evaluating the competition. Although we all like to believe we have the best product to sell in our product class, this may lull us into a false sense of security and the competition can move by us very quickly. The marketing plan must be truly objective and realistic about the products evaluated for the best results. After making a list of all of the competitors for your product, you will need the information shown in Exhibit 5-2.

competition Anyone competing for the same customer with the same or a similar product or a reasonable alternative that the customer has a reasonable opportunity to purchase at the same time and in the same context.

Exhibit 5-2 Competitive Information Needed in the Marketing Plan

DESCRIPTION

A brief description is needed of the physical attributes of the competing hotel or restaurant or lounge (or, for marketing tourism destinations, countries, states, or cities). Examine strengths as well as weaknesses. Determine such things as when the product was last renovated, plans for upgrading in the near future, physical facilities, and all features that compete with yours—that is, the product/service mix. The description should include both tangible and intangible features, relative quality, personnel, procedures, management, reservation systems, distribution networks, marketing efforts and successes and failures, promotions, market share, image, positioning, chain advantages and disadvantages, and so forth. All of these items will be important in the final analysis. A physical inventory and description—number of rooms, meeting space, F&B outlets, and so on—is simply not enough. All strengths and weaknesses need to be defined.

CUSTOMER BASE

Who are your competitors' customers? Why do these customers go there? Are they potentially your customers? Part of the marketing plan will focus on creating demand for your product. Much of the plan will focus on attracting customers away from your competitors. It will be difficult to take customers from your competition if you do not know who their customers are. In a restaurant situation, for example, do your competitors have a high volume of senior citizens eating at traditionally quiet times, a group that you

desire? Does their lounge have a successful happy hour that you could augment for your lounge, and, if so, what type of people go there? Does a competitive hotel have a higher percentage of transient guests than your own? What particular market segments does the competition attract?

PRICE STRUCTURE

Where is your competition in relation to price? Although food and beverage prices are relatively easy to obtain, the product delivered for the price is also important. Is their $6.95 chef's salad as good as yours for $8.95? When analyzing prices, you must compare apples with apples. Published guest room prices are relatively easy to discover. Negotiated prices with volume producers take a little more effort but usually can be obtained from purchasers or directories made available to the public.

FUTURE SUPPLY

It is important to determine whether any new projects will affect your competitive environment in the future. This information can normally be obtained from the chamber of commerce or other local sources. The fact that a new 300-room hotel is scheduled to break ground soon will be very important when developing your marketing plan. Likewise, if the building that houses a major food and beverage competitor is scheduled for demolition to make way for a new office park, this could also influence your decision-making process for the following year.

Once again, keep in mind that competition is all relative. Traditional boundaries of location may no longer apply. For a restaurant in New York City, the competition may encompass a three-block radius that is less than one quarter square mile. For a five-star resort, the competition might be located thousands of miles away. When determining who your competition is, ask the question, Where else do or might my customers or potential customers go?

If you do not know who your competition is or care to validate your assumptions, just ask your customers. They will tell you what other hotels they prefer to patronize in the destination. They will tell you the other restaurants in which they dine when they are not in your establishment.

In the development of the marketing plan, it is also critical to keep in mind that you want new customers and that you are looking for opportunities to attract them. This means that sometimes you have to break the "rules" of competition. For example, a Hilton property might normally be positioned against Sheraton, Hyatt, and Westin in good times. However, when occupancies are low—as after the terrorist attacks in the United States—the Hilton hotel might consider customers that it could capture at a profit from other competitors. If rooms are going vacant, a "normal" Holiday Inn customer might be a target of the marketing plan for the Hilton. A Holiday Inn customer paying $95 for a room that is offered at $125 but only costs $25 to clean may be a good customer to have when the room might otherwise be vacant. In addition, there might be a longer-term benefit—retaining this person as a regular customer. On the other hand, the Red Roof Inn customer who only wants a room at $59 would not be considered a target.

Internal Environment

The third area of collection is that of internal data. One hopes that accurate and adequate records have been kept and much of this information will be readily at hand. Once you have prepared your first marketing plan, you will have said, at least a dozen times, "I wish I knew that." Thus, you will have set up procedures so that next year you will know "that."

Hotels and restaurants should have current data at all times on occupied rooms by market segment—occupancy percent, fair market share, actual market share, revenues, average rate (total and by market segment), REVPAR (revenue per available room), market segments served, restaurant covers, seat turnovers, check average, food-to-beverage ratios (total and for each outlet), and ratios as a percentage of gross revenue. This is also the place to identify market segments and target markets—past, present and future. These are "hard" data and the easiest to obtain, but don't stop there. Now list what you know about the markets:

- Who are they?
- What do they like?
- What are their needs and wants?
- Why do they come here?
- Where would they go if they didn't come here?
- What are their complaints?
- What are their characteristics, attitudes, opinions, and preferences?
- What is the market's perception and awareness?

If you are unable to answer these questions, it is time to start doing some research. At a minimum, start talking to your customers. Have personnel in every department keep logs on all customer comments—good, bad, and otherwise.

Formal research is even better. A basic tenet of all effective marketing is that you must know your market. Yet it is surprising how few hospitality establishments do. This is why so many marketing plans, rather than addressing what they will do for the customer, deal with bricks and mortar, physical facilities, inaccurate definitions of the competition, too-broad market segments (e.g., the business traveler), vague budgets and forecasts, and unfocused advertising.

The second category of internal data collection is the objective listing of resource strengths and weaknesses. The resources include not only the staff, but the physical condition of the property. What is the condition of the property? Where is it weak and where is it strong? How can or should it be improved? What does it offer in terms of facilities? How attractive is the location?

Exhibit 5-3	Data Collection Abstracts from a Paris Hotel Marketing Plan

ECONOMY

The economy in France is wrestling with the creation of the post-war socialist state, the European Union (EU), and the euro. The latter, especially, is having multiple effects on pricing and operating costs. High unemployment has put pressure on the government to reduce benefits to workers to increase jobs. Neighboring European states, as well as China and India, are producing quality goods with one-third of the fixed labor costs. In addition, there is unrest within a segment of the population who do not feel as if they have been treated fairly. Riots captured world attention, and potential tourists and potential new businesses considered other locations within Europe.

EXTERNAL IMPACTS

Although Paris remains a strong destination in the worldwide marketplace, pricing has become an issue. Five-star hotel rooms are more expensive in Paris than in Los Angeles, New York, and London. The result is occupancy that is way down, making most hotels unprofitable. Customers are coming to Paris from other European destinations for only one- and two-day trips, leaving rooms empty almost half the time.

Internal impacts—while service levels remain high, workers share a growing anxiety about the French economy. At this writing, the country has a high unemployment rate and Euro Disney continues to lose money. Will jobs be combined or eliminated to meet owners' needs? Will the customers return after the renovations or be absorbed by another hotel in the area?

FUTURE OF OUR MARKETS

The European Union made travel among citizens of member countries much easier. The single currency has been a success for the participating countries. Overall, the economy of Europe itself is growing, led by the United Kingdom, Paris' largest feeder market. Tourists, however, are more prone to short holidays and weekends and packages, especially in summer, our busiest season. Company individual bookings have also declined.

Next comes the hardest part—how strong is management? The marketing team? Personnel training, experience, and attitude? How are guests being treated? What do complaints look like? How successful have marketing efforts been in the past? What is the customers' image of the property? What is the property's position in the marketplace? This is the time for realistic objectivity, not glossing over or wishful thinking. Finally, make a list of what you do not know—that is, what additional research may be needed.

To give an idea of how all this comes together, we show some extracts from the data collection portion of the actual marketing plan of a five-star hotel in Paris in Exhibit 5-3.*

DATA ANALYSIS

Thus far, we have been engaged only in the collection of data. It is wise to complete this stage first without attempting any data analysis, because you want to obtain the complete picture. Analyzing factors in isolation can be misleading.

Data analysis follows the same flow as data collection. Essentially, we want to draw some conclusions about market position, market segments, customer behavior, environmental impacts, growth potential, strengths and weaknesses, threats and opportunities, performance trends, customer satisfaction, resource needs and limitations and other factors that will be pertinent to the marketing plan.

 **Web Browsing Exercise**

Use your favorite search engine and look up "dining trends." How, if at all, have dining trends changed since the publication of this book? How would you recommend that a restaurant incorporate these trends into its marketing plan? Be prepared to discuss in class.

Environmental and Market Trend Analysis

The first data to look at are environmental and market trends. Are they positive or negative? How will or might they affect us? How can we take advantage of or compensate for them? What are our alternatives? How long will they last? What courses of action are possible and feasible? How do these fit together?

---

*Note that the entire marketing plan is 78 pages, so we can only show excerpts.

Competitive and Demand Analysis

What are the potentials and opportunities in the marketplace? This requires a close analysis of all the demand factors, various market segments, and target markets, such as the following:

- What are the strongest market segments?
- What is their potential for further growth:
    Steady?
    Growing?
    In decline?
- What is their contribution in room nights?
- What is their contribution in covers?
- What is their contribution in revenue?
- What can be done to accelerate a growing trend?
- What can be done to begin growth in a steady trend?
- What can be done to reverse the direction of a declining trend?
- What other segments are there, perhaps untouched, that could be developed?
- How do these segments affect our market mix?
- Are they compatible?
- Can they be expanded to fill gaps such as seasonal or day-of-week fluctuations?
- What types of business would complement these segments?
- What types of action could be taken to attract more business during low-occupancy periods?
- How does the competitive situation affect all these factors?

Property Needs Analysis

Here, we have added a new category. A **property needs analysis** is an analysis of major profit areas to see what gaps have to be filled. These gaps could be in occupancy, market share, average room rate, market segment mix, food sales, beverage sales, seasonal needs, and many other areas. In other words, instead of looking at where we can cut costs, we want to look at where we can increase revenues. When we have done that, we can match property needs with market needs to determine target markets and how to reach and serve them.

Needs analysis also means identifying other marketing problems. For example, there might be marketing strategies that are not working, image changes that are needed, ineffective advertising or promotion, pricing problems, a loss of business to a particular competitor (perhaps because of a new facility, new product or service, or even better marketing), or changing needs of a market segment that we cannot meet.

In short, property needs analysis is an identification of problems to be overcome. It makes the case clearer if we can apply some quantitative measurements to our analysis, which are no more than best estimates based on all of the data assembled. To demonstrate this, we will use a simplified case to determine what the overall increase or decrease for the product will be for the forthcoming year. Ideally, this would be done by market segment.

property needs analysis An analysis of major profit areas to see what gaps (which can be in occupancy, market share, average room rate, market segment mix, food sales, beverage sales, seasonal needs, etc.) have to be filled.

Exhibit 5-5   This advertisement by the Fairmont Hotel is an example of a promotion to create new business.

Source: Fairmont Hotels and Resorts, Bermuda. Used by permission.

Exhibit 5-4   Hypothetical Competitive Universe of ABC Hotel

Hotel	# Rooms	Available/ year	%Rooms FMS*	Rooms Occupied	Sold	AMS**	Variance	Rank
ABC	200	73,000	20%	67	48,910	20%	0	3
Westin	350	127,750	35%	73	93,258	38.2%	3.2	1
Hyatt	250	91,250	25%	72	65,700	26.9%	1.9	2
Hilton	200	73,000	20%	50	36,500	14.9%	(5.1)	4
Total	1,000	365,000	100%	67	244,368	100%		

*FMS (fair market share) is the number of available rooms per hotel divided by the total number of available rooms.

**AMS (actual market share) is the number of rooms sold per hotel divided by the total number of rooms sold.

In this example, we will say that we are anticipating an increase of 2 percent in the demand for both group and transient hotel rooms in the product class category. From the data collected, a competitive universe can be compiled as shown in Exhibit 5-4.

Assume, also, for the purpose of this discussion, that a Crowne Plaza of 200 rooms is opening next year with a projected occupancy of 55 percent. Its forecasted market mix is 50 percent group and 50 percent transient.

Now, for the purpose of developing the marketing plan, we have some quantitative data with which to work. One thing is immediately obvious: ABC Hotel has a relatively low occupancy and is barely achieving its fair market share. After all the data collected in the situational analysis have been analyzed, two main areas of concentration must be addressed: creating new business and capturing competitors' business.

Creating New Business.   Given the current situation, what plans can be developed to create new demand for the product? McDonald's created new demand for its product by opening for breakfast. Package weekends have created new demand for hotel products in the past. Spas are the latest amenity driving demand for many

hotels and resorts. Creating new demand in the hospitality industry, however, may be the toughest part of marketing. The important point to remember is that we are creating demand that until now did not exist for a product. This usually means creating a new use.

The advertisement by the Fairmont Hotels in Exhibit 5-5 is an example of a promotion to create new business. In this case, the target market is people who want to spoil themselves. The advertisement constitutes direct marketing against the competition, rather than creating a new use for the product. Other parts of the marketing plan will carry out and specify the implementation of this promotion in terms of the specific target market.

Capturing Competitors' Business. Most marketing plan program executions are concentrated in this area. Specifically, let's return to the competitive universe shown in Exhibit 5-4. ABC Hotel's main competitors are Westin, Hyatt, and Hilton, plus the new Crowne Plaza being built. A demand analysis for these five hotels, two of which are capturing more than fair market share while ABC and Hilton are not, might appear as shown in Exhibit 5-6.

A red flag should be raised with this scenario. Although the forecast reveals an increase in demand for the hotel product, the increase in supply will be greater than the increase in demand. Each hotel will now be fighting for a smaller piece of the pie. If ABC Hotel does everything the same as the year before, it will be drawing on a smaller pool of rooms and occupancy will drop even further. In fact, ABC and its four competitors are now competing for 209,105 rooms versus 244,368 in the previous year, after the new Crowne Plaza takes its share.

Exhibit 5-6   Hypothetical Demand Analysis

	Total	Group Segment	Transient Segment	New ABC FMS	@20% AMS
Rooms sold previous year	244,368	146,231	98,137		
Next year projection with 2% increase in demand	249,255	149,156	100,099		
New supply from Crowne Plaza	40,150	20,075	20,075	16.7%	49,851

ABC's marketing team can now see the task that lies before it. Just to maintain the occupancy of the year before, it will have to create new demand for the product, aggressively attack competitors for new business, and maintain its own customer base, which the competition will be trying to lure away with their marketing plans. It will also have to exceed its new fair market share by 2.95 percentage points, something it hasn't been able to do in the past. ABC's strength may be as a transient hotel, whereas this may be a weakness of the other properties. In this case, ABC might choose to direct its major marketing effort toward that market.

Another possibility is that ABC has neglected the group market and needs to concentrate greater effort in that direction. Of course, it may have to make major efforts in both directions. Let us assume, for the sake of argument and because it is easier to demonstrate, that there is a high degree of price sensitivity in the market within either one or both of these segments. In either case, specific marketing plans must be made to attack the competitive hotels to capture rooms from them. The plans might call for lowering prices for specific days of the week or during times of the year when ABC's occupancy suffers the most.

The ABC example is clearly an oversimplified one. There are many other factors affecting any similar situation and a number of approaches. In fact, we haven't even mentioned the customer in this discussion, and that database would be the first one to consider! The point we want to make is that there is an absolute need for complete and adequate data and information followed by a thorough analysis of all possible considerations. It is only through such methods that workable, realistic, and effective marketing plans are developed.

Internal Analysis

We now turn to the internal analysis. Using the realistic and objective data we have gathered, we start by asking questions such as those shown in Exhibit 5-7. We would do this by segment and target market. Strategies by segment for a hotel might include efforts directed against group sales, national sales, local corporate, weekend, or international markets. Catering might include segments such as freestanding, local corporate, evening, and social markets. Descriptions of how each market should be addressed should be included in your internal analysis.

The list shown in Exhibit 5-7 could go on indefinitely. Once again, we have to state that workable, effective, and realistic marketing plans can be developed only through the gathering of complete and adequate information and its thorough and objective analysis.

Exhibit 5-7   Internal Analysis Questions

- What is the gap between what your customers want and need, what you promise them, and the product/service you provide?
- How well do you meet or exceed customer expectations?
- How does the market's estimation of your product/service agree with yours? What makes you think so?
- What items, product improvements, or services are needed to improve customer satisfaction?
- Are you actually delivering what you think you are?
- What patterns are appearing in guest comments? What types of problems seem to recur? What areas seem to need improvement?
- Do you have the proper organization to accomplish what you are trying to? For instance, although the manager is a strong operations person, does she understand the customer?
- Do you reward your staff strictly on bottom-line results? If so, does it show up in matters affecting the customers?
- Do you know, identify, and deal with your real strengths and weaknesses?

Market Analysis

Our final step in analysis focuses on the market itself, the customer. Because this entire book is about the hospitality and tourism customer, it would be redundant to repeat here all that we will say about this ever-changing individual who is the reason for the existence of any hospitality enterprise. For purposes of developing the marketing plan, this step means determining where the gaps are, where needs are unfulfilled, where problems are not being solved, and what niches the competition is not filling.

The market analysis must be matched with the environmental trends, the competitive and demand analysis, the property needs analysis, and the internal analysis. We would, of course, combine all of these analyses by segment and target market. We are then ready to develop a mission statement for the property, determine opportunities, establish objectives, and begin preparing the actual marketing plan, which will include a plan and course of action for each segment or target market.

MISSION AND MARKETING POSITION STATEMENT

The mission statement at the beginning of the marketing plan flows from the strategic mission statement and from the corporate mission statement. It is different from the former, however, in that it is a broad statement of objectives at the basic level. The general guideline of the corporate mission statement is a good starting point.

"WE'VE JUST BEEN GIVEN A NEW MISSION STATEMENT, AND IT DOESN'T SAY ANYTHING ABOUT CUSTOMERS, SO SHOVE OFF."

In multilevel organizations, however, there can be a lot of variety. Many chains have diversified products selling in diversified markets for diversified uses. Corporate strategies established in corporate headquarters in Atlanta, Chicago, New York, London, Paris, or Tokyo do not necessarily fit the situation in India, Germany, Kuwait, New York City, Minneapolis, or Los Angeles.

The situational analysis of the marketing plan provides the test of the strategy and requires rewriting the local mission statement. Therefore, only after the situational analysis has been completed do we recommend writing the marketing plan mission statement and, if necessary, adjusting the strategic mission statement. Recall, moreover, that the latter is the long-term mission; the marketing plan mission is set forth one year at a time. This mission statement will have specific objectives such as "to be the business traveler's hotel of choice in the city." More specific objectives will be in the statement of objectives, such as, "by increasing our ratio of business customers to pleasure travelers from 55 to 65 percent."

Further, the intended "position" in the eyes of customers will be identified. This statement flows from the strengths and weaknesses of the competition as well as the property for which the plan is written. For example, our Paris hotel example would be positioned, at minimum, as one of the top five luxury hotels in Paris. Other attributes, such as a "corporate luxury hotel," might also be part of the positioning to that market segment.

OPPORTUNITY ANALYSIS

If we have done a thorough job of data collection and analysis, we should now be able to determine the opportunities available. The section heading is self-explanatory and can best be discussed by example. Therefore, we take information again from the marketing plan of the Paris hotel previously mentioned, as shown in Exhibit 5-8.

The opportunities in Exhibit 5-8, although perhaps too general, have been developed after an analysis of the market, market segments, the competition, trends, the needs of the property, and so forth. Its brief form contradicts the groundwork that goes into identifying the opportunities. Sometimes this groundwork is not done—that is, someone says something like, "How about the incentive market? We don't have any of that business. That's an opportunity! Let's put it down." Of course, a thorough study of

---

Exhibit 5-8 Marketing Plan Opportunity Analysis of Paris Hotel

A. Market segments relating to existing customer mix
  1. Corporate groups
  2. Corporate individual bookings
  3. Leisure travelers from abroad
  4. Weekend packages with special features
  5. Special off-season incentives
B. New markets
  1. Packages for individual travelers
  2. U.S. upscale travel agencies

  3. Incentive market—London, New York
  4. High-ranking government officials
C. Image
  1. More professional and colorful F&B promotions
  2. Improved reputation of service and cuisine
  3. Provide better background information about city, emphasize price/value
  4. Professional advertising to improve hotel image

the incentive market, its needs and wants and the organization's ability to serve them, is necessary first. Opportunities, in the true sense, are not just something that's "out there"; they are, instead, a match between customer needs and an organization's capabilities and, one hopes, a drop in the competition.

OBJECTIVES AND METHODS

The next step in the marketing plan is to establish the objectives and how they will be accomplished. The objectives listed need to be specific and fairly typical of hotel marketing plans. Many are directly measurable. Action plans will be designed to carry out each one and are discussed next.

Marketing Action Plans

Marketing action plans address how the marketing plan will be carried out. They assign specific responsibility to individuals and dates for accomplishment. Marketing action plans include detailed lists of the action steps necessary for carrying out the strategies and tactics for reaching each objective. One format for an action plan is shown in Exhibit 5-9, but there can be numerous variations on the theme.

Action plans deal with the various parts of the marketing mix, which, of course, result in the use of the marketing plan. For example, the action plan for the communications mix might include advertising (both offline and online), direct mail, personal sales efforts, promotions, merchandising, and public relations campaigns. These are brought together in order to achieve the maximum impact of the desired strategies.

marketing action plans Plans of action set forth in the marketing plan that include time frames and who will implement the plan. (Sales action plans are part of the marketing plan, too.)

Exhibit 5-9	Marketing Action Plan

XYZ HOTEL MARKETING ACTION PLAN

NAME: Quarter:

BOOKING GOALS

1,000/MONTH   250/WEEK   50/DAY

NEW ACCOUNTS OPENED

20/MONTH   5/WEEK   1/DAY

ACTION PLAN BY WEEK: PERSON RESPONSIBLE:

Week 1_____	Begin advertising campaign, corporate group	_____
Week 2_____	Trade show schedule, third quarter	_____
Week 3_____	Direct mail, corporate transient	_____
Week 4_____	Good accounts function, associations	_____
Week 5_____	Public relations for catering	_____
Week 6_____	Public relations for catering	_____
Week 7_____	Focus groups, meeting planners	_____
Week 8_____	Focus groups, travel agents	_____
Week 9_____	Image advertising campaign begins	_____
Week 10_____	Strategy session, tour and travel	_____
Week 11_____	Direct mail, past users	_____
Week 12_____	Develop comarketing partners	_____

The action plan should be developed for a full year and updated quarterly for all products and actions, consistent with the stated performance goals by market segment and time of year. Advertising support may be necessary in certain months to create awareness and accommodate requests for more information.

Yearly schedules for other support elements of the communications mix are needed to organize the entire plan. A direct mail campaign might be used in conjunction with the advertising for the promotion to generate the best response. Without action plans, too many things are forgotten too often or are done too late to be effective.

There are other concerns as well. The communications mix is expensive to carry out. The savvy marketing executive will constantly be looking for ways to maximize the impact of communications dollars. Cooperatively funded advertising is possible with related travel partners such as American Express. Airlines are increasingly willing to work with hotels to create business through collective advertising and direct mail. Credit card companies are doing dual promotions with restaurants and lounges on a consistent basis to differentiate their products and combine resources. The Internet now requires resources for individual hotels to allocate money and personnel to this growing channel of business. All of these efforts require a lot of advance planning and specific actions completed on time.

Except for the final forecast and component budgets, the marketing plan is now complete. Remember, this should be a "fluid" document, ready to be changed with shifts of the marketplace. This does not mean that the entire marketing plan should be rewritten every time there is a major change in the business or competitive environment; if the situational analysis was done properly, the conclusions drawn should not change quickly or dramatically.

MARKETING FORECAST

marketing forecast
A component of the marketing plan that projects revenues to be generated such as guest room sales, food and beverage sales, and others for the approaching year.

Making accurate **marketing forecasts** is one of the most difficult tasks when developing the marketing plan. Regardless, the best attempt possible is essential. Forecasts are trips into the unknown that are subject to any number of peculiar situations in the marketplace. Accuracy is ensured only by access to the best information available, thorough analysis of that information, and the learned judgment of the forecaster.

Many hotel marketing plan formats require the projection of room nights for every day of the approaching year to forecast, by segment and day of the week, the upcoming year's business. It is not uncommon for forecasters to use some figure—say, 5 percent—as the projected increase in sales over the previous year. Such a method is usually random and may have no basis in market fact. It is better to start with a zero base each year and build the forecast according to the strategies set forth in the marketing plan. In this way, room nights, covers, and other projected sources of revenue are based on performance goals that have been realistically established. Monetary amounts, such as average room rate per segment; average breakfast, lunch, and dinner check; and so on, are used as the multipliers to forecast revenue.

MARKETING BUDGET

The industry-wide average for marketing expenditures, referred to as the **marketing budget,** for average U.S. hotels falls between 4.5 and 5.5 percent of gross revenue (rooms, food, beverage, and miscellaneous). There are no reported averages for restaurants except a general figure of 2 to 3 percent of revenue spent on advertising for an individual sit-down, mid-to-upper-scale operation. As a rule of thumb, the marketing payroll expenses of a hotel are normally one half of the total marketing budget, although this varies depending on the importance of group business to the property and the corresponding size of the sales force. Traditionally, resorts have slightly higher marketing budgets as a percentage of gross revenue because of the seasonal nature of their business, as do new properties that are relatively new or find themselves battling a number of new or aggressive competitors. The overall trend in the industry has been toward increasing the marketing budget as a total percentage of revenue as the cost of implementing various aspects of the marketing program has continued to rise. For example, marketing expenses increased 6.1 percent in 2004.

> **marketing budget** A component of the marketing plan that projects marketing expenses to be incurred for the forthcoming year.

The marketing budget should be a natural extension of the marketing plan—no more and no less. Once a strategy has been developed to create and keep customers, enough funds need to be distributed to be sure of success.

The budget will normally include the following components, regardless of the size or type of the operation, even if, for example, the manager of a restaurant performs all the marketing and sales duties for that restaurant. Parts of that person's salary and expenses should be allocated to the marketing budget.

- Payroll for all sales and administrative staff, plus any secretarial or related work
- Communication, including all advertising, promotion, direct mail, public relations, collateral, and related items
- Travel, including all marketing-related travel
- Office expenses, including related telephone and office supplies
- Entertainment, including that of clients or prospective clients both in-house and off premise
- Agency fees and expenses, including professional services purchased from outside marketing service suppliers

These are broad categories of expenditure. A further breakdown depends on the needs of the operation. It is important that marketing expenses be clearly and correctly assigned.

The budget should be carefully prepared, not done at the last minute or by guesswork. If you are not your own boss, you will probably have to have it approved by someone. In that case, you may have to justify each cost item as one that will produce tangible results.

The marketing budget should also be a fluid tool, responding to changes in the marketing plan. It is critical to protect the integrity of the budget and plan throughout the planning year. The plan and budget should be changed if results are falling short of forecasts. For example, the "Sex in the City Weekend" might be considered cost effective if it produced 50 rooms for a given Friday and Saturday night. If, after three or four weekends, the demand never exceeds 35 rooms a night, the responsive marketing team will have another look at the value of the package. The decision might be to revise the components of the package, try the promotion again at a later date, or scrap it altogether and send the funds elsewhere.

Problems occur when managers think only in terms of short-term response (i.e., improving short-term financial performance by cutting costs) rather than using longer-term strategies to increase and keep customers. This type of situation occurs frequently in the careers of sales and marketing professionals. Although there is no clear-cut answer to the problem, the need to create and keep customers should be the first consideration of any successful organization. Short-term rewards are put first too many times at the expense of future business.

MARKETING CONTROLS

A final and important phase in the development of the marketing plan is to monitor the performance of the plan throughout the year and at the end of the year. The first step, of course, is to continuously match performance against the desired results and to detect when and where changes occur. The extent of each change should be measured and the worst ones addressed. The cause of the change should be determined and dealt with either by bringing it into line or adjusting the plan.

Benchmark measurements are established in advance. These could include any of the following as well as others:

- Actual market share versus fair market share
- Occupancy (both achieved and matched against the direct competitors)
- Covers served
- Seat turnovers
- Check averages
- F&B ratios
- Revenue per available guest room
- Average room rate by segment
- Business mix by segment
- Advance bookings
- Advertising inquiries
- Website traffic
- Return per marketing dollar
- Customer satisfaction
- Complaints and compliments
- Repeat business
- Revenue
- Profit

A feedback system should be established to synchronize with these benchmarks. You should be able to answer questions such as the following:

- Is the product or service meeting the needs of the segment(s)?
- Is the segment growing, static, or declining?
- Is the segment profitable?
- Is customer perception what you thought it would be?
- Is your positioning correct?
- How are you doing in comparison to the competition?
- Are weaknesses showing?
- Are strengths being used to their fullest potential?
- Is there price resistance?
- What are the reasons for the changes?

You may have to make changes where necessary or move to emergency plans. You may have to rethink your strategy or your plan or perform a new situational analysis. Marketing plans are not static, but dynamic; they are executed under dynamic conditions and must be monitored in the same way.

A final word: The marketing-driven organization must not permit the demands of short-term performance to dictate decisions and actions that may result in the loss of customers. The marketing budget and plan should be adjusted to show the needs of the customers, not the accountants. To do otherwise is like getting rid of maintenance to improve short-term bottom line figures, then having to buy new equipment at some inconvenient time in the future. Nevertheless, accountants must have their say. Thus, plans and budgets must ultimately stand the test of cost-effective results and proven revenues.

Chapter Summary

The marketing plan and the marketing budget are fluid tools designed to create, capture, and retain customers. Their development is based on grounded and realistic situational analysis, which requires good data collection, primary research where necessary, and thorough analysis. Instead of relying on traditional methods to deal with unique situations, the marketing team needs to develop new strategies based on insightful information. The funding made available to use these strategies must then be enough to get the job done. The marketing plan and budget also need to take into account new trends. Exhibit 5–11 illustrates a current trend of the everyday traveler: bringing pets along on the trip.

The marketing forecast needs to be as realistic as possible to ensure that the budget is appropriate for the upcoming year. Marketing controls are put into place to monitor performance throughout the year. A marketing plan serves no purpose if it sits on a shelf collecting nothing but dust!

Exhibit 5-10 is a template that may be used to develop a marketing plan. The chapter fills in many of the empty spaces with explanations, but the actual data must come from those who work the plan.

Exhibit 5-10 Marketing Plan Template

1. Overall mission statement
   Broad-based as guideline, subject to revision later
2. Situational analysis
   The situational analysis describes the business climate of the hospitality entity. This portion of the plan gives an overview of the business, recapping what worked well in the current year and what needs to be accomplished for the upcoming year. It also indicates environmental and industry trends.
   a. Recap of past year, what worked well, what didn't, and broad needs for next year
   b. Environmental trends
      Possible major shifts that affect customers or operations: political, economic, social, technological, ecological, regulatory
   c. Market trends
      Provide the foundation for the marketing plan, including possible major or minor shifts that affect customers: international, national, regional, local. Statistics are gathered for analysis. Market potential is an important part of this. Macro statistics can be obtained from sources such as the local visitors and convention bureau or the National Restaurant Association. Micro statistics such as local trends can be obtained from Smith Travel Research for rooms or Fasttrack for food and beverage.
   d. Competitive trends and forces—current and forecast
      Needed to anticipate competitive moves and strengths and weaknesses, both current and forecasted. A competitive review needs to be completed at least yearly. Current supply of hotel rooms or restaurants should be documented with not only the obvious statistics, such as number of guest rooms, seats, or square footage of function space, but also a determination of the direction of each competitor in terms of the overall marketplace. Strengths and weaknesses of each should be documented. Any new or lost supply of rooms or restaurants should be identified.
3. Internal data
   Provides all pertinent data and statistics on the operation, both financial and customer information, including segments and target markets.
4. Data analysis
   Involves analyzing the data collected in steps 2 and 3. We add, however, one new category called *property needs analysis*. In this category we analyze the internal gaps to be filled so that we can match them with the opportunities revealed by analyses of the other stages.
5. Mission and market position statement
   The mission and objectives for the forthcoming year are detailed. Although this is not done until all the previous work is done, it will be inserted at the beginning of the marketing plan right after the broad mission statement.
   The market positioning strategy follows. With the statistics gathered and the competitors reviewed, the hospitality entity has to define clearly where it belongs in the constellation of competitive brands. For example, Mary's Restaurant will be positioned just below Cathy's Restaurant, directly in competition with Jenny's Restaurant, and above Meagan's Restaurant. The market position statement gives direction for the marketer as well as the employees as to what the hospitality entity expects of itself.

6. Opportunity analysis
   Involves identifying all the opportunities in the marketplace revealed by all the previous analyses that are consistent with the mission and market position adopted. Once the positioning has been established, a strategy for each market segment needs to be developed. This can be as simple as breakfast and lunch for Mary's Restaurant or 22 segments of customers at the Grand Hyatt in New York City. Each segment has to be defined and addressed in the marketing program.
7. Objectives and methods—objectives are measurable where appropriate
   Define precisely where we want to go and how we plan to get there.
8. Action plans
   Establish specific responsibilities for every member of the team and some who are not directly part of the marketing team. For example, a chef may be brought in here, if not before, to get involved in the action.
9. Marketing communications
   Each segment responds to different communications. For example, the corporate group market has to be called on by salespeople. The "two-fers" for a restaurant read newspaper ads or listen to the radio. The marketing communications portion of the marketing plan establishes how we are going to reach our customers and tell them what we have to offer.
   Clearly, multiple communication vehicles are available to reach each market segment. The corporate customer not only responds to a salesperson, but also reads the business press, corporate directories, and other periodicals.
10. Market forecast and revenue projections
    The revenue forecast takes all of the assumptions of the marketing program and establishes a financial goal for the upcoming year. In many cases revenue goals by market segment are determined to track the progress of the marketing programs.
11. Marketing budget
    With the data collection in place and strategies outlined by segment, the allocation of marketing resources such as advertising, public relations, direct sales, and database marketing may begin to support the strategies.
12. Marketing controls
    The old expression "What gets measured gets done" applies here. This is the follow-up to stay on track, see what is working and what is not, and decide whether any changes or new directions are needed.

Exhibit 5-11   A new trend for many travelers is the practice of taking their pets with them while they travel. Here is Belle, resting during one such trip

Key Terms

competition, p. 129

marketing action plans, p. 139

marketing budget, p. 141

marketing forecast, p. 140

marketing plan, p. 123

property needs analysis, p. 133

Discussion Questions

1. How would you define a marketing plan?

2. What is meant by internal and external data collection?

3. What are marketing action plans? How do they fit into the overall marketing plan?

4. Preparing accurate marketing forecasts is one of the most difficult tasks when developing a marketing plan. Why?

5. The marketing budget is an important piece of the marketing plan. Briefly describe some of the parts of this budget.

6. How would you define marketing controls as they relate to the marketing plan? Give some examples of benchmarks used when using controls to monitor the performance of the marketing plan.

Endnote

1. Hayes, R. H. (1986, April 20). Why strategic planning goes awry. *New York Times*, Sec. 3, p 2.

.... case study

Emir's Delicacies Lebanese Restaurant

Discussion Questions

1. What is the current product line offered by Emir's Delicacies?

2. Who are the primary competitors for each product line?

3. Which of the product lines appear to be in the growth stage of the product life cycle? In the mature stage of the product life cycle?

4. If you were to develop a marketing plan for Emir's Delicacies, who would you suggest makes up its primary market? Its secondary market? Who makes up the primary market for the restaurant side of the business? The catering side of the business?

5. What do you think should be the positioning of Emir's Delicacies? Primarily an ethnic restaurant that also offers catering? Or a catering company that also has an ethnic restaurant? (Should they consider closing the restaurant and becoming solely a catering company?)

6. Can Rabih continue to be so actively involved in the many facets of the business? Should the family consider hiring professional or experienced staff to help grow the business? If so, in what area(s) would you recommend?

"Hey Mom, please keep Dad away from the restaurant until around 11:00 or so this morning," Rabih asked his mother, Nicole, as he started to head out the door at 7:00 A.M. "I have to finish up last week's accounting before we open today, and he just gets in the way!" Nicole shrugged and said she would do her best. "But you know your father, and telling him what to do is a challenge at best. Nonetheless, I will try to keep Roger busy here at home for a while and we will be over in a couple of hours to start final prep for today." Rabih thanked his mom and went over to Emir's to work quietly.

FAMILY BACKGROUND

Roger and Nicole Abouhalka emigrated from Lebanon in the late 1980s with their three young sons—Rabih, Chadi, and Marwan. The civil war and political unrest in Lebanon was not abating,

and they wanted to start a new life for their young and growing family in North America. Roger and Nicole initially came to the United States as visitors to Menomonie, Wisconsin. This area was home to the University of Wisconsin–Stout, which was well known for its hotel and restaurant management program. Their eldest son, Rabih, was interested in studying the hospitality business and enrolled in the program shortly after they arrived.

Once settled, Roger and Nicole ardently explored immigration alternatives as they clearly did not want to rely on refugee status. What they had learned was that, at the time, Canada offered perhaps the best opportunities for becoming landed immigrants in North America. Quebec (the French speaking province) was particularly inviting in that the entire family was fairly fluent in the French language. (They also had excellent command of the English language.) The family visited Montreal where they had friends and found it appealing and to their liking. It was after this trip and much discussion that Roger and Nicole made the decision to seek landed immigrant status in Canada as entrepreneurs. As part of the required process for entrepreneurial applications, they developed a business plan for a franchised restaurant operation. Though the government initially lost their application for nearly 14 months, in time, and after several hurdles, the application was approved in December 1988. During this period Rabih continued his course work into his third year at Wisconsin–Stout. The family had moved to Montreal in June 1987.

Neither Roger nor Nicole had been in the restaurant business, yet their professional experiences in Lebanon were in somewhat related fields. Roger, an entrepreneur, was an import/export trader. Many of his clientele were suppliers to the food service industry including Kraft, Best Foods, Amstel Beer, Heineken Beer, Nestle, and Mazola. Because the nature of his import/export business included being a customs broker, Roger became quite knowledgeable about the specific ingredients and quality of the products he bought and sold for his customers. Roger also dealt with food service equipment products including ovens, refrigerators, china, and cutlery. Nicole was an executive secretary for senior management at Nestle (of Switzerland) in their Beirut headquarters. She then held an

This case was contributed by Margaret Shaw, PhD, School of Hospitality and Tourism Management, University of Guelph, Ontario, Canada. All rights reserved. Used by permission.

administrative position at a major Beirut hospital in Lebanon. In addition to French and English, Roger and Nicole were fluent in their native tongue, which was Arabic.

While enrolled at Wisconsin–Stout, Rabih had worked 20 to 30 hours per week in the university's food service division. This operation served 6,000+ students daily. Rabih worked in several positions including assistant cook, cashier, dishwasher, steward, and, on occasion, wait staff for catered events. He was starting to enjoy his decision to pursue a career in the hospitality industry. There was so much to learn.

Shortly after the family arrived in Montreal, the franchise company with which they had wanted to sign an agreement had gone bankrupt. Now what? At this juncture, and after much thought, they decided to open an independent restaurant specializing in Lebanese cuisine. But where? Learning more about Canada, Quebec, and Ontario in particular, they chose Guelph, Ontario, as the location for their new venture. This location was about an hour's drive west of Toronto. Guelph had much to offer including being a cosmopolitan yet small town (with a population of 80,000 at that time) and home to the University of Guelph. At the university was one of Canada's leading hospitality programs, the School of Hospitality and Tourism Management. Perhaps Rabih could continue his studies once the restaurant was open and well established, and the two younger sons were starting to show an interest in hospitality as well. The Abouhalka family moved to Guelph in June 1989 and opened Emir's in June 1990.

EMIR'S DELICACIES

In Ontario, Guelph had come to be known as the "Royal City." To identify with this royalty image, Roger and Nicole chose "Emir" as the name for their restaurant. An emir, by definition, is an independent prince or commander in the Middle East. And, for Roger and Nicole, the name Emir's meant the freedom to become successful entrepreneurs in their newly adopted homeland serving first-choice, quality cuisine worthy of a prince. They were willing to work hard, which is what it would take to successfully launch their new Lebanese cuisine restaurant concept. And the boys were now old enough to help out as well.

The health food and vegetarian trends were picking up in the early 1990s, and Lebanon was well known for its excellent cuisine including the use of fresh ingredients as a mainstay for all food preparation. The multicultural environment at the university made it a natural target for their restaurant clientele. Lebanese favorites such as falafel, tabbouleh, and baklawa were very much a part of Emir's menu offerings. Shown in Exhibit 1 is the main course menu for the 30-seat table service restaurant. The menu also included appetizers, dips, salads, soups, sandwiches, and dessert items. Take-out was popular, for which there was a separate menu, which also offered side orders by the portion size such as tabbouleh for $4.75 per 8-ounce, $6.50 per 16-ounce, and $12.00 per 32-ounce servings.

Emir's was located in a small plaza, a mini shopping mall area, in close proximity (about 2 kilometers) to the university. The restaurant was situated next to a major health club and tanning salon as well (another interesting target for their healthy cuisine). Free parking was available for guests, which was not always the case for the competition in the downtown core. The lease agreement for their facilities was also less expensive in this location. Emir's application for a license to serve wine and beer was finally approved by the local authorities in 1992.

Also in 1992, Roger and Nicole decided to expand into the catering business. In truth, Rabih was the instigator for this endeavor. It became an immediate success and prospered quickly. As well, Emir's was the only caterer in the local area to offer kosher style food for clients who preferred this type of preparation. Events they catered included weddings, anniversaries, funerals, Christmas and New Year's parties, bar mitzvahs, and so forth. Though somewhat limited because of on-campus hospitality food services, catering opportunities at the University of Guelph included departmental gatherings and special occasion events.

In addition to the local area (see Exhibit 2 for geographic location), Emir's catered to various events in the nearby cities of Cambridge, Kitchener, and Waterloo. They also had clients in the Greater Toronto area, Ottawa, Montreal, and Owen Sound in eastern Canada. In western Canada, locations included Winnipeg, Manitoba, and Whistler, British Columbia. Across the border in the United States, they had catered events in New York City, Chicago, Minneapolis, and Orange County, California. Most of this business was a result of positive word of mouth.

Emir's did not offer alcoholic beverages at catered events because the additional insurance cost to do so was prohibitive. Most clients, however, served their own alcoholic beverages of choice. By 2002, 50 percent of revenue for Emir's came from the catering side of the business. Restaurant table service comprised 40 percent, and take-out comprised 10 percent of total food sales.

MAIN COURSES

Masbaha

A mixture of fresh egg plant, Lebanese zucchini & potatoes simmered
in fresh tomato sauce served on a bed of Lebanese style rice. $12.75

Okra

Simmered in a tomato & onion sauce, the Okra is served on a bed
of Lebanese style rice. $12.75

Chicken Mishwy

Boneless skinless breast chicken, barbecued, shaved & served with saffron
rice & a garlic dip. $12.75

Kafta

The kafta, a mixture of extra lean ground beef, parsley & spices served on
a bed of Lebanese style rice, topped with tomato sauce. $12.75

Chicken Shawarma

Boneless skinless chicken breast, marinated in Shawarma spices & table
wine, barbecued, shaved & served with saffron rice, spiced onions mixed
with shaved parsley. $12.75

Beef Shawarma

Sirloin steak marinated in table wine & spices, barbecued, shaved & served
with saffron rice, spiced onions & tahineh mixed with shaved parsley. $12.75

Special

Please don't hesitate to ask your host about this phenomenal special dish
of the week if you haven't been informed about it first.   Price Varies

Fish Kibbeh

Fresh Grouper mixed with cracked wheat & green vegetables, baked on a
bed of onions & pine nuts, served with saffron rice & tahineh. Upon
availability of fresh fish. $18.25

(Taxes and Gratuity are not included in the prices)

Exhibit 1   Emir's main course menu lists Lebanese favorites, ideal for the university's multicultural environment.

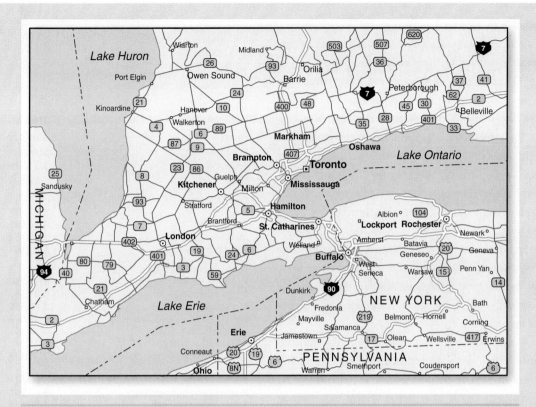

Exhibit 2   Map of southern Ontario.

FACILITY LAYOUT AND OPERATIONS

Emir's was actually quite small. The kitchen prep and adjacent service and cashier area was 10 feet by 20 feet. The table service dining area comprised 320 square feet. Including the bathrooms and a small storage area, total square footage at the restaurant was 780 square feet. Because of limited space in the kitchen, the upright rotisserie equipment used for broiling chicken (and other meats) was placed on a counter in the service area near the entrance of the restaurant. In some respects this became a "show kitchen" for Emir's.

Within the first two years of operations, quality black-and-white laminated professional photos of Lebanon were mounted on the interior walls in the dining area. Drapes, linen tablecloths, and soft lighting were introduced. Potted plants were set on the sills of the large bay window stretching across the front of the restaurant along with hanging baskets to complete the decor. Background music added to the pleasant ambiance in the comfortable surroundings. Their goal was to create a pleasing Lebanese atmosphere for patrons to relax and enjoy their midday or evening meal.

Hours of operations at the restaurant were 11:00 A.M. to 9:00 P.M., Monday through Saturday. Emir's was closed on Sundays. On request, though infrequently, private parties were catered on premise in the evenings after 8:00 P.M. or Sundays. Off-premise catering was done throughout the week and weekends, including Sunday. Rabih and the two younger sons did most of the catering service. Roger and Nicole managed the restaurant, with part-time help, when catering events took place during the regular operating hours. Rabih did much of the routine maintenance for the restaurant and outsourced repairs as need be. The family closed the restaurant for two weeks in mid-August each year when the family usually returned to Lebanon. This was a quiet period in Guelph especially with the university between semesters and much of the community on holiday.

During the week, lunch and dinner averaged a three-quarter turn for each meal period. A "turn," in this case, means the number of patrons served in a given period relative to the seating capacity of the restaurant. For example, at Emir's, one turn at lunch would mean 30 customers were served during that particular lunch period. Friday and Saturday evenings ranged from one to two turns. When demand was high on weekend evenings, they would extend the hours of operation to accommodate their guests. Saturday luncheon patrons came in throughout the day averaging one turn between the hours of 11:30 A.M and 2:30 P.M.

The number of catering events and guests served each week varied widely. In any given week the number of parties catered ranged from 0 to 5 and the size ranged from 4 to 1,300. The party of four, a golden wedding anniversary, was a nine-course meal including 12 appetizers, 3 salads, 3 main courses, 2 digestives, and 2 desserts. The 1,300-person event was a private New Year's Eve party catered in the nearby city of Cambridge. In general, however, a typical-sized event ranged from 30 to 150 guests.

Spring and summer were the busiest times for their catering business. Fall, winter, and spring were the high-volume periods for the restaurant business. Prime time for take-out was early evening during the fall and winter semesters at the university.

The marketing budget for Emir's was low. They did little advertising and had few traditional special promotions. The focus was on direct or one-on-one marketing by offering patrons an appetizer, salad, or dessert as a "treat" to accompany a meal. Rabih did not attempt to keep track of these expenses. Rather, he simply let them be part of the overall food cost of the operation. Roger, Nicole, and Rabih felt that patrons appreciated this gesture, and sales of signature items "promoted" did increase as a result of this effort. This was especially true for appetizers and desserts.

Another marketing effort was Emir's involvement with local fundraising events. There were four major events each year in which they participated. These included the Wine Gala in March supporting the local music foundation, the Wyndham House fundraiser in April for homeless teenagers, the Taste of Guelph fundraiser in September for hospitals in the community, and the MacDonald-Stewart Art Center (located at the university) fundraiser for the arts in November. A number of restaurants and caterers, including Emir's, would donate and serve appetizer-type food at these half-day events in return for menu, flyer, promotional material, business card, and signage displays.

Emir's continued to earn a profit, albeit a modest one. Shown in Exhibit 3 is a summary statement of income and expenses for the previous year. Rabih made every effort to follow the uniform system of accounts for restaurants. Sales for the restaurant, take-out, and catering were easily identified. However, separating food cost expenses accurately for each proved to be laborious and difficult. He knew that catering was less costly and had higher contribution margins.

Exhibit 3   Summary Statement of Income and Expenses

For Previous Year Ending December 31

Sales:	
Food restaurant	$184,500
Food catering	$230,625
Food take-out	$ 46,125
Beverage	$ 27,675
Total sales	$488,925
Cost of Sales	
Food	$163,477
Beverage	$ 4,981
Total cost of sales	$168,458
Gross profit	$320,467
Controllable Expenses	
Part-time payroll	$ 56,226
Direct operating	$ 19,557
Marketing	$ 2,455
Energy	$ 18,943
Administration	$ 14,862
Repairs and maintenance	$ 13,571
Total controllable expenses	$125,614
Occupation Costs	
Rent	$ 28,624
Property and business tax	$ 4,901
Total occupation costs	$ 33,525
Net income before income tax depreciation and debt service	$161,328

Yet Rabih decided it was not worth his time to guess the numbers to itemize these expenses separately for accounting purposes. "It is all there," he thought to himself, "and my time is probably better spent finding new customers!"

GUELPH COMMUNITY AND EMIR'S CLIENTELE

As previously noted, Guelph is home to the University of Guelph. By 2002 enrollment had grown to 12,000+ students at the undergraduate level and 3,000+ students at the graduate level. The university offered several programs and was especially well known for its College of Agriculture, Veterinary College, School of Engineering, and School of Hospitality and Tourism Management. In the mid-1990s, the Ontario Department of Agriculture made the decision to open a major research center adjacent to the university. This was a welcomed boost to the prestige and overall economy for the Guelph community.

Guelph's population had grown to 115,000+ by 2006, and continued growth was anticipated. Housing developments in the area were on the rise (including low-income accommodations). The downtown core had a new theater and sports arena, both of first-class quality. Guelph was also home to Sleeman's Brewery, which had gained national recognition as a boutique brewery. Annual community events included the Winterfest in February, the Spring Festival in May, the Multicultural Festival in June, Festival Italiano in July, the Jazz Festival in September, and the Christmas Festival of Lights in December. These and other events reflected the strong sense of community involvement for residents of Guelph.

Because of the university and Guelph's central location in southern Ontario with its close proximity to Toronto, the local population had a high international presence. Asian, European, and Middle Eastern representation was particularly strong. As a result, ethnic restaurants abounded in the community and throughout the region.

Emir's clientele were from many walks of life. These included locals who had grown up in the city and university faculty, staff, administrators, visiting faculty, visiting scholars, graduate students, and parents of students from near and far. Some patrons from the surrounding area came to Emir's because they had a particular zest for authentic Lebanese cuisine. Others came because they simply enjoyed ethnic cuisine of all types. Nonetheless, Roger and Nicole felt that the major sources of their clientele were those directly (or indirectly) affiliated with the university. It was still difficult to discern, however, exactly who were the restaurant and take-out customers versus the catering customers. Another question they pondered was whether people first discovered the restaurant and then became catering customers. Or, perhaps, was it through positive word of mouth from either the restaurant or catering customers that their catering business grew? The catering business continued at a strong pace, yet the restaurant business became stagnant. Other questions arose: If they grew the restaurant business, would that in turn foster growth in the more lucrative catering business? Their advertising and promotion budget was limited. What direction should they take?

COMPETITION

Overall competition in Guelph was strong with 100+ restaurants in the area including upscale fine dining, casual fast food, pubs, sports bars, taverns, cafes, and coffeehouses. Ethnic and international establishments offered Indian, Mediterranean, Asian, Italian, Greek, Latin American, and Middle Eastern cuisine. However, Emir's was the only restaurant offering authentic Lebanese cuisine. Within a 5- to 10-block radius of their surrounding area they were the only restaurant offering ethnic cuisine.

In the nearby downtown core area Rabih identified three direct competitors: Café Insomnia & Deli featuring Italian cuisine, Latino's offering Chilean cuisine, and the Greek Garden. Each of these competitors was somewhat larger in its seating capacity, especially Latino's, which accommodated 75+ persons. Latino's also offered all homemade and all natural ingredients and had a fairly extensive wine list with a well-trained staff knowledgeable on the types and regions of the wine they served. The general price ranges of these three competitors were similar to that of Emir's.

There were 12 catering companies in Guelph, each specializing in various cuisines. Appetizingly Yours and Bounissimo's were solely in the catering business with each offering European and North American fare. The Essence of Italy Specialty Catering was also a dedicated caterer focusing on Italian cuisine. The Black Mustard Bistro Lounge and Sorbaro's Fine Dining were upscale restaurants that offered catering services but not to the extent of Emir's. Latino's and Café Insomnia, their direct restaurant competitors, did extensive catering. The University of Guelph's Hospitality Services offered on-campus catering, albeit at lower prices, and not at the quality level of Emir's, Latino's, and Café Insomnia.

FAMILY MEETING

Nicole felt it was time to call a family meeting to address a number of issues they currently faced regarding the future of Emir's. For the past year the two younger sons, Chadi and Marwan, were based in the Whistler, British Columbia, ski resort area teaching snowboarding, at which they had become quite skilled. In a recent phone conversation they had with the family, however, the boys let them know about an opportunity that had arisen to take over management of a local café. It would be a full-time job for both of them in that the restaurant was in disrepair and needed much attention to rebuild and renew its loyal customer base. Chadi and Marwan were excited about the challenge, and they had grown to like this area in western Canada.

Meanwhile, Rabih's cousin from Lebanon, Souheil, had come to Guelph to enroll as a graduate student in the MBA program at the School of Hospitality and Tourism Management. He took up residence with the family and helped out tremendously at the restaurant, especially with the catering. Roger, Nicole, and Rabih were delighted to have him join the team. The four of them decided that the best time for their meeting would be on a Sunday afternoon in the kitchen of their home. The following dialogue captures the essence of what transpired at the meeting.

Nicole: Well, here we are, and perhaps I'll take lead role as the family matriarch that I am. (Forthright that she was, Nicole also had a lovely sense of humor.) For starters, Chadi and Marwan are doing well. I am so glad we encouraged them to explore Canada and venture out on their own while they are still young. And Souheil, it is wonderful that you have joined us, but be sure to keep up with your studies. Remember, your graduate degree comes first, and we are grateful that you can assist at Emir's as time allows.

Souheil: It is so great to be here. I enjoy my studies, yet I am learning so much from you all, too.

Rabih: Hey, Mom's right, Souheil. But it sure is great to have you on board.

Nicole: Okay, now it is time we get down to business. The boys are on the West Coast having fun and working hard. Here at home, Emir's continues to move along, but we have much room for growth. Your father's recent mild heart attack, Rabih, set us back a bit, but we are getting on track again—

Roger: I feel fine!

Nicole: May I continue? Thank you. Rabih, please report and give us your thoughts on the catering side of the business.

Rabih: We continue to do pretty well, Mom. And we are starting to get more business from the Kitchener/Waterloo area. Here in Guelph, Latino's, in particular, continues to be our main competition. So I have lowered our prices somewhat to be more competitive. They sure are tough, and they have a good product.

Roger: Rabih, you spend so much time on catering. And you try to keep up with the bookkeeping. I think the restaurant and take-out business is suffering as a result. We need to be strong for this side of the business as it helps to get exposure from retail to garner new business for catering. You also spend a lot of time going to Toronto each week to purchase much of our fresh ingredients.

Rabih: I know, Dad, but the best quality and price value for items such as seasonings, vegetables, bread, and imported specialty products are in Toronto. Sure, I could run to Kitchener or Waterloo for some of these, but the quality is just not as good. And you insist on quality!

Roger: Okay, okay, but you can't do everything. We need to decide where to focus our efforts. Are you aware that the retail space next door to Emir's may become available for leasing soon? We could double our capacity at the restaurant, create a larger kitchen we very much need, and, perhaps, better compete with Latino's. Aha!

Rabih: You are dreaming, Dad. It is all we can do to manage what we have and do so effectively. And I want to keep working on my Chef de Cuisine Certificate (CCC) credentials, too. The next exam is coming up soon.

Nicole: Maybe we should bring in a new full-time position. Right now we only use part-time help when needed. Both Roger and I are energetic, but we are not getting any younger. I know it has always been a family business, yet maybe it is time to rethink that philosophy.

Souheil: When looking over the books, I notice two things: First, your advertising expenses are really low, and second, turnovers in the restaurant could be higher. Thus, shouldn't you build these two areas before delving into doubling the size of the restaurant?

Nicole: Thank you, Souheil; you are making a good point.

Roger: But you have to spend money to make money, that I know. And you have to take risks now and then, too. We do have some capital funds available, so let's just do it. Advertising can wait. We have good word of mouth, so why waste the money? And we could have a really grand reopening to let the town know we are on the move!

Rabih: I know you are bright, Dad, and ran a very successful import/export business in Lebanon. But we are now in the food service business and a very competitive one at that. Latino's advertises like crazy.

Roger: Latino's has over twice our current capacity and a prime downtown location. Look at all the exposure they are getting. What I have learned is that restaurant goers want to see a lively place, one that is active but not boisterous. I must confess, Latino's really is a great place to relax, enjoy a good meal, and have a pleasant evening. Right now we close up at 9:00 P.M except when there is a crowd that we did not anticipate. No wonder their catering business is doing well. People just feel good about Latino's. Emir's may be too small to create that kind of atmosphere, and, hence, some locals may just not think of us when it comes time to look for a caterer for that special event.

Nicole: That is all well and good, Roger, and you bring out an interesting perspective. I have another idea. Let's franchise! We could open a second Emir's in the Kitchener/ Waterloo area. I am sure we can find some young, energetic entrepreneurs who might be interested.

Remember, Roger, being an entrepreneur was so important to you. And when we first came to Canada our plan was to franchise with a known brand ourselves. Emir's could become the Lebanese restaurant of Southern Ontario . . . and I could retire! (with a smile of her face)

Roger: Nicole, you are nuts.

Rabih: Dad, she has an idea that just might have merit. Remember, we are having this meeting to explore possibilities of where we might be headed next.

Souheil: I have what may be an even better idea. I have gotten to know several of the professors pretty well at the university. Some are pretty good. In one of our courses we actually develop a case study on a hospitality establishment in the local area and then analyze the situation that the management of the operation is facing. The students get excited about the project. So maybe we could get some graduate students involved

with Emir's and see what direction they think we should take. Anyway, just a thought.

Rabih: Hey, that is a good idea. Why not?

Nicole: Sounds pretty good to me. Hopefully the professor who teaches the course could get involved, too.

Souheil: I think she would be happy to do so.

Roger: Okay, but I still like my idea best. Nonetheless, getting some outside opinions will not hurt. And perhaps it might even help us to see our situation a little more clearly. Too bad Chadi and Marwan are missing all the fun! Ah, but they are enjoying their independence. And, Rabih, I know you really want your CCC. Somehow we will make this all work if we just stay determined.

Nicole: Super. This was a productive meeting. Souheil, would you contact the professor who teaches the course? Thanks. We could invite her to lunch at the restaurant so she can see what we are all about. I look forward to meeting her.

Strategic Marketing

overview

Strategy and the importance of developing a strategic plan before developing a marketing plan is the subject of this chapter. Specifically, this chapter begins with a discussion of strategic marketing versus marketing management and the concept of strategy. This leads to the mission statement followed by a full discourse of the strategic marketing system, including SWOT analysis and feedback loops to ensure the success of strategic planning.

chapter 6

Strategic Marketing

learning objectives

After reading this chapter, you should be able to:

1. Develop marketing strategies for a place of business.

2. Discuss the basic differences between marketing strategy and marketing management.

3. Develop mission statements and important criteria to make them effective.

4. Explain the strategic marketing system model, what it means, and how it works.

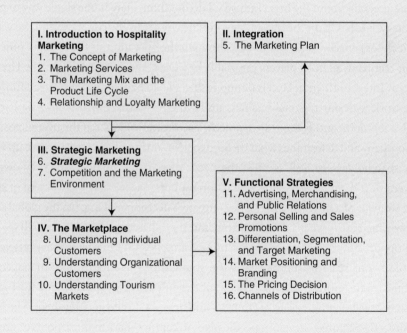

I. Introduction to Hospitality Marketing
1. The Concept of Marketing
2. Marketing Services
3. The Marketing Mix and the Product Life Cycle
4. Relationship and Loyalty Marketing

II. Integration
5. The Marketing Plan

III. Strategic Marketing
6. *Strategic Marketing*
7. Competition and the Marketing Environment

IV. The Marketplace
8. Understanding Individual Customers
9. Understanding Organizational Customers
10. Understanding Tourism Markets

V. Functional Strategies
11. Advertising, Merchandising, and Public Relations
12. Personal Selling and Sales Promotions
13. Differentiation, Segmentation, and Target Marketing
14. Market Positioning and Branding
15. The Pricing Decision
16. Channels of Distribution

Christian Hempell
Vice President, Strategy, InterContinental Hotels Group

Christian Hempell is vice president, strategy, for InterContinental Hotels Group (IHG), PLC, the world's most global hotel company, with more than 3,500 properties under seven brands in nearly 100 countries. He joined in the strategy function in 2003 and is based in London. Hempell is responsible for supporting the development of the company's global strategy and managing the internal strategic planning process. Prior to joining IHG, Hempell served as practice manager for Andersen's Hospitality Consulting Group in the United States. He holds a BS degree from the School of Hospitality Management at Cornell University and an MBA from Harvard Business School.

 Marketing in Action
....

Christian Hempell
Vice President, Strategy, InterContinental Hotels Group
....

How does IHG practice strategy, and who is involved in the strategic planning in this organization?

Speaking generally about the hotel industry, I do not think there is one single way to practice strategy. It is based on the capabilities and the culture of the company and the way in which it chooses to compete. Therefore, although the operation of the strategy function varies by company, all hotel players invariably practice strategy very deliberately. The way it works at InterContinental Hotels Group is this: We do have a central strategy function and we work with our regional business units (the Americas, EMEA, and Asia Pacific) through a *top-down* and *bottom-up approach*. The top-down looks at the overall position of the company and determines what we need to focus on at a global level to drive up value for our shareholders, as well as meet the needs of our stakeholders—be they owners, employees, or consumers. The central function supports the overall development of strategy and strategic objectives that drive the company's decision making. On the global scale, we know which areas we need to focus on, and that's based on a series of analyses into market opportunities, consumer trends, our existing position in various markets versus competitors, our scale in supply share, the strength of our brands, and management capabilities on the ground that are needed for execution. We know which are the biggest markets—we call them the *pools of growth*—and how we should go after them. The business units—the regions—develop specific plans to capture share in those pools of growth from the bottom up. So they develop specifics on how many hotels we could add into our system and which consumer brands would be most successful in each market.

The real trick, in my opinion, is the iteration that happens among the different trade-offs that you have to make across business units. For example, a business unit may find an attractive opportunity to invest and grow, but they don't know if that's a better return relative to something in another part of the world. So, they submit these opportunities, and then it's the job of the center and the business units collectively to pick and choose which ones are the most meaningful for the company. Strategy is about tradeoffs. It's about saying no more often than saying yes, so it has to be done in a culturally vibrant way. There are many good growth opportunities, but some are better than others. Picking up these opportunities on a relative basis to one another allows you to make your best choices.

Every company, regardless of how big it is, has limited resources, and one of the scariest resources is management focus. Capital is important, and the number of staff, employees, and so forth, is important; but really, it's about having the top managers of the company focusing on a handful of things that are really going to drive the biggest profit for the company. A lot of things are worth doing, but they won't have the same return or impact on achieving your strategic objectives. Time is limited and focus is finite, so companies needs to pick the big drivers that really matter.

How do you measure or track the success of your strategy?

One of the most important functions of our strategy team is to track the progress and success of the corporate strategy. We have two methods of measuring how our strategy is performing. One is the level of total shareholder return, which is essentially how the shareholders benefit from the actions we take. A share price is simply the earnings per share times the price/earnings multiple. We have direct control of our earnings per share through our operations, and we can influence (but not control) the multiple based on management credibility, a good track record on delivery, and a lot of future growth opportunities. This, however, is only the financial measure of our strategy.

We also track our strategic progress by taking our relationships with stakeholders—owners, consumers, and employees—into account. We monitor how many new contracts we have signed with owners and how many renew their agreements. We measure our brand premiums and loyalty program to ensure long-term relevance and preference with consumers. And we assess the talent base of the company to develop a deep pipeline of management capability and a high level of employee engagement. These measures describe more of our future earnings growth and longterm viability as a company.

How many people tend to work in the strategy department in your field?

There is a difference between the strategy department and the strategy function. The actual department is a rather small group of individuals who are available to do analysis on corporate-level initiatives and on big strategic decisions that arise. The actual strategy function involves putting together ideas for growth opportunities, looking at your competitors, assessing your capabilities and what you can execute upon, and being very clear on how they work together. That requires a much larger group of people.

Whether it is our head of franchising, head of finance, head of management operations, or the president of a region, the entire senior management team is involved in the strategy function. The strategy process is balanced and integrated with our budget and operations plans for the next year. However, each company will carry out the process differently based on how they're structured, the role of the corporate office, and where decisions are made.

Strategy is both an art and a science. The science part is what people usually think about with regard to the branding, the financial modeling, and competitors—the analytical side. Then there is the art side, which is where people get together and come up with different ideas that seem equally plausible and, at the end of the day, the chief executive and executive committee decide how best to grow the company. There is not necessarily one clear answer, and this is where the art comes in. Being able to pick and integrate the best strategy overall involves not only numbers or a column in a spreadsheet, but it also involves creativity to come up with the management team's best assessment about choosing and achieving your strategic objectives.

Used by premission from Christian Hempell.

STRATEGIC MARKETING, MARKETING MANAGEMENT, AND MARKETING EFFECTIVENESS

strategic marketing The overall view of marketing that includes defining marketing, setting marketing goals, and allocating the resources necessary to reach those goals.

marketing management The day-to-day management of the marketing process. The development of the annual marketing plan is part of marketing management.

strategic business unit (SBU) A division of a company that may operate separately from the main organization. For example, InterContinental Hotels Group (IHG) is the organization, whereas Holiday Inn is the SBU.

Strategy takes a long-range view of an organization. Creating a strategy is about matching the strengths of the organization with the external environment to develop a competitive advantage. Once the firm has established its objectives, it must develop the strategies for obtaining those objectives. Strategy must not be confused with tactics. The strategy is the plan that guides the day-to-day activities, whereas the tactics are the specific steps or activities undertaken by the firm to execute the strategy.

Many marketing students find it hard to differentiate between **strategic marketing** and **marketing management**. This is because some authors place strategy largely at the corporate or **strategic business unit (SBU)** level and marketing management at the local level (e.g., individual hotel). In contrast, we believe that true marketing leadership develops from the practice of strategy and marketing management at both the corporate level and the local level.

It is necessary to distinguish between strategy and management effectiveness (the term *management* here refers not only to marketing management, but also to management of all areas), because too often management tools have replaced strategy. Management effectiveness refers to activities that allow a company to operate more smoothly and produce better results. It includes things such as employee motivation, total quality management (TQM), better salesmanship, revenue management, global distribution, and employee empowerment, among others. For example, consider Southwest Airlines. While other airlines struggle, Southwest's profits continue to grow. Management effectiveness keeps employees happy and loyal to the airline. This is in contrast to many other carriers that experience antagonistic relationships between employees and management.

Management effectiveness is necessary for a company to produce better results, but it is not enough. In other words, management effectiveness alone will not lead to better results. A solid, well-thought-out, and well-executed strategy is equally important. Too often, management effectiveness is emphasized instead of strategic thinking, which is the long range view. Strategic effectiveness means performing activities that are different from competitors' or performing similar activities in more effective or meaningful ways. A company can outperform others only if it can preserve an established difference. Members of the target market must also see this difference and find it of value.

Southwest Airlines' strategy of flying point to point to "secondary airports" instead of the traditional hub and spoke system favored by many "legacy" airlines and its emphasis on keeping operating costs to a minimum enabled the airline to survive higher fuel prices and the economic impact of 9-11, when other airlines could not. Southwest's strategy is to perform things differently and to perform different things— no assigned seating, no meals, and flights to secondary airports that still provide convenient access to major destination cities. Southwest does what it does well because it sticks to what it does best and it delivers on its promises to customers.

Web Browsing Exercise

Visit Southwest Airlines' website (www.southwest. com/about_swa/). Which areas on this website show management effectivness? How? Which areas show strategic effectiveness? How? Compare and contrast Southwest's management effectiveness and its strategic effectiveness.

Strategy

Competitive strategy is about being different. Strategic competition is the process of perceiving new market opportunities that bring new customers from competitors, or bring new customers into the market. Often these opportunities open up, or close, because of environmental change or because they have been given up by competitors. Consider, for example, the way Starbucks changed how Americans drink coffee and how the Internet changed the way people book airlines and hotels.

What, then, is strategy?[1]

1. Strategy is taking advantage of opportunities to create/design a product/service to serve customers. If there were only one ideal way to do this, there would be no need for strategy. In reality, there are many ways a firm can conduct its business. The idea is to choose the products, services, and methods of promotion that are different from competitors and valued by the targeted market.

2. Strategy is making trade-offs. Companies cannot be all things to all customers, nor can they afford to pursue every opportunity. An essential part of strategy is not only choosing what to do, but choosing what not to do.

3. Strategy is creating "fit," or directing all activities that occur in an organization toward the same goal. Consider again the case of Starbucks Coffee. Its competitive advantage comes from the way its activities fit and complement one another. Fit locks out competitors by creating a chain that is as strong as its weakest link. The best fits are strategy specific because they enhance the company's uniqueness. The whole matters more than any individual part, and sustainable competitive advantage grows out of the entire system of activities. Prior to Starbucks, getting a cup of coffee was a task, not an experience. Starbucks offered not only many varieties of coffee, but also a great place to sit and enjoy the coffee. Because of these two differences, Starbucks has higher perstore revenues than the local coffee shop does.

Unfortunately, many companies do not have marketing strategies, nor do they make strategic choices. They also let their strategies disintegrate and disappear. Although these situations may be due partially to external causes (e.g., changes in technology or competition), more often they originate within the organization, from a misguided view of competition, organizational failure, or the undisciplined desire to grow "bigger and better."

Managers are under increasing pressure to deliver measurable performance growth. Managerial effectiveness becomes the goal, and strategic thinking is avoided. Companies imitate one another in a kind of herd behavior. This has been especially true in the hotel industry. Consider the hotels in Las Vegas. After Steve Wynn built the Mirage with its volcano and Siegfried and Roy magic show, other companies copied this idea. Although each hotel has a different theme, the strategies of the hotels are essentially the same. (The difference was Harrah's, which focused on customer loyalty and the use of technology to better serve the customer.)

Then there is the desire to grow with steps that blur a company's strategic position. Attempts to compete on many fronts can create confusion and cause organizational chaos. The goal is for more revenue even if profits fall. Meanwhile, uniqueness is fuzzy, compromises are made, fit is reduced, and competitive advantage is lost. Rather than being deepened, the strategic position is broadened and compromised. Consider Darden Restaurants, operators of Red Lobster and Olive Garden. They tried to grow through the introduction of a Chinese food concept called China Coast. After opening 55 units across the United States, they shut them all down and abandoned the concept because China Coast lacked strategic fit and competitive advantage. Another chain, P.F. Chang's, has been very successful with Chinese cuisine in multiple locations. P.F. Chang's had a better strategic fit and a strong competitive advantage. The Chinese food was good with some signature dishes, and it was served in a hip environment with a great local bar scene.

Success, ironically, can also be one of the greatest threats to survival and future success because it may invoke the attitude, We can do no wrong. Examples of firms that suffered from hubris include IBM, General Motors, Mirage Resorts (now part of MGM Mirage), Caesars Entertainment (now part of Harrah's Entertainment), Digital Equipment (now part of Hewitt Packard after it first became part of Compaq), Sheraton (now part of Starwood), and Planet Hollywood. Some of these companies were able to turn things around; others were purchased by other companies. Bill Gates of Microsoft, an astute observer of business strategy, has vowed never to be afflicted by the same mentality.

Strategic Leadership

Strategic leadership is the fundamental element necessary to establish or reestablish a clear strategy. This requires strong leaders willing to make choices. The well-known Michael Porter of Harvard Business School describes his view of strategic leadership in Exhibit 6-1.

Strategic leadership is the ability of the leader to express a strategic vision for the company, or a division of the company, and to motivate others to buy into that vision. Characteristics of this type of leadership also include commitment, being well informed, a willingness to delegate and empower, and smart use of power.

Exhibit 6-1   Michael Porter's View of Strategic Leadership

General management is more than the stewardship of individual functions. Its core is strategy: defining and communicating the company's unique position, making trade-offs, and forging fit among activities. The leader must provide the discipline to decide which industry changes and the customer needs to which the company will respond, while avoiding organizational distractions and maintaining the company's distinctiveness. . . . One of the leader's jobs is to teach others in the organization about strategy—and to say no when appropriate.

Strategy renders choices about what not to do as important as choices about what to do. . . . Deciding which target group of customers, varieties, and needs the company should serve is fundamental to developing a strategy. But so is deciding not to serve other customers or needs and not to offer certain features or services. Thus strategy requires constant discipline and clear communication. Indeed, one of the most important functions of an explicit, communicated strategy is to guide employees in making choices that arise because of trade-offs in their individual activities and day-to-day decisions.

. . . A company may have to change its strategy if there are major structural changes in its industry. In fact, new strategic positions often arise because of industry changes and new entrants unencumbered by history often can exploit them more easily. However, a company's choice of a new position must be driven by the ability to find new trade-offs and leverage a new system of complementary activities into a sustainable advantage.

Source: Reprinted by permission of *Harvard Business Review.* From "What is strategy?" by Michael Porter, Nov-Dec 1996, 77–78. Copyright © 1996 by the Harvard Business School Publishing Corporation; all rights reserved.

Exhibit 6-2   Major Differences between Strategic Marketing and Marketing Management

---

Point of Difference	Strategic Marketing	Marketing Management
Timeframe	Long-range; i.e., decisions have long-term implications	Day-to-day; i.e., decisions have relevance in a given financial year
Orientation	Inductive and intuitive	Deductive and analytical
Decision process	Primarily bottom-up	Mainly top-down
Relationship with environment	Environment considered ever-changing and dynamic	Environment considered constant with occasional disturbances
Opportunity sensitivity	Ongoing to seek new opportunities	Ad hoc search for a new opportunity
Organizational behavior	Achieve synergy between different components of the organization, both horizontally and vertically	Pursue interests of the decentralized unit
Nature of job	Requires high degree of creativity and originality	Requires maturity, experience, and control orientation
Leadership style	Requires proactive perspective	Requires reactive perspective
Mission	Deals with what business to emphasize	Deals with running a delineated business

---

Source: Marketing Planning and Strategy, 5th ed., by Jain, 1997. Reprinted with permission of South-Western, a division of Thomson Learning: www.thomsonrights.com. Fax 800 730-2215.

Strategic Marketing

We now examine the differences between strategic marketing and marketing management. Strategic marketing takes an overall, big-picture view, distributing resources and setting objectives after defining the market; marketing management develops the product or service, prices it, tells the customer about it, and gets it to the customer. Thus, strategy must come before management. A hotel's restaurant, for example cannot be appropriately designed without first correctly selecting the market it is to serve. Exhibit 6-2 delineates the differences between strategic marketing and marketing management.

CONCEPT OF STRATEGY

We begin with the standard textbook definitions of strategy and tactics, which come directly from the military: Tactics are the way to win the battle; strategy is the way to win the war. In a simplistic example, we could demonstrate this as follows:

Objective: Surround the enemy.
Strategy: Take one area at a time.
Tactic: Use armored tank divisions.

Actually, marketing is not much different. The objective is to increase revenues. Strategy is the way to gain and keep customers; tactics are the step-by-step procedure of how to do it. For example:

Objective: Increase revenues 6 percent during the next fiscal year by being perceived as the hotel of choice.
Strategy: Always give customers better value.
Tactics: Always have their reservations and guest rooms (tables) ready; call them by name; make sure they receive their wake-up call and have full-length mirrors and good bathroom lighting in their rooms; offer fresh-brewed coffee as soon as they sit down for breakfast; have room service delivered on time; provide complimentary Internet access from guest rooms; have the print on the menu large enough to read; offer a selection for those who are light eaters; and so forth.

This example shows that tactics flow from strategy. That means that the first thing we have to do is develop an appropriate strategy. The strategy drives the firm and specifies the direction in which it is going.

STRATEGIC MARKETING SYSTEM MODEL

strategic marketing system model A model that shows the process by which marketing strategies and substrategies are developed.

mission statement A statement that defines the purpose of a business and that may include many ways to achieve that purpose in terms of all stakeholders. It should drive all subsets of the business.

constituents Those who have an interest in the business entity, including customers, employees, owners, financial backers, and the local community. Constituents are also referred to as **stakeholders**.

Exhibit 6-3 shows the **strategic marketing system model**. We will guide you through this model as we proceed through this chapter. Notice that the model starts with the firm's **mission statement**. This is true whether you are Hilton Hotels or an independent restaurant. Both Hilton and the independent restaurant have specific objectives and missions. Hilton's may be put together by an executive committee of senior vice presidents, while an independent restaurant may carry their objectives and mission around in their head, and if you asked either of them, they might be unable to express them. It doesn't matter; they are still there and they will drive the operation of the restaurant, for better or for worse, every bit as much as Hilton's will guide the operations of that multinational corporation. The mission statement usually includes the broad, long-term goals of an organization.

The mission statement defines the purpose of a business. It states why the firm exists, who the firm competes against, who the target market is, and how to serve the **constituents**—those who have an interest in what the firm does. These include customers, employees, owners, financial backers, and the local community, all of whom are commonly called **stakeholders**.

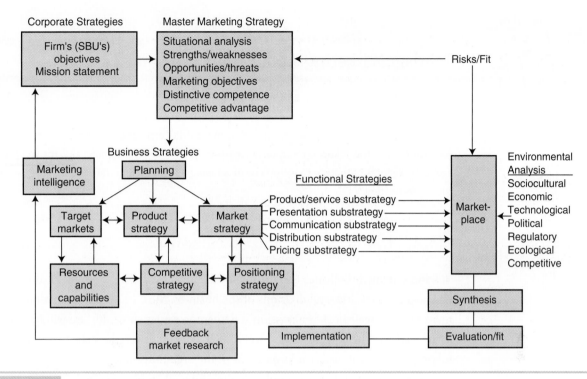

Exhibit 6-3 The Strategic Marketing System Model

Mission statements exist not only at the corporate level, but also at the level of every strategic business unit (SBU) within the firm. Recall that an SBU is a unit of a business that serves a clearly defined market segment with its own strategy (but in a manner consistent with the overall corporate strategy), its own mission, and its own competitors.

As Marriott Hotels and Resorts has a mission, so to does the Courtyard by Marriott division, the Fairfield Inns division, and the Residence Inn division. The same is true of the Marriott Long Wharf and the Marriott Copley Place, both in Boston. By the same token, each of the restaurants and lounges at the Marriott Copley Place has its own objectives, mission, and competitors.

All mission statements in the organization flow from the corporate mission statement, as shown in Exhibit 6-3. This strategic marketing system model, can also be applied to any strategic business unit. Thus, in the case of Marriott Copley Place, we could replace "Firm's objectives" with "Hotel's objectives." By the same token, we could replace "Hotel's objectives" with "Restaurant's objectives." Strategic planning occurs at every level at which a strategic business unit exists. We should bear this in mind as we continue to discuss the strategic marketing system model.

Tourism destinations often have mission statements, too. As discussed in the Tourism Marketing Application, Victoria, a pristine island on the west coast of Canada, has a distinct, yet simplified mission statement developed by the Tourism Victoria Corporation.

A firm's (or SBU's) objectives may include growth, return on investment, profit, leadership, industry position, or other factors. These are included in the mission statement. Thus, developing the mission statement is a crucial assignment. Because the

Web Browsing Exercise

Search the World Wide Web for restaurants, travel and tourism companies, or hotels. Find two companies that have different positions in the marketplace and evaluate their mission statements. How effective are these mission statements? Next, compare and contrast the mission statements. In what ways are they similar? In what ways are they different?

Tourism Marketing Application

Mission Statement for Victoria, British Columbia, Canada

Tourist destinations can also have mission statements. For instance, consider Victoria, an island located on the west coast of Canada. The Tourism Victoria Corporation, which promotes tourism for Victoria, has the following mission statement: "Maximize employment and the long-term economic benefits of tourism to Victoria by developing and marketing the State as a competitive tourist destination."

Source: www.tourismvictoria.com.au/index.php?option=displaypage&Itemid=126&op=page (accessed March 5, 2006).

mission statement indicates the purpose of the business and is a statement of why the business exists, it drives all divisions of the business. Most important, a mission statement must be realistic. For a mission statement to say, as some do, "we will be known as the leader in the hotel industry" when such a possibility is not realistic or meaningful only leads to confusion at lower levels of the organization.

The mission statement should be something all employees can believe in. It sets goals, and it urges everyone in the organization to meet those goals. Properly, it is communicated throughout the organization for all to follow. When the response at lower levels is, "Who are they kidding?" the entire effort becomes a meaningless and self-defeating effort. Although some mission statements are quite brief, an effective mission statement, which is nothing less than an overall strategy statement, should fulfill the criteria shown in Exhibit 6-4, either implicitly or explicitly.

Three things should be equally represented in the mission statement: employees, customers, and shareholders. Too often, mission statements reflect the needs of the owners: more profits. Profits can only be achieved with satisfied customers and satisfied employees. Conversely, if customers enjoy prices that are too low or employees are paid too much, the owners will not be satisfied. All three entities must be represented in any successful mission statement.

Consider the mission statement of Ritz-Carlton, probably one of the most complete ones in hospitality, in Exhibit 6-5.

MASTER MARKETING STRATEGY

Developing the master marketing strategy is the next stage in the strategic marketing system, as shown in Exhibit 6-3. The master strategy is designed to be long term, not short term. This does not mean that it will never change; if market conditions change, then so too should the master strategy.

The intention, however, of master strategies is that they will endure for some time. This means that they take a long-range view of the environment as opposed to the short-range perspective of the marketing plan, even though many of the issues are the same.

Exhibit 6-4	Criteria of an Effective Mission Statement

1. **It states what business the company (or SBU) is in or will be in.** This goes considerably beyond being in the hotel, restaurant, or food business. Instead, it is more specific and states how we serve our customers and specifies who they are. For example: Hotel XYZ is in the business of providing the traveling and price-sensitive public with modern, comfortable, and clean accommodations at a very reasonable price. Accordingly, Hotel XYZ recognizes the basic needs of travelers as well as the need for a pleasant and hassle-free experience, but without the amenities for which this market is unwilling to pay. Hotel XYZ wants to be known as the best buy at the moderate price level, satisfying all essential needs for the motoring public.

   You can see that this statement has numerous ramifications such as how, what, when, and where. These are enumerated later in the strategic plan, but the answers will be driven by the mission statement.

2. **It identifies the special competency of the firm and how it will be unique in the marketplace.** Hotel XYZ is and will continue to be a leader in its field because of its special identification with the budget-minded traveling public and its needs. By continuous communication with its market and regularly adaptating to the changing needs of that market, Hotel XYZ will maintain its position as the hotel of choice of its customers.

   Again, the mission statement has committed the firm to a definite course of action. Its competency and uniqueness is special knowledge of the target market and a commitment to maintain and implement that knowledge.

3. **In a market position statement it defines who the competition will be—that is, it actually chooses whom it will compete against and does not leave this to chance.** Hotel XYZ's niche in the market will be between the full economy, highly price-sensitive market that chooses accommodations almost solely by price, and the middle-tier market that will pay $20 more for additional amenities and services. Accordingly, XYZ competes only tangentially against ABC and DEF, on the one hand, and GHI and JKL, on the other hand. XYZ competes directly for the same market against MNO and PQR as well as other companies that choose to enter this market.

   As Burger King knows it has to beat McDonald's and Pepsi-Cola knows it has to beat Coca-Cola, XYZ knows it has to beat MNO and PQR and will watch these competitors very closely.

4. **It identifies the needs of its constituents.**
   a. **Customers:** XYZ will conduct ongoing research of its customers needs, both at the corporate and unit levels. It will continuously seek to satisfy those needs within the constraints of its mission.
   b. **Employees:** XYZ recognizes all employees as internal customers with their own varying needs and wants. Accordingly, it will attend to those needs and wants with the same attitude it holds toward its paying customers and will maintain an open line of communication for that purpose.
   c. **Community:** XYZ recognizes its position in the economic, political, and social communities. Thus, it will maintain a role of good citizenship in all endeavors and efforts.
   d. **Owners:** XYZ has committed itself to a 15 percent ROI for its investors as well as a positive image of which they can be proud. XYZ will function both in the marketplace and in its operations to maintain these commitments.

5. **It identifies the future.** Hotel XYZ will develop and expand through controlled growth in suitable locations. Its strategy will be to develop regional strength as a gradual development toward national strength, with the objective of reaching that goal by the year 2005.

The master marketing strategy shapes goals after developing and weighing options. It specifies where the firm is going and provides the framework for the entire marketing effort. Derived from the mission statement and objectives, the master marketing strategy turns to the marketing emphasis to fulfill those missions.

The mission statement of the hypothetical hotel company XYZ in Exhibit 6-4 noted that it wanted to be perceived as the best buy at the moderate price level. The master marketing strategy, then, should address that goal to make it happen.

SITUATIONAL ANALYSIS

The master marketing strategy begins with a **situational analysis**, often referred to as a **SWOT analysis**, of strengths and weaknesses, again asking the questions, Where are we now? and Where do we want to go? It is the "where" of the strategic marketing system that shapes objectives and sets the stage for all decisions. A master marketing strategy deals with the generic stages we have just discussed such as positions, trade-offs, and fit. It also deals with issues such as new markets, growth sectors, customer loyalty, repeat business, quantity versus quality, cheap versus expensive, best versus biggest,

situational analysis/SWOT analysis An analysis of the current internal strengths and weaknesses and external opportunities and threats of a business entity.

THE RITZ-CARLTON®
HOTEL COMPANY, LLC

MISSION STATEMENT

The Ritz-Carlton Hotel Company will be regarded as the quality and market leader
of the hotel industry worldwide.

We are responsible for creating exceptional, profitable results with the investments
entrusted to us by efficiently satisfying customers.

The Ritz-Carlton Hotels will be the clear choice of discriminating business and leisure travelers,
meeting planners, travel industry partners, owners, partners and the travel agent community.

Founded on the principles of providing a high level of genuine, caring, personal service; cleanliness;
beauty; and comfort, we will consistently provide all customers with their ultimate expectation,
a memorable experience and exceptional value. Every employee will be empowered to
provide immediate corrective action should customer problems occur.

Meeting planners will favor The Ritz-Carlton Hotels. Empowered sales staff will know their own product and
will always be familiar with each customer's business. The transition of customer requirements from Sales to
Conference Services will be seamless. Conference Services will be a partner to the meeting planner, with
General Managers showing interest through their presence and participation. Any potential problem will be
solved instantly and with ease for the planner. All billing will be clear, accurate and timely. All of this will create
a memorable, positive experience for the meeting planner and the meeting participants.

Key account customers will receive individualized attention,
products and services in support of their organization's objectives.

All guests and customers will know we fully appreciate their loyalty.

The Ritz-Carlton Hotels will be the first choice for important and social business events
and will be the social centers in each community. Through creativity, detailed planning,
and communication, banquets and conferences will be memorable.

Our restaurants and lounges will be the first choice of
the local community and will be patronized on a regular basis.

The Ritz-Carlton Hotels will be known as positive,
supportive members of their community and will be sensitive to the environment.

The relationships we have with our suppliers will be one of mutual confidence and teamwork.

We will always select employees who share our values. We will strive to meet individual needs because our
success depends on the satisfaction, effort and commitment of each employee. Our leaders will constantly
support and energize all employees to continuously improve productivity and customer satisfaction. This will be
accomplished by creating an environment of genuine care, trust, respect, fairness and teamwork through
training, education, empowerment, participation, recognition, rewards and career opportunities.

Exhibit 6-5   The Ritz-Carlton mission statement

Source: Ritz-Carlton. Used by permission.

high or low markups, quick turnover, product/service range, building brand name, consumer awareness and perception, and a host of other things that will guide the business strategies. Marketing objectives are identified in the master marketing strategy in these contexts.

To address these issues, a firm must first define its current state. We continue with an environmental analysis (see Exhibit 6-3), looking especially at the long-range trends and effects. These trends could be economic (the state of the international, national, or local economy); sociocultural (the graying of America, generation X); lifestyle (interest in fitness and health); legal (laws pertaining to employees such as minimum wage requirements); ecological (a greater awareness of environmental concerns); political (from room taxes to terrorism), technological (Internet reservations); and competitive (industry consolidation). The major purpose of an environmental analysis is to identify external opportunities and threats to the organization.

An opportunity is a favorable trend in the environment such as an emerging market segment or a need or demand for certain specialized services. On the other hand, a threat is an unfavorable trend such as reduced demand or new competition. External opportunities and threats emerge when various questions are asked about the organization, along with its internal strengths and weaknesses, in a situational analysis. Elements of a situational or SWOT analysis are shown in Exhibit 6-6. We discuss trends in more detail in the next chapter.

The main objective of a SWOT analysis is to identify strategies that fit a company's resources and capabilities to the demands of the environment in which the company operates. The purpose of the strategic alternatives developed by a SWOT analysis is to build on a company's strengths by exploiting the opportunities, countering the threats, and overcoming the weaknesses, thus developing distinctive capabilities and a competitive advantage.

The distinctive ability of an organization is more than what it can do; it is what it can do particularly well. It often takes a great deal of self-analysis to understand this and to abide by it. Objective situational analysis is the tool to lay bare the facts. Following it is sometimes more difficult. Marriott learned this the hard way when it diversified into different businesses in which it lacked strength or special skill. It fixed these mistakes by

---

Exhibit 6-6   Some Elements of a Situational (SWOT) Analysis

INTERNAL—STRENGTHS AND WEAKNESSES

- **Brand Demand:** Who is our customer? Why? What is our position? Who are our market segments and target markets? To which do we appeal the most? What use do they make of our product/service? What benefits do we offer? What problems do we solve or not solve? What are the levels of brand awareness, preference?
- **Customer Profile:** What do our customers look like—demographically, psychographically, behaviorally? Are they heavy users or light users? How do they make the decision? What influences them? How do they perceive us? What do they use us for? Where else do they go? What needs and wants do we fulfill? What are their expectations?
- **Organizational Values:** What are the values that guide us? What is the corporate culture? What drives us in a real sense? Do these limit alternatives?
- **Resources:** What are our distinctive capabilities and strengths? What do we do particularly well? How do these compare to the competition? What are our physical resources? Are there any conflicts among our resources, our values, and our objectives? Do we have a good fit?
- **Product/Service:** What is our product/service? What benefits does it offer or problems does it solve? How is it perceived, positioned? What are the tangibles/intangibles? What are our complementary lines? What are our strengths and weaknesses?
- **Objectives:** Where do we want to go? What do we want to accomplish? Are our objectives quantifiable and measurable? How

do we want to be perceived? What are the long-range and short-range considerations? What trade-offs do we need to make?
- **Policies:** What rules do we have now? How do we operate? What guides us? Are any rules conflicting?
- **Organization:** How are resources, authority, and responsibility organized and implemented? Do we become proactive rather than reactive? Does the organization enhance the strategy, or does the organization need to be changed?

EXTERNAL—OPPORTUNITIES AND THREATS

- **Generic Demand:** How are we positioned? Why do people come here, and why do they use this product? Where else do they go? What do they need, want, demand? Are there unmet needs? What do users and nonusers look like? What are the segments for this product category? What are the alternatives? What are the trend patterns—cyclical, seasonal, or fashion?
- **Competition:** Who are our competitors? Where are they? What do they look like? How are they positioned against us? In what market segments are they stronger or weaker? Why do people go there? What do they do better or poorer than we do? What is their market share? What are their strengths and weaknesses? What are their expectations? What are their strategies? What are they doing, and where are they going? Are there new ones coming?
- **Environment:** What are the impacts of technology and sociocultural, economic, political, and regulatory trends?

Tourism Marketing Application

The Bulgarian Association for Alternative Tourism prepared a SWOT analysis for alternative tourism in Bulgaria. This organization defines alternative tourism as travel that is "personal and authentic and encourages interaction with the local environment, people, and communities"—in contrast to mass tourism. The organization believes that by 2007 Bulgaria will be a leader in alternative tourism. Their report details the SWOT analysis of this type of tourism.

Source: www.alternative-tourism.org/english/index.php?page 15 (accessed March 18, 2006).

selling a number of the businesses including its restaurant division, which included Allie's and the Roy Rogers fast-food chain. In many cases, failure to recognize strengths and weaknesses results in targeting the wrong markets. Strategically speaking, a firm should do only what it has the experiences and resources to do well. Ignoring this fact may result in a huge strategic error.

Often a SWOT analysis primarily provides indications of past performance that are unlikely to produce assessments of future opportunities. Following are some actions to take to avoid this mistake:

- Involve the managers who will make the final strategic choices.
- Test alternative strategies against strengths and weaknesses.
- Evaluate strengths and weaknesses in terms of the future and their strategic significance, and relative to the competition.
- Separate weaknesses from simple problems to be overcome.

The final output should be a list of the most significant strengths, on which the future should be planned, and the most important weaknesses, which should be targeted for solution and avoided as underpinnings of strategy.

Solid strength and weakness analysis may be the most neglected phase of strategic planning in the hospitality industry. Without a doubt, this lack was a major contributor to the failure of Howard Johnson in the 1980s. In less than 10 years Howard Johnson went from a top company to a bankrupt company. Eventually, the company was broken up and sold off in pieces, all because management failed to understand its strengths and weaknesses or to see its opportunities and threats. The good news is that a "new" Howard Johnson emerged in the 1990s under new management and strategic leadership. Much of the growth of Howard Johnson is in Eastern Europe.

USING THE STRATEGIC MARKETING SYSTEM MODEL

To help illustrate the strategic marketing system model, we will examine its various stages with excerpts from the actual strategic marketing plan of an international hotel that we will call the International. The mission of this hotel was to be the top upscale hotel of choice for international travelers to its destination. Without showing the full situational analysis of this firm, we have provided some key issues presented by managers in their plan. Three years after this strategic plan was drawn, this hotel lost much business to new competition. Examine the analysis and strategic plan to see whether you can understand why the hotel was not successful. We have added questions in italics to help guide the discussion.

Objectives and Master Strategies

Marketing Objective: To be perceived as a premier super deluxe hotel marketed to the connoisseur consumer.

Master Marketing Strategy: To create an image of exclusivity and uniqueness with premium quality facilities and services.

Strengths:

- Personalized and professional service
- Prime strategic location
- Part of a chain that has already made its mark
- High standards of food and service
- Newly refurnished outlets
- Renowned shopping arcade on premise
- Wide variety of excellently appointed suites

Key Question: Do these strengths represent unique competitive differences perceived by the customer that build defenses against competitive forces or find niche positions in the market? What is the hotel's distinctive competency?

Weaknesses:

- Higher room and F&B rates make it difficult to secure international conference business
- Market sensitivity that we are more pro-foreigner and have less identification with local community
- Lower percentage of national clientele
- Marketing is more product oriented than customer oriented
- Lack of exclusive executive club
- Absence of well-located properties in chain that reduces chain utilization

Key Question: Are these weaknesses, or problems that need to be solved?

Opportunities

- The commercial market in the city is very active, and our location is strategic.
- Development in this area is strong and has a strong affiliation with our hotel.
- The entrepreneurial market is growing, and most businesses are locating to this area.

Key Question: Is this a matching of strengths and competencies to opportunity?

Threats:

- Foreign traffic will be dependent on the political stability of the country.
- Corporations are developing their own facilities to encourage privacy and reduce expenditures.
- The biggest competitor has renovated rooms.
- Some corporations are moving to the suburbs.

Key Question: Are these threats caused by weaknesses? Can they be avoided? Can resources be more effectively deployed?

Business Strategies

Referring back to Exhibit 6-3, you can see that the next stage in strategic marketing is the planning stage. It consists of both operational- and business-level strategies. These strategies are the "how" of strategic marketing—that is, *how* we're going to get from *here* to *where* we want to go. Strategies at this point are more easily measurable and may have time and performance requirements.

This is the stage at which the organization acts in advance, by planning for change. That is, it plans rather than reacts. It is here that the organization shapes its own destiny. At this stage the company attempts to minimize risk, maintain control, and assign resources to keep in focus and reach its goals.

The planning stage is also the stage of specific matching of the product to the market, of understanding where the business is going to come from, of developing new products and services, and of influencing demand. You should take special note of the interrelationship among the various elements of the business strategies.

Target Market Strategy. Target market strategy clearly depends on, among other things, resources and capabilities. To target a market with similar needs and wants is insufficient, if not fatal, when the resources and competencies are not there to serve that market. The appropriate strategy is to target not just markets that appear to have the most opportunity, but also those that the firm can serve best and, one hopes, better than the competition.

A common failing in this respect may be observed among hotels that target the upscale market and price accordingly, but do not have the resources or capabilities to sustain an advantage against this market segment. The hotel then has to accept lower-rated business while management continues to vehemently maintain that it is serving the upscale market. The result is a confused image and failure to fulfill potential. Such strategies are often built on wishful thinking rather than on objective analysis.

Another potential peril derives from targeting too many markets—a strategy of providing something for everyone, that lacks focus and results in confusion for all.

Target market strategy means defining the right target market within the broader market segment. The strategy of the International Hotel we have been discussing is to target the following market:

Age: 35 plus
Income: High

"My new marketing strategy is to sell stuff to you two."

Lifestyle: Results-oriented, professional businessperson, aristocratic with a modern outlook on life, respected in the community, voices an opinion, a leader, and an active socializer

Desired Consumer Response:

- *Rational:* I like staying here because the rooms are spacious and beautiful. I like the computerized telephone exchange with its automatic wake-up call and wireless Internet. The executive club with computers and fax machines is timesaving, smooth, and trouble-free. Check-in/check-out is fast and efficient. Because the hotel is so exclusive, I don't encounter undesirable people. Service is smooth, courteous, and efficient.
- *Emotional:* I like staying here because everyone knows me and takes care of me. I feel very much at home with the room service and restaurants. They know my likes and dislikes and make it a point to remember. It is so exclusive; I like to be seen here.

Product Strategy. Product strategy is concerned with the offering of various products and services to satisfy market needs. It deals with the benefits the product provides, the problems it solves, and how it differentiates from the competition. Product strategies should be based on opportunities in the environment and customer needs rather than just owners' or management's concept of what the product should be. For example, it is quite common in Southeast Asia for upscale hotels to have as many as five formal dining rooms. These will inevitably be Chinese, Japanese, and French, plus one native to the country. The other is likely to be Italian or American. The reasoning, of course, is that all these geographic markets are served by the hotel. Each room usually seats 100 or more, and in most cases is fortunate if it is 50 percent occupied.

The low patronage does not occur because there is no market need. Demand exists for all these ethnic foods, but at varying levels. Further, numerous freestanding restaurants in the city are also filling these needs—at least the Chinese, Japanese, and native. Regardless, the product strategy is to have something for everyone instead of defining the specific needs of the target market.

The essence of marketing is to design the product to fit the market. Sometimes, however, the product, such as a hotel, already exists and the situation is reversed: The market must be found that fits the product. Such a case might exist when the market changes or new competition takes it away.

Competitive Strategy. In developing a competitive strategy, the firm actually chooses its competition and when and where it will compete, as well as whether it will be a low-cost producer, a differentiator, a focuser, or some combination of these. This is realistic provided the choice is realistic, and if it is based on an objective situational analysis. Take the case of Wendy's restaurants, which used a focus strategy when it started its first restaurant in Columbus, Ohio, on November 15, 1969:

> In the late 1960s, when McDonald's and Burger King were already well established, industry experts did not think that there was room for another hamburger chain to enter the market and grow to any substantial size. However, in March 1978 Wendy's had grown to a chain of 1,000 units. A year later, it opened its 1,500th unit. Wendy's

did not compete head to head with McDonald's or Burger King, but rather it focused on a special niche in the crowded hamburger market. For example, it became the first fast-food chain to introduce the salad bar. It went after the baby boomers who were young adults in their 20s and 30s. Surveys showed that over 80 percent of Wendy's business came from those over 25 years old. Compare this with McDonald's, which derived 35 percent of its business from those under 19 years.

The secret to a successful competitive marketing strategy is to find a market where there is a clear advantage or a niche in the market that can be defended. The trick, then, is to match the firm's product strengths with the market. It does not matter whether this position occurs in the high or low end of the market, and it is sensible to consider examples on both ends of the spectrum.

Ritz-Carlton uses a differentiation strategy and is positioned at the top of the market. Its product is "5 star," and it is almost never compromised. After acquiring Ritz-Carlton, Marriott clearly indicated it would never put the Marriott name on the Ritz-Carlton brand. On the other hand, Red Roof Inn in the United States and Formule1 in Europe have maintained their position in the budget segment as low-cost leaders. Both of these companies chose their competition, stuck to it in the marketplace, and were realistic about the choice. This is the essence of clear and compelling competitive strategy. These are single product companies that compete in a single position.

Other companies, such as Choice International, choose their competition in different roles with different brand lines (Clarion, Quality, Comfort, and Sleep Inns) and market them together, giving the customer a choice. Groupe Accor of France has 13 brand lines, as shown in Exhibit 6-7. Each of the brands is marketed separately.

In some companies with widely varied products carrying the same name, each property has to choose its own competition. This is sometimes difficult when the brand name is carried on all products. For example, prior to being purchased by Starwood, Sheraton operated the five-star Sheraton St. Regis in New York City; the convention hotel Sheraton

Exhibit 6-7   Formule1 has proven that cost leadership strategies work

Source: Retrieved October 11, 2005, from www.accor.com/gb/groupe/activities/hotellerie/marques/formule1.asp. Used by permission.

New York a few blocks away; tiny Sheraton Russell on Park Avenue; and franchises in Bordentown, New Jersey (50 miles away), and Westchester County, a half an hour from New York City. Each of these Sheratons had markedly different competition and customers, but all were marketed together under the Sheraton flag. To correct this situation, Starwood rebranded many of the properties by creating different classifications of hotels. For example, it established the Four Points (by Sheraton) hotel brand for lower-tier properties. Other brand names now include St. Regis, W, Sheraton, Westin, and the Luxury Collection.

Market Strategy.

Market strategy is founded on the idea that the firm needs to reach the right market with the product in order to survive. In the final analysis, if you can't reach the market, the best product and the most well-defined strategy will fail. For the hospitality industry, reaching the market can be looked at in two ways. The first is taking the product to the market; the second is bringing the market to the product. We examine these strategies in more detail in Chapter 16. By contrast with manufactured goods, taking the product to the market is a major commitment and, in some cases, a major capital investment. For multi-unit hotel and restaurant companies, taking the product to the market is part of the distribution system. This is the area where location becomes a major factor. The strategy involved concerns the appropriate markets to enter.

For multi-unit companies that seek growth, the case is multiplied many times. When McDonald's saw its growth limited to freestanding, drive-up stores, it changed its market strategy. Soon McDonald's appeared in inner-city locations, office buildings, universities, and almost anywhere else one looked. It then headed overseas to both the European and Asian markets. In Singapore, on the main road of the city and right next to the Hilton International, sits what became the highest-grossing McDonald's in the world, later supplanted by the one in Red Square, Moscow. Market strategy has been a major factor in McDonald's success.

Hilton International's market strategy was to be in major capital cities throughout the world. Intercontinental was developed for cities where Pan American Airlines, its former owner, flew. Le Meridien Hotels chose to enter primary cities such as Boston, New York, and San Francisco when it expanded into the United States. Marriott likes to saturate an area with multiple units, as it has done in Boston, Washington, D.C., Atlanta, Dallas, and other cities.

Getting the market to the product (or making the market aware of the product) involves a new or different set of strategies. When resources are scarce, as they usually are, the market strategy must designate where to use those resources. A restaurant may choose the surrounding neighborhood and concentrate on word of mouth to help get the market to the product. McDonald's, on the other hand, uses national television to

Web Browsing Exercise

Visit Starwood's website (www.starwoodhotels.com/), as well as the websites for its various brands (you can link to these from the preceding web address). How do the brands' individual websites highlight their different competitive strategies? Compare and contrast the brands' individual websites. How are they similar? How are they different?

Tourism Marketing Application

The city of Dublin, in Ireland, presents its marketing strategy and its plans to reach its goals on the website www.trade.visitdublin.com/trade/marketing/default.asp. This website is a nice example of concepts discussed in this chapter. (Accessed March 18, 2006)

cover the entire United States, as well as other countries. The Internet, of course, has become a major force in getting the market to the product. Again, we discuss this subject in more detail in Chapter 16.

Positioning Strategy. The last, but by no means least, of the business strategies is the positioning strategy. We define *positioning* as the way the product is defined by consumers — the place the product occupies in consumers' minds. The goal of positioning strategy is the creation or enhancement of a specific brand image.

The strategic plan for positioning the International Hotel to its market is as follows:

> The hotel will be positioned as a super-deluxe property for the "up" market. It will be positioned to image-conscious elitists and high-flying business executives. All marketing will be geared to the top-brass higher-echelon bracket of both the social and business circles for which facilities, specialties, and personalized attention are the main criteria for selection. The exclusive executive club; the businessman's club with business equipment; and the rooms with antiques, objects d'art, and special butler service will symbolize luxury plus.

Functional Strategies

Functional strategies (refer again to Exhibit 6-3) are the "what" of the strategic system—that is, *what* we are going to do to get *where* we want to go. The important thing to remember is that these are still strategies, not tactics, which come immediately afterward. This set of strategies flows directly to the consumer in the form of the value chain. For example, in the International Hotel situation the communication strategy might be to portray luxury; the presentation strategy to price exclusively with luxurious rooms; the product/service strategy to render personal attention (e.g., butler service); and the distribution strategy to use exclusive referral systems and select travel agents. The functional strategies represent the substrategy implementation of the business strategies.

Product/Service Substrategy. The product/service mix is defined as the combination of products and services, whether free or for sale, aimed at satisfying the needs of the target market.[2]

In the better known four Ps (product, price, place, and promotion) developed for goods marketing, this is the product. For the seven Ps of services marketing, it is also the product. For the thirteen Cs of marketing, this is the categories of offerings. It can also refer to customer care, and customization, especially if tangible products are used. Ritz-Carlton Hotels has a top-of-the-line product/service strategy at the master and business strategy levels. At the functional strategy level, strategic decisions must be made regarding the level of service to offer and when and how to offer it. The same criteria, of course, apply: What is important to the target market? What does the target market expect? What problems does the target market have?

Let's say the product/service substrategy is to provide luxury. This would naturally derive from the master strategy and the business strategies. The question is how to put it into practice. These are the tactics. Consider terry-cloth bathrobes in each room—is this important to the market? Does the market expect it? Does it solve a problem for the customer? For Ritz-Carlton the answer may be yes, and the customer is willing to pay the additional cost. For most other hotels, the answer may be no.

The Oberoi Hotel chain in India once changed its master strategy and decided to aim at the super-luxury market. Into the rooms went antique desks, personalized stationery, beautiful brass ashtrays, and terry-cloth bathrobes, among other things. The rooms themselves weren't much different; it was the symbols of luxury that made the difference and the product/service strategy had to change. Many other changes were also made in the hotel's marketing mix to carry out this strategy.

These examples demonstrate that product/service functional strategies concern the level of product and service offered consistent with higher-level strategies. Higher-level strategies must be built around the target markets and the product, as should the functional strategies that flow from them and the tactics that are implemented. As shown in Exhibit 6-8, this is sometimes not the case in practice.

Presentation Substrategy. The presentation mix is defined as all elements used by the firm to increase the tangibility of the product/service mix in the perception of the target market at the right time and place.[3] These elements include the physical plant, atmospherics, employees, customers, location, and price. This mix has no true counterpart in the four Ps, but includes price. For the seventh P, it is process. In terms of the thirteen Cs, the presentation substrategy includes control of the process (for example, including RATER system in interaction and collaboration within).

Physical plant and atmosphere must be consistent with the product/service strategy. This means they shouldn't be overdone or underdone.

Employees must be hired and trained accordingly. Certainly we expect a bigger smile and quieter maids at a four-star than at a two-star hotel, and better service at a three-star than at a one-star restaurant. In either case, we expect an emphasis on the customer rather than on the service. This difference, in fact, is why Ritz-Carlton does so well at what it does.

Exhibit 6-8   Tactics That Failed to Support Strategies

Hyatt Hotels once had a policy that every dish that went out of its restaurant must have fresh fruit on it. Strawberries showed up in the strangest places, but the tactic, at least, was consistent with the strategy of fresh quality. This was also the communications strategy at that time, and ads portrayed fresh fruit (tactic). In other situations, this was not the case. The Sheraton Boston Hotel (formerly the Sheraton Towers Hotel) had a product/service strategy of exclusivity and provided bathrobes, but didn't open the pool until 9:00 A.M., and all the lightbulbs in the rooms were only 67 watts. Marriott's Courtyards didn't open the pool until 10:00 A.M., in spite of people trying to get in at 8:00 A.M. The Fairmont Hamilton Princess in Bermuda (formerly the Southampton Princess Hotel) emphasized service and convenience and told you that its coffee shop was open until 1:00 A.M., but closed it if no one happened to be there at 10:00 P.M. It also closed its lobby restrooms and waterfront pool and bar in the slow season to keep costs down, but maintained expensive bathroom amenities and high room rates in the largely empty rooms.

The Grosvenor Hotel in Orlando (formerly the Americana Dutch Inn) once offered nightly dancing, but only disco, and only locals went. The hotel was full of families, its target market. The Crowne Plaza in Kuala Lumpur offered fresh orange juice, but not before 10:00 A.M. (because "the juicer is in the bar") and targeted Americans. The Asia hotel in Bangkok, catering to an American and European market, had minibars in the rooms, but no wine in them. The Westin Harbour Castle in Toronto had Do Not Disturb signs to hang on your doorknob. The maid knocked at 8:00 A.M. anyway.

The Marriott Copley Place in Boston had drapes that didn't close all the way to shut out the morning light, no pull-out clotheslines in the bathroom because they're "too much trouble," and a restaurant with coffee shop decor, appointments, and service, but fine dining room prices. The Walaker Hotel in Sognefjord, Norway, billed itself as the finest hotel in Norway, charged $200 for a double room, but had tiny soap bars (albeit in fancy boxes) reminiscent of the early American motels.

All of these examples show how tactics executed at the hotel level may not support the mission statement at the corporate office.

The reverse is also true: At McDonald's we expect service to be consistent with the product strategy; for example, McDonald's expects you to clear your own tray when finished!

The *customer* strategy is very important. In some deluxe hotels in Paris and London men don't get in the door without a coat and tie. At other places you may be an "oddball" if you have them on. There is a basic strategy here that really applies in almost all cases: Don't mix incompatible markets if you can possibly help it; if you have to deal with incompatible markets, keep them separated in both time (e.g., seasonally) and space (e.g., separate dining rooms).

Location strategy means being where the customer can get to you or you can get to the customer. Again, McDonald's, in its popularity, is a prime example of this strategy in practice with stores located in just about every conceivable facility or location.

Pricing strategy, again, should be consistent with the other functional strategies. In too many cases, in fact, there seems to be no strategy at all. Prices sometimes seem to be set totally independent of all other strategies and without regard to their interrelationship. Price creates many expectations, which is why we consider it part of the presentation mix, as well as a separate part of the marketing mix.

Consider the airline passenger who pays $2,000 to fly first class versus the one who pays $600 in economy. They leave and arrive at the same time and travel at the same speed. What does the $1,400 difference in price tell you? That's an easy one: leg room, good food, personal service, movies, and so on. What does $200 for a hotel room tell you? Or $100, or $75, or $35? In Toronto we found a hotel that charged $150 for a room service imperial quart bottle (40 ounces) of liquor, plus tax and tip, that sold in the liquor store for under $30. The room cost $119. Although the customer is the same for both, there is clearly no relationship between the pricing strategies.

Pricing Substrategy.
The pricing mix is the combination of prices used by the firm to represent the value of the offering. The pricing mix is how the customer values what is being offered and what is received.

The following points were considered when developing the pricing strategy for the International Hotel example (take particular note of the last line):

- Special features of the product
- Spending power of the market
- Traffic movement of the market
- Possibility of losing regular users of high-rate rooms to lower-rate rooms
- Pricing of the competition
- Management policy to avoid discounted business, group business, and any upgrading to the new rooms
- Rates will be raised in three months

As Jain stated,

Increase [in price] should be considered for its effect on long-term profitability, demand elasticity, and competitive moves. Although a higher price may mean higher profits in the short run, the long-run effect of a price increase may be disastrous. The increase may encourage new entrants to flock to the industry and competition from substitutes. Thus, before a price increase strategy is implemented, its long-term effect should be thoroughly examined. Further, an increase in price may lead to shifts in demand that could be detrimental.[4]

All of the possibilities mentioned by Jain have happened in the hotel and restaurant industries in recent years, because of overpricing in the short run. We will spend all of Chapter 15 on pricing, as a separate part of the marketing mix, so it will not be belabored further here. Suffice it to say that pricing is both a powerful and a dangerous strategic tool.

Communication Substrategy.
The communications mix is defined as follows:

> All communications between the firm and the target market that increase the tangibility of the product/service mix, that establish or monitor consumer expectations, or that persuade customers to purchase.[5]

The communication mix replaces promotion in the four Ps and seven Ps. For the thirteen Cs, it is, obviously, communication. The issue here is obviously the strategy to be used to communicate all of the preceding to the marketplace. The strategic issue is what to say, not how to say it. The "how to say it" requires exceptional creativity in many cases and is often best left to those with that kind of expertise. The "what to say," however, is a strategic management decision and should not be left for advertising agencies to decide without extensive consultation.

Management's failure to clarify its strategy will not stop the agency from being creative. But it could, and too often does, result in advertising that does not clearly communicate the desired or appropriate message. The finished ads, the "how to say it," should always be filtered through the strategy to be certain that is what they are really saying.

An example that happens quite frequently in practice is advertising copy that positions a hotel or restaurant at a higher level than its strategy calls for. The property may be at a three-star level and aimed at the corresponding target market. The "creative" agency, however, gets inflated with terms such as *luxurious and elegant.* The appropriate target market believes it cannot afford it, and the upscale market, which is attracted, is disappointed. The result is a net loss for everyone. This is Gap 4, which was discussed previously.

Advertising, of course, is not the only part of the communication strategy. The strategy should consider all methods of communication. This is likely to include some combination of advertising, personal selling, public relations, promotion (including frequent guest programs), merchandising, direct mail, e-mail marketing, web blogs, and the like. The strategy will dictate where the emphasis and proportion of the budget should be placed on each.

Distribution Substrategy.
The distribution mix is made up of all channels available between the firm and the target market that increase the probability of getting the customer to the product and the product to the customer. This is *place* in the four Ps and seven Ps and *chain of relationships* in the thirteen Cs.

Strategies for distribution deal with channels and, in the case of most hospitality services, how to "move" the customer to the product. These include travel agents, tour brokers, wholesalers, referral services, reservations systems, websites, airlines, travel clubs, third-party online intermediaries, and so forth. Strategies involve the emphasis placed on each (or none) as well as the particular channels used. For getting the product to the customer, strategies include franchising and management contracts. We discuss distribution in more detail in Chapter 16.

Destination hotels and resorts place special emphasis on using these channels. Because distribution systems have become increasingly complex in the hotel industry, they require far more attention today than they did even three years ago.

The International Hotel in our example belongs to a consortium that represents many hotels and hotel chains in the world in a similar product class. This enables the International to benefit from international advertising that it otherwise could not afford.

Restaurants are also involved in distribution channels. In the aftermath of Hurricane Katrina, restaurants in New Orleans are seeking to regain heavy convention levels and tourist traffic, and are working closely with tour operators and incentive travel planners to bring in customers. There are other special cases, too. Many restaurants use the services of concierges at hotels to make recommendations to out-of-town guests. This distribution channel in many cases is worked every day, with financial rewards to the concierges that send the most business to certain restaurants.

In fact, destination management companies (DMCs) are increasingly becoming strong channels of distribution. Originally designed to handle the land transportation needs of groups upon arrival, the DMCs quickly recognized the ability to steer potential customers to restaurants, catering facilities, attractions, and other related hospitality providers.

Feedback Loops

feedback loop An important part of the strategic marketing system model, this refers to issues that need to be studied both before and after implementing a marketing strategy, including the potential risks of the strategy, the strategic fit to the operation, how or whether goals and objectives are being met, and so forth.

There are two **feedback loops** in the strategic marketing system model in Exhibit 6-3. One is the risk/fit loop. Feeding back to the master strategy, this loop questions the risks if the strategy is pursued and the strategic fit. Some of the critical risk questions that must be asked are, What can happen? Will it work? What if it doesn't? How will competitors react? What are the economics? Does it meet objectives? and Is there a fit between the marketplace and the master strategy? If answers are negative, reevaluation must take place. This is far better than following hunches that might end in failure.

The second loop starts with a synthesis of all the analysis that has been done. Through analysis we learn to break a problem into its many parts such as the marketing, financial, organizational, and environmental components. Many students and managers are good at this, but what they often do not do is put the pieces back together again. Too often the ability to analyze is valued over the ability to synthesize. Miller stated it this way:

> Analytical skills are fine for delving into problems, but they are inadequate for generating the insight needed for a workable solution. Analysis requires systematic probing, thoroughness, and logic. Synthesis, on the other hand, calls for artful pattern recognition, receptiveness, and magical insight — traits much neglected in the western world.[6]

In other words, synthesis means restating the important elements in a concise, clear summation that considers the needs of the firm, the needs of its customers, and the challenges of its competitors. It means identifying a theme or a vision for a configuration that is durable, defensible, and possible.

Synthesis is followed by evaluation/fit. Here we ask some of the same questions that we asked in the risk loop. We also make value judgments about whether the strategy matches capabilities and whether the organization can support it; that is, can it be successfully implemented, and if so, what will it take? Evaluation is the summation of the upside and the downside.

Once the strategy is planned and approved, of course, it has to be executed. Unfortunately, this is sometimes where it all falls apart, especially if the previous planning stages haven't been analyzed thoroughly. Implementation may require a number of events.

First, there is the organization. Strategy is put into action through organizational design. This means creating an organizational structure that will spot its own weaknesses and make the strategy work. Too often a strategy is put in place with the existing organizational structure, which doesn't allow it to work, or, worse, the strategy is designed around the existing structure. Structure follows strategy. Employees' activities must be coordinated, and employees must be motivated to make the strategy work — to create value and obtain competitive advantage. The organization must also be designed to have an effective control system that compares actual performance against established targets, evaluates the results, and takes action if necessary.

Because this is a marketing text, we won't elaborate further on organizational structure and implementing change. These issues, however, are not to be taken lightly and should be considered in any strategic plan.

The second feedback loop continues with the feedback on whether the strategy is working once it is in place. Marketing research is fed into the marketing intelligence system. This is the control that warns management to act before the system gets out of control.

STRATEGY SELECTION

As we have progressed through the strategic marketing system model illustrated in Exhibit 6-3, we have provided a framework on which we can later expand. That is because there is no single right marketing strategy for any situation; there are simply right alternatives. The situational analysis, if done objectively, should show the facts. The environmental analysis provides the bases for assumptions. From these sources, the strategic planner develops alternative courses of action.

Which action should be chosen? That is a simple question that has no simple answer. When you consider that there are also alternatives at every step of the strategic planning process, you find that you have dozens, perhaps hundreds, of choices to make. That seems like a difficult task, and it may well be. Some are better at it than others. Good common sense, wisdom, judgment, and intuition still have their place. Interpreting information, although objective, is not mechanical. The functions of strategic planning are to define objectives in terms other than profit, to plan ahead, to influence and not just react to change, and to inspire organizational commitment. Once your strategy has been formulated, it should also be evaluated for content. Exhibit 6-9 is a checklist for that purpose. Strategy selection should also include understanding the customer's

Exhibit 6-9	Strategy Checklist

- Is it identifiable and clear in words and practice?
- Does it fully exploit opportunity?
- Is it consistent with competence and resources?
- Is it internally consistent, synergistic?
- Is it a feasible risk in economic and personal terms?

- Is it appropriate to personal values and aspirations?
- Does it provide stimulus to organizational effort and commitment?
- Are there indications of responsiveness of the market?
- Is it based on reality to the customer?
- Is it workable?

role in the process. This emphasis on the customer, one of the thirteen Cs, is illustrated in the following quote:

> A business . . . is defined by the want the customer satisfies when he buys a product or service . . . To the customer, no product or service, and certainly no company, is of much importance . . . The customer only wants to know what the product or service will do for him tomorrow. All he is interested in are his own values, his own wants, and his own reality. For this reason alone, any serious attempt to state "what our business is" must start with the customer, his realities, his situation, his behavior, his expectations, and his values.[7]

Chapter Summary

Strategic planning is a difficult but necessary process. At the highest level of the firm, it drives the firm. At the lowest operational level, it drives day-to-day activities. It is an essential phase of marketing and management leadership. In the short term, it is the annual marketing plan.

Good strategic planning rests on knowing where you are now and where you want to go, and finding the best way to get there. Its success rests on objective analysis, knowing what business you are in, understanding markets, integrating within the firm, and creating an organizational structure that will facilitate the implementation. There is no substitute for strategic planning and execution in today's competitive environment.

At the same time, all strategic thinking and planning need not take place only at the corporate or higher levels of management. Unit managers have to be involved in strategic planning. We have given numerous examples of what happens when strategic planning is not done, is done poorly, or is poorly executed. At the least, every manager should be thinking strategically at every level.

Strategic planning occurs at the functional level following the strategies set forth at the higher levels. It may occur for a 60-seat coffee shop or a 20-unit motel. Regardless, it is strategy that drives tactics and that, when done and executed properly, will produce the best marketing performance. The functions of strategic planning are to define objectives in terms other than profit, to plan ahead, to influence and not just react to change, and to inspire organizational commitment.

Key Terms

constituents, p. 164
feedback loop, p. 180
marketing management, p. 160

Discussion Questions

1. Describe in your own words the concept of strategy.

2. How does marketing management differ from marketing strategy development? Give a brief example to support your answer.

3. What is the strategic marketing system model? How does it work? (This is a big question.)

4. What are mission statements? Who are they for? Why are they important when developing a marketing strategy?

5. Within the strategic marketing system model, what is meant by business strategies? How do they relate to functional strategies? Which comes first? Give an example showing how these two concepts relate to each other.

6. Within the strategic marketing system model, describe what is meant by feedback loops. Why are they an important aspect of the system model?

Endnotes

1. We have borrowed some ideas here and in other parts of this section from Porter, M. (1985). *Competitive advantage*. New York: The Free Press; and Porter, M. (1996, November–December). What is strategy? *Harvard Business Review*, pp. 61–78.

2. Renaghan, L. M. (1981, April). A new marketing mix for the hospitality industry. *Cornell Hotel and Restaurant Administration Quarterly*, p. 32.

3. Ibid.

4. Jain, S. C. (1997). *Marketing planning and strategy*, (5th ed). South-Western Publishing, 410.

5. Renaghan.

6. Miller, D. (1990). *The Icarus paradox: How exceptional companies bring about their own downfall*. New York: Harper Business, 208.

7. Drucker, P. F. (1974). *Management: Tasks, responsibilities, practices*. New York: Harper & Row, 79–80.

The Marketing of "Little England"

Discussion Questions

1. Based on the information in the case, conduct a SWOT analysis for the Barbados Tourism Authority. What do you consider the strengths, weaknesses, opportunities, and threats for this tourism island in the Caribbean?
2. Who should Barbados identify as its primary markets? Why?
3. What should the overall product strategy be? Why?
4. Is Barbados trying too hard to be a little bit of everything for everyone?

Earlyn Shuffler, president of the Barbados Tourism Authority (BTA), reviewed his speech for the last time. He still had a few minutes before the start of the annual BTA board meeting, at which he was presenting a review of activities for the past year and recommendations for the future. On his way to the boardroom, he reflected on the events of the previous year.

Last September the appointment of a new three-member BTA executive committee, with Allan Batson retaining the chairmanship, signaled the end of a long-standing controversy surrounding the government's management of the island's tourism industry, which had seen various structures and numerous directors coming and going. The "executive," as the new committee was commonly called, consisted of Earlyn Shuffler, president; Thomas Hill, vice president sales and marketing; and Errol Griffith, vice president finance and corporate affairs. Each of these men had extensive experience in his respective field and was well equipped to face the daunting task that lay ahead. The members of this newly appointed team, soon nicknamed the "Dream Team" by the media, were viewed as saviors of the flagging tourism industry, with the ability to recapture lost ground, increase visitor arrivals and spending, and develop an effective marketing plan for the island.

CARIBBEAN TOURISM

The entire Caribbean region, including some not truly Caribbean islands such as the Bahamas and Bermuda, received over 14 million stay-over visitors in the previous year. Of these, approximately 60 percent originated from the United States, 18 percent from Europe, 16 percent from other markets, and 6 percent from Canada. There were over 165,000 hotel rooms in the region, and tourism receipts were almost US$15 billion. The Caribbean region is shown in Exhibit 1.

Each island of the Caribbean conducted its own marketing campaign, and each was responsible for attracting visitors to its shores. Continued efforts by the Caribbean Tourism Organization, the Caribbean Hotel Association, and a number of private sector entities, however, resulted in a number of the islands collaborating on a joint marketing campaign. The outcome of this effort was the Caribbean Vacation Planner, a full-color book that contained basic tourist information about 34 islands in the region and aided travelers in planning their Caribbean vacations. This was the only instance of cooperation and collaboration in Caribbean tourism marketing.

BARBADOS

Barbados, affectionately known to visitors and locals alike as "Little England," was the most easterly of the islands of the Caribbean, 166 square miles in size, with a population of slightly more than 254,000. Barbados gained its independence from Britain in 1966 and was considered to be one of the best governed and most politically stable islands in the region. The island earned its nickname from its resemblance to the English countryside and the adoption of certain characteristically English traditions, such as afternoon high tea.

Barbados is bordered by the Caribbean Sea on its west coast and the Atlantic Ocean on its east coast. The relative calm of the Caribbean Sea made the west coast of the island ideal for swimming, sailing, and snorkeling, while the turbulence of the Atlantic Ocean transformed the east coast to a haven for surfers. A wealth of attractions was offered to visitors to the island, ranging from tours of old plantation houses, to a day at the races; a wander through the coolness of underground sea caves, to a canter

This case was contributed by Lisa M. Jebodhsingh, developed under the supervision of Dr. Margaret Shaw, Ph.D., School of Hospitality and Tourism Management, University of Guelph, Ontario, Canada. All rights reserved. Used by permission.

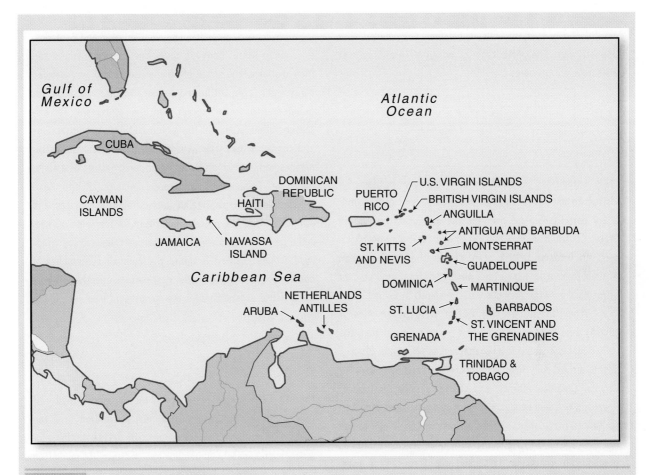

Exhibit 1   The Caribbean region

through the countryside; or a day of frolic cruising on the popular pirate ships, to an elegant dinner at one of the many top restaurants. The more adventurous visitors tried parasailing or deep sea fishing, while others preferred to relax on one of the many beaches.

In the past Barbados had been viewed as an elite destination, catering to the upper income end of the tourism market. According to Allan Batson, chairman of the BTA, however,

> Barbados has rededicated itself to providing a well-rounded product, from economy hotels and villa accommodations, to world class luxury resorts, offering value at all three price ranges — budget, mid-range and luxury. We now target travelers in each of these three categories with offerings to appeal to their individual sense of value and style.

In recent years, tourism had become the mainstay of the economy, replacing sugar as the largest foreign exchange earner, and was the second largest employer. The island had close to 6,000 hotel rooms, including apartment hotels. The winter tourist season ran from mid-December to mid-April, and the summer season from mid-April to mid-December. Traditionally, the winter season was the peak season for arrivals with visitors from North America and Europe flocking to the island to escape the cold. In recent years, however, peak arrivals had been experienced in the months of July and August during the annual Crop Over Festival. This monthlong festival had gained international recognition and signified the end of the harvesting of the sugar crop, hence the name "Crop Over." It culminated in a day of costumed revelry, food, and entertainment enjoyed by visitors and locals.

In the year just passed, Barbados received over a million visitors, and tourist expenditure was estimated at US$ 850 million, making a net contribution to the island's gross domestic product of about US$291 million. Of the total number of visitors, approximately 482,000 were stay-over visitors, an increase of 3.9 percent over the previous year. These visitors spent at least 24 hours on the island, and between 30 and 40 percent of them had visited Barbados at least once before. Stay-over visitor arrivals for January to May, and July, were lower than the corresponding months the year before. In June, and August to December, however, monthly stay-over arrivals consistently exceeded the previous year. The percentage by month of total stay-over visitors is shown here:

Jan	Feb	Mar	Apr	May	Jun	Jul
8.1	8.3	8.9	8.6	7.0	6.8	10.0

Aug	Sep	Oct	Nov	Dec	Total
9.2	6.9	7.6	8.5	10.1	100%

Europe had provided the largest number of stay-over visitors to the island, contributing 44.3 percent of the total, a 2.9 percent increase over the previous year. The United States contributed 25.3 percent of the total stay-over visitors, an increase of 2.7 percent over the previous year. Arrivals from other areas, their percent of total, and percent of increase over the previous year are shown in Exhibit 2.

American Airlines and British West Indian Airways (BWIA) were the two main carriers providing airlift from the United States; there were no charter services from that country. Flights arrived daily from New York, Miami, and San Juan. Air Canada, Canadian Airlines, and charter airlines Air Transat, Royal Airlines, and Canada 3000 provided scheduled and charter service from Canada. British Airways was the major carrier from the United Kingdom, along with numerous charter airlines providing air access from the UK and other European countries.

Cruise passengers on land trips accounted for about 485,000 of the total visitors, a 5.5 percent increase. These visitors spend less than 24 hours on the island and spent approximately US$56 million of the total visitor expenditure. It was estimated that stay-over visitors spent approximately US$1,635 each during their visit, while cruise passengers spent on average US$113 each while on the island.

COMPETITION AND ACCOMMODATIONS

In a macro sense, all warm-weather destinations, especially islands similar to the Caribbean Islands (for example, Hawaii,

Tahiti, Fiji) were competition for the Caribbean and Barbados. In a micro sense, however, given the present markets and their distance, the scope of competition could be narrowed considerably even within the Caribbean region. Bermuda, for example, was north enough that the winter was its off-season. Other islands in the region had different attractions and characteristics. The Bahamas, San Juan, and the Lesser Antilles had large chain hotels and offered gaming as a major incentive. Cuba, Haiti, Dominican Republic, Puerto Rico, Trinidad, and Jamaica were all quite different in character and ethnicity. Thus, the prime competition for the same markets could be narrowed down to the Lesser Antilles and the Windward Islands. Exhibit 3 shows the approximate number of hotel rooms in each of these destinations and the approximate visitor expenditure for the most recent year.

THE BARBADOS TOURISM AUTHORITY (BTA)

The BTA was a statutory agency of the Barbados Ministry of Tourism. It was funded by an annual subvention from the central government, most recently of US$18 million, of which approximately 70 percent was dedicated to marketing costs and the remaining 30 percent to administrative costs. Total actual expenditures were about $US20 million. The authority also managed overseas offices in the United States, Canada, United Kingdom, and Germany. A global advertising agency was contracted by the BTA to create and execute its advertising campaigns, and a global public relations firm was contracted to support these advertising campaigns.

The primary role of the BTA was the marketing of Barbados' tourism product. The official role, as dictated by the government and the Barbados Ministry of Tourism, consisted of the following:

▪ To promote, assist, and facilitate the development of tourism in Barbados.

Exhibit 2	Past Year Stay-Over Visitors to Barbados Area		
	Stay-over visitors	Percent of total	Percent increase
United Kingdom	126,621	28.6%	2.6%
Other Europe	69,237	15.7	4.4
Canada	53,373	12.1	2.1
United States	111,983	25.3	2.7
Other	80,898	18.3	9.4
Total	442,107	100	3.9

Exhibit 3	Hotel Rooms and Visitor Expenditures of Major Competition		
Islands	Hotel Rooms	US$ Expenditures per Available Room	Expenditures (Millions US$)
Anguilla	978	58	59,304
Antigua/Barbuda	3,317	448	135,061
Barbados	5,685*	680	119,613
British Virgin Islands	1,224	214	174,837
Dominica	757	35	46,235
Grenada	1,428	67	46,919
Guadeloupe	7,798	443	56,809
Martinique	7,220	431	56,695
Montserrat	710	21	29,577
St. Kitts/Nevis	1,593	87	54,614
St. Lucia	2,954	255	86,324
Saint Maarten	3,710	464	125,068
St. Vincent/Grenadines	1,215	57	46,913
U.S. Virgin Islands	5,461	1,045	191,357

*Luxury 24.8%, A class 5.4%, B class 3.4%, guest houses 1.5%, apartment hotels 64.9%

■ To design and encourage marketing strategies for the effective promotion of the tourism industry in Barbados.

■ To seek to enhance the provision of adequate and suitable air and sea passenger transport services to and from Barbados.

In partial fulfillment of these duties, the BTA also engaged in sports promotion, hotel and restaurant registration, licensing and classification, public relations and public awareness programs, media relations, press releases, crime/damage control, international media tours, facilitation of tourism partners, travel agent familiarization visits, and cooperation with private sector partners. An annual marketing plan was also prepared by the BTA as part of its annual report, which outlined the strategies and projections to maintain Barbados as a desirable tourist destination. With the appointment of the new executive, this marketing plan became confidential, and was no longer made available to the public.

In recent years, Barbados' position in the marketplace had deteriorated, as reflected by stagnation in the level of stay-over visitor arrivals. Reasons cited for this included the strength of the Barbadian dollar; the inability to compete on cost with other warm-weather destinations (e.g., Mexico and Cuba); mild winters in North America; well-publicized instances of crimes against visitors; and hurricane devastation, which reduced visitor arrivals throughout the region. These difficulties, however, were also faced by other Caribbean destinations that experienced surges in arrivals substantially greater than those in Barbados.

There was an apparent lack of recognition of the island in the marketplace due to a recent history of lackluster marketing. This lack included travel agents and tour operators in the island's main markets. The controversy that surrounded the BTA from its inception resulted in a perceived absence of leadership in the Barbados tourism industry. Efforts at promoting the island and attracting visitors were hindered by this perception, and, consequently, there was a general lack of confidence in the island by the recipients of these efforts. In addition, marketing efforts were not limited to the BTA, the official marketing body of the Barbados tourism product.

There were also instances of advertising and promotion by other stakeholders in the industry, for example, the Barbados Hotel and Tourism Association, St. James Beach Resorts, and the Elegant Resorts of Barbados. These also served to undermine the authority of the BTA. According to the minister of tourism, Billie Miller:

An analysis of the past performance of the management of Barbados' [tourism] . . . abroad reveal there has been wastage, ineffective use, duplication of effort, limited sharing of information and a lack of complementarity.

Earlyn Shuffler, president of the BTA, saw his role as a facilitator for the industry:

I am to provide leadership by example; to build a strong management team and be the primary facilitator in bringing all sectors of tourism together to make Barbados the preferred warm-weather destination in the Caribbean.

Prior to the appointment of the new executive there were a number of government-backed marketing strategies already in place for the island (See Exhibit 4). In order to achieve some of their goals, the new executive of the BTA developed a number of additional strategies. Their main obstacle, however, was the lack of current, comprehensive market information. Thomas Hill, vice-president sales and marketing of the BTA believed:

The product is here, it has always been here, it is an excellent product and improves as time goes on. But what we have to determine now is how do we take that to the rest of the world and convince them that this is the place to visit.

As a result, the BTA launched a US$2.3 million advertising campaign in the United States in an attempt to recapture some of their lost market share. The campaign focused on the theme "Imagine Yourself in Barbados" and depicted scenes of visitors enjoying themselves in Barbados. The BTA also branched into new markets and sought entry into the South American market through linkages with Brazil. This venture failed to reach its full potential with the loss of airlift between Brazil and Barbados. British West Indian Airways (BWIA), the airline that provided transportation from Brazil to Barbados, canceled this route shortly after its introduction due to lack of profitability for the airline and underusage of the route.

"Barbados is concerned about market share," said Jean Holder, secretary-general of the Caribbean Tourism Organization. "During the past decade there has been a significant loss of market share [from] important markets [for Barbados]." Barbados was a traditional warm-weather destination, offering doses of "sun, sea and sand." In an attempt to diversify and recapture some of its lost market share, a number of product offerings was expanded with the intention of capturing specific niches, which are shown in Exhibit 5. With these

Exhibit 4   Marketing Strategies in Place before the Hiring of the New BTA Executive

UNITED STATES

- A super value package
- Barbados welcomes VISA in collaboration with VISA International—VISA cardholders received promotional material with their monthly statements
- Increased advertising and public relations
- Educational tours to the island of 800 travel agents by year end
- Strategic alliances with British West Indies Airways (BWIA) and American Airlines

UNITED KINGDOM

- Promotional tour to the UK in November
- Participation at World Travel Market in November
- Road shows with Kuoni and British Airways

OTHER EUROPE

- Educational tours in Barbados for travel agents
- Participation in Travel Trade Workshop, Switzerland
- Participation in an all-Caribbean travel fair in Paris
- Radio advertising in Germany
- Spot advertising in media

CANADA

- Promotional tour to Montreal, Toronto, Calgary, Edmonton, Vancouver
- Sponsorship agreement with Hamilton Tiger Cats football team
- Negotiating with the Toronto Raptors basketball team to use Barbados as their official destination
- Program with Regent Holidays (tour operator) to bring visitors from Vancouver, Quebec, and Toronto

Exhibit 5   Specific Niche Markets Targeted

Niche Market	Activities
Sports:	Cricket, field hockey, horse racing, wind surfing, golf, yachting, soccer, body surfing, bridge, chess, surfing, running, cycling, swimming
Culture:	Paint It Jazz Festival (past performers included Roberta Flack), Holetown Festival, Holders Opera Season (past performers include Luciano Pavarotti), Oistins Fish Festival, De Malibu Congaline Carnival, Gospelfest, Crop Over Festival
Weddings/ honeymoons:	In the UK, Barbados was voted the top wedding destination. There were no residency requirements, and couples could be married on the day of their arrival on the island.
Heritage:	Old slave huts, plantation houses, churches, and sugar mills formed part of heritage tours.
Diving:	Natural reefs and sunken wrecks provided underwater attractions.
Film making:	A Film Credential Industry brochure was being compiled for use as a marketing tool. This followed the filming of several episodes of various movies and the US$1.5 million Malibu Coconut Rum ad that was filmed on location on the island.
Ecotourism:	Harrison's Cave, Turner Hall Woods, and Welchman Hall Gully were touted as part of the ecoproduct.

products, the island was able to offer a wide range of activities throughout the year, shown in Exhibit 6, catering to a variety of tastes.

"Was this enough? Do we really have a strategy to move the island forward?" wondered Mr. Shuffler. Reports so far for the year indicated an increase in arrivals to the island. Was this due to the efforts of the BTA? He wondered what decision the board would reach after he presented his report: Who should we target, and what should we offer? Can we really support so many different products? What should the plan of action be for the future? Who should we target, and what should we offer?

Exhibit 6   Monthly Activity Schedule

JANUARY
- Wind Surfing World Championships
- Paint It Jazz (jazz festival)
- Mount Gay International Regatta (sailing)
- Barbados National Trust (heritage tours, January–April)

FEBRUARY
- Flower Show
- Holetown Festival (cultural festival)

MARCH
- Cockspur Gold Cup Race (horse racing)

APRIL
- Holders Opera Season (opera and Shakespeare season)
- Oistins Fish Festival (cultural festival)
- Caribbean Atlantic Cricket Cup

MAY
- Gospelfest (International Gospel week of activities)

JUNE
- Aqua Splash (aquatic-based competition and show)
- Shell June Rally (international car rally)

JULY
- Carnival/Crop Over (July–August; Carnival)
- Caribbean Storytelling Festival

AUGUST
- International Schools Netball Festival
- Banks International Hockey Festival
- Tulip Rally (international car rally)

OCTOBER
- Sir Garfield Sobers Seniors Cricket Festival
- Sun, Sea, Slam International Bridge Festival
- International Triathlon
- Pro Am Cricket Festival

NOVEMBER
- Sprite Caribbean Surfing Championships

DECEMBER
- Run Barbados Road Series
- United Barbados Open Golf Tournament
- Red Stripe Series Regional Cricket

overview

We begin this chapter with an overall look at the marketing environment in which all firms operate. Environmental impacts on hospitality and tourism enterprises are accelerating at a fast pace and changing the competitive scene. We then address how to study the competitive environment, including how to identify the different types of competition; how to defend against it; and how to identify, compare, and measure the competitive environment.

chapter 7

Competition and the Marketing Environment

learning objectives

After reading this chapter, you should be able to:

1. Explain environmental scanning and undertake an environmental scanning analysis.

2. List the various types of environments that affect hospitality and tourism, and explain why an understanding of these environments is important for strategy development.

3. Analyze the competitive environment and explain how to choose the right competition.

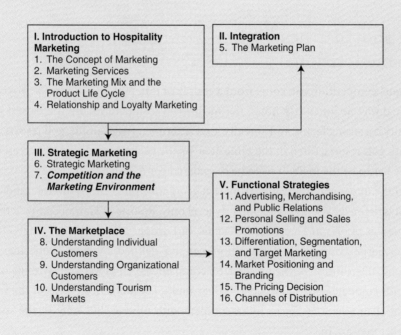

I. Introduction to Hospitality Marketing
1. The Concept of Marketing
2. Marketing Services
3. The Marketing Mix and the Product Life Cycle
4. Relationship and Loyalty Marketing

II. Integration
5. The Marketing Plan

III. Strategic Marketing
6. Strategic Marketing
7. *Competition and the Marketing Environment*

IV. The Marketplace
8. Understanding Individual Customers
9. Understanding Organizational Customers
10. Understanding Tourism Markets

V. Functional Strategies
11. Advertising, Merchandising, and Public Relations
12. Personal Selling and Sales Promotions
13. Differentiation, Segmentation, and Target Marketing
14. Market Positioning and Branding
15. The Pricing Decision
16. Channels of Distribution

Jan Freitag
Vice President, Global Development, Smith Travel Research

Jan Freitag is a recognized leader in the U.S. lodging industry. He is a frequent public speaker and is quoted in trade publications and the general news media such as the *Wall Street Journal, New York Times*, Associated Press, Reuters, Forbes, etc. He also writes a monthly column for *Lodging* magazine, the official magazine of the American Hotel and Lodging Association.

At Smith Travel Research (STR) Freitag oversees a variety of projects and departments, all charged with the accumulation and interpretation of U.S. and global lodging data. Prior to joining STR, Freitag was the director of content integrity at hotelreports.com in upstate New York and a hospitality consultant with Ernst & Young in Phoenix. Freitag holds a bachelor's degree, with distinction, from the School of Hotel Administration, Cornell University, and received his executive MBA, with honors, from Vanderbilt University. He resides in Nashville, Tennessee.

 Marketing in Action
....

Jan Freitag
Vice President, Global Development, Smith Travel Research
....

How should a hospitality firm define the competition?

For a hotel competition can come from a couple of different angles. One is geography. For a highway or inner-city hotel, the most likely competitors are the closest ones. Proximity is a strong factor, and amenity level is another. For example, golf resorts compete with golf resorts, often in the same area. As the geographic distance increases, the amenity has to be the main draw to be considered competitive.

At the upper end of the luxury market geography plays almost no role, and competition is defined by service and amenity level only. Resorts in Hawaii compete for the same guest as resorts in Thailand, the Pacific, or Europe.

For large companies, competition can be defined by brand offerings and geographic distribution of the product. So companies such as Marriott with its multiple brands compete with Starwood and their multiple brands in all areas of the world. Regional chains such as Taj in India compete with Oberoi, but only in the specific submarket they are in.

How competitive is the hospitality industry in Europe, Asia, the Middle East, and the United States? If there are differences, what are they?

Following the life cycle of the hospitality industry in those regions can give you a good insight into the competitiveness. The most developed market is the United States, with brands in all segments of the marketplace—from economy to luxury. Other regions of the world (e.g., Asia) are just now starting to develop middle-tier brands, and some have yet to develop strong regional midscale brands at all (e.g., Africa, excluding South Africa). Europe and Asia have a strong presence of independent midmarket hotels, and brands trying to compete in this market have to educate the public about the merits of buying a brand. Over time we see that the multinational brands are expanding their global reach to capitalize on economies of scale. But independent hotels with a strong regional "flavor" will likely be in a good position to compete with the multinational brands.

What do you see as the major differences among the European, Asian, Middle Eastern and American methods of defining and addressing the competition?

I'm not sure there are differences. Addressing competition is up to the motivated GM in each hotel. With the right tools and staff, the way to deliver memorable guest service probably does not differ that much from country to country.

How should a firm go about gaining information on the competition? Has anything proven successful or unsuccessful?

Luckily a variety of tools are now available that allow firms and hotels to gain access to good competitive data. Rate-scraping companies give an idea of what competing hotels charge with OTAs. The Bench (www.thebench.com) and Smith Travel Research (www.str.biz) provide benchmarking data for a self-selected competitive set and show you ADR, occupancy, and REVPAR for last week, last month, and last year. In a lot of cities GMs meet for regular round tables, facilitated through trade associations such as HSMAI or their CVB, and discuss topics that have an impact on the wider market.

How critical are measures such as REVPAR (revenue per available room), yield index, and so forth, to hotel firms?

As an operator, not knowing how your competitive marketplace performs is akin to driving blind. You need to understand not only what drives your performance but also whether your competitors are doing better or worse to see whether your marketing and sales efforts work. Our client hotels give bonuses to their GM's and director's of sales on the yield index, so those people would probably argue that those measures are hyper-critical. But on a larger scale, you cannot manage what you cannot measure, so access to good data helps you perform better because you substitute gut feeling with real data and can be held accountable for your actions.

What advice do you have for students interested in a marketing career?

Get involved with HSMAI in your city or college. Meet with local directors of sales and marketing to understand the nuts and bolts of the sales process. Find internships with larger regional or national companies to learn about the tools they deploy to measure the effectiveness of their marketing efforts. Read the trade press. Ask a lot of questions. *Used by permission from Jan Freitag.*

Every organization is part of a larger environment that includes many forces that are beyond the control of any one firm. A careful assessment of these environmental forces is necessary before any strategic plans are used. As the world's economies become more intertwined, the environment of a company becomes larger and more complex than ever before. Events in faraway places have impacts close to home. Customer needs and wants are also ever changing, creating an environment that is always changing and demanding.

Environmental impacts such as the North American recession of the early 1990s changed the buying habits of consumers just as the robust economy of the late 1990s did. The customer today is looking for "value added" in every purchase, from cars to homes to hotel guest rooms to fast food. As one hotel executive stated, off the record, "During the 1980s we just provided the product. The 1990s is a service decade. Unless we are able to provide the service, we cannot compete." This executive could have added: The new millennium is shaping up to be the value decade. Customers want more for their money, and they are willing to look at all their options (i.e., your competitors) to get it. The 2000s are also about giving the customer choice and customizing, as discussed in Chapter 4.

Customer demands create opportunities to serve them in new and better ways. There is no other choice. The right choice means reading the environment, understanding that change is constant, and developing the products and services that anticipate customers' problems, as well as fulfill their needs and fantasies. Yesterday's success is tomorrow's failure without constant adaptation to the environment. This calls for continuous and careful environmental scanning and analysis.

ENVIRONMENTAL SCANNING

environmental scanning Scanning the external environment to learn of potential opportunities and threats that could affect a business.

Marketing leadership means planning for the future. Trying to determine what the future holds in store has come to be known as **environmental scanning**. This simply means asking the question, What is going on out there in the environment that is likely to affect our business? What this also means is having an awareness of the need to be proactive, rather than reactive, and the need to constantly see what is happening and to anticipate what will happen. Environmental scanning is an essential leadership tool. All of this is relative to the competition, which is also an environmental force. We discuss types of environments first followed by the competitive environment.

TYPES OF ENVIRONMENTS

Macro environmental forces can be broadly classified into technological, political, economic, sociocultural, regulatory, and ecological/natural environment. We will discuss some of the impacts of these changing environments on the hospitality industry.

Technological Environments

Scanning the **technological environment,** from a marketing viewpoint, involves considering the impact of technology on both the organization and the customer. In terms of the organization, one of the major impacts has been the decrease in the cost of computers and computer memory. Much of customer relationship management, which we discussed in Chapter 4, would not have occurred if firms had not been able to store and manipulate a massive amount of customer information. Newer integrated guest history databases connect the central reservations office to the front office property management system (PMS). Some advantages of database systems are shown in Exhibit 7-1.

According to recent research, technology is shaping up to be one of the biggest competitive advantages a hospitality firm can have. Technology will define and direct customer service, information management, and hotel design as well as create alternatives to existing products and services. The new use of technology-based management information and decision support systems, property management, yield management, and database marketing has improved management control and, in turn, its efficiency and effectiveness. Given the speed of change, firms cannot wait. They must invest today, not only in the opportunities presented, but also in systems that help them monitor changes and their impact on the evolution of customer expectations.[1]

Technology and travel are natural partners. The rapid growth of travel and tourism in the last 20 years has been encouraged by technological developments, such

technological environment
Technology advances that can or will affect a business from both a consumer and competitive perspective.

Exhibit 7-1   Advantages of Customer Database Systems

- Databases speed the taking of reservations.
- A guest's record can be located quickly.
- Guest history reports can be reviewed and rooms prepared accordingly. A guest room can be selected and/or stocked to reflect the guest's preferences.
- Guest preferences in the aggregate can be reviewed to see whether guests have a common need that the hotel is not providing.
- Profiles of frequent guests can be established. Mailing lists of names that match the profile can be purchased.

- Sales and marketing can track the effectiveness of specific advertising campaigns and promotional rates.
- Hotels in a chain can fulfill guests' needs on the first visit when that guest has stayed at other hotels in the chain.
- Profiles of guests' demographics and spending patterns can be used to "fingerprint" trends of current customers. This fingerprint may then be used to obtain lists of customers with similar demographics and spending patterns to be used to prospect for new customers.

Go to your favorite search engine and type in the words "technological impacts" or "impact of technology." Choose one of the sites listed, read the information, and be prepared to report to the class on what you discovered and the implications for the hospitality industry.

as in air transport through the development of the jet engine, more sophisticated aircraft design, and improved systems management. The original Boeing 747 had to make a refueling stop at an intermediary location in making a flight from Los Angeles to Tokyo. Newer aircraft make the same flight nonstop. This capability affected many destination points, both pro and con. A new plane being manufactured by Airbus, the A380, is capable of carrying over 800 passengers, although most are expected to be configured to accommodate 555 passengers.

Although technological advances have brought other benefits to organizations such as faster elevators, self-service kiosks, immediate credit card approval, electronic door locks, and many other amenities, it is the impact of technology on consumer behavior with which hospitality marketing executives need to be most concerned. The rise of the Internet as a distribution channel is perhaps the most obvious example of the impact of technology on consumer behavior. According to Yesawich, Pepperdine, Brown & Russell/Yankelovich Partners 2005 *National Travel MONITOR*, almost 7 out of 10 business travelers now use the Internet to plan some aspect of their business trips. The comparable percentage for leisure travelers is just over 60 percent. The Internet has provided consumers with an infinite number of choices, as well as access to large amounts of information. This has led to more knowledgeable consumers, which in turn has led to much more demanding customers.

Increases in guest room technology have also helped the guest. Not only has TV been used for on-demand movies, checkout, and a message operator, but it is also being used as a high-tech concierge. The in-room television will ultimately become a monitor for Internet surfing. Guests at a number of hotels can watch video channels describing local attractions, shops, and restaurants and, for example, see what the restaurant looks like and hear about its specialties. By pressing another button they can learn not only the address, but also what the cab fare should be for the ride over. Another channel provides information on special exhibits at museums and whether tickets are hard to get. Guests can order breakfast in advance in six different languages. Some hotels have a system through which guests can buy theater tickets and shop at local merchants directly through their in-room televisions.

Marketing executives must determine whether a technological solution can be used to improve the nontechnological amenity and increase customer satisfaction. For example, when the guest states that she would like a phone on the desk, she is really stating that she would like a phone that is more accessible. Perhaps putting a portable in the room would better suit her needs as opposed to two traditional phones. Similarly, the additional data line accessible to the desk can best be met by a wireless solution. As one customer told one of the authors, "I want wireless in my room so I can use my computer while in bed. I do not want to be tied to the corner to be plugged into the Internet." Finally, the desire for express check-in and checkout is currently being met by self-service kiosks, as evidenced in Exhibit 7-2.

Political Environment

Political impacts on the hospitality industry vary both with the stability of government and with the interest of destinations in developing tourism. In the United States there has been relatively little interest at the federal level in funding the development of

| Exhibit 7-2 | Hilton Hawaiian Offers Guests Self-Service Hotel Check-In at Honolulu International Airport |

Hilton Hawaiian Village® Beach Resort & Spa and the Honolulu International Airport partnered in 2004 to become the first hotel and airport in the United States to offer self-service kiosks that allow Hilton guests to check into the hotel and get their room keys—before they even claim their baggage and leave the airport.

Hilton installed four kiosks at the airport, two each in Baggage Claim areas "G" and "H," which serve United, Continental, Northwest and American Airlines. The kiosks are readily identifiable with Hilton signage.

Hilton and IBM developed the kiosk hardware and software and began testing it in lobbies of selected hotels on the U.S. mainland in January 2004. Hilton installed 100 kiosks in 45 hotels in 2004. However, Honolulu International was the first airport in the nation with hotel self-service kiosks. Hilton Hawaiian Village also installed three kiosks in its main lobby to provide guests with an alternative to the high touch service associated with a traditional front desk check-in. The kiosks may also be used for check-out or as a private check-in solution for large groups.

"We are delighted to work with the team of the Honolulu International Airport and Hawaii Department of Transportation on this important leap forward in the travel and tourism industry," said Gerhart Seirbert, area vice president and managing director of Hilton Hotels Corporation-Hawaii. "Cooperative efforts such as this are yet another example of the commitment by the state and the tourism industry to keep Hawaii at the forefront of customer service and technology."

The kiosks function in much the same way as airline self-service kiosks for air travelers using e-tickets. After inserting a credit card for identification purposes, guests follow a set of simple on-screen instructions and utilize the touch screens to check into the hotel.

The kiosk displays the traveler's reservation information, offers a room based on the customer's known preferences, which the customer can accept or change, issues a room key and provides printed room directions and information. The kiosks can also offer guests the opportunity to upgrade to more premium accommodations than originally reserved, should the guest desire.

Hilton guest service agents are on-hand at the airport to answer questions and assist guests in the check-in process. Hilton's long-term commitment to personal service and a warm welcome adds to the convenience, control, and efficiency the kiosk check-in provides. Guest service agents also have access to Hilton's entire technology platform OnQ via Xybernaut Atigo wireless, handheld computers.

At the end of the stay, the traveler can checkout at a kiosk in the same fashion by reviewing and confirming their bill and printing out a receipt for their records. At checkout customers can also change their payment credit card, enter Hilton HHonors® and airline frequent flier account numbers, and request an email copy of their receipt.

"This is the trend of the future," said Seibert. "Seasoned travelers, whether on business or vacation, value time and convenience. At Hilton, we continue to explore new technologies to meet their needs and we hope to roll out this technology in other locations around the country in the future."

"These kiosks are an exciting addition to an array of high-tech services we already provide our guests at Hilton Hawaiian Village," continued Seibert. "With high-speed internet access in place in all of our guest rooms, wireless internet access in many of our meeting and public areas, and an impressive array of technology-based services throughout the resort, even the most tech-savvy guest can stay connected at the Village."

Source: gohawaii.about.com/od/oahulodging/a/hiltonhv090804a.htm. Used by permission of Hilton Hotels.

tourism. The U.S. federal government usually ranks about 20th in the world, behind such countries as Korea, Malaysia, and Greece, in the amount of money allocated to market the country to international visitors. This has made it harder for the states and private enterprise to develop international trade. In other destinations, such as Singapore and Bermuda, government interest and investment in developing tourism has been a tremendous help to the industry. Even some unstable governments, such as the Philippines, make strong political efforts to boost tourism.

Political uprisings, of course, can have a negative effect on tourism. On the other hand, the growth of political stability such as in the former Soviet Union can greatly boost tourism. Political differences between countries, such as the ongoing airline route conflicts between the United States and the United Kingdom, can also hurt tourism. Positive agreements, on the other hand, such as the agreements between the United States and China and between Canada and Cuba, can help it.

The **political environment** is particularly important for large and multinational companies in many areas of the world. The opening of the People's Republic of China to trade led to many opportunities for both hotel and fast-food companies. In 2004 there were more than 1,000 KFC restaurants in China, and the number is increasing at an annual rate of 200 per year.[2]

political environment
Government movements and positions at the local, state, provincial, and federal levels that affect consumers and businesses, such as feminism, discrimination, the funding of tourism, border crossings, and business ownership.

Globalization has become so predominant in the hospitality industry, in fact, that no company can remain distant from international politics. Formerly American companies such as Holiday Inns (now part of InterContinental Hotel Group), Omni, Westin (now part of Starwood), and Motel 6 (now part of Accor) were bought by British, Hong Kong, Japanese (then Mexican, then American again), and French companies.

Consider Hyatt Hotels. This family-owned company operates Hyatt hotels and resorts around the world. The recently acquired Amerisuites brand gives Hyatt another 100 hotels in the midmarket position. Hyatt must understand and deal with political conditions in all of these countries.

India now allows foreign corporations to hold more than 50 percent ownership in a business with prior approval of the finance minister. In 2005 there were over 100 KFC units in India; by 2014 KFC would like to see 1,000 restaurants.

Other political impacts occur within a nation's borders and are either to be taken advantage of or worked against. Within the United States, both the National Restaurant Association and the American Hotel & Lodging Association maintain lobbyists in Washington, D.C., to fight taxes and minimum wage laws and lobby for a greater emphasis on tourism. Florida (United States), Ontario (Canada), the Loire Valley (France), Algarve (Portugal), Costa del Sol (Spain), and the Bosporus (Turkey) are examples of locations where state and local politics have had major impacts in helping or hindering tourism.

No operator can afford to ignore local politics. When the state of Louisiana licensed a casino in New Orleans, the local restaurant association lobbied to prevent the casino from having any substantial food and beverage services that would compete with existing local restaurants. At the even more local level—town, village, or city—politics can control such things as liquor licenses, zoning variances, building permits, and hours of operation.

Government interest in tourism has stemmed primarily from its economic significance, particularly tax earning and employment potentials. Tourism demand, however, is also largely influenced by legislative actions at various levels of government and intergovernment agencies (e.g., the World Tourism Organization [WTO] and the International Air Transport Association [IATA]). International politics play a significant role in the volume of travel and tourism business, especially when there are warring factions and terrorist activity.

The air transport industry was revolutionized in most tourist-generating countries in the 1980s, although the European Union (EU) did not give full freedom to carriers to operate intra-Union flights until 1997, which increased competition and reduced fares. The deregulation of the airline industry in North America further fueled an increase in intercontinental flights, which, in turn, positively contributed to the growth of world tourism. The adoption of an open sky policy in Asian countries resulted in a substantial increase of air traffic within Asia and helped the introduction of new carriers such as Eva Airways (Taiwan) and Asian Airlines (South Korea).

Relaxed travel restrictions as well as the increased leisure time and income of residents in newly industrialized countries also contribute to the growth of tourism within, to, and from Asia. In the past, both the Taiwanese and South Korean governments restricted or limited overseas travel by their citizens. With rapid economic growth and an increase in consumer disposable income, the concept of leisure travel became widespread in these countries and their governments gradually lifted overseas travel bans. Once, for example, the Korean government prevented its citizens from obtaining a passport for "sightseeing" purposes or pleasure travel; at one time an applicant had to be at least 50 years old. Eventually, the government eliminated all age restrictions on the issuance of passports to its citizens. The Taiwanese government followed the same course. The number of outbound tourists from Taiwan increased more than threefold in four years. The number of South Korean outbound travelers increased fivefold during the same period.

The European Union eased restrictions on intracountry travel in the late 1990s. Prior to lifting the restrictions travelers had to produce documentation and both enter and leave each country through local customs. Now travel between EU countries is like traveling between states in the United States.

After 9-11 the U.S. government created numerous rules and regulations to protect its borders. One such regulation is that all Americans must now have passports to enter the United States from Canada and the Caribbean. Travel to the United States for noncitizens has become so cumbersome that many tourists who would like to visit the country find it not worth the trouble.

Economic Environment

Many factors in the **economic environment** affect any business: recessions, inflation, employment levels, interest rates, personal discretionary income, and so forth. One that has a particularly important impact on international travel is the change in currency values. For example, if the U.S. dollar is strong relative to the Japanese yen, Japanese tourists cannot purchase as many dollars with their yen, thus making the United States a less attractive place to visit. In the late 1990s, with the Canadian dollar at $1.50 to the U.S. dollar, Americans flocked to Canada for the bargains on everything from ski vacations to Canadian fashions, while many Canadians decided to "visit" their own country. The so-called "Asian contagion" of the late 1990s, when many Southeast Asian economies and currencies collapsed, caused that area also to become a bargain for travelers from other parts of the world, but stopped many Asians from traveling abroad. British Columbia, Canada, heavily dependent on this market, went into official recession while eastern Canada was doing quite well.

economic environment Economic realities that affect the hospitality industry, including recessions, inflation, employment levels, personal discretionary income, and foreign exchange rates.

Tourism Marketing Application

Tourism experts project that the requirement of a passport to enter the United States from Canada and Mexico starting January 1, 2008, will have a huge impact on tourism in the United States because the majority of residents of those countries do not have passports. The expectation is that, rather than going through the aggravation of obtaining a passport, they will simply stay in their own countries. Cruise operators are also worried that U.S. passengers will stay home rather than go through the hassle of getting a passport—which they will be required to have to reenter the United States.

In 2007 the U.S. dollar was at an all-time low against the euro. This exchange rate made the United States a very inexpensive place for Europeans to visit. In contrast, Europe had suddenly become very expensive for Americans. The Argentinean peso is very inexpensive relative to the U.S. dollar, so instead of going to Europe, Americans can go to South America. To take advantage of this favorable exchange rate, many hotels in Buenos Aires are promoting their city as a great price value, relative to European cities.

Among other economic factors affecting the hospitality industry is price resistance, which is strong in many areas of the world. The expense account customers whom many hoteliers had classified as "non-price-sensitive" are now resisting higher prices. Corporate controllers have forced cutbacks in expense accounts, and organizational travel planner and buyers are seeking reduced-price contracts with both airlines and hotels. Corporate travel planners reduce airline costs by negotiating a flat per mile charge per ticket used. Under such arrangements tickets cost the company the same per mile from New York to Paris as from New York to Boston. Hotel costs are reduced by the same travel managers, who choose two or three hotels in a given travel destination. Once rates are negotiated with these properties, they are placed on a directory for use by company travelers who can use only the selected properties if they want to be reimbursed for their hotel expenses.

In the United States, recent tax laws have further decreased the deductibility of meals from corporate expense accounts. Guest room city tax increases have deterred travel to some destinations. In Sweden, the Swedish Hotel Association is actively lobbying its government to reduce the heavy tax burden paid by restaurants and hotels, which greatly affects their overall profitability.

As the economies of many countries grew stronger, particularly those of developing nations that had not previously introduced their own upscale hotel chains, two things happened. The first was that foreign developers began to invest. Many at first brought in North American companies to manage their properties. Eventually, they developed their own chains, primarily in the upscale market.

The second thing that happened was that the traveler changed. People with lower incomes or not on expense accounts began to travel both abroad and in their own countries. These people resisted the high prices of the upscale hotels and searched for alternatives. The result was that many hotel companies had to learn to adjust to the local economy.

Hotel companies in countries other than the United States are also likely to enter foreign markets at the upper end of the scale. An exception was Accor of France. This company took its two- and three-star concepts into neighboring European countries and Asia, as well as the United States, under the names of Ibis and Novotel, then bought Motel 6 and Red Roof Inns, both U.S. budget chains. Accor's strategy, as it turned out, was a preview of the future as economic conditions continued to change. Days Inn and Choice International budget properties are now worldwide; Holiday Inn has Garden Courts in Europe and Africa and Holiday Express properties in India. Exhibit 7-3 shows the global brands of Accor.

Many countries are still without sufficient middle-tier hotel accommodations to serve the market. India is one example, with a middle class of 200 million people.

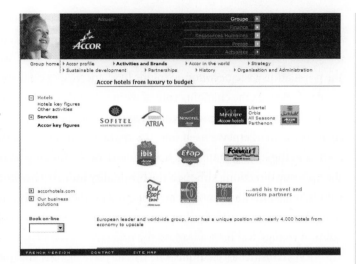

Exhibit 7-3 Groupe Accor of France has 13 brand lines

Source: Retrieved October 11, 2005, from www.accor.com/gbgroupe/activitites/hotelleri/marques/formule1.asp. Courtesy of Groupe Accor. Used by permission of AccorHotels.com.

Residents of these countries, as well as international travelers who visit them in increasing numbers, must often choose between top-rated hotels or less-than-desirable facilities. Many companies are moving to take advantage of these opportunities. Microtel, for example, a U.S. budget chain, is franchising in South America.

Asia is a special case because it was economically less developed for so long. Strong economic growth along with greater political stability in many Asian countries was the major reason for the strong growth surge. Higher disposable incomes, increased leisure time, and relaxation of travel restrictions added to these promising trends until the 1998 economic collapse. The collapse is now over, and Asia is once again a booming market.

Of all the environmental concerns, the economic environment may be the most universally important to the company doing business in foreign lands. The economic environment opens doors to opportunities, and also closes them. This is because countries differ greatly in growth rate, consumer consumption, level of economic development, and discretionary income.

Sociocultural Environment

Although the economic environment will probably have the greatest impact on major international marketing decisions, the **sociocultural environment** affects market behavior the most. Culture is the common set of values shared by most citizens in any country or region. This includes personal beliefs and goals, values and attitudes, opinions and lifestyles, interpersonal relationships, religion, and social structure.

The hospitality business is extremely vulnerable to social and cultural change. Jain stated, "The ultimate test of a business is its social relevance. This is particularly true in a society where survival needs are already met. It therefore behooves the strategic planner to be familiar with emerging social trends and concerns."[3] Two-income families; later marriages; higher divorce rates; AIDS; fewer children; female careerism; greater sophistication; an increasing interest in physical fitness and well-being, escaping from boredom, and returning to nature; and many other social changes worldwide have

sociocultural environment Cultural values, customs, habits, lifestyles, personal beliefs, demographic changes, and so on, that affect consumer behavior.

affected the hospitality industry in recent years. Many hotels and restaurants have reflected these changes—some sooner, some later.

The sociocultural environment includes demographic trends (e.g., the aging of the population), socioeconomic trends (e.g., increasing dual-income households), cultural values (e.g., the changing role of women), and consumerism (e.g., certain "rights" such as full information, safety, and ecology). Contained in the sociocultural environment is the marketplace itself and the characteristics of society. Many of these trends started in the United States and have moved, or are moving, abroad. Others came from the opposite direction. Although the hospitality industry has not been unaware of these trends, it is sometimes slow in catching up with them. This is not surprising, because so many have come so quickly, but the organization that is constantly alert and adapts to cultural change will have a lead on the others.

Let's consider an example. The female traveler, now approaching 50 percent of the U.S. business travel market, initially said, "I just want to be treated the same way as men." But her wants also included special hangers, full-length mirrors, nicer bathroom amenities, better lighting around mirrors (never fluorescent!), irons, softer colors, and other items that often do not occur to men. Electronic door lock security has become critical, as have lobby lounges that are out in the open and room service to avoid going to restaurants. Exhibit 7-4 elaborates on food service trends.

Consider the baby boomers born during the 20-year period following World War II (1946 through 1964). Some of them are now in their 60s. They won't tolerate old products or poor service. They are taking many short vacations, are eating out frequently, are more sophisticated than their parents, have more choices, and are more demanding. They want five-star standards in four-star hotels at three-star prices. They want personal service, and they don't want excuses for poor performance. They want added value—a superior product and better service at a reasonable price. Peter Warren of Warren Kramer Advertising talked about the current attraction to "hip" upscale boutique hotels and questioned the true difference between the needs and wants of baby boomers and generation Xers. In other words, there is more going on out there in the marketplace than just age differences.

Many baby boomers were categorized as "yuppies," meaning young, urban professionals with high incomes and on the fast track, who never worried about tomorrow but spent for today. Now we have the "DINKS," (dual income, no kids) with high

Tourism Marketing Application

The *Journal of Tourism and Cultural Change* is an academic journal that examines the impact of sociocultural changes on tourism destinations. The website is www.multilingual-matters.net/default.htm (accessed February 27, 2005). In addition, the Travel Industry Association presents a Minority Travel Report. Results publicized in this report indicate that minorities exhibit a greater likelihood than travelers in general of choosing group tours, engaging in gambling and nightlife activities, and visiting theme parks and amusement parks. Minorities generate about 17 percent of all U.S. travel expenditures.

Source: Travel Industry Association of America report. (2005). *Travel Insights, 1,* 7.

Exhibit 7-4   Overview of Food Service Trends

To understand food service trends, it is important to first understand the consumer megatrends, social trends, and current industry trends. Time, health, trust, and emotion are all crucial megatrends that restaurateurs need to be aware of. By 2010, 53 percent of every food dollar will be spent outside of the home; consumers will demand greater individual customization of menu items, convenience, and food that has perceived health benefits (e.g., organic foods). According to the National Restaurant Association (2004), current industry trends include the following:

- Continued expansion
- Interest in health and nutrition
- Intensified government impact
- Embracement of diversity
- Greater productivity through technology

- Importance of tourism
- Energy-cost management
- Focus on service
- Heightened competition

The current trend in European and American quick service restaurants (QSRs) is toward "fast casual" or "quick casual" dining, which in part is due to the "get healthy" megatrend. In fact, Wendy's is in the process of becoming a fast casual restaurant, rather than a QSR. Another trend that is similar in both the United States and Europe is the increased placement of branded restaurants in European hotels. This allows customers to conveniently eat at a hotel restaurant that they already trust. These branded restaurants have a much higher success rate and enjoy better profitability than nonbranded hotel restaurants.

Source: www.ats.agr.gc.ca/US/3666_e.htm#Overview (accessed March 19, 2006).

---

discretionary dollars and the DEWKS (dual earnings with kids). DEWKS, who don't get to see much of their children, now take their kids with them on vacation but still want escape. Club Med, other all-inclusive resorts, and the cruise industry have all capitalized on their needs. They market themselves to parents but offer "kid concierges" and other activities during the stay to give the parents a break.

Generation Xers (adults born between 1964 and 1982) want "answers" and high tech. There are also the echo boomers. Members of this group were born between late 1982 and 1995. There are nearly 80 million of them, and they spend approximately $170 billion a year. They are called echo boomers because they are children of the baby boomers. Other names for this group include millennial generation (millennials) and generation Y (gen Y).[4] They are also the most diverse generation ever: 35 percent are nonwhite. Word of mouth or "buzz" is very important to members of this group.[5]

The sociocultural environment facing the hospitality industry continues to change. The "can you top this" policies of coupons, two for one, concierge floors, frequent traveler giveaways, pillow mints, and extended bathroom amenities are not going to address these social changes or take the place of better price–value relationships. The hospitality industry works hard at increasing customer expectations, but it has to deliver. Michael Diamond, then senior vice president of marketing at the Boca Raton Hotel and Club in Florida, stated it this way:

> [W]hatever their needs, travelers want those needs met. You can have all the amenities you want, but unless there's a friendly, knowledgeable, caring person to make those amenities work, you have nothing. It's knowing what travelers need and want in a hotel and then providing it that keeps us in the travel business. If we lose sight of that, then we might as well forget the concierge floors, special menus and technologies. We're here to serve our customers' needs.[6]

Regulatory Environment

The **regulatory environment** tells restaurateurs to whom they can sell liquor and when. Regulations can order hoteliers to collect certain information from guests; in many countries this includes passport numbers, where they came from, where they are going,

regulatory environment
Government regulations or ordinances at the local, state, provincial, and federal levels that affect consumers and businesses, such as tax codes, truth in menu regulations, minimum wage laws, and smoking bans in restaurants.

how and when they are traveling, and a multitude of other details. Regulations tell us how much tax to add to a bill, what we can say on the menu, how much we have to pay employees, what to do with our waste, where to smoke (and where not to), and whom we must accept as a customer. This is not to mention the mass of paperwork required just to comply with city, state, and national government information requirements, or the taxes we have to pay.

Costs and profits can sometimes be affected almost as much by regulations as by management decisions or customers' preferences. This, of course, is why hospitality professional associations have lobbyists in Washington, as well as in state capitals. In many countries there is no such luxury: If the government decides it wants to do something, it does it.

Aside from fighting proposed regulations that will affect a business, scanning the regulatory environment means preparing for the event if and when it materializes. Regulatory impacts are bound to be with us for a long time. There is no way they can be successfully ignored. The marketer's task is to be aware of them, prepare for them, and develop an emergency plan before they occur.

A good example is KFC's response to the New York City department of health's plan to ban all trans fats in New York City restaurants. Prior to the department's meeting, KFC held a press conference in which it announced it would stop using trans fats in all its restaurants across the United States.

Clearly, the regulatory and legal environments of other countries differ from those of your own, based on economic and cultural differences. Many times it is largely who you know and who you pay "extra" that will determine when, or whether, it will be done. It may take economic influence to get the political influence to get around the regulatory barriers. Sometimes competitors know someone in a higher position of authority than you do. Problems of political and regulatory environments are not limited to developing countries; they occur in almost all countries. All of these problems affect marketing strategies and tactics.

Consider the requirement in the European Union that firms cannot mail information to consumers unless they have the customer's permission to do so. It is no longer possible to buy e-mail addresses and send out mass mailings. Doing so could invoke huge penalties. In the United States there is the National Do Not Call List, with which consumers can register to avoid unwanted telephone solicitations. People on this list cannot be contacted by telemarketers except under clearly stated circumstances such as the existence of a prior business relationship with the supplier. If people on this list are called, fines are assessed to the firm making the calls.

Despite perceived difficulties, companies from all countries with expanding economies will increasingly go abroad and find new ways to get their products and services to the consumer.

ecological/natural environment Ecological or natural environmental issues and concerns such as pollution, waste disposal, and recycling that affect consumers and hospitality tourism businesses alike.

Ecological/Natural Environment

Issues relating to the **ecological/natural environment** have risen to the forefront as consumers become more aware of the fragility of our natural environment. In fact, one of the growing segments of the travel industry is ecotourism. Belize, a small country nestled between Mexico and Guatemala, has created an image of an ecotourism paradise. With many unspoiled natural resources ranging from tropical mountains to the second largest barrier reef in the world, Belize is approaching tourism with caution and has positioned itself to attract the growing number of environmentally conscious travelers.

Among environmental concerns, waste disposal, recycling, and pollution are attracting attention not only from customers but regulators as well. Cruise ships are no longer allowed to dump their wastes into the sea, and some even have biodegradable golf balls so that their customers can practice from an on-board driving range without polluting the sea. A long time ago McDonald's Corporation was committed to the use of Styrofoam containers; when it realized that customers' attitudes had changed, it switched to paper wrappers. Golf courses are looking for new strains of grass to minimize the use of pesticides, and hotels are slowly moving toward recycling of solid wastes, not to mention asking you to reuse your towel and sheets. In Germany citizens are so concerned about ecological issues that McDonald's totally revised its waste handling. Increasingly, the public expects the hospitality industry to incorporate ecological concerns into its decision making. Some companies have already started and even found it profitable.

Much of the progress in ecological management in hotels and restaurants has been made outside the United States. Steigenberger Hotels of Germany now places unwrapped soap in guest rooms and saves 50 percent of the corresponding cost. When it eliminated individual portion packs for butter and jams, it estimated the savings at 40 percent. The Crowne Plaza in Wiesbaden uses low-energy lightbulbs and has a central switch in guest rooms to make it easy to turn them all off. The Thai Wah Group of Thailand won the International Hotel Association Environmental Award for converting an abandoned former tin mining area into the luxurious Laguna Beach resort on Phuket Island with a total commitment to the physical, cultural, and social environment. All organic waste is composted. Treated sewage is cycled into a chemically treated system and recycled into the gardens.

Highlighted in the Tourism Marketing Application, the Tourism Commission for the province of KwaZulu-Natal in South Africa indeed uses both the cultural and ecological strengths of this region to promote the uniqueness of traveling to the destination.

Tourism Marketing Application

Marketing KwaZulu-Natal, South Africa

The province of KwaZulu-Natal is located in South Africa. The goal of the Tourism Commission for this region is for the region to be recognized as Africa's premier tourism destination. To reach this goal, the commission has created a website (http://zululand.kzn.org.za) that provides information about the various destinations in the region. Each destination is positioned on different features including cultural attractions and natural wildlife reserves.

Restaurants, of course, are also involved in environmental practices. According to a press release issued by Norcal Waste Systems, Inc, all restaurants and food service vendors at the Oakland International Airport separate kitchen trimmings and plate scraping for composting. In addition, bottles, cans, paper, and cardboard from all airline supplies are also recycled. In 2003 the airport was recycling 9 tons a month. In 2004 it was recycling 54 tons a month.[7] In San Francisco nearly 2,000 businesses participate in the Food Scrap Compost Program started by Norcal Waste Systems (www.norcalwaste.com). Anything that can be made into compost is accepted; for example, half-eaten hamburgers, coffee grinds, stale bread, and cheesy pizza boxes. Robert Reed of Norcal states that the compost of this material is high in nitrogen and other nutrients. This compost is called Four Course Compost (www.fourcourse.com) and is favored by commercial growers who apply the compost to orchards, farms, and vineyards.[8]

So far, we have discussed the components of the environment and how they affect hospitality and tourism. However, it must be noted that these components are dynamic and often tied to each other. The major task of environmental scanning is not only to identify those elements that will affect the firm but also to assess the nature of the effect. A favorable effect is an opportunity, whereas an unfavorable effect is often a threat. Exhibit 7-5 demonstrates the environmental forces we have discussed thus far. It also picks up on competitive forces, which we discuss next.

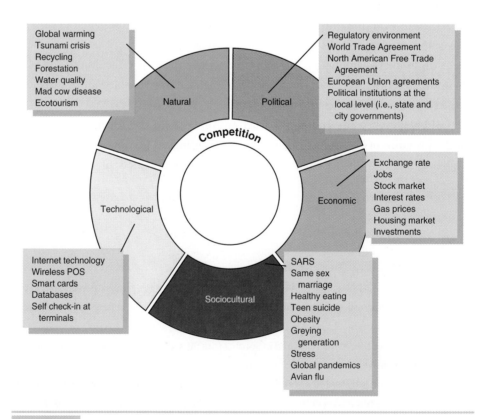

Exhibit 7-5   Types of environments

Source: Van W. McVitty, Jr. Used by permission.

COMPETITION

In marketing, opportunity and understanding competition go hand in hand. This is because opportunity may be defined as being where the competition isn't, or being where the competition is weak, or being the first to market with a new or better idea. Someone once said that marketing opportunity is "the niche that cries out to be filled." That statement may be a little melodramatic, but it does make the point. The only problem is that most opportunities don't cry out; in fact, they can be very well hidden.

In marketing, competition is the enemy. To outmaneuver the enemy, you have to know its strengths and weaknesses and what it does and does not do well; you must understand the customers it serves, why they go there, and what they do when they get there. You also want to know something about customer loyalty and dissatisfaction, what needs and wants are not being fulfilled, and what problems customers may have with competitors' brands. This calls for a great deal of marketing intelligence.

Competitive analysis is not limited to investigating bricks and mortar, the number and size of meeting rooms, decor, what grade beef is bought, what size drink is poured, or what prices are charged. That would amount to a product orientation, and you don't want to be any more product oriented in analyzing the competition than you do in analyzing yourself. Of course, if the competition itself is product oriented, you must know that as well. It may reveal your marketing opportunity.

If, on the other hand, the competition is marketing oriented, you must first determine precisely how marketing oriented they are. Are they creative? Do they adapt quickly? Will they accept short-term losses for long-term gains? Do they worry about the customer? How soon will they copy or react to what you do (or, how short-lived will any advantage you gain be)? In other words, you have to get to know the competition. Of course, you have to know something about their facilities, product, services, and resources, but here's a basic saying that is often ignored: The competition is not simply other hotels and restaurants; it is also the people who manage and operate those hotels and restaurants, the strategies they employ, and the tactics with which they carry them out. Understanding the competition means understanding these people. As Jain stated:

> Most firms define competition in crude, simplistic, and unrealistic terms. Some firms fail to identify the true sources of competition; others underestimate the capabilities and reactions of their competitors. When the business climate is stable, a shallow outlook toward the competition might work, but in the current environment, business strategies must be competitively oriented.[9]

Consider the Japanese. When Toyota wanted to learn what Americans preferred in a small, imported car, it didn't ask the people who owned Chevrolets and Pontiacs (as General Motors did); it asked the owners of Volkswagens what they liked or disliked about the Beetle. It identified the true competition and then addressed consumers' problems.

Understanding competition is a vital part of environmental analysis. But first you have to know who the competition is. It is easy to identify the wrong competition or to fail to identify the right competition. Second, you have to decide what information is important when profiling competitors, and then obtain it. Third, you have to anticipate competitors' future moves and reactions. Finally, you need to determine what you can do to gain a competitive advantage, now and in the future. We discuss these topics next.

Macrocompetition

How do we define competition? There are actually two broad forms of competition, macrocompetition, or industry competition, and microcompetition, or product class competition.

macrocompetition
Industry competition; refers to organizations that compete for the same consumer dollar yet are not necessarily considered direct competitors.

Macrocompetition is any other organization that competes for the same consumer's dollar. This means that any restaurant represents competition to any other restaurant, at least in the same geographic area, and any hotel is theoretically competition for any other hotel. We can carry it even further: We can say that a local supermarket is competition for a nearby local restaurant, or that a new car is competition for a two-week cruise, even though it is satisfying a different need and solving different problems. The question is, Where does the consumer want to spend her money? What are her priorities? Will she make a down payment on a new car this year or go on that cruise she has so much wanted to take?

Upscale restaurant operators are threatened by the competition they are getting not only from high-end supermarkets (especially those that sell prepared dinners), but also from take-out services, catering services, and casual dining. Hotel and motel operators are threatened by the competition they are getting from campers, recreational vehicles, and the hospitality of friends and relatives. Time shares, or vacation ownership, is growing competition for the resort industry. Indeed, some resorts, including Marriott, Hyatt, and Fairmont, participate in the vacation ownership business to compete head-on. These examples represent environmental changes in which the macrocompetition is moving in to fulfill a need.

Why are these other businesses competition? Because they are both satisfying the same needs or solving the same problems—although in a different way. The customer wants it cheaper, quicker, easier, or more conveniently, with less hassle. The hospitality industry is not, of course, totally unaware of this. McDonald's, Burger King, Wendy's, and others have gone back to drive-up windows. Even traditional sit-down restaurants now offer carry-out service and provide convenient short-term parking for order pickup. Many tablecloth restaurants, especially in hotels, have gone into catering and casual dining. Marriott, Choice International, and Groupe Accor of France have bought or built accommodations in almost every level of the hotel industry. All-suite and luxury budget hotels have become commonplace. All of these started out as opportunities in a changing environment. Someone saw an opportunity in the marketplace left empty by the competition as the environment changed.

Microcompetition

microcompetition
Businesses that compete for the same customers in the same product class at the same point in time.

We define **microcompetition** as any business that competes for the same customers in the same product class at the same point in time—in other words, a business that is a direct competitor with a similar product in a similar context. According to this definition, the gourmet restaurant does not compete with fast food restaurants, and upscale hotels do not compete with budget motels.

Caution is necessary to avoid overgeneralizing these contrasts, such as gourmet restaurant versus fast-food restaurant. An alternative in a different product class can easily become a competitor if the one product class is not fulfilling consumers' needs or if the environment changes. For example, we would not normally think that a three-star hotel would be a direct competitor of a four-star hotel. The situation changes, however, if economic forces change and consumers turn to lower priced alternatives to save money. This may lead to upscale hotels with full services offering rates almost as low as limited-service properties. This is called the *slide down effect,* in which five-star hotels price themselves at the four-star level, four-star hotels at the three-star level, and so on. The initial result may be an increase in business. But the cost of providing more service for less money does not last very long, and something (either the rate or the service offerings) has to match for the business to be viable.

One also needs to be careful about designing competition as those properties that actually represent different product classes. This failing can lead to major strategic errors. For example, if the operator of a French restaurant says his competition is the Red Lobster across the street, he is probably basing his statement on geographic proximity rather than product class or customer needs, wants, and demand. We will analyze this point in some detail to be certain that it is clear.

The people who eat at the gourmet French restaurant may also eat at the Red Lobster. They may even eat there more often, but this does not make the two restaurants direct competitors. The reason is that customers are fulfilling different needs and wants at the two restaurants. Except in rare cases, one of these restaurants would not be an acceptable alternative to the other.

Suppose these are the only two restaurants within 100 miles. Does the situation change? If the Red Lobster disappeared, would patronage increase at the French restaurant? Probably very little. On those occasions when people would have gone to the Red Lobster, they may now be more likely to stay home. Suppose, however, that the Red Lobster is doing volume business and the French restaurant is doing poorly. Now does the situation change? Yes, but the Red Lobster is not the competition; it is that the French restaurant, assuming it is a good operation, has misjudged the market and is not catering to the needs and wants of the marketplace. These examples assume that prices are consistent with the product, but even at the same price level there would be some noncompetitive separation of needs and wants.

CHOOSING THE RIGHT COMPETITION

Choosing the right competition is very important when conducting a competitive analysis because it has a huge impact on the marketing strategy and tactics of any hospitality operation. Choosing the "wrong" competition is an error that can be illustrated with the following restaurant example.

Management of the only French restaurant in a small city was unhappy with the volume they were doing. They were at a loss to explain this, because they received very favorable customer comments. In fact, quick research showed that they were overwhelmingly rated the best restaurant in the city in terms of food, service, and atmosphere. In final desperation, after scouting all the "competition," management decided to put in a prime rib buffet, added steaks and chops to the menu, and did other things that other restaurants in the area were doing. A little less than a year later business had fallen to the point at which ownership sold out.

What happened? Two things led to the restaurant's undoing. One, the new menu had alienated the old clientele. Two, the restaurant failed to attract a new clientele that still perceived it as an expensive French restaurant. The failing was not recognizing that its major competition was not in the same city; the competition was in neighboring cities where this restaurant lacked awareness. The appropriate strategy would have been to more fully develop its strong position and pull in more customers from out of town and the surrounding area who would have been willing to drive a great distance for fine French food, excellent service, and wonderful atmosphere.

With all the possible alternatives to consider, then, how does a firm effectively compete? Unfortunately, there is no simple answer to that question. The answer requires thorough analysis of any given situation. We can suggest, however, two launching points. First the firm needs to determine (1) which other firms in the same product class are pursuing the same customers (2) at the same point in time.

First, deliberately choose with whom you want, and can, compete. Rarely do markets simply appear out of nowhere; most of the time you have to lure them away from a competitor. As Michael Porter pointed out in his extensive writings on competition, choosing with whom you want to compete is one of the first decisions that has to be made in developing a product or business.[10] Of course, as we discussed in the beginning of the book, the first decision that has to be made is which consumer need you are trying to fill or which consumer problem you are trying to solve.

The second way to determine how you compete is to ask customers where they would be if they were not at your property. This is easy to do as guests are checking into your property. After welcoming the guest, first ask, "If you had not come here, where would you have stayed?" Then ask why. Or, if you're developing a new product, research the market. Investigate what problems customers currently have. Attempt to determine their needs. Then investigate where they go now to solve the problems or fulfill their needs. The answers to these questions will tell you, at least, who the market perceives to be your competition. The answers will also reveal what you have to compete against in terms of attributes and services. If customers mention properties other than those against which you have chosen to compete, it is clear that your perception differs from the markets. It may be necessary to rethink your competitive strategy.

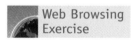

Web Browsing Exercise

Use your favorite web browser and type in the words "competition tourism destination." You will find many articles on this topic. Read one of the articles and be prepared to discuss it in class.

COMPETITIVE INTELLIGENCE

As in war, one always wants to know what the enemy is doing, its position and intentions and strengths and weaknesses, where it is most vulnerable and least vulnerable, and where the best place is to attack. In marketing, we refer to this as competitive intelligence. There are a number of ways to get this information, and it is well worth getting. It goes beyond physical property descriptions. The 2006 Oscar winning film *The Departed* taught us one thing about this: Be close to your friends, but closer to your enemies.

How do we get this information? First, there is public information. The media, articles in the trade press, information available through trade associations, annual reports for publicly traded firms, company brochures, flyers and advertisements, publicity releases, and so forth, are examples of some sources. Then there is trade gossip—information from vendors and others who deal with the competition, such as consulting and accounting firms, universities, local convention and visitors bureaus, and local hotel and restaurant associations.

Another technique for getting information is simply to talk to your competitors. You might do this one on one or at industry conferences and trade shows. However, you cannot talk to them about what the hotel rates should be (this is against the law). You can also talk to your customers, who might have been their customers. You can talk to your employees, who might have been their employees, or at least might know some of their employees. Finally, there is always value in visiting or using the competitor's product. Don't forget, of course, that while you are doing this, so is the competition. Finally, there are measurable differences that can be determined, which are outlined in the following sections.

Market Share

In some areas, hotels exchange room occupancy percentages and average rate figures nightly by mutual arrangement. In addition, firms such as Smith Travel Research and TravelCLICK provide competitive information on a daily basis. Many restaurant operators do likewise. These arrangements are beneficial because they tell you how you are doing relative to the others. Refusing to share this information, or lying, is generally self-defeating. It still remains to discover why you are doing better or worse.

Comparison figures of occupancy and restaurant covers are called *market share figures* and are used to compare **actual market share** with **fair market share,** or the amount of business you could expect if you received your proportional share of the total business conducted by the properties that make up the competitive set. In calculating fair market share, it is important to be certain that you are comparing properties that are in the same competitive set (those that are directly competitive in the same product class and that are competing for the same customers). Management needs to be realistic about identifying the proper competitive set. This means that you will often make this calculation multiple times. For example, you would make calculations for the leisure travel segment, for the group segment, and the like. To calculate your fair market share, divide your capacity by the total capacity in the product class. This is the share you should get if every competitor is performing equally well. To compute actual market share, divide your actual occupancy (or covers) by total competitive set occupancy. Then compare actual to fair market share as a measure of how well you are doing relative to the competition. The goal is for your actual market share to exceed your fair market share.

actual market share
The amount of business a firm actually receives in relative proportion to its direct competitors in the same product class.

fair market share The amount of business a firm expects to receive in relative proportion to its direct competitors in the same product class.

Exhibit 7-6	Hypothetical Example of Market Share				
Hotel	Actual Rooms	Rooms Sold	Occupancy %	Fair Share %	Actual Share %
Upper-Tier Hotels					
A	300	220	73.3	11.5	15.5
B	500	350	70.0	19.2	24.7
C	1,200	500	41.7	46.2	35.2
Yours	600	350	58.3	23.1	24.7
Total	2,600	1,420	54.6	100.0	100.0
Middle-Tier Hotels					
E	275	220	80.0	31.3	39.6
F	425	360	84.7	48.3	50.0
G	180	140	77.8	20.4	19.4
Total	880	720	81.8	100.0	100.0

Consider the hypothetical example shown in Exhibit 7-6 for one city area for one night. All the participants in the analysis are not in the same product class. This does not mean that you are not interested in their occupancy. For example, it would be worthwhile to know why middle-level properties are running at higher occupancy than upper-level properties. It might indicate that the upper-level properties are pricing themselves out of the market, or it could mean something entirely different, for instance, concerning the type of business that was in town last night.

Now consider the market shares of the properties in your competitive set. Hotel C's actual share is considerably lower than its fair share. But look at the size of this hotel; it is still filling more rooms than any of the others in the analysis. Perhaps this is primarily a convention hotel with widely changing occupancies; perhaps it should not be included in the same competitive set. What this means is that one has to interpret these figures with good judgment before making a decision.

As you can see, your hotel is getting slightly more than its fair market share (actual share is 24.7 and fair share is 23.1) and would not be doing even that if Hotel C's occupancy were up. Hotels A and B, however, are substantially exceeding their fair share (15.5 versus 11.5 for Hotel A and 24.7 versus 19.2 for Hotel B for actual share and fair share). What, you might ask, are they doing right? Or, what are you doing wrong? This calls for an examination of their segments and marketing strategies.

REVPAR

revenue per available room (REVPAR) Guest room revenue generated over a given period of time divided by the total number of guest rooms available for sale for the same period of time.

Market share is one method of measuring relative performance in the marketplace; the calculation of **revenue per available room (REVPAR)** is another method commonly employed in the hotel industry. REVPAR is calculated by dividing the room revenue by the number of rooms available for sale. It can also be calculatedd by multiplying the average daily rate (ADR) by the occupancy percentage. The fallacy of market share is that a competitor can gain actual share in the market at the expense of room rates. By dropping its rates $10, more people may book at that hotel. REVPAR measures the revenue generated per available room and essentially

Exhibit 7-7 Hypothetical Example of REVPAR

Hotel	Actual Rooms	Rooms Sold	Occupancy %	Average Daily Rate (ADR)	Revenue	REVPAR	Yield Index
Upper-Tier Hotels							
A	300	220	73.3	$120	$26,400	$88.00	1.17
B	500	350	70.0	130	45,500	91.00	1.21
C	1,200	500	41.7	150	75,000	62.50	0.83
Yours	600	350	58.3	140	49,000	81.67	1.08
Total	2,600	1,420	54.6	137.96	195,900	75.35	1.00
Middle-Tier Hotels							
E	275	220	80.0	$110	$24,200	$88.00	1.06
F	425	360	84.7	100	36,000	84.70	1.02
G	180	140	77.8	90	12,600	70.00	0.85
Total	880	720	81.8	101.11	72,800	82.70	1.00

controls for pricing decisions. REVPAR is the method most widely used in the industry today.

REVPAR more accurately measures the balance of marketing efforts as shown in Exhibit 7-7. The middle-tier hotels indicate better asset management. Their REVPARs are comparable to the upper-tier hotels, or better, and they undoubtedly are lower-cost producers. One possible conclusion is that they are stealing business from the upper-level hotels with lower rates. Another is that, on this particular day, the upper-level hotels had booked lower-rate conference groups. In any case, the reason needs to be examined on a regular basis. Calculations of trends are more enlightening than calculations performed to evaluate single days, and other conclusions might be drawn.

Restaurant Comparisons

Restaurant market share revenue per square foot (RSQFT) and revenue per available seat (REVPAS) are two very new methods developed for restaurants. Both calculate competitive restaurant share by measuring one's own position against that of the competition in a similar manner to that of market share analysis for the lodging industry.

Other Measures Used to Analyze Competition

Measures such as market share and REVPAR show the end result of customer choice. They do not show why the customer made the choice. To determine this information, we need to ask the customer. One way be to this is shown in Exhibit 7-8.

Perceptual Mapping

Another way to analyze competitive intelligence is through the construction of **perceptual maps** that plot customers' perceptions of your property versus those of your competitors. This technique often involves simple plotting (or sophisticated statistical methods known as multidimensional scaling or discriminant analysis). The technique may be used to evaluate a single property or multiple properties. Information used in perceptual mapping comes from the questions shown in Exhibit 7-8.

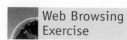
Web Browsing Exercise

Go to the websites for TravelCLICK (www.travelclick.net), the Daily Bench (www.thedailybench.com), and Smith Travel Research (www.smithtravelresearch.com). Compare and contrast the different offerings. What other information do they provide that can be helpful to the marketing executive? Be sure to read their sample reports.

perceptual map A visual map, or "plot," of customer perceptions of an establishment (or brand) relative to the competition based on features and benefits important to the targeted audience such as price, location, services offered, and the like.

| Exhibit 7-8 | Questions Used to Determine Importance and Performance |

IMPORTANCE QUESTION

Next, please think for a moment about the reason for visiting a specific legalized gambling establishment in Las Vegas. Please tell me how important each reason is for you in your decision to choose one specific property over another. Please use a 1 to 10 scale where a 1 means the reason *is not at all important* and a 10 means the reason *is very important* in your decision to choose one legalized gambling establishment over another. You may use any number on this 1 to 10 scale. Do you understand how this 1 to 10 scale works? How important is _____ in your decision to choose one place to visit over another?

PERFORMANCE QUESTION

Now I am going to read you a list of features that may or may not describe some of the casinos in the Las Vegas area. We'll use a 1 to 10 scale where 1 means it *does not describe the casino at all* and 10 means it *describes the casino perfectly*. If you have not been to the casino personally, please base your answers on what you have heard or what you believe to be true. The first feature is _____. How well does this feature describe casino _____?

| Exhibit 7-9 | Calculation of Customer Satisfaction Index, AKA Competitive Index |

	Importance	Brand A		Brand B	
		Rating	Score	Rating	Score
Column	A	B	C	D	E
Feature	Scale: 1–10	Scale: 1–10	A × B	Scale: 1–10	A × B
It is a place friends like to go	7.30	7.60	55.48	6.40	46.72
Atmosphere is very pleasant	8.80	7.70	67.76	7.60	66.88
One place seems to have better odds	7.40	6.80	50.32	6.00	44.40
Slot machines filled in a timely manner	7.50	6.80	51.00	6.80	51.00
Types of promotions offered	7.40	7.70	56.98	6.80	50.32
Total	384.00		281.54		259.32
Index			73.32		67.53

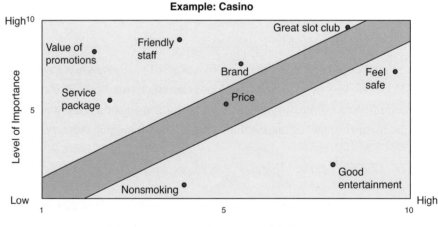

Example: Casino

For example, one of the authors conducted a study of hotels in Las Vegas. Visitors to Las Vegas were asked what was important in their decision to visit a specific property over another. Customers were also asked to rate the importance of criteria described for different properties. This information was then plotted for each hotel individually, as shown in Exhibit 7-9.

The feature "Good entertainment" is of very low importance to the customer, as shown in Exhibit 7-9. However, customers believe that the casino offers "good entertainment." This is in contrast to the feature "Value of promotions." This feature is very important to the customer, but customers do not believe the casino delivers on this feature. The feature "Great slot club" is important to the customer and customers believe the casino has a "Great slot club." The same interpretation can be made for the feature "Feel safe." The result is that management should spend less money on entertainment and more on creating high-value promotions.

USING COMPETITIVE INTELLIGENCE

The purpose of competitive intelligence, of course, is to use it to your best advantage. If you are behind, you need to seek and increase competitive advantage. If you are ahead, you need to sustain and increase competitive advantage. In the first case you need to overcome **barriers** to move ahead. In the second case you need to erect barriers to stay ahead.

Raising Barriers

Strong **economies of scale,** for example, may provide an unbeatable cost advantage. In the hotel industry, this is especially true today in the use and cost of the latest technology. This technology also gives a firm preferred access to customers. This helps to explain at least one reason the industry has been in such a consolidation stage.

Barriers may also be raised based on cost differences or service differences. A successful barrier returns higher margins if it is sustainable and unreachable by the competition; that is, it must cost the competition more to overcome it than it costs the firm to defend it. For example, Southwest Airlines, Easy Jet, and Ryan Air have created high barriers by reducing costs to the minimum. The service and culture offered by Four Seasons is a barrier because the culture has been built over the years by promoting employees from within. It is difficult to copy this. When something is done by a firm that is difficult to copy, we say the firm has a *sustainable competitive advantage.*

A sustainable competitive advantage in marketing strategy happens only when (1) customers perceive a consistent difference in important attributes between one firm or property and its competition; (2) the difference is the result of a capability gap between the firm and its competitors; and (3) both the difference in important attributes and the capability gap can endure over time.[11]

The essence of opportunity is beating the competition. Intense competition in an industry is neither coincidence nor bad luck. The competitive objective is to find the most vulnerable position that, if broken, would eliminate a competitive barrier, or one that would enable you to erect the best defense barriers. This means finding what makes the competition vulnerable. Attacking vulnerability, as in the military sense, means attacking the weaknesses and avoiding the strengths in the line. The latter is as important as the former. Wendy's is one fast-food operator that has proven this point. Wendy's attacked where McDonald's was weak in two areas. One was the area of "adult" hamburgers. Wendy's saw that McDonald's was not really serving this market, so it set out to carve its own position. The second area was in the product. Wendy's saw a dislike in the market for the frozen, precooked hamburger, especially among a portion of the adult market, and offered fresh hamburgers. Wendy's has survived where many others that copied McDonald's failed.

barriers Impediments that make reaching a desired goal difficult. To increase its competitive advantage, a firm needs to overcome barriers to move ahead and create barriers to stay ahead.

economies of scale Spreading costs across multiple units or within a large establishment of an organization to save money.

Avoid the competition's strengths, at least until you are strong enough and have the resources to challenge them with meaningful differentiation. Look for the weaknesses. These may be in the product line, positioning, value, segmentation and target markets, capacity, resources, cost disadvantages, product differentiation, customer loyalty, or distribution channels.

COMPETITIVE MARKETING

Porter suggested three strategies for beating the competition: positioning to provide the best defense, influencing the balance by taking the offense, and exploiting industry change.[12]

Defensive positioning means matching strengths and weaknesses against the competition by finding positions where it is the weakest and developing strengths where the competition is least vulnerable. Four Seasons Hotels has accomplished this by maintaining its level of service at all costs.

Influencing the balance by taking the offensive, or being proactive, means attempting to alter the industry structure and its causes. It calls for marketing innovation, establishing brand identity, or otherwise differentiating the product. Darden is a restaurant company example with its Olive Garden, Red Lobster, and Bahama Breeze brands. This has also happened with all-suite hotels, conference center hotels, and, especially, "luxury budget" hotels, and also with Marriott entering every product class and buying the Ritz-Carlton and Renaissance brands.

Exploiting industry change means anticipating shifts in the environment, forecasting the effect, constructing a combination for the future, and positioning accordingly. Taco Bell and Domino's pizza delivery are two examples. Robert Hazard accomplished this in his successful metamorphosis of Quality Inns (now Choice International) in the early 1980s, when he offered three product levels to the marketplace. All-suites have done the same thing, as has SAS Hotels in its strategic alliance with Radisson.

A successful company must look beyond today's competitors to those that may become competitors tomorrow (e.g., convenience stores and take-out versus the fast-food industry). It must also watch out for new entries in the race (e.g., condominium and private home rentals, conference center hotels) and the threat of substitute products (e.g., supermarket "make your own meal" bars).

The key to growth, even survival, is to obtain a position that is less vulnerable to direct attack, old or new, and less vulnerable to consumer manipulation and substitute products. This may be done through relationship marketing, actual or psychological product differentiation, and constant and foresighted competitive awareness and analysis.

Chapter Summary

Today's complex and dynamic marketing environment requires hospitality firms to constantly evaluate new and evolving opportunities and threats. Environmental scanning has become increasingly important for all businesses in this fast-moving and fast-changing world. Without it, a company can only react to what happens in the environment. The basis of hospitality marketing, which was presented in the first part of this book, and the application of hospitality marketing, which will be presented in the remainder of the book, lie in what is happening in the environment.

Environmental scanning looks at the big picture, the broad view, and the long range. Marketing builds on this by approaching it through a narrower perspective, resting ultimately on the individual consumer.

Environmental scanning, both in textbooks and the real world, has been used primarily at the corporate level and in strategic planning. By definition, environmental scanning presents a macro view. We believe, however, that this macro view can be given a micro perspective. In other words, we believe that every individual or unit operator, as well as corporate and higher-level management, must be conscious of and continuously analyzing the environment and forecasting its likely impact on each unit operation. The hospitality industry today is too broad based, too diversified, and too much operated in multiple micro environments to ignore this technique at any level.

Understanding all facets of competitors' businesses, and the actors who make them work, requires constant analysis. Without this understanding, and often good research to explain it, marketing is doomed to ignore opportunity. Sustainable competitive advantage is hard to come by in the hospitality and tourism industry. Regardless, it is an essential competitive element.

Key Terms

actual market share, p. 211

barriers, p. 215

ecological/natural environment, p. 204

economic environment, p. 199

economies of scale, p. 215

environmental scanning, p. 194

fair market share, p. 211

macrocompetition, p. 208

microcompetition, p. 208

perceptual map, p. 213

political environment, p. 197

regulatory environment, p. 203

revenue per available room (REVPAR), p. 212

sociocultural environment, p. 201

technological environment, p. 195

Discussion Questions

1. Overall, and in your own words, describe what is meant by environmental scanning.
2. There are essentially six broad classifications of the external environment: technological, political, economic, sociocultural, regulatory, and ecological/natural environment. Give an example of each from a hospitality or tourism perspective.

3. What is the difference between macrocompetition and microcompetition?

4. What are some of the ways you can learn about direct competition?

5. Describe what is meant by fair market share versus actual market share. Why is it critical that you compare "apples to apples" and "oranges to oranges" when putting together and analyzing this type of information?

6. Explain what the figures mean in Exhibit 7-6, Hypothetical Example of Market Share. Which hotels are doing well and which hotels are not doing so well from a market share perspective?

7. What is REVPAR?

8. What is perceptual mapping?

Endnotes

1. Olsen, Michael, D., & Cassee, Ewout. (1997). The international hotel industry in the new millennium: Visioning the future. In Into the new millennium, a white paper on the Global Hospitality Industry. Paris: International Hotel Association, 58–60.

2. Retrieved December 1, 2006, from www.chinatoday.com.cn/English/e2004/e200406/p26.htm

3. Jain, S. C. (1997). *Marketing planning and strategy* (5th ed.). Cincinnati: South-Western, 134.

4. Retrieved December 1, 2006, from www.worldwidewords.org/turnsofphrase/tp-echl.htm.

5. Retrieved December 1, 2006, from www.cbsnews.com/stories/2004/10/01/60minutes/main646890.shtml.

6. Presentation at World Hospitality Congress III, March 9, 1987, Boston.

7. Retrieved December 1, 2006, from //www.sunsetscavenger.com/prrestaurantre cycling.htm.

8. Ibid.

9. Jain, 70.

10. Porter, M. E. (1980). *Competitive strategy: Techniques for analyzing industries and competitors.* New York: Free Press.

11. Jain, 96–98.

12. Porter, M. E. (1975). Note on the structural analysis of industries. Harvard Business School Case Services, 22.

.... case study

The Beefsteak Steakhouse

Discussion Questions

1. How would you describe the product concept of the Beefsteak Steakhouse, including its history, layout, and design?

2. What is the customer base for this steakhouse located in the main lobby of the Hilton in Atlanta, Georgia?

3. How would you describe the competitive environment, including primary and secondary competition, that Leslie faced in this downtown location of Atlanta?

4. How would you assess the marketing situation for the Beefsteak Steakhouse? Who was responsible for what? Does this approach to marketing make sense to you?

5. What advice would you give to Leslie to increase the customer base and improve the sales volume, especially given the competitive environment in which this steakhouse operates? In other words, can Leslie turn the situation around, and if so, how?

Leslie, the general manager of the Beefsteak Steakhouse, Hilton Center, Atlanta, was sitting in her office looking at the restaurant's statistics. She wondered what had gone wrong and how it happened. The past few years had been a rollercoaster. Everything was going perfectly before she went on her maternity leave almost two years ago. While she was away, the director of operations for the company took over her responsibilities. When she came back to work, after a shortened maternity leave, too many changes had happened in her absence: new policies, new employees, and a new menu. Then, three months later, the Keg Steakhouse opened its doors just around the corner. Ever since, her job had become harder, with more pressure and intensity, and the numbers had become more alarming.

THE BEEFSTEAK STEAKHOUSE

The Beefsteak had been established 25 years ago. It originated in Dallas, Texas, as a butcher shop and ever since the company had been expanding and become known as the "Legendary Masters of Steak." The first Steakhouse restaurant was opened on the west side of Dallas in 1981. Today there were five locations in Texas, four in Georgia (three in Atlanta), and a sister company Steakhouse in Florida with up to eight locations.

The Hilton Center location, in downtown Atlanta, opened its doors in February 2000. It was open 364 days a year serving lunch and dinner. The restaurant was *fully* licensed and offered full dine-in service and take-out. The hours of operation were Sunday through Thursday 11:30 A.M. to 10:00 P.M., and Friday and Saturday 11:30 A.M. to 11:00 P.M.

LAYOUT, DECOR, AND ATMOSPHERE

The Beefsteak Steakhouse was located in the main lobby of Hilton Center, Atlanta. It had three entrances: from streets on both sides and from the lobby of the hotel. Both streets were major ones in Atlanta that gave opportunity for more exposure. The location of the restaurant in the Hilton Centre was considered strategic because the Beefsteak was located in a major hotel and convention center in the heart of downtown Atlanta, and was the main dining facility in the hotel. The restaurant had a seating capacity of approximately 400. The bar accommodated an additional 20 guests. Furthermore, there were two available rooms for private functions. The restaurant decor was one of a traditional steakhouse, with dark pastel colors, electric fireplaces, and mirrors and pictures on the walls. The atmosphere was casual.

CUSTOMER BASE

The customers of the downtown location were usually referred to as people from all walks of life. The Beefsteak targeted the business crowd and hotel guests for lunch, cocktail hour, and dinner; families for dinner and weekends; as well as private

This case was contributed by Olia Galeva and David Erlich, developed under the supervision of Dr. Margaret Shaw, PhD, School of Hospitality and Tourism Management, University of Guelph, Ontario, Canada. Names and places have been disguised. All rights reserved. Used by permission.

(*Continued*)

functions that were offered in the two available private rooms. The lunch had been fairly successful, catering to the various business clientele in the area. The restaurant had a well-established repeat customer base during the lunch hour. The breakdown of the customer base, according to the overall estimation of the restaurant's management, was 45 percent business clientele, 30 percent families, and 25 percent hotel guests.

COMPETITION

Located in Atlanta's downtown financial and theater districts, the Beefsteak was surrounded by many competitors. The steakhouses in the vicinity were Hy's and Ruth's Chris, which were not considered primary competitors by the management of the Beefsteak. Their menu prices were much higher, and they were targeting different, higher-end markets. Other secondary competitors included food courts and independent establishments. Management did not consider that it had any primary competitors during the first few years of operation. The environment, however, had changed. The Keg Steakhouse opened its doors on York Street, just around the corner. The Keg was considered a primary rival competitor by the Beefsteak management, due to the fact that their prices were comparable. The new establishment offered an exciting and contemporary interior, an atrium lounge area, real wood fireplaces, big-screen flat TVs, a stylish main dining room, and a cozy patio. The atmosphere was unquestionably inviting. In addition, the Keg was heavily advertising during peak TV time in hotels. Direct mail and ads emphasized the experience. Furthermore, it offered higher and consistent quality of steaks in a trendy ambiance, compared to the Beefsteak. In addition, the Keg was considered to provide a unique experience and excitement that definitely was not a salient attribute related of the Beefsteak. Other steakhouses, which were in the vicinity, included Morton's in Marriott Center, Baton Rouge, and Canyon Creek on Front Street.

CURRENT SITUATION AND STAFFING

The restaurant had been open for five years, and Leslie had been GM for four of those years. She had been with the company for more than 10 years. She loved her job and her staff. Currently she had a staff of 90 people. Until recently, the

employee turnover had been low. Nearly 80 percent of the staff had been working in the store since it opened its doors in 2000. The front of the house personnel had had no turnover until quite recently. The turnover rate was higher in the back of the house environment. Long hours, high volume, lack of promotion, and negligible pay increase were some of the reasons people gave for leaving.

Recently, however, Leslie had noticed that more and more people were leaving, and the turnover rate increased in the front of the house as well. Servers were trying to get jobs at different establishments, because business had been slow. She found herself looking at resumes quite frequently to offset the increasing resignations. She wondered if the restaurant would be able to survive until the new opera opened its doors next September. Everyone had great expectations about the volume of customers that the opera would bring in.

All Leslie's decisions had to be approved by the director of operations and one of the major shareholders, Charles. Charles was also a working manager at the downtown location. He was about 70 years old, yet still very active for his age, with a sharp mind and an extensive economic background.

After the Keg opening Charles decided to renovate the bar area and compete with the lounge area of the Keg. The funds invested were rather modest and sufficient for a new marble bar top, three wall mirrors, bar lights, and two 15-inch TVs. The initial idea of a full renovation was cut short, and the bar area was not brought up to the initially intended standards to compete with the Keg's lounge.

Leslie was speculating how she could address the issue of constantly decreasing sales and turn it all around. The Keg, on a regular Thursday, its busiest day, was breaking $34,000 in sales by early evening. That was the sales volume her restaurant used to reach on a regular basis until the Keg opened its doors. Nowadays, she was hoping to reach $8,000 sales on a weekday and $10,000 on a Friday or Saturday. Yearly sales had reached their peak of $7.9 million in 2002, and had decreased to $4 million in 2004 as shown in Exhibit 1.

MARKETING

The Beefsteak Steakhouse had positioned itself as "The Biggest Steakhouse in Town." Since the beginning of 2005, however, the company had dropped the motto, and was left with just its name. The main office in Dallas handled the major decisions

Exhibit 1	Sales for the Beefsteak Hilton Center for 2000–2004				
	2000	2001	2002	2003	2004
Sales	3,125,965	4,995,423	7,881,367	5,269,752	4,023,235
Food cost	1,131,568	1,648,489	2,491,951	1,844,413	1,649,526
Beverage cost	375,116	599,451	945,764	632,370	482,788
Labor cost	937,790	1,498,627	2,364,410	1,580,926	1,206,971
Utilities	1,100,000	1,100,000	1,100,000	1,100,000	1,100,000
Gross income	**(418,508)**	**148,856**	**979,242**	**112,043**	**(416,050)**

concerning operations and marketing. There was, however, a marketing director for each city with more than three locations. Two years ago the marketing director for the three Atlanta locations was released, and the marketing responsibilities were now handled by an appointed functioning manager at each location. Frank, a functioning manager in the downtown location, was assigned the additional marketing responsibilities. Frank lacked formal marketing training and experience. He was under the strict supervision of Charles, who was the ultimate decision maker for each marketing initiative. Frank's main duty was to get familiar with all the concierges in the downtown area and obtain a referral business from them. He had developed an efficient system to track the numbers of guest referred to the restaurant by each concierge and reward them individually for their effort. Furthermore, he was in charge of Christmas party booking and was selling door-to-door to the businesses in the area.

The head office in Dallas was advertising weekly in the *Atlanta Constitution*. Ads were published once or twice a week. They usually promoted weekly specials and were located in the comic section. More often than not, individual restaurants were not informed of the upcoming weekly specials, which left them to find out about them from the newspaper.

The restaurant also catered to tour groups. These sales were handled in the head office. The Beefsteak had the highest group visitations percentage during the summer months, which in the past had helped offset the low summer season sales, due to the fact that the restaurant lacked a patio. Central bus tour reservations had been profitable in the past, but this year the restaurant was barely breaking even from

these tours. Prices had not been increased for the past three years. The majority of groups were charged $12 per person, including a three-course meal, coffee or tea, and taxes and gratuity. Leslie raised the issue a number of times with the director of operations, as well as with the head office in Dallas; however, no measures had been taken to date. She invested additional time and resources in serving the tour groups from which she could barely cover her costs. At the same time she had been constantly pressed to cut production and labor costs. Cutting costs seemed to be the major issue with the shareholders, and she was reminded of this on a regular basis.

In addition, occasionally surveys were distributed to the guests, per Charles's request, in order to evaluate customer satisfaction. The surveys, after being collected, were put in the bottom drawer in the office. Sporadically, Charles would look at a few. The establishment was also collecting business cards for a monthly free lunch draw. The destiny of the business cards was the same as that of the surveys.

THE FUTURE?

Leslie was sitting in her office trying to figure out what strategy she needed to develop that would appeal to the top decision makers, in order to increase the customer base and improve the sales volume. She knew that if her plan for action were approved, the funds allocated to implement it would be limited. Financial constraints were a considerable issue in the company. What plan, she wondered, did she need to present to the director of operations and to Charles that would increase her sales?

The Marketplace

overview

This chapter reviews the basics of consumer behavior; specifically, how consumers make choices and decisions. This knowledge helps marketing professionals meet the needs and wants of customers. We place a special emphasis on hospitality purchases. The second part of the chapter examines these behaviors for two broad-based segments: business travelers and pleasure travelers.

chapter 8

Understanding Individual Customers

learning objectives

After reading this chapter, you should be able to:

1. List the basic principles of consumer behavior.

2. Apply the basic principles of consumer behavior to hospitality and tourism customers.

3. Identify the various types of individual customers in hospitality and tourism and then apply this information to the development of successful marketing strategies.

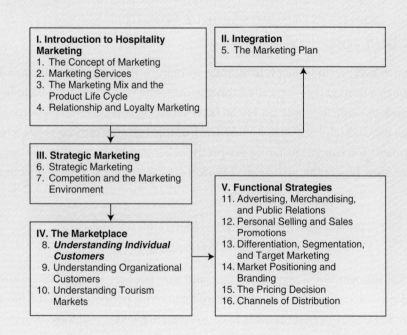

I. Introduction to Hospitality Marketing
1. The Concept of Marketing
2. Marketing Services
3. The Marketing Mix and the Product Life Cycle
4. Relationship and Loyalty Marketing

II. Integration
5. The Marketing Plan

III. Strategic Marketing
6. Strategic Marketing
7. Competition and the Marketing Environment

IV. The Marketplace
8. *Understanding Individual Customers*
9. Understanding Organizational Customers
10. Understanding Tourism Markets

V. Functional Strategies
11. Advertising, Merchandising, and Public Relations
12. Personal Selling and Sales Promotions
13. Differentiation, Segmentation, and Target Marketing
14. Market Positioning and Branding
15. The Pricing Decision
16. Channels of Distribution

Thomas Storey

Executive Vice President, Development, Fairmont Hotels & Resorts

Tom Storey was appointed executive vice president, development, of Fairmont Hotels & Resorts in June 2004. He is responsible for defining and executing the growth plan for the firm and also oversees new business development and Fairmont Heritage Place.

Storey joined the company as executive vice president, business development and strategy, in February 2001. Prior to joining Fairmont Hotels & Resorts, he was with Promus Hotels, a hotel management and ownership company, as executive vice president, strategic planning and venture operations, from 1998 until 2000, and executive vice president, marketing, from 1997 to 1998. Prior to joining Promus Hotels, Storey was executive vice president, sales and marketing, at Doubletree Hotels.

 Marketing in Action

Thomas Storey

Executive Vice President, Development, Fairmont Hotels & Resorts

How do you segment your customers (e.g., business, leisure), and would this be typical of all hotels?

At a macro level, we do segments in business and leisure. We also segment them as direct or through a third party—typically a travel agent—and we also segment them by how they book us—whether they go to the hotel direct through the reservations center, through the GDS [global distribution system], or through an e-commerce site. Plus, we do segmentation from a clustering perspective relative to their spend levels, their stay patterns, their geographic source, etc. I think it is typical of most hotel companies.

Are business travelers one segment or many? How are they different?

The ideal segment is the segment of one, and that is where the business is moving, toward customerization* by individual. From a marketing perspective, it's still more difficult to attract one person than it is to cluster people together and look for commonalities that you can market to. In the business travel segment, people would segment

---

Customerization can be defined as a "buyer centric company strategy combining mass customization with customized marketing." For more information, visit www.smeal.psu.edu/ebrc/publications/res_papers/1999_06.pdf.

from the perspective of whether they're booking themselves or booking through an intermediary. They would also probably segment by some sort of product purchase intent. For example, we have a Gold product, which is a higher service product bundle that is within our hotel, so people who purchase that product will be different from people who are just purchasing the standard Fairmont room. Hotel companies have multiple segments within business travel that either relate to the types of products they offer, to the price point that they offer them at, or again, in terms of how their decision is ultimately made.

What are the needs of business travelers today by segment? How have these needs changed over time?

They're always evolving. The key evolution in the past five years has been the increased use of technology—whether that's high-speed Internet in the room, wireless networks, plasma televisions, the fact that many consumers now are carrying their own content, whether they are doing that through their own computers, memory sticks, or iPods, or some other format. For a while, the content was resonant in the TV, then the content was resonant in the traveler's laptop, then it became resonant on the network, and now it's really in all of the above. Being able to access multiple forms of content through multiple technologies is probably the latest need of the business segment. In a hotel, we need to make sure that, if a person is carrying their information on a JumpDrive, that there is a place in the hotel where they can access that JumpDrive. If a customer accidentally forgets the power cord for their iPod, we need to make sure that we have those available at the front desk or an in-room stereo system that they can plug that equipment into. There are multiple ways of looking at it. The comfort level with technology that business travelers have today is dramatically different than what it was in the past.

Are leisure travelers one segment or many? How are they different?

They are the same as business travelers, but they have a wider array of segments because leisure travelers have multiple trip occasions. They could be traveling with family or friends, they could be attending an event, it could be personal leisure, or it could be vacation. I think that any one leisure traveler could have multiple reasons for booking a leisure trip, which could actually cause them to fall into different segments.

How do hotel firms attract the leisure segment?

Promotions and price are the two big ones, with price being more for short-term demand and promotion more typically being used when a person is looking for some type of value added type of booking. In terms of how you reach leisure travelers, it is largely the same tools as you would use to reach business travelers. I think for long-haul destination leisure travels, travel agents play an even greater role. Travel agents come more into play the more expensive the trip gets, because if somebody just wants to book a weekend someplace, it's less likely that they're going to use a travel agent. If they're booking a week, 10-day, or perhaps two-week vacation—especially if it's overseas, because there's a lot of risk in that vacation—they might research it online, but still talk to a travel agent to give them a bit of a security blanket. Additionally, it's still relatively difficult to do a multisegment booking online—using air, car rental, and multiple hotels, for example. It's often easier to have a travel agent put something like that together, and obviously the suppliers are paying the agents, not the traveler.

Do the needs of travelers change depending on their country of residence? If so, how?

One key difference is language. It may not seem like a big thing, but it is when you're taking people from Asia to the Rockies or from some parts of Europe to someplace like Bermuda—language is key.

Food is also key. You need to be able to offer people food that is consistent with their typical diet. Whenever you have people that are traveling in larger groups such as tour groups, you need to be able to help them coordinate all the ground handling activities, whether that is transport or sightseeing or any sort of destination type product that they might experience beyond the hotel. People like to be able to experience the food in the culture they are visiting, but at the same time, they like to see types of food that they are used to back home as a sort of fallback. When we bring Asians to Banff, we typically try to have Asian food like sushi, rice, etc., that would fit a traveler coming from Asia. It's not that it replaces the indigenous foods, but rather is a complement to them.

What advice can you give to those studying marketing?

I think there is a continuing evolution toward customization, so I think that students need to focus more on the intricacies of market research, both qualitative and quantitative, to really understand how to develop hypotheses, how to test hypotheses, and how to monitor behavior. Those are all critical components in being an effective transient marketer.

I think that gaining a high level of comfort with technology and how to use technology is also important—again, to do the research, but also to deliver a customized product. I would carry that through to operations research from the standpoint of being able to create service bundles that truly are differentiated and customized.

The hotel industry still operates within a lot of paradigms around the way the service is delivered, and it tends to be from the hotel out, rather than from the customer in. I think the hotelier of tomorrow is going to be even better at identifying and recognizing somebody as an individual, providing them with a customized set of products and services, and then maintaining a dialogue with them over time so that it becomes a community of one, rather than trying to dump them into segments. Fairmont is continuing to go in that direction. It's more challenging in larger hotels when you're dealing across multiple customer segments. Some customer segments are better served by being able to handle large numbers of people and some customer segments—transient in particular, especially high-end transient—are more oriented toward intimacy and personalization. Blending the two of these together is somewhat of a challenge, especially doing it in a portfolio where you probably have upwards of one and a half million a year going through the hotel. Luckily, technology has come a long way, so it's easier than it used to be, but it's still not an easy task in general.

Used by permission from Thomas Storey.

---

*I*f the first step in marketing is to recognize customers' needs, wants, and problems, then it is obvious that we must understand how and why customers behave the way they do, as well as what leads them to that behavior. This is no small task.

There are no easy and clear-cut answers to these questions. Instead, many theories, concepts, and models have been developed to explain the customer. These have been developed from many disciplines such as sociology, psychology, social psychology, anthropology, philosophy, and economics; these approaches must be integrated before we can approach even a limited understanding. Our ultimate goal, of course, is to be able to influence buyer behavior. We may fall far short of that goal in its full sense, but we will learn, at least, to understand some "hows" and "whys" and their causes.

It is important to begin with some basic and generally agreed upon principles of **consumer behavior,** because effective marketing must be based on these ideas. Managerial decisions that ignore these principles tend to lead to marketing failures.

> ■ **Principle 1:** Consumer behavior is purposeful and goal oriented. What may appear to be completely irrational to the outside observer is, nevertheless, the action that an individual views as the most appropriate at the time. To assume otherwise is to underestimate the consumer.
> ■ **Principle 2:** The consumer has free choice. Messages and choices are processed selectively. The frequency of these messages is increasing daily. Those that are not felt to be relevant are ignored, disregarded, or forgotten.
> ■ **Principle 3:** Consumer behavior is a process. The specific act of buying is only an intermediate stage in that process. There are many influences on consumer behavior both before and after purchase.
> ■ **Principle 4:** Consumer behavior can be influenced, but only if we address perceived problems and potential needs and wants.
> ■ **Principle 5:** There is a need for consumer education. In all their wisdom and purposeful behavior, consumers may still behave unwisely, against their own interests. Marketers have a responsibility in this effort.

consumer behavior A process a consumer undertakes (either consciously or unconsciously) when making a purchase decision. Marketers must recognize that consumers are goal oriented, have choices, and can be influenced.

CHARACTERISTICS OF CUSTOMERS

Needs and Wants

Abraham Maslow was a psychologist who wanted to explain how people are motivated. What he learned was that motivations are based on different needs in different contexts. He labeled his theory of motivation the hierarchy of needs.[1] **Maslow's hierarchy of**

Maslow's hierarchy of needs Psychologist Abraham Maslow's premise that people are motivated by differing levels of needs and that lower-level needs, such as hunger and thirst, need to be satisfied before higher-level needs, such as self-esteem, can be met.

needs model has stood the test of time and is the basis of much of what we know about human behavior. The model is shown in Exhibit 8-1.

The basic idea of Maslow's hierarchy is that lower-level needs have to be met before the higher-level needs become important. Thus, until the physiological needs of hunger and thirst are satisfied, they remain primary in human motivation. Once these needs are satisfied, our safety needs of security and protection become primary, and so forth on up the pyramid. Of course we will not all act in exactly the same manner, but it has been shown, in a general sense, that the order succeeds.

Maslow *did not claim that the hierarchy was completely rigid or necessarily exclusive*. In fact, it should be noted that we may seek to satisfy two or more diverse needs at the same time; for instance, reserving a hotel suite instead of just a room might be an attempt to satisfy needs at opposite ends of the hierarchy. Or, in another marketing sense, we might *need* a room but *want* a suite—the guest room satisfying a lower-level need and the suite satisfying a higher-level want (if we can afford it!).

Application of the Theories

Do we have to know a psychological theory to know that when we say we are "starving," the first thing we want to do is eat; or that we have higher-level needs of belonging, esteem, and self-actualization? The reason the answer is yes will become clearer when we get to Chapter 13's discussion of segmentation and target marketing, both of which mean pinpointing particular customers. In the meantime, let us consider the need–context relationship.

Businesspeople who travel have the need to sleep, shower, work in the evening, change their clothes, and perhaps watch a little TV. These are basic needs, so almost any hotel will satisfy them. But they also may have the need to reward themselves and will select a hotel to fulfill that need. For instance, one of the authors of this textbook was teaching in Singapore and staying at a very nice three-star property where the lectures took place. When the teaching finished, he checked himself into the InterContinental Hotel in town to sleep in a comfortable bed. (He enjoyed his pillow so much that he bought one.) The nice hotel was the reward for the hard work.

At the same time, people will have other needs, such as a desk to write at, good lighting to read by, convenient Internet service, a on time wake-up call, and the ability

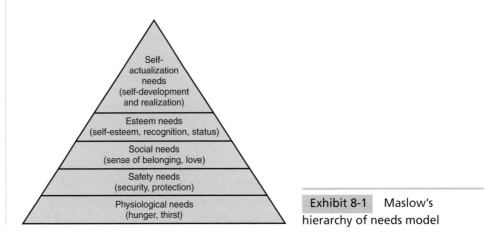

Exhibit 8-1 Maslow's hierarchy of needs model

to order room service. These are not needs in the sense of Maslow's hierarchy; they are needs, however, in the sense that the consumer has a problem that needs to be fulfilled. Consumers seek solutions to these problems and are willing to pay for them (or have their companies pay).

Beside these needs, consumers have wants. They want the bed to be comfortable, the room to be large, a comfortable chair to sit in; they want to be able to see the TV while lying in bed, to be able to have breakfast in the room, and the front desk to have their reservation so they can check in and check out without any hassle.

They definitely do not want to stand in line, to wait an hour for breakfast to be delivered, to have the telephone ring 15 times before the operator answers, and to hear the housekeepers yelling at each other in the hallway. In this sense, customers have "do not wants," which really means that their needs are not being satisfied.

Consider the businessperson who arrives home on Friday after spending a hectic week meeting with clients out of town at a hotel. She says to her significant other, "Let's go away for the weekend. I just need to relax." Her basic needs haven't changed; she still needs to sleep, shower, and change clothes, but the context has changed. Now price could be a factor, but it could be a high cost or a low cost. She may want a low-cost room because she is saving for a specific item or does not plan to spend time in the hotel. Instead she plans to be hiking and visiting museums. There is no need or less need for a desk because she is on holiday, not working. She still doesn't want housekeepers yelling in the hall. The hotel restaurant that was perfect for entertaining her clients is now too expensive and the service is too slow for her.

She may be prepared to pay a high cost because she wants to pamper herself. She wants the property to have a spa, fitness facilities, and a nice restaurant with a nice wine collection. In short, the needs are the same—a place to sleep, shower, and eat. What have changed are the wants and problems to be solved. In the first case the problem was to find a hotel near museums and hiking trails, and in the second case the problem is too little "me" time, which can be cured by being pampered.

Clearly, management must understand the needs hierarchy, the wants that go with each level of the hierarchy, the "problems" of given people, and the context in which they will consume the hospitality product. Still, needs and wants cannot be generalized across the entire population and, especially, the international market. Consider, then, what hoteliers and restaurateurs might like to know of the needs and wants of women versus those of men, of tour operator planners, and of self-employed business travelers, just to mention a few of the broader possibilities.

Maslow's hierarchy of needs is a critical foundation of human behavior. At the same time it is only a foundation on which we must build. Motives activate people's behavior, but perceptions determine the course of that behavior.

THE BUYING DECISION PROCESS

We stated at the beginning of this chapter that consumer behavior is a process and many influences affect this process both before and after the act of purchase. We will now

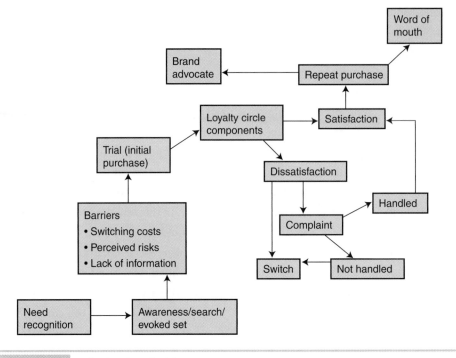

Exhibit 8-2   The buying decision process

examine the buying decision process as shown in Exhibit 8-2 in some detail. Only by understanding the process can the marketer hope to influence it.

Needs, Wants, and Problem Recognition

The buying decision process normally begins with needs, wants, and problem recognition or identification. Sometimes the need, want, or problem recognition will come as a response to **stimuli,** for example, a TV commercial, an Internet advertisement, or a billboard on the highway.

At the beginning of the buying decision process, a, consumer knows, or thinks, he has a problem and begins a search for a solution. This may be very subtle. The consumer does not suddenly jump up and shout, "I have a problem." In fact, he may not even think of it as a problem; he might simply say,

> "I'm hungry [problem].
> Let's eat [solution]."

On the other hand, he might think,

> "I'd really like to go to a quiet [want]
> place for a good [want]
> dinner [need] tonight.
> Where should we go [problem]?"

stimuli In a marketing context, this refers to advertisements, reference groups, and so forth, that might incite or quicken the actions, feelings, or thoughts of a potential or repeat customer.

Search Process

Having determined a need, want, or problem, the consumer begins the search for a solution. He may simply search his memory, he may ask others, he may look to the newspaper or the telephone directory, he may search the web, or he may do any number of other things to obtain either new information or additional information. He may do this in a split second if there is a suitable restaurant right around the corner, or he may take a year to do it, as in planning an annual vacation. He may give the task to someone else, such as an assistant, a secretary, a spouse, an airline, or a travel agency. The set of places that initially come to mind through the research is called the *evoked set*.

In the preceding example the consumer has recognized the problem without help. Marketing has had little, if any, impact on that recognition. Once the problem has been recognized, however, marketing can begin to take an active role. Where consumers go for information, what they read or hear there, whether the information was already in their memory, for whatever reason—marketing can have an influence on this part of the process.

At other times the problem may arise not through recognition, but through identification. In this case, the consumer is unaware that she has a problem until it is presented to her and identified for her. Now marketing has a role right from the beginning. Let us say that this consumer plans, as usual, to eat at home tonight. She is reading the newspaper or watching television when suddenly she sees an ad for what appears to be a nice, quiet restaurant. She thinks, "Boy, wouldn't that be nice for a change." Her wants and problem have been identified for her. Marketing has not created a need—that was already there. Marketing has created a want and caused a problem that needs a solution.

These simplistic examples can be applied to the beginning of any purchase decision process, whether it is for an ice cream cone or a yearlong trip around the world. For simple purchases the process may be totally subconscious, or parts of it may be skipped. For similar decisions that have been made many times before, the process may be instantaneous because of what has become a learned reaction. Of course, the process could also end in a decision not to purchase. This might be the case when marketing has not done its job adequately. Regardless, the role of marketing is apparent even at this early stage. It is also apparent that if marketers want to affect the process at this stage, they must be aware of the complexities of the decision and the influences that will modify it.

Stimuli Selection

Sometimes consumers determine their own problem; other times their problem recognition is caused by a stimulus. The degree of impact of this stimulus, as well as the intensity of the entire search process, is determined by the level of involvement the consumer has with the purchase decision. There are two levels of involvement: high involvement and low involvement.

High-involvement cases are those in which the decision has high personal importance or significance to the consumer, such as a high risk of making a wrong decision, a high effect on self-image, or a high effect on health. High involvement does necessarily mean high cost. Sugar is a very low cost product, but for those with a weight problem whether to eat it is a high-involvement decision. Obviously, the impact and intensity is greater in cases of high involvement.

Low-involvement cases are those in which the decision has little personal importance or significance to the consumer. With low-involvement decisions, the process proceeds very quickly and some stages may be skipped, especially when information is readily at hand. Regardless of high- or low-involvement decisions, consumers are affected only by stimuli that they selectively choose.

It should be noted that the same purchase choice may be of high risk to one person, but low risk to another.

Selectivity

The process of selective choice represents a hierarchy. That is, the steps are taken in sequence, or dropped at any point, as follows:

> Selective Attention: We attend only to what interests us. Advertisers may use graphics or headlines to get this attention.
> Selective Comprehension: We try to comprehend what is still of interest.
> Selective Acceptance: We accept or reject what we comprehend.
> Selective Retention: We retain in memory what we want to remember.

Because consumers selectively attend, comprehend, accept, and retain messages, we should be aware that much of what we direct at them does not sink in. Unless we can bombard them, as in the case of McDonald's, we need to be certain that what we want them to select is directed in a manner that appeals to their needs, wants, and problems. This saying applies not only in advertising but also in personal selling, in-house merchandising, public relations, and any other way consumers gather information or marketers communicate it.

Perceptions

perceptions Meanings we assign to what we see, hear, and sense around us, which are influenced by sociocultural and psychological forces. Perceptions are images we have, and images influence purchase behavior.

Perceptions are meanings we assign to what we see, hear, and sense around us. Sociocultural and psychological forces, as well as our needs, heavily influence our perceptions. Perceptions are selective. We cannot possibly perceive all the stimulus objects that are

Tourism Marketing Application

After the tsunami disaster of 2004, many tourists stayed away from Indian Ocean destinations, even though not all the destinations were damaged. The belief, however, was that all destinations were destroyed. Firms in nonimpacted areas had to change people's beliefs so that much-needed tourist currency could flow into the region.

presented to us, so we select what we want to perceive. If you are looking for a honeymoon spot, you select to perceive the beach, the patio, the quiet, and the moon from the brochures, ads, or materials presented you. If you are looking for a spot for your company's next sales meeting, you select to perceive the meeting rooms, the banquet facilities, and the sports facilities from the very same brochures, ads, and materials. When you are at a hotel, you may select to perceive the decor, elevator service, bar, golf course, or anything that you felt was promised you. Perceptions are images, and images influence purchase behavior.

Sociocultural forces that influence our perceptions include the culture of society, social class, and small reference groups, among others. A **reference group** may be defined as people who influence a person's attitudes, opinions, and values, such as family, friends, and business associates. Reference groups are especially important in the purchase of hospitality services because word of mouth recommendations play a major role in the buying decision. Psychological forces that influence consumer behavior generally come from within a person and include learning experiences, personality, and self-image.

reference group People who influence a person's attitudes, opinions, and values, such as families, friends, and business associates.

For the consumer, perception is reality. Perhaps one of the greatest mistakes we can make as marketers is thinking that what we perceive is also what the customer perceives. If the customer doesn't perceive it, it doesn't exist. If the customer does perceive it, it does exist. You cannot make something what it is not by simply saying so; you have to make sure that the product matches the perceptions you are trying to create. If it does not match, we have Gap 4, discussed in Chapter 2.

Hotels and restaurants that brag about how great they are must be able to live up to their boasts. Exhibit 8-3 is an example. When they promise the moon and then fail to deliver the moon (in fact, are bewildered when the customer asks for the moon), they have defeated their own purpose: They have alienated a customer. Expectations arise from initial perceptions and may be disconfirmed by subsequent perceptions.

Initial perceptions depend on stimulus factors. This is the area of traditional marketing. Consider a resort hotel brochure that illustrates an indefinitely long (as far as the eye can see) stretch of white sandy beach, a quiet remote setting, elegant dining on your own private patio overlooking the ocean, and a romantic full moon. These are stimulus factors and your perception is, What a perfect place for a honeymoon! You book your honeymoon with great expectations. When you arrive at the resort, you find that the beach is the size of a postage stamp. The beach in the brochure is on the other side of the island. When you check in, you find that a 300-room convention is checking in ahead of you. The remote setting is in the flight path of the airport. The private patio overlooking the ocean is in the $800 suite, which you didn't reserve. You can eat in your room or in the enclosed dining room without a view.

Now, reality is perception; expectations were not fulfilled and your perceptions are negative. You go home and spread negative word of mouth. These differences in perceptions create many problems for service marketers. Therefore, they must deal very acutely with perceptions. Marketers must create images with the stimuli related to the specific target market they are trying to attract. They must use stimuli that are appropriate for that market, and they must be certain that reality equals, or almost equals, expectation, so that reality doesn't negatively influence perception. Failure to do this will

| Exhibit 8-3 | The hotel provides a "happy ending" to its guests |

Source: Outrigger Hotels & Resorts.

create a dissatisfied customer and negative word of mouth. Recall that it is not enough just to create a customer; it is necessary, as well, to keep the customer.

Alternative Evaluation

Rarely is there only one possible solution; the consumer often has to evaluate alternative solutions. Reference groups and other evaluative criteria have strong impact at this stage. More important, marketers must ask themselves how well they have done their job. Has the case been well presented? Does the solution seem practical? Is the risk worth it? Is the price–value relationship appropriate? Does it cover the necessary needs and wants? What is the word-of-mouth reputation? Is it different or better than the other alternatives, and if so, why? These are the thoughts that are going through the mind of the consumer. Again, the higher the involvement, the longer and more deliberate the process and the greater the search for more information.

The marketer's most important impact is probably at this stage, at least for high-involvement purchases. The level of involvement will vary with the individual. Companies want you to have high involvement with them.

As the service element becomes more important, the level of involvement increases. But the involvement level in choosing a hotel or upscale restaurant is always relatively high, because the product is consumed as purchased, unlike a good, which can

be returned if it doesn't work. It is also because the entire hotel experience is personal; if it's a bad experience, it affects the user personally.

Hospitality purchase choices often include many elements. There are the obvious elements such as price, location, accessibility, reputation, and quality. There are also the less obvious elements such as service, ambience, attitude, newness, and other clientele. Researchers have developed different models to explain how consumers compare alternative choices. One assumption is that consumers will make trade-offs of one characteristic for another; that is, a weakness in one attribute can be made up by a strength in another. For example, we might select a hotel on the outskirts of town (weak in location) because the price is very reasonable (strong in price). Another model would have the consumer establish a minimum level on only one or a few attributes—for example, price. These choice models and others require that the marketer perform consumer research to determine the target market's choice process.

Beliefs

Beliefs are thoughts we think are facts; they derive from perceptions and are the beginning of the **belief→attitude→intention trilogy.** We attach a belief to an object. An object could be a restaurant, and a belief could be that it is expensive. Whether or not the restaurant is expensive is secondary to the belief. Beliefs exist in the mind regardless of where or whom they came from. If beliefs are accurate (if the restaurant is expensive) and we want consumers to have that belief, then we do not need to change the belief.

Sometimes, however, marketers want to change or create beliefs. The restaurant is really not expensive, we say. But how can we say that? For some people it may be very expensive, and for others it may be quite inexpensive. The solution lies in the definition

belief A thought we think is fact, which we derive from perceptions.

belief→attitude→ intention trilogy A process consumers go through once they have formed perceptions of a product but have not actually purchased it. A belief is a thought the consumer thinks is fact; attitude refers to emotional feelings about that fact; and intention is the plan to make (or not make) the purchase.

---

* Note that this trilogy is often stated "belief → attitude → intention," but in many cases this is not so. For example, we may intend to do something and then develop beliefs and attitudes about doing it.

of the target market. These are the people we want as customers; what are their beliefs? We have to learn this before we decide whether we want to change those beliefs.

The same is true if we want to create beliefs, as for a new restaurant. Creating beliefs, however, is much easier than changing them. If there are no beliefs, there is essentially a blank slate and all that needs to be done is to fill it. When we want to change a belief, we have to both get rid of the old one and replace it with a new one. This is why it is important that we create the right belief in the first place.

Attitudes

attitude How we judge and react to beliefs; emotional feelings toward beliefs.

Attitudes are the emotional component of the belief→attitude→intention trilogy that consumers often follow. They are the subjective feelings toward the belief. Attitudes are tendencies to respond to beliefs. If you believe that a restaurant is expensive, how do you actually feel about going there? In a sense, this is the application of our beliefs; this is how we judge our beliefs and how we react to them.

Let's assume that our restaurant is expensive and our target market believes this. There is no point here in trying to change the belief, because it is true. Yet people are not coming to the restaurant because of their response to their belief that it's too expensive. Unless, of course, we want to change the restaurant to lower prices (and research may show that to be the only possible course of action, given this market), what we have to change is consumers' attitudes. One way to do this might be to try to persuade people that the restaurant is expensive but worth the price. If we could succeed in this effort, we will have changed consumers' attitudes toward the restaurant while they maintain their belief.

Research conducted by Coca-Cola found that a significant majority of the 40,000 people who taste-tested "new" Coke against "old" Coke preferred the new variety. When they switched their formula to the new Coke, the market revolted. What Coca-Cola's research did was to measure beliefs and ignore behaviors and attitudes. People believed that new Coke was better, but not better enough to make the change because their attitude toward the old Coke was so positive. Changing to a new product just did not seem right.

Intention

intention A consumer's plan to make or not make a purchase.

The final stage of the belief→attitude→intention trilogy* is the stage when the consumer has the **intention** to purchase (or not purchase) a product. This is not actual behavior, but it may be as close to behavior as we can get, as will be discussed shortly. There is no way we can positively be assured of behavior until after it happens. Failing this, we want to know what people intend to do.

Let us assume that we believe a certain restaurant is expensive, but worth it. We have a positive disposition toward this restaurant. Do we intend to go there? No! We can't afford it. Our positive attitude thus turns out to mean nothing. Of course, the specific context may change: Would you intend to go there on your 10th wedding anniversary next week? Perhaps now the answer is yes. Clearly, context can change behavior, or at least intended behavior. Also, you can see that simply asking people what they intend to do can be very misleading without also measuring belief, attitude, and context.

Barriers to Purchase

Intention is followed by actual behavior (which could include doing nothing). As shown in Exhibit 8-2, before making an actual purchase, the consumer needs to work through the potential barriers to purchase. Marketers attempt to reduce the barriers to purchase of their product and raise the barriers to purchase of competitors' products. For instance, familiarization trips with meeting planners are a way to introduce the product and service to meeting planners before they bring their groups. Starwood's 100 percent guarantee is another way to reduce the perceived risks to purchase. Information provided on the Internet is an additional way to reduce barriers to purchase. Restaurants often post their menus outside their restaurants as a way to reduce the risk of customers realizing they do not like the menu until they are seated.

One barrier to purchase is **cognitive dissonance,** which is a state of mind in which attitudes and behaviors don't mesh.

> "According to cognitive dissonance theory, there is a tendency for individuals to seek consistency among their cognitions (i.e., beliefs, opinions). When there is an inconsistency between attitudes or behaviors (dissonance), something must change to eliminate the dissonance. In the case of a discrepancy between attitudes and behavior, it is most likely that the attitude will change to accommodate the behavior.[2]

For example, suppose we book a hotel room at an expensive hotel (behavior), but we believe we should not spend the money for the hotel (belief). This state causes us to have second thoughts or doubts about the choice we made. This is especially true when the choice was an important one psychologically, or financially, or both, and when there were alternative choices with a number of favorable features. To remove the cognitive dissonance, we convince ourselves that it is okay to spend the money because we deserve a nice hotel; or, we reserve at a different hotel. Because a gap usually exists between the time the guest makes the reservation and the time he checks in at the front desk, there is a lot of opportunity for the guest to change his behavior. The role of marketing is to make the guest believe he has made the correct decision. This can be done any number of ways, such as sending an e-mail reassuring them that they made the right decision.

Research has shown that people try to reduce cognitive dissonance by seeking or choosing to perceive information that supports the correctness of the decision, by finding fault with the alternatives so that they look less favorable, and by downplaying the negative aspects of the choice and enhancing the positive elements. Advertising that supports the choice or personal communication that commends the wisdom of the choice has been found to be helpful in reducing cognitive dissonance and increasing loyalty.

cognitive dissonance
A state of mind in which attitudes and behaviors don't mesh; a customer's potential uncertainty after making a purchase.

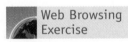 Web Browsing Exercise

Go to www. learningandteaching.info/ learning/dissonance.htm to learn more about cognitive dissonance. Who first investigated this theory, and under what applications? How can cognitive dissonance apply to learning in school? How can we overcome cognitive dissonance?

Outcomes—Satisfaction or Dissatisfaction

Once the choice is made and the purchase takes place, an outcome will occur. What is the outcome? Does performance match expectation? Is perception changed? We have managed to create a customer, but have we managed to keep one? Will she come back? Will she tell others? Is the new customer satisfied, dissatisfied, or just so-so? All the components of the loyalty circle that were discussed in Chapter 4 come into play here. This initial trial is the opportunity to create a loyal customer.

We could never hope to know whether every customer left satisfied or dissatisfied, but we certainly should have a good idea of their overall level of satisfaction. With individuals, we should randomly sample to find this out; with groups and large parties, we should have some contact with some members of each. Remember, we want to keep these new customers. As much as we can, let's follow up with them. If they are satisfied, let's find out why. Maybe it will teach us something.

If our customers are not satisfied, why? What can we do about it? Can we still get them to come back? Can we correct the problem? In Chapter 4, we discussed customer complaints and service recovery. The strategies and tactics discussed in that chapter come into play at this point of the service. How well we execute service recovery will depend on whether the customer continues up the loyalty ladder or jumps off.

Now, let us apply the consumer behavior process to the hotel industry. The process shown in Exhibit 8-4 is one of choosing a vacation destination but it is easily adapted to other choices such as for a restaurant. A look through the steps will show how they fit the elements of consumer behavior we have discussed thus far. This analysis will make the theory more practical. You might even attempt to fit the model to your own mental process on a recent or proposed purchase. Better yet, think of how marketing could affect each stage of the process. You will find advertisements (only one phase of marketing) throughout this book that address each step.

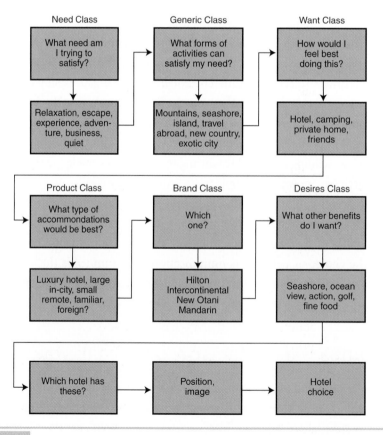

Exhibit 8-4   Consumer mental evaluation process (conscious and unconscious)

Exhibit 8-4 is an oversimplification of a very complex process. In fact, this process is so complex that only one's own mind can process it for oneself, as it is full of many different variables. The process in an overall consumer behavior sense, however, is important and marketers should understand it. Here, we tie them together to understand how they are interrelated. So far we have discussed hospitality customers and their purchase behavior in a general sense. Now we will look at some specific types of customers, as commonly defined in the hospitality industry.

TYPES OF HOSPITALITY CUSTOMERS

There are various ways of grouping hospitality customers based on common needs and wants. This will be discussed in greater detail in Chapter 13 and 14 addressing market segmentation and target marketing. Here we will try to understand these customers in broad category types so that we can make an effort to influence their purchase behavior. In the Marketing Executive Profile, Tom Storey, vice president, marketing for Fairmont Hotels, nicely articulates how this company broadly separates its individual customers into business and leisure categories.

Business Travelers

The business traveler market segment is one of the most desirable for the hospitality marketer. This market consists of over 50 million travelers a year in the United States alone. It is not only the largest major segment, but it is also considered the least price-sensitive market available. The **business traveler** is defined as a customer who is using the product because of a need to conduct business in a particular area. Although the hotel or restaurant facility may be used during business, the facility is not the sole reason for the buy. Purposes of the business trip include company-related business, consulting, sales, personal business, and the need to fulfill managerial functions. Most of these travelers spend between two and three nights away from home on each trip.

> **business traveler** A customer who purchases hospitality products or services because of a need to conduct business in a particular area.

On the surface, the needs of this market are simple; in practice, they are far more complex and not so simple to deliver. One thing is certain: The business traveler group contains the greatest "demanders," which is best explained by considering the nature or purpose of their travels. They would rather be home, they may have had a bad flight or business dealing, they are quickly in and out, and they want everything to run smoothly.

Business Traveler Needs
Whereas business travelers (like others) once complained loudly about the small towels and small bars of soap, they now have other things to complain about. The industry has changed radically and so has the customer. In the past, a hotel provided a place to sleep while a customer was on a business trip. Today, a hotel has to provide the services for a successful business trip, and may just happen to be a place to sleep as well. Research has shown that when business travelers are asked what their first consideration is when selecting a hotel, convenience of location receives the highest response. This is followed by reputation and price. It is useful to go through the actual decision process.

Because business travelers consider location first, many hotels emphasize it in their advertisements. If the location is inappropriate, it's out of the running. This rarely happens because they look at location first, and only then at what hotels are situated within that location. Many hotel companies try to circumvent the location issue by offering strong "bonus awards" to their customers in an attempt to influence the location choice of the business traveler.

Business travelers look next at rate ranges because of company rules or personal limitations. They choose by product class, assuming that hotels within the product class are in the appropriate rate range, be it the upscale, middle level, or budget. If the product class desired is middle-level (Ramada), and the only other available choice is upscale (Four Seasons), then price may be the most important factor in the decision, including location.

Many hotels have what are called corporate rates. To get them, you have to work for an organization that has contracted a certain number of rooms for a specific rate. These are not necessarily the lowest rates but often are better rooms that are better furnished for the business traveler at a discount from the rack rate. Some are on concierge floors, where, at a higher price, special services, a lounge, and complimentary continental breakfast are available. The concept is that the business traveler will pay more for less hassle.

Other concerns for the business traveler are check-in lines, employee attitudes, deferential treatment, lighting, skirt hangers, mirrors, security, type of clientele, coffee makers, business services, noise (some business travelers avoid convention, atrium-lobby hotels and prefer more boutique, smaller properties), operational efficiency, Internet access, limousine service to the airport, and a host of other things.

On the whole, business travelers who are choosing a hotel do not consider bathroom amenities, shoe polishers, bathrobes, turn-down service, chocolates on the pillow, and other such factors, except, perhaps, in luxury hotels where they are expected. These may be nice "extras" but not critical, although customers have come to expect certain amenities, such as a decent size bar of soap. Goat's milk shampoo, herbal soap, and bubble bath are mostly take-home items. Even when some are used, their absence wouldn't be considered serious. These travelers are more concerned with how the shower works.

For many hotels superfluous amenities have become a cost they can no longer afford to provide at the prices travelers are willing to pay. A better way, perhaps, is the approach now taken by some hotels to provide amenities only when really needed, as shown in Exhibit 8-5. It appears the clear winners in the future will be hotel management which is truly committed to customer satisfaction, value, and consistent provision of the basic lodging fundamentals," plus the availability of things that certain business travelers want. Many hotels, for example, now offer special "business plan" rooms as shown in Exhibit 8-6. Further, simply having a swift, friendly check-in, with all of the information correct the first time, solves many problems and greatly improves customer satisfaction.

Most business travelers visiting cities do not consider hotels' restaurants as a determinant factor, simply because there are usually numerous alternatives available. A good breakfast room is assumed, and a quick and easy "grazing" restaurant open all hours is desired; having other restaurants in a hotel is considered convenient, sometimes, but not totally necessary. A majority of city hotel customers, in most developed countries, eat out for lunch and dinner. This somewhat contradicts the idea of convenient location, which tends to reappear when staying at a roadside hotel. (This does not

Exhibit 8-5 Renaissance Hotels & Resorts provides amenities that customers really want

Source: Marriott International, Inc. Used by permission.

mean that an upscale hotel should not have good restaurants, but that they are seldom the determinant in the choice of hotel.) These are generalizations. As we have said, each hospitality establishment has to know its own target market.

Exhibit 8-7 shows what business travelers are looking for in a hotel/motel experience and some differences between men and women. Note that basic cleanliness, friendly and efficient service, and safety are at the top of the list.

Dealing with the Business Segment While the corporate office is saying, "Raise the rates," the local marketing team frequently believes that lower pricing is the way to capture new customers or keep existing customers. Naturally, if a hotel is significantly higher or lower in price than its competition, a choice may be made on price, but this alone is not the answer—and there is no one answer. What it is necessary to know is the appropriate price range for existing market conditions. This is not a random decision, as is sometimes assumed. A hidden conflict often exists in this regard between the local sales department and the corporate office of hotel chains. Consider the following example:

A 1,400-room hotel was undergoing a three-year, $100 million renovation program. This had caused considerable customer discontent and some serious loss of

| Exhibit 8-6 | Westin shows that a hotel room is much more than a bed—it's a place to work out |

Source: Westin Hotels & Resorts. Used by permission.

| Exhibit 8-7 | What Business Travelers Are Looking for in a Hotel or Motel Experience |

Gender Differences' 2004	Men %ᶜ	Women %ᵇ
Attributes Considered Extremely/Very Desirable:		
Mastery of Basics:		
Clean/well-maintained guest rooms	97	99
Friendly and efficient service	93	96
Safe place to stay	84	96[a]
Pricing/Value:		
Free local phone calls	71	83[a]
No long-distance access charges	69	75
Complimentary shuttle service to and from the airport	63	74[a]
Complimentary area shuttle service to places other than the airport	60	75[a]
Complimentary breakfast included with the nightly room rate	52	67[a]
Provides complimentary newspaper daily	54	59
Offers free premium movie channels like HBO and Cinemax	53	55
Streamlining/Simplification:		
Offers express check-in and checkout	79	87[a]
Offers an interactive system through the guest room television set that provides information about local events, activities, dining, entertainment, and news	38	45

business at the same time that the economy had turned down. Although hotel management believed there would be no problem selling the 600 renovated rooms at a higher rate, there was a problem with the rest of the hotel and the disrupting construction. The corporate rate for this hotel was $144. The corporate rate of the

Exhibit 8-7	What Business Travelers Are Looking for in a Hotel or Motel Experience

Gender Differences[c] 2004	Men %[a]	Women %[b]
Business Services/Command Center Concept:		
Business services (e.g., copying, faxing, etc.)	NI	NI
Wireless Internet access in the guest rooms	53	62
Free high-speed access to the Internet from guest rooms	48	52
Computer data ports in guest rooms	49	47
Personal service enabling you to send and receive faxes from your laptop	43	50
Wireless Internet access in public areas such as the lobby or pool	32	40
Fax in room	27	30
High-speed access to the Internet from guest rooms at a cost of $10 or less per night	25	26
Multiline telephone in guest rooms	25	20
Creature Comforts/Ambiance:		
Casual atmosphere, three-meal restaurant	66	71
Diversity of restaurants on premises	60	65
Casually elegant atmosphere and décor	56	69[a]
Fine-dining restaurant	50	53
Formal and elegant atmosphere and décor	47	50
24-hour room service	45	48
Exercise facilities	41	50
Delicatessen on premises	36	43
Spa services	30	43[a]
Concierge or executive floor	31	37
Bar or lounge	34	28
Sports bar	37	20[a]
CD stereo system in room	28	29
Small "boutique" hotel with unusual architecture and décor	20	31[a]
VCR in room	18	29[a]
Live entertainment	18	19
Minibar in guest rooms	16	14
Perks:		
Hotel frequent-guest points	45	53
Airline frequent-flyer points	44	48

a. *statistically significant difference from men.*

b. *Asked in versions of the questionnaire. Interview base varies.*

c. *Top two box score on a scale of 1 to 5 where 1 equals not at all desirable and 5 equals extremely desirable.*

NI *Not included due to insufficient n size.*

Source: Yesawich, Pepperdine, Brown & Russell. (2004). *National Business Travel MONITOR,* 97–98.

major competitor was $135. The corporate office of the hotel under renovation, in the midst of this situation, mandated that the property increase its corporate transient room nights sold from 60,000 to 65,000. At the same time, it mandated a corporate rate increase to $158 *while* the disrupting renovations continued. This mandate left the property sales staff in an impossible and noncompetitive situation and caused the eventual loss of a considerable amount of business, much of which was never regained.

For the restaurant industry, business travelers mean expense account travelers. Like hotels, a sizable number of restaurants would not be in business today were it not for these customers. Although restaurant meals became only 50 percent tax deductible in 1993, this issue is not the primary factor in the eating-out decision.

Restaurants that serve the needs of this market well will prosper even if the tax laws are changed. No logical businessperson would decide not to take a good customer to lunch for lack of a $25 tax deduction. On the other hand, restaurants have needed to adjust menus and prices with more creativity to charge lower prices, create greater value perception, or both.

The hospitality industry worldwide cherishes the business travel segment, which, sometimes or in some cities, is simply not large enough to go around. The property that gets its fair market share, and more, will be the one that truly understands the needs, wants, and problems of this market. The business traveler segment is not homogeneous. All of its members do want convenience of location and cleanliness. The irony is that some want price, some want service, some want room appointments (such as a large desk with good lighting), and some want a number of other things. Separating the "somes" is the essence of target marketing discussed in later chapters.

pleasure traveler A customer who purchases hospitality products or services for leisure or other nonbusiness purposes.

Pleasure Travelers   The leisure market is comprised of **pleasure travelers** that individually, in couples, in families, or in small groups visit hotels and restaurants for nonbusiness purposes. They may be traveling on vacation, but often are not. Many, of course, are weekend or other package users. Others travel to cities for shopping, visiting friends, going to the theater, "just for a change," personal business, and other purposes.

Increasingly, many pleasure travelers and business travelers are becoming the same person. Many businesspeople extend their trips to see the sights of a destination or just plain relax. One survey of a Wall Street (New York City's financial district) hotel's weekend guests found that 95 percent were originally there because of business.

In the restaurant business, this segment includes those who eat out just for pleasure. In many cities, both large and small, this is a powerful segment with many diverse needs and wants. Not as constrained as the business customer by having to "get back to the office," the pleasure diner tends to be more relaxed and casual. At the same time, because the primary purpose of being there is to eat and socialize, these diners have more time to be critically conscious of the product or service delivery.

Some pleasure travelers use the best hotels and visit the best restaurants. In recent years, however, there has been a growing trend toward short pleasure trips and frequent dining out at less expensive properties by those with limited budgets. This has considerable impact on the hospitality industry both in expanding the industry and in creating the need to better serve this market.

When traveling to some tourism destinations, such as the islands in the Caribbean, some leisure customers want more than sun, sand, and sea. The island of Barbados has tried to capitalize on this, as shown in the Tourism Marketing Application, "Marketing the Island of Barbados in the Caribbean."

Many destinations have recognized the value and significance of tourism, and there is intense competition among countries and states to attract the pleasure traveler. For example, residents of Louisiana are exposed to advertising campaigns from the states of Texas, Arkansas, Mississippi, and Tennessee, all of which compete heavily for visitors from

Tourism Marketing Application

Marketing the Island of Barbados in the Caribbean

Gary Leopold, President,
Irma Mann Associates, ISM Marketing

Our agency is currently managing the marketing for the country of Babados. We came up with the tag line "authentic Caribbean experience." Our research showed that the consumer had very little knowledge of any Caribbean island's actual location, that all the islands were perceived as having somewhat the same sun and sand. Or, that Barbados is a destination perceived as far away and expensive. There's a certain truth to that, and a lot of what we have done in our marketing is turn that into an advantage—that this is a destination that takes a little longer to get to and it's farther away, but that is also what has allowed it to stay so authentically Caribbean. It attracts a certain type of traveler, so it is not your typical Americanized destination, like the Bahamas. Not all customers want McDonald's, Gold's Gym, and other American icons at their Caribbean destination—not that this is a bad thing. Some customers want a different experience, so we began to think about the island of Barbados as hard to get to and hard to leave. We discovered that the oldest synagogue in the Western Hemisphere is on this island. "God's trail" gives visitors tours of the ancient churches and houses of worship on the island.

We work with a triangle approach. Imagine three circles: customer needs, product offering, and guest experience. At some point, these circles intersect, and that's what we market. We develop a list of all of the rational reasons for choosing the destination and then make a list of all of the emotional reasons for visiting. Many times they are not the same.

---

that area. Malaysia seeks Singaporeans and vice versa. Advertising campaigns have raised customers' awareness of their many vacation choices such as that shown in Exhibit 8-8. Thus, demand for hospitality services is being created and spurred on by foreign, state, and local governments, which reap their share from taxes levied on visitors. The international tourism market has grown huge and is still growing with many different needs and wants. We discuss tourism in more detail in the next chapter.

The pleasure market is a high-growth-potential market. While the business market remains relatively stable, with businesspeople traveling and eating out when they have to, a large portion of the pleasure market has yet to be developed. This is even truer in countries other than the United States. Many countries have only recently seen a large growth in the so-called middle class with more disposable income. Because they are not "big spenders," however, they are often closed out of a market that caters and prices to the expense account customer. Lower-cost hotels and restaurants in some countries have expanded this market.

A major part of the pleasure market is made up of family travelers. Even in tough economic times a family vacation has become an essential part of many lifestyles. This market is more price sensitive than the business segment and is more fickle about choices of destinations and hotels. Just as hotels must learn the needs of business travelers, however, they must also determine the underlying reasons and needs of pleasure travel. Exhibit 8-9 shows some research results on what people are looking for in a pleasure

Web Browsing Exercise

Use your favorite search engine and type in phrases such as "gay travel," "lesbian travel," "multigenerational travel," and "pet travel." What do you find? What might you recommend to those in the hospitality industry?

Exhibit 8-9   What People Are Looking for in a Leisure Travel Experience

Age Differences[b] 2004	Echo-Boomers %[*]	Xers %[*]	Boomers %[*]	Matures %[*]
Attributes Considered Extremely/Very Desirable:[f]				
Experimentation/Fantasy/Ambiance:				
Beautiful scenery	82	91	87	88
A place I have never visited before	83	87[c,d]	78	76
A beach experience	71	74	69	49[a,b c]
An opportunity to eat different and unusual cuisines	58	66	57	53
A hotel with casually elegant atmosphere and décor	52	55	54	40[c]
Nightlife and live entertainment	73[b,c,d]	56	45	47
The option of scheduling vacation activities in advance of arrival	54	56	49	51
A hotel with a historical atmosphere and décor	43	50	42	43
Going to theme parks	63	58	40[a,b]	18[a,b,c]

Exhibit 8-9 What People Are Looking for in a Leisure Travel Experience

Age Differences[h] 2004	Echo-Boomers %[e]	Xers %[e]	Boomers %[e]	Matures %[e]
A destination that is remote and untouched	38	65[a,c,d]	44[d]	23
A hotel or resort with a distinctive theme or atmosphere	47	45	42	34[a,b,c]
A hotel with a formal and elegant atmosphere and décor	43[c,d]	38	33	31
Going to a spa	47	39	32[a,b]	24[a,b,c]
A small "boutique" hotel with unusual atmosphere and décor	34	29	33	32
Learning a new skill or activity	41	29	31	24[a]
Being able to gamble	29	25	25	23
Physical Activities:				
Getting exercise	61[b,c,d]	42	45	39
Hiking and outdoor adventure	52	52	42[a,b]	25[a,b,c]
Snorkeling or scuba diving	58	48	37[a]	20[a,b,c]
Participation in water sports	50	42	35[a]	9[a,b,c]
Whitewater rafting	38	37	28	4[a,b,c]
Bicycling trips through the countryside	24	29	28	19
Snow skiing	32	23[a]	13[a,b]	6[a,b,c]
Playing golf	15	14	16	17
Mountain biking	29	27	8[a,b]	8[a,b]
Snowboarding	27	15[a]	7[a,b]	1[a,b,c]
Playing tennis	11	5	8	4
Other Activities:				
Participating in activities with children while on vacation[g]	62	77[a,c]	63	39[a,b,c]
Visiting arts/architecture/historical sites	47	51	54	54
A hotel having a kids' club or organized family activities[g]	43	44	36	35
Shopping	42	44	36	35
Familiarity/Control:				
Safety of hotel or motel	75	85	86	83
Safety of destination	69	81	83[a,d]	72
A place I have visited before	54	55	54	60
Having a separate children's/teen program	37	36	28	26
Having access to the Internet or an online service to stay in touch with the home or office	40	27	30	17[a,c]
Pricing:				
An all-inclusive vacation price (one that includes air transportation, accommodations, food, transfer to the hotel or resort, and some recreation)	70	69	64	56[a,b,c]
An all-inclusive resort price (one that includes my accommodations, food, beverage, and recreation)	76[b,c]	67	62	49[a,b,c]

a = statistically significant difference from Echo-Boomers, b = significant difference from Xers, c = significant difference from Boomers,

d = significant difference from Matures.

e Asked in versions of the questionnaire. Interview base varies.

f Top two box scores on a scale of 1 to 5 where 1 equals not at all desirable and 5 equals extremely desirable.

g Asked among respondents who have taken or plan to take one or more leisure trips with children.

h Echo-Boomers = those adult consumers born since 1979, Xers = those adult consumers born 1965 through 1978, Boomers = those adult consumers born from 1946 through 1964, and Matures = those adult consumers born before 1946.

Source: Yesawich, Pepperdine, Brown & Russell. (2004). National Business Travel MONITOR, 97–98.

travel hotel experience. You can contrast this with Exhibit 8-7 to see some of the differences between business and pleasure travel needs for both men and women.

Another important pleasure market is made up of people traveling to visit friends and relatives. Although many of these travelers stay with friends and relatives at their final destinations, they often seek out lodging accommodations along the way. This is generally a value-conscious market that is attracted to budget hotels and eating places such as McDonald's and family restaurants. In these lower-level markets, pleasure travelers are actually less demanding than customers in almost any other market. One reason for this is the lack of experience. Travelers may not realize just what is available, or they may simply not know how to demand. (The exceptions, of course, are the business travelers now turned pleasure travelers.) They do, however, have long-term memories. These customers are likely to simply walk out of a bad experience without complaining, never to return. They also are very likely to spread negative word of mouth. They will become, however, more demanding as their travel experience increases.

Package Market

This increasingly popular method of attracting customers during low-demand periods is becoming more crowded with offerings every day. In the *New York Times* Sunday Travel Section, hotels from the upscale Ritz-Carlton on Central Park and the Carlyle on the Upper East Side, to the Waldorf-Astoria on Park Avenue, to the convention-type Hotel Pennsylvania in the garment district, and the downscale Milford Plaza on Broadway, are all offering weekend packages. The same is true in major cities throughout the United States; at resorts; and in London, Paris, Rome, Athens, Singapore, Bangkok, and just about any other place you look. Packages have also become quite important to online retail travel agents. Some include airfare. Another popular version of packages is the "escape" or "getaway" theme. One of each of these is shown in Exhibit 8-10.

package market
Consumers who purchase a hospitality product offering that includes a combination of services for an all-inclusive price.

The hotel **package market** is defined as consumers who purchase a combination of room and amenities for an inclusive price. Although normally these packages are designed to boost occupancy during low-demand periods, such as weekends and off-seasons, some packages are used to maximize revenues at all times.

Tourism Marketing Application

It is important to understand the motivations for vacations and trips. Although the outcomes of these motivations are reflected in what people are looking for, the understanding may lead to other offerings that are not currently available. Motivations for travel include: (1) to observe the lifestyle of exotic peoples, such as a trip to India to study yoga and meditation; (2) to experience or participate in festivals or living museums, such as those found in Williamsburg, Virginia; (3) to visit historic sites, such as the Coliseum in Rome; (4) to visit natural and environmental attractions, such as national parks; (5) to participate in sports, such as skiing or swimming; and (6) to attend business meetings, conventions, or other functions.

Source: Valene Smith. 1977. *Hosts and Guests.* Philadelphia: University of Pennsylvania Press, pp. 2–3.

Exhibit 8-10   This advertisement for Starwood exhibits an "escape" package
Source: Starwood Hotels Hawaii. Used by permission.

An example of this might be a resort package that includes three nights' accommodations and breakfast and dinner daily. The purpose of this combined package is to ensure that while the hotel is full, the guests are required to make use of the food and beverage facilities. Also, the three nights are sold at once, ensuring their occupancy over the period. If sold individually, one night might sell out before the others, eliminating longer, more desirable bookings. Naturally, the hotel would have to forecast some significant demand to be able to force the customer to purchase this type of package.

We define a package as a bundling of goods and services, be it food and beverage, coupons to a nearby retailer, or a welcome gift upon arrival. Often the term is misused to describe obvious discounting. Offering a guest room at a significant discount is nothing more than that; it certainly does not package anything for the consumer.

An example of this bundling of services is the joint marketing promotions of the New York Palace hotel and the famous retailer, Saks Fifth Avenue. The result was satisfied customers and significant business increases for both parties.

There is an important warning about packages that too often is violated: Provide what you promise in the package! This advice is obvious, but is not always followed, resulting in very negative feedback for the property. For example, we know one small city hotel that offered the usual weekend package. The main appeal of this package in the winter in New England was the indoor pool and lounge area. People from only a few miles away bought

---

the package for that reason. More often than not, however, the hotel had a wedding party function on many, if not most, Saturday afternoons that was held by the pool. Including setup and breakdown time, package customers could not use the pool for much of their stay.

The main reasons hotels do not deliver on the promises made with their packages seem to be that they do not plan for packages and consider them secondary, low-rated business. This results in extremely negative word of mouth. Research on customer complaints has revealed a disproportionate number of complaints about package 'promises.'

Mature Travelers

mature traveler A hospitality customer who is 55 years old or older. This market is generally considered a subsegment of the pleasure traveler market.

The **mature traveler** market, actually a subsegment of the pleasure market, is another important growth segment for both the hotel and restaurant marketer. Usually defined as age 55 and over, this market's size is on the increase as people tend to live longer and better. This segment is important to the hospitality industry not just for its size but for other reasons. People in this segment travel extensively, spending over 50 percent more of their time away from home than the younger pleasure segments. Members of this market today live longer, healthier, and more vigorous lives; are better educated; and have a wider range of interests and activities than their counterparts in earlier generations. Their children are grown; their mortgages are paid; and they have the time, energy, and inclination to travel.

The needs and wants of the mature market are different from those of other segments discussed in this chapter. Studies have reported that "to visit new places" was the number one reason for trips taken by mature travelers, followed by "to visit friends and family." Many mature travelers are price sensitive, and getting a discount is an important attraction. Because they have the flexibility to plan trips any time, they can take advantage of the lowest prices. These travelers use hotels of all price ranges, from luxury to budget, but hotels must be able to provide attributes that are important to this segment, such as increased security, well-lit public areas, readable signage, nonsmoking rooms, easily maneuverable door handles, grab bars and supports in bathrooms, and wide doorways to accommodate wheelchairs and walkers.

The mature market is not homogeneous. This market can be segmented in a variety of ways. Travel habits of mature travelers differ depending on retirement status and travelers' life stages as they grow older and encounter physical restrictions. Some mature travelers prefer to travel as part of a group, whereas others travel in pairs.

Many hotel chains are aggressively pursuing this market. Choice Hotels has long featured famous but active seniors in its television advertising. Hilton has a Seniors HHonors program in which members can receive up to a 50 percent discount on room rates. Radisson SAS once gave discounts by age starting at age 65. Best Western provides the following advertising guidelines to hotel members who want to target this market:

- Keep advertisements and collateral pieces upbeat and positive.
- Always depict older customers as active, healthy, and involved.
- Use language that is sensitive to mature audiences.
- Emphasize convenience.
- Show price–value relationship.
- Stress service, reliability, and savings.

Restaurants also tap into this market. Active senior citizens spend a large proportion of their food budgets on food away from home, and most prefer midscale restaurants. Many are bargain hunters who are conservative in their eating habits. Restaurants should have good lighting to avoid safety hazards, and menus should be easily readable with enough variety to satisfy senior citizens' nutritional needs. Service staffs should be trained to recognize changes in vision and hearing so that people with special needs can be provided with better service without calling attention to their impairments.

The restaurateur can fill some seats early in the evening because seniors tend to eat dinner earlier. In fact, "sunset dinners" or "early-bird specials" have become quite popular in attracting diners from 5:30 to 7:30 P.M., before regular patrons arrive. These menu offerings normally include beverage and dessert at an attractive price.

The needs of the senior citizen are basically simple. They are not, as a group, demanding. They want rooms close to the lobby, they want help with luggage, and they want information. Like most customers, they want clean rooms, convenient location, and value. They do not want to be publicly singled out for service, but at the same time hospitality employees must recognize their special needs and provide them in a subtle way. Seniors tend not to rush through their stays the way businesspeople do.

Senior citizens tend to travel outside traditional patterns, such as the businessperson's Monday through Thursday, the weekend package guest's Friday and Saturday, or the busier times of the year. They are also more flexible in rearranging their schedules. Senior travelers can often check in on a Thursday and stay through Monday, making their stays attractive to hoteliers.

As the baby boom generation in the United States matures, it is quite possible that the needs of this market will further evolve and change, as they have in only the past 10 years. It is up to hospitality corporations to research these needs as they evolve so that this market can be better served.

International Travelers

Tourism is already the world's largest retail industry, and travel between nations is expected to continue to grow. Although international tourism was on the rise before September 11, 2001, **international traveler** spending in 2003 consisted of only 11.6 percent of total U.S. tourism spending. However, international tourism is expected to increase in the upcoming years.

international traveler A person who travels and visits outside his or her own country for business, personal, or pleasure purposes.

Canada and Mexico provide the most tourists to the United States, and vice versa, because of their close borders. Overseas visitors are led by the Japanese, followed by Europeans from the United Kingdom, Germany, France, and Italy. Growth markets in the future are visitors from Argentina and South Korea, which have shown huge increases. Singapore targets Australians and Japanese. Thailand targets Germans. Portugal, Spain, and Turkey target the British. And so it goes.

The international market is staggering in its size and complexities. Over 400 million people travel outside their own countries every year. This market is obviously not homogeneous, and hospitality marketers must be sensitive to the cultural differences of

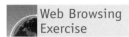

Web Browsing Exercise

Visit the website www.tia.org/PowWow. Read about the Pow Wow and the list of attendees and answer the following questions:

1. What is TIA's International Pow Wow?

2. How is this meeting different from traditional trade shows?

3. Where are the future meetings being held?

visitors from other nations. Because it is expensive and risky to try to directly market to individual international visitors, hospitality operators often seek out an intermediary, such as a consortium, reservation system, referral network, or tour operator with which to establish marketing relationships.

International trade shows such as the Travel Industry Association's Annual Pow Wow are also essential for reaching this market. This show brings together tour operators from all over the world who meet with hospitality industry representatives to conduct business. The tour operators account for over 70 percent of all international tourist arrivals to the United States. As the number of international travelers has increased, hospitality corporations and tourist destinations have become more user-friendly. However, much can be done.

The basic principles of marketing to the international traveler are no different wherever you go: They always involve the needs, wants, and solutions of customers' problems. Likewise, the concepts of positioning, segmentation, and marketing planning or strategy are no different. What changes, of course, are the consumers. International marketing does not involve changes in marketing concepts; instead, it involves understanding the changes in consumers.

Worldwide, the customer is looking for the same things: to establish relationships and to be sure they are taken care of on arrival. Communicating the message that a property will do this may vary by country of origin, but the meaning will be the same. Small offerings such as a Japanese breakfast or directional arrows to Mecca for Arabs are like mints on the pillow; they are nice to have, but without trust and a relationship, they become nonissues.

When McDonald's first opened in Moscow, it served 35,000 customers a day. Russians, however, ate with utensils and were not accustomed to picking up food with their hands. So McDonald's created brochures and tray liners explaining *how* to eat a hamburger, not *why* to buy one. Burger King had similar experiences in Venezuela, where hamburger buns no longer have sesame seeds (the Venezuelans kept brushing them off), the catsup and milkshakes are sweeter, the menu includes ice cream (everyone's favorite dessert there), and the outlets stay open as late as 1:30 A.M., because Venezuelans eat late.

Burger King also realized that mere adaptation to cultural differences does not mean that one gets to know the market. Burger King originally served wine in its restaurants in France, but customers tended to linger longer over glasses of wine. This slowed table turnover, so wine was removed from the menu. Conversely, Disney did not allow alcoholic beverages to be served at Euro Disney. The French were outraged, and Disney now sells wine.

But what do you do when your customer mix originates in 20 or more countries as is the case for many hotels worldwide in major destination areas, and some not so major? This is the challenge that faces hospitality firms that have an international focus.

Free Independent Travelers (FIT)

free independent traveler (FIT) An independent traveler who is not affiliated with an organized travel group and is difficult to identify within a well-defined market segment.

A final category of travelers is somewhat of a catchall for everyone else—the **free independent traveler(FIT).** In fact, in many hotels' segment breakdowns of their customer base, this may be quite a substantial proportion. That is because everyone who is not known to fit some other category will fall into this one.

The FIT traveler is a "nonorganized" visitor who does not belong to a group. Although these travelers may well participate in tours during their visit, they essentially come on their own and do as they please. Unidentified business travelers will also be lumped into this category. Hotels catering to the FIT market will usually set aside a block of rooms a year in advance and fill them in as reservations are made. The lead time may be three to six months in advance. The hotel releases the unused blocked space according to its buy-time schedule.

Both wholesalers and retail agents handle the FIT. This segment is normally willing to pay higher rates than group customers. However, a conflict arises with this situation. Although the FIT is willing to pay a higher rate because of a lack of volume, the wholesaler and retailer are able to negotiate large discounts as a result of aggregate FIT bookings.

The resulting savings are not always passed on to the traveler. Therefore, the guest may pay a high price while the hotel receives a relatively low room rate. Often the FIT booked by an intermediary may get the poorest room in the house based on the rate being paid to the hotel. The traveler is at a disadvantage in these situations and is surprised at the accommodations. This can hurt the hotel that is caught in the middle.

Incidentally, the term FIT is also used by some to designate "free individual traveler," or "free international traveler."

Consumers Who Are Members of Private Clubs*

We would remiss if we did not discuss consumers who belong to private clubs—whether country clubs, city clubs, yacht clubs, tennis clubs, or military clubs, to name a few. Consumers who join private clubs do so for a variety of reasons. These reasons include: (1) to meet people of similar social status or a status to which the member aspires; (2) to partake in recreational activities; (3) to entertain past and future clients; (4) to visit for lunch and dinner in a social setting; and (5) to have a place to hold private meetings.

Like most businesses, clubs require a strong marketing presence to attract and keep members. Many clubs rely extensively on word-of-mouth to attract members. However, clubs also use traditional forms of communication (e.g., advertising, direct mail, newsletters). All of these activities need to be coordinated through the marketing department.

Consumers who join clubs are looking for customization of their experience. They want to be recognized by name, they want staff to know what their preferred drink is, and they want luxury service. In other words, they want the type of service one would find in a high-end hotel, only better. Hali Freitus, a student at the University of Houston who works at River Oaks Country Club, said it best when he described the wants and needs of this group of consumers: "They want to know that the word *no* is not in the vocabulary." They want all their needs taken care of.

Raj Vijayakumar works at the University Club of Washington, DC, and describes his experiences in Exhibit 8-11.

*This section is based on conversations with Hali Freitus, a student at the University of Houston, April 2007.

| Exhibit 8-11 | Mini-Profile on Working in a Club versus Working in a Hotel |

Raj Vijaykumar began his career with the Swissotel, Boston, as a guest services associate and was then promoted to assistant front office manager. At Sheraton Hotels in Boston and Washington, DC, Raj worked in the rooms department. The University Club of the City of Washington, District of Columbia, a prestigious 104-year-old institution, gave Raj the ability to learn the hospitality of private social clubs. He began as front office manager in 2002, was promoted to rooms manager, and is currently the assistant general manager and director of rooms. In this position, Raj is responsible for the overall marketing and operations of the rooms department of the club.

BENEFITS OF WORKING IN PRIVATE CLUBS

I have been with the University Club since July 2002 and have excelled in my position. I had a great mentor, Mr. Albert Armstrong, CCM, (Certified Club Manager), whom I credit for grooming and training me in the club industry. It took a while for me to get adjusted to the club experience because it was a more relaxed environment. I was more used to always being on your toes in hotels, ready to react. While members do expect a high level of service and recognition, employment in clubs is less stressful. We foster the "home away from home" concept to cater to their personal needs. I have become accredited in several different areas in rooms management due to the scholarship programs in place for employees at the club. Continuing educational programs are encouraged and sometimes rewarded with scholarship money for

employees. While this does not always hold true for each individual club, their benefits usually outweigh those of hotels.

POTENTIAL SETBACKS OF WORKING IN PRIVATE CLUBS

The club is an independent property; members are equity owners of the club. It is a great place for someone right out of college to get acquainted with the hospitality business. As a private club, the revenues must be generated from the members and their guests. When it comes to capital improvement of the club, the resources are limited and there are committees that have to review and approve certain projects (i.e., replacing carpet for the lobby). The house committee must approve the reason for purchase and its design, and then the board of governors must approve of the project before it is implemented. In a hotel setting, the general manager consults with corporate or the owners and is able to get the job done much more quickly. The employee tenure in clubs is also much longer than in a hotel. There is very little turnover in management; some of the employees at the University Club have been here over 20 years, and some over 40 years. This leads to some discomfort if one is seeking to move up in a company more quickly.

Used by permission from Raj Vijaykumar, Assistant General Manager, The University Club.

Chapter Summary

There is a tremendous amount of research on the topic of consumer behavior, and it is impossible to review all of it in any one chapter. However, we have tried to show how some of the theories of consumer behavior can be applied to understand the behavior of hospitality customers. This chapter has also shown that this can be very difficult, because we cannot be sure what goes on in a person's mind. Maslow's hierarchy of needs forms a foundation, but perceptions and expectations play an important role. Differences between perceptions and expectations create many challenges for hospitality marketers, as seen in the gap model.

Perceptions lead to beliefs, which in turn affect attitudes, and much of marketing deals with attitudes and the changing of attitudes. Positive attitudes toward a product or service are required before customers will include it among their choices. Consumer behavior is a complex process, and the different stages are need or problem recognition, search, stimuli selection, alternative evaluation, alternative comparison, and choice.

In the latter part of this chapter we reviewed the most common broad individual market segments, business and pleasure, that are encountered in the marketing of hospitality. There are numerous other segments, as well as more specifically defined target markets. The most important point to remember is that market segments represent groupings of customers with similar needs and problems. Ideally, the scenario would be to operate a hotel or restaurant that catered to one market segment year-round. Unfortunately, this is rarely possible. Different segments will often stay in a hotel at the same time, making service and execution of the experience difficult. An example is the wedding held by the pool, which meant that the pool was closed to the leisure traveler. The marketing-oriented team responds to this challenge by truly understanding the needs of the customer and communicating these needs to the staff that will deliver the product promised. When all is said and done, relationship marketing provides the tie that binds.

Key Terms

attitude, p. 238

belief, p. 237

belief→attitude→intention trilogy, p. 237

business traveler, p. 241

cognitive dissonance, p. 239

consumer behavior, p. 229

free independent traveler (FIT), p. 254

intention, p. 238

international traveler, p. 253

Maslow's hierarchy of needs, p. 229

mature traveler, p. 252

package market, p. 250

perceptions, p. 234

pleasure traveler, p. 246

reference groups, p. 235

stimuli, p. 232

Discussion Questions

1. What are the five basic principles of consumer behavior?
2. What is meant by the belief→attitude→intention trilogy?
3. How does this trilogy relate to the overall buying decision process?
4. Briefly describe the types of customers in hospitality and tourism.
5. How are individual business travelers similar to and different from individual pleasure travelers?

Endnotes

1. Maslow, A.H. (1954). *Motivation and Personality*. New York: Harper and Row.
2. This discussion of consumer information processing is quite limited. Those who would like to explore it further are encouraged to read Ajzen, I., & Fishbein, M. (1980). *Understanding Attitudes and Predicting Social Behavior*, Upper Saddle River, NJ: Prentice Hall.

.... case study

Yore Heroes: Tap and Grill

Discussion Questions

1. Describe the types of customers that went to Yore Heroes Tap and Grill over the past five years.

2. What were the primary needs and wants of these customers?

3. Do you think the perceptions of the restaurant changed across these segments over the past five years?

4. Describe as best you can the likely change in beliefs, attitudes, and intentions toward patronizing Yore Heroes for each of the customer types you identified in question 1.

5. If you became the new manager of Yore Heroes during the 30-day closing period, what marketing strategy might you put into place prior to reopening the restaurant? Does it really have a chance to become a successful operation? In other words, does Yore Heroes Tap and Grill have a future?

Mike Monroe was troubled and angry as he locked the doors of Yore Heroes Tap and Grill for what he feared could be the last time. Yore Heroes would not be open for the next 30 days because of the recent charges of serving alcohol to minors and exceeding maximum capacity. "It figures this would happen at the worst time imaginable," Mike thought to himself. Over the past year and a half business had been declining steadily, and they were struggling to break even. Mike wasn't sure the restaurant would be able to survive 30 days of no business activity whatsoever. As he picked up his coat and left to go home, Mike thought back to when Yore Heroes had first opened five and a half years earlier. Then, the house had always been full. People used to line up and wait for an hour to get a table. He had worked so hard to please people and felt he had something for everyone at his establishment. What could have gone wrong? Why was patronage so low? What could he do to save his restaurant? Even more important, could his restaurant be saved at all?

This case was contributed by Erin O'Brien, developed under the supervision of Margaret Shaw, PhD, School of Hospitality and Tourism Management, University of Guelph, Ontario, Canada. Names and places have been disguised. All rights reserved. Used by permission. All Monetary amounts are in Canadian dollars. For conversion use 1$C/0.85$US.

THE RESTAURANT

Yore Heroes Tap and Grill was owned by Mike Monroe and his wife, Barbara Brooks, in the college town of Prince George. Mike was a graduate of the College of New Caledonia's Hotel and Food Administration Program. In addition, Mike and Barb each had many years of experience within the restaurant industry, both as minimum wage employees in the front and back of the house, and as managers. However, Yore Heroes was the first restaurant the two had owned and operated.

The city of Prince George was located in central British Columbia, Canada, and had a population of about 70,000. The College of New Caledonia had an enrollment of over 10,000 students in the fall and winter semesters, and 4,000 during the summer.

There were only two malls within the city of Prince George: Westwood Mall and Chesire Road Mall. Yore Heroes was located in Chesire Road Mall, the largest of the two by far. People would travel from all areas of Prince George not only to shop at this location, but also to attend the Cineplex Odeon movie theater located on the second floor of the mall (one of the two movie theaters in Prince George).

The mall was situated close to the college. It was only about a 20-minute walk to the mall from campus, and it was also easily accessible by bus. Buses left from both the college center and from the central downtown square. The mall had operating hours of Monday–Saturday, 9:00 A.M.–9:30 P.M., and Sunday 12:00–5:00 P.M. Heroes was not required to adhere to the same operating hours. It was open for business Monday–Saturday, 11:00–1:00 A.M., and Sunday, 11:00 A.M.–8:00 P.M.

Heroes was located on the second floor of the mall, directly across from Cineplex Odeon. The restaurant had a chalkboard at the entrance on which they would display a list of the movies currently playing in the theater, and their start times. Mike wanted the restaurant to have a very relaxed atmosphere. The interior was woodsy with wine-colored covering on the seats, brass railings, and a large number of booths. Pictures of popular "Heroes of Yore," such as Marilyn Monroe, Clark Gable, Elvis, Joe DiMaggio, Lou Gehrig, Joe Namath,

Bobby Orr, and Ken Dryden, and even "heroes" like Bugsy Siegal and Al Capone, covered the walls of the restaurant. Although most of the students, when first patronizing Heroes, had no clue who most of these people were, lively contests ensued on a regular basis. Once names were identified, there were always new queries and bets about what they did, sports statistics, and so forth.

The restaurant had a seating capacity of 200, which included a "boardroom" facility. The boardroom was separated from the restaurant by beautiful glass doors and could be rented out for parties, meetings, and occasion dinners. When this room was not rented out, it was used as part of the restaurant during regular business hours. Potential existed to expand the capacity of the restaurant during the summer months by means of a patio. However, it was, at the time, in need of much work, and the owners felt it would be best to put off the improvement/use of the patio until a later date.

There was a large upraised rectangular bar situated at the rear of the restaurant. Televisions were placed in the four corners around the bar so they could be viewed from any angle. The sports channel was shown on the televisions at all times, but the sound was kept off unless there was a big game or match. Lively upbeat music played throughout the restaurant, but customers could also choose a selection of music from the jukebox found directly across from the bar. There was also a pool table, dart board, shuffle board, dance floor, and DJ booth all located in the rear of the restaurant near the bar.

THE CLIENTELE

Within just a few months of opening, the restaurant was experiencing great success, at both the lunch and dinner hour, without any advertising. Mike and Barb found that their clientele appeared to be primarily families shopping within the mall, as well as a large number of movie goers, stopping in either before or after the show. They didn't have as many students frequent the restaurant as they had expected, but they were quite pleased with their success and felt it best to focus on the customers they had.

They had also managed to obtain a regular bar crowd. This consisted of men who would often gather in the afternoons or evenings to watch sports or take advantage of the happy hour, as well as mall staff who would often come up to Heroes after work for dinner and/or drinks.

THE MENU

The Heroes menu was printed on a brown paper bag. There was a wide selection of items offered. These ranged from inexpensive appetizers and sandwiches to full entrée dinners offering a choice of soup or salad, and starch (rice, baked potato, house pasta, or fries). Some of the most popular menu features were all-you-can-eat soup and salad priced at $8.99, make-your-own pasta, and make-your-own pizza.

A kids' menu was also available for families with children 10 years of age or younger. This menu was printed on a separate page, which was covered with cartoon characters for the kids to color (crayons were supplied to each child with a menu). When items were ordered from this menu, the child would receive a free beverage, free ice cream for dessert, and a free balloon on exit.

A variety of specials were offered on a regular basis. Each day there was a pizza of the day, soup of the day, appetizer of the day, pasta of the day, lunch special, and nightly special. On Friday and Saturday evenings two nightly specials were available. There were also daily specials available from Monday to Thursday. Following are some examples:

Monday: Two-for-one 16-ounce T-bone steaks (full entrée) for $19.99

Tuesday: All-you-can-eat pasta for $8.99 (has to be the pasta of the day)

Wednesday: All-you-can-eat chicken wings for $8.99

Thursday: Brontosaurus ribs (full entrée) for $16.99

THE BAR

A happy hour was available at the bar daily from 3:30 until 5:00 P.M., when customers were given a free four-slice pizza, with any three toppings, for every pitcher of draft beer purchased. There were also bar snacks, and two-for-one appetizers (available from 9:00 to 11:00 P.M.), offered at the bar but not available on the restaurant menu.

COMPETITION

Competition was not a major concern of Mike's when Heroes Tap and Grill was first opened. There were no other restaurants within the mall, only the fast-food outlets found in the mall's center. In addition there were very few restaurants in the area surrounding Cheshire Road Mall, only an East Side Mario's and a Pizza Hut, which were located in two separate strip malls across the street. Mike had considered one potential threat, the Potted Pigeon. While it was not located directly near Heroes, it was a similar restaurant in its relaxed style and in the types of food it served. It was situated closer to the college and was also the only restaurant to accept the college's meal card. However, seeing as the majority of their customers were not students, at the moment anyway, it was not a major issue. Besides, Mike thought having some competition is healthy. "It keeps you on your toes," he said.

THE FIRST TWO YEARS

While business remained steady throughout the first couple of years, Mike noticed some bothersome trends that were occurring in his business. The most disturbing of these was the incredible slow period from January to April. Although this was common for most restaurants, due to the tightening of purse strings following the Christmas season, Mike felt that his situation was even more severe because of his location in the mall. The number of customers in the mall (Mike's main client base) was very low, as was the number of people going to see movies. Mall customers dropped by 40 percent and, at times, as much as 50 percent below the norm for September to January.

To counteract this trend Mike decided to implement a number of changes. First, he allowed the two-for-one appetizer special, previously only offered to the bar patrons, to be available to all customers of Heroes (between 6 and 9 P.M.). He also introduced a two-for-one dessert special: Any customer who produced a movie stub (for that night) could receive two desserts for the price of one with the purchase of any beverage. Mike already had a number of regulars who came into Heroes to watch sports or play pool, so he felt there was opportunity to expand his bar business. To do so, he invested in a large-screen TV on which to show hockey, football, and other sports games.

Mike also observed over these two years that more and more students were coming into the restaurant. He recognized that it was primarily to take advantage of the all-you-can-eat specials: the soup and salad deal, the all-you-can-eat pasta (Tuesday nights), and the all-you-can-eat

wings (Wednesday nights). While the majority of customers still came from the mall crowd, Mike wondered if the students couldn't be the solution to his problem from January to April. He began to advertise Heroes in the two student newspapers, the *Crest* and the *Caledonian*. Mike then applied to obtain a meal card contract with the college's Express Centre and was granted a three-year contract to begin the first of the school year.

THE EXPRESS CARD

At the College of New Caledonia, students living in any of the on-campus residences were required to purchase a meal plan. Off-campus students were not required to purchase the plan but were given the opportunity to do so. Students chose from meal plan options ranging from a light meal plan ($500) to a full meal plan ($1,100) per semester. When the card was used to purchase food, it was swiped at the cash register and a number of points were deducted, like a debit card. This meal card was accepted in all of the campus cafeterias and until this time had been accepted in only one restaurant off campus, the Potted Pigeon. The meal card could not be used to purchase any alcoholic beverages.

For restaurants interested, the meal card contract was obtainable (by application) from the Express Centre, a department of the hospitality services at the college. When reviewing the potential clients, the Express Centre evaluated a number of factors, including the following:

1. The "fit" of the restaurant in terms of suiting and meeting students needs
2. Whether the restaurant attained enough of its business from students to warrant the use of meal cards
3. Whether the restaurant would be able to make money for the Centre

If the preceding factors were met, the restaurant in question would then offer a bid (a percentage of sales they were willing to pay back to the Centre). Heroes was found to fit the requirements and was granted a three-year contract, agreeing to pay the Centre 9 percent of Express card sales each year.

A further benefit of having the meal card was the free advertising done by the Express Centre for those restaurants accepting the meal card. For example, table cards promoting

Yore Heroes Tap and Grill were displayed on the tables in all cafeterias. Restaurants accepting the meal card were also listed in brochures that were mailed to students both on and off campus, and in the packages for new incoming students. In addition, for a minimal monthly fee, advertisements for the restaurant would also be run in a monthly publication released by the college's hospitality services department.

THE NEXT YEAR

Mike had been unprepared for the impact the meal card would have on his business. He had expected patronage by students to increase, of course, but he did not realize by how much. The first September with the card was the busiest month Heroes had yet seen. Not only was the mall busy with "back to school" shoppers, but the college students were pouring in. Having five or six nightly reservations for groups of 40 to 50 students was not uncommon. Unfortunately, not all ran smoothly. There were numerous complaints of slow service, long waits for food, and food coming at inconsistent times. The restaurant was often understaffed, and the majority of the kitchen help were young, inexperienced teenagers. This led to frequent mix-ups in the kitchen and the inability of wait staff to offer excellent service.

Business toned down to a more manageable level in October, but Mike was excited by the success and felt he had found his new market. There were, however, a few things that concerned Mike. The first was the fact that students could not purchase alcohol on their cards. He had noticed that, for the most part, the students coming in were not spending their money at the bar, but more often they took advantage of the unlimited refills offered on the soft drinks.

Another major concern was the staff. Mike's customers had always commented on the pleasantness and promptness of his staff, and he believed this was what brought many of his return customers back. The majority of servers had worked at Heroes since it opened and had come to know the regulars very well. It was not uncommon for the wait and bar staff to greet customers by name, but now servers often grumbled about having to serve the large groups of students. The most common complaint was the lack of tipping by the students. Tipping on the meal card was not permitted, and often students would come in with

no cash because all they needed to pay for their meal was their card. Mike feared the staff did not give the students the same level of service they did to other customers and also wondered what impact the situation would have on their morale. He had noticed staff scheduling off Tuesdays and Wednesdays, the two nights most frequented by students, saying they were not available to work those evenings.

Another issue of concern arose when the next summer turned out to be much slower than the summers of the three previous years. Mike attributed the lag to the students' going home for the summer months and decided it would be beneficial to focus on reaching more of the mall/family crowd during these months. A number of things were done in an effort to do just that.

First, Mike introduced a new summer menu that offered various items that were not available at other times of the year, such as a fresh fruit plate, a strawberry spinach salad, and a mango and kiwi chicken dish. In addition, he began once more to focus on increasing bar business. He began implementing promotions within the mall in hopes of getting in a drinking crowd during the summer months, consisting of both mall staff and mall patrons. He developed a plan to host theme night "parties" at Heroes throughout the summer. One example was a Hawaiian luau night. On these nights the restaurant would be decorated accordingly, drink and food specials would be offered, and a DJ would be brought in. To promote the night, stores in the mall were asked to display posters. These posters would also be displayed within the restaurant about a week in advance of the "big night." In addition, two staff would walk throughout the mall for about an hour every day, about a week in advance, handing out free Hawaiian leis and informing people of what the "party" was about and the type of drink and food specials that would be available. The nights were somewhat successful, attracting mainly mall employees and Heroes staff who were off and would bring in their friends.

THE FOURTH AND FIFTH YEARS

The next two years proved to be a tough period for Heroes. Competition increased dramatically. Three new restaurants and a new coffee pub went up in the immediate area: Up the

Exhibit 1   The Competition

Restaurant	Type	Segment	Independent or Chain	Meal Card?	Other Amenities	Price Range
Potted Pigeon	American food	Casual theme, popular with students and a younger crowd	Independent	Yes	Yes—two pool tables;	$8.99–$15.99
Pizza Hut	Pizza joint	Young families, price conscious, casual theme	Chain	No	No	$6.99–$14.99
East Side Mario's	Italian/American	Family restaurant, midscale, fun/boisterous atmosphere	Chain	Yes—next Sept	No	$10.99–$20.99
Swiss Chalet	Primarily spit-roasted chicken	Family, restaurant, casual theme, as the name suggests—like a Swiss Chalet	Chain	No	No	$7.99–$12.99
Red Rooster	BBQ food, ribs, chicken, beef	Midscale, open concept kitchen, caters to mid 20s to late 30s age group	Chain	No	No	$10.99–$20.99
Walters Coffee Pub	Coffee, light lunch, desserts	Quick service, casual theme, varied clientele from late teens to seniors	Chain	No	No	$3.99–$9.99
Up the Creek	American food	Casual theme, fireplace, camping atmosphere, early 20s to middle age	Chain	Yes—next Sept	No	$8.99–$18.99

Creek, the Red Rooster, Swiss Chalet, and Walter's Coffee Pub, not a full-service restaurant but one that served light food dishes as well as a variety of specialty deserts. More detailed descriptions are given in Exhibit 1. Other problems were also arising. Staff turnover was incredibly high, with 50 percent of Mike's former staff leaving to work at the Red Rooster restaurant. Mike was faced with the problem of finding not only new staff, but also the time and money to train them.

Customers, and subsequently sales, began to decline dramatically. Students were now the main clientele, about 75 percent, but their numbers were decreasing. Mike knew he would lose even more of this business in the future as East Side Mario's and Up the Creek were expected to have meal card contracts by September. Few families frequented the establishment. Those that did come were mainly seen only at peak shopping periods (i.e., September and November/December). Mike was at a loss. He wasn't sure how he could or should compete with this new competition.

Mike knew he needed an edge. His first decision was to introduce a delivery service to students living in residence (so they could still use their meal card) because none of the other establishments offered this service. The delivery service was a success and very well liked by the students; however, it was not making the money Mike needed to turn things around.

The second decision was to introduce Wacky Wednesdays, a new concept in an attempt to bring in the students, not to eat, but to get them spending their money at the bar. Mike felt Wednesday was appropriate because students were still coming in for the all-you-can-eat wings. The new Wacky Wednesday special was a 32-ounce draft beer for the price of a 16-ounce one, or a double bar shot for the price of one, starting at 9:00 P.M. It took some time for the idea to catch on. At first it attracted mainly mall employees and staff. Then Mike brought in a DJ on Wednesday nights and began promoting the special in the student newspapers. Eventually, more and more students began pouring in; some nights there would be lineups at 9:00 waiting to get in. The dance floor in the restaurant was packed, and it was not uncommon to meet the maximum legal capacity of the restaurant. Mike raised the price of the drinks to $4.25 for either a 32-ounce beer or a double shot, and the students kept coming.

Staff had commented to Mike a number of times concerning the regulars feeling alienated by this new atmosphere. More than one server had heard the line, "It's just not like it used to be." Not only did regulars complain, but other customers did as well. On more than one occasion, a table seated at 8:30 on a Wednesday night would become frustrated and annoyed at 9:00 when the DJ came on and dance music began to blare. Yet, as the night became more and more popular, Mike decided to take advantage of this and, in the summer, invested in improving the old patio attached to the restaurant. He put in a separate bar outside so that more customers could be fit into the restaurant on Wednesday nights.

Unfortunately, Wednesday evenings were now the only nights when the restaurant was busy. The rest of the week the restaurant was practically empty. Mike couldn't figure it out. Nothing had really changed at Heroes except for the implementation of Wacky Wednesday. Yet, not only were the families and mall patrons not coming in, the students were no longer coming in to eat either. Even the before-and after-movie business had dwindled. Something needed to be done and soon, but what? Mike had 30 days to figure it out.

overview

The organizational customer is described as the purchaser of hospitality products for a group or an organization that has a common purpose. Following are several subsegments of the organizational customer some of which are described in detail in this chapter:

- Generic organizational market
- Corporate travel market
- Corporate meetings market
- Incentive market
- Association, convention, and trade show market

continued on pg 266

chapter 9

Understanding Organizational Customers

learning objectives

After reading this chapter, you should be able to:

1. Explain what organizational customers are all about.

2. Identify the types of organizational customers in hospitality and tourism.

3. Discuss convention centers and the trade show market.

4. Describe conference centers specifically designed for the meetings market.

5. Explain the role of convention and visitors bureaus in hospitality and tourism marketing.

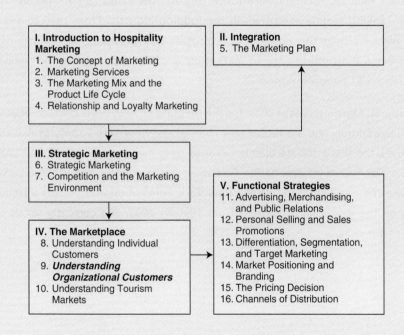

I. Introduction to Hospitality Marketing
1. The Concept of Marketing
2. Marketing Services
3. The Marketing Mix and the Product Life Cycle
4. Relationship and Loyalty Marketing

II. Integration
5. The Marketing Plan

III. Strategic Marketing
6. Strategic Marketing
7. Competition and the Marketing Environment

IV. The Marketplace
8. Understanding Individual Customers
9. *Understanding Organizational Customers*
10. Understanding Tourism Markets

V. Functional Strategies
11. Advertising, Merchandising, and Public Relations
12. Personal Selling and Sales Promotions
13. Differentiation, Segmentation, and Target Marketing
14. Market Positioning and Branding
15. The Pricing Decision
16. Channels of Distribution

overview
continued

- Airline crew market
- Social, military, education, religious, and fraternal (SMERF) markets
- Government market
- Group tour and travel market
- Meetings, incentives, conventions, and expositions (MICE) market

Marketing Executive Profile

Charlotte St. Martin
President and CEO, St. Martin Enterprises

After 28 years with Loews Hotels, Charlotte St. Martin left Loews to start her own company, St. Martin Enterprises. The company specializes in marketing, branding, and operations for the hospitality industry.

St. Martin served as executive vice president of marketing and sales for Loews Hotels, where she was responsible for all of sales and marketing for the chain's 20 hotels and resorts. Since joining the company in 1977 as director of sales and marketing for the former Loews Anatole Hotel, where she also served as president and CEO from 1989 to 1995, St. Martin rose through the ranks to become one of the highest-ranking female executives in the lodging industry. From 1990 to 1996 she simultaneously held the position of executive vice president of operations, but turned her full attention back to marketing in 1996 as a result of the chain's unprecedented $1 billion expansion.

Currently, St. Martin serves as chair for the Meeting Professionals International Foundation. In addition, she is a recent past chair of the New York Society of Association Executives, the first associate member ever to serve as a chair not only of NYSAE, but of any major Society of Association Executives in the country. She has also served as both vice chair and treasurer of the New York Convention and Visitors Bureau and served on the executive committee from 1990 to 2003. She currently serves on its board of directors.

Marketing in Action

Charlotte St. Martin

President and CEO, St. Martin Enterprises

What is the customer actually purchasing when "buying" a meeting? How do hoteliers provide this? Are they providing it now?

The customer is buying the services of a group of professionals who provide a site that, ideally, is problem free so that the meeting is all about achieving the desired results for the organization. Hotels are providing it now, although we can continue to improve our understanding of the purpose of the meeting so that we can deliver that problem-free meeting. For example, if we know that one of the goals of the meeting is to get an interactive dialogue going, we would not recommend a theater or school-room setup. If we simply "follow orders," we aren't helping to ensure the success of a meeting.

What do you think meeting planners are looking for these days in hotel services?

I think they are looking for more bells and whistles online and they are looking for no reduction in services at the hotel level. They want 360-degree views of ballrooms, meeting layouts that they can access online, online menus, RSVPs online—essentially they want everything made easier for them. They want you to have the very best technological capabilities to make their lives easier. And when they arrive at the hotel, they want to have everything exactly the same as it was online. They want 24-hour room service. They want all the bells and whistles. It has certainly created an interesting relationship for hotels and has increased the capital budget because it is difficult to do it all.

You have been involved in the meeting business for many years. How have you seen it change over time, and where do you see it moving in the future?

Although relationships have always been important—and still are—the technology that has been added has provided a degree of pace and efficiency that was not a factor years ago. We as an industry went through a period several years ago during which relationships were put on a back burner. Online RFPs were replacing the personal nature of the business, but the meeting planners and their suppliers soon learned that, although technology is terrific and helps us a great deal, we need to have personal interaction to do the best job for our organizations. I don't see this changing in the future. While procurement will be involved in the meetings industry increasingly in the future, personal relationships are still important to providing the best meetings possible.

I think that the biggest trend will be the continuation of outsourcing. But I honestly believe that the last trait that I just mentioned—developing the ability to bring an ROI component to the meeting—will be the factor that enables meeting planners to keep their jobs and be very valuable. Otherwise, companies would simply outsource.

You are very involved in MPI and other organizations. Can you indicate how these organizations have affected your career? The industry?

MPI, ASAE, PCMA and CIC are all organizations in which I have held leadership positions. I believe that volunteering and getting involved in them is probably the single biggest factor in my success in the early stages of my career. I held leadership roles in organizations prior to these, and they helped me learn things that I hadn't yet learned in my paid positions, enabling me to gain the respect of my bosses for being more accomplished than the average person at that point in my career. Also, being in sales and marketing, being visible, certainly helped our brand. As president and CEO of the then-Loews Anatole Hotel, I once analyzed our local business and found that over 75 percent of our local customers came about as a result of a relationship I had developed in one of my leadership roles. Additionally, as I noted earlier, these roles helped me understand my customer, which is so crucial in being successful in the sales/marketing arena. The industry has raised the visibility of meetings and has made work in the meetings industry a "real career." Although there is still much to be done, we wouldn't be where we are without them.

You have been very successful in this industry. What lessons have you learned that can be passed on to those reading this text?

The most important thing I've learned is that listening to your customer is critical for long-term success. Truly understanding their business, their needs, and what you can do to make them more successful is such an important thing that I did during my career. There is no substitute for caring about the customer. I attend industry organization meetings, learn about their concerns, challenges, needs, and wants; and then I do my best to meet those needs and challenges. Volunteering for leadership roles in their organizations has been the cornerstone of my success, both in the meetings industry and within my own company. Integrity follows you throughout your career. Not having it will kill you in the long term.

What types of skill sets would be needed by a student interested into going into the meeting planner side of the business?

Actually, I would recommend they have two opposite skills: the first is to be extraordinarily detail oriented because a meeting is all about the details. But the meeting planners who are getting the bigger jobs with higher pay and bigger incentives are the ones who understand that their jobs are much more than just details. They are much more strategic. They are able to actually develop a strategy for the meeting with their client and are able to measure the results of that meeting. They build that into their processing, so at the end of the meeting they can go to the CEO or the VP of marketing or whomever and discuss the outcome of the meeting: how the meeting ranked, how much money was made, or how much they saved. They make themselves much more valuable in that process.
Used by permission from Charlotte St. Martin.

*T*he **organizational customer** *is defined as the purchaser of hospitality products and services for a group or organization that has a common purpose. These customers are particularly important to hotels and resorts that offer extensive meeting facilities and, in many cases, represent the majority of such properties' annual occupancy. This customer's needs are somewhat different from those of the individual customers described in Chapter 8. Although all of the basic principles for the organizational customer are the same as those for the individual customer (stimuli, search, perceptions, beliefs, attitudes, etc.), the organizational customer is typically the purchaser for the end users. Although* **meeting planners, travel managers,** *and actual users are all organizational customers, we will use the term* user *or* end user *for the person who actually stays in the room. In other words, the end user is not usually the decision maker for the purchase. We will use the term* planner *or* manager *for the decision maker.*

organizational customer A buyer of hospitality products and services for a group or organization that has a common purpose for the purchase.

Planners and managers act as go-betweens to satisfy the needs and wants of the users as a group. In a way, they "sell" to the organizational customer, just as travel agents, tour operators, and incentive travel planners do; or, as is the case with meetings and convention planners, they may organize and plan meetings at the request of the organizational customer. Travel managers "manage" the travel arrangements for their organizational customers.

meeting planner An organizational customer who plans and executes off-site meetings and is the primary decision maker for hospitality and related purchases. Related purchases include airline seats, ground transportation, and the like.

Although there are a number of target market categories in the organizational market, we classify them into six major segments: the corporate travel market; the corporate meetings market; the incentive market; the association, convention, and trade show market; the social, military, education, religious, and fraternal (SMERF) and government markets; and the group tour and travel market. An acronym for these markets is **MICE** (meetings, incentives, conferences, and exhibitions).

travel manager An organizational customer who helps plan individual travel itineraries for members of an organization and makes hospitality purchase decisions for those members.

MICE An acronym for meetings, incentives, conventions and expositions customers.

GENERIC ORGANIZATIONAL MARKET

When a couple books a hotel room on a weekend package, they know what their expectations are. Similarly, business travelers often choose to be close to their places of business for the next day, sometimes at the expense of comfort. The needs and purposes of these customers are different. The meeting planner or travel manager, however, intends to satisfy the needs of multiple travelers at that same time. Although the group may have a common purpose, such as a business meeting of a corporation, a computer industry convention, or an incentive trip for insurance salespeople, each member of each group may have somewhat different needs. This makes the overall task somewhat more difficult for the planner or manager. The similarity, of course, is that with both individual and group travelers, if expectations are not met, the customer may go somewhere else the next time.

The challenge for planners has increased in recent years because of the corporate trend toward downsizing. Large corporations have decreased their workforces by tens of thousands of employees. Many meeting planners and travel managers have been victims of downsizing, leaving these tasks to administrative assistants who lack experience in this field or to travel agents or travel planners who don't know the organization as well. This gives the hotel marketer an even more critical task and need to understand the buyer. This has also led to the growth of more independent professional meeting planners, entrepreneurs who plan meetings for numerous corporate clients.[1]

Specifically, the planner or manager must try to predict the needs of the group, as well as select the proper facilities to accomplish the group's common purpose. For example, the meeting planner of a corporation may be given the task of planning a sales conference for the international division. The planner must understand the needs of that particular department within the company, with which he or she normally has very little contact, as well as the needs of its individual members.

At times, planners or managers may not even visit the hotels or restaurants to which they send their organizational groups. Thus, to make the right decision, they need to rely on a different set of stimuli from those used by other customers. Yesawich, Pepperdine, Brown & Russell revealed in their 2004 *Portrait of North American Meeting Planners* that if planners do not make a site visit (the most important source of information when selecting a site), then they rely heavily on their correspondence with the property sales staff. Word-of-mouth referrals from fellow planners or managers are also very influential factors when choosing a facility. About 4 out of 10 professional meeting planners rely on information in brochures or other collateral, regional hotel/resort sales staff, and hotel/resort representation company sales staff. Advertising in meeting trade directories, the business press, the meeting trade press, and direct mail all rank far down the list on the reliance hierarchy,[2] thereby underscoring the importance of a property or destination's sales staff in providing this type of business.

MEETING PLANNERS

Planners rely heavily on hotel salespeople, unlike individual customers. Also, conference service managers of hotels, who handle the details during an event, become extremely important in the decision to book and to rebook after the event is over. Even the chef, who is going to be serving perhaps 300 attendees three meals a day, becomes critical. The organizational planner is at far more risk from a bad meal than is the weekend package customer who is not pleased with room service.

As planners gain more experience on the job, however, they are less influenced by salespeople. Corporate meeting planners are far more likely to rely on site inspections than on sales staff when selecting meeting destinations.[3] These people, and many other planners, want to see for themselves and will often visit the property before booking it. All planners, no matter what their level of experience, are most concerned that the hotel and its staff perform so that their meetings are successful. Quite often a planner's promotion—or even her job—is on the line, and hotels often reflect this in their advertising (Exhibit 9-1). Even if the hotel is entirely at fault, it is ultimately the responsibility of the planner who chose the wrong site for the meeting.

The leading concerns of meeting planners, both corporate and association, are shown in Exhibit 9-2.

One planner explained many years ago (and it is still true today), what customers need:

> You can have the most gorgeous facility in the world.... I still need professional staff to augment what I do.... I often follow the same people as they move from hotel to hotel. The people I do meetings for like to be pampered a little bit. A property may be less than desirable, but if they can provide service and if the food is good we can overlook the other things. What's important to me is ... that everything I've ordered is there. Problems occur when hotels don't deliver what they say they can deliver.[4]

Exhibit 9-3 shows the highest 18 site selection criteria generally agreed on by association and corporate meeting planners.

To begin to understand the needs of the organizational planner, it is important to see how the planning process should go for a meeting or function. Understanding this, the sales and operations departments of a hotel can prepare for problems before they happen, perhaps preserving the success of an entire meeting. We next discuss important issues related to working with the organizational planner.

Buy Time

Each segment of customers has different **buy times** (also called **purchase cycles** or **lead times**) for purchasing the hotel product. A corporate traveler may make reservations one week in advance of an upcoming trip. According to a study undertaken by PKF Consulting on behalf of *Convention South* magazine, the average booking lead times for large events moved from 21.6 months in 2002 to 23.4 months in 2004.[5] The tour operator will have routes calculated a year or more in advance. For smaller meetings, the lead time is less. For example, Yesawich, Pepperdine, Brown & Russell reported in their 2004 *Portrait of North American Meeting Planners* that for meetings of fewer than 100 people the average lead time is 6.3 months; for 100 to 200 attendees, it is 9.5 months; for 201 to 400, it is 13.2 months; for 401 to 600, it is 15.2 months; for 601 to 800, it is 16.6 months; and for 801 to 1,000 attendees, the average lead time is 18.2 months.[6]

Knowing the timing of the purchase is important in selecting potential market segments because it will determine the scheduling of sales, advertising, and related

buy time How far in advance a buyer of hospitality services makes the decision to purchase the product and book the reservation accordingly. Also referred to as **buy time (lead time, purchase cycle.)**

Exhibit 9-1   The Tarrytown House focuses on tradition on entice meeting planners to consider them for their next meeting

Source: Destination Hotels and Resorts. Used by permission.

marketing activities. To maximize revenues, the ideal business mix of segments may include a variety of group customers. With the different room rate potential of each customer grouping, managing the inventory becomes critical.

For example, a 400-room hotel may have an opportunity to sell out to a midweek convention three years in advance at a rate of $100 per night. At first glance, this might appear to be a good sale; the sales department can spend its time trying to fill other, less busy time periods. More careful analysis, however, might show that this hotel has an average of 300 rooms per night occupied by business travelers during the week. The rate this year is $125 per night, and in three years it is expected to be at $150.

Exhibit 9-2	Top 12 Factors with Which Meeting Planners Are Extremely or Very Concerned		

Factors of Concern	Corporate Planners	Association Planners
Making the meeting agenda relevant	1	1
Convention services staff	2	3
Room rates	3	2
Accessibility of the destination by air	4	6
Cost of food, beverage, and entertainment at the destination	5	5
Cost of flying to the destination	6	7
Availability of low-cost air carrier service to the destination	7	9
Hotel or resort security services	8	N/A
Internet access from all meeting rooms	9	N/A
Meeting attendance projections	10	4
AV company services	11	N/A
Adequacy of the high-speed Internet	12	N/A
Popularity of the destination	N/A	8
Availability of new/interesting speakers	N/A	10
Popularity of the hotel or resort	N/A	11
Accessibility of the destination by car	N/A	12

Source: Yesawich, Pepperdine, Brown & Russell, (2004). *Portrait of North American meeting planners,* 42.

Exhibit 9-3	Hotel or Resort Selection Criteria of Association Meeting Planners and Corporate Meeting Planners (% stating extremely important or very important)		

	Corporate Planners	Association Planners
Small meeting rooms	82%	75%
Internet access from guest rooms	82%	57%
Free Internet access from guest rooms	79%	57%
Ability of "headquartered hotel" to accommodate all delegates together	76%	81%
Ballroom	61%	68%
Bar or lounge on premises	63%	62%
Internet access my attendees can trust is secure	56%	76%
On-site convention service manager	74%	76%
Working desks in guest rooms	74%	46%
Business Center	72%	63%
Hotel/resort brand's reputation in the meetings industry	70%	59%
Casual, three-meal restaurant on premise	68%	68%
Complimentary transportation to/from airport	67%	69%
Ability of on-site technical support for high speed Internet in the meeting rooms	67%	58%
Internet access from the meeting rooms	67%	54%
Professional on-premise support for Internet/AV	66%	67%
Secure high-speed Internet to protect data	64%	45%
Fine dining restaurant on premises	64%	44%

Source: Yesawich, Pepperdine, Brown & Russell (2004). *Portrait of North American meeting planners,* 33.

Few, if any, business travelers plan business trips three years in advance. These travelers call the hotel a few days in advance for their room reservations, unaware of the convention that is being held there at that time. If all patterns hold, the hotel will "lose" $50 per room on the 300 rooms that it could have held for the segment that pays a higher rate, but books the shortest lead time. Mistakes like this are very subtle, because the hotel is sold out, yet the room revenues are decreased by $5,000 a night. The hotel may also push away some regular customers who cannot get rooms when they need them. It does not take many miscalculations like this to understand the importance of the lead time on profitability.* Managing this forecast revenue is the goal of the revenue manager. We discuss this issue in more detail in Chapter 15.

Another buy time variable is the use of a property at different periods of time. City hotels generally target business travelers and conventions during the week and pleasure travelers on weekends. The same variation may occur between summer and winter. Thus, many hotels offer "package" meetings at special rates during slow periods just as they do for individual travelers. Resort hotels have similar situations depending on the season of the year. At one time, many resorts simply closed during the off-season. Now most stay open year-round but seek a different mix of occupied rooms that includes more discounted meetings and conventions during the slower periods.

Assessing Needs

Each collection of people with a common purpose has different needs. For example, the Elks Club, a fraternal organization, convention certainly has different reasons for a meeting than does the new product development team for Eastman Kodak, yet both of these organizations may meet in the same hotel at the same time of the year. Both the planner and the hotel employees must understand the purpose of each meeting. If, in fact, the meeting is purely a social one, theme parties, golf outings, fashion shows, and so on, are expected and welcomed. If, on the other hand, the purpose of the meeting is to think of strategies that will bring a corporation out of bankruptcy, the entire schedule and tone of the meeting will be changed accordingly. These are obvious differences; there are many far more subtle ones.

The most common complaint planners have about hotel salespeople does not relate to high-pressure selling. Rather, it is the salesperson who has not taken enough time to find out about their business. They may be pitched by a property unsuited to their needs and resent the fact that their time is being wasted by someone who didn't make enough effort to find out what they were like.

Resolving Conflicts

Planners have to work with both the hotel and their own organizations to anticipate and resolve potential problems. Although planning may alleviate possible conflicts, the hotel may be only half of the problem. The organization itself presents problems that must be addressed before the function occurs. There may be a hierarchy of attendees within the organization that needs suites, first-class travel, and seats at the head table. Failing

---

*With convention: 400 rooms @ $100 = $40,000. Without convention: 300 rooms @ $150 = $45,000. This calculation, of course, ignores F&B revenues from the convention, which could change the picture. Nevertheless, the question of alienation remains.

to accommodate these needs can cause conflicts that ruin the meeting through no fault of the hotel. A hotel staff can anticipate these needs by asking to review the VIP list and discussing its needs.

There are numerous other potential issues. Nonsmoking guest rooms and meeting rooms are entering into the spectrum of worries. Individual special meals during a banquet are no longer limited to just kosher or vegetarian meals. Today guests have many different dietary needs and restrictions, and hotels must work hard to cater to these to satisfy the needs of attendees.

The best way to avoid conflicts is to have a preconference meeting. The term *preconference* is generic and can be applied to incentive trips as well as to corporate meetings. At this meeting, the planner reviews the details of the meeting with each department to ensure that communications have not been distorted through the conference service manager. The front office, housekeeping, banquet managers, and general manager, if the situation warrants, should be in attendance with the salesperson and conference service manager to ensure that all potential conflicts are discussed and remedied before the function ensues.

Executing the Meeting

Executing the meeting may be the simplest phase of the planner's job if all the previous steps were followed and done well. If they were not, this is certainly the hardest portion of the process. The execution of the meeting could occur without the planner being in attendance. The needs of the planner are now being transposed onto the group.

Sometimes, even if the organizational planner is on site, the end users' needs are not met. For example, the association planner may want the general session set up theater style, with the room having chairs that face the podium for a guest speaker. The guest speaker might demand that the room be set up classroom style, with each chair having a desk in front of it so that participants can take notes during the presentation. One of the authors attended a function in Chicago in which the meeting specifications called for a podium on the platform. At the last minute it was decided that a sit-down panel format would be more appropriate. The flustered setup man was clearly annoyed at the last-minute change.

These are classic examples of how the planner is not the end user and how the needs of the group can change right up to the last minute. The hotel that adjusts accordingly will be the one that receives the future business. There are no right and wrong sides to this scenario. The task must be completed to satisfy the needs of both the end user and the organizational planner. It really does not matter how many times a group changes the setup of a room. The hotel is responsible for making the changes. This is what marketing is all about—giving customers what they want at a time and place of their choosing.

Evaluating the Results

Based on the goals of the organization, was the meeting a success? The hotel should be as interested in the results as the planner is. The evaluation process can take place in a postconference meeting held shortly after the conclusion of the function. Department heads and the planner can review face-to-face all the things that went right, as well as those that went wrong. The marketing-oriented organization will take immediate steps to correct the malfunctions and to reinforce the positive aspects.

The evaluation process is also critical for the planner. When these customers are the buyers, but may not be present at the actual event, it may be difficult for them to understand exactly what took place. Even when the hotel delivered as promised, the organization may not have accomplished its goals. The planner will need to assess the results before starting to plan the next similar function and should be made aware of the problem areas by the hotel that wants to recapture the business.

CORPORATE TRAVEL MARKET

The corporate travel manager or coordinator plans travel and entertainment for a company's employees. Corporate travel managers are different from corporate meeting planners in that they plan individual travel schedules. A common purpose may still exist, because the corporate unit is relatively the same, but people at different levels of the organization will be traveling on different missions. In some organizations, the travel manager and the meeting planner are the same person. Corporate meeting planning will be discussed in the next section.

The size of the corporate travel market is very large, running into tens of millions of business travelers worldwide. Behind salaries and technology, travel and entertainment costs are the third largest controllable expense of private sector companies in the United States[7]. About half of these end users are directed or influenced by the corporate travel manager who plans, controls, mediates, negotiates, evaluates, and/or approves travel expenditures.[8] In 2003 the typical travel manager managed $4.2 million in U.S. booked air volume and $5.1 million in travel and entertainment. This travel manager was responsible for on average 5,190 U.S. and foreign travelers.[9] This market is very desirable for hotels because it tends to pay good rates, is large, and delivers business consistently throughout most of the year.

The corporate travel manager needs to find the correct products and services for the entire group of corporate travelers. Once the product is identified, the best rates are negotiated. The supplier needs to understand the culture of the organization to fulfill its needs. For example, some companies go to the top of the line for their hospitality and service needs. From first-class airplane seats, to limousines for ground transport, to the best hotel in the area, some companies spare no expense when entertaining themselves or their customers.

Some corporate cultures are just the opposite. They use hotel rooms sparingly, have meetings in their own offices, and use cabs or airport shuttles to reach hotels. Negotiations with a large corporate planner at the Embassy Suites Hotel in Boca Raton revealed this. After renovation, the hotel approached its major customer for a rate increase. During the sales call, the customer did not disagree that the hotel was new and worth more money. The customer simply said that the company at this time would not pay more than $125 per night for a hotel room, even if it was the Ritz-Carlton. The Embassy Suites lost this business. Most companies are somewhere in between.

Typically, corporate executives get the best treatment and company trainees get the least. It mostly comes down to examining the purpose of travel, who is traveling, and their position in the corporate hierarchy.

Many companies have come to realize the extent of their travel and entertainment budgets. In some cases, this can be as much as 25 percent of an organization's costs. Thus, many corporations are tightening the screws on travel costs. As one corporate travel manager told us, "You can't believe what $5 a night means to this company over a year's time."

The way corporations manage their travel and entertainment dollars significantly affects the revenues of the hospitality industry. The emergence of corporate travel buyers is a result of this cost control effort. Essentially, their task is to control these costs without losing the quality of the product. Corporate travel buyers first figure out the level and service of product that the organization is willing to accept, and then they negotiate the prices.

Knowing the Volume

It is difficult to negotiate anything without knowing the restrictions both parties are dealing with. A hotel might give a discount based on expected volume, only to find that the volume never happens. A corporation, on the other hand, might underestimate its true room requirements at a destination and be paying more than it could negotiate at the actual volume. The same is true with airline travel, where companies can often negotiate volume discounts. Most major airlines even have a special website for corporate travel managers.

If hotel rack rates are high, the travel manager has come to expect a discount no matter what the volume. One of the authors once received a call from a corporate travel department asking for a discounted rate. The company, which happened to make shoes, claimed its volume would be about 100 room nights annually. The hotel happened to enjoy high occupancies and rarely discounted rooms, even for 1,500 room nights a year. The shoe company planner was not convinced that his perception of volume did not apply in this case. Finally, the author asked if he could get a discount on shoes if he bought three pairs a year. The response was, "Of course not! You have to be a big retailer to command a discount!" The point was finally made.

Hotel room rates are negotiated initially from the published or rack rates. Rarely today do customers pay the rack rate unless they are uneducated enough not to ask for one of the many other rates available, or are traveling during peak demand periods. From rack rates come corporate or commercial rates, usually at least 10 to 15 percent lower than the rack rate. We discuss pricing in more detail in Chapter 15.

Hotels now negotiate individual corporate rates with individual corporate customers. Volume corporate customers recognize the wide-scale availability of corporate rates for anyone and demand their own corporate rate relative to their volume. These rates can run 15 to 35 percent below the rack rate. This, of course, makes the rack rate a ridiculous requirement, so hotels may raise their rack rates, say 10 percent, to raise the corporate and volume rates.

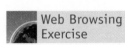

Web Browsing Exercise

Visit the website of Northwest Airlines for corporate travel managers (www.corporate.nwa.com). Compare and contrast this with their regular website (www.nwa.com). What differences, if any, are there? What makes this website unique for the target audience? If you were a corporate planner, would it make you want to join the program?

consortia Large travel agencies that have substantial volume purchasing power that enables them to command highly discounted rates from hotels, airlines, car rental companies, and the like.

Large travel agencies, or **consortia,** provide volume purchasing power for their customers. Woodside Travel in Boston, American Express in New Jersey, and Rosenbluth Travel in Philadelphia are examples of consortia. They are consolidated purchasers under which smaller agencies can access the technology and buying power of larger entities by combining the volume of a number of corporations to get the best fares and rates for all.

Corporate travel managers also like hotel chain representatives to negotiate rates that will apply chainwide. A chain that provides the convenience of "one-stop shopping," or one place where corporate travel managers can negotiate discounted room rate agreements for all hotels in the chain, would have a competitive advantage in this market.[10]

Understanding Travel Patterns

The corporate travel manager uses a knowledge of corporate travel patterns to negotiate with hotel suppliers; the suppliers respond in kind. For example, if the corporation has people traveling to a given city mainly when occupancy is already high, the manager will have far greater difficulty negotiating preferred rates. On the other hand, if travel can be planned during low-occupancy periods, the manager may obtain not only discounted rates but also preferred availability during periods of high occupancy.

CORPORATE MEETINGS MARKET

The corporate meetings market covers a wide range of organizational customers. In 2003 this market spent $44.7 billion dollars over 1,058,800 meetings attended by 84.6 million people.[11] The average corporate meeting expenditure was $262,000 in 2003, with hotel and F&B costs representing up to 56 percent of this figure.[12] Some hotel companies aim to specifically attract this business (See Exhibit 9-4).

The most common type of corporate meeting is a management meeting, wherein executives gather to discuss company business. These tend to be small gatherings with largely senior-level managers attending. An additional type of corporate meeting is the sales meeting, which is usually organized once or more a year to discuss and review company sales goals and strategies and "pump up" the sales team. Another common type of corporate meeting is the training meeting or seminar. These provide corporations avenues to exchange information and improve the performance of their personnel. They are often new employees, and the goal is to introduce them to the corporate culture. The corporate planner is responsible for all three types of meetings, and any others. Some incentive meetings are also in this situation, but this unique type of meeting is usually handled by specialized incentive companies.

To understand the needs of the corporate meeting planner, one must review all the components of the organizational customer. In a nutshell, meeting planners need to "look good." They need to look good to their boss, to the person whose meeting they organize, and also to the hotel if they want to continue to look good to the first two people. Exhibit 9-5 shows an ad for a hotel company that appeals to this need of meeting planners.

What meeting planners do not need is for hotels to mislead them with regard to the capabilities of the physical plant and the personnel. The sometimes short-term

Productive meetings arise from a harmonious setting.

222 MASON STREET · SAN FRANCISCO, CA 94102 · 415.394.1111· FAX 415.421.0455
HOTELNIKKOSF.COM · In Partnership with Le Meridien Hotels & Resorts

hotel nikko san francisco
Meetings

Exhibit 9-4   Some companies, such as Hotel Nikko, specifically pursue the corporate meetings market

Source: © 2005 Hotel Nikko San Francisco. Used by permission.

Exhibit 9-5   Hotel Sofitel is appealing to meeting planners

Source: Sofitel. Used by permission.

thinking of the hotel sales staff may lend itself to misrepresentation and the eventual loss of customers. With as much as a 70 percent annual turnover in many hotel sales offices, and bonuses based on room nights sold, the reward system basically decides how fast you can make your quota to increase your income or get promoted. Sales personnel are often told to "book it, not cook it," meaning to get the signed contract and go find more business. Many experienced planners, however, have little need for the salesperson, instead requiring the attention of a professional conference service manager, who works for the hotel, to service the meeting. A conference service manager is the on-site need fulfiller. All the detail work will be done through this person, who, like the salesperson, may also be on a similar "fast track," often leaving the meeting planner in the hands of inexperienced new people.

Meeting planners need meeting rooms that will suit the purpose of their event. They also want quiet rooms. Often, hotel ballrooms are divided by thin, movable walls that allow noise from the meeting next door to filter through. One of the authors recently gave an all-day seminar on research methods directly across a narrow hallway from a national convention of gospel singers. Guess what?

The meeting planner also needs an efficient front desk that will assign rooms to the right people: the VIPs in the suites and the attendees in the regular rooms. The billing needs to be right: Some rooms may be billed to the organization, while some attendees may have to pay for their own. The meeting planner needs meeting rooms to be set up on time and coffee breaks to arrive when ordered. The audiovisuals need to be in the meeting room at the right time and in working order. The spare bulb for the projector should be on the cart, not locked in a closet at the other end of the building.

Meeting planners expect all of the details to be handled absolutely professionally. If a hotel is able to provide planners not only with what they think they need, but also with what they don't realize they need, the planners will return. All of the previous concerns, and much else, fall on the shoulders of the conference service manager assigned to service a group.

CONFERENCE CENTERS

conference centers
Carefully designed lodging facilities that target and meet the specific needs of the small-to-medium-size meetings market.

Today there is a variety of hotels from which meeting planners may choose. With this additional supply in most marketplaces, the need to attract meeting planners' business has grown. Some hotels attempt this by claiming to be **conference centers** by adding the words to their name, believing they can establish a new identity (i.e., XYZ Motor Inn becomes XYZ Motor Inn and Conference Center). Howard Johnson properties have, for example, in many areas, adopted the practice of adding the term *conference center* to their signs. These properties are not, however, conference centers in the true sense. In fact, many are far from it and may, in the long run, be hurting themselves with this pretension.

One way the industry has responded to the unique needs of the meeting planner is by developing "dedicated" conference centers that offer carefully designed facilities and services that are reserved mainly for meetings (and not social functions) and do an excellent job specializing in this market. Pure conference centers cater almost exclusively to meetings. Scanticon Conference Center in New Jersey and Arrowwood Conference Center in Westchester, New York, for example, are true conference centers as defined by the sole trade association that represents the category, the **IACC (International Association of Conference Centers)**. This organization has established a number of criteria that must be met by a facility to use the "conference center" label, including a requirement that at least 60 percent of available meeting space be dedicated, single-purpose conference space. These rooms must be separated from living and leisure areas and made available to clients on a 24-hour basis for material storage. In addition, a minimum of 60 percent of total revenue from guest rooms, meeting space, food and beverage, conference technology, and conference services must be conference related. The average group size must be 75 people or smaller.[13] The needs of conference attendees are very different from those of leisure guests or business travelers. Exhibit 9-6 shows an example of a dedicated resort conference center, the website for ARAMARK Harrison Lodging.

According to the IACC, the big difference between a combined hotel and conference center and a true conference center is not just technical services but human services. The IACC claims that the business of the typical hotel is temporary, limiting the attention and service it can give to every meeting. A conference center has a greater commitment to managing conferences because they are essentially the only market it serves.

In some cases, the combination of hotel and conference center works well when the markets are separated by day of the week or season. In other cases conflicts are created that can be damaging. The ballroom that is ideal for weddings or for trade shows may be entirely inappropriate for meetings. For example, in the first case, wedding

International Association of Conference Centers (IACC) A trade association that represents and serves the needs of conference centers. The membership of the association includes owners and mangers of conference centers, professional meeting planners, and others who have an interest in this type of facility.

 Web Browsing Exercise

Visit the website for the International Association of Conference Centers (www.iacconline.org/). How does this organization help move the planner through the buying process model? Why would an organization want to belong to this association? Why would a planner want to use this organization? Why would a planner not want to use this organization? If you work for a conference center that is not part of this organization, how do you compete?

Exhibit 9-6 The website for ARAMARK Harrison Lodging, which is a division of Aramark Services Inc.

Source: Retrieved September 15, 2005, from www.aramarkharrisonlodging.com/home/flash.php. ARAMARK Harrison Lodging is a division of Aramark Services, Inc. Information submitted 1/5/06. Used by permission.

guests would probably not hear noise from the kitchen. During a meeting sales presentation, however, those noises can break the concentration of the speaker and ruin the meeting. The dividing walls of the same ballroom may be ideal for the separation of a cocktail reception and a dinner, whereas too thin for the hosting of two meetings simultaneously. Very few facilities are ideal for all markets.

True conference centers attempt to serve one market only, the conference meetings market, and do so in a controlled environment. With soundproof meeting rooms, dedicated audiovisual rooms with state-of-the-art equipment, and conference service managers whose only job is to facilitate the needs of the meetings, these properties offer a serious environment in which to conduct meetings. Most are located away from major cities so that recreational and other distractions are also held to a minimum.

INCENTIVE MARKET

The Society of Incentive Travel Executives, SITE, defines *incentive travel* as follows: "Incentive travel is a modern management tool that motivates salespeople, dealers, distributors, customers, and internal employees by offering rewards in the form of travel for participation in the achievement of goals and objectives."[14] Exhibit 9-7 reveals the results of a study undertaken by *Incentive* magazine. This table shows how many incentive programs are run annually, the objectives of incentive travel programs, the annual expenditure for incentive travel programs, and the amount spent on trip delivery per participant.

incentive travel Trips that are taken by employees of an organization as rewards for their superior performance over a given period of time. Incentive travel is sometimes offered to customers, distributors, dealers, and others for motivational purposes.

The **incentive travel** meeting planner has a unique need when compared with other customers of the hospitality product. The incentive planner has to plan for not only the meeting and sleeping room requirements of a group, but also for the group's idea of "fun," such as entertainment, golf, sightseeing, and other activities. This is a difficult task. When you think of your own idea of a good time, it is probably different from that of some other people you know. This problem of disparity challenges the incentive meeting planner.

The incentive meeting planner organizes travel as a reward for superior performance within a group. As stated by the Sales Marketing Network, "what distinguishes incentive or motivational travel from traditional travel is the focus on creating an extraordinary experience for the winner, or one that builds morale, communicates the corporate message, or promotes improved communications between employees and/or the company and its customers."[15] For example, the sales team of a computer manufacturer may have exceeded its sales quota by 30 percent. The reward is a trip to the Caribbean for a week, with spouses. Managers of a retail store chain may be eligible for travel incentives if their profit margins are above a certain quota.

Travel certainly is not the only form of incentive reward, but it is one that projects an image of excitement and relaxation away from the job. When this is done in the group format, teamwork and morale increase with the sense of accomplishment. Merchandise rewards, such as televisions, stereos, and cash bonuses, are the competition to travel rewards. Travel rewards are preferred by many companies, however, and managing that travel becomes an important task.

Exhibit 9-7	2003 Incentive Travel Facts

Number of Programs Run Annually	
1	22%
2	24%
3	2%
4	8%
5	7%
6 or more	37%

Objectives of Incentive Travel Programs (35% or more mentions)	
Increase sales	86%
Recognize performance	74%
Build morale	63%
Increase market share	54%
Improve employee loyalty	49%
Sell new accounts	48%
Build customer loyalty	40%
Create new markets	36%
Foster teamwork	35%

Annual Expenditure for Incentive Travel Programs	
Under $25,000	20%
$25,000–$49,999	9%
$50,000–$99,999	7%
$100,000–$149,999	7%
$150,000–$199,999	6%
$200,000–$249,999	7%
$250,000–$499,999	10%
$500,000–$999,999	15%
$1 million or more	19%

Amount Spent on Trip Delivery per Participant	
Under $1,000	19%
$1,000–$1,999	28%
$2,000–$2,999	21%
$3,000–$3,999	20%
$4,000 or more	12%

Source: Retrieved November 13, 2005, from www.incentivemag.com/incentive/reports_analysis/index/jsp. Used by permission from American Express Incentive Services.

The popularity of incentive travel has led to the growth of **incentive houses,** companies that provide professional incentive planning services and hope to ensure no-hassle, successful, and satisfying trips. The incentive planner often becomes involved in the development of criteria for incentive success. To have winners to send on trips, the framework of the incentive must be established. To motivate successfully, the reward must be worth the extra effort required to achieve it.

Once the framework of the incentive has been developed, the incentive planner must figure out the appropriate travel prizes. Even if incentive planners do not specifically design the promotion, they must have a full understanding of the composition of the group and its achievements. Incentive planners need to establish the perceived level of incentive and plan accordingly.

The actual incentive trip can take three forms: pure incentive, incentive plus, and incentive weekends. The pure incentive trip is dedicated to having a good time without any business-related activities. The incentive plus is a more popular form of incentive

incentive houses
Companies that provide professional incentive planning services for corporations that prefer to have this complex activity handled by an external firm.

trip and represents over 70 percent of the trips taken. Incentive plus trips combine pleasure with some form of meeting or new product introduction. In this way, companies maximize the use of their incentive travel dollars. The company can distribute valuable information without having another sales meeting elsewhere. Incentive weekends are increasingly being used as rewards for good, but less than superior, performance. Companies recognize that although incentive trips are productive, they also take time away from the workplace. Three-day weekend incentives are more cost effective from a time management viewpoint than a trip made in the middle of a work week.

The incentive planner has a multifaceted job when planning the actual trip. Specifically, all phases of the excursion must be carefully planned to enhance the end user experience. This is different from planning corporate or association meetings or conventions. In those cases, the planner plans the functions but leaves it to the individuals to get there and participate. The incentive planner, on the other hand, arranges for literally everything: air and land travel, hotel, food, excursions, sightseeing, entertainment, sports, and anything else that might take place during the trip. Each of these categories can be critical to the success of the trip.

Many companies are not large enough or skilled enough to develop incentive trips internally. A company may have a full-time corporate travel manager and a meeting planner, but the complexities of the incentive purchase are entirely different. For example, staying familiar with destination areas and necessary ground arrangements is incredibly time consuming. Incentive houses are a popular go-between for the companies that need the dedicated attention of a professional. The incentive house is more than a travel agent; professional incentive planners help in all phases of incentive management.

Overall, the incentive organizational customer has a unique job among hotel customers. The "fun" aspect of the planning can be anything but that. Hotels that want a greater share of the incentive market must be extremely flexible in their approach to this marketplace. Standardized approaches to capturing this market are likely not to be fruitful.

ASSOCIATION, CONVENTION, AND TRADE SHOW MARKETS

The association, convention, and trade show markets overlap. **Association** and **convention markets** have similar needs, although they are somewhat different types of groups. Both tend to have large guest room and meeting/function space requirements. An association meeting can comprise a group of people convening on a social basis to elect officers, have social functions, and organize activities on a regional or a national basis. This category of organizational customer also tends to meet throughout the year in smaller groups, and social contacts are a major reason for attendance. There are, of course, many professional (e.g., the American Medical Association) and business associations (e.g., the National Association of Manufacturers) that meet both regionally and nationally to present papers, have board meetings, and set policy.

Convention planners are more focused on annual activities, such as annual meetings of delegates for a political caucus. Other examples are union gatherings to decide policies for the coming year, or a commercial fishermen's convention to plan lobbying

association market
Groups that need guest rooms, food service, and function space facilities to accommodate their meeting needs and convene at the local, regional, national, and international levels. Membership in associations and attendance at their annual meetings are normally voluntary (excluding officers and elected officials of the organization).

convention market
Large corporate or association groups that typically meet on an annual basis.

efforts. The participants may or may not meet throughout the year, and distribution of information, more than social contacts, is the primary goal.

Finally, the main purpose of **trade shows** is to showcase and sell products. This requires wide open space as advertised in Exhibit 9-8. The hotel's task in booking trade shows is to provide the space, ease of access for products to be brought in, and the facilities (e.g., electric power and lighting,) to display the products. This requires a great deal of work, which can be disruptive to other guests. In addition, the hotel sells rooms and meals to exhibitors and those who attend. Exhibitors also make wide use of "hospitality suites" where they entertain customers. This puts heavy pressure on the hotel's room service division, although at high cost to the exhibitors.

Although association, convention, and trade show planners have different reasons for purchasing the hospitality product, their needs are similar and sometimes interchangeable. For example, an association may meet as a convention in connection with a trade show. At times, an entire facility will be booked for a two- or three-day period. Usually, the planner arranges for guest rooms to be held, but reservations are made individually by the participants. The organizer will have a list of VIPs, but the majority of guest rooms are booked by direct contact with the hotel or through the use of reservation forms.

Reservations forms are basically order forms that are provided by the hotel and are designed specifically for the use of attendees. Attendees, of course, are always free to stay somewhere else if they prefer. Thus, the hotel sales department tries to make it convenient for them to stay there. Handling reservations in this manner can make coordination difficult. The hotel must be flexible about meeting the needs of the attendees, many of whom are buying the hotel sight unseen. Strict inventory control is necessary. If, for example, the hotel accepts more king-size bedroom requests than it can accommodate, it may have many unhappy customers. In addition, the sales department needs to coordinate with reservations and revenue management to ensure that attendees cannot book a room at the same hotel less expensively by going through a different distribution channel such as the Internet.

trade shows Exhibitions whose primary purpose is to display and sell products. Trade shows typically take place in large exhibit halls, such as convention centers.

Exhibit 9-8   Atria solicits trade show business on its website

Source: Retrieved September 16, 2005, from www.accor.com/bg/groupe/activities/hotellerie/marques/atria.asp. Used by permission of AccorHotels.com.

Food and beverage is also a unique proposition for hotels in these markets. The organizational buyer tries to be as precise as possible when forecasting the number of people who will attend meal functions, but the actual attendance can vary widely. If there are alternatives, as in a large city, many attendees will go out for meals. Attendances at meal functions can vary widely even within the same meeting. The first night's award banquet might have close to 100 percent attendance. The following night might have a boring speaker scheduled, and half the attendees may choose to go elsewhere.

Association, convention, and trade show planners need extremely good convention service managers within the hotel to execute all phases of the event. These managers are more important than the salespeople in delivery of the final product. Rutherford and Umbreit found that convention service managers of hotels had the greatest number of encounters with meeting planners during the process of planning and executing an event than any other personnel.[16] This may be true for a meeting of any size, but it is especially true for large, complex ones. Technical details such as the voltage in the main ballroom, the delivery space for exhibits, and the audiovisual support for the speakers are all critical to the success of the meeting.

Delegates to these kinds of functions often will not stay for the duration of the meeting. They may book for three nights yet stay two nights and not give any notice of the early checkout (although hotels are beginning to penalize this type of behavior). Many are small business people who cannot make definite plans for the future; others will simply decide they've had enough and leave.

Delegates to these functions also tend to be quite price sensitive. The organizer, who wants to keep the delegates happy, looks for low rates and for low-cost or free meeting space. All three of these markets are tough to sell and tough to service, but they can represent profitable business, especially if booked during slow business periods.

CONVENTION CENTERS AND CONVENTION AND VISITORS BUREAUS

convention center
A freestanding large exhibit hall where trade shows are typically housed. Unlike most conference centers, convention centers do not have lodging accommodations (though hotels are often adjacent to or near the facility).

Two external bodies often closely involved in the handling of association, convention, and trade show marketing need brief mention. The first is the freestanding **convention center,** sometimes called a conference center, but generally meaning a freestanding, independent property. Most major cities in the world and many secondary and tertiary cities, especially in the United States, have such convention centers. In these cases, the main event takes place in the convention center, which is usually publicly owned but privately operated. The trade show or convention itself, on the other hand, may be handled by a private organizing firm.

The National Restaurant Association (NRA) annual trade show in Chicago and the annual Consumer Electronics Show in Las Vegas are examples of trade shows held in convention centers. Either booths or space are sold to purveyors, and attendees inspect the offerings under one roof. Although informational seminars may be given during the show, the main purpose of the event is to display products and take orders. The trade show organizer makes money from the booth or space sales. In turn, the purveyors hope to write enough business to make their expenses worthwhile.

The American Hotel & Lodging Association (AH&LA) annual trade show is in New York City each year. The Tourism Marketing Application highlights how the convention center, the AH&LA trade show planners, and local hospitality and tourism operators work together to market the event.

Many trade shows are arranged, at least initially, through **convention and visitors bureaus (CVBs).** These organizations are publicly and privately supported by those they serve: convention centers, hotels, restaurants, merchants, theaters, airlines, and so forth.

CVB organizations are nonprofit, serving their constituents, who pay annual fees. They exist in both large and small cities. Their mission includes promoting the city as a destination area, assisting groups with meeting preparations, providing promotional material to encourage attendance, and working with hotels to coordinate room blocks. Hotels and restaurants usually work very closely with CVBs. If the CVB can sell the city to a group, it then provides information on hotel accommodations, restaurants, and other attractions that are part and parcel of the overall enticement.

Exhibit 9-9 details what every meeting planner should know about CVBs. Clearly, CVBs offer many services.

convention and visitors bureaus (CVBs) Local, nonprofit organizations that help market a city. CVBs are supported by and work closely with hotels, restaurants, and local attractions to foster tourism business for the area.

SMERF AND GOVERNMENT MARKETS

The **SMERF market** (social, military, education, religious, and fraternal) is not solicited by many major hotels because their inventory is primarily targeted to more upscale corporate customers and associations. Yet during weak supply and demand times of the year, SMERF customers are being called on by many hotels. The SMERF market is now considered a "segment" by the Professional Conference Managers Association (PCMA). SMERF customers include all organizational customers that do not fit into the other categories; hence, they are considered to be part of a catchall market. Major submarkets of SMERF—social, military, education religious, and fraternal—cover most of this market.

SMERF market Price-sensitive, not-for-profit organizations including social, military, education, religious, and fraternal.

Tourism Marketing Application

Marketing the Annual New York Hotel Show

To help market the annual New York hotel show each year, several entities come together in a joint effort to enhance marketing for the trade show. The Jacob Javits Convention Center (where the show takes place), AH&LA (the main sponsor of the event), local hospitality establishments, and other service providers take part in this endeavor. Collateral is sent out by the show operator, which includes locations and guest room rates of selected hotels in Manhattan, schedules of free bus service put together specifically for attendees of the show, and restaurant and theater advertisements to make it easier for attendees to plan their stays. The Jacob Javits Convention Center is located on the west side of the city, somewhat out of the mainstream; thus, attendees find this information quite helpful.

Exhibit 9-9	What Every Meeting Planner Needs to Know about CVBs

Misconception #1: CVBs solely book hotel rooms and convention space.
Fact: CVBs represent the gamut of visitor-related businesses, from restaurants and retail to rental cars and racetracks. Therefore, they are responsible for introducing planners to a full range of meeting-related products and services the city has to offer. Basically, they match needs to a city's resources.

Misconception #2. CVBs only work with large groups.
Fact: More than half of all meetings involve less than 200 people. These meetings are just as important to a CVB as larger ones. In fact, larger bureaus often have staff members specifically dedicated to small meetings.

Misconception #3: Bureaus own and/or run the convention center.
Fact: Only 5 percent of CVBs run the convention center in their location. Nevertheless, CVBs work closely with local convention centers and can assist planners in getting what they need from convention center staff.

Misconception #4: Planners have to pay CVBs for their services.
Fact: In truth, the services of a CVB are free. Michael Gehrisch, president of the International Association of Convention & Visitor Bureaus (IACVB), points out, "Convention bureaus are both a hotel's and a meeting planner's best friend. They don't charge either one, but book business for the hotel without a fee and provide the same service, for free, to planners." Most bureaus are primarily funded through hotel occupancy taxes. Some bureaus also charge membership fees.

Some may question the need to work through a CVB when planning a meeting, particularly in cases where the bulk of an event takes place at one hotel or at the convention center. The bureau can help you work with those entities and can help fill out the convention schedule with off-site activities (including spouse tours and pre- and post-conference tours). An objective resource, the bureau can direct planners to products and services that will work best to accommodate their needs and budgets. In summary, a CVB acts as a mediator, matching meeting needs to the products, services and speakers available in a community.

Why use a CVB? CVBs make planning and implementing a meeting less time-consuming and more streamlined. They give meeting planners access to a range of services and packages.

Before a meeting begins, CVB sales professionals can help locate meeting space, check hotel availability and arrange for site inspections. CVBs can also link planners with the suppliers, from motorcoach companies and caterers to off-site entertainment venues, helping meet the prerequisites of any event.

What are some of the specific services CVBs offer planners?

- CVBs can offer unbiased information about services and facilities in the destination.
- CVBs serve as a vast information database and a one-stop shop, thus saving planners time, energy and money in the development of a meeting.
- CVBs act as a liaison between the planner and the community. For example, CVBs are aware of community events that may beneficially coincide with your meeting (like festivals or sporting events). They can also work with city government to get special permits and to cut through red tape.
- CVBs can help meeting attendees maximize their free time through the creation of pre- and post-conference activities, spouse tours and hosting of special evening events.
- CVBs can provide hotel room counts and meeting space statistics and will keep a convention/meetings/events calendar in order to help planners avoid conflicts and/or space shortages.
- CVBs can match properties to specific meeting requirements and budgets.

Other services provided to planners include:

- Collateral material
- Help with on-site logistics, including registration
- Housing bureaus
- Auxiliary services, such as production companies, catering and transportation
- Site inspections/familiarization tours and site selection
- Speakers and local educational opportunities
- Security
- Access to special venues

The overall job of a CVB is to market and sell a destination. A CVB wants every single client to be happy. It will do everything it can to match every client with the perfect setting and services for their meetings. The bottom line—the CVB is working for you.

Source: Retrieved November 13, 2005, from www.pema.org/resources/convene/archives/displayArticle.asp?ARTICLE_ID=4622. Used by permission. Reprinted with permission of *Convene*, the magazine of the Professional Convention Management Association. © 2006. www.pcma.org.

So what is the SMERF market? Essentially, it is a price-sensitive, nonprofit organization market. All social-related group business is considered SMERF. Wedding parties needing overnight accommodations, rehearsal dinner parties, society events, fundraisers, and so forth, are all considered part of this market. So are gospel singers and military customers who use hotel rooms for reunions and travel on business. The education subsegment consists of groups such as faculty and school sports groups. Religious groups include large Baptist conventions filling cities or the Order of the Rising Star meeting in hotels. Finally, fraternal orders such as Elks Clubs and the Benevolent Order of Moose fall into the SMERF market.

Although the SMERF market has the reputation of being low rated, the customers nevertheless fill guest rooms, ballrooms, and local restaurants, especially during slow periods. The Head Buffalo of an Elks group can be no less important a customer to some hotels than a corporate meeting planner.

The **government market** is also low rated, but in the United States it is a $12 billion market. It is large in other countries as well. Again, economic slumps in bookings may lead hoteliers to see government as an attractive market. This market is a reliable source of revenue for many budget and midlevel properties. Upscale properties also cannot ignore the upper end of this market: higher-level government officials.

Government at all levels is engaged in many activities that tend to be travel intensive: research, regulation, investigation, enforcement, oversight, litigation, education, and coordination. Although government employees may be end users, they may not be the customers to whom to make the sale. Government travel planning is an official affair. A program manager or travel coordinator is often responsible for their reservations, and per diem rates (maximum daily expense allowance) are set by state or national law. Hotels often target this market for weak demand periods. Similar to the SMERF market, the government market tends to be fairly price sensitive and regularly looks for discounted rates.

> **government market**
> Those who travel for official government business at either the local, state, provincial, or federal level. This could be for individual or group purposes.

> **Web Browsing Exercise**
> Go to your favorite search engine and search the word "SMERF". What entries do you find? In addition, look up different fraternal organizations such as Free Masons, Elks Club, and Rotary International. What is the purpose of these organizations? How can hospitality firms help these organizations reach their objectives? Do you believe such organizations will be around in 10 or 15 years? Why or why not?

GROUP TOUR AND TRAVEL MARKET

This market is defined as leisure travelers who travel in groups, with or without an escort. This is a wide-ranging market that has changed dramatically in recent times and is no longer characterized by hordes of ignorant travelers visiting five countries in four days. Tours may range from trekking in the Himalayas to whale watching in the Pacific; from a ladies garden club tour of Japan to a high school senior trip to Spain. Group tour travelers have different motivations for selecting this form of travel, the most important of which is generally the convenience of having all arrangements made for them. Other motivations include companionship (especially among mature travelers), lower travel costs, and planned itineraries that will ensure that travelers do not miss the "must-see" places. There are over 2,000 tour operators in the United State and many more around the world. Tour operators usually belong to trade associations such as the CrossSphere (formerly known as the National Tour Association), the American Bus Association, or the United Bus Owners of America. Member directories provide useful information and are a good starting point for hospitality marketers interested in pursuing this type of business. Although there are many kinds of tours, a common type is the escorted **motorcoach tour** or the group inclusive tour (GIT) arranged by wholesale tour brokers.

> **motorcoach tour** When five or more travelers arrive at a hospitality establishment by motorcoach as part of an organized leisure travel tour package.

Motorcoach Tour Travelers[17]

In the United States, the motorcoach tour market segment has traditionally consisted of older travelers. In other parts of the world, however, where owning cars is not as prevalent, motorcoach tours have long been a popular mode of sightseeing and vacationing. Things are changing in the United States as well. Many younger travelers are

| Exhibit 9-10 | The Tauck Tours Difference |

TAUCK TOURS MISSION STATEMENT

Our mission is to offer enriching travel experiences that enhance people's lives by broadening their knowledge and fulfilling their dreams. To that end, we strive to create unique and imaginative travel itineraries.

By conducting each Tauck vacation with pride and enthusiasm, we bring to our guests the best possible experience in every destination. It's a difference that's uniquely Tauck.

CAREFREE TRAVEL, BOTH RELAXED AND REFINED

An important benefit of "Tauck Style" is the freedom to relax and travel as an individual because your needs and expectations are anticipated. Our carefully designed and choreographed itineraries are a refreshing mix of sightseeing, leisure, adventure and culture. You'll experience spectacular scenery and exciting events other travelers only dream about. Every detail is handled seamlessly. Imagine . . . luggage usually in your hotel room before you arrive and preferred seating at shows! Your time is truly your own to enjoy.

UNIQUE AND DISTINGUISHED HOTELS IN THE BEST LOCATIONS

When you travel with Tauck, you enjoy some of the best rooms at the world's finest hotels. With Tauck, you may stay in oceanfront rooms at the Royal Hawaii Hotel, lake view rooms at Chateau Lake Louise and you'll be on the Grand Canyon rim at Kachina Lodge. Many of our hotels, such as the Raffles Hotel in Singapore, the Grand Hotel in Paris and the Westin Excelsior Hotel in Rome, are rated among the world's finest.

EXCEPTIONAL DINING AND FREEDOM OF CHOICE

The enjoyment of dining is a delightful and integral part of travel. That's why Tauck selects restaurants that offer a true "taste" of each region you visit. You have the freedom to dine from regular restaurant menus—whether you're in the mood for a sophisticated Chateaubriand in a hotel's "signature" restaurant or a specialty salad in the café. You even have the freedom to dine anytime and with anyone you wish. You simply sign the check "Tauck World Discovery."

INSIGHTFUL TOUR DIRECTIONS

During your travels, you have the advantage of being with a Tauck Tour Director who is fully knowledgeable about the destination, its history and its culture. You can relax knowing that you are accompanied by a professional who manages the flow of events . . . a problem-solver should one arise . . . an insider who saves you precious time . . . and a friend who will help transform your trip into a truly memorable journey.

EVERY TRAVEL EXPENSE IS INCLUDED

When you travel with Tauck, you will save up to 40 percent from the cost of traveling the same itinerary on your own. Virtually every travel expense is included . . . hotels, meals, cruises, sightseeing, entertainment—even taxes and gratuities. Guests are delightfully surprised how much is included—you can practically leave your wallet at home! A Tauck vacation offers you both superb quality and extraordinary value.

FREQUENTLY ASKED QUESTIONS

How do you describe Tauck World Discovery and how is your brand different?

For three generations we have focused on providing travel experiences that truly enhance people's lives. We create journeys that go beyond the ordinary; they are innovative, authentic and unique experiences that our guests couldn't get "on their own." Our company is different in many ways; predominantly in our highly knowledgeable and passionate staff that helps others to broaden their horizons and see a destination in a new way. We are very different from mass market "tours" in that we offer "in-depth experiences" in a variety of different ways, such as luxury small-ship cruising, Heli-hiking and rail journeys. We include the finest accommodations available, our tours are all-inclusive and we take care of all the details. This is what makes our brand different!

What new travel patterns have emerged for 2004/2005?

The upscale travel market is on a fast rebound and Tauck is experiencing some significant new trends in travel. Most notably, travel to Europe is up 75 percent, with incredible growth in interest to Eastern Europe, Scandinavia, Russia, Italy, Great Britain and Ireland. Family travel remains strong, including family reunions and three-generation family travel, as evidenced by a 150 percent growth in Tauck's family-focused sister company, Tauck Bridges. Tauck's small ship cruises are seeing significant growth as the market seeks a higher quality, intimate cruise experience to fantastic, remote destinations. For the long term there continues to be growing interest in exotic destinations, including the Galapagos Islands, Costa Rica, Australia, New Zealand and South Africa. More than ever guests are seeking the finest "all-inclusive," guest protection and "done for you" services as they venture abroad once again.

For many people, vacations are becoming shorter (such as one-week tours to Europe) and multi-generational family travel is very popular. Europe is a treasure trove of experiences and there is growing interest in year-round travel to Europe—by train, riverboat and lightly traveled roads—all unique ways to explore cultural "hidden gems" of the Old World.

Source: Tauck World Discovery website. Retrieved January 3, 2006, from www.tauck.com

using motorcoach tours to see domestic sights inexpensively. Exhibit 9-10 shows website information for Tauck Tours, a global leader in motorcoach tours.

The motorcoach tour market for hotels and restaurants may be defined as five or more travelers arriving at a hospitality establishment by motorcoach, as part of an organized leisure tour package. This market really has to be separated from other travelers arriving at the hotel by bus, simply because of their reason for the purchase. A group of corporate business people could arrive at a hotel by bus, yet their main purpose for the visit would be to attend a corporate meeting, making them a corporate group. A convention could have an entire delegation from a similar geographic area arrive by bus, but again the main reason would be to attend the convention, not to visit local attractions.

When soliciting the motorcoach market, hoteliers should respond to its specific needs. Tours employ tour leaders responsible for the well-being and satisfaction of the group. Tour leaders are also, in essence, sales representatives for the tour company. The hotel salesperson sells the hotel to the tour company, but the tour leader is the one who travels with the group and ensures their satisfaction. As with most other products, many similar tours are available to the consumer, and often the tour leader develops a following of repeat customers.

Although the average age of motorcoach tour travelers is dropping into the 50s, older travelers typically prefer guest rooms that do not require the use of stairs. They prefer rooms with views, they like being close to each other, and they want the correct bedding configuration—all requirements that may result in some misunderstanding. A weekend package user might ask for a double room and be completely satisfied with a queen bed assignment. The same double room request by the motorcoach guest may indicate a need for two beds in the same room—and a roll-away is not an acceptable substitute.

Motorcoach tour groups are a practical market for many hotels; in fact, some hotels survive on them. The warning here is the one we have mentioned before: These customers may not mix well with some other market segments, and special care is needed to see that this mix is not a problem. Unlike other travelers, the needs of these travelers arise all at once. An average busload of 40 means 40 bags all at once (maybe 80), 40 luncheons all at once, and 40 breakfasts all at once. Staffing to handle this is important, especially when these customers normally do not tip well and employees are not especially eager to serve them.

Chapter Summary

Organizational meeting planners and travel managers are unique to the hospitality industry in that they often represent the purchasers but not the users. These people are responsible to the organizations they represent and have to prepare for the wide variety of needs of the members of the organization.

Overall, organizational planners and travel managers are better educated about, and have more experience in, the hospitality industry than the individual purchasers of hotel and restaurant products. The most important factor in their destination and hotel decision-making process remains the word-of-mouth endorsement of their fellow professionals. References from someone within their organization also help steer this customer toward a specific hotel or resort. And the conference service manager at the host hotel probably plays a more significant role than the salesperson in creating and keeping customers.

Conference centers are lodging establishments that are specifically designed to meet the needs of the meetings market and compete head-on with the hotel industry to capture this business.

The association, convention, and trade show markets constitute large groups that have extensive guest room and function room space requirements. Trade shows, in particular, often use convention centers for their function space needs with nearby hotels providing overnight lodging accommodations.

The incentive, SMERF, and government markets are also major organizational buyers in the hospitality industry. The SMERF and government markets tend to be price conscious, whereas the incentive market is usually top-level.

Convention and visitors bureaus (CVBs) promote the cities (and areas) they represent to all of the preceding organizational buyers. They work in concert with hotels, restaurants, motorcoach tours, local attractions, and the community at large to help market and sell the destination.

Key Terms

association market, p. 284

buy time (lead time or purchase cycle), p. 271

conference centers, p. 280

consortia, p. 278

convention center, p. 286

convention market, p. 284

convention and visitors bureaus (CVBs), p. 287

government market, p. 289

incentive houses, p. 283

incentive travel, p.282

International Association of Conference Centers (IACC), p. 281

meeting planner, p. 269

MICE, p. 269

motorcoach tour, p. 289

organizational customer, p. 269

SMERF market, p. 287

trade shows, p. 285

travel manager, p. 269

Discussion Questions

1. What are organizational customers, and how they are similar to and different from individual customers?

2. Who are some of the major group market players in hospitality and tourism?

3. What is meant by "buy time" (sometimes referred to as "lead time" or "purchase cycle")?

4. How is the job of the meeting planner somewhat different from that of the travel manager?

5. How do conference centers differ from convention centers?

6. What are convention and visitors bureaus, and what is their role in hospitality and tourism?

Endnotes

1. In the 2003 "State of the Industry" report prepared by *Successful Meetings,* the average number of full-time meeting planners in an organization had dropped from seven in 2000 to four in 2002. In the 2003 "State of the Industry" report, the number of participants who identified themselves as corporate planners was 33.7 percent. Another 27.4 percent claimed they were association planners, and 13.1 percent claimed to be independent planners. Retrieved January 3, 2005, from www. successmtgs.com/successmtgs/images/pdf/2003-01-coverstory.pdf. see also www.successmtgs.com/successmtgs/search/search_display.jsp?vnu_content_id=1957337.

2. Yesawich, Pepperdine, Brown, & Russell (2004). *Portrait of North American meeting planners,* 5.

3. Ibid., 59.

4. Seal, K. (1987, July 20). Staff, service, top priorities for planners. *Hotel & Motel Management,* 40–43.

5. Mandelbaum, R. (2004, December). Understanding the recovery occurring in the meeting's market—Special report. *Hotel Online.* Retrieved January 3, 2005, from www.hotel-online.com/News/PR2004_4th/Dec04_PKFConventions.html.

6. Yesawich, Pepperdine, Brown, & Russell, 23.

7. American Express 1994 Survey of Business Travel Management. New York: American Express Travel Related Services Company. Inc.

8. Ibid.

9. Warcholak, E. S. (2004, August 2). Travel manager salary and attitude survey. *Business Travel News.* Retrieved January 4, 2005, from www.btnmag.com/businesstravelnews/headlines/breaking_news.jsp.

10. Bell, R. A. (1993, April 1). Corporate travel-management trends and hotel-marketing strategies. *Cornell Hotel and Restaurant Administration Quarterly,* 31–39.

11. *Meetings and Conventions.* 2004 Meetings market report. Retrieved January 3, 2005, from www.mcmag.com/mmr2004/index.html.

12. Ibid.

13. IACC North America website. Retrieved January 3, 2005, from www.iaccnorthamerica.org/about/index.cfm?fuseaction=memcrit.

14. Shaw, M. (1985). *The group market: What it is and how to sell it.* Washington DC: The Hotel Sales and Marketing Association International Foundation, 45.

15. Sales Marketing Network website. Incentive travel overview: Resources & statistics [no. 4010]. Retrieved January 3, 2005, from www.info-now.com/moshow/article55#stat.

16. Rutherford, D. G., & Umbriet, W. T. (1993, February). Improving interactions between meeting planners and hotel employees. *Cornell Hotel and Restaurant Administration Quarterly,* 68–80.

17. To see how one independent hotel captured the senior citizen bus tour market, see Shoemaker, S. (1984). Marketing to older travelers. *Cornell Hotel and Restaurant Administration Quarterly, 25* (2), 84–91.

Bridgeport Inn & Conference Center

Discussion Questions

1. Shown in Exhibit 1, what are the typical types of meetings and major sources of business for executive conference centers in general?

2. Describe what Bridgeport Inn & Conference Center is all about. Who are the major clientele for this hospitality organization?

3. What is unique about the Bridgeport Inn in Bridgeport Village?

4. What is the pricing strategy for its conference business?

5. What types of competition did the Bridgeport Inn & Conference Center face?

6. Describe in your own words the essence of the occupancy statistics shown in Exhibit 2 for the conference versus the social markets.

7. Though at first somewhat confusing, how does the marketing function work at the Bridgeport Inn in Bridgeport Village? What is the relationship/communication link with corporate headquarters?

8. What advice do you have for Mr. Biden, the general manager? Should he be given more decision-making authority at the local level given the unique situation of the Bridgeport Village property?

9. Relatedly, should Mr. Biden seek approval to hire on-site sales people even though this would certainly increase his marketing budget and is not done at the other Bridgeport Conference Services properties? Does it make sense?

According to John Phillips, CEO and president of Bridgeport Conference Services, typical meeting planners devote only 20 percent of their time to planning meetings. In most cases, the individual responsible for the organization of the meeting is also a participant who finds it difficult to both host the event and make a suitable contribution. "They aren't 'pros,'" explains Phillips, "so we try to take the worry away from them to let them concentrate on the business at hand. Our aim is to make them look like a hero."

A consulting firm profiled this industry segment, based on a small sampling of executive conference centers, as shown in Exhibit 1.

This case was contributed by Scott Flagel and Karl Grover and developed under the supervision of Robert C. Lewis. Names and places have been disguised. All rights reserved. Used by permission.

Exhibit 1	Profile of Executive Conference Centers

- **Meeting Types:** Executive conference centers were most often used for training sessions (44% of all meetings, with a mean group size of 28), management planning (27% of meetings with a mean of 23), and sales meetings (16% of meetings with a mean of 44).

- **Sources of Business:** Most conference center meetings were sponsored by business organizations (82%), although trade and professional associations accounted for 9% of the business at participating centers, and academic institutions and government bodies occasionally met at the centers.

- **Occupancy:** Although some conference centers did play host to transient traffic in order to boost business during slow periods, average annual occupancy among the centers was still low at 59%. December and January were the slowest months; March and October the busiest.

- **Traffic Patterns:** Conference centers reported that meeting participants most often arrived on Sunday or Wednesday and stayed through Friday. An earlier trend toward weekend meetings appeared to be reversing, except in resort areas where a Friday arrival was common. Conference centers handled an average of 5 to 12 meetings each week.

- **Recreational Facilities:** Most executive conference centers offered tennis, swimming, golf, an exercise room, and a game room; some also boasted facilities for bowling and horseback riding. (Although most centers allowed meeting participants to have unlimited access to these facilities, some derived additional revenue from recreation.)

- **Operating Statistics:** Revenues and operating expenses of executive conference centers compared with those of convention hotels found most operating ratios to be similar. At 8.5% of sales, however, marketing costs at executive conference centers were almost twice those of convention hotels (4.9%)—presumably reflecting the need to communicate both the existence of a new facility and a fairly new concept to prospective clients.

BRIDGEPORT CONFERENCE SERVICES

Bridgeport Conference Services, Inc., was a large and highly developed marketer of meetings at conference centers. It operated nine properties around the country. The emphasis at Bridgeport Conference Centers was on self-contained meeting packages with exceptional services to enhance the productivity of meetings. As opposed to a hotel in a major metropolitan area complete with all the distractions, Bridgeport Conference Centers were located in natural settings away from the hustle and bustle of the city.

The centers offered extensive recreational activities, which invigorated the body as well as the mind. The advantage of the self-contained facility was that meeting participants tended to discuss related business matters while enjoying the leisure activities, thus increasing the overall productivity of the meeting. Bridgeport Conference Services catered to Fortune 500 companies that normally demanded first-class dining, accommodations, and amenities and looked toward executive conference centers to improve the productivity of their meetings.

Bridgeport Conference Services performed most external marketing functions at the regional and national levels, rather than at the individual properties. Bridgeport Conference Services believed this was beneficial for the company because interested parties could contact one central office to obtain information on any or all of the properties.

All of the centers operated by Bridgeport Conference Services were closed to transient guests, with the exception of the Bridgeport Inn at Bridgeport Village. This facility was open 365 days a year to the general public whenever rooms were available and not used by conferences.

THE BRIDGEPORT INN

The Bridgeport Inn at Bridgeport Village was conveniently located close to the major metropolitan areas of the Northeast, yet completely removed from the noise and distractions of city life. The Inn itself was located in Bridgeport Village, a self-contained community that also boasted a 3000-unit condominium complex, a shopping bazaar, a professional building, and a bank. The physical structure of the inn reinforced the natural atmosphere.

The meeting rooms of the inn and the very heart of its business were fixed in size, not having the retractable walls that allow for greater flexibility. The missing thin, retractable

walls, trademarks of many downtown convention hotels, were replaced with soundproof walls, which did not allow the outside world to enter. The meeting rooms were equipped with comfortable chairs, table space, screens, and state-of-the-art audiovisual equipment which, along with coffee breaks, were included in the price of the room.

The 121 guest rooms were also fitted in the rustic mold, with a camplike firmness to the beds and custom furnishings in the bright and airy rooms. Most of the rooms had two twin beds, although more and more were being converted to include queen-size beds and private sitting areas in response to customer requests. Deluxe facilities at the inn included a golf course, outdoor swimming pool, tennis courts, health club with saunas, billiard room with connecting pub, and a beautiful restaurant and lounge called Woody's.

The Bridgeport Inn was owned by the Bridgeport Village Developers and operated by Bridgeport Conference Services. Joseph Biden, the general manager of the property, believed that there were problems inherent in this type of arrangement. For example, plans for an expansion of 80 rooms and an indoor fitness facility consisting of a swimming pool, racquetball courts, and an improved health club had been discussed again and again with no decision being made. Biden felt that the expansion would be not only beneficial in the long run, but actually essential if they intended to maintain their competitive position in the years to come.

The pricing strategy of the hotel was to offer an all-inclusive price that included a room, three meals a day, coffee breaks, meeting facilities, and the use of all recreational facilities with the exception of the golf course, where there was a minimal greens fee. The conference rack rate was $275 per person double occupancy and $345 per single occupancy, which was comparable to city hotels offering the same services.

The hotel split its markets into two categories: conferences, which had a 50 percent repeat factor, and transient social customers, who returned 60 percent of the time. The conference markets comprised almost three fourths of all business and were very diverse demographically. The conference market also generated the greatest percentage of the inn's revenue.

Organization

The conference manager of the inn served as the liaison between the regional sales office and the property's clients. There was no sales department for conferences on the premises. When arranging with a future meeting planner, the conference representatives acted as consultants to the company planner. This was beneficial to the client because the representatives were very experienced in responding to particular meeting to requests.

The operations manager had created strategies to encourage better use of the food and beverage outlets, such as promoting sports-oriented theme nights in the lounge to the mostly male guests. Woody's restaurant catered to the local market and also those social guests not on a meal plan, in addition to the American plan conference guests. Business was brisk on the weekend nights, and Sunday brunch was also popular with the locals.

The general manager, Mr. Biden, was the final member of the executive staff. He was pleased with the performance of the hotel, but would have liked to add an on-site sales position to the budget. All executive committee members agreed that a salesperson who was based at the inn would be better equipped to make a sale than one who was geographically distant.

Competition

Because of the success of hotel concepts similar to the Bridgeport Conference Centers, there had been an increase in competition from two sectors. The first form of competition was from hotels that had shifted their marketing efforts toward executive conferences when they recognized the potential for revenue. The second was from companies invading the niche that Bridgeport had carved out. An example of this was the Arrowwood facility, which was located not far away in Rye, New York. Arrowwood was built at an average room cost of $250,000 and boasted the latest in audiovisual equipment and indoor recreational facilities, although at a higher price than Bridgeport. More competition like this was expected as more and more companies attempted to jump onto the executive conference bandwagon.

Demand

The occupancy rate for the Bridgeport Inn at Bridgeport Village had hit a peak two years before at 69 percent. Mr. Biden believed that the occupancy rate was still well above the industry average, but was nonetheless concerned with the negative trend in occupancy. He was convinced that decisions by the developers had triggered this downward spiral, stating

that "not adequately renovating the existing facility and not proceeding with the expansion/addition has contributed to the occupancy problem. The addition of 80 rooms would help meet the demand of weekday requests for conferences, and the indoor facility would attract more winter guests."

For the conference market, January and February had the lowest weekday occupancy, but the remaining months were consistently higher. Weekend occupancies, however, fluctuated drastically as a result of special packages and other incentive plans. The impressive weekend occupancy rates for January and April were a little misleading because room rates were discounted considerably in those months.

For the social market, the biggest demand fluctuations occurred between summer and winter months and between weekdays and weekends. The hotel had much more to offer the social guest in the summer. Although overall weekend occupancy had improved over the last five years, it still ran a poor second to weekday occupancies. Occupancy information for the previous year is shown in Exhibit 2.

THE MARKETING FUNCTION

Mr. Biden was concerned about the negative trend in occupancy. He had done what he felt was necessary at the local level, but any long-range actions were subject to approval of both Bridgeport Village and Bridgeport Conference Services.

A problem, however, existed among Bridgeport Village Developers, Bridgeport Conference Services, and the management of Bridgeport Inn in terms of short-range versus long-range objectives. Management had felt constraints in having to set objectives that adhered to short-term goals while also keeping long-range objectives in mind.

The marketing function of Bridgeport Inn was somewhat diversified. First, there was the corporate marketing department that handled all but local group sales, advertising, and corporate strategy. In fact, 88 percent of conference bookings went through the main sales office at corporate, which handled all but on-the-spot arrangements. This was designated corporate procedure even to the point that when the inn received inquiries, it was required to refer them to the corporate office.

Then, there was the network of interactions between the corporate marketing department and the inn's catering and conference sales force, which could at times become somewhat tense. Finally, operations management had the

| Exhibit 2 | Occupancy Information* |

Social Market

Month	Weekday	Weekend	Total
January	8.0	23.5	12.8
February	9.9	27.3	16.1
March	9.8	37.7	17.9
April	13.3	26.9	18.3
May	17.7	49.7	28.0
June	7.3	56.8	20.5
July	17.3	59.6	33.6
August	13.8	44.7	21.8
September	15.7	53.1	30.4
October	9.8	50.4	25.5
November	8.9	47.4	19.0
December	5.1	22.4	18.6
Average	**11.4**	**41.6**	**21.9**

Conference Market

Month	Weekday	Weekend	Total
January	57.0	33.3	49.4
February	49.6	6.6	34.2
March	61.5	8.2	46.0
April	61.2	19.5	45.9
May	61.9	5.1	43.6
June	71.8	13.8	56.4
July	65.1	5.6	42.1
August	59.8	3.1	45.2
September	74.0	5.0	54.6
October	83.4	10.0	59.8
November	68.5	9.5	46.2
December	72.0	2.2	37.6
Average	**65.5**	**10.2**	**46.8**

*Weekday occupancy is Sunday night through Thursday night, and weekend occupancy is Friday and Saturday night.

responsibility of marketing the restaurant and lounge both internally to in-house guests and conferees and externally to the local community. The overall marketing function was covered by all three of these divisions, but no one person was involved and responsible for the mass marketing activities of the inn.

Mr. Biden noted that, in spite of the decline, the customer repeat rate was over 50 percent of the total business.

He wondered if this was good or bad; was the inn losing customers, was it losing potential customers, was the overall market declining, or was it a combination of all three? He summed up the situation by saying, "Our bottom line is fantastic in terms of profit, but we have to correct the problems at hand. Only then will we be able to make meeting planners look like heroes."

overview

In this chapter we focus on the importance of travel and of local residents' support for the success of a tourist destination and the types of marketing activities that national tourism organizations (NTOs) perform in the marketing of countries as tourist destinations. To explain how public and private sectors collaborate to market a destination, we provide an interview with the VP of marketing of the Hilton Waikoloa to show how a specific hotel works with a destination to promote itself and the destination.

continued on pg 300

chapter 10

Understanding Tourism Markets

learning objectives

After reading this chapter, you should be able to:

1. Identify current trends in international tourism.

2. Define national tourism organizations (NTOs) and destination marketing.

3. Explain how individual hospitality and tourism organizations work with NTOs to market their destination together.

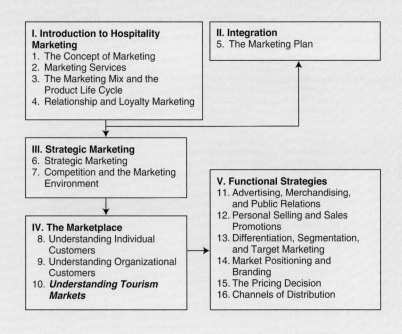

I. Introduction to Hospitality Marketing
1. The Concept of Marketing
2. Marketing Services
3. The Marketing Mix and the Product Life Cycle
4. Relationship and Loyalty Marketing

II. Integration
5. The Marketing Plan

III. Strategic Marketing
6. Strategic Marketing
7. Competition and the Marketing Environment

IV. The Marketplace
8. Understanding Individual Customers
9. Understanding Organizational Customers
10. *Understanding Tourism Markets*

V. Functional Strategies
11. Advertising, Merchandising, and Public Relations
12. Personal Selling and Sales Promotions
13. Differentiation, Segmentation, and Target Marketing
14. Market Positioning and Branding
15. The Pricing Decision
16. Channels of Distribution

This chapter also examines the current macro environmental trends and ways to identify these trends. Finally, we examine the types of tourists (e.g., one who visits friends and families versus one who travels for rest and relaxation) and how to communicate with them.

Marketing Executive Profile

Vincent Vanderpool-Wallace
Secretary General, Caribbean Tourism Organization

Before being selected as secretary general of the Caribbean Tourism Organization, Vincent Vanderpool-Wallace was director-general of the Bahamas Ministry of Tourism. Prior to that he had been involved in the tourism industry for more than 15 years, holding senior management positions in both the private and public sectors.

Born in Nassau, Bahamas, Vanderpool-Wallace graduated from Harvard University and then worked for the Ministry of Education and Culture from 1975 to 1977. Joining the Ministry of Tourism's marketing department, he rose to the position of deputy general manager of marketing, which he held from 1979 to 1982. In 1982 Vanderpool-Wallace joined the staff of Resorts International (Bahamas) Ltd., where he held various managerial positions including senior vice president in the Office of the President.

 Marketing in Action

Vincent Vanderpool-Wallace
Secretary General, Caribbean Tourism Organization

Please briefly explain the role of the Caribbean Tourism Association.

Right now we have 32 member states within the Caribbean, from Barbados to Bermuda. We all get together—recognizing that tourism in this area is the dominant industry—and try to find ways to collaborate for the greater good of tourism development for the entire region. Each member country pays dues that are proportionate to the total number of visitors that they receive. It has been in existence for over 50 years.

What are the key challenges in marketing a destination such as the Caribbean?

Each member country is an individual sovereign nation that has a right to develop tourism in any way, shape, or form they wish. So, although we have established certain policies that we think would make their development of tourism travel more efficient, they don't have any obligation to follow our lead in anything that we're doing. We are always trying to allow enough leeway for our members to do what is necessary in their individual interests while at the same time advancing the cause for the Caribbean as a whole.

The other issue that we have been trying to grapple with is that you have different markets that the individual members are going after and also individual products within each of those segments. It is a peculiar situation in that we are trying to market diversity, which is very important, we think, to the long-term success of the Caribbean, and many people see this variety of products and variety of markets as a problem.

How does your organization identify key market segments and key competitors?

We've got some emerging trends that we can clue our numbers onto, so that we can prepare a long time in advance for them. These trends are in terms of the types of products and services that need to be delivered in any kind of product you have or any market you're going after. We are always on the lookout for the best piece for a particular market so we can learn the lesson of being more efficient—not inventing anything and certainly avoiding any mistakes that any other member states make within our league towards a particular market or any mistakes somebody not in our league has made in a particular market.

Our key competitors are really anyone taking a vacation anywhere. In the United States, for instance, we compete with Orlando, Las Vegas, or other vacations people may pick within the United States. Our job is always to get people to leave the country of the United States and come to our part of the world. In New York, for example, the United Kingdom and the Caribbean are equidistant, so our competition, a lot of times, really depends on where the customer is coming from. We recognize that any other place a customer can chose as a destination is our competitor.

What factors influence consumers' decisions to visit the Caribbean? Have these factors changed over time? How do you monitor these changes?

It certainly has changed over time. One of the things that you will notice is that there used to be a very sharply defined season—of people escaping the cold. The market would really die after that. Then, as people began looking for and selling the "experience," we got more than the people who were merely escaping the cold. The focus has now changed more towards what the experience is that we can give our customers, and the customers are looking for great value. This has probably been the most significant change over the years and means a lot more business on a more year-round sort of basis.

Some of our members have the most sophisticated data-gathering techniques and technologies in the world, such as the reservation card, which collects data from every single piece of data about people coming into the country— where they come from, why they are coming, e-mail addresses, the flight they came on, and which hotel they're staying in. There's a piece of the card that they hand back in on their way out of the country that has a questionnaire on the back of it asking what they did enjoy and what they didn't enjoy. That is used for guidance in terms of seeing trends, seeing what is happening, and looking at who is coming through travel agents versus who is booking online.

We also have some fairly sophisticated data-gathering techniques that give information about what people thought about their vacation and what you need to fix in order to increase their level of satisfaction. One of the things that we want to do is aggregate the information that is being taken for each country so we can look at trends on a global basis—especially where people are coming from.

What advice can you give to students who want to be involved in destination marketing?

There is no doubt in my mind that it is the most exciting area of any career I could ever think of. The one thing about the tourism business is that it is broader in depth and scope than any other business on earth because it affects and is affected by so many areas. If you think you are somebody who can think in very broad terms and also see the large numbers and information and choose a course of action from that, then this is a great field for you. In our constituency, it is the people who live and work in the destination that you rely on to treat the customers very well and very fairly. You have to satisfy the tourists and locals equally, which is one of the peculiarities of destination marketing that many people forget and often run into trouble with because they focus on the local constituencies and you need more than that for long-term success.

Used by permission from Vincent Vanderpool-Wallace.

IMPORTANCE OF TRAVEL AND TOURISM

United Nations World Tourism Organization
A division of the United Nations responsible for tracking tourism visitor statistics on a worldwide basis, including country of origin, length of stay, and so on.

According to the **United Nations World Tourism Organization (UNWTO),** tourism comprises the activities of people traveling to and staying in places outside their usual environment for not more than one consecutive year for leisure, business, and other purposes. Tourism can also be defined as the processes, activities, and outcomes arising from the relationships and interactions among tourists, tourism suppliers, host governments, host communities, and surrounding environments that are involved in the attracting and hosting of visitors. Both of these definitions suggest that tourism is made up of a number of tangible and intangible components, often taken to include the tourist, the tourist-generating region, the transportation system, the tourist destination, hospitality services, and the tourism industry.

All of these components are highly interrelated, and they are very sensitive to changes in environmental trends. Any small change in any of the environments is likely to have some impact on all or most of the tourism components because they are all connected to each other. This makes the whole industry very unstable and vulnerable to external factors (e.g., terrorism). Despite this vulnerability, many countries and destinations depend heavily on the money spent by tourists.

Tourism affects the economy of every country and of every city and local community in the world. According to the UNWTO, tourism is the number one industry in many countries and the fastest-growing economic sector in terms of foreign exchange earnings and job creation. Tourists from other countries are often the major source of external money for most nations. If one were to add up all the money earned from tourists who do not reside in the visited country, tourism would rank as the number one source of income. For tourist-receiving destinations, it has become one of the most important sources of employment and an enormous reason for investment in buildings, roads, and communication technologies. It is "one of the most remarkable economic and social phenomena of the past century."[1]

The expenditures of international travelers can have a tremendous effect on host country economies. The impact is not limited to direct traveler spending. The **multiplier effect,** whereby a tourist dollar is spent and respent throughout the economy, plays an important role in measuring tourism's contribution to the gross domestic product (GDP). The Australian Bureau of Tourism Research studied this indirect contribution of tourism from 1997 to 2001 and found it to be slightly higher than direct spending, thus doubling tourism's contribution to the Australian economy.

multiplier effect The impact the tourist dollar has on a destination's economy in that wage earners working in hospitality and tourism also spend much of their earnings in the local community.

Exhibit 10-1 provides an overview of the multiplier effect and how it would work for a hotel purchase.

Tourism's enormous growth and influence has spread widely since World War II, particularly since 1960. International recessions in 1982–1983 and 1991 caused slight declines, but recovery was swift and strong. From 2001 to 2003, the worldwide tourism industry contracted as the result of a series of events. The attacks in the United States on September 11, 2001, and subsequent tightening of travel restrictions, the SARS outbreak in Asia, expanded military operations in the Middle East, and an increase in terrorist attacks worldwide further weakened the global economy and contributed to an overall decrease in international travel. There are indicators, however, that global travel is overcoming these challenges. For example, the total estimated visitors to the United States increased 4.5% from 2005 to 2006, an estimated increase of 2.1 million visitors.[2] According to the UNWTO, international tourist arrivals (the term used to indicate that travelers leave their home country and visit another) increased at a phenomenal average annual growth rate of 6.5 percent from 1950 to 2004.

The fastest-growing destinations in recent years have been outside the traditionally economically stronger continents of North America and Europe. While the Americas have just begun to see a small increase in the number of visitors in the last couple of years, after years of decline, Asia and the Pacific have grown by an average of 13 percent per year, and destinations in the Middle East have grown by 10 percent a year. Explanations for the rapid growth of tourism in regions outside of North America and Europe include the following:

- Robust economic growth
- Relatively stable political environments

Exhibit 10-1   The Multiplier Effect

Tourism not only creates jobs in the tertiary sector, it also encourages growth in the primary and secondary sectors of industry. This is known as the multiplier effect, which in its simplest form is how many times money spent by a tourist circulates through a country's economy.

Money spent in a hotel helps to create jobs directly in the hotel, but it also creates jobs indirectly elsewhere in the economy. The hotel, for example, has to buy food from local farmers, who may spend some of this money on fertilizer or clothes. The demand for local products increases as tourists often buy souvenirs, which increases secondary employment.

The multiplier effect continues until the money eventually "leaks" from the economy through imports—the purchase of goods form other countries. A study of tourism "leakage" in Thailand estimated that 70% of all money spent by tourists ended up leaving Thailand (via foreign-owned tour operators, airlines, hotels, imported drinks and food, etc.). Estimates for other Third World countries range form 80% in the Caribbean to 40% in India. A chart of the tourist multiplier effect is shown below.

The tourist multiplier effect.

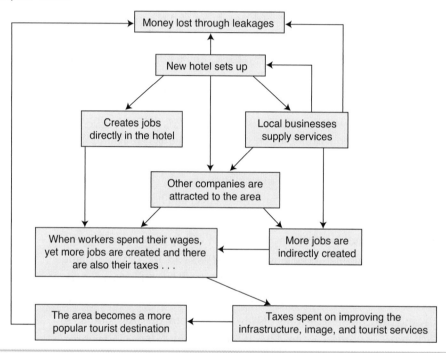

Source: Courtesy of Barcelona Field Studies Centre S.L. Used by permission. Retrieved January 3, 2006, from www.geographyfieldwork.com/TouristMultiplierhtm.

- Relaxed visitation regulations
- Ability to finance infrastructure development
- Aggressive marketing efforts by national tourism organizations (NTOs) of host countries
- Increase of tourism demand worldwide

Even though tourism is the world's fastest-growing industry and the number of travelers is likely to reach a record in the next decade, communities that are planning on developing tourist destinations and existing tourist destinations need to understand that competition among the destinations is fierce, travelers needs and wants constantly change, and tourism is susceptible to changes in the macro environment. To be successful in today's environment and sustain success, tourist destinations need to thoroughly understand local residents' attitudes toward tourism, the type and level of tourism products and services offered, their competitors, and the changing needs and wants of the current and prospective markets and their characteristics.

LOCAL RESIDENTS' ATTITUDES TOWARD TOURISM

Understanding local residents' attitudes toward tourism development is crucial for local governments, policymakers, and businesses because the success and sustainability of any sort of tourism development depends on the goodwill of the local residents and their active support. Once a community becomes a tourist destination, the quality of life of the local residents is affected by the consequences of tourism development. These include an increased number of visitors, increased use of roads, and various economic and employment-based effects. The success of any tourism development project is threatened to the extent that the development is planned and constructed without the knowledge and support of the local residents.

Although successful tourism depends on attractions and services, it requires the hospitality of local residents. Anger, lack of interest, or mistrust on the part of the local population will ultimately be conveyed to the tourists and is likely to result in reluctance on the part of the tourists to visit. In addition, active opposition can lead to delays, legal action, and abandonment of projects if they become financially unfeasible. The importance of local residents' support has been widely recognized by both planners and businesses.

Several factors affect the level of local residents' support for tourism development. Residents are likely to support tourism development as long as they believe that the expected benefits of development exceed the cost of the development. Because the most visible benefits of tourism are its contribution to the local economy and the employment opportunities created, some local residents are likely to see tourism as an economic development tool, and therefore, they are likely to support tourism in their community. However, not all residents place such an importance on the economic benefits of tourism.[3]

Several factors are likely to influence how locals view tourism and their support for tourism. One of those factors is the state of the local economy. Many communities faced with a narrow resource base embraced tourism as a cure for their economic distress. If the economy of the community is depressed, residents are likely to place more importance on the perceived benefits of tourism and consequently support tourism development even though they are aware of the potential for tourism to result in negative impacts.[4]

To be successful over time, destination managers need to make sure they have local residents' support and endorsement for tourism development. However, having that support alone is not enough. Destination managers also need to know their situation and must be in tune with their external environment and changing customer needs and wants. There must be a fit between what the market wants and what the destination has to offer. This requires a marketing strategy that is endorsed and supported by all stakeholders to guide the product development and all marketing activities. However, it should be remembered that in most destinations, public agencies such as **national tourism organizations (NTOs)** and convention and visitors bureaus

destination marketing A collective and coordinated effort among various constituencies to market a destination such as a country, a regional area, a state, a province, or even a city, such as the successful "I Love New York" campaign.

national tourism organization (NTO) A department or agency of a national government responsible for marketing and promoting tourism for its country at the international level.

(CVBs) are likely to promote the tourist business for the whole destination or country, whereas the private businesses such as hotels and restaurants are likely to promote their own businesses.

NATIONAL TOURISM ORGANIZATIONS

At the national level, governments promote their countries in the international tourism market through NTOs. In some countries, such as Mexico and Canada, government leaders have recognized the importance of tourism to the nation's economy and have elevated the status of the NTO to cabinet level, instituting a Ministry of Tourism. Regardless of their position in governments, NTOs have similar objectives. They promote their countries through the following:

- Publicity campaigns
- Research
- Plans for destinations

As a result of increased competition in world tourism, NTOs today spend much more on their tourism marketing budget than at any other time. A significant part is spent to publicize a country as a destination through outlets in major source countries to create public awareness and to promote positive images.

After the Indian Ocean tsunami at the end of 2004, several Southeast and South Asian countries increased their tourism budgets to educate travelers about the geography of the region, promoting locations that were not directly affected by the tragedy and publicizing the recovery of impacted areas.

The U.S. Travel and Tourism Administration (USTTA), under the Department of Commerce, functioned as the NTO for the United States until 1996. At that time its funding was cut, leaving the United States as the only major country without a federally funded NTO. The Travel Industry Association of America (TIA), established in 1941 as a consortium of industry trade organizations and private companies, has attempted to fill this gap by promoting and facilitating increased travel to and within the United States. In 2005, TIA teamed up with the Travel Business Roundtable to further strengthen its influence with government lawmakers and to promote travel-friendly regulation in the United States.

Historically, the principal marketing roles of NTOs have been fairly narrow in scope: creating and communicating overall appealing destination images and messages to the target market. However, NTO functions are changing as today's international tourism industry becomes more competitive and tourists become increasingly more sophisticated in their destination choice behavior. Because tourism industry leaders in many countries recognize the importance of collaboration between the public and private sectors, they launch and implement various collaborative marketing programs. The Greater Mekong (river) Subregion (GMS) is being marketed as a single destination; it includes the Southeast Asian countries of Myanmar, Laos, Thailand, Cambodia, Vietnam, and southeastern China.

In 2003 tourism to East Asian countries was severely affected by incidences of the SARS virus. Even countries with no record of SARS cases were negatively affected. In one month Thailand experienced a 46 percent drop in arrivals compared to the prior year.[5] Tourism marketing budgets in the region were boosted by the millions in a post-SARS strategy to attract tourists. By October 2004 UNWTO reported that Northeast and Southeast Asian countries had "bounced back stunningly from the losses suffered in 2003 due to SARS."[6] However, 2005 and news of a looming threat of an avian flu epidemic originating in Asia had the World Tourism Organization warning that an overreaction could once again damage international tourism.[7]

The marketing activities of an NTO are mainly centered on the promotion of the country as a whole. Subsets, however, are common. States, provinces, regions, areas, cities, and other small parts of a country participate in similar activities to promote their unique destinations. Advertisements for some of these locations are shown in Exhibits 10-2, 10-3, 10-4, 10-8, and 10-9.

NTOs also play a facilitation role that typically includes the following:

- Collecting, analyzing, and decimating market research data
- Establishing a representation in the markets of origin

Web Browsing Exercise

Go to the Peru Tourism Bureau website (www.visitperu.com). What major aspects of Peruvian culture are highlighted to attract tourists to the country? Is it possible to make a hotel booking from the Visit Peru website?

Exhibit 10-2   This image illustrates the promotion of states

Source: Vermont Department of Tourism & Marketing and Stowe/Smugglers' Notch Region. Used by permission.

The Doctor Knows Best.
Dr. Beach Rates Fort De Soto Park
AMERICA'S #1 BEACH.

ORLANDO
ST. PETERSBURG/
CLEARWATER AREA

Just 90 minutes from Orlando.

Best sun. Best sand. Best beach. Fort De Soto Park brings you everything you'd expect from America's best beach. Come enjoy it. Together. Visit FloridasBeach.com for reservations. Or call 877.352.3224.

ST. PETERSBURG/CLEARWATER
FloridasBeach.com

Exhibit 10-3   This image illustrates the promotion of regions
Source: St. Petersburg/Clearwater Convention & Visitors Bureau. Used by permission.

- Participating in trade shows
- Organizing and coordinating familiarization trips
- Supporting the private sector in the production and distribution of literature

Noting that "a growing economy is one of the strongest indicators of tourism growth,"[8] TIA has put additional resources into studying the potential of outbound travel to the United States by four emerging markets: China (PRC), India, Russia, and Poland. All have been identified as having more affluent populations with growing middle-class segments that have the resources for international travel. These countries are targeted in addition to the traditional U.S. markets—the UK and Japan.

Because the United States has no national tourism offices, the Nevada Commission on Tourism was able to obtain certification by the China National Tourism Administration (CNTA) to operate an office in China. Nevada is now the only state tourism office in China, putting it in a unique position to capture Chinese travelers to the United States.

Web Browsing Exercise

Go to the Nevada Commission on Tourism website (www. travelnevada.com). Besides China, in which other countries will you find a Nevada tourism promotion office?

LUXURY

China World Hotel
BEIJING
AT CHINA WORLD TRADE CENTER
A SHANGRI-LA HOTEL

LOCATION

China World Hotel, Beijing.
What will your reason be?

For some, it's the inspired setting
in the city's commercial and diplomatic heart.
For others, it's the legendary Shangri-La hospitality.
But perhaps most of all, it's the luxurious accommodations
unmatched anywhere else in the city, if not the world.
www.shangri-la.com

Exhibit 10-4   This image illustrates the promotion of countries

Source: Developed for Shangri-la Hotels and Resorts. Used by permission.

HOW HOTELS AND TOURIST DESTINATIONS WORK TOGETHER

To provide an example of how the marketing executive of a specific hotel might work with a tourism office to promote both the hotel and the destination, we conducted an interview with Leanne Pletcher, who is the VP of marketing for the Hilton Waikoloa on Hawai'i—the Big Island. Tourism is the number one source of revenue for the state of Hawaii; as such, much money is used to promote Hawaii as a destination. A critical issue for a hotel in such a destination as Hawai'i is to maintain its identity and not get lost in the promotion of the destination. That is, Leanne Pletcher wants the traveler to say, "I am going to the Hilton Waikoloa in Hawai'i" rather then, "I am going to Hawai'i."

Pletcher explains how this is done in Exhibit 10-5. The advertisement in Exhibit 10-6 promotes the destination, whereas the advertisement in Exhibit 10-7 promotes the individual property. Both of these advertisements would appear together in a magazine or newspaper, just as they do in this book.

Exhibit 10-5   Interview with Leanne Pletcher of the Hilton Waikoloa on Hawai'i

DO THE HILTONS IN HAWAII PROMOTE THE DESTINATION FIRST AND THE HILTONS SECOND, OR IS IT THE OTHER WAY AROUND?

Whether the promotion carries destination versus property-specific information depends on the nature of the message and the size and scope of the target market. For a broader appeal—if the guest is coming from a further distance and more apt to spend a longer amount of time on their trip—the message will incorporate more of a destination message for the Hilton Hawaii property: Hilton Waikoloa Village, Hilton Hawaiian Village, or Doubletree as the main attraction. Because the three properties have their own unique characteristics, each has its own separate marketing plan, as well as joint advertising messages for Hilton Hawaii.

Hilton Waikoloa Village recognizes the importance of promoting not only the resort, but also the destination. When a visitor is making the commitment to travel to Hawaii, they should be informed of all there is to offer at our resort and the Big Island.

Hilton Waikoloa Village promotions typically incorporate a message about the property along with the destination, as the two go hand in hand. Especially on the Big Island, we recognize the importance of conveying to the visitor that Hawaii's Big Island is part of the Hilton Waikoloa Village experience. It is also important to sell the destination to groups because they are inclined to do activities both on and off the property.

With our local or "kama'aina" markets, which can encompass on-island residents or visitors from the outer islands, Hilton Waikoloa Village focuses promotional efforts on special events, dining, spa, and room packages. Kama'aina business complements our group and FIT business, especially during the, "shoulder" periods. Kama'aina promotions include radio and newspaper advertising, pitching local media writers, and e-mail communications.

WHAT SORT OF ACTIVITIES DOES A PROPERTY IN A DESTINATION LIKE HAWAII DO TO GET NOTICED OR TO STAND OUT?

A resort like no other, Hilton Waikoloa Village is really a destination in itself. Hilton Waikoloa Village stands out amongst other resorts and properties in Hawaii due to the following:

- Set on 62 ocean-front acres on the sunny Kohala Coast of the Big Island with tropical gardens, abundant wildlife, nine international restaurants, world-class shopping, art and culture, golf and tennis.
- Dolphin Quest Learning Lagoon—offering guests the unique experience of a dolphin encounter.
- Largest amount of indoor/outdoor meeting space combined with room availability on the outer islands.
- Special events like Dolphin Days Summer Fest, Return to Paradise, and Big Island Festival, which showcase the unique culture, agriculture, art and foods of Hawaii.
- Guests can explore the resort by air-conditioned trams; take a leisurely stroll on flagstone walkways with $7 million in Polynesian, European, and Asian artwork; or cruise on mahogany canal boats along waterways throughout the property.

Being owned and managed by Hilton, Hilton Waikoloa Village has the advantage of Hilton Hotels Corporation's brand marketing campaign, collateral pieces, and electronic media vehicles. Hilton Hawaii also partners with strong marketing companies, like American Express, on strategic mailings.

HOW DOES A PROPERTY OVERCOME THE LENGTH OF TRAVEL TIME NEEDED TO REACH THE DESTINATION?

While the length of time certainly varies with the origination of the traveler, the number of direct flights continues to increase, making it easier to travel to Hawaii, and especially to Hawaii's Big Island. Marketing and public relations efforts are targeted to the cities and regions that provide the direct lift.

One marketing strategy is to work with the airline partners of the Hilton HHonors program to cross-promote the resort and the airline to the target markets. Hilton Waikoloa Village also works with Big Island Visitors Bureau (BIVB) on special promotions and events that are in conjunction with new flights/airlines coming to Keahole-Kona International Airport.

Given the amount of activities available and the time commitment for travel, we also try to recommend at least a week stay at the resort.

WHAT SORT OF ACTIVITIES ARE UNDERTAKEN TO PROMOTE THE HILTON AND HAWAII IN YOUR FEEDER MARKETS?

- Direct sales efforts to groups, companies, organizations through Hilton Waikoloa Village–based national sales managers, and Hilton Corporation–based national sales team on the mainland.

Exhibit 10-5   Continued

- Media/marketing plan incorporates targeted advertising in publications that have distribution and frequency in the feeder markets. Direct mail pieces are targeted to the same areas. Efforts are also made to capture repeat visitors in these markets.
- Strong website presence and e-communication programs in place.
- Providing information, packages, and specials to travel agents and wholesalers.
- Targeted press releases to major metro markets.
- Invitations to media based in feeder markets to take individual or group press trips to the resort.
- Media blitzes several times annually to feeder markets.

HOW DOES AN INDIVIDUAL HOTEL WORK WITH THE TOURIST AND VISITORS CONVENTION BUREAU TO PROMOTE ITS PROPERTY?

Hilton Waikoloa Village works with Big Island Visitors Bureau on a number of levels:

- Co-op advertising opportunities.
- Representation on media trips to top market destinations in order to pitch stories (articles published in top magazines).
- Offering accommodations to Big Island Visitors Bureau–sponsored media/writers visiting the Big Island. The Hawaii Visitors and Convention Bureau is also a major partner, as well as the Kohala Coast Resort Association. Both groups also sponsor media/writers.
- Participation in Big Island Visitors Bureau–sponsored trade shows and special events on the mainland or other target markets.
- Participation on the board level to give input on marketing plans and sales efforts.
- Being a part of "aloha" greeting committee to new flights/passengers coming into the Keahole-Kona International Airport.
- Promotions, sweepstakes, and other partnership opportunities with the bureaus.

HOW DO INTERACTIONS WITH TRAVEL WRITERS FIT INTO YOUR OVERALL PROMOTIONAL STRATEGIES?

Interactions with travel writers are key to the success of a public relations program. Relationships with trust must be developed with writers so that they can call on a public relations professional at any given time for information. Speedy, accurate response to writers' requests will guarantee optimum coverage.

Travel writers provide the unique opportunity to convey our messages through magazines, newspapers, website, and broadcast media. Articles can capture more detail and give the reader the essence of a true testimonial. With the more seasoned and better-known writers, the believability factor increases. We refer to these writers as the "A List," while the more novice writers might be on the "B List." If the publication has a strong circulation and targets the right audience, receiving "ink" can be an extremely effective promotional tool. Some may view this type of coverage as a "free" advertisement; however, it is important to understand that there is a cost associated with recruiting the writers to visit the property and providing accommodations, meals, or activities. The risk is that there is less control over what is presented in the article than what is placed in a paid advertisement.

"Ink" that is given to Hilton Waikoloa Village is sometimes valued at the space-equivalent price of an ad. A third party endorsement, such as that by a reputable journalist, is oftentimes given a value of three times the publication's advertising rate. Clipping services provide copies of articles that mention the Hilton Waikoloa Village.

If Hilton Waikoloa Village is focusing its marketing efforts on the incentive markets, public relations will work in tandem with marketing to proactively pursue writers for incentive magazines so that articles will appear in conjunction with or as a complement to paid advertising.

IS HAWAII CONSIDERED A FAVORITE AMONG INCENTIVE COMPANIES? IF SO, HOW DOES A PROPERTY LIKE THE HILTON MARKET ITSELF TO THIS SEGMENT?

With the warm sunshine, abundance of outdoor activities and beautiful tropical scenery, and stunning sunset, Hawaii is considered a favorite and top destination of incentive groups. It is considered to be a "prize" for the hard work that brings the "winners" of the incentive programs to Hawaii.

Direct sales calls through the Hilton Waikoloa Village–based national sales team and the Hilton Corporation national sales team help to reinforce Hilton Waikoloa Village as an incentive destination. Advertising is also placed in trade magazines that target companies, meeting planners, and incentive houses. The public relations team works in conjunction with marketing to recruit writers for these publications to come to Hilton Waikoloa Village and produce articles as well.

ARE THERE ANY OTHER UNIQUE CHALLENGES THAT I HAVE NOT MENTIONED?

Hawaii's Big Island is experiencing unprecedented growth. Consequently, the competition among places to stay for visitors is escalating. Visitors now have the option of resorts, hotels, condominiums, and time shares for accommodations. Hilton Waikoloa Village has been undergoing an aggressive capital campaign to keep the property fresh and new. The enhancements include guest room and suite renovations, expanding the lower lobby, updating pools, and renovations to the Kohala Sports Club & Spa.

Source: Used by permission from Leanne Pletcher and Hilton Waikoloa Village.

Exhibit 10-6   An advertisement that promotes a destination

Source: Hilton Hotels. Used by permission.

and the Spectacular Resorts of Hilton Hawaii.

HILTON WAIKOLOA VILLAGE

A lush oceanfront oasis, cradled by miles of untamed coastline. You have found paradise on Hawaii's BIG ISLAND.

Hilton Waikoloa Village® is a spectacular destination resort set on over 62 acres on the sunny and exclusive Kohala Coast of Hawaii's Big Island. Recently ranked as a Departures Magazine readers favorite hotel in Hawaii.

- Explore aboard mahogany boats that cruise tranquil waterways or in Swiss-made trams
- Enjoy award-winning dining, shopping, Hawaiian arts and culture
- Snorkel or kayak in a private ocean-fed lagoon
- Experience an interactive experience with dolphins*
- Play 36 holes of championship golf or renew body and spirit at the Kohala Sports Club & Spa

For a complete resort tour, visit
HiltonWaikoloaVillage.com

Independently owned and operated.

CHAMPIONSHIP GOLF

DOLPHIN QUEST LEARNING LAGOON*

TAKE ME TO

HAWAII

TO A PARADISE I'VE DREAMED ABOUT.

TAKE ME TO THE HILTON™

HILTON WAIKOLOA VILLAGE

At this stunning destination resort on the sunny Kohala Coast of Hawaii's Big Island, Hilton Waikoloa® Village is set on over 62 oceanfront acres of breathtaking scenery surrounded by 36 holes of championship golf. Enjoy endless activities and experience Hawaii art and cultural treasures at this spectacular destination resort. Truly a world-class resort with service to match—a resort like no other.

For reservations visit
Hilton.com
or call **1-800-HILTONS**.

Hilton
Waikoloa Village®
Travel should take you places™

425 Waikoloa Beach Drive
Waikoloa, Hawaii 96738
Phone 808-886-1234
HiltonWaikoloaVillage.com

Exhibit 10-7   An advertisement that promotes an individual property

Source: Hilton Hotels. Used by permission.

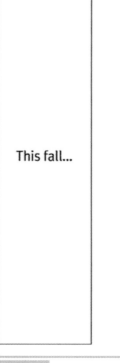  

This fall...

watch Mother Nature light up the night.

The Outer Banks
Click now for your free Getaway Card.

Exhibit 10-8 This image illustrates the promotion of areas

Source: Outer Banks Visitors Bureau. Used by permission.

SPECTACULAR LOCATION. EXCEPTIONAL SERVICE.

TAKE YOUR MEETING ON
A VACATION TO MONTEREY

174,000
square feet of meeting space

10,000
guest rooms within 10 minutes

80
flights daily at Monterey Peninsula Airport

5
minutes from airport to hotels and attractions

0
other destinations like Monterey

MONTEREY
CONVENTION AUTHORITY
Groups: 800-749-8091
Leisure: 877-666-8373
Fill out our online RFP at
www.montereyconventionauthority.com

Exhibit 10-9 This image illustrates the promotion of cities

Source: Monterey Convention Authority. Used by permission.

YOU'LL BE TEMPTED TO CALL THE GAME DUE TO WEATHER.

Sure, we've got incredible sports facilities here. And yes, we do host every type of tournament from baseball to soccer to volleyball, and every sort of sport from triathlons to gymnastics to cheerleading. But, when all is said and done, what really makes holding a sports event here a hit is the beach. Our beach. And that's just the beginning. Call 1-800-PCBEACH or visit thesportsloversbeach.com. And discover all the other ways to play at Panama City Beach.

PANAMA CITY BEACH
The Sports Lover's Beach
thesportsloversbeach.com

Exhibit 10-10   Tourism officials have had to make efforts to overcome the negative images of Florida as a hurricane target

Source: Panama City Beach Convention & Visitor Bureau, Inc. Used by permission.

DESTINATION MARKETING STRATEGY

Before a destination can begin to formulate a marketing strategy, management must understand the external environment to identify possible opportunities and threats. Destination managers are expected to be aware of major environmental factors likely to affect the industry, the destination, and current and future markets to consider their possible impact on marketing. They must also be ready to respond quickly and intelligently to new events and trends. Exhibit 10-10 is an illustration for Panama City Beach showing that the hurricanes that struck Florida did not affect these beach areas.

Macro Environment

The macro environment consists of several major forces including economic, technological, political/legal, ecological, sociocultural, demographic, and competitive forces. Most of the time, an organization has limited or no control over these forces. Therefore, the organization needs to monitor and respond to these forces. Managers should engage in environmental scanning. When scanning the environment, managers should be able to identify the factors in each environment that are likely to influence the way they run their business, their ability to attract visitors, and the potential market worth of services and products offered in the destination in the future. The analysis of a destination's macro environment should include a brief review of major forces such as market acceptance, social perceptions, visitation levels, consumer trends, economic changes, competitive activity, and technology advancements. The following sections discuss in greater detail the general environmental factors that are likely to influence the success and growth of a destination.

Economic Environment

International tourism is largely dependent on the economic conditions in the market of the countries or regions in which prospective visitors live. Developed and growing economies sustain large numbers of trips away from home for business purposes of all kinds. Business meetings, attendance at conferences and trade shows, and travel on government business are all important parts of the travel and tourism industry. The influence of economic conditions is even more obvious in leisure travel, where, in many counties with advanced and developed economies, average disposable income per capita has grown to a size large enough to enable a majority of the population to take vacation trips in foreign lands. This expanding middle class has the means and desire to travel, making an impact on destinations worldwide.

Technological Environment

Technology and travel are natural partners. The rapid growth of travel and tourism in the last 20 years has been fostered by technological developments—principally in air transport through the development and refinement of the jet engine, more sophisticated aircraft design, and improved infrastructure that permits more landings and takeoffs per hour and better control of planes in the sky.

Wide use of computer technology has facilitated another leap forward by the travel industry. Computer software has been designed to cover a wide range of activities undertaken by the travel trade (e.g., information retrieval, reservations, ticketing, invoicing, etc.). The advent of the Internet and the World Wide Web has further fueled growth in international travel by making distribution channels directly and quickly accessible to the consumer. Destination marketers are able to tap into these sophisticated technologies, enabling international travelers to "experience" the destination via virtual tours, interactive discussions, and even live webcams.

Political/Legal Environment

Government interest in tourism has stemmed primarily from its economic significance, particularly employment earnings and tax potential. Tourism demand, however, is also largely influenced by legislative actions at various levels of government and intergovernment agencies (e.g., the World Tourism Organization [UNWTO] and the International Air Transport Association [IATA]). International politics also play a significant role in the volume of travel and tourism business. After the terrorist attacks in the United States on September 11, 2001, international travel to the United States experienced a 16 percent drop—the worst decline in the history of international arrivals. Part of this decrease was a result of the travel restrictions put in place by the U.S. government. For example, the Western Hemisphere Travel Initiative will require all travelers to and from the Americas, the Caribbean, and Bermuda to have a passport or other accepted document to enter or reenter the United States. The implementation schedule is as follows:

▪ December 31, 2005: All travel to or from the Caribbean, Bermuda, Central and South America

Web Browsing Exercise

Use your favorite web browser to gather information on Boeing, Airbus, and the new A380 from Airbus. Be prepared to discuss the history of Airbus, Boeing, and the new A380. In addition to this information, also be prepared to discuss new planes on the drawing board for each company. After reading the information on the websites, what predictions can you make about the future of international travel? What support can you use for these predictions?

- December 31, 2006: All air and sea travel to or from Mexico and Canada
- December 31, 2007: All air, sea, and land border crossings.[9]

Conversely, nearly a quarter million Americans have visited Vietnam since the two countries normalized relations in 1995.

The air transport industry was liberalized in most tourist-generating countries in the 1980s. The deregulation of the airline industry in North America generated a significant increase in intercontinental flights, which, in turn, positively contributed to the growth of world tourism. The adoption of an "open sky" policy in Asian countries resulted in a substantial increase in air traffic within Asia and fostered the introduction of new carriers such as Eva Airways (Taiwan) and Asiana Airlines (South Korea). The advent of low-cost carriers in Europe—for example, Air Berlin, Easy Jet, and Ryan Air—have also contributed to the increase in tourism among countries.

Relaxed travel restrictions and increasing leisure time and income of residents in newly industrialized countries contributed significantly to a growth of tourism within and from Asia. In the past, both the Taiwanese and South Korean governments restricted or limited overseas travel by their citizens. With rapid economic growth and an increase in consumer disposable income, the concept of leisure travel became widespread in these countries, and their governments gradually lifted overseas travel bans. In the past, for example, the Korean government prevented its citizens from obtaining passports for "sightseeing" purposes. Previously, to obtain a passport for pleasure travel, an applicant had to be at least 50 years old in 1983, 40 in 1987, and 30 in 1988. In 1989 the government eliminated all age restrictions on the issuance of passports to its citizens. The Taiwanese government followed the same course. The number of outbound tourists from Taiwan increased more than threefold in four years. The number of South Korean outbound travelers increased fivefold during the same period.

In 2004 the Chinese government increased the number of "approved" countries its citizens are able to visit to 90. This in turn has spawned considerable activity in tourism-related organizations in the approved countries to attract and accommodate Chinese visitors. Buoyed by the sheer number of potential visitors, *BusinessWeek* magazine dubbed 2005 the "Year of the Chinese Tourist" for Europe because more than 25 European countries were added to the approved destination list.[10]

Ecological Environment

A growing awareness of planet Earth's finite resources and the impact of travel on a destination's ecology have spurred new consciousness on the part of some international travelers and host communities alike. In a landmark survey of American travelers by TIA and the National Geographic Society, nearly three quarters claimed that it is "important to them that their visit not damage the environment."[11] Host governments of popular tourism destinations—for example, the Great Barrier Reef in Australia and the Galapagos Islands of Ecuador—have taken steps to manage the carrying capacities of these destinations. It is hoped that measures such as limiting the number of tour operators and increasing visitor fees will minimize the ecological impact of tourism.

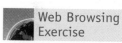

Web Browsing Exercise

Use your favorite search engine to find the most up-to-date travel requirements for inbound passengers to the United States. What requirements exist? Do these requirements vary by the passport carried? That is, are travel restrictions different for those who carry Chinese passports than they are for those who carry UK passports? For comparison, choose another country and examine its travel restrictions. How do these restrictions compare to those of the United States?

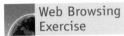

Web Browsing Exercise

Go to www.unep.net and search the topic "tourism." Be prepared to discuss the type of information on this site. Review one of the pieces of information in detail and be prepared to discussion it in class.

Web Browsing Exercise

Go to the World Travel & Tourism Council website (www.wttc.org). Click on "Blueprint for New Tourism." What are the three fundamental conditions for the Blueprint for New Tourism? After looking at the "Case Studies of the Blueprint for New Tourism in Practice," which initiative is of most interest to you and why?

In late 2005 the government of Mexico received a $200.5 million loan from the World Bank "to promote sustainable development by balancing socioeconomic development with sound environmental management." The money is to be used in part to promote sustainable action plans in selected tourist destinations and to make improvements in areas such as wastewater and solid waste disposal.[12]

Sociocultural Environment Social and cultural considerations involve the beliefs, values, attitudes, opinions, and lifestyles of those in the market environment, as developed from their cultural, ecological, demographic, religious, educational, and ethnic conditioning.

A key element in the tourism marketing process is the significant demographic shifts affecting the population, particularly in selecting target markets. In 1970, 24 percent of the U.S. population was between 25 and 45 years of age. By 1990 this age group accounted for 33 percent of the population, a 38 percent increase. A major turning point in the United States will be in 2010, when the majority of the population will be 45 years and older. Another continuing phenomenon is the growth rate of the over-60 segment, a group that likes to travel. Other countries have had similar demographic shifts. Moreover, as social attitudes change, so too do the leisure patterns of consumers. The popularity of "ecotourism" in recent years is one good example.

Demographic Environment Although demographic change is a constant process, it has received a lot of public attention in recent years because it is seen as one of the important drivers for new trends in consumer behavior. Global demographic trends are likely to have far-reaching consequences for the future of destinations. One of the major demographic trends that is likely to have a significant impact on tourism and the future of destinations is the rapidly aging population of the developed world. For example, many baby boomers in the United States are at the peaks of their careers and possess the highest earning power of their lives, resulting in the highest level of discretionary income. They are reaching their retirement age rapidly, but even after retirement they are likely to stay active and retain their independence for a long time. Research has shown that people do not change their travel behavior just because they turn 60 or 65, or because they retire. In most cases they stick to the travel patterns acquired until the middle of their lives. This fact allows for predictions of the tourist behavior of future senior generations.

The new senior citizens in 5 or 15 years' time will be different from the present senior citizens when it comes to travel behavior. While senior travelers today are already relatively active, the senior generations to come are more than likely to surpass them. The effects of demographic change (more and a bigger share of older people) and consumer behavior patterns (sticking to learned travel patterns) will show up as more senior trips with different preferences. For example, it is estimated that in Germany within 15 years the number of tourists in the age group 70 to 80 years will rise by more than 50 percent (from 4.2 million in 2003 to 6.6 million in 2018) with more than two thirds choosing destinations abroad. Studies also suggest that participation in cultural and heritage activities increases through middle age, peaks between 45 and 65, and subsequently falls off, which makes this group a prime target for these kinds of activities. People in this group also have more available time than they had previously, and those with older children choose to expose them to enriching educational experiences. In addition, this group is likely to be one of the best target markets for tourism focusing or health, spas, and keeping fit.

Another demographic trend that is likely to influence the future of tourism destinations is the shrinking population of the developed world as a result of lower fertility rates in many industrial countries. This combined with the dissolution of traditional family patterns may require destinations to develop products and marketing strategies that will attract visitors from developing countries.

Another important demographic trend is rising education levels. This is likely to result in consumers with unprecedented sophistication and depth. Destinations may need to develop products that can satisfy their sophisticated needs and wants.

The increasing economic role of women worldwide is another demographic trend that may influence the future of tourism and destinations. More women are working, and they are earning more money and controlling more discretionary income. Women typically make the decisions regarding the educational experiences of their children and set vacation plans. They also account for a large majority of bus tour passengers, trip planners, and elementary schoolteachers who make decisions on field trip destinations for their students. Women account for 60 to 65 percent of museum attendance and are more likely than men to support and participate in heritage and cultural activities. As more women move into positions of power and influence, funding for these interests will tend to increase. These demographic trends suggest that tourism will have the largest, wealthiest, and best-educated market for the next 20 years.

Competitors: Rivalry among Destinations

In today's global marketplace, hundreds of destinations are competing for the market share. To maintain a competitive position in the market, a destination must be able to accurately evaluate its competition. To do this, those in destination management must have a sound understanding of the market in which they operate: their customers, market boundaries, market conditions, and competitors.

In the tourism industry, destinations are mutually dependent. A competitive move by one destination can be expected to have a noticeable effect on its competitors. For example, if one destination starts offering deeply discounted packages or rates to all leisure customers, chances are good that other destinations will develop discounted packages to maintain their market share.

Identifying competitors is especially crucial because it influences all marketing decisions such as pricing, promotion, distribution, products development, attraction design, and positioning. It also provides the basis for competitive analysis, which assesses the industry structure, market scope, and focus of competitive advantage. Destinations must also understand the dimensions on which they compete when assessing and selecting competitive strategies.

industry/supply-based approach to identifying competition This approach defines competitors as those brands that operate in the same industry offering similar products and services to similar customers. This is also called the supply-based approach because it requires identifying competitive destinations through the similarity of their offerings, resources, and strategies.

market or demand-based approach to identifying competition This approach defines competitors as those satisfying the same customer need. This method is also called a demand-based approach because it requires identifying competitors by the similarity of their customers (i.e., their attitudes, behaviors, and motivations).

A destination can identify its competitors using one of two approaches: an **industry/supply-based approach** or a **market/demand-based approach.** The industry approach defines competitors as those brands that operate in the same industry offering similar products and services to similar customers. This is also called the supply-based approach because it requires identifying competitive destinations through the similarity of their offerings, resources, and strategies.

The market approach classifies competitive destinations as those satisfying the same customer need. This method is also called a demand-based approach because it requires identifying competitors by the similarity of their customers (i.e., their attitudes, behaviors, and motivations).

Research indicates that managers tend to use an industry (or supply-based) approach more often than a market (or demand-based) approach because of its ease. Basically, managers identify destinations that appear outwardly similar—for example, Malta and Cyprus. This approach ignores the fact that customers choose between competitive brands. Competition takes place in the minds of customers, which makes it very important for destinations to create and clearly communicate a strong image of the destination. In addition to creating and communicating a strong image, a destination must be able to distinguish itself from the competition and promote its superiority and relevance to target customers. By providing a point of distinction to its target customers, the destination may be able to entice the target customer to choose that destination over others.

The context and situation in which destinations are considered travel options can be influenced by a variety of factors such as the customer's cultural orientation, knowledge level, available time, discretionary income, and perception of safety, as well as major catastrophic events, terrorism, and so on. For example, tourists from an English-speaking country may seek holiday destinations where people also speak English. Similarly, people from Muslim countries may purposely limit their destination choice to other Muslim destinations. Available time and discretionary income are likely to play significant roles in determining which destinations are likely to be included in a tourist's consideration set. If a person has limited time or limited money to travel, that person's choice is likely to be restricted to cheaper, closer-to-home destinations rather than expensive and or long-distance destinations.

SEGMENTING THE TOURIST MARKET

Like any product, destinations cannot be everything to everyone. It is almost impossible for a destination to offer travel and tourism products that will satisfy the needs of everyone. Travel markets consist of travelers who may differ in their wants, needs, resources, locations, buying attitudes, and buying processes. Because travelers have unique needs and wants, each traveler can potentially be classified as an individual market. However, modifying the marketing strategy and developing customized product for each traveler may not be feasible for destinations because of their limited resources. Therefore, destinations may need to identify groups of travelers who have common interests and share common values. For example, students who travel for spring break purposes are likely to have similar interests and share similar values. The same may be true for those who travel for the purpose of visiting ancient Roman ruins. Destinations can identify their target markets in two ways. One is to gather information about current visitors to develop profiles of the current market segments. What are they? Where do they come from? Why do they visit the destination? What are their demographic characteristics? How do they make their vacation decisions? Basically, the destination uses the market segmentation approach to develop profiles of its current visitors using some or all of the bases for market segmentation.

Another way to identify target markets is to inventory the kind of attractions, services, and facilities the destination offers and then identify market segments that may be interested in what the destination has to offer. This approach enables a destination to identify underused attractions, services, and facilities and provides an opportunity to increase the visitation levels by going after the segments that may be interested in those underused attractions, services, and facilities. While developing an inventory of what the destination has to offer, managers should also identify the variables, or internal strategic factors, within their destination that may be important to the operation. These internal strategic factors are likely to determine whether a destination can take advantage of external opportunities.

Experts in the area suggest that differences in performance among destinations in the same market may be explained best through the differences in destinations' capabilities and resources and their application. To gain and sustain a competitive advantage, destination managers need to understand the destination's capabilities and use those capabilities to maintain current visitation levels and attract new segments. If competitors also have similar capabilities, the manager needs to make sure that the operation uses those capabilities and resources better than competitors do. Otherwise, competitors are likely to start stealing customers.

To identify capabilities, destination managers list and describe the service(s) the destination offers. For each offering, managers identify the main points, including what the attraction, service, and facilities are; how much they cost; what sorts of customers make purchases; and why. What customer need does each service fill? It is always a good idea to think in terms of customer needs and customer benefits, rather than thinking of the destination side of the equation, such as how a destination generates the service. After identifying the attractions, services, and facilities it offers, the destination needs to determine what it does better than its competitors. These are the destination's distinctive, or core, competencies, the products and services the destination provides better than any other destination in the marketplace. As a destination lists and describes its attractions, facilities, and services, it may run into one of the unexpected benefits of good planning, which is generating new ideas. Describing the offerings in terms of customer types and customer needs may enable managers to discover new needs to fill and new kinds of market segments to target.

After identifying the current and prospective market segment, destination managers need to conduct marketing research to find out who those people are, where they are located, and so on. For example, a destination may identify several market segments from Germany and France. However, different segments or groups of travelers from

Germany and France may visit the destination for different purposes. To identify segments that visit or that may be interested in a destination, destination marketers need to know who comes to the destination and why. In other words, destination marketers may need to develop a profile for each segment, making sure they include the location of the segment and the reasons they travel.

Most destinations have limited marketing budgets and are under constant pressure to maximize the rate of return on every marketing dollar they spend. It is always financially more feasible to go after market segments that are heavily concentrated in fewer locations. If most of the members of a market segment are located in a few major cities, the destination can develop very cost-effective localized promotion strategies to attract those customers. If the market segment members are located all over the country, it may be cost prohibitive to go after that segment.

How to Describe Markets

Markets can be described in terms of geographic, demographic, psychographic, and behavioral attributes. Market geographies address where the members of each segment are physically located. Destination managers are likely to attempt to identify regions and cities where most customers are located (for each product) in order to develop cost-effective promotion strategies and communication materials.

Market demographics refer to travelers' wants and preferences and the frequency of their purchases because they are often associated with travelers' demographic characteristics. Travelers' demographics include information about their age, gender, nationality, education level, household composition, occupation, and income.

Market psychographics describes the market segments in terms of psychographic information. It is more challenging than the previous categories because it is less quantifiable and more subjective. Psychographics categorize people on the basis of their lifestyle or personality attributes. For example, the lifestyles and personality attributes of people in a large metropolitan area are likely to be very different from those of people living in a small agriculture-based community.

Market behaviors categorize travelers based on their knowledge, attitude, use, motivation, or response to an attraction, facility, or service. These behavioral variables may include the occasions that stimulate a visit, the benefits they realize, the status of users, their usage rate, their loyalty, their buyer-readiness stage, and their attitude toward the attractions, facilities, and services a destination offers.

For example, a market segment may be described in these terms:

Geographic: They are located in the suburbs of Seattle, Chicago, and Los Angeles that have populations of 65,000.

Demographic: The average incomes of this predominately couple group is $70,000 or more, most have attended college, they are between the ages of 35 and 50, and they have children at or out of the home.

Psychographic: They consider time their most limited resource, and security—both physical and financial—is important.

Behaviors: After checking with people they know and trust, they choose destinations that offer unique and authentic cultural heritage attractions. They are likely to keep visiting such destinations. However, although they may be loyal to a destination, they may eventually stop visiting the destination because of their desire to visit someplace new.

After developing profiles of each market segments and identifying where they are located, destination marketers identify the segments that are likely to yield the highest return on each marketing dollar spent to attract those travelers. Once the most profitable segments are identified, destination marketing managers develop promotional campaigns and communications materials to communicate with each of those segments.

COMMUNICATING WITH THE TOURIST MARKET

In today's dynamic global environment, understanding how travelers acquire information is important for making marketing decisions, designing effective marketing communications campaigns, and delivering services. Before designing communication materials for specific target market segments, destination marketers need to understand what kind of image the destination travelers have in their minds. This is because a destination image is one of the factors most likely to influence travelers' destination selection.

Importance of Image Promotion

Every communication related to a destination helps consumers form an image of that place; websites, books, movies, television, postcards, songs, photographs, news stories, and advertising all contribute. Favorable images, of course, greatly improve the chance of increasing tourist traffic. *The Lord of the Rings* movie trilogy spawned a number of operators in New Zealand to offer tours to the movie locations. Similarly, two of America's popular television series, *Survivor* and *The Amazing Race,* increased Americans' interest in adventure travel destinations around the world.

On the other hand, negative images can also have a profound impact by increasing the challenge for NTO marketers to promote a country's tourism overseas. The book and movie *Midnight Express,* which depict the story of the convicted drug smuggler Billy Hayes' treatment in a Turkish prison before he escaped after 5 years (he was sentenced to 30 years as an example to others), hurt tourism to Turkey. Turkey was portrayed as having a corrupt and inhumane government, and as a result, people were afraid to visit. In another example, India had difficulty promoting itself as a tourist destination because of the image of poverty that people associated with the country. The scenic beauty and many cultural attractions of India were overshadowed by the negative scenes of starving people and squalid living conditions. In recent years, however, India has been able to overcome these challenges and has seen double-digit increases in foreign tourist arrivals since 2003.[13] In the United States, tourism in the entire state of Florida suffered following the massive hurricanes in 2004 and 2005. These forced tourism officials to make efforts to overcome the negative images of Florida as a hurricane target, showing that the hurricanes that struck Florida did not affect certain beach areas.

Images of a destination are so important that states and countries spend millions of dollars to build positive images of their destinations. Some researchers postulate that a tourist's experience is nothing but a constant modification of the destination image. A tourist makes a destination choice based on a previously held image of the destination. The tourist's actual experience in the destination provides comparison with the previous image—a "reality check"—and determines the tourist's level of satisfaction or dissatisfaction with the overall experience.

Choosing a Destination

The following example illustrates the process of choosing a destination: Ken is a 21-year-old college student from Pittsburgh who decides to take a vacation over spring break and reads brochures for spring travel packages to Cancún, Mexico. He has never been to Mexico and is excited about the idea of going to Cancún. At this point, Ken's image of Cancún is based mainly on three things: the written information from the tour package brochure; "Visit Cancún" websites; and the knowledge about Mexico he acquired from books, mass media, and friends. Ken's image of the destination is most important here because his expectations of Cancún (and also Mexico in general) are based on his images of the area.

Spring break comes and Ken travels to Cancún. He participates in water sports and also meets new friends. When he returns home, he will go through a "recollection" stage in which he evaluates his overall experience, including a comparison of his expectations and actual experiences. If the actual experiences lived up to his expectations (based on his images of the destination), he will be satisfied; if the actual experiences did not live up to his expectations, he will be dissatisfied. Depending on his level of satisfaction or dissatisfaction, Ken will decide whether to return to Cancún, as well as other places in Mexico, in the future. More important, he'll talk about his experiences with his friends, which will, in turn, help his friends form images of Cancún and Mexico.

TRAVELERS' INFORMATION SEARCH BEHAVIOR

As with the purchase of any product, when it comes to making a vacation decision, travelers are likely to go through a decision-making process, which includes an information search. Understanding the information search behavior of key current and prospective markets can help destination managers and marketers develop effective communications. Travelers' information source utilization patterns can be used as either a segmentation base or descriptor, which can help marketers focus on positioning and media selection.

Recent hospitality and tourism studies and marketing and consumer behavior literature suggest that a consumer's prior product knowledge comprises two components: familiarity and expertise. Familiarity represents the early stages of learning and expertise represents the later stages of learning. As consumers' familiarity with the product increases, their expertise with the product increases as well.

Expertise

Expertise can be defined as product-related experiences such as advertising exposures, information search, interactions with salespersons, purchasing, and product usage in various situations. The term *consumer expertise* is also used in a very broad sense to include both beliefs about product attributes and decision rules for acting on those beliefs.

At the most basic level, mere exposure to a brand name may help the consumer recognize the brand during a later visual search. Repeated exposure to a single brand or attribute may lead to easy retrieval of information about that single brand or attribute. Wider experience results in the accumulation of more information, which enables consumers to include more brands in their memory and to recall and use more attributes in their decision making. When decisions are based on internal information, knowledge may offer an expert consumer an opportunity to use processing decision strategies that are very different from the ones used by the consumer who is low in expertise. When a consumer who is high in expertise and a consumer who is low in expertise learn the same information and later must make a decision, the expert consumer may be able to rely on memory, whereas the consumer who is low in expertise may again need to engage in an external search to avoid making an incorrect decision.[14]

Certainly, understanding external information source utilization can help marketers effectively tailor the promotional mix. In addition, understanding the similarities and differences in familiar travelers' and expert travelers' external information search behavior and identifying which information sources are most likely to be used by familiar travelers and expert travelers can help marketing managers design effective marketing programs and communication strategies.

Past research in the area of information search has focused on nearly 60 variables that are likely to influence external information searches. These include several aspects of the environment (e.g., the difficulty of the choice task, the number of alternatives, and the complexity of the alternatives), situational variables (e.g., previous satisfaction, time constraints, perceived risk, and the composition of the traveling party), consumer characteristics (e.g., education, prior product knowledge, involvement, family life cycle, and socioeconomic status), and product characteristics (e.g., the purpose of the trip and mode of travel).

Travelers' involvement is also likely to have a positive effect on familiarity and expertise. Highly involved travelers are likely to be more familiar with the product, remember the product information, develop better category structures, analyze the information in more detail, elaborate on it, and make automatic decisions. A traveler's involvement may also positively influence intentional learning. A traveler who is highly involved is likely to pay more attention to incoming information such as commercials about the destination.

Familiarity

Familiarity with a product category has been recognized as an important factor in consumer decision making. Consumers' familiarity with a product category is measured as a continuous variable that reflects their direct and indirect knowledge of a product category. Familiarity has been defined as the consumer's perception of how much he or she knows about the attributes of various choice alternatives being considered. However, several researchers suggest that what people think they know and what they actually know often do not correspond because familiarity represents a traveler's subjective knowledge of the destination, whereas the traveler's expertise represents his or her objective knowledge.

Because familiarity represents early stages of learning, consumers are likely to gain knowledge and, therefore, familiarity through an ongoing information search such as reading guidebooks or other related books, seeing advertising and write-ups in newspapers and magazines, watching advertisements on TV, listening to advertising on the radio, and talking to friends and relatives. Studies show that product familiarity has a direct impact on consumers' information search behavior. In both familiar and unfamiliar product categories, consumers first search their memory for some information to help guide them to make decisions. Consumers' familiarity with a product category is likely to lead to direct acquisition of available information from memory. If the consumer has sufficient information, there may be no need to search for additional information; the consumer can make a decision based on the information he already has.

A traveler who has been to the destination before is likely to have more familiarity with and expertise on the destination than a traveler who has never been to the destination. The previous visits may also have a positive influence on a traveler's involvement. Previous studies suggest that as the number of visits to a specific destination increases, a traveler's involvement is likely to increase as well.

Learning is also likely to influence a traveler's information search behavior. Studies suggest that travelers' learning has two dimensions: **intentional learning** and **incidental learning.**

Intentional learning is likely to increase a traveler's expertise and familiarity, whereas incidental learning is likely to increase a traveler's familiarity. Travelers who gather information through intentional learning are likely to pay more attention to incoming information and process the information thoroughly and therefore increase their objective knowledge and expertise. On the other hand, travelers who learn through incidental learning are not likely to process information thoroughly. However, because incidental learners have some information about the destination and its attractions, their learning is likely to increase their subjective knowledge and therefore their familiarity with the destination and its attractions.

It is crucial for destination managers to understand the importance of the perceived cost of an information search. The negative relationship between the perceived costs of external information and the amount of information received should cause marketers to take steps to make external searches as inexpensive and time efficient as possible. This is often not the case in travel marketing. For example, a perusal of destination websites quickly reveals sites that are difficult to navigate, take a long time to load, and are linked to empty sites and incomplete information. This increase in time cost to acquire information can cause travelers to look elsewhere for information. Destination marketers should also be aware of the fact that the more information there is available about a destination, the more likely travelers are to increase both incidental and intentional learning. These two factors are likely to lead to increased familiarity and expertise, which, in turn, decrease information search costs, reduce the necessity for an extensive external search, and help focus the search on specific attributes rather than on general information.

intentional learning
Learning that occurs through the active cognitive processing of information. An example of intentional learning is when you read this book in preparation for class or an exam.

incidental learning
Learning that occurs unintentionally. It can occur by completing tasks, watching others, doing the same thing over and over again, or talking to experts. A professor who makes the class fun and interesting fosters incidental learning.

Marketers recognize the value of actual visitation to a site for improving marketing outcomes. Indeed, previous visits may positively affect involvement with a destination while increasing familiarity and expertise, which lead to the outcomes discussed earlier. Both familiar and expert travelers are likely to use external information sources to varying degrees. However, travelers' use of external information is likely to be influenced by their perception of the cost of the information search. Therefore, marketers and advertisers need to develop different communication strategies for familiar and expert travelers. Because unfamiliar travelers are likely to have a hard time examining the information gathered from external sources because of their limited processing ability, they may require a different communication strategy than expert travelers. Marketers communicating with unfamiliar travelers should provide simple information about the overall destination. They may also need to include a comparison of the destination with other destinations that target the same market to make it easier for the traveler to digest the information.

In other words, communication materials should clearly identify the unique selling propositions of the destination to differentiate the destination from competitors and to make positioning of the destination easier for unfamiliar travelers. Establishing a good and understandable communication with unfamiliar travelers is critical in convincing them to choose a destination over other destinations because low familiarity is associated with higher perceived importance of, and receptivity to, new information.

Another method of communicating with unfamiliar travelers is through word of mouth. Unfamiliar travelers may have a hard time comprehending and evaluating product-related information because of their inferior ability to comprehend and evaluate product-related facts. Because of their limited ability to process the product-related information, unfamiliar travelers are more likely to sample the opinions of others such as their friends and family. Because positive word of mouth is the result of satisfaction, special attention needs to be given to customer satisfaction and complaint handling. Customer satisfaction should be constantly monitored in order to identify the problem areas and to make necessary modifications to enhance customer satisfaction. In addition, customers' complaints should be handled delicately and quickly to ensure satisfaction and positive word of mouth.

Communication materials developed for expert travelers should include detailed information about the attributes that are important to the target market. These attributes can easily be identified by conducting formal or informal research. In addition, destinations also need to monitor changing consumer needs and wants that may shift the importance placed on attributes. Destinations can design surveys or conduct focus groups to find out and monitor what attributes are most important to expert travelers. Managers may also identify the important attributes by just talking to their existing customers. Destinations need to pay special attention to identifying expert travelers. If destination managers and marketers fail to ask the right questions to the right audience, they may end up making the wrong conclusions and developing ineffective communication strategies.

After the important attributes are identified, destinations need to communicate them to expert travelers. Expert travelers are more likely to search for detailed information. Therefore, destinations need to develop communication materials (i.e., brochures, direct mailing materials, etc.) that provide detailed information about the destination and its important attributes. These materials need to be modified as expert travelers' needs and wants change. Destination managers and marketers should understand that different travelers have different information needs. While unfamiliar travelers need simple, understandable, and general information, expert travelers need detailed information about the destination and attributes to make their vacation decisions. Destination managers and marketers can use travelers' levels of product knowledge (familiarity and expertise) as a segmentation tool to develop communication strategies that are most appropriate for each segment.

Travel marketers must have an overall picture of how travelers acquire information. They must also know the major components of the search process and how they fit together. With this understanding, marketers can design communication strategies aimed specifically at different stages in the information search process, which will lead to the efficient use of resources and more success in attracting tourists to their specific destinations.

Chapter Summary

Marketing destinations is no different from marketing hospitality products. External macro and micro environmental factors that are likely to influence the marketing efforts of any hospitality organization are likely to influence any destination marketing efforts. The only difference between hospitality marketing and destination marketing is that most destination marketing activities are likely to be undertaken by public agencies such as NTOs with the support and help of private hospitality organizations.

A factor that is likely to influence the success of any tourist destination is local residents' attitudes toward tourists and tourism. Any tourism development in a destination and any marketing activity that is not supported by local residents is likely to fail. To make sound marketing decisions, an NTO or its subset needs to constantly monitor any changes occurring in the environment and travelers' behaviors and take a proactive posture in its marketing programs. Because of the importance of building positive destination images, in the future, NTOs and their subsets will continue to play an important image building and image communication role in the marketplace. However, to communicate the desired image, an NTO needs to understand the factors that are likely to influence travelers' information search behavior and develop communication strategies that are aligned with their search behavior. At the same time, NTOs will play a greater role as facilitators for market research and as collaborators of market efforts by the private sector of the tourism industry.

Key Terms

destination marketing, p. 305

incidental learning, p. 326

industry/supply-based approach to identifying competition, p. 320

intentional learning, p. 326

market or demand-based approach to identifying competition, p. 320

multiplier effect, p. 303

national tourism organization (NTO), p.305

United Nations World Tourism Organization, p. 302

Discussion Questions

1. Briefly describe the United Nations World Tourism Organization (UNWTO) and its involvement with international tourism marketing.

2. Discuss how the economic environment of a destination or an area affects international tourism today.

3. What impact does technology have on international tourism today?

4. How do concerns for the ecological environment affect international tourism today?

5. What are NTOs? What do they do?

6. Describe what is meant by destination marketing and give an example to demonstrate what it is all about.

Endnotes

1. World Tourism Organization. Historical perspective of world tourism. Retrieved November 1, 2005, from www.world-tourism.org/facts/menu.html

2. www.tia.org/Travel/forecasts.asp, accessed April 10, 2007.

3. Gursoy, D., & Rutherford, D. (2004). Host attitudes toward tourism: An improved structural model. *Annals of Tourism Research, 31*(3): 495–516.

4. Gursoy, D., Jurowski, C., & Uysal, M. (2002). Resident's attitudes: A structural modeling approach. *Annals of Tourism Research, 29* (1): 79–105.

5. East Asian countries launch campaigns to revive tourism lost to SARS. (2003, June 9). *Travel Agent*, 83–84.

6. UNWTO World Tourism Barometer. (2004 October), 2(3), 6.

7. World Tourism Organization. (2005, October 18). Avian flu: Overreaction could damage tourism industry, says WTO [press release]. Retrieved November 29, 2005, from www.worldtourism.org/newsroom/menu.htm

8. Travel Industry Association of America. (2004). *Emerging international tourism markets: Trends and insights.*

9. Retrieved January 3, 2006, from www.allstays.com/Services/2005/04/new-us-travel-restrictions.htm

10. Tiplady, R. (2004, December 13). The year of the Chinese tourist. *BusinessWeek Online.* Retrieved November 8, 2005, from www.businessweek.com/magazine/content/ 04_50/b3912081_mz054.htm

11. Travel Industry Association of America. (2003). *Geotourism: The new trend in travel. Executive Summary.*

12. The World Bank. (2005, September 6). Mexico: World Bank approves $200.5 million for sustainable development [press release]. No. 2006/06/LAC.

13. Tourism statistics for India. (2005, October 4). *Federation of Hotel & Restaurant Associations of India (FHRAI) Magazine.* Retrieved November 8, 2005, from www.fhrai.com/magnews/magTourismStatisticsIndia.asp.

14. Gursoy, D., & McCleary, K. W. (2004). An integrative model of tourist's information search behavior. *Annals of Tourism Research, 31*(2): 353–373.

Eco Paraiso

Discussion Questions

1. Briefly describe the Eco Paraiso ecolodge and its surrounding environment.

2. What are some of the key points raised in this case regarding ecotourism and the attention it is receiving in Mexico in particular?

3. How are "casual" and "dedicated" ecotourist segments distinguished in this case?

4. Based on her experience, how does Ms. Gerber distinguish between European and the North American leisure travelers?

5. What percentage of business at Eco Paraiso comes from travel agencies and tour groups?

6. As discussed in the case and shown in Exhibit 1, what are the various ways Ms. Gerber promotes the property?

7. Looking at your responses to the preceding questions and other information given in the case, put together a SWOT analysis for the Eco Paraiso lodge. In particular, what are the opportunities that look most promising for this ecolodge in the Yucatan region of Mexico?

Ms. Verena Gerber, managing director and primary owner of Eco Paraiso, was wondering how to market and operate her nature-oriented, upscale, ecolodge in Yucatán, Mexico. Although the lodge was five years old, it still had not broken even in any fiscal year. Her co-investors indicated they were getting impatient with the lack of profits and were beginning to express doubts about the financial viability of ecotourism. The terrorist attacks of September 11, 2001, had made matters even more difficult. International tourism across the world registered dramatic declines after the attacks, but bookings at her hotel, as well as at others, were beginning to rebound slowly. Her recently hired manager, Max Voggensperger, an experienced hotelier from Switzerland, was optimistic about the potential of ecotourism in Mexico and believed Eco Paraiso offered some unique benefits to a variety of tourists. But Ms. Gerber was not sure about what to do.

This case was written by Gregory Osland, PhD, Robert MacKoy, PhD, and Daniel McQuiston, PhD, Butler University, Indianapolis, Indiana. All rights reserved. Used by permission.

ECOTOURISM

International tourism, both a cause and effect of globalization, increased dramatically over the last four decades. In the most recent decade, 1991 to 2000, international arrivals increased 4.3 percent a year to 697 million. Tourism was Mexico's third largest industry, contributing US$54 billion to its GDP in 2001. Major contributors to this figure were the 20 million international tourists who visit the nation each year. Although Mexico's tourism industry is best known for its "sand-and-surf" destinations, such as Cancún and Acapulco, an alternative kind of tourism called ecotourism was attracting increased attention from governments, nongovernmental organizations (NGOs), tourists, and the tourism industry.

Ecotourism is a sustainable form of nature-based travel that is managed to conserve the physical environment on which it depends, to provide economic benefits to the local community and the owners, and to educate and satisfy the tourists. The combination of the nature-based context and these multiple perspectives and goals distinguishes ecotourism from general, or mass, tourism. Ecotourism has been embraced by many nongovernmental and some governmental organizations because it is a promising means to both conserve natural areas and foster sustainable economic development (e.g., Conservation International, Nature Conservancy, United Nations World Ecotourism Summit, 2002). The United Nations declared 2002 the "International Year of Ecotourism," and the Mexican Tourism Ministry committed 256 million pesos to promote this type of tourism in 2002.

Tourists are increasingly attracted to ecotourism as a means to learn about and experience relatively undisturbed natural areas and charismatic fauna such as flamingos, monkeys, and whales. As high-quality natural areas deteriorate and disappear around the globe, there appears to be more interest among many travelers to see endangered species and natural habitats "before they are gone." Finding and watching birds in natural settings is the most popular ecotourism activity. The intense, high-stress lifestyle of many professionals in developed nations is also leading a growing number to seek out vacations in remote areas where they can "get away from it all" and relax in a peaceful, natural setting. Another relevant trend is the increased desire of "empty nesters" and retirees for alternative vacations that involve learning and education. Ecotourism's growing popularity among tourists has motivated countless tour operators and accommodation providers to start businesses or position their businesses around ecotourism themes. An illustration of this trend is Planeta.com, a successful web-based network of more than a thousand ecotour operators, ecolodge owners, and researchers that focuses its attention on nature-oriented travel in Mexico and other Latin American countries.

HISTORY OF THE HOTEL

Verena Gerber, owner of a successful hotel south of Cancún, on the Caribbean coast of the Yucatán Peninsula of Mexico, decided to build a lodge in a remote natural area on the western side of Yucatán to serve people who enjoy vacationing in nature. She had become disillusioned with the changes in the flavor of tourism along the Caribbean, as her previously tranquil lodge became surrounded by intense development. With strongly held environmental ideals and goals, she initiated the purchase of 530 hectares of mangrove swamps, coastal dunes, and a coconut palm plantation on the undeveloped Gulf of Mexico coast of the Yucatán Peninsula.

Her land adjoined Ría Celestún, a protected natural area owned by the Mexican government, which attracts the world's largest colony (20,000) of greater flamingos during the winter and over 350 other species of birds. Three species of endangered sea turtles nest on the quiet beaches of this area.

Ms. Gerber opened an ecological hotel on her property, naming it Eco Paraiso (Ecological Paradise). To better serve nature-oriented tourists, she hired one of the best local naturalists to guide guests on a variety of nature tours and employed a biologist to manage the lodge. His special strengths were adhering to ecotourism ideals and educating tourists. He developed a natural history museum in a room in the reception center and worked at Eco Paraiso for three years. The hotel has consistently employed eight people full-time, including one expatriate manager and seven nationals.

The occupancy rate of Eco Paraiso increased slowly, from 9.5 percent to 22 percent. In the past year an average of 3.3 rooms were occupied per night. The lodge's direct and indirect variable costs comprised about 40 percent of the rooms revenue, consistent with the hotel industry's typical margin of approximately 60 percent. The "high season" was November through April, when occupancy rates averaged about 34 percent. During the "low season" from May to October, occupancy rates averaged around 14 percent. Ms. Gerber calculated her breakeven occupancy rate at 55 percent.

TARGET MARKETS/CUSTOMERS

When Eco Paraiso opened, the Swiss-Mexican owner targeted fairly wealthy, nature-oriented Europeans as the primary market for her ecolodge. Ms. Gerber selected this group based on her vision for the lodge, her cultural background, and previous success with Europeans at her Caribbean resort. Europeans tend to have long vacations and travel throughout the year. North Americans seemed to be a less promising segment; she believed they were less willing to travel "off the beaten track" in Mexico than were Europeans. "I have had experience with North Americans at my Caribbean resort, and I know those people would not enjoy a vacation here at Eco Paraiso," stated Ms. Gerber.

Academic research has found that there are two primary segments of ecotourists: dedicated ecotourists and casual ecotourists. Dedicated ecotourists visit natural areas with the primary purpose of observing nature, such as the flamingos and other endemic species of plants and wildlife. They often spend most of their day in natural habitats with a guide. Dedicated ecotourists are a small, but growing segment of tourists and are characterized as middle-aged to retired, with education and income levels that are higher than the average tourist's. January to March is a prime time for dedicated ecotourists to visit Eco Paraiso. The following quote from a guest at Eco Paraiso exemplifies this kind of tourist:

Together with another couple we spent five days and four nights (March 9–13) in your lovely resort. We had an unforgettable time in this pristine part of the world. The cottages we stayed in were very attractive and the service of the staff was A-1. We took three excursions to the Celestún Reserve and admired the beautiful nature, the bright orange flamingos, the great number of other birds, and during our night trip, the crocodiles. Your guide, Alejandro Zib, was superb. Not only is this young man extremely knowledgeable, but in addition he was just great company. We believe that to really see the bird life well, one has to start at the hotel not

*later than 6 A.M. which we fortunately did. The estuary
at that time is just mind boggling and so is the fauna.
We also made very long walks on the extensive beach.
All in all it was a great experience which we enjoyed to
the fullest. Hopefully this beautiful setting will be
preserved forever.*

—Helene and Ben, Switzerland

Casual ecotourists are a much larger segment that
appears to be increasing in numbers, as well. People in this
group seek a natural setting, but in contrast to dedicated
ecotourists, spend a smaller portion of their time observing
nature. They often expect more amenities and have less tolerance for physical discomfort than dedicated ecotourists.
Furthermore, they typically spend more time lying on the
beach or in hammocks than looking for wildlife on a trail.
This group often has little specific knowledge of nature, but
enjoys seeing dramatic natural spectacles, such as the large
concentrations of flamingos in the estuary near Eco Paraiso.
The tranquility and pristine character of Eco Paraiso would be
attractive to most casual ecotourists.

Ecotour operators are important intermediaries in
serving the dedicated ecotourist market. Birding groups,
such as Wings, Victor Emanuel Nature Tours (VENT), and
Field Guides, design one- to two-week tours to attractive
natural areas around the world and charge groups of 10 to
15 ecotourists a package rate that averages $200 to $300
per person per day. These groups may stay one or two
nights at a hotel before moving on to other natural areas
in the region. Eco Paraiso has been marketed to these kinds
of groups.

Ms. Gerber also sought out other target markets.
"Esoterics" are a group of potential customers looking for
a spiritual retreat and/or meditation in quiet, remote settings. These include New Age, yoga, and religious groups.
Honeymooners may also be attracted to this kind of isolated
setting. Lastly, corporations and associations that would
use the facilities for small conferences, retreats and work-
shops are a target market. Ms. Gerber believed that with 15
rooms and a capacity of 30 (at double occupancy), the
lodge was too small to attract most tour groups, or the
interest of most travel agencies who operate on a high-
volume basis. Because she primarily targeted international
guests, she was surprised that the percentage of national
guests increased each year, to almost half of all guest
nights. Upper-middle-class Mexican families came to the
lodge for quiet beach vacations, most frequently in late
summer.

FACILITIES

The hotel was conscientiously built to be environmentally
sensitive and self-sustaining and to provide guests with an
authentic natural experience. Fifteen tastefully designed bungalows, each with a porch and a hammock, were placed well
behind the beach, with each offering private views of the Gulf
of Mexico. The Clubhouse, in the center of the complex, contained an elegant, but casual restaurant that seated 40, a
well-stocked bar, a reading/satellite TV room, and a game
room. The reception center included a natural history museum
and two administrative offices. Guests could also swim in an
attractive freshwater pool, paddle sea kayaks in the Gulf, ride
bicycles, and climb a 17-meter-high observation tower that
provided stunning views of sunsets and the surrounding
wilderness. The hotel offered a wide array of guided natural
history and cultural tours that ranged from an hour to a full
day. Eco Paraiso offered many combinations of its services and
had tried to emphasize discounted packages that included
meals, tours, and three to six nights of lodging.

Ms. Gerber had developed and implemented a rather
costly facility operations system that was consistent with her
environmental ideals. Well water and sewage were treated
with state-of-the-art "green" technologies, including a reverse-
osmosis system, while organic waste was composted and recycled. Nonrecyclable waste was hauled to a garbage site two
hours away. Some of the food was grown in gardens on the property, while most of the food supplies are purchased in the city
of Mérida, a one-and-one-half to two-hour trip to the east. In
contrast, most of the other ecotourism lodges in Mexico had not
taken as "green" an approach in hotel management operations.

The areas around the buildings had been landscaped
with native shrubs and flowers, providing excellent habitat for
local species of butterflies, birds, and small reptiles. Eco Paraiso
generated its own electricity and had installed a digital telephone system to communicate via phone, fax, and Internet.
Guests could also use these telecommunication services.

LOCATION

Cancún was the gateway for most international, and some national, guests to Eco Paraiso. The large Cancún international
airport on the opposite side of the Yucatán Peninsula was a
seven-hour drive to Eco Paraiso. Guests could also fly to
Mérida, the capital city of Yucatán, which was a two-hour
drive to the hotel. Continental and American Airlines offered
direct flights to Mérida from the United States. Some guests

paid a ground transportation fee to be picked up at the Mérida airport. A few took a second-class bus which went to the nearest town, Celestún. This small fishing village was the access point to the Ría Celestún Biosphere Reserve, where the flamingos fed and roosted.

Almost all of the 10,000 visitors per year who rode on guided boats to access the estuary of the reserve came from Mérida on group day tours. There were five one-star and two-star hotels in the town, primarily targeted for Mexicans. Eco Paraiso guests endured a 20-minute ride from Celestún on a poorly maintained sand-and-gravel, and then sand, road to get to the remote "ecological paradise," where there was virtually no air, water, light, or noise pollution.

Eco Paraiso was located on the edge of biologically rich breeding grounds of the Gulf of Mexico that teemed with many species of fish and other marine life, such as octopus. Magnificent frigatebirds and brown pelicans continuously plied the waters for food. Located in the tropics, Eco Paraiso experienced hot and humid weather from April to October. In the dry "winter" season from December to March, temperatures often cooled off to the low 80s (Fahrenheit) for highs, and the mid-60s for lows. Near the beach was a narrow coastal dune habitat of grasses and shrubs, while much of the property was composed of spiny thickets of low-growing trees and shrubs. The east edge of Eco Paraiso bordered mangrove swamps and the estuary of the Ria Celestún Biosphere Reserve.

Insects were abundant, with mosquitoes and sand flies present in significant numbers for most of the year. The hotel used a mild, environmentally sensitive pesticide to reduce the number of mosquitoes around the rooms at night. Ms. Gerber chose to not use stronger pesticides that would nearly eliminate mosquitoes and sand flies, because these would harm other insects and wildlife as well.

The primary distribution channel was a direct one— only 9.5 percent of guests arrived through travel agencies and tour groups. Fifty percent of the guests made reservations, and 50 percent were "walk-ups"—an extraordinarily high percentage for an upscale ecolodge in a remote location.

COMMUNICATIONS

Ms. Gerber promoted the lodge extensively. She advertised the lodge on her own website (www.ecoparaiso.com/ecopa_e/navi/navi_frame_e.html), on six other websites, and in 14 publications, primarily in tourist-oriented newspapers and magazines in Mexico. Advertising also included the production and distribution of brochures about Eco Paraiso and several billboards and signs in the Celestún area. She personally sold the lodge to travel agencies and tour operators at eight trade shows in Mexico and abroad. Approximately 20 percent of her annual promotion budget was spent on free nights of lodging for travel agents, tour operators, conference leaders, and travel and business writers. She considered the publicity generated from a *BusinessWeek* article on ecotourism and her lodge as the most effective single promotion of the past five years. Ms. Gerber believed she needed to reevaluate her promotion strategy. For example, she wondered whether her promotion efforts should try to get people interested in the Celestún area first, and then to the hotel, or if she should just try to attract visitors to the hotel directly. Exhibit 1 shows previous promotion methods and costs. Twenty-five

Exhibit 1	Previous Promotion Methods and Costs (in pesos)					
Item type		Year 1	Year 2	Year 3	Jan-May Year 4	Total Cost
Free rooms (to the trade)	prom	28,571.20	25,576.00	42,098.40	10,387.60	106,633.20
Yucatan Today	publ	6,108.00	9,900.00	13,095.00	4,744.00	33,847.00
Hotels Directory	publ		7,500.00	9,500.00	12,650.00	29,650.00
MITT magazine	publ			4,200.00	1,400.00	5,600.00
Yucatan newspaper	publ	3,846.00		9,132.00		12,978.00
Peninsula map	publ			1,800.00		1,800.00
MID map	publ			3,500.00		3,500.00
Tourist stamp	publ			6,300.00		6,300.00
Phonebook	publ			3,000.00		3,000.00
CD	publ		8,123.00			8,123.00
Specialty Travel magazine	publ	7,200.00	4,250.00	2,018.75		13,468.75
Discover Mexico magazine	publ		2,905.00	2,880.00	2,773.00	8,558.00
Transitions abroad	publ		2,500.00		2,350.00	4,850.00

Exhibit 1   Continued

Item type		Year 1	Year 2	Year 3	Jan-May Year 4	Total Cost
Footprint Tourism guide	publ			13,492.40		13,492.40
Canada Hotel Directory	publ				2,820.00	2,820.00
Flyers	paper	11,460.00	6,673.00	880.00	4,340.00	23,353.00
Brochures	paper	14,500.00	20,700.00			35,200.00
Business cards &misc.	paper	1,500.00	3,582.00	473.00	240.00	5,795.00
Photos and transparencies	paper	15,124.00	7,199.00	20,817.00		43,140.00
Postal cards	paper			4,520.00	4,860.00	9,380.00
Tianguis Acapulco	fair	11,500.00	550.00	19,838.00	575.00	32,463.00
Chicago	fair		18,237.00			18,237.00
Let's Take Vacations	fair		7,026.00			7,026.00
Kihuic Merida	fair			6,089.00	9,305.00	15,394.00
Expoadventure	fair		13,654.00			13,654.00
Speleology/diving	fair		3,000.00			3,000.00
Exotic Escapes	fair		4,000.00			4,000.00

Exhibit 2   Sources of Information on Eco Paraiso

Source	# of Rooms	% of Rooms
Internet	79	24.9%
Miscellaneous	58	18.3%
Tourist guide	43	13.6%
Celestún billboard	32	10.1%
Travel agencies	30	9.5%
Repetitive	23	7.3%
Recommendation	16	5.0%
Family	11	3.5%
Great Plan (AeroMex)	10	3.2%
Yucatan Today	8	2.5%
Swiss tour company	7	2.2%
Total	317	100.0%

percent of the guests learned of the hotel from independent Internet communication. Exhibit 2 shows other sources from which guests learned of Eco Paraiso.

PRICE

Hotel pricing is rather complex and variable, with European pricing plans and American pricing plans. Eco Paraiso used the American plan, with a rack rate per person for a double occupancy bungalow, two meals, tips, and taxes in the previous year of US$96. (American plan pricing typically includes all of the preceding items in one price. The European plan highlights the basic room charge, does not include meals and tips, and adds on the taxes. Rack rates are the highest retail prices.) At Eco Paraiso children under six were not charged; kids 6 to 12 in the same room were charged US$31 each on the American plan. The rate for one adult in a bungalow was US$154. During summer (the low season) the retail rates were reduced 15 percent.

To induce longer stays and to provide a greater sense of value, Ms. Gerber designed a number of packages. One of the most popular was the "Ti Ma'an." For US$381 per person in the winter, including tips and taxes, a couple received three nights' lodging, three meals per day, nonalcoholic beverages, a half-day guided boat tour in the estuary, and unlimited use of bicycles and kayaks. Alternatively, for seven days and six nights, and all of the above, the price was US$732 per person, double occupancy. This package also included two additional half-day and one extra day tour in the region. Most of the individual reservations were sold at these rack prices; however, walk-up guests often negotiated lower prices. In the previous year the average rate charged per room for independent guests was US$147.30, including two meals, tips, and taxes. The average per person rate was $72.60. Government sales and room taxes were 17 percent, which were included in these average rates. Tour operators' and travel agencies' prices on all the aforementioned products were approximately 20 percent less than the rates for individuals. However, these intermediaries could also, at any time of the year, book a minimum of 12 of the 15 bungalows for seven days and six nights, including all meals, tips, and taxes, for $497 per person and then sell these wholesaled rooms at whatever rates they wished.

CUSTOMER EVALUATIONS

While checking out of the hotel, guests were asked to complete a satisfaction survey regarding 11 aspects of the lodge. Customers had rated Eco Paraiso as "very good" to "excellent"

Exhibit 3	Customer Evaluations of Eco Paraiso*

	Nov.	Dec.	Jan.	Feb.	Mar.	Apr.	May	Jun.	Jul.	Aug.	Sep.	Oct.
Reception	9.5	9.6	9.5	9.6	9.6	9.4	9.4	9.3	9.2	9.2	9.5	9.6
Rooms	9.0	9.6	9.4	9.4	9.4	9.0	9.1	9.2	9.1	9.2	9.6	9.5
Breakfast	9.4	9.2	9.3	9.3	9.5	9.3	9.3	9.7	9.3	9.3	9.7	9.4
Lunch	9.2	9.2	9.4	9.5	9.6	9.1	9.7	9.1	9.2	9.0	9.6	9.6
Dinner	9.2	9.2	9.0	9.1	9.4	9.3	9.2	9.7	9.3	9.3	9.6	9.7
Clubhouse	9.0	9.1	8.9	9.1	9.3	9.0	9.0	9.3	9.3	8.8	9.3	9.2
Library	9.1	8.9	8.4	8.7	8.9	9.1	9.3	8.9	9.2	8.8	8.9	8.7
Pool	9.1	8.9	8.9	9.0	9.1	8.7	8.9	9.5	8.9	8.9	9.4	9.4
Green hotel	9.6	9.5	9.4	9.5	9.3	9.5	9.6	9.5	9.5	9.5	9.7	9.6
Ambience	9.5	9.6	9.5	9.5	9.8	9.5	9.6	9.5	9.5	9.3	9.9	9.8
Average	9.26	9.28	9.17	9.27	9.39	9.19	9.31	9.37	9.25	9.13	9.50	9.40
Price												
Very high	5%	0	2	2	14	13	3	0	10	13	0	5
High	47%	38	36	36	36	29	26	42	33	54	41	20
Right	47%	62	60	62	50	58	71	58	56	33	59	75
Low	0%	0	2	0	0	0	0	0	0	0	0	0
Too low	0%	0	0	0	0	0	0	0	0	0	0	0
Average price	US$76	80	77	87	74	72	69	75	71	66	68	66
# of surveys	45	42	52	49	43	56	44	12	41	42	17	20
% foreign	58	83	71	92	74	52	75	67	37	50	59	60
% nationals	42	17	29	8	26	48	25	33	63	50	41	40

*10-point scale where 10 is excellent, 9 is very good, etc.

on physical facilities, meals, ambience, reception, and ecological elements. About 60 percent of the guests in the winter season felt that the price–value ratio "seemed right," whereas approximately 40 percent felt the prices were "too high." In August, the traditional vacation month of Mexicans when the guests are mostly nationals, 67 percent of the guests felt the prices were too high. Exhibit 3 shows some customer evaluations of Eco Paraiso.

THE QUANDARY

"I don't know what to do," sighed Ms. Gerber. "Our guests really enjoy the lodge, but I don't know how to get more people to come here. If I expanded the number of cabins, or even added tent platforms, we could probably get more groups and more support from travel agencies for the high season. But we would still have periods when almost no one wants to come to the Yucatán. I would like to shut down the hotel in three months of summer, but nature would take over and it would cost too much to reopen. Ecotourism seems to be a tougher sell than beach tourism. Maybe I should just try to find a buyer for this property."

"But where else can people find such a quiet and pristine area?" asked the new manager, Mr. Voggensperger. "Maybe we are just a few years ahead of the market for this area. It will be successful eventually."

part V

Functional Strategies

overview

This chapter begins with a review of the marketing communications mix first introduced in Chapter 3. Once strategic marketing decisions, the mission statement, situational analysis, and business strategies are in place, communications mix strategies come into the forefront. In this chapter we take a closer look at three marketing communications mix tools: advertising, merchandising, and public relations. In the next chapter we continue with the communications mix, focusing on personal selling and sales promotions.

Advertising, Merchandising, and Public Relations

learning objectives

After reading this chapter, you should be able to:

1. Explain how the marketing communications mix is used to attract the right customer.

2. Develop an advertising plan for a hospitality enterprise.

3. Develop a public relations campaign.

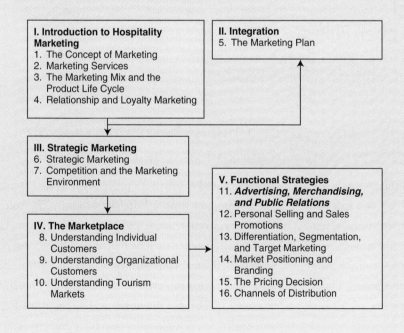

I. Introduction to Hospitality Marketing
1. The Concept of Marketing
2. Marketing Services
3. The Marketing Mix and the Product Life Cycle
4. Relationship and Loyalty Marketing

II. Integration
5. The Marketing Plan

III. Strategic Marketing
6. Strategic Marketing
7. Competition and the Marketing Environment

IV. The Marketplace
8. Understanding Individual Customers
9. Understanding Organizational Customers
10. Understanding Tourism Markets

V. Functional Strategies
11. *Advertising, Merchandising, and Public Relations*
12. Personal Selling and Sales Promotions
13. Differentiation, Segmentation, and Target Marketing
14. Market Positioning and Branding
15. The Pricing Decision
16. Channels of Distribution

Jennifer Ploszaj

Director of Global Brand Communications and Public Relations,
InterContinental Hotels & Resorts

Jennifer Ploszaj currently serves as director of global brand communications and PR for InterContinental Hotels & Resorts, the luxury brand of InterContinental Hotels Group. In this role she has overall responsibility for communications and public relations for the 140 InterContinental hotels around the world.

Ploszaj's career spans more than 15 years, with a focus in communications and marketing in the luxury hotel segment. Prior to 2005 Jennifer was vice president of communications for Noble House Hotels & Resorts, a privately held boutique luxury resort company based in Seattle. In that role, she managed all communications including advertising and marketing and led the company to win more than 20 HSMAI awards for advertising and marketing excellence. Jennifer also spent time working in San Francisco during the dot-com boom and led the communications efforts for a NASDAQ-listed online marketing company.

 ## Marketing in Action

Jennifer Ploszaj

Director of Global Brand Communications and Public Relations,
InterContinental Hotels & Resorts

What is the role of a public relations specialist at the property level and at the corporate level?

We'll start with the property level because I think that's probably the most tactical area in which you can make the biggest impact from a hotel standpoint. The role of the public relations specialist or director is to drive local and regional awareness. The role of PR directors in each individual location, whether it be Beirut, San Francisco, or Hong Kong, is to drive awareness within their local community, as well as within their key source markets. For example, the PR director in Paris focuses largely on French media; U.S. media, which is also a big source market for the hotel; and German media, which is another big source market for the hotel. [The local property approach] is the same approach as marketing, so it's very tactical; it's getting people in the door, getting people to try, getting people to write about the hotel within the regions in which they can drive business. The local approach is about driving awareness of the hotel in the local and regional markets, but also driving trial of the property.

The corporate or global role is to drive awareness for the brand across the board and across the globe. This is where a PR director is focused on brand positioning and key messages. The global role is more of an awareness approach as opposed to a tactical approach, which is what the local PR directors embark upon. We approach PR as the fifth P in the marketing mix, so our role is to support and supplement the overall brand marketing plan. The other piece of corporate PR is to ensure that individual staff in the local and regional markets are on message—that what they're doing and what they're communicating supports the overall brand positioning.

How is PR practiced at a firm like InterContinental Hotels Group?

In a large, global organization, we face challenges relating to the levels of sophistication with regard to the discipline in various markets around the world. For example, in the United States the practice has evolved to a highly sophisticated level—results are critically measured and planning is aligned with the marketing efforts. In Asia, where PR is still very much a young, growing field, there are still many basic principles that are being practiced at the hotel level. However, we see very positive progress and view our corporate communications and PR teams as extremely advanced and highly strategic. Therefore, working from a global standpoint is quite interesting because you are working on significantly different platforms around the world. It is very hard to build a single policy that applies to all markets in the world. As markets mature and businesses mature in each of those markets, then the discipline matures along with it.

Our corporate team approach is to try to create "best practices" around the world. Additionally, it's important that we create processes and resources to support the regions. As with any business around the world, we must adapt our strategies to fit the needs of the local market.

What would you say has been the biggest challenge that you have faced as a PR executive?

For me, the biggest challenge is trying to educate people about the nature of the discipline itself and how it can add value as a part of their marketing team. People have such varied opinions and views on what public relations is. In our business in particular, there are organizations that remain very traditional and see the PR director of their hotel as someone who plans parties and may host VIPs or plan events, but they do not see PR as a strategic discipline.

If we as practitioners can demonstrate the importance of PR on a global basis and if the organization thinks enough to create a top-level PR position, then PR becomes visible. It then becomes evident to teams in the field that PR is an important discipline and they begin to think, "Maybe I should think about PR and maybe I should talk to them [PR director] and find out how I can apply these strategies to my hotel."

How do you feel the PR role fits into the entire area of marketing? In other words, what role does PR play in the overall marketing strategy of the firm?

I think that PR should be treated at the fifth P in the marketing mix. Everything done from a public relations standpoint strategically as well as tactically should support the overall marketing plan. PR should be treated as an additional channel to achieve the business goals of the hotel or the organization.

At a hotel level, much of what PR directors do is tactical and promotion driven. The marketing department is responsible for creating the promotion, pricing the promotion, and communicating the promotion by advertising. Public relations, on the other hand, is responsible for communicating the promotion to the media audience. I think from a local level, the key is to drive immediate call to action and drive trial through the promotion.

I read a quote somewhere that "advertising is the nail and PR is the hammer." If we look at traditional advertising effectiveness, we know that cutting-edge creativity and frequency is what makes advertising work. The more times a consumer sees a message, the better the chances are that the message will be retained. Consumers today are more savvy and understand the difference between advertising and editorial. Advertising is subjective and editorial is objective. Therefore, most consumers place greater value on editorial because they see this as a nonbiased communication. We can, therefore, conclude from this statement that advertising and PR must work in tandem. The more that PR supports marketing and the more that PR supports advertising, the more effective the overall brand message will be to the consumer.

If one were to choose a career in PR, what advice would you give them?

I would advise them to obtain a marketing degree with an emphasis or a minor in communications. I believe the key to being a successful PR practitioner is having a strong brand communications background. One has to understand the entire marketing mix as well as gain important communications skills and be able to apply the specific tactics of public relations to a much larger brand strategy.

Used by permission from Jennifer Ploszaj.

The communications mix is what we have come to know as all communications between the firm and the target market that increase the tangibility of the product/service mix, that establish or monitor consumer expectations, or that persuade or induce customers to purchase. The communications mix contains five elements:

- Advertising
- Sales promotions
- Merchandising

■ Public relations and publicity

■ Personal selling

Some elements of this definition need further explanation. Note the phrase *between the firm and the target market.* This tells us that communication is a two-way street. It does not consist simply of how or what the firm communicates; it also includes the feedback from the target market that tells the firm how well it is communicating and how well it is providing the services promised.

Second, the definition says that communications "increase the tangibility of the product/service mix." As we have seen, the presentation mix does the same thing. The difference is that the presentation mix does this with tangible physical evidence of the product. The communications mix does it with words and pictures, not the product itself.

Third, communications "establish or monitor consumer expectations." Not only do communications create expectations, but they should also signal the firm when expectations change or are not being met. This occurs by continuously monitoring consumer complaints, attitudes, and behaviors.

Finally, marketing communications "persuade customers to purchase"—we hope. Although communications, particularly in advertising and public relations, may be crafted to achieve such nonbehavioral outcomes as the creation of awareness or the modification of a brand image, the ultimate goal of most marketing communications is to induce purchase and increase "buzz" or word of mouth.

COMMUNICATIONS STRATEGY

Communications strategies are concerned with the planning, usage, and control of persuasive communication with customers. The strategy is the plan, and tactics represent the actions. This is an important distinction because it is very easy when using marketing communications to get bogged down in the tactics. When this happens, communications are often not consistent with strategic objectives.

For example, in personal selling we might call on a client hoping to convince him to book his next group meeting at our hotel. Knowing when his next meeting will be held, we might try to persuade him to book that period. That is a tactic. But if the client has already reserved at another hotel for that meeting, the result is no sale, and we will have to go through the same process for his subsequent meeting.

Instead, we might use strategic persuasion. Our strategic objective is to persuade the client that our hotel, of all hotels, can best serve his meeting needs. We don't mention dates, and we don't "sell" our product; we address the client's needs. Instead of a "no sale," we receive this response: "I've already booked our next meeting, but I'll get in touch with you for the one after that." If our persuasion has been successful, he will.

In advertising, the same concept applies. The first step in the development of a communications strategy is to decide what our objectives are and what we hope to accomplish. These are broad objectives that will serve as an umbrella for all our communications efforts; that is, these objectives will cover all advertising, personal selling, sales promotions, merchandising, and public relations efforts. We may have more than one objective at a time. We want the objectives to be congruent and not in conflict with each other.

There are many possible main objectives. Here, we list just a few:

■ Create or change an image
■ Position (both objectively and subjectively)
■ Provide benefits
■ Offer solutions to problems
■ Create awareness
■ Create belief
■ Stir emotions
■ Change attitudes
■ Create expectations
■ Stimulate action
■ Create buzz or word of mouth

These are all strategic objectives, and one or more may guide the communications process.

The communications process has six broad stages, which are shown in Exhibit 11-1. The first of these stages is "to whom to say it." This stage sets the guidelines in terms of featured attributes, positioning, benefits offered, promises made, and so forth. Consider the

Exhibit 11-1	Six Stages of a Communications Strategy
Stage	**Possible Strategic Element**
1. To whom to say it	The target market(s)—those who either use our product or who we want to persuade to use it.
2. Why to say it	Prior users—persuade to use again or more often: • Offering new benefits. • Offering specials at slow periods. • Show improvements. • Develop relationships. • Recapture, reposition. • Adopt as first choice. Nonusers: • Get interested, get attention. • Make part of evoked set. • Position. • Arouse desire. • Provide more information to evaluate, explain features. • Persuade to use.
3. What to say	• Awareness, to desire, to buy. • Awareness, interest, evaluation, trial. • Logos, pathos, ethos. • Cognitive, affective, conative
4. How to say it	Humor, sex, cost/value, bargain, slice of life, lifestyle, mood, atmospherics, testimonial, service. quality, action, etc
5. How often to say it	Depends a lot on budget, reach, effectiveness.
6. Where to say it	Selecting the media or personal selling that would most likely reach the target market, most effectively and efficiently.

advertisements for two hotels (Exhibits 11-2 and 11-3) and two restaurants (Exhibits 11-4 and 11-5) for the purpose of relating the communications process to an actual strategy. Each advertisement has a different strategic objective.

To Whom to Say It

The first stage of a communications strategy is to define the target market. The appropriate research should be done, and the needs and wants of the target market should be clearly identified. The target markets for both Exhibits 11-2 and 11-3 are vacationers, but that's too simplistic. Do the target markets for the properties in the advertisements have the same needs and wants? Not really. The advertisement for the Fairmont Hotels in Bermuda (Exhibit 11-2) focuses on the sophisticated traveler, as evidenced by the use of the terms *precious destination, glorious greens,* and *brilliant gems.* In addition, the "$92 million dollar renovation" also suggests that properties must be very nice and expensive. Notice that nowhere is price mentioned.

The advertisement for Ohana Hotels and Resorts (Exhibit 11-3), on the other hand, also focuses on the vacationer. However, the emphasis is not at all on the sophisticated traveler, but instead on the more casual family/friends traveler. This is evidenced by the pictures of the people in very casual clothes, the focus on "you're among friends," and the mention

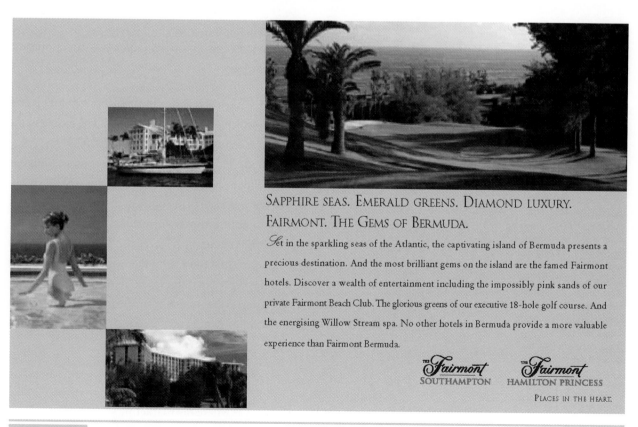

Exhibit 11-2 This advertisement for the Fairmont Hotels in Bermuda focuses on the use of exclusive gems to make tangible the high quality of the resort

Source: Fairmont Hotels, Bermuda. Used by permission.

Exhibit 11-3   This advertisement for Ohana Hotels and Resorts focuses on words such as *family* and *friends* and is crafted to suggest it is a property where you can bring kids
Source: Outrigger Hotels & Resorts. Used by permission.

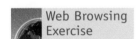

Web Browsing Exercise

As this book is getting ready to go to print, the latest marketing trend is podcasting. Use your favorite search engine and type in the word "podcasting." Be prepared to discuss the definition of podcasting, its origins, its growth over the last couple of years, and its future. Also be sure to address how a hospitality organization (restaurant, hotel, or tourist destination) might use podcasting as part of its advertising strategy.

of the "real island experience." The advertisement also shows the price per night. This is in contrast to the previous advertisement in which price is not mentioned. The feeling in the Fairmont advertisement is "if you have to ask the price, you cannot afford it."

The advertisement for Legal Sea Foods (Exhibit 11-4) is for the consumer who likes fresh seafood and casual, relaxed dining. This is suggested by the descriptions of the meals. They are promoting the product, not the experience. In contrast, the advertisement for the Newport Room (Exhibit 11-5) is promoting a more formal experience. This is evidenced by the emphasis on the fact that the restaurant is "So elegant, it even comes with diamonds." Rather that showing an example of a food item, the reader is left to ponder what the restaurant might be like and the experience one would have eating there. Here one would imagine that users are couples who dress more formally for dinner, like really fine food and service, drink wine, and want romance and peace and quiet.

Why Say It

This stage in the communications strategy is where the marketing strategy comes in. At this stage we are concerned with what effect we expect the communication to have; that is, what we want to accomplish. Some of these purposes are shown in Exhibit 11-1.

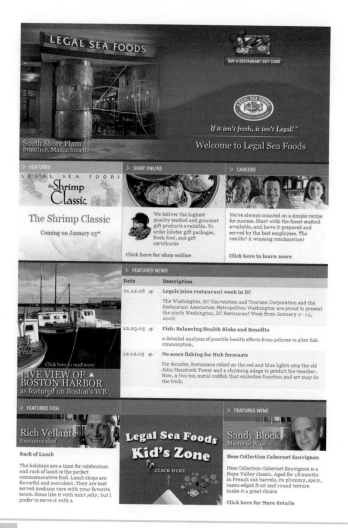

Exhibit 11-4   Legal Sea Foods delivers its food to customers

Source: Legal Sea Foods, Inc. Used by permission.

The strategy for the Fairmont is to position the resort as a quiet, sophisticated alternative to other destination resorts in Bermuda and to tell the target market, "We are one of the gems of Bermuda." The advertisements are designed to communicate how the Fairmont hotel differs from the numerous other resorts on the island of Bermuda.

The strategy for Ohana is to position it as a family resort with a variety of facilities and recreational services for the whole family or group of friends, at a reasonable price. Notice the emphasis on relaxation, with an "inside track to the islands" and a suggestion to relax in "real OHANA comfort" and get "real Hawaiian values."

In Exhibit 11-4, the strategy is to position Legal Sea Foods as the place for fish. The food is fresh but not pretentious, the restaurant offers a relaxed atmosphere.

In Exhibit 11-5 the strategy is to position the Newport Room as a fine dining restaurant, evidenced by the use of five diamonds and the jacket and tie requirement. It hopefully makes the restaurant the first thing that comes to mind the next time someone wants that perfect romantic dining experience.

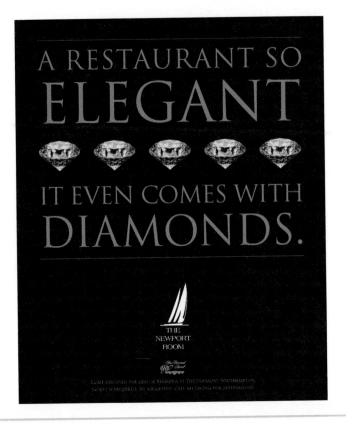

Exhibit 11-5   This advertisement for the Newport Room positions the restaurant as a fine dining establishment

Source: Fairmont Hotels & Resorts, Bermuda. Used by permission.

Tourism Marketing Application

India and Singapore provide an example of "what to say." In August 2005 the tourism boards of India and Singapore announced that they planned to advertise a seamless, attractive tourist package that included Kolkata (Calcutta), India, and Singapore. The target market was business tourists from Western countries. The distance between these two areas is 2,870 kilometers. As a point of perspective, the distance between New York and London is 5,585 kilometers. *Source:* www.bridgesingapore.com/4bsnews_7oct.htm (accessed April 2, 2006).

consumer adoption process model A model drawn from consumer behavior theory suggesting that the "adoption," or purchase, of a product is a process undertaken by the consumer. This process essentially entails awareness → interest → evaluation → trial → adoption. The model also contends that the consumer may or may not be aware that he is going through this process.

What to Say

This stage evolves from Stage 2 (why say it). It deals with the method chosen to achieve the strategic objective. In other words, what are we going to say that will effectively reach the consumer depending on where she is in the buying process? One particular model drawn from consumer behavior theory helps us better understand this stage. It is called the **consumer adoption process model**.

The consumer adoption process model contends that adoption, or the purchase, of a product is a process. In its essential form, this process essentially entails awareness → interest → evaluation → trial → adoption.

Marketers should know the stage of their target market before developing marketing communications. This knowledge can help influence the development of the communications strategy and related objectives. Strategies can and will change depending on where the consumer is in the consumer adoption process.

How to Say It

Once the first three stages of a communications strategy have been considered, the creative juices of everyone tend to focus on the advertising copy and its appeal to the customer; this is called the execution stage. Of the many creative options, the search is for the one that will work best, the one that most accurately accomplishes the goals consistent with the identified target market. Some of the elements used to generate appeal include humor, sex, surprise, slice of life, life style, and testimonial.

The creative strategy for the Fairmont is to emphasize the product's beautiful seas, physical ambience, high quality, and uniqueness. For Ohana, it is fun, family, friends, relaxing, and food. The creative strategy at Legal Sea Foods is to relax and enjoy (lifestyle). And for the Newport Room, it is elegance, quality, romance, and atmospherics.

Now consider all four advertisements in terms of the definition of the communications mix:

- Do they establish communications between the firm and the target market?
- Do they increase the tangibility of the product/service mix?
- Do they establish or monitor consumer expectations?
- Do they persuade or induce consumers to purchase?

These are the things we want to accomplish for our communications strategies to be successful.

How Often to Say It

The next decision is both a consumer-driven and a budget-driven one. Repetition has been shown to help build awareness and enhance product recall over time. Perhaps ironically, continued repetition of a message also produces an undesirable outcome called "wear out" or the gradual decline in the ability of the advertising execution to capture and retain the attention of the intended audience. Therefore, careful thought and planning are required to determine the number of executions that will be used in the campaign, as well as how often to use them.

Where to Say It

Where, as used here, applies to the various components of the communications mix. If we use advertising, then we must select the appropriate media: offline or online, electronic or print, and "in home" or "out of home." Electronic media include television, radio, and now many online services. Print media include newspapers, magazines, and direct mail. Out-of-home media include billboards, bulletins, transit displays, and the like. The demography, lifestyle, and social values of the target audience are all evaluated when composing a media plan to ensure that the media selected reaches the target both effectively and efficiently.

Push/Pull Strategies

There is one more important element in developing the communications mix. This is especially true in an industry that deals so heavily with other customer providers or go-betweens, such as travel agents, tour operators, and external reservation systems.

Using a **push strategy** means "pushing" the communications mix down through the distribution channels. For example, a hotel company calls on travel agents, advertises in travel agent publications, uses media available on the computer screens of travel agents, provides travel agent bonuses, and so forth. This is intended to get their cooperation in sending customers.

Using a **pull strategy** means going directly to customers, who then go through the distribution channels to book reservations with an idea of what they want. Advertisements that say, "call your travel agent" are using this method; that is, the company is "pulling" the customer up through the distribution channel.

Both push and pull strategies are common in the hospitality industry, and they are often used simultaneously. The communication question is: Who are you targeting—the customer or the go-between? The answer to that question will guide the communications message and the placement of that message.

WORD-OF-MOUTH COMMUNICATION

The most powerful form of communication, especially in the hospitality industry, is word of mouth (WOM). The reason is that hospitality products are considered credence goods. These are products or services that typically cannot be tested before purchase, so consumers are forced to seek outside advice on whether to purchase the service.

Elements of the communications mix can, of course, influence word-of-mouth behavior. In fact, creating WOM is a critical outcome of the communications mix. We may see an ad, read or hear publicity, or talk to a salesperson and from any one of those experiences develop a perception and expectation. We may then communicate that perception to someone else via word of mouth even though we really have no actual experience with the product. In this sense, the communications mix affects word of mouth and, indirectly, may persuade someone to purchase or not to purchase.

Word-of-mouth behavior originates from an actual experience with the product or the word of mouth of others who have had an actual experience. Thus, we control behavior more by what we do (relationship marketing) than by what we say. A strong foundation for good word-of-mouth communication is built by fulfilling the needs and expectations of customers. Failing this, we can recapture our reputation by responding appropriately to customer requests and complaints.

Impact of Word of Mouth[1]

Janet Johnson, the vice president of marketing communications at Marqui, has written extensively on the various types of word-of-mouth advertising and how to create WOM.

push strategy Directing marketing communication efforts to intermediaries such as travel agents, who then help "push" the product to their customer base.

pull strategy Directing marketing communication efforts directly to the consumer, who in turn (hopefully) purchases the product directly or through an intermediary such as a travel agent or an online booking agency.

Web Browsing Exercise

Use your favorite search engine and type in the phrase "undercover marketing" or "stealth marketing." Read some of the sites listed and be prepared to discuss undercover marketing or stealth marketing in more detail. What are the pros? What are the cons? Do you think this is a good idea for hotels and restaurants? How would you implement such a plan for hotels and restaurants?

Her firm works with a variety of tourism companies, including motorcoach tour operators, destination marketing organizations, and cruise lines, as well as other businesses across multiple industries. Following are types of WOM available to organizations:

Buzz Marketing: Using high-profile entertainment or news to get people to talk about your brand.

Viral Marketing: Creating entertaining or informative messages designed to be passed along in an exponential fashion, often electronically or by e-mail.

Community Marketing: Forming or supporting niche communities that are likely to share interests about the brand (such as user groups, fan clubs, and discussion forums); providing tools, content, and information to support those communities.

Grassroots Marketing: Organizing and motivating volunteers to engage in personal or local outreach.

Evangelist Marketing: Cultivating evangelists, advocates, or volunteers who are encouraged to take a leadership role in actively spreading the word on your behalf.

Product Seeding: Placing the right product into the right hands at the right time, providing information or samples to influential individuals.

Influencer Marketing: Identifying key communities and opinion leaders who are likely to talk about products and have the ability to influence the opinions of others.

Cause Marketing: Supporting social causes to earn respect and support from people who feel strongly about the cause.

Conversation Creation: Using interesting or fun advertising, e-mails, catch phrases, entertainment, or promotions to start word-of-mouth activity.

Brand Blogging: Creating blogs and participating in the blogosphere in the spirit of open, transparent communications.

Referral Programs: Creating tools that enable satisfied customers to refer their friends.

According to Johnson, to create WOM and encourage communications, the firm should give people something interesting to talk about. For example, in Johnson's work with a cruise line, she has included tips on the firm's website about wine education, environmental tourism, and safety and security.

A second recommendation for creating WOM is creating communities and networks to connect people. Johnson's company has enabled guests taking a motorcoach tour to send electronic postcards to friends and relatives. They have also enabled newfound friends to stay connected on a secure website.

A third suggestion is to work with influential communities. One example is the Conrad Punta del Este Resort and Casino, in Uraguay, which has created a program wherein local children come to the resort for after-school meals and entertainment. In addition to positive results for the community, this program has also generated positive WOM for the resort.

A fourth idea is to create evangelist or advocate programs, enlisting people to be evangelists while they're taking advantage of your services. Starwood Resorts benefited from an employee evangelist called the Starwood Lurker, who was well known for solving problems with Starwood frequent traveler programs. The Lurker's full-time job is to monitor Internet discussion boards such as FlyerTalk (www.flyertalk.com).

Another method for generating WOM is the use of blogs. A weblog, or blog, is a journal (or newsletter) that is frequently updated and intended for general public consumption. Blogs often have hyperlinks, which enable the user to search other similar information or connect with others reading the same blog. Johnson stated that 27 percent of adults who go online in the United States read blogs and that more than 28 percent of journalists now rely on blogs for reporting and research, according to a survey by EURO RSCG Magnet and Columbia University.

Measuring Word of Mouth

Roger Hallowell and Abby Hansen of Harvard Business School stated that to calculate the value of word of mouth, one needs to know (1) the likelihood that the customer will refer the property, (2) the number of people to whom the recommendation will be made, (3) the percentage of referrals that are empathetic (i.e., may act on what they hear), (4) the probability that those who are empathetic will buy the service, and (5) the lifetime value of the customer.[2]

Given this information, the formula for WOM is: WOM = $(a \times b \times c \times d)$.
The lifetime value of the incremental customer (LVIC) is calculated as:

a = Gross profit on an average purchase
b = Average number of purchases a customer makes each year
c = Average number of years customer will continue to purchase
d = Probability that customer will continue to purchase

LVIC = $(a \times b) + (a \times b \times c \times d)$, where $(a \times b)$ is the profit for the year.

BUDGETING THE COMMUNICATIONS MIX

The amount that a company or an individual property may spend on its total communications effort is not easy to determine. There are no universally accepted standards of how much should be spent in a given product or market situation. This is because the situation is compounded by a complex set of circumstances that is never constant within or among properties or companies. Our focus here will be on individual properties, not on entire firms such as McDonald's or Burger King, which spend huge amounts and use lot of television, or Marriott and Hilton, which are heavy users of national print media.

What does the budget consist of? Let's start with some common practices. Independent restaurants typically do not have marketing departments. They may or may not fund advertising campaigns. If they do, the amount spent on them is most likely based on "gut feel." How good is business, how well are they known, and how far is their reach? Chances are, those who advertise will spend 2 to 3 percent of sales on advertising. In a city where advertising costs are higher, they may spend more. Restaurants that are part of a chain or a franchise will pay up to 4 percent of revenue to the parent company that does national or regional advertising for all units in the chain. Any local advertising that they may do on their own comes under the 2 to 3 percent category.

In the hotel industry similar rules apply, only now reservation costs are added to the equation. A franchised Days Inn, for example, may have no marketing department, do zero local advertising, but pay the franchisor a set fee per guest room sold or a percentage of revenue for both advertising support and the reservation system. Again, any local advertising they do will be similar to that of independent restaurants. Also, if they have meeting space, they may have salespeople who also represent part of the operation's total marketing cost.

Most major hotels have marketing departments, so this is the more complex case. We'll deal with common practice, which, of course, will vary among properties. A major hotel will usually base its marketing budget on forecasted total sales (rooms, food, and beverages). This includes the salaries of the marketing and sales staff and fees paid to the brand affiliation (or flag) for national or regional advertising and for the reservation system. These fees may be based on percentages, rooms, or, in the case of advertising, total revenue. Major hotels will also pay the brand a fee to participate in its frequent traveler program. (These are usually forecasted as a separate item outside the marketing budget.) The cost of any local advertising is typically borne by the property.

What does all this amount to? A good rule of thumb is 5 to 6 percent of forecasted total revenue for the marketing budget including all sales payroll and related expenses, advertising, brochure production, direct marketing materials, website costs, and fees paid to advertising or public relations agencies. Brand-name hotels tend more toward the 5 percent figure because of the economies of scale with a large channel of distribution. Independent hotels and resorts tend more toward the 6 percent level. Also, the percentage allocated to marketing by commercial hotels is typically lower than that allocated by resorts.

How the money is allocated across the communications mix is another issue. Other than contractual fees, which are fixed, a property has considerable latitude when deciding whether to spend on advertising, promotions, collateral, research, or the sales force. A hotel with a large proportion of meetings business will have a larger sales staff to support. The director of marketing will negotiate with management for the department's budget and how and where it will be spent, with one choice clearly affecting the other. Assuming a 5 percent budget, a usual split would be as follows:

2.0% salaries and benefits
1.5% advertising
1.0% travel/entertainment
0.5% collateral/miscellaneous

The Internet has presented a new challenge to hospitality marketers. Websites, banner advertisements, pay per click advertising, and other online communications vehicles all cost money. Unfortunately, the marketing budget in most cases remains at 5 percent of sales, forcing the director of marketing to reduce expenses in other areas to fund Internet marketing. In this regard, it is interesting to note that 2005 was the first year that Google and Yahoo's U.S. advertising revenues exceeded the combined advertising revenues of the biggest three broadcast television networks stations: ABC, CBS, and NBC.

Although it is generally accepted that the effects of advertising and other forms of promotion may last over a long period, there is no certainty about the duration of the benefits. The total effect depends on the loyalty of customers, their frequency of purchase, and competitive efforts, each of which may be influenced by a different set of variables. Further, promotions may tempt competitors to react, but it is hard to tell just what that reaction will be until it happens. When it does, the strength of the response may require additional expenditures to meet it.

Following are other methods, or rules of thumb, for determining marketing budgets:

- **Zero-Based Budgets:** Budgets set with respect to specific performance goals rather than projected gross revenues or previous levels of expenditure.
- **Competitive-Level Budgets:** Budgets set with reference to the level of marketing expenditure by competitors.
- **Whatever's Left Over Method:** This method is perilous and is typically used by sole proprietors.
- **Return on Investment Method:** This method compares the expected return with the desired return and treats the expenditure as an investment to be recouped over the years.

Some promotional costs, of course, produce immediate results. In both cases, however, it may be very difficult to determine the outcome.

In practice, it is not easy to pinpoint the separate roles of advertising, sales departments, and sales promotions because these three methods almost always overlap. A further difficulty occurs when something happens that hasn't been accounted for in the budget, such as a group suddenly deciding to come to the city and seeking competitive bids that may include expensive promises and great sales efforts. In the final analysis, a hotel may randomly set a figure such as 5 percent of sales as a benchmark, develop the marketing plan, and then adjust up or down as need be. New openings, special events, and special situations will affect the final budget.

ADVERTISING

advertising A tool used in the marketing communications mix to reach identified target markets. Advertising is paid mass communication typically placed in media publications such as newspapers, radio, and television, and on the Internet.

Advertising is mass communication that is paid for. It is the most visible element of the communications mix. It has the broadest potential reach of all the components of the communications mix; that is, it can reach the largest number of prospects and do so very quickly. It can also be the most expensive component. The question is, How effective is it? That is a question that researchers, especially those connected with advertising agencies, have been trying to answer for years. It is an especially important question in the hospitality industry, where word of mouth is such a potent force.

We include in advertising all media that are part of the messages consumers receive daily through newspapers, magazines, television, radio, transit displays, outdoor boards, and online. We also include collateral, such as hotel brochures, flyers and pamphlets, and direct mail.

Again, the Internet has made the advertising choices for hospitality marketing professionals more complex. Banner advertising, skyscraper ads, pay per click ads (see Exhibit 11-6), text links, and many other forms of communication online are available to marketers today. With hundred of thousands of travel-related Internet sites to choose from, all claiming thousands of visitors monthly, the choices seem unlimited.

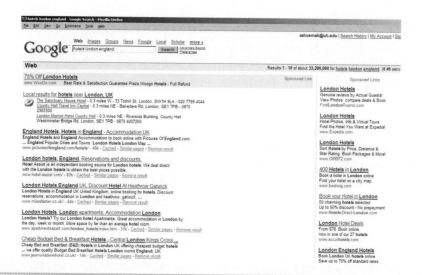

Exhibit 11-6 Many search engines, such as Google, exhibit pay per click advertisements on their search results pages. The pay per click ads are the one at the top of the page shaded in blue and the listings on the right-hand side of the browser

Source: Google. Used by permission.

Tourism Marketing Application

Many states provide opportunities for co-op advertising, in which attractions and destinations can get together to advertise the region. For example, *USA Today* ran a section on North Carolina in the weekend edition on April 30, 2006. The cost for an advertisement started at $4,500. The projected circulation was 3,700,000, and the median age of the reader was 48 with a median income of around $53,000.

Source: (www.commerce.state.nc.us/tourism/advertising/2006/print/default.asp (accessed April 2, 2006).

Role of Advertising

Advertising, of course, is intended to perform the same general role as all marketing communications: to inform, create awareness in, and attempt to persuade new customers and reinforce the buying behavior of present customers. It also can play a major role in positioning, as we have shown. Advertising development is subject to the same guidelines that we discussed in the first part of this chapter; the major difference is that it is paid mass communication.

For the hospitality industry, the most important objective function of advertising may be to create and maintain awareness of the company or the property, or to promote a particular component such as a new addition or a new service. The most important individual function is to support the position of the property or the company, as revealed in Exhibits 11-2 to 11-5.

What Advertising Should Accomplish

Major advertising campaigns in the hospitality industry are conducted only by very large companies with significant resources. In the restaurant industry, we are all familiar with the television commercials from McDonald's (one of the largest advertisers in the country), Burger King, and other fast-food chains.

On the other end of the scale are the individual restaurants or motor inns that do almost no advertising. Between these two extremes lies a vast group of hospitality operators who do limited advertising on very limited budgets. For these operators, the "more bang for the buck" principle is especially appropriate: Advertising dollars have to be carefully allocated to where they will do the most good.

To do the most good, the ideal hospitality advertisement will accomplish five things:

1. Tangibilize the service element so the reader can mentally grasp what is offered
2. Promise a benefit that can be delivered and/or provide solutions to problems
3. Differentiate the property from that of the competition
4. Have positive effects on employees who must execute the promises
5. Capitalize on word of mouth

We can demonstrate many of these accomplishments by referring to the Panama City Beach advertisement in Exhibit 11-7. The same principles apply if you are advertising only in your local newspaper or are a national advertiser.

BEFORE HURRICANE DENNIS.   AFTER HURRICANE DENNIS.

Same Beach. Same Vacation. Only Now You Get Your 4th Night Free.*

Exhibit 11-7 This Panama City Beach advertisement has accomplished the five goals necessary for a successful hospitality ad

Source: Panama City Beach Convention & Visitors Bureau, Inc. Used by permission.

The Panama City Beach advertisement was launched after Hurricane Dennis swept through Florida and captured the news media's, and hence the general population's, attention. The news media coverage implied that all of Florida was destroyed, yet most of Florida remained untouched. The problem for Panama City Beach was how to convey this information to vacationers. Exhibit 11-7 does this by showing the beach before and after the hurricane. Clearly there is no difference. We next discuss how this advertisement illustrates four of the five features the ideal hospitality advertisement should have.

First, the use of the beach chairs signifies relaxation. Because relaxation is intangible, the advertisement needs to somehow tangibilize this feeling. This is accomplished by the lounge chairs, which by themselves signify relaxation. However, the advertisement goes beyond this by using old-style wooden chairs instead of modern plastic chairs. This conveys the feeling of tradition and stability, which helps define Panama City Beach. Notice how this is also reinforced with the words "Thankfully, some things never change."

The promised benefit is that Panama City Beach has remained unchanged after the hurricane; it offers the same great golf courses and the same great restaurants. The other benefit is that vacationers get the fourth night free. The solution to the problem of being stressed out is to come to Panama City Beach to relax and take a vacation. The third component of an ideal advertisement is the ability to differentiate the property or destination from other competing properties or destinations. The Panama City Beach advertisement addresses this component in two ways:

1. The use of the positioning statement, "The beach lover's beach"
2. The implication that, unlike other beaches, Panama City beaches are undamaged

This advertisement doesn't have the fourth component of an ideal advertisement—that it will have positive effects on employees who must execute the promise.

The fifth component is that it will capitalize on word of mouth. For example, look at the two pictures and the headings, "Before Hurricane Dennis" and "After Hurricane Dennis." The goal of these ads is to counter the beliefs created by TV and radio coverage, which tend to make things look worse than reality.

It is seldom easy to get all these elements into one advertisement; usually we have to settle for less. Even then, however, one should strive to differentiate with something other than grand claims that characterize some hotel and restaurant advertisements. Unless there is something truly unique about the property, the "grandiose" approach does not meet the preceding requirements and will not help the property gain more customers.

Use of Advertising Today

If you cannot make an impact on the market through advertising, other than to create awareness and provide information, it might be better to save your dollars and put them to better use elsewhere (e.g., in the product itself or in lower prices, which will generate positive word of mouth, a far more powerful force than most advertising). This is not to say that advertising agencies are not creative. It is just that they don't always develop advertising in accord with the guidelines we have presented so far.

The consumer today is constantly bombarded with advertising messages from all directions (over 400 different messages each day by some estimates). The human mind is not capable of paying attention to all these messages. Instead, the mind will selectively perceive, attend to, comprehend, accept, and retain only those to which it is most responsive.

Hospitality properties and services are very similar in the same product class; some would say that they have even reached commodity status. The competition is selling the same thing, unique niches are harder and harder to find, services are easy to copy, and aggressive competitors are using new positioning strategies. This means that it is difficult to gain a competitive advertising advantage solely through advertising. In many cases, it may be too expensive to achieve effective awareness and persuasion levels just through advertising.

These factors demonstrate that advertising must be managed with extreme care. Successful advertising is not just copy and graphics, nor even just clever copy and graphics, but derives from a well-thought-out and well-planned strategy. However, there is a strong tendency to just look at the execution (the copy) and ignore the strategy. Many copy decisions in advertising are based on what someone likes rather than on how it affects the customer. It is no wonder the famed retailer John Wannamaker once said, "I know that only half of my advertising is working; I just don't know which half."

Collateral

collateral Promotional material such as brochures, flyers, and direct mail, that are used to inform customers and create interest. "Electronic brochures" are often posted on websites.

Brochures, direct mail, and other forms of advertising are commonly referred to as **collateral** in the industry. Much hospitality and tourism print advertising has grown beyond the stereotypical advertisements of yesterday. Hotel collateral, on the other hand, has progressed only in rare instances and generally consists of relatively uninspiring visuals and copy. In at least 75 percent of all hotel brochures you couldn't tell one hotel from another. Websites have provided a new outlet for creativity, however, and have become mandatory "brochures" on the Internet.

Because collateral is a frequently used form of advertising for hotels, we recommend that you evaluate the website shown in Exhibit 11-8. Do the rules apply in this example?

Merchandising

merchandising A tool used in the marketing communications mix to reach identified target markets. Merchandising is primarily an in-house marketing technique used to stimulate sales of additional products or services on premise.

Merchandising is primarily an in-house marketing technique designed to stimulate immediate purchase behavior through means other than personal selling or the purchase of time or space in media. In a sense, merchandising is marketing to the captive customer once the customer comes into the hotel or restaurant to purchase a room or a meal. Many customers will buy nothing other than the basic product. The goal of merchandising is to provide opportunities for customers to purchase related or supplementary products and services.

The goal of merchandising, however, should not be just to stimulate sales; it should also have a more long-term goal of increasing customer satisfaction. When the dessert tray is presented at the end of the meal, the goal is to have customers order dessert and thereby increase the check average. It is also to have customers feel even more satisfied because they have finished their meal in a very pleasing manner. If hotel guests order room service, they add to their overall bill. Also, we hope, their stay has been made just a little bit better and we have a few more satisfied customers.

Exhibit 11-8 Do the collateral rules apply to this website for the Victoria Jungfrau Hotel?

Source: Victoria-Jungfrau. Used by permission.

Like everything else, we approach merchandising from a marketing perspective—fulfilling customers' needs and wants and solving their problems. It may also be to provide a unique or exciting experience, as in the case of the casino. If we are able to do this, the higher check averages and the larger bills will follow. If, instead, we put all the emphasis on the increased revenue, we are likely to fall into the same old trap of forgetting about the customer.

Basic Rules of Merchandising

The opportunities for merchandising in a hotel or restaurant are almost endless and are limited only by the imagination. A few helpful rules affect all merchandising, including its purpose, compatibility and consistency with the overall marketing effort, practicality, visibility, and simplicity. Following is a brief discussion of each.

Purpose. All merchandising should have a purpose. The commonly expressed purpose—to increase sales—is true, but not sufficient. Instead, let's say that the overall purpose is to increase customer satisfaction and loyalty. Of course, we could also say that the purpose is to fulfill needs and wants and solve problems. Much of merchandising does that, but in this case we go beyond the basic marketing concept.

Sometimes just knowing that something is available and can be had if wanted will establish the need or want or increase satisfaction—even when the customer does not consciously need or want that thing. A good example is the year-round swimming pool in an urban hotel. Comparatively, very few guests use these pools, but research has shown that they like the idea that the pools are there to use if they wish. A positive, however, is turned into a negative when the pool is not open at reasonable times that people want to use it because of operational convenience. The same is true of fitness areas. It is human nature to want to know that we can have something if we want it; merchandising creates that feeling and increases satisfaction and loyalty.

Marketing can stimulate basic needs, but it can also create wants. Restaurant diners might feel a need for chocolate after dinner (might even want it, in fact), but limit that need because of the concern about the number of calories in the dessert. However, along comes the pastry cart with all those goodies, and now, instead of not ordering any dessert, customers at the table may order a couple of desserts to share. Restaurants have many merchandising opportunities. The most powerful one, sometimes neglected, is the menu itself, which can range from boring and unoriginal to exciting and provocative.

By the same token, hotel guests do need to eat; merchandising can make them want to eat in one of the hotel's restaurants. Cards can be put up in the elevators, and signs may be displayed in the lobby and in promotional flyers that are placed on desks in the guest rooms. In many hotels, guests also see and hear about the in-house restaurants on the televisions in their rooms. In many European hotels, merchandising is practiced upon check-in, when the desk clerk asks whether the guests would like a dinner reservation made for them. The in-room dining menu is a great opportunity for selling in-room dining, but only if the menu is clean, looks appealing, and does not show the previous guest's order. Unfortunately, all too often such menus do not fit these criteria, and the hotel misses a great merchandizing opportunity.

All merchandising tactics need to have their purpose understood. One purpose, as we mentioned, is to create the feeling, "If you want it, we have it." Another might be to create excitement, as with an exotic drink, a flambé dessert, or a "spinning salad bowl," which made Don Roth's Blackhawk Restaurant in Chicago famous. Another purpose might be entertainment, such as that provided by slot tournaments held by casinos. Other possible purposes are convenience (room service), relaxation (aperitifs), contentment (after-dinner cognac), or information (in-room directories).

A merchandising technique used by Marriott in some of its hotels, which can also be used in restaurants, occurs when customers are first seated. A waiter or waitress immediately approaches with a basket of house wines, offered by the glass. The customer has immediate service and a need identified by the customer has been met. Even if there is a delay in ordering the meal, instant satisfaction has been created. Some hotels also use the same approach at breakfast: Customers are immediately greeted by a server with a pot of coffee in one hand and a pitcher of fresh orange juice in the other.

Compatibility and Consistency. Merchandising efforts should be well matched and consistent with the rest of the marketing effort in terms of quality, style, tone, class, and price. They should reinforce the basic product/service mix, because these efforts are part of the enhanced product. Hotels that have an eye on the growing family vacation travel market should consider opening a child care center where parents can leave their children with trained, licensed, professional staff. Some resorts offer 24-hour child care service. However, this market and this service may not be compatible in a hotel with a strong transient business traveler base.

Practicality.   The rule here is: If you can't do it right, don't do it. Failure to follow this rule results in lost customers, not satisfied ones. The child care center is an example where serious problems can result if the service is not offered in a professional way.

Visibility.   Let the customer know about it and how to get it. Elevator cards merchandising restaurants often fail to say where the restaurants are or what hours they are open. In today's modern hotels, where restaurants might be anywhere, it can be a mind-boggling experience to find one. One mistake management often makes is to forget that for many customers this may be their first time in the hotel. In-room directories sometimes are so confusing that the guest either turns to the telephone or gives up. We have even seen directories with full pages on the swimming pool and health club facilities but no indication of how to get there or what to wear on the way. Many people don't use pools simply because they are too embarrassed to go there in a bathing suit and do not want to change in dressing rooms. The Royal Garden Hotel in Trondheim, Norway, solved this customer problem by identifying a swimming pool/health club elevator specifically for that purpose.

On the other hand, visibility doesn't mean total clutter. Some restaurant tables or hotel desktops have so many table tents, flyers, and brochures on them that there isn't room for anything else and it is too confusing to find what you want. They also can get in the way of the necessary work space that many business travelers need. Some people just remove them without looking so they can put down their own things.

Clutter can also occur on the casino floor or, more important, hanging from the casino ceiling. Some casinos have so much merchandizing hanging from the ceiling that the consumer does not see anything through the clutter. It is always good to put yourself in the shoes of the customer.

Simplicity.   Make it easy to understand and easy to obtain. Make it clear how much it will cost, how long it will take, and when it is available and provide any other information that will make it unnecessary for the customer to have to make additional inquiries. Customers tend to give up when they have to go through too much effort to purchase a service. Placing a star next to an "extra hot" menu items provides quick information to customers and increases sales of those items for those who are looking for something very spicy. It also ensures that those who do not like spicy foods do not order them.

Examples of Good Merchandising

Examples of good merchandising techniques can be found everywhere. One case in point is the business centers that are frequently found in hotels that cater to business travelers. These business centers offer a variety of administrative support services such as typing and dictation, together with copying, fax machines, Internet access, and computer terminals. Business centers are usually located off the lobby and have a separate room in which to work. Guests pay for these services, so hotels can make a profit on them. More important, they fill a need of the traveling businessperson and create a better guest experience, especially if they are priced fairly.

The emergence of pizza on finer hotels' room service menus is a merchandising opportunity that fills a need of many customers. Many people do not want a full, heavy meal in their guest room. Some just want to watch television and have something "fun," as if they were at home. The pizza (merchandised often with beer) fills the need of the customer while putting money into the hotels' cash registers. Courtyard hotels recently redesigned their lobbies to include 24-hour "mini-markets" that sell food products and other items that travelers may require. This type of merchandising can only increase revenues. Price does not become the deciding factor; instead, the product (and its ease of accessibility) becomes the reason for the purchase. Those customers who really wanted pizza in the first place might have called for a delivery from outside or gone out of the hotel; either way, the money would have been spent outside the hotel or not at all. More important, once again, the hotel has satisfied a customer by fulfilling a need.

The inclusion of minibars in guest rooms is satisfying to customers and increases hotel profits. Minibars are self-contained units that have beer, wine, liquor, juices, and soft drinks together with snacks. An inventory is taken of the unit's contents before the guest checks in, and all items consumed are posted to the bill upon checkout. A guest is unlikely to call room service for only one beer. With a minibar in the room, customers can lean over while watching television or reading and open a beer at their convenience. Minibars now feature many nonfood items such as playing cards and disposable cameras. Again, however, improper merchandising can lose customers. Too many minibar contents are overpriced, and customers are not particularly pleased when they have to pay the high prices. Because of the lack of items that guests want and the overpricing of these items, many buy outside and use minibars as refrigerators. This is an opportunity lost. Smart hoteliers are talking with their guests before their arrival and asking what items they would like to see in the minibar. This way, the food items are things the guest will like and will be more likely to order. Some are also including the minibar items in the price of the room or pricing the items fairly.

Merchandising is marketing to the "captured" customer. Do not translate *captured* as *captivity*. Instead, translate it as *opportunity:* Here's an opportunity to make the customer even more satisfied.

PUBLIC RELATIONS AND PUBLICITY

public relations (PR)
A tool used in the marketing communications mix to reach identified target markets. Public relations is planned management of the media's and the community's perception of a hospitality or tourism enterprise. Press releases are often used to aid in this process.

publicity News related to a hospitality or tourism enterprise that is made available to the public-at-large. It may be positive or negative.

Public relations (often referred to simply as **PR**) is used to present the product or service to the media and the community in the best possible light. Positive **publicity** is the desired outcome. Public relations and publicity are grouped together because of their commonality, which is the "free" use of the media to present management's view to the community at large. Instead of buying space in a newspaper or time on a radio station to get the firm's information out, the organization obtains it for free—provided the media think the organization is newsworthy or of interest. Usually, though, the organization does not control the actual placement and appearance of the information.

Every organization exists in a community that influences its success. It wants to have a positive image in the community and to be seen as a contributor to the overall well-being of society. Although publicity can come from public relations, the difference is that publicity represents only the information the media freely and without influence chooses to use. Thus, publicity may be positive or negative. Public relations, on the other hand, is an attempt to manage publicity and to "plant" information in the press or to create a favorable image for reasons other than its formal product. In politics this is called "spin control," a phrase that is used quite frequently.

Public relations, as well as publicity, also occurs through word of mouth. Although much of this may be started by the media, other aspects may be spontaneous. For example, a restaurant makes a special effort to employ people with special challenges. This fact may never strike the media, but the word gets around and the restaurant is looked at as a "do-gooder." This reflects positively on other aspects of the restaurant.

To the public, public relations and publicity may be the most believable forms of the communications mix. A salesperson pitching a product or a slick advertising campaign may be subject to skepticism from consumers. When an independent source, such as a newspaper, writes about the product in a seemingly objective setting, credibility unmatched by any other media format is given to the message. A potential customer for a restaurant is more likely to try the veal special recommended by a restaurant reviewer than to try the same dish shown on a full-page advertisement proclaiming its excellence.

Public Relations

Public relations is the planned management of the media's and the community's perception of the hospitality establishment or tourism attraction. Although the press certainly cannot be told what to publish, a public relations effort can steer the story toward the best features of the product and away from negative images. Public relations efforts are designed to create stories that capture writers' attention with the hope that the writers will, in turn, communicate "the good news" to the desired readers or target market. Exhibit 11-9 is an example of a public relations **press release.**

press release A document prepared for release to selected media containing information or "news" about a hospitality or tourism enterprise. Press releases normally include the name of the public relations professional who created the release and how she or he can be reached for further information.

Tourism Marketing Application

The World Tourism Organization (WTO) sponsored two conferences in 2006 focusing on the best practices of tourism communications. Called TOURCOM, the meetings were held in Bamako, Mali, for Africa, and in Rosario, Argentina, for the Americas. WTO believes that public relations plays a key role in the international tourism process. As one person stated at a TOURCOM meeting in Amman, Jordan, in 2005: "The media are an equally important partner of destinations to tour operators and airlines."

Source: www.world-tourism.org/newsroom/Releases/2006/february/tourcom.htm (accessed April 3, 2006).

FOR IMMEDIATE RELEASE

Contact: Jennifer Ploszaj
           InterContinental Hotels & Resorts

DO YOU LIVE AN INTERCONTINENTAL LIFE?
InterContinental Hotels & Resorts Launches Global Advertising Campaign

LONDON (September 22, 2005)—InterContinental Hotels & Resorts announced today it will launch a new global brand advertising campaign on September 26.

InterContinental's new advertising campaign is the result of positioning work that has been ongoing since late 2004. The brand's new tagline challenges its audience to answer the question, "Do you live an InterContinental Life?"

"At a time when other hotel brands are working to keep people in a 'bubble', InterContinental wants to provide our guests with memorable and unique experiences that will enrich their lives and broaden their outlook," said Jenifer Zeigler, senior vice president, global brand management, InterContinental Hotels & Results. "We believe the new social currency is about being 'in the know.' And we deliver that to our guests through great travel experiences."

The campaign launch consists of a television commercial filmed in Sydney, Australia, featuring Australia's Challenge Yacht, named Spirit, built for the 1992 Americas Cup in San Diego. The campaign's first print executions feature photography shot on the beaches and in the local markets of Bali, Indonesia.

"At InterContinental, we believe travel is a great thing," said Zeigler. "Launched in 1946, InterContinental Hotels became the symbol of glamour, sophistication and success that years later, continue to define international travel."

The television spot will air on CNN International as well as in-flight programming on British Airways, United, American, Emirates and Singapore airlines. The print media schedule includes insertions in *The Wall Street Journal, The New York Times, Newsweek, Time Magazine, Forbes, The Financial Times, Economist, The Times* and *Business Week* as well as major in-flight publications.

The campaign launches on Monday, September 26 and continues through the 2005 business travel season.

###

Note to Editors:

InterContinental Hotels Group PLC of the United Kingdom [LON:IHG, NYSE:IHG (ASRs)] is the world's largest hotel group by number of rooms. InterContinental Hotels Group owns, manages, leases or franchises, through various subsidiaries, more than 3,500 hotels and over 537,000 guest rooms in nearly 100 countries and territories around the world. The Group owns a portfolio of well recognized and respected hotel brands including InterContinental® Hotels & Resorts, Crowne Plaza® Hotels & Resorts, Holiday Inn® Hotels and Resorts, Holiday Inn Express®, Staybridge Suites®, Candlewood Suites® and Hotel Indigo™, and also manages the world's largest hotel loyalty program, Priority Club® Rewards, with over 26 million members worldwide. In addition to this, InterContinental Hotels Group has a 47.5% interest in Britvic, one of the two leading manufacturers of soft drinks, by value and volume, in Great Britain.

InterContinental Hotels Group offers information and online reservations for all its hotel brands at www.ichotelsgroup.com and information for the Priority Club Rewards program at www.priorityclub.com.

For the latest news from InterContinental Hotels Group, visit our online Press Office at www.ihgplc.com/media.

Exhibit 11-9   Publicity helps with promotions

Source: InterContinental Hotels Group. Used by permission.

We can demonstrate these points with some examples. After Hurricane Katrina hit the Gulf Coast region, hoteliers in the region were faced with running their businesses while showing compassion. Area hotels provided food and shelter for evacuees and relief workers, helped coordinate medical care, organized fundraisers, and relaxed policies for pets and the number of people per room. At the corporate level, response was widespread and swift, aimed at helping both evacuees and employees of affected properties. Immediately after the hurricane struck, staff members of the Millennium

Hotels and Resorts across the United States held a Basic Essentials Drive that collected over 12,000 donations of toiletries, food, clothes, medicine, paper products, and baby items for distribution to those in need. In addition, Millennium helped provide long-term assistance to displaced hospitality industry members by posting nationwide employment opportunities to the Travel Industry Association of America's (TIA) job bank.[3] Starwood ran a weeklong, worldwide online auction on Yahoo! to raise money for the Starwood Relief Fund benefiting Starwood associates affected by Katrina, offering honeymoon, spa, and gold packages at Starwood properties around the world, celebrity packages, and other auction items, representing a combined retail value of over $1 million.[4] Marriott developed a program for Marriott Rewards members to convert their rewards points into monetary donations to the Red Cross.[5]

Another example is that of McDonald's, a company widely acclaimed for its public relations efforts. For McDonald's, public relations is an important part of the brand marketing strategy. Ronald McDonald homes for families of sick children at nearby hospitals are nationally famous. When disaster strikes anywhere near a McDonald's, some of the first people on the scene are McDonald's employees with coffee and hamburgers for the victims and workers on the scene. When a man went into a McDonald's in California and shot and killed customers, McDonald's immediately closed the store and provided financial aid to the victims' families. When the company wanted to reopen the store a few months later, the townspeople strongly opposed it. McDonald's quickly complied by closing the store permanently. A great deal of positive publicity came from its public relations efforts to manage this tragedy.

In such cases, and more often in less serious situations, public relations are used to create an image in the consumer's mind of what the company or product represents. Public relations–engendered publicity enabled McDonald's to capitalize on a possible negative image. Hosting Ronald McDonald homes has nothing to do with the production of hamburgers. The story is "created": The company cares for children and that is the perception McDonald's wants to portray.

Undertaking Public Relations.

Public relations efforts are not only for dealing with negative happenings or simply creating positive happenings; instead, they represent an ongoing task and are an important part of marketing planning by doing the following:

> Improving awareness, projecting credibility, combating competition, evaluating new markets, creating direct sales leads, reinforcing the effectiveness of sales promotion and advertising, motivating the sales force, introducing new products, building brand loyalty, dealing with consumer issues and in many other ways.[6]

Public relations happens at both the corporate level and the property or unit level. Large companies or properties usually have their own public relations firms, which are hired on a monthly retainer to develop and maintain favorable publicity for the organization. But smaller companies that cannot afford PR agencies, as they are called, must practice public relations in-house on an ongoing basis. Doing this involves managing employee relations. It also involves maintaining good relationships with taxi drivers and local police, the press, competitors, members of the distribution channels (such as

| Exhibit 11-10 | Sample Timetable for Preopening Public Relations for a Hotel |

This schedule begins six months before the hotel opening, at which time the announcement of construction plans and the groundbreaking ceremony will have been completed.

150–180 days before opening
1. Hold meeting to define objectives and to coordinate public relations effort with advertising; establish timetable in accordance with scheduled completion date.
2. Prepare press kit (printed and electronic form).
3. Order photographs and renderings.
4. Begin preparation of mailings and develop media lists.
5. Contact all prospective beneficiaries of opening events.
6. Reserve dates for press conferences at off-site facilities.
7. Create a special "press room" on the website.

120–150 days before opening
1. Send announcement with photograph or rendering to all media.
2. Send first progress bulletin to agents and media (as well as corporate clients, if desired).
3. Begin production of permanent brochure.
4. Make final plans for opening events including commitment to beneficiaries.

90–120 days before opening
1. Launch publicity campaign to national media.
2. Send mailings to media.
3. Send second progress bulletin.
4. Arrange exclusive trade interviews and features in conjunction with ongoing trade campaign.
5. Begin trade announcement.
6. Post all press releases in the online press room.

60–90 days before opening
1. Launch campaign to local media and other media with a short lead time; emphasize hotels' contribution to the community, announcement of donations and beneficiaries, etc.
2. Send third and final progress bulletin with finished brochure.
3. Commence "behind-the-scenes" public tours.
4. Hold "hard hat" luncheons for travel writers.
5. Set up model unit for tours.

30–60 days before opening
1. Send preopening newsletter (to be continued on a quarterly basis).
2. Hold soft opening and ribbon-cutting ceremony.
3. Hold press event to announce opening.
4. Establish final plans for opening gala.

The month of opening
1. Begin broadside mailing to agents.
2. Hold openings festivities.
3. Conduct orientation press trips.

Source: Adapted from the work of Aaron D. Allen, Founder/CEO of Quantified Marketing Group (www.quantifiedmarketing.com).

airlines, travel agencies, tour operators), purveyors (who can be excellent carriers of good tidings), shareholders, bankers, and other publics with which the firm interacts.

Hotel and restaurant managers should belong to the appropriate community and public service organizations such as the Rotary, Chamber of Commerce, community task forces, and other groups. One could almost say that everything management does contributes in some manner to the public relations program. Even the employees of the firm may be excellent public relations ambassadors; in fact, for some firms they may be the most important of all. What your employees say about you and the way you operate reflects heavily on the image that will be created in the public's mind. Public relations efforts may serve well in times of need as a defensive weapon; but more important, as a continuous and ongoing offensive weapon.

Planning Public Relations. Rules that govern the proper planning of the communications mix also apply to public relations, including identifying its purpose and target markets accordingly. The purpose of a specific public relations program effort should be established before any further planning occurs. For example, a restaurant might be under a new management that has to overcome a perception in the marketplace of slow service. In this case, an advertising campaign would be unlikely to convince anyone that the service is now better. Improving the customers' perception of the restaurant's service would be the purpose of the public relations campaign. This measurement, as in advertising, would show positive shifts in local opinions about the restaurant's service standards and, ultimately, an increased number of covers.

A hotel might have an image of being too expensive for the local customers and might thus be avoided by them. The purpose of the public relations effort would be to eliminate this perception by improving the price–value relationship image in the local marketplace. The success of this program may be measured by increased usage of guest rooms by local customers or in increased restaurant or lounge business outside of usual occupancy trends.

When planning public relations, one must consider the benefit to the customer in the target market. Choosing a target market for a public relations effort is as important as choosing the correct market for any marketing communications effort. You must ask, How will the target market be influenced by the communications? This involves not only short-term benefits, but long-term ones as well, because hotels and restaurants are a major part of the community in which they exist. They are the most public of all commercial enterprises, so much so that they often become public places where people meet. These same people, will answer questions from out-of-towners such as, "Where should I stay?" or "Where's a good place to eat?" Public relations can influence local responses to these questions even when the people themselves have never stayed or eaten at the property. Public relations efforts create an image in the mind of the consumer and reinforce that image in many ways.

Along with identifying a target audience comes the task of reaching these prospects. Although the geographic location of the customers needs to be understood, the correct media to reach that geographic area must be analyzed as well. Although a computer trade journal may appear to be a good place to advertise for a corporate meeting, this is probably not where a potential vacationer would be reading an article on the benefits of staying in a hotel.

Although "selling stories" may sound unusual, good public relations experts will have a network of editors to whom they can do just that by calling on them personally. This relationship with editors and writers can be critical to breaking a story. For this reason, the discipline of public relations is becoming more of a science and less of an art form.

The public relations expert will push a story much as a salesperson sells a product. Calls are made to the editors, they are wined and dined, and thank-you notes and flowers are sent in appreciation of the placement of a story or press release. A press release is a document that contains the message or story the hospitality enterprise wishes to communicate that is prepared in a manner that is consistent with the expectations of the media. A press release always contains the contact name and number of the public relations professional who wrote the story, background information on the facility, and the body copy of the story. It is then "pitched," or "sold," to the media.

Personal contacts with the media make the difference between a good public relations firm and a poor one. Anyone can write stories and send them to papers and broadcast media, but only a true professional has the contacts to follow up until the article is published or the story makes the nightly news.

Developing Public Relations Tactics.

Before the public relations program is employed, it is important to begin to develop stories on the product itself. Good starting points in a public relations campaign include looking at personnel, customers, and history.

Numerous personnel stories can be developed and submitted based on the employees who work every day in a hotel or restaurant. The Clarion Hotel in New Orleans received much media attention when an off-duty bellman chased and apprehended the attacker of a foreign tourist who had ventured into an unsafe area of the city. For restaurants, the background of the chef can provide an interesting story. If the chef has won any awards or trained outside of the country, the local media are often willing to convey the story to their readers.

Sometimes customers become stories themselves. A honeymoon couple from 30 years ago checking into the same room can generate tremendous interest in the press. A customer who dines regularly in a restaurant conveys an image of contentment that might cause readers to try the product. And when celebrities or politicians dine in a restaurant or stay in a hotel, the public has a natural curiosity. Positioning also becomes an important element in using a customer for a lead story for a hotel or restaurant. Be sure, however, that the customer being featured is representative of the desired target market.

From a history perspective, a story line developed about the building, the neighborhood, or the owner's or manager's background can also interest the public. Examples include any of the hotels developed by the likes of Donald Trump, Steve Wynn, or Ian Schwager.

Public Relations Guidelines.

Rod Caborn, executive vice president/public relations for the firm Yesawich, Pepperdine, Brown & Russell suggested additional guidelines for public relations.[7]

1. The most common mistake hotels make is not budgeting for PR expenses. In order to get results, you need to spend some money.
2. Be careful who you hire, as PR titles are often bestowed on people who have no training or experience in public relations. It is best to use reputable PR firms.
3. Like a good marketing plan, it is imperative to have a written PR plan. Without such a plan, PR will not happen.
4. PR people must understand your marketing plan. You can't expect results unless you let them in on your plans and objectives. Make sure they understand that PR is part of the marketing mix.
5. A consistent, ongoing PR program should provide consistent, ongoing results.
6. It takes innovative ideas to get deserved PR coverage.
7. Remember: Great public relations depend on creative management.

Publicity

Once "natural" stories like those generated in public relations have been fully developed, other methods need to be employed to keep the press interested in the hospitality enterprise. Publicity now needs to be "created" so that editors will continue to have something to write about.

The creation of publicity events is not as simple as it may sound. The purpose of the event needs to be established together with a target medium, and an evaluation of the event needs to follow. Publicity, in this sense, is like promotion, except that publicity is specifically aimed at the media to produce more public relations. Promotions can be held without publicity; publicity helps with promotions.

Publicity begins with inviting the appropriate editors and radio or TV station managers to the property for a specific event. Again, personal relationships developed by the public relations manager are critical for successful attendance by the right people.

The event must be organized so that everything goes perfectly. If an event is not executed well, the hotel or restaurant may be the object of negative publicity. In addition, any other potential customers who hear of it by word of mouth many be turned off. Although this is exasperating, it is nothing compared to the potential lost business that one editor could produce by writing a critical review in a newspaper with a circulation of 300,000 readers.

At the event, press releases with background information should be made available to the press. A prepared press release will answer questions such as the number of seats available in the restaurant, the name of the manager, and so on. The public relations professional will "work the event" by attending and "pitching" the points personally to the attendees from the media. The end of the actual promotion signals the beginning of the placement work for the public relations effort. Thank-you notes, special commemorative gifts, and/or flowers are typically sent to remind the attendees of the importance of the event. Follow-up calls are made to convince the writers to place the story in the best light and to request favorable placement from the editors. Having a story placed in a newspaper or on the radio or television is not the only measure of success. The physical placement of the story in the medium (i.e., its position relative to other stories) is as important as getting the story into the media.

After all of this work is finished, the last stage of the public relations/publicity effort is program evaluation. Have more customers been generated? Was the perception in the marketplace altered to the satisfaction of the management team? The evaluation process is as important as any other phase of the effort. Restaurant covers and rooms sold can be tracked at the property, but changes in customer perceptions are more difficult to measure. These measures are typically taken through consumer research that measures both pre- and post/campaign images.

Effective public relations programs always include a provision for the unexpected—the negative publicity that can follow a natural disaster or otherwise unfortunate event. When Hurricane Wilma devastated the resort community of Cancún, local hoteliers had an immediate need to communicate quickly and accurately with the worldwide media

Tourism Marketing Application

Crisis Management Center in Nepal

In an article published in the *Kathmandu Post*, Basanta Raj Misras stated "Tourism in Nepal has been hit not only by the insurgency but also by the wrong perception created by the inflated news and other types of misinformation campaign. The misconception has kept significant number of tourists off Nepal, despite improved security situation in the country." To counteract the negative publicity, the government of Nepal set up a crisis management center comprised of high-level officials throughout different parts of the government. The goal of this center is to "to disseminate correct and authentic news, fight against negative publicity by establishing strong relationship with foreign media, adopting appropriate marketing strategy, suggesting policy measures and handling seen and unforeseen incidents."

Source: www.strategypage.com/messageboards/message/76-270.asp (accessed April 3, 2006).

covering the story. Marriott managed the tragedy masterfully, chartering buses to take guests to a regional airport several hundred miles away, then picking up the tab to fly them home. Other properties that were less well prepared were the object of repeated negative stories that featured disgruntled guests who had no food, water, or clean shelter for days. Accordingly, every public relations plan must include a provision for "crisis communications," setting forth policies on such critical issues as who will serve as the official spokesperson to the press, what arrangements will be made to get timely and accurate information to the press, how often the press will be briefed, and so on.

Crisis management teams are often put together for tourism destinations as well. Security issues, for example, have become very important for today's international traveler. The Tourism Marketing Application tells of the government of Nepal setting up a crisis management center to combat negative publicity concerning security and safety in this region.

Chapter Summary

In this chapter we discussed the foundations of the communications mix and three of its major components–advertising, merchandising, and public relations. The foundations apply to all aspects of communications, and successful implementation depends on a comprehensive knowledge of the market.

Word-of-mouth communications are a strong force in the hospitality industry; recommendations from those who have personally experienced a product or service play an important role in the selection process. Complaints and praise are ways that customers choose to communicate. These can be just as effective as any advertising campaign, or more effective, yet equally destructive if negative.

Although it is easy to employ advertising techniques, it is not so easy to do so successfully. Success depends on the advertisement's ability to address the needs of the customer. Advertising should have clear goals and should be developed only after asking the right questions.

Merchandising is primarily in-house marketing. The planning process includes assessing the needs of the customer and then providing the product to the customer in a cost-effective manner. With a strong planning process and a proper evaluation mechanism in place, both revenues and customer satisfaction will be enhanced.

The public relations component of a hospitality marketing plan is a very effective element of the communications mix. It may be the most effective element because it is the generally the most believable. A person is more likely to be influenced by reading or hearing a third party's praise for a product than by an advertising campaign. Publicity, a division of the public relations umbrella, is used after "natural" stories have been highlighted by press coverage. Publicity, in a way, is similar to sales promotions except that publicity is aimed specifically at the media to generate more good public relations.

Key Terms

advertising, p. 354
collateral, p. 358
consumer adoption process model, p. 348
merchandising, p. 358
press release, p. 363
public relations, p. 362
publicity, p. 362
pull strategy, p. 350
push strategy, p. 350

Discussion Questions

1. In your own words, describe what the marketing communications mix is all about.

2. What are some of the broad objectives of a communications strategy?

3. Briefly describe each of the six stages of a communications strategy. You may find using a hospitality or tourism example helpful in responding to this question.

4. What is the consumer adoption process model? Why is it helpful when developing communications strategies?

5. What are push/pull strategies? Can you give a hospitality or tourism example for each?

6. Describe the role of advertising in the marketing communications strategy. In other words, is advertising the most important tool in the communications mix?

7. What is merchandising? What is the intended audience of a merchandising plan?

8. What is the role of public relations in the marketing communications mix? How does publicity fit into this picture?

Endnotes

1. Retrieved August 16, 2006, from www.hotelmarketing.com/indexphp/contentarticle/ 060413_starwood_laundches_blog_the.lobbycom.

2. Hallowell, R., & Hansen, A. (1999). *A taste of Frankenmuth: A town in Michigan thinks about word-of-mouth referral*. Cambridge, MA: Harvard Business School Publishing, Product number 9-800-029.

3. Retrieved November 2, 2005, from www2.millenniumhotels.com/MCIL.nsf/ unidlookup/E742B59BD6400B1548257098000C43A6?opendocument.

4. Retrieved November 2, 2005, from www.starwoodhotels.com/luxury/about/news_ release_detail.html?obj_id=0900c7b9804db82b.

5. Retrieved November 2, 2005, from http://marriott.com/news/detail.mi?marrArticle =102037.

6. Haywood, R. (1984). *All about PR*. London: McGraw-Hill. Excerpted from Buttle, F. A. (1986). *Hotel and/food service marketing*. London: Holt, Rinehart and Winston, 400.

7. Conversation with Rod Caborn, executive vice president/public relations of Yesawich, Pepperdine, Brown & Russell, November 2005. Updated from Adams, J. (1987, June 8). Good P.R. plan can be potent marketing tool for hotels. *Hotel & Motel Management*, 60.

.... case study

The Lacandon Rain Forest, Chiapas, Mexico

Discussion Questions

1. What is the mission of the Na Bolom Institute in Chiapas, Mexico?

2. Briefly describe the Lacandon people.

3. What are some of the cultural impacts, both positive and negative, that tourism is having on the Lacandon people?

4. Describe each of the four types of Lacandon tours presented in the case, including ecological, cultural, Mayan history, and specialty.

5. Do you agree with the proposed measures for visitor accommodations, meals, and other amenities presented in the case?

6. Who is the "right" kind of tourist for the Lacandon rain forest? How can the Na Bolom Institute effectively reach out to this market given the very low marketing budget constraint?

In 1951 Frans and Gertrude Blom founded Na Bolom Institute, whose purpose was to preserve the Lacandon rain forest and the culture of the Lacandon peoples. A modern web page describes the Bloms as "pioneers in the research of the Rain Forest and the Lacandon" (The True Men, www.burn.ucsd).

Na Bolom means "house of jaguar" in Tzotzil, a Mayan dialect. Na Bolom remains today a nonprofit organization that offers limited lodging facilities, an extensive library, and a photo archive of Chiapas. Recently, Na Bolom has expanded into the areas of ecotourism to the Lacandon rain forest and cultural tours of the Lacandon Indians. The staff at Na Bolom is concerned that ecotourism may help save the rain forest but destroy the culture of the Lacandon.

The Lacandon are the last remaining direct descendants of the Maya. Because the Spaniards never conquered the Lacandon, the Lacandon provide the best insights into the great Mayan culture. Their culture is intertwined with the

This case was written by Eddie Dry and Stephanie Duncan of the Anderson Schools of Management, University of New Mexico, Albuquerque. The authors thank Caitlin Thomas and Victor H. Gallegos Coutiño of the Na Bolom Institute for their concern for the people and the rain forest.

rain forest, as their food shelter and clothing have been made or grown in the jungle for centuries.

Years of tradition and isolation have left the Lacandon defenseless against the influx of tourists and Western ideas. Realizing that total isolation is no longer a possibility, Na Bolom is developing transition strategies that would provide a key to the Lacandon survival.

LOCATION AND PEOPLE

The Lacandon rain forest covers much of the eastern portion of the state of Chiapas, Mexico. The Lacandon, the indigenous people who have occupied the Lacandon rain forest for hundreds of years, are direct descendents of the Maya who built the great cities of Palenque and Bonampak.

The Lacandon rain forest is located approximately four hours or 80 miles by van from San Cristóbal de Las Casas, Chiapas, Mexico. Chiapas, Mexico's southernmost state, borders Guatemala and Belize. The state of Chiapas is home to hundreds of Mayan ruins and is an ideal adventure destination. See http://www.travelchiapas.com/map/map-2.php for a map of Chiapas and the location of Lacandon in the southeast. Na Bolom Institute is located in San Cristóbal de Las Casas, Chiapas.

PROBLEMS FACING THE LACANDON

The basic subsistence crop of the Lacandon is corn. As a result of NAFTA (The North American Free Trade Agreement), crop prices have fallen drastically to the point that growing corn no longer serves as a viable means of income. With the growing popularity of ecotourism, the Lacandon are shifting their efforts to arts and crafts to sell to tourists. The Lacandon see many opportunities in arts and crafts, lodging, food, guiding services, and other tourism activities.

Unfortunately, the Lacandon are unaware of the tremendous cultural impact that tourism can inflict on traditional cultures. The isolation of these people has helped them maintain their culture ever since they successfully resisted the Spanish in the early 16th century. The modern world, however, has created in the Lacandon the desire for material goods and services that are far beyond their capacity to purchase, thus creating a frustration and a desire to turn away from a culture that has existed for centuries.

Other undesirable impacts from tourism are turning children into beggars, producing arts and crafts for public consumption instead of a religious expression, losing their Mayan languages, creating a desire to move to the city to earn "real" money, and facing the problems of living in a city with no modern city survival skills.

Another problem is the slash-and-burn agriculture practiced by the Lacandon for centuries. Rain forest topsoil is very thin; new immigrants as well as the Lacandon slash the rain forest, burn it, farm it for a few years, deplete the soil, move on to another plot, and then begin the process again. The Lacandon rain forest is the last vestige of a great rain forest that once covered most of southern Mexico.

A political problem that the Lacandons face involves the Mexican government. The government has allotted the Lacandon a great deal of land through the historic ejido land system, a system similar to that of the Indian reservations in the United States. However, with the tremendous demands by thousands of immigrants into Chiapas, the government needs huge amounts of new land to maintain political stability. The historic ejido land grants are an easy target; therefore, the government is tempting the Lacandon to relinquish their land. Many of the poverty-stricken Lacandon are falling for the government's offer.

The answers are not simple; there are no perfect solutions. The hope is that the Lacandon can generate an ecotourism industry without suffering the many social problems that tourism can bring.

BACKGROUND

Some people believe ecotourism is only a label for vacation packages to symbolize adventure and to ease the minds of wealthy travelers, who take comfort in knowing that a part of the money that they pay is given to indigenous communities. These people also believe that ecotourism often damages more than it helps to protect fragile areas.

One website focusing on world social issues criticizes village ecotourism especially as "the first step in the process of appropriation, where the physical environment, the human societies within it, and any historical remains, subtly become universal property" (www.georg.unm.edu). This website goes on to wonder if, in fact, village ecotourism is a "Trojan horse" that penetrates into the personal space of residents and causes political and cultural changes that are irreversible. Does village tourism help preserve artifacts of culture only to destroy the spirit that created them? Do villagers adopt city or Western ways, moving from traditional hospitality to cash transactions to acculturation, and then to manufacturing solely for tourist consumption? In the development of a village ecotourism program, how will these negative impacts be addressed, and how will they be avoided in the future? The answer to these questions lie in the definition of ecotourism. How do tourists around the world define this extremely important word?

WHAT IS ECOTOURISM?

Ecotourism is a nature-based form of travel that is defined by the Ecotourism Society as "responsible travel to natural areas, which conserves the environment and sustains the well-being of local people." (Ecotourism Society, 1998) Ecotourism then seems like the perfect solution to the "problems" in Chiapas: saving the rain forest and at the same time preserving the culture of the Lacandon. It is great in theory, and it just may be realistic in practice.[*]

Another popular definition of ecotourism (of which the authors could not find the source) is: "Ecotourism is ecologically responsible and culturally sensitive tourism that has a low impact on the environment and local cultures; meanwhile creating jobs and raising funds to conserve wildlife and vegetation."

There is no doubt that there is ecotourism, but are there ecotourists out there in sufficient numbers to generate significant impact? The World Tourism Organization (WTO) estimates that nature tourism generates 7 percent of all international travel expenditures.

Numerous authors are reporting annual growth rates of ecotourism of 10 to 30 percent. For Na Bolom, however, ecotourism is not so much a term to define as it is a philosophy and belief system that each person involved will understand and support. A definition of ecotourism is important, but is the definition clearly embraced by all involved?

NEGATIVE IMPACTS OF TOURISM

Tourism can easily destroy (and often does) the very things that the tourists are seeking: the Lacandon rain forest, the Lacandon culture, the streams, and the ancient Mayan ruins. Examples of tourism destruction are pollution, begging children, and the demonstration effect.

[*]*www.ecotourism.org/webmodules/webarticlesnet/templates/eco_template_news.aspx?articleid=12&zoneid=25, accessed April 10, 2007.*

The destructive nature of pollution is easy to understand. Many thoughtless tourists think nothing of throwing trash on the trails when proper receptacles are not readily available. The trash results in disappointed tourists who travel hundreds of miles to visit ancient ruins. Tourists lose themselves in their imagination of years lost, only to be rudely awakened by a candy wrapper.

Preserving the cultural integrity of the Lacandon and, more important, the innocence of their children is a foremost concern. Begging is a crucial issue, especially with children. Begging not only takes away from the tourists' experience, but also begins a downward cycle leading the children further away from their heritage. A child quickly learns how to manipulate the sympathies of the tourists with a sad smile, or even worse, a gesture toward his little sister who is holding their baby brother. A clever child can earn more in an hour by begging than his parents can make in a day. Tourists must be carefully educated about the very harmful impacts of their giving money or gifts to begging children.

The demonstration effect may have produced more cultural destruction than will ever be known. When wealthy tourists flaunt their portable radios, cameras, expensive clothes, and cars, the impact on a people without any of these conveniences is immense. The indigenous people want the conveniences, and they want them now! This process is called the demonstration effect and is widely thought to be a major negative impact of tourism. Again, tourists must be carefully educated about the demonstration effect and asked to leave many of these conveniences at Na Bolom for safe-keeping.

Social interactions between the Lacandon and Westerners is another sensitive issue. Naturally, some interaction is necessary, but it should be limited to tour and accommodation information and craft sale transactions. Children should not be permitted to interact with tourists for extended periods of time, and then, only with adult supervision.

TYPES OF TOURS

Lacandon tours could be broken down into four categories: ecological, cultural, Mayan history, and specialty. Each category has unique target markets and marketing techniques. All, however, share the basic organizational framework. Ecological tours would provide essential information on conservation and teach low impact on the ecosystems while hiking through the rain forest. One key issue to focus on is the slash-and-burn agriculture, its importance, and alternatives for the Lacandon. While most agricultural experts decry slash-and-burn agriculture, development experts need to be prepared to provide the resources to teach other forms of agriculture, or even alternatives, such as tourism. Other ecological tour issues are the complex ecosystems of the rain forest, the huge number of species in the rain forest, herbal medicines, and global warming.

Botanists and horticulturists would be attracted to a program of classification and identification of Lacandon flora. One may begin the week with an introduction to Tzeltal plants and their role in daily Tzeltal life, instructed by a Tzeltal ethnobotanist. Throughout the week tourists will be in the rain forest identifying and classifying native flora and photographing the species for a future exhibit. Experts on the tours can help identify unknown species.

Alternative medicine has a surprisingly large following internationally. With the possibilities of undiscovered remedies in the midst of the Lacandon rain forest, its mystery is quite appealing. Herbalists could be invited to research potential remedies. Na Bolom could form partnerships with U.S. and Mexican medical schools to send their students to study and research alternative medicine.

Cultural tours would provide tourists with an opportunity to browse through the photography library, listen to stories of the villagers, and even live the life of a Maya for a day. A day in the life of a Maya would involve Lacandons teaching tourists to work in a milpa (ejidos are broken into milpas for each village family); cook in a hut; and make crafts such as baskets, pottery, flutes, and bows.

Mayan history tours would provide travelers with the background history of each site, including information on rulers, festivals, wars, marriages, and any other known facts. Tourists would have a choice of one-to-five-day tours; bus to sites such as Palenque, Bonampak, Tonina; or boat down the Rio Usumacinta to the Mayan ruins of Yaxchilan. Time permitting, visitors could also opt to visit the breathtaking natural water sites such as Misol-Ha, Agua Azul, or Lagos de Montebello.

Na Bolom currently provides tourists staying in San Cristóbal de Las Casas with day tours to nearby villages. San Juan Chamula and Zinacantan are both within a half-hour's drive on dirt roads. These communities are extremely different from each other economically and culturally. Tenejapa and Amatenango del Valle are two other Tzeltal villages that offer tours where Na Bolom's assistance could be used.

Na Bolom could create a photography tour of Chiapas and the Lacandon. During the tour a variety of photography opportunities could be provided offering various levels of instruction. The first day's subject would be Mexican lifestyles in San Cristóbal de Las Casas; following days could focus on the Lacandon rain forest and its inhabitants. Tourists would also be encouraged to send back their photos, which would preserve part of the Lacandon forever; pictures could be displayed in a new gallery dedicated to Gertrude Blom (Trudi) for her years of dedication.

Avi-tourism is the largest growing sector in eco-tourism and an amazing opening to attract new visitors. Birders enjoy observing, an extremely low-impact activity, and can start the Lacandon Bird Sighting List. An impressive sighting list will draw tourists from around the globe. Many bird watchers have developed an extremely important personal document, the life list, which documents their sightings for their entire lives! A marketing strategy would be to appeal to those serious birders through e-mail, bird club newsletters, press releases, and a press kit sent to writers and editors.

These specialty tours would target market niches such as photographers, botanists, horticulturists, avi-tourists, and herbal medicine specialists. Each tour would concentrate on the individual group's interests and also provide them with exceptional photography opportunities, or Lacandon insight into their rain forest. The tourists could not only provide income and jobs to the Lacandon, but they could also give back to the Lacandon through friendships, photos, and a sense of caring. Very importantly, these tourists could "carry the word to the world" of the importance of preserving the rain forest, protecting cultures, and being a bit more cautious in international trade agreements, such as NAFTA.

PROPOSED MEASURES

Accommodations

A visitor's complex located at least a half a mile from the village would reduce unsupervised and unnecessary social interactions between tourists and the Lacandon. The complex would consist of 10 thatched sleeping huts, a large dining hut, restrooms and bathing facilities, and cleaning and cooking facilities for the Lacandon. Construction would be done with native materials and tools in Mayan architecture and decor. The labor for construction could be provided by ecotourists who actually pay for the experience of working alongside the Lacandon.

Meals

Village meals, prepared in the cooking hut next to the dining hut, would consist of Mayan staples such as maize, rice, squash, and chicken. The Lacandon would offer three meals daily, and service would be in executed in traditional attire and etiquette. To further enhance the experience, tourists could harvest their own food from the rain forest and the milpas. These experiences would provide the Lacandon the opportunity to interact with outsiders in a controlled environment.

Other Amenities

The Lacandon Ecological/Cultural Tours would require all the elements necessary in any tour. Besides the accommodations and meals, other necessary amenities are transfers, guide services, and shopping opportunities.

The transfers would be from the airport in Tuxtla Gutierrez to Na Bolom Institute in San Cristóbal de Las Casas and from Na Bolom Institute to the village of Lancanja. Because of the small size of the tour groups (fewer than eight to minimize the impacts), vans would be the best means of transportation. Initially, these vans could be rented from any of several places in San Cristóbal de Las Casas. Hopefully, in the future a Lacandon business could own the vans.

Currently, Victor H. Gallegos Coutiño, a guide at Na Bolom, typifies the characteristics of an ecologically and culturally sensitive guide. Victor speaks the Mayan dialect. At every stage of the tour, he uses local Lacandon to prepare the food, guide the hikes, and provide the canoes. Victor knows many of the Lacandon by their Mayan names and stops frequently on jungle roads to talk with Lacandons that he knows. He provides news from their friends and relatives who are staying at Na Bolom for medical reasons. He carried news back to these friends and relatives. Victor has compassion, an understanding of his role, and an appreciation

of the fact that someday (hopefully) he will turn over the business to the Lacandon.

Competition

Because of Na Bolom's long, unselfish, and dedicated service to the Lacandon rain forest and the Lacandon, the current tours have few real competitors. Other tour companies offer similar tours, but none have the access and trust of the Lacandon. Another competitive advantage is that Na Bolom is known throughout the world and is written up in all guidebooks that focus on Chiapas.

Marketing

Because the Lacandon have limited funds to work with in promoting ecotourism, low-cost marketing is essential. Ultimate success is dependent on the Lacandons' frame of mind and attitude and their ability to keep in tune with emerging markets.

The first step might be to form a marketing committee to implement these strategies and periodically reevaluate them. The marketing committee could include Sectur (the national Mexican tourism marketing department), Mundo Maya (a multinational association promoting Mayan tourism), and the Chiapas office of Sectur. Others that might be included are students and professors at the National Autonomous University of Chiapas as well as the Superior University of Chiapas. Industry representatives could include the airlines, hotels, bus companies, and tour guide organizations. After the committee has a preliminary course of action, the next step is making people aware of their ecotours.

E-mail-based newsletters are an effective low-cost strategy for reaching target markets, especially associations. The newsletter can be synergistically marketed in conjunction with the Maya Peace Park (a joint organization of preserves in Mexico, Guatemala, and Belize). A familiarization tour (or FAM) is an inexpensive means of acquainting key industry professionals to the experience of the Lacandon. The invitation list could include travel writers and environmental magazine editors. The tour would provide a firsthand experience of the uniqueness of the Lacandon exploration.

The next step is to write a press release and create a press kit. A press kit should include several press releases each on a different tour or topic. A press kit should include pictures related to the press releases (preferably slides), articles written about Chiapas and/or the Lacandon rain forest, and literature on Na Bolom. It could be mailed in an eye-catching envelope. The press kits should be sent out to travel writers, travel editors, and conservation magazines.

This project promoting ecologically/culturally sensitive tourism to the Lacandon rain forest is a special sort of tourism. The impacts could be beneficial or devastating, depending on the types of tourists attracted to the Lacandon. Therefore, marketers take on a very different responsibility. Some criteria for marketing the tours will help attract the "right" kind of tourists:

■ The marketing message should emphasize the uniqueness of the Lacandon culture.

■ The marketing messages should promote the responsibility of the tourists to protect the culture and the ecology.

■ Marketing should attract educated, culturally sensitive, and environmentally aware tourists.

overview

This chapter continues our discussion of the tools of the marketing communications mix by focusing on personal selling and sales promotions. These tools are as important to the communications mix as advertising, merchandising, and public relations (addressed in Chapter 11).

Personal selling is direct, face-to-face communication with prospective buyers and regular clientele who have needs or wants for the product both now and in the future. Personal selling is commonly used for complex purchases such as meetings, conventions, and group tour business.

continued on pg 380

chapter 12

Personal Selling and Sales Promotions

learning objectives

After reading this chapter, you should be able to:

1. Define personal selling and sales promotions, two more tools of the marketing communications mix.

2. Explain why personal selling is a particularly effective tool of the communications mix for complex purchases of hospitality and tourism product offerings. You will be able to use this knowledge to develop a sales plan call sheet.

3. Explain how sales promotions can stimulate sales on a short-term basis and bring in new business from first-time customers. You will be able to use this knowledge to develop effective sales promotions.

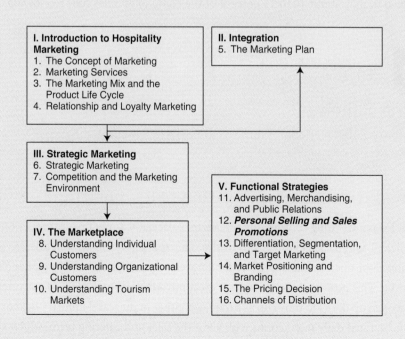

I. Introduction to Hospitality Marketing
1. The Concept of Marketing
2. Marketing Services
3. The Marketing Mix and the Product Life Cycle
4. Relationship and Loyalty Marketing

II. Integration
5. The Marketing Plan

III. Strategic Marketing
6. Strategic Marketing
7. Competition and the Marketing Environment

IV. The Marketplace
8. Understanding Individual Customers
9. Understanding Organizational Customers
10. Understanding Tourism Markets

V. Functional Strategies
11. Advertising, Merchandising, and Public Relations
12. *Personal Selling and Sales Promotions*
13. Differentiation, Segmentation, and Target Marketing
14. Market Positioning and Branding
15. The Pricing Decision
16. Channels of Distribution

Sales promotions are short-term product offerings that are used to stimulate sales on a short-term basis. They can also be effective to stimulate trial purchases, especially for buyers new to the product. Sales promotions, as well, are often used to bring in business during slow periods of demand or to help reduce excess inventory.

Marketing Executive Profile

Carrie Ballew
Corporate Sales Manager, Four Seasons Hotel Silicon Valley

Carrie Ballew was born and raised in Las Vegas and attended the University of Nevada, Las Vegas. During her first two years out of college she worked in the restaurant field and then moved over to the hotel industry as a front desk agent at Caesars Palace. She remained at Caesars Palace for nearly five years as a front desk agent, wedding coordinator, travel industry sales coordinator, and sales manager. After this, she began work at the Four Seasons Hotel as a corporate and travel industry manager, where after some time she transferred to Northern California and opened a new hotel for the Four Seasons in Silicon Valley in January 2006. Ms. Ballew achieved the President Club award in 2006. This award is given to sales managers who achieve over 110 percent of their goals and is awarded to only about 10 to 15 percent of the companies' sales staff worldwide.

 ## Marketing in Action

Carrie Ballew
Corporate Sales Manager, Four Seasons Hotel Silicon Valley

How did you first get into a sales position?

When I was a wedding coordinator at Caesars Palace, it was my first introduction to the sales field. I sold packages directly to brides. This sparked my interest in sales, so when a position came open in the sales office for a travel industry sales coordinator, I applied for a transfer. This position was known as a great introduction to the sales department that would open up doors to move up as a manager either in the travel industry field or group sales field. I had daily interactions with all of the sales and catering team, which allowed me to observe and understand how the whole process takes place.

What do you particularly like about being a sales manager? What do you dislike?

What I love most is the interaction that I get to have with clients and the relationships that I build with them over the years. A goal for me is to build a great rapport with clients, where they know they can call me for anything, and that I am here to make their lives easier. Although the relationships you have with your clients are a very big part of the sales process, the ultimate goal is to have these clients bring in business. So, on the flip side of enjoying the sales process and clients, your ultimate responsibility is to produce revenue for the hotel. The business you are booking into the hotel is feeding all other departments of the hotel. If you aren't booking business, then your fellow employees in other departments are not getting all the hours they could be if the hotel were busy. Sometimes the pressure of "making your numbers," or the budgeted amount of revenue that your particular segment is assigned, can take away from the things that you enjoy about sales. In the sales field, there are good months and bad months, and keeping a positive attitude will keep you going through the bad months, when you might not have as many bookings.

What is the typical career path for a person in sales and marketing?

All companies have different structures, but from a sales manager position there might be a step up to a senior sales manager or national sales manager. After that one would move to an associate director of sales or director of sales position. Then from a director of sales position you could branch off to a director of marketing or VP of sales and marketing.

Can you describe a typical workday for a person in sales?

The great thing about sales is that every day is different. A typical day includes answering incoming calls from clients inquiring about availability, as well as sending out collateral, proposals, and contracts. A salesperson is also making outgoing calls to drum up new business. During the day, some effort also goes towards interdepartmental items, such as communicating client needs to the reservations department, front desk, room services, or housekeeping, and having meetings within your department to make sure everyone is on the same page. Other items include planning and going on sales trips to visit your clients face to face or to industry trade shows.

How do you see personal selling changing in the next five years?

I do not see the elements of personal selling, such as face-to-face sales appointments, trade shows, and client events, changing, but instead becoming more important. As this world becomes more and more high-tech, hard collateral is being phased out and replaced with electronic forms of communication that can be tailored to each client's needs. I believe that the personal side of selling will become even more important. Although electronic selling will always be efficient, face-to-face interaction with your client will almost always have the most impact.

Can you talk about research that is being done targeting the specific market needs that sales promotions can satisfy? How do you envision such research methods improving over time?

One research method that we do in different market segments and business segments is roundtable events, where we gather a handful of clients together with other hoteliers and have open discussions on key topics that the clients are challenged with in the travel/meeting industry. This is a great way to combine the client relationship development with an educational element. From these events we have direct client feedback on the hotel and corporate levels to improve our relationships, to understand what our clients want, and to examine how we can market to new clientele.

What advice would you give to a student who wants to get into sales?

You have to have sales experience to be a sales manager, so although you may have graduated with a hospitality background, this field requires hands-on experience. The best way to acquire that is to get into the sales department in a support staff role such as an administrative assistant where you work with a group of sales managers supporting them and their sales efforts. This way you get to see and understand how the sales process works. You would then be a great candidate for a coordinator position. A sales coordinator is a great step toward becoming a manager for a person looking for more interaction with clients and a broader scope of the sales process to prepare for becoming a manager. I find that some people look down on starting positions like these when really they should see them as a great way to get their foot in the door and show everyone their solid work ethic and determination to learn the sales process.

Used by permission from Carrie Ballew.

---

personal selling Direct, face-to-face interaction between a seller and a buyer for the purpose of making a sale.

salesperson A person within a hospitality or tourism organization given direct responsibility for personal selling to identified target markets (also referred to as a **sales representative** or **sales manager**).

Personal selling is the direct interaction between a seller and a prospective buyer for the purpose of making a sale. Personal selling may be one of the most challenging aspects of the communications mix. Whereas public relations communicates through stories and the media, advertising communicates through copy and artwork, and merchandising communicates through in-house promotions, the **salesperson** *(also referred to as a* **sales representative** *or* **sales manager**) *communicates through direct oral presentations to the customer.*

Obviously, every employee should be a "salesperson" for his or her organization. In this chapter, however, we will discuss selling from the perspective of hospitality organizations that specifically designate people to carry out the direct sales function.

Organized personal selling is not universally used in the hospitality industry. Rarely will you meet a salesperson from the local Pizza Hut, McDonald's, or even Motel 6. On the other hand, full-service hotels and restaurants with large catering facilities for groups use sales representatives as an essential part of their communications mix.

Whether personal selling is used by an organization depends on several factors, including the following:

◼ Targeted source of revenue
◼ Complexity of the products and services offered
◼ Quantities in which they are purchased
◼ Price that is paid

In the fast-food or budget hotel case, the products and services are relatively simple, customers know pretty much what they are purchasing, individuals or small groups are typical buyers, they purchase in small quantities, and the price is low. For the buyer, it is a low risk, low-involvement purchase. Hiring a salesperson would not be cost effective or productive. The interaction between the buyer and seller is easy and straightforward.

Contrast this with the 1,600-room Hilton Americas in Houston. One salesperson may book a group of 500 rooms for three nights at $150 per room night. This is $225,000 worth of room revenue alone, negotiated between the salesperson and a meeting planner. In addition, there are food and beverage functions, hospitality suites, a general session, and breakout rooms, representing possibly another $100,000, and dozens of other details to be worked out. This is a high-risk purchase with high involvement on the part of the buyer, who wants every detail to go perfectly. The products and services are complex and need a lot of explanation, negotiation, and confirmation before a contract is signed. A good salesperson will decrease the risk factor by offering a guarantee and providing examples.

Besides sales managers at the unit level, chains such as Marriott, Hyatt, and Hilton employ national and international sales managers to represent the entire chain to accounts that have ongoing needs for many hotels in many locations. Personal selling has been found to be most appropriate in the following situations:

◼ The product requires that the customer receive assistance, perhaps personal demonstration and trial, and the purchase decision requires a major commitment on the buyer's part.
◼ The final price is negotiated, not fixed, and the final price and quantity purchased allow an adequate margin to support selling expenses.
◼ Information sought by potential customers cannot be provided thoroughly through advertising.
◼ The market sees personal selling as an essential part of the product.

The hospitality product and organizational markets clearly lend themselves to the personal selling component of the communications mix. Buying the product requires a major commitment on the purchaser's part. Even a small meeting of only 30 people in a suburban hotel for three days can easily exceed a $10,000 expenditure. Assistance is necessary in application; that is, understanding customers' goals and helping them achieve them is very important. Personal demonstrations and trial are common and include site inspections, trial stays, and booking a small meeting before a large one. Pricing for meetings, group bookings, and corporate accounts are normally negotiated. Advertising will not work and is too expensive to reach and fulfill the needs of the buyer and explain the benefits. Finally, the marketplace sees the salesperson as an important part of the product. Because many hotels in the same product class are quite similar, salespeople can develop relationships that provide a competitive edge. In other words, a good salesperson can become part of the augmented product.

Following are several advantages to using personal selling:

1. Selling is really about solving customers' problems. Customer problems become needs.
2. Personal selling can be used to make services tangible and describe products and services in greater detail.
3. The sales presentation can be tailored to customers' needs. Solutions to specific customer's problems can be offered.
4. Prospective buyers can be identified and qualified before engaging personal selling so that overall communications mix dollars may be more effectively spent.
5. Personal selling can reduce risk and is more effective in getting customers to close the deal and sign the contract.
6. Personal selling is the only part of the communications mix that permits direct feedback from the customer.
7. Personal selling provides an excellent opportunity for relationship marketing.

THE SALES PROCESS

The steps in the personal sales process are as follows:

1. Prospecting
2. Qualifying prospects
3. Sales approach
4. Probing
5. Benefits and features
6. Handling objections
7. Closing the sale
8. Following up

Following is more detailed discussion of each of these steps.

Prospecting

prospecting The initial stage of looking for potential customers to purchase a product.

Prospecting is the term used for finding new customers. The goal of prospecting is to convert names of potential customers into sales leads. Sales leads turn into making sales calls, in person or by phone, to qualified customers who are not currently using the product. Prospecting is more difficult than calling on existing customers because new customers do not necessarily know the product, although they may certainly have some perception of it. New customers need to be convinced that the product they are currently using does not satisfy their needs as well as your product could. One axiom goes like this: If you want to sell your product to our company, be sure your product is accompanied by a plan that will help our business so that we will be more anxious to buy than you are to sell.

It is highly unlikely that a meeting planner will "create" a meeting just because of your facility. The meeting either already exists, having occurred before at a competitor's hotel or in another area, or has been partially developed and is waiting to be placed in a property. In direct sales, the most common way to get new customers is to take them away from competitors. Prospecting has evolved over the past decade as the real challenge for selling in a competitive marketplace.

"Cold calling" (calling on a prospect without notice) used to be the main method to create new business. One technique was for a sales team to "blitz" an area or office building by making calls, unannounced, on companies within the buildings. This method is still in place in some organizations, but it is not generally recommended because of its limited effectiveness. Few like to have a salesperson walk in "cold" and ask to speak to the person who books meetings or banquets. Many salespeople also do not like cold calling and the risk of rejection that comes with it. On the other hand, some salespeople use cold calling, and not necessarily a blitz, to successfully set up appointments and obtain pertinent information.[1]

Improved methods of generating leads or prospects have emerged. The Pacific Area Travel Association (PATA) sets up travel marts as described in the Tourism Marketing Application. At travel marts tour wholesalers and the like can explore

Tourism Marketing Application

Pacific Area Travel Association (PATA) Travel Mart

One of the major methods of prospecting used by tourist destinations is regional travel marts, such as the one sponsored by the Pacific Area Travel Association. The Travel Mart 2007 was held in Bali. "The Director General for Indonesian Culture and Tourism Marketing, Mr. Thamrin Bachri, said that Indonesia would promote special interest tours including golf, community-based tourism, eco-tourism, village tours and marine tourism at the Mart. The PATA Bali and Nusa Tenggara Chapter see the successful bid as 'a real breakthrough in boosting the declining visitor arrivals at Bali.'"

Former chapter chairman and Bali Discovery Tours president and director John M. Daniels said that "PATA's decision to select Bali as the venue for PATA Travel Mart 2007 represents excellent timing. The event will bring top travel industry members from around the world to see first hand the enduring charms of our tropical island and the significant new investments over the past few years."

Source: www.tourismindianoline.com/highlightpages/highlights2.html#Bali%20To%20Host%2PATA%20Travel%20Mart%202007 (accessed April 6, 2006).

firsthand a destination for possible group bookings and other travel plans. In other words, the prospect comes to the potential venue instead of the salesperson calling on the prospect.

Many sales directors are recognizing the cost of sales calls and realize that sending salespeople out on calls without appointments is an expensive way to do business. Depending on the experience of the salesperson, the location, the account, and other variables, a sales call can cost $55 to $500 or more, after salary, benefits, office space, secretarial support, collateral, and travel are factored in as part of the cost of doing business. At even $50 per call, the salesperson becomes an expensive resource, not to be used without a well-devised plan. Direct mail, e-mail campaigns, telemarketing, or sometimes a combination of both can be used effectively to set up sales calls in advance. Once a face-to-face relationship is established, salespeople often use the telephone and the Internet for follow-up as much as possible, because of the time and costs associated with face-to-face calls.

Qualifying Prospects

In this step, the salesperson determines whether prospects are qualified to make the purchase by asking the following questions:

- Can they afford it?
- Do they have business in this destination?
- Do they have the authority to make the decision?
- How serious are they about using our facilities, or are they using us for leverage against another property?
- Do they use properties in our product class?

qualifying (a prospect) Determining whether a prospect has an interest in and ability to purchase a product.

sales leads Sources of prospective customers.

Qualifying is done during prospecting, telemarketing, or as a follow-up to direct mail or advertising responses. The qualifying process turns hundreds of names into a few **sales leads.** One way to generate sales leads is through direct mail, either through the post office or the Internet. A mailing list or e-mail list is purchased, and a mailing piece is developed and mailed with a response request or card and, often, some kind of incentive. For example, a facility may want to increase its share of a certain market, such as medical meetings. Certain parameters that fit the property, such as size of meeting and geographic preference, are established. A creative direct mail piece is created to generate a response for more information from a prospective buyer. A sales manager or telemarketer would then follow up the lead to better determine a potential customer's specific needs. This method is effective in qualifying prospects because only those who are genuinely interested in the service will bother to respond.

Advertising to meeting planners is also common practice for large hotels, resorts, and hotel chains. Media that reach professional meeting planners include such trade publications as *Meetings & Conventions, Successful Meetings, Corporate Meetings and Incentives, The Meeting Professional, Convene,* and a host of others. It is also possible to advertise on the Meetings Industry Megasite (www.mimegasite.com), as shown

| Exhibit 12-1 | Advertising on websites is another way to reach a target market |

Source: Retrieved from www.mimegasite.com—VNU Business Media – USA. Used by permission

in Exhibit 12-1. The business press, including newspapers and magazines, is frequently used as well. Advertising appearing in these publications may provide special incentives or direct solicitations of meeting planners. Responses are followed up with more information; for example, sending brochures, meeting room specifications, facilities, and so forth. Some properties also send out DVDs to provide more graphic and direct presentations. Eventually, a personal sales call or a telephone call followed by an invitation to make a "site inspection" will be in order.

Sales Approach

The sales approach, which includes communicating personally with customers, is not an easy skill to master. In past years, when the demand for guest rooms exceeded supply, the selling process was simple—order taking. A salesperson would simply answer the telephone or call back customers who had telephoned earlier and take their orders. The sales process is now very different. In most markets, supply exceeds demand and business does not come automatically. An effective sales process is what generates business.

There is a right way to sell and a wrong way to sell, yet it is still a very personal skill. Many salespeople in the hospitality industry think they are selling by knocking on office doors without appointments (cold calling) and leaving behind brochures. Today, good salespeople get "inside the buyer's head." The sales process has evolved to where good salespeople are good problem solvers.

In the past, and perhaps today, the seller tried to unload onto a buyer what the seller decided to offer. Over time, selling has progressed to the point that the seller must learn about buyers' needs, desires, and fears, and then design and supply the product in all its forms to solve these problems. Instead of trying to get the buyer to purchase what the seller has, the seller makes sure she has what the buyer wants. The "product" is no longer just an item; it is a whole bundle of benefits that provide solutions to problems and provide overall customer satisfaction.

Today successful selling is more about long-term relationships between sellers and qualified buyers. The point is not just keeping customers. It is about delivering what the buyer wants. The buyer wants a seller who will keep in contact and deliver on the promises made. Thus, an interdependence develops between the seller and the buyer.

Web Browsing Exercise

Use your favorite search engine and look up the term "sales process." Examine two or three of the sites. What did you learn? If possible, find a site that offers an article explaining the sales process in detail.

The Ritz-Carlton Hotel Company learned this lesson when it hired a research firm to study its relationships with meeting planners. The research revealed that Ritz-Carlton was seen as lacking during the period after the sale and before the event, a period of sometimes a year or more. Buyers felt, "Now that they've got my business, they don't care about me," when there was no contact from the hotel until just before the event. Buyers wanted an ongoing relationship during the period that assured them that their business was appreciated.

sales account managers Salespeople assigned to specific accounts with only a few clients that have large volumes of business. This is often a sales position at the national level for large hospitality and tourism organizations.

Similarly, some of the larger hospitality companies, such as Marriott and Hyatt, have specific **sales account managers** who work with only a few major accounts such as IBM and GE. These accounts have a large number of buyers for many properties in the chain at the national (and sometimes international) level. These salespeople "live" with these companies so as to get inside their culture and fully understand their needs when planning a meeting or event. This is a good example of relationship selling.

The art of selling begins, like all marketing, by understanding the needs of the customers. This is critical for a successful sale. Once again, the idea is to solve a customer's problems. First, one must determine what they are. Sometimes this calls for an interpretation. For example, a customer might say, "I need a good hotel close to the airport for my meeting." This means that one of the customer's requirements is convenience. The salesperson's property might not be near the airport, and the sale might be lost. However, offering transportation (overcoming the distance by offering convenience) might save the sale. In any event, the salesperson must determine the real problem in order to address it.

Sometimes a customer will express an opportunity, not a need. The salesperson has to understand the difference. Whereas a need is a customer want or desire that can be satisfied by the hospitality entity, an opportunity is a statement of a problem without the expressed desire to solve the problem. Let's look at the difference:

- "I need a hotel with a location near the airport." (need)
- "The hotel we use now is too far from the airport." (opportunity)

In the first statement, the problem is obvious. The first statement expresses the desire to solve the problem. The second statement is an opportunity that calls for further interpretation. Suppose the salesperson responds by telling the customer how close her hotel is to the airport. The customer then might reply: "That's nice, but it doesn't matter. My boss lives on the other side of town and doesn't want to travel all the way to the airport. I personally think the hotel we use now is out of the way, but what can you do?"

What the salesperson saw as an opportunity was not an actual need. By addressing a perceived opportunity with a benefit not important to the customer, the salesperson lost the sale. The location of the hotel was a problem for the customer, but not one that he had a desire to solve. It is not uncommon for a salesperson to sell to an opportunity, rather than to a real need; however, it is important to know the difference.

Probing

probing Asking a prospective customer questions to learn more about his or her real needs and wants.

How do you ensure that you are not selling to an opportunity? By asking the right questions. Asking the customer questions is defined as **probing.** Probing comes in two forms, open probes and closed probes. Open probes encourage customers to speak freely and to elaborate on their problems. Closed probes limit the customer response to a yes or no answer or a limited range provided by the salesperson. An example of an open probe might be "Tell me about what is important to you when you select a restaurant?" or "What is the nature of your conference?"

In both cases, the customer is encouraged to freely discuss her feelings (and hopefully reveal some needs). On the other hand, a closed probe may sound like this: "Is location important to you?" or "Do you prefer chain or independently run restaurants?" The customer can answer yes or no to the first question and has a limited option for the second question.

Let's go back to differentiating between needs and opportunities. To confirm that an opportunity is a need, a closed probe is appropriate. Let's review the selling example already presented:

> *Salesperson:* "Tell me a little bit about your meetings." (open probe)
> *Customer:* "Oh, we have had many lately, but the hotel we use now is too far from the airport." (opportunity)
> *Salesperson:* "Is an airport location important to you in choosing a hotel?" (closed probe)
> *Customer:* "Not really, my boss lives on the other side of town and doesn't want to change now."

The salesperson has avoided talking about something the customer does not need. The salesperson would then open probe further, until a customer need or a new opportunity was identified. In some cases, however, the salesperson might address the objective by further probes (why the boss won't change) and suggest a potential solution before going on to new opportunities. Once the need has surfaced, the salesperson has to support the need. Supporting is done in two stages:

1. Acknowledging the need
2. Introducing the appropriate benefits and features to the customer

Tourism Marketing Application

Helen C. Broadus of the Africa Travel Association recommends, in order to sell Africa more effectively as a destination, that salespeople ask the following probing questions of their prospects:

- Have they taken a trip to Africa before?
- How physically active are they?
- What kinds of tourism activities do they like?
- Are they interested in wildlife viewing, exploring local cultures, or educational/ historical venues?
- Would clients be comfortable on trips where they are up close and personal with the people and/or the wildlife, or do they prefer more luxurious accommodations in a relaxed atmosphere?

Source: www.africa-ata.org/ata_tips.htm (accessed April 6, 2006).

Acknowledging the need tells the customer that the salesperson understands the problem to be solved. The salesperson then introduces the solution (benefit and feature). An acknowledging statement by the salesperson might be: "I understand your need for a large ballroom, because a screen projection can take up quite a bit of room. However, there will be plenty of room for all, so no one will feel cramped or crowded."

Once the salesperson has uncovered and supported a customer need, he looks for other needs to which he can respond by further probing. When he has uncovered all the needs and responded to them, it may be time to close the sale.

Benefits and Features

Once the needs of the customer have been established through a series of probes, the customer is introduced to the product benefits and features. A feature is a tangible or intangible part of the product the customer will buy. It is a characteristic of the service being offered. It is also important to recognize those features that differentiate the product from the rest of the competition. These distinctive features should be especially emphasized if they are important to the customer.

As we discussed in the beginning of this book, the most important thing to remember is that customers do not buy features; they buy benefits. A benefit is the value of the feature to the customer and should be mentioned first to get attention. Unless the benefit is clearly explained, the customer may not understand why the feature is important. A feature might be a ballroom with high ceilings; the benefit to the customer might be that he can produce a high-tech show because the function room's high ceilings can accommodate complicated audiovisual requirements. A feature might be a good location; the benefit to the customer might be that the attendees of the meeting do not have far to drive from the office.

To encourage the potential customer to buy the product, the good salesperson will attempt to match the benefits and features to the customer's objectives and needs. Features that do not provide any benefit to the customer should be left out of the presentation. A typical mistake made in sales is not knowing the needs of the customer and presenting features and benefits of the product that are unimportant.

One of the authors was training a salesperson on the selling process and encountered the following scenario:

Salesperson: "Tell me what is important to you when you choose a hotel for your meetings." (open probe)

Customer: "I want a hotel that can handle a large checkout all at once. The last hotel we went to took one and a half hours for our guests to pay their bills!" (need)

Salesperson: "Was the problem the billing or the time to get through the line?" (closed probe)

Customer: "The time to get through the line."

Salesperson: "I can see why it is important for the group to check out effortlessly. Like most hotels now, we are able to check out the guest without the guest coming to the front desk by using video checkout. Unlike most hotels, however, we are able to provide a zero balance due on the bill we slip under their door on the last night of their stay."

In this case, the salesperson was doing fine (satisfying the needs of the customer), until she introduced a feature (video checkout) without clearly stating how the benefit worked. The customer thought that the hotel had a video camera in the lobby to record checkouts. The customer was confused as to how this would ease the checkout process. The correct presentation would have been:

> *Salesperson:* "I can see why it is important for the group to check out effortlessly. At our hotel, there is no waiting in line at all (benefit). Attendees can call up their bill on their in-room television and check out right there! (advantage). We have video checkout on the television to make leaving the hotel easier" (feature).

Once the feature was translated into the value for the customer, the customer accepted the benefit.

Salespeople must sometimes translate the features of the facility into the benefits because, although the customer may have some idea of the feature being sold, she may be skeptical, misunderstand, or have the wrong impression from previous experience. For example, a health club may not seem to be in need of translation into benefits, but the quality of health clubs in hotels differs greatly. Some "health clubs" are rooms with two or three treadmills, whereas others offer space with spas, Jacuzzis, state-of-the-art training equipment, personal instructors, and so on. If a health club is relevant to the customer, then the benefits should be explained.

The same holds true for other generic hotel features. A concierge in a Le Meridien Hotel may be sophisticated, multilingual, and resourceful. A concierge in a Ramada Inn may be the manager's secretary. Both are marketed as concierge services, with very different benefits to customers. Convention services in one hotel may consist of a staff of 10, in another, the bell person may also set the room. Translating what the features do for the customer is important in the sales process once you have determined what is important to the customer.

Handling Objections

Objection occurs when there is a problem with your product offering that cannot be changed. If the customer wants a hotel near the airport and yours is not, then you have an objection. If the customer wants a restaurant with a private dining area for a group and yours does not have one, then you have an objection. The objection that cannot be changed may be on the customer's side. For example, corporate headquarters may mandate where a meeting is to take place.

Objections are hard to overcome if one is not prepared to provide counterarguments. Think about getting your parents to agree to something that you knew in advance they might not want you to do. The strategy for overcoming the objection is to present benefits already accepted that outweigh the objection, or to learn why the objection is important and figure out an alternative. Although the customer may want an airport location, the fact that your hotel is newly renovated, has a better pool, and has a more flexible meeting space may outweigh the location objection.

A restaurant customer desiring a private room for a meal may be presented with more parking facilities, better food, and billing privileges. After all, customers make trade-offs between the products or services to choose the one that offers the best bundle of benefits. After reviewing the entire buy decision, the restaurant customer may be convinced to choose the restaurant without the private dining room. The focus, once again, should be on the objective and needs of the customer. You can overcome the objection either by providing a solution acceptable to the customer or by agreeing that it cannot be overcome and hoping that the benefits of using your facility will outweigh the disadvantages.

Closing the Sale

closing the sale Confirmation (preferably in writing) that a potential customer has now agreed to purchase the product.

The close, or signed contract, should come naturally without having to ask for it. If you have done a good job selling and if you have handled all objections and solved all the problems, the client may tell you he is ready to sign. This is commonly referred to as **closing the sale.** Asking for a close prematurely puts too much pressure on the buyer and may lose the sale. Closing a sales call entails asking the customer for a commitment. Hence, a sales call close might involve meeting the decision maker's boss on the next visit, having the customer visit the hotel, or making a presentation to the board of directors that will make the decision.

The ultimate closing is when the salesperson "asks for the business." (This entire process does not necessarily take place in one meeting.) At this point, the salesperson summarizes all of the benefits accepted by the customer and then asks for the customer's commitment.

It should be apparent that the sales approach is a difficult one at best. With many new facilities in each market, the selling process is more competitive than ever. The salesperson who has the correct selling skills and uses them on the sales call will close on a larger portion of business.

Tourism Marketing Application

Tourist destinations can be perceived poorly because media tend to report only tragic events, thus giving the impression that a whole country suffers the same calamity. Africa is one such country where this occurs. Rick Taylor, Johannesburg-based head of South Africa Tourism's National Convention Bureau and an at-large member of MPI (Meeting Planners International), stated: "Our growth strategy is to eliminate seasonality and get geographical spread, promoting the entire country—building on our asset base like a corporation or sports team. The image problem does exist. People come here with preconceived notions of starving children and wild animals walking the streets, but when they arrive they're gob-smacked by the sophisticated nature of the destination. We have to convey the message that not all of Africa fits the CNN image—that we're open for business, and the time is now. South Africa's one of the world's top value destinations. I estimate about a three-to-one value compared to the United States."

Source: (www.mpiweb.org/CMS/mpiweb/mpicontent.aspx?id2940 (accessed April 6, 2006).

Following Up

Follow-up entails regularly contacting the customer and keeping in touch after the close until the event takes place. One of the most common complaints of meeting planners in the hospitality industry is that the person who made the sale is not around when the services are actually being performed. A convention or meeting may be booked in a hotel two to five years in advance. The promises made by a salesperson who may no longer be at the property or who may be out trying to sell to another account should still be kept when the event takes place. Too often, operations says, belatedly, "We can't do that." Also, human error in incorrectly tracing a file can cause chaos. To create good word-of-mouth communications and to get customers to come back, follow-up is extremely important.

The role of the professional salesperson is evolving. The "order taker" of the 1990s has become the "order getter" in the early 2000s. Clearly the events of September 11, 2001, changed the selling scenario drastically. Suddenly there were very few buyers. Security became a need that was not discussed before the terrorist attacks. Price became a serious issue as all companies cut back spending on travel until events became more stabilized. Today, successful salespeople have to be relationship managers. This is especially important in service businesses where trust, credibility, and confidence that promises will be kept form essential parts of the relationship between the buyer and the sales representative. Of course, the working relationship of the salesperson with operations is also critical. Customers sometimes ask for the impossible. However, sometimes the impossible can be done with the right knowledge and solid working relationship within the hospitality organization itself.

SALES ACTION PLAN

The **sales action plan** sets out detailed tasks for a sales manager to accomplish over a given period of time. A typical sales action plan is shown in Exhibit 12-2. This plan includes both daily and weekly goals, sources for prospecting, and weekly action plans for the 12-week period. The sales action plan shows the salesperson clearly what needs to be done on a monthly, weekly, and daily basis.

The sales action plan is purposely formatted to accommodate 12 weeks of work. Each quarter, a new sales action plan is written to reflect the next 12-week period of work. This planning process allows the sales team to be flexible and to change activities to reflect the possibility of changing market conditions, the entry of new competitors, and so forth.

sales action plan A salesperson's plan of action for the forthcoming quarter to reach identified goals and objectives.

DEVELOPMENT OF SALES PERSONNEL

The development of personnel is critical to the success of a sales organization. If the wrong people are hired, business will be lost. If good salespeople leave to go to a competitor for better opportunities, which frequently happens, there is an additional opportunity to lose business. If a position remains open for any length of time, necessary sales

Exhibit 12-2   Sales Action Plan

Name: Mary Jones
Quarter: 3rd

- Referrals from current customers
- Newspaper leads, etc.

SALES EQUATION/SALES CALLS

PAST CUSTOMERS	+	PROSPECTING	=	GOALS
100	+	60	=	160 per month
25	+	15	=	40 per week
5	+	3	=	8 per day

BOOKING GOALS

480 per month   120 per week   30 per day

NEW ACCOUNTS OPENED

20 per month   5 per week   1 per day

PROSPECTING RESOURCES

- Meeting planner directories
- Convention and visitors bureau leads

ACTION PLAN BY WEEK

Week 1	Sales calls to Hartford
Week 2	Make appointments for trade show
Week 3	Attend trade show
Week 4	Develop direct mail for groups
Week 5	Sales calls to Boston
Week 6	Send direct mail to planners
Week 7	Local sales calls
Week 8	Local sales calls
Week 9	Follow up leads from direct mail
Week 10	Follow up leads from direct mail
Week 11	Follow up leads from direct mail
Week 12	Develop next quarter sales action plan

calls to existing and new customers will not be made. Companies need to be conscious of this problem and address the reasons for it. Through better selection and training, Ritz-Carlton reduced its sales force turnover from 40 to 10 percent once it identified the problem that salespeople well trained by other hotel chains did not necessarily fit into the Ritz-Carlton model.

Another reason for high turnover is the "move up and out" philosophy. Promotions often involve relocation. If a salesperson is unwilling to relocate, the only other way to further a career is to move to another hotel. Also, salespeople are highly visible, not only to their clients but also in networking functions. This high exposure provides increased opportunities for other positions.

Development of an effective sales staff begins with recruitment. There should be an ongoing effort to locate and know the best salespeople in the marketplace. Although new talent can be found at the college graduate level, there is still a gap at the experienced salesperson level. Organizations such as Hospitality Sales and Marketing Association International (HSMAI) are good forums for getting to know the best salespeople in an area.

Training is critical to the development of salespeople. Although there are many existing sales training programs, the challenge is to use them. At least one month of training is necessary for new salespeople to minimally learn the product specifics and understand the needs of customers in the target markets. Even seasoned salespeople need to be constantly trained through role-playing and sales meetings to keep their skills sharp. The extent of training also shows the level of commitment that the company has toward the development of individual salespeople's careers.

SALES AND OPERATIONS

"Sales sells and operations provides" is an expression that describes what is often seen as the relationship between sales and operations. As previously mentioned, that relationship is a critical one and needs further explanation here, because a conflict between sales and operations can be incredibly damaging to a hotel's relationship marketing effort. In fact, this situation represents a real need for internal relationship marketing.

Knowledge of the product and the capabilities of the organization is essential to successful selling. Constant and continuous communication between sales and operations is important for the effective marketing of hospitality and tourism products and services. If what the salesperson sells cannot be delivered, the hotel will in most cases eventually lose the customer. It is natural for a salesperson to want to make promises to close the sale.

Salespeople have two difficulties in this regard, which they need to overcome. One is perception. Because operations people see salespeople largely when they are in-house entertaining clients (for example, having lunch with customers in the restaurant or giving tours), they may see their jobs as "cushy" ones. The second difficulty is that salespeople have no direct authority over operations people, yet they need to make certain that their promises are executed properly. This can cause conflict in the lines of authority needed to get the job done and keep promises to customers.

Knowledge of the product and the capabilities of the organization will go a long way toward keeping these promises from creating unreasonable expectations for the customer. If the salesperson is not sure that the hotel can deliver, she should confirm with operations before making the promise. This not only provides a confirmation; but it also gets operations into the act so that there is more likelihood that someone will follow through. The services manager position in large hotels helps avoid these problems. This person works closely with the salesperson and also personally attends to the function when it takes place.

If both parties are truly tuned in to solving the customer's problems and each party fully understands the problems of the other, a satisfactory resolution is almost always possible. This is both internal and relationship marketing at their best. The marketing and management leadership of the property or the company sets the tone and

Tourism Marketing Application

The city of London's Tourist Board and Convention Bureau (now called Visit London) changed its business model in 2005 after attending a meeting where members were exposed to different ways of running a sales organization. Initially the team consisted of people with tourism backgrounds who were reactive rather than proactive and focused primarily on the leisure market. After receiving major funding, they now sell, market, promote, and act as the voice of the tourism industry for London. They promote London not only to Londoners and the rest of the United Kingdom, but to overseas markets as well. Instead of being only a reactive team, they split into two groups: a sales team and an account management or client services team. The sales team feeds leads to the account management team, which follows up with the actual bid. They also changed the focus of their public relations team to look after both leisure and business travelers. In terms of measuring their success, they focus on the economic benefit to the city of their efforts versus focusing on room nights driven.

Source: (www.mpiweb.org/CMS/mpiweb/mpicontent.aspx?id 2940 (accessed April 6, 2006).

should make sure customers are satisfied. The salesperson must go back to dissatisfied customers and ask for their business again.

PRINCIPLES AND PRACTICES OF SALES PROMOTIONS

sales promotions
Marketing communications that serve as incentives to stimulate sales on a short-term basis.

Sales promotions are marketing communications that serve specifically as incentives to stimulate sales on a short-term basis. In addition, sales promotions can also be used to stimulate trial purchases. In hospitality, they are frequently used to bring in business during periods of slow demand and corresponding low occupancy. In most of these cases, the lure is tied to some form of discounting or the bundling of products and service at one price that gives the perception of a price discount. Marriott's Two-For-Breakfast weekend package is an example of such bundling, as are the combination of travel, rooms, meals, sightseeing, and so forth, in one all-inclusive price. Exhibit 12-3 shows a package for Bali that includes airfare, transfers to and from the hotel to the airport, and lodging accommodations.

An example of a sales promotion in the restaurant business was when McDonald's offered reduced prices on Teenie Beanie Babies with the purchase of a Happy Meal. It is also a sales promotion when restaurants offer discount coupons or two entrees for the price of one. Sales promotion involves the development of creative ideas aimed at producing new customers or driving more frequent purchases in support of the total marketing effort. Sales promotions must be in tune with overall goals and must work with other elements of both the communications mix and the marketing mix. Sales promotions, by definition, although they should provide customer satisfaction, are not likely to build long-term customer loyalty. The only exception is when they are used to reward loyal guests.

Sales promotions are typically oriented toward the short term. If they are perceived to be offered all the time, they rarely succeed in the long term. The reason is that the promotion becomes part of the product; that is, it no longer performs as originally intended. The promotion becomes something you are forced to give customers as part of the customary transaction because the customers have come to expect the offer. In fact, they may not buy unless the offer is in place.

Web Browsing Exercise

Use your favorite web browser and search the words "restaurant promotions." What did you find? What did you learn about such promotions?

GUIDELINES FOR SALES PROMOTIONS

A sales promotions is designed to fulfill a specific marketing need. The first thing to be done when developing a sales promotion is to define that need. There are many reasons for the development of promotions. To create new business, to create awareness and to create trial purchase are typical. Other reasons are to increase demand in slow periods, to take business from the competition, or to meet the competition in its own promotional efforts. Whatever the reason, there is one major warning with regard to promotions: They should be tied to something positive such as a new or better facility, a new product, a special time or offering, or a trial.

Promotions tied to negative concerns—for instance, lack of business when it is expected to be good—tend to backfire. An example of this is restaurant two-for-one promotions offered by such companies as the Entertainment Guide. These promotions are designed to generate business by bringing in new customers. In the best situations, they succeed in doing this, but the customers they bring in may not be from the designated target market, and few of them may ever return. Although there may be a temporary increase in business, it is obtained at a cost: If food cost percentage is normally 35 percent, it is now 70 percent. At the same time, regular customers who would normally pay the full price are also dining at half price. The net gain is minimal, if not negative. As Feltenstein stated:

> The trick is to discount in such a way that you do not sabotage the integrity of your menu. Disguise the lure so that it's perceived as something other than an attempt to discount mainline items.... In the consumer's mind, there is always a correlation between product and price.... But over time, discounting is bound to raise questions in the consumer's mind about the integrity of your pricing structure.... If you must discount ... [and] there are times when discounting is a sound promotional technique—then put together a separate package to your regular offering, that will engender no recognizable negative effect on your customer's perception of the value and price of your menu ... [once] you get the customer in the store, remember it is going to take more than a cents-off coupon to bring him or her back.[2]

HOW TO CREATE SUCCESSFUL SALES PROMOTIONS

The following sections contain some general guidelines for promotions that should apply to most cases.

Take a dream vacation in exotic Bali
From $1,154*
- Round-trip Economy Class airfare on Singapore Airlines
- 5 nights at the 5-star Nusa Dua Beach Hotel & Spa, with daily breakfast
- Airport-hotel transfers
- Full-day Kintamani tour with lunch

Experience world-renowned luxury at a Four Seasons Resort
From $3,328*
- Round-trip *Executive Economy Class* airfare on Singapore Airlines
- 5 nights in a one-bedroom villa at the Four Seasons–Jimbaran Bay or Four Seasons–Sayan, with daily breakfast
- Airport-hotel transfers

Extend your trip – visit an *Ubud Hideaway*
Add $495**
- 3 nights in a poolside villa at the new Royal Pita Maha Resort
- Transfers to/from Ubud

1-888-742-3443
www.asianaffairholidays.com

*Rate is per person, double occupancy, based on departure from LAX. Valid for travel from 1/6/06 – 3/31/06. Price does not include government mandated taxes and fees, including international departure tax, customs user fees, $2.50 Sept. 11th Security Fee and passenger facility charges, all totaling up to $72. Fares subject to change without notice. **Ubud Hideaway extension must be purchased in conjunction with the Bali Deluxe or Four Seasons package.

Exhibit 12-3   An advertisment that offers an all-inclusive package

Source: Singapore Airlines. Used by permission.

Exhibit 12-4 This advertisement for Singapore Airlines is an example of single-mindedness

Source: Singapore Airlines. Used by permission.

Be Single-Minded

It is well to remain single-minded and not try to accomplish too many things at one time. Is the purpose to create new business, awareness, or trial? To increase demand during a slow period? To take away business from the competition? To compete head-on with the competition? Or to sell specialty items? Trying to do more than one or two of these things tends to diffuse the promotion, confuse the market, and accomplish none of them. Exhibit 12-4 is an example of an advertisement that reflects single-mindedness. Specifically, it focuses on the comfort of economy class.

Define the Target Market

Is the identified target market first-time users, heavy users, nonusers? What benefits do they seek? What are their demographic and psychographic characteristics? The promotion must specifically focus on the needs of the target market you are trying to reach.

Decide Specifically What You Want to Promote

What you want to promote is not necessarily the tangible item that you may be promoting. For example, you may want to promote a new decor or atmosphere in the lounge, but the tangible item could be a special drink. A hotel might want to promote its guest rooms on weekends; the tangible promotional feature could be free breakfast in bed with champagne.

Decide on the Best Way to Promote It

It is not always necessary to give something away or to charge a lower price when creating promotions. The alternative is typically referred to as a premium program. Rather than emphasize discounts, premium programs focus on the addition of a specific "premium" that accompanies the offer. An example might be to include a $100 food and beverage credit in the purchase of an upscale weekend package. Another example might be to include complimentary high-speed Internet access as part of a nondiscounted guest room rate for business travelers.

Make Sure You Can Fulfill the Demand

This is a critical point. Many customers are lost forever—the opposite intention of the promotion—by failure to deliver on the promotion. If you are promoting lobster dinner specials, don't run out of lobsters even at the risk of having to let some spoil. If you're offering weekend packages, provide the rooms even if you have to upgrade. The worst thing that can happen is that you'll have a happier customer. At a minimum, do as the retail stores do and provide rain checks that customers can use at a later date; then, when customers collect on them, give something better, just to compensate for the inconvenience. All too often promotions lose customers rather than winning them, because management forgets why it is having the promotion in the first place.

Communicate Clearly All Related Aspects of the Promotion

Some promotional literature or advertisements are confusing, hide the conditions of the promotion in fine print at the bottom of an advertisement, or assume that customer's have so much knowledge that they ignore the advertisement or get irritated. Specify clearly all important information about the promotion, such as price, quality, procedure, place, dates, time, and any other necessary details. When customers ask for "it," give it to them; don't play games and say, "There are no more of those left." If you are promoting "children free in the same room," do not hassle with the customer as to whether the child is under 16 or charge $15 for a cot—both create no-win situations.

Communicate the Promotion to Your Employees

Employee knowledge of a special promotion is critical. Too often promotions break down over failure to do this. Management may run an advertisement in the local paper about a promotion and assume that the employees who have to execute the promotion will know exactly what it's all about and how to handle it. A restaurant we know once ran a promotion of free movie tickets with certain special dinners. None of the waiters or waitresses knew anything about it. When diners asked for their tickets, management was "out of town." The result was a disaster. Guest service agents at hotels have the same problem with weekend specials, not to mention reservationists at toll-free 800 numbers, who often don't know what they are supposed to be promoting.

Measure the Results

Be sure to measure the results of the promotion, and not just in terms of bodies or dollars. Did the promotion meet its objectives? What were the benefits, gains, losses? Specifically, what was the total cost relative to the actual return? Did the promotion at least pay for itself? Will it work again? If it did not work, why not? Will there be a lasting effect, or was it a one-shot deal? Some of the best promotional results are nothing more than goodwill, which will pay off in the future.

DEVELOPING SALES PROMOTIONS

The use of sales promotions in the hospitality industry centers on the creation of demand. A promotion is the development and execution of an event outside the normal day-to-day business. The goal of a sales promotion is twofold: to increase the satisfaction of the guest while increasing revenues for the hospitality establishment. If the guest is extremely satisfied with a promotion but the costs are so high that money is lost, then the promotion is unsuccessful. On the other hand, if the hotel or restaurant makes a great deal of money but the customer feels slighted, then the promotion is equally unsuccessful.

Normally, there are two types of sales promotions, those centered on established events and those created entirely on their own. A promotion created around an established event might be a Mother's Day brunch, a Bastille Day food offering, a hotel package for Valentine's Day, Christmas shopping, or, in the case of casinos, a slot tournament or a blackjack tournament. In these cases, hospitality establishments have an opportunity to create excitement for customers and to build volume. Participation can vary from flying in a French chef to cook for Bastille Day to placing a corned beef sandwich on the menu for St. Patrick's Day.

Web Browsing Exercise

Using your favorite web browser and search engine, type in the words "sales promotion" What is listed? What can you learn from this listing? Choose one of the listings and be prepared to discuss it in class.

The second type of sales promotion—created independent of an established event—is more difficult to develop and execute. Good examples of creative sales promotions reported in the industry press are shown in Exhibit 12-5.

Identify the Gap

One purpose of a sales promotion from the perspective of management is to increase revenues. Many promotions are designed to build revenues during slack times or sell products that are traditionally in low demand.

The New York Palace completed an extensive renovation in 1997 and repositioned itself at the top of the New York City luxury market. Specifically, the Towers section of the hotel competed head on with the established St. Regis and Four Seasons hotels. Although the physical product was five-star quality, the customers' perception of the hotel remained confused. Customer research indicated that the St. Regis and Four Seasons were well liked by travelers, and changing their behavior

Exhibit 12-5	Some Uncommon Promotional Ideas
Kids Rule	The 205-room hotel in New York City, 70 Park Avenue, offers a special package for all families. This package offers a balance of activities for both the parents and the children. The children are given cooking/baking lessons with a member of the Silverleaf Tavern's culinary team, while the parents receive a two-hour chauffer-driven limo for a shopping spree. That evening the children enjoy an in-room pajama party with pizza, ice cream, and a babysitter while the parents enjoy a romantic dinner at Silverleaf Tavern.
It's a Wonderful Life Package	The Hotel Burnham, located in Chicago, offers a special Christmas time package during the slow Christmas season, between November 1 and December 31. This offer includes a two-night stay in one of our Executive Suites, luxury town car service for up to four hours, Christmas stockings with $50 gift certificates from Neiman Marcus, Nordstrom and Marshall Field's, and a complimentary copy of the film, *It's a Wonderful Life*. In addition, guests will indulge in hot toddies or hot chocolate the evening they arrive. And on the second night of their stay, they will enjoy a one-hour, in-room massage.
No Monkey Business Package	The Onyx Hotel in Boston understands how important a good night's sleep and excellent technical support is to the business traveler, They have created a special program to encourage business travelers to stay at their hotel. They offer 24-hour on-call technical support, executive car service to the financial district, in-room printing services, and a banana upon arrival to energize the business traveler.
So Hip It Hurts	The Hotel Triton in San Francisco focuses on giving their guests one-of-a-kind experiences. They have developed a tattoo and piercing package that gives the guest a $65 credit to get a tattoo or piercing at Mom's Body Shop, and the general manager will escort you to the parlor with smelling salt in case you faint. This program has helped Hotel Triton establish itself as a trendy place to stay.
Downtown Divas	The Alexis Hotel in Seattle, Washington, understand the needs of the woman traveler. To increase their business among women, they have created a program centered completely around the woman traveler. This program offers a guest room for two, a makeup consultation with an Aveda consultant, a $50 perfume credit at Nasreen, a Nordstrom personal shopper at your service, and a pint of Haagan Dazs ice cream at turn-down.
PAWS Package	The Hotel Monaco chain has a pet friendly policy. They encourage guests to bring their dogs and cats with them on vacation. They provide a VIP check-in for all pets, a welcome letter for your dog or cat, a puppy bed and bowl, a puppy bone for your dogs at turn-down, and puppy massages for your pets. If you are not able to bring your pet with you on your trip, the hotel will provide a complimentary goldfish for your stay.

Source: Retrieved November 6, 2005, from www.70parkave.com; www.burnhamhotel.com; www.onyxhotel.com; www.hoteltriton.com; www.alexishotel.com; and www.hotelmonaco.com.

would prove difficult. The problem was that the competition was doing a good job and there were no real reasons for their customers to look elsewhere. Palace management was convinced, however, that once customers tried the Towers they would return. Getting trial usage became the marketing problem to be solved. A sales promotion was created to give the customer a reason to try the Towers. The promotion offered a round-trip airline ticket on Delta Airlines to anyone staying in the Towers for one night.

The tickets were purchased through a special program offered by Delta for $249. Rates in the Towers began at $525. With an average stay of two nights, each booking paid for the sales promotion. Future stays created incremental revenue and guest loyalty.

Design the Sales Promotion

One important aspect in the design of a promotion is the customer. Normally, the customer should be considered before putting any type of promotion together. However, management might design a promotion because of excess inventory. For example, perhaps some wine was bought in too large a quantity and needs to be sold. A wine promotion is created, regardless of the needs of a customer, but the promotion itself is designed to satisfy the property's needs. The promotion must be consistent with the positioning of the restaurant or hotel. A disco promotion at the Ritz is not in keeping with the positioning of the hotel. Similarly, a caviar promotion at a family restaurant is equally inappropriate.

The second important aspect in the design of the promotion is the timing and planning. It takes time to plan a promotion. Promotional materials need to be prepared, and supplies need to be ordered to guarantee that when guests arrive at the facility the promotional item is available and employees are aware of the promotion.

The proper delivery of a promotion includes the use of a variety of items in the communications mix. Advertising, merchandising, and public relations all need time to be coordinated. Promotions that do not have the proper timing and planning are usually failures.

Throughout the design of the promotion, a clear and concise message must be put forth to the customer. This may not be as necessary for promotions centered on established events, but promotions that are attempting to present a new concept must be clear. A St. Patrick's Day promotion can be easily understood by most customers because the event is expected; however, a novel promotion may have to be explained to customers and all employees.

Analyze the Competition

Competition should be analyzed before a sales promotion is developed. If all of the restaurants in town are offering a turkey dinner for Thanksgiving, what will make this promotion different? A close watch on competitive activity can give the promotion designer a head start on potential problems.

Establish Goals

How should success be judged? If a sales promotion is to satisfy both the customer and management, how many extra rooms, covers, or cases of wine can be reasonably expected to sell? What level of sales must be achieved simply to cover the cost of the promotion? Goals should be set in advance for evaluating the promotion at the conclusion of the event. Goals also need to be realistic, and a measurement form should exist before the promotion takes place.

Execute the Sales Promotion

Execution includes delivery of the product to the customer in the framework of the created expectation. Promotion delivery is more important than normal delivery because the customer is excited. The promotion has created a demand. Demand has created a special reason to use the product, and customer expectations are unusually high.

Proper execution includes employee participation. The entire staff needs to understand the promotion and its specific involvement. When a bartender shows up for work in the middle of an Oktoberfest without knowing the service steps involved, problems can occur. Employee involvement, perhaps even in the design stages of the promotion, will increase the chances for better delivery of the correct product.

Execution also means maintaining the proper inventory of goods to be sold. If the restaurant runs out of bratwurst during the Oktoberfest and has to substitute hamburgers, the customers' expectations will not be met. Part of the planning process of the promotion is the development of goals. Purchasing should be based on the attainment of these goals, at a minimum. It is more desirable to have some waste than to not fulfill expectations.

Evaluate the Sales Promotion

All sales promotions should have an evaluation mechanism installed that includes asking the following questions:

- Were the goals met?
- Would the customers have come anyway?
- Did it generate revenues sufficient to cover the costs?

Although these questions are certainly relevant and necessary, they constitute only half the equation for success. The second half consists of the following questions:

- Were the customers satisfied?
- Were there any unusual complaints?
- Do comments reflect any information that might be useful for future promotions?

All of these questions should be addressed in the evaluation process to allow a total assessment of the event.

A good and true evaluation of a promotion is one of the most difficult analyses to undertake. Judd Goldfeder of the Customer Connection, a firm specializing in restaurant loyalty programs, stated:

> While we try to do all we can to objectively and accurately measure the sales generated by a frequent diner program, no analysis can provide absolute evidence that any program produces a definitive amount of incremental sales. Therefore, the best we can do is make some subjective assumptions, temper them with common sense and good business judgment and reach a "comfort zone" regarding what portions of sales were generated as a direct result of the program versus guest patronage that would have occurred anyway.[3]

Despite the preceding comments, there are two tests management can undertake to help understand whether "customers would have come anyway." Exhibit 12-6 shows one such test, and Exhibit 12-7 shows a second.

The first test shown is a traditional pretest/posttest with a control group. In this test, the target population is divided into two similar groups based on such characteristics as buying behavior, loyalty toward the firm, and the like. Notice that in Exhibit 12-6, groups are determined by their **RFM** (recency, frequency, monetary value). These are the rows in the table. Each group is then split into a test group and a control group. The test group receives the sales promotion offer, while the second group receives nothing. The purchase behavior of the two groups is tracked after the promotion. To be successful, the purchase behavior of the test group should be much higher than the purchase behavior of the control group. If this is the case, then the promotion affected behavior. If the purchase behavior is the same for the two groups, then the promotion had no impact. Obviously, the difference between the two groups should exceed the cost of the promotion. The calculation appears in the Exhibit 12-6.

A second way to judge the success of a promotion is to use what is called "they would have come anyway analysis." In this analysis, management makes basic assumptions about the percentage of customers who came because of the promotion and those who would have come anyway. By examining the percentages of each group and the cost of the promotion, management can gain an estimate of whether the promotion was a success. Obviously, this analysis can be undertaken before the promotion begins. This analysis is shown in Exhibit 12-7. When all feedback has been analyzed, the final stage is formulating the next promotion. What other promotions can be developed to fill in gap periods or to sell slower-moving products or time periods? The process of promotional development begins all over again.

Exhibit 12-8 provides ideas for restaurant promotions from the Quantified Marketing Group, a restaurant consulting company.

An honest effort to assess the success of a promotion will help in the development of future promotions.

RFM An acronym for recency (How recently have you been to the establishment?), frequency (How often do you visit the establishment?), and monetary value (How much are you worth to the organization?). RFM is used in database marketing to understand customers and to show that some customers are worth more to the firm than others.

Exhibit 12-6	Tests to Determine the Success of Promotions

BASIC FORM OF PREPROMOTION/POSTPROMOTION WITH CONTROL GROUP

Code	Group	Number of Visits
01	Prepromotion test group (e.g., receives promotion)	5.3
02	Postpromotion test group (e.g., receives promotion)	7.8
03	Prepromotion control group (e.g., does not receive promotion)	5.4
04	Postpromotion control group (e.g., does not receive promotion)	5.6

Effect of the promotion = (02–01) − (04–03)

(7.8–5.3) − (5.6–5.4)

(2.5–0.2)

2.3 increase in number of visits because of promotion

POPULATION DIVIDED BY RFM (ROWS) AND TEST CONTROL GROUP (COLUMNS)

Count of Account	R (Recency)	F (Frequency)	M (Monetary Value)	Test Group	Control Group
90	5	5	3	45	45
99	5	5	4	50	49
106	5	5	2	53	53
123	5	5	5	62	61
281	5	5	1	140	141
Subtotal: 699				350	349
105	5	4	1	53	51
135	5	4	2	66	68
139	5	4	3	70	69
154	5	4	4	77	77
167	5	4	5	84	85
Subtotal: 700				350	350
152	5	3	5	76	76
156	5	3	4	78	78
146	5	3	3	73	73
159	5	3	2	80	79
86	5	5	1	43	43
Subtotal: 699				350	349
125	5	2	5	63	62
145	5	2	4	73	72
160	5	2	3	80	80
148	5	2	2	74	74
122	5	2	1	61	61
Subtotal: 700				351	349
132	5	1	5	66	66
146	5	1	4	73	73
164	5	1	3	82	82
152	5	1	2	76	76
106	5	1	1	53	53
Subtotal: 700				350	350
Total: 3,498				1,749	1,749

Chapter Summary

The sales process is becoming more complex in the competitive marketplace of today. The selling process is not unlike the marketing process; understanding the needs of the customer is the primary focus of the sales organization and is the foundation of effective selling. The selling process involves the skilled use of probing, supporting, and closing to manage the sale. Overcoming objections is typically part of this process, too. All three need to be handled with professional selling skills.

Exhibit 12-7	They Would Have Come Anyway Analysis*	
% Who Would Have Come Anyway	Total Sales of Those Who Would Have Come Anyway	Incremental Revenue Less Cost of Program
5%	$69,720	$1,263,688
	(.05 × 1,394,408)	(.95 × 1,394,408–$61,000)
10%	$139,441	$1,193,967
15%	$209,161	$1,124,247
20%	$278,882	$1,054,526
25%	$348,602	$984,806
:**		
50%	$697,204	$636,204
:		
75%	$1,045,806	$287,602
:		
90%	$1,254,967	$78,441
95%	$1,324,688	$8,720
100%	$1,394,408	$−61,000

*Cost of the promotion is $61,000, and total revenue of time period is $1,394,408.
**The ":" indicates that numbers no longer go in 5% increments. Obviously, a spreadsheet could be created for each percentage from 0 to 100.

Exhibit 12-8	Restaurant Promotions from the Quantified Marketing Group

10 TACTICS FOR DRIVING F&B SALES

Restaurant Promotions Tactic 1: Publicity Stunts

Stunt is a word with negative connotations for restaurant owners, but I wanted to use a word that conjured up images that are different than traditional press relations efforts. Sending a standard press release about a new menu may result in a small write-up. To cut through the clutter and generate extensive exposure, you need a newsworthy angle. Something like a celebrity chef cook-off, really unique contest or other major event. Think beyond typical events like golf tournaments and simple fundraisers. Challenge your staff or marketing firm to think what you'd have to do to make it into the Guinness Book of World Records. Challenge them to think much bigger and come up with ideas that tie in to what your club stands for but also have potential for national exposure. If you create events that have only local appeal, you'll be limited with your media exposure potential and may not even make the local paper. If you think much larger, you won't have to worry about getting coverage. A well-constructed publicity stunt can be worth its weight in gold in terms of positive exposure for your restaurant. And everybody wants to be associated with a winner.

Restaurant Promotions Tactic 2: Public Relations

Public relations has been called advertising that you don't have to pay for. If you have a successful public and media relations program, you'll get increased exposure and prestige without spending a fortune. For this to work, though, you'll need to create and publicize newsworthy stories. Hiring a new chef isn't always enough to garner the kind of attention you deserve. Create other angles that are unique and make your restaurant stand out. Also, review your restaurant's marketing and advertising expenses over the last three years. Then determine the percentage that was spent on traditional advertising compared to public relations. It's worthwhile to spend 15–30 percent of your budget on a solid public relations program. Find a firm that has creativity and excitement about your restaurant. If that firm doesn't seem genuinely curious and interested in your restaurant and what it has to offer, it'll have a hard time creating interest with the media.

Some higher-end restaurants are understandably concerned about publicity stunts and other marketing activities that seem to fly in the face of the exclusivity of their establishment. My answer to that is simple—these tactics won't be appropriate for everyone. That being said, if you are one of the restaurant owners that cringes at the thought of creating buzz in the community at large, I urge you to think about your position.

Everyone wants to be associated with a winner. For some of your regulars the whole reason they belong in the first place is because it's exclusive and their being a part of that is an extension of their self-brand and identity. Creating buzz won't distract from that; it will reinforce it in many cases. The key is how the publicity comes across. If done correctly, it supports your position in the market, exclusivity and prestige.

Restaurant Promotions Tactic 3: Bouncebacks

This is an underutilized tool that bounces guests from peak times to off-peak times and can also work to encourage frequency in your food and beverage operations. While simple in theory and execution, this tactic can produce far more in revenues per dollar invested than traditional advertising. All you do is offer incentives at the point of purchase on popular services to encourage the guest to try your restaurant another time. For instance, if you're busy for lunch and need to drive sales for dinner, offer bounce-back certificates that can only be redeemed during dinner hours. Test different offers and delivery vehicles and track response rates for each to hone in on what works best with your clientele.

Exhibit 12-8 Continued

Restaurant Promotions Tactic 4: Stop Discounting

Discounting tells your customers and prospective customers. "We don't deserve full price, so we'll be happy to lower our rates to make up for the difference." This point was driven home to me during my tenure with The Breakers of Palm Beach, a lavish resort whose guests spend a small fortune to walk the halls. Discounting the price would be to discount the 105 years spent building a brand. Instead of discounting, consider no strings offers that do not rely on percentages. Examples include value-added perks such as free valet parking, complimentary services, merchandise, etc. And, in a related topic, never offer coupons, only offer certificates. There is a big difference in perception.

Restaurant Promotions Tactic 5: Business Socials

A no brainer, right? Well, you'd be surprised how unreceptive or apathetic some restaurant owners are to hosting business socials with outside organizations at their establishment. However, if you select the right group to partner with, you can leverage their resources to promote your restaurant, and you can also target your core audience. Host socials where the food is center stage. Arrange photo opportunities that include your displays in the background and submit to local media. Partnering with a business or charitable organization works on many levels and can help you stretch your marketing budget while still delivering higher returns on investment than can be achieved with traditional advertising.

Restaurant Promotions Tactic 6: Sampling

Tasting is believing and if you would grade your food a B minus or above, you need to get it in potential customers' mouths. That's the best way to build recognition and it is more effective and less expensive than advertising. Every public event that draws your core audience is an opportunity to offer samples of your product. Pick the best 2–3 items on your menu that can be easily transported and get some solid representatives of your restaurant out to meet and greet at these off-property functions

Restaurant Promotions Tactic 7: Host Food Events

Hosting food events such as the "Taste of (insert your town)" is a great way to position your restaurant as a center of the food scene in your market. It allows you to leverage the reputation, profile and credibility of all of the other participants, and it can also help you share the expense of holding the event. Hosting an event also provides your restaurant with the opportunity to recruit additional manpower and resources for promoting the event and gives that added edge with garnering local publicity.

Restaurant Promotions Tactic 8: Toss Up Tuesdays

Promote this program through your next newsletter and other internal marketing vehicles to your existing customer base. Pick Tuesdays (or your slowest food day) and flip for the food tab. Guests will have a 50 percent chance of getting their food bill paid by the restaurant. This attracts your guests' attention much more than a "buy one get one free" restaurant promotion. Guests are also more likely to have higher check averages than normal because there is a chance they won't have to pay. It creates a tremendous attention among your core guest base.

Restaurant Promotions Tactic 9: Menu Bingo

This is a great tactic for encouraging frequency and getting members to try different items on the menu. You simply create bingo cards that have different menu items in boxes. Have the cards designed with five columns and five rows. You can also promote other non-food items such as merchandise, cookbooks, and gift certificates. Guests have an allotted period of time—60 days for example—to complete a connection just as they would with a bingo card. Once they try five items in any direction, they receive a free gift basket or other incentive that is roughly equal to one of the items purchased.

Restaurant Promotions Tactic 10: Birthday Program

Research shows that 50 percent of all Americans eat out on their birthday. This presents an opportunity for establishments with solid birthday programs. So why don't restaurateurs do more to take advantage of this? You've got me, but it does offer a chance for you to swoop in and capture your increased share of the market. A birthday program can be executed through new automated tools like those that are available through e-mail marketing service providers. You simply plug in the birthday and e-mail address of your members, and a secure and nicely designed e-mail is sent to them at a time you determine in advance. The system knows who and when to send the e-mail to and also tracks view rates for reporting that allows you to know how well your program is working. You can also have the e-mail include a redemption code that will allow you to track what percentage of the e-mails are bringing in guests and calculate a return on investment. Recent research has shown that retention based e-mail marketing is 300 to 400 percent higher than traditional vehicles such as direct mail and faxes. It's a great way to communicate and manage your club's birthday program.

The restaurant industry has been conditioned to believe that only traditional marketing efforts can be applied to grow sales because it's what everyone else is doing. Fact is, the restaurant industry is getting more competitive and will continue to do so. In the face of increased competition, the most effective strategy is to differentiate your restaurant from the others and create excitement in a way that reinforces your positioning strategy. Again, restaurant promotions are only gimmicky if they are created that way; it is entirely possible to execute these restaurant promotions in a way that is completely in alignment with the image of your restaurant no matter how exclusive.

Remember, differentiation and exciting tactics like the ones described here are particularly potent for your food and beverage operations.

Smart marketing is best achieved through non-traditional techniques that are executed inside your restaurant and among your existing customer base. Opportunities abound if you look at your situation through the right lens. Use the preceding ideas to spark your own thinking of similar underutilized programs in your own operation and reap the rewards as other successful restaurants are around the country.

Source: Adapted from the work of Aaron D. Allen, founder/CEO of Quantified Marketing Group (www.quantifiedmarketing.com).

Having the ability to sell is only half of the selling process. Planning the sales function is also important. Tools such as an account management system and a sales action plan assist the sales manager in focusing on what is important. Maintaining a balance of resources to call on past, present, and new customers is critical to the success of the entire organization.

The sales personnel have to be knowledgeable and consistent about their goals and be prepared to sell the customer with appropriate features and benefits. The sales office that carefully organizes the sales team and develops and motivates its people effectively will be the most productive. Sales and operations need to continuously work together to keep the customer coming back.

Sales promotions are special promotions that serve as incentives to stimulate sales on a short-term basis. They should be developed with reference to the interests of specific target markets, and employees as well as customers need to be aware of the product or service being promoted. Different products are promoted in different ways, and results should always be measured in some way to determine the success of the promotion.

Key Terms

closing the sale, p. 392

personal selling, p. 382

probing, p. 388

prospecting, p. 384

qualifying (a prospect), p. 386

RFM, p. 404

sales account managers, p. 388

sales action plan, p. 393

sales leads, p. 386

sales manager, See *Salesperson.* p. 382

sales promotions, p. 396

salesperson, p. 382

sales representative, See *Salesperson.* p. 382

Discussion Questions

1. Taking time to review the marketing communications mix, briefly describe different ways to reach out to current and potential new customers for hospitality and tourism product offerings.

2. Why is personal selling a good communications tool for the more complex purchases such as meetings, conventions, banquets, and group tour sales?

3. What is the personal selling sales process all about? What are the steps involved in this process?

4. Describe in your own words what is meant by sales management. What is being "managed"?

5. Why are sales promotions an important part of the marketing communications mix?

6. What makes a good sales promotion?

Endnotes

1. For a positive view on cold calling and how to do it successfully, see Gitomer, J. H. (1994). *The sales bible.* New York: William Morrow, 94–108. Gitomer's recommended opening line is, "Can you help me?"
2. Feltenstein, T. (1987, November 9). How to discount your product without sabotaging your image. *Nation's Restaurant News*, F20.
3. Personal correspondence with Judd Goldfeder.

.... case study

Castle Spa

Discussion Questions

1. In your own words describe the vision and mission of Castle Spa.
2. What is the definition of a spa?
3. Who are the primary, secondary, and tertiary markets for Castle Spa? Do you feel these markets as described in the case are well defined?
4. Briefly describe the Castle Spa pricing strategy? In your opinion is this a good pricing strategy? Is there any room for improvement?
5. How would you assess the competitive environment in which Castle Spa operates?
6. Should Sonja consider personal selling (which is somewhat expensive) as part of her communications mix strategy to reach out to the incoming residents of the new multimillion-dollar luxury condominium? What do you think of her idea to focus on sales promotions to boost nonpeak period sales in this new potential market?
7. How can Sonja go about building her loyal customer base? Is there evidence in the case that Sonja truly understands who her customer base really is?

Sonja Drury, CEO and owner of Castle Spa Inc., sat in Goody's restaurant in late September sipping a ginger tea and staring out the expansive floor-to-ceiling windows onto the historic Downwind market of downtown Vancouver. Directly across the street, the framework of a new multimillion-dollar luxury condominium complex was beginning to take shape.

In six months' time, the condominium complex was due to be completed, and Sonja wanted to make sure that Castle Spa captured this market before its competition, the Devon Spa, which was located a mere three blocks away.

In anticipation of an influx of new clients that match Castle Spa's target market, the spa management team was busily brainstorming ways to make sure the experience that these potential new clients had on their first visit would ensure that they would become loyal clients in the future. But how to ensure that their needs were best met? How might their booking demand differ from Castle's current client base, given their proximity to the Devon Spa and unusually high income bracket?

Sonja was concerned the condominium clients were likely to increase sales in periods that were already booked, such as high-demand Saturdays and evenings in the latter part of the week. If this were the case, all spa clients were likely to become frustrated with overcrowded lounges, a decrease in service quality due to service providers having to rush through treatments to accommodate back-to-back bookings, and an overall decline in the ambiance of relaxation and personalized service that Castle strived to uphold. How could Castle Spa develop a marketing strategy to maneuver the condo business into nonpeak periods, to ensure

This case was contributed by Ana Yuristy and Cheryl Pylypiuk, and developed under the supervision of Margaret Shaw, PhD, School of Hospitality and Tourism Management, University of Guelph, Ontario, Canada. Names and places have been disguised. All monetary figures are in Canadian dollars. For conversion, use 1$C/0.85$U.S. All rights reserved. Used by permission.

that their experience upheld the five-star quality that kept clients coming back to Castle? Sonja felt the beginnings of a headache coming on with the thought of all that was at stake. Perhaps she needed a massage herself.

COMPANY BACKGROUND

Castle Spa was located in downtown Vancouver in the Downwind market area, adjacent to many major office buildings. It began as a private fitness club (the Essex Club) in 1987 and had been transformed over the years into a five-star health and beauty spa. Castle Spa was 5,000 square feet spread over two floors and employed over 60 full- and part-time esthetics staff, massage therapists, and hair stylists/colorists. Castle Spa's mission was to be on the cutting edge of new treatments and technologies, providing world-class service to its clients. The owner also owned Goody's, an Asian-fusion restaurant located on the reception floor of the same building, which operated under separate management.

COMPANY VISION

Castle Spa was committed to nurturing the physical, mental, and spiritual well-being of its guests (See Exhibit 1). The management team, under the vision of Sonja Drury, owner and CEO, actively sought out new trends in the world health market. As a consequence, Castle Spa treatments and services were at the cutting edge. This required a top-notch external communication system, as clients had to be educated about the benefits of the products and services. For example, Castle Spa was the first spa in Vancouver to bring in Thai- and Asian-influenced body treatments; was the first to offer now common treatments such as Arizona Hot Stone Massage and Shiatsu Massage; and was now promoting detox treatments, adding detoxifying algae wraps and body scrubs to its menu of services. It had even incorporated a line of herbal and mineral supplements in its retail line. This required a knowledgeable reception staff as well as esthetics, body treatment, and massage staff.

INDUSTRY PROFILE

By definition, a spa is a curative mineral spring. Therapeutic spa treatments, using products of the sea, were practiced by the Romans in the south of France as early as 450 B.C. Throughout the ages cultures have used nature's resources to treat various ailments of the skin, body, and mind. Europeans, Asians, and North American native populations soaked in thermal springs and mud pools. The spa, or use of nature's elements to invoke healing and health maintenance, had been a part of every culture for centuries.

In recent years, interest in spas and the services and the experiences they offer had been significant (see Exhibit 2). The cultural motivation to visit spas had changed from mere curiosity to need. The spa industry had grown as well. Suddenly spas were opening on many city corners. Establishments that had hot tubs or tanning beds or that offered manicures and pedicures were starting to identify themselves as spas.

Exhibit 1   Castle's Philosophy: The Essentials to Health and Wellness

PHILOSOPHY OF BALANCE

Stepping in from the heart of historic downtown Vancouver, your journey toward serenity begins. The spa's natural elegance both relaxes and inspires.

Our team of professionals is committed to nurturing the physical, mental, and spiritual well-being of our guests, utilizing a unique blend of the finest products, spa therapies, and exotic healing traditions from around the world.

HEALING RETREAT

Sensuous, calming, and fragrant treatment rooms, where a soothing massage, luxurious body treatment, or signature skin care are performed with the utmost intimacy. All our spa guests are invited to relax in our beautiful Orchid Lounge,

experience a rejuvenating eucalyptus steam, or enjoy the tantalizing cuisine offered by our own Goody's restaurant.

A European edge on innovative spa treatments, the purest natural products available, and highly trained and certified therapists. Castle Spa has all the essentials to health and wellness, and our goal is to exceed your expectations.

Experience excellence
Spa boutique
Orchid Lounge
Turkish hummum
Massage clinic
Anti-aging services
Medical/advanced skin care

SPA SERVICES

Atmosphere

Castle Spa, by design, epitomized luxury and relaxation by appealing to all the senses. Candles, aromatherapy essences, and orchids imported directly from Thailand were placed throughout the spa to help create an atmosphere of serenity and a relaxing and exotic escape from the stresses of the outside world (see Exhibit 3). All guests, whether they were there simply for a manicure or for a full day of relaxation, were offered coffee or herbal tea by the receptionist upon arrival and were escorted to the nearest lounge while they awaited their treatment. Guests to the spa had use of the lockers and shower facilities and were offered the opportunity to change into plush Castle Spa bathrobes for optimal comfort.

Spa guests were encouraged to order food and beverages from Goody's restaurant during their visit. Champagne, fruit smoothies, and appetizers could be enjoyed in any of the lounges, or even during treatments (depending on the treatment type).

Services and Packages

Spa services (See Exhibit 4) ranged in price from $35 for a basic Signature manicure to massage, advanced facial, and body treatments in the $100 to $200 range. Castle Spa also had a number of price-bundled packages. The holiday season was a popular time for gift certificate purchases, and customers could choose from among a dozen or more "Spa Getaway" certificates, including the Thai Spa Getaway and the Supreme Day at the Spa—a full day at the spa that included a catered lunch.

Exhibit 2   International Spa Association 2001/2002 Industry Study

With all the attention focused these days on holistic remedies, organic produce, and botanical beauty products, it's no surprise that interest in spa treatments was on the rise. Everyone from harried executives to soccer moms—and baby boomers in particular—is eager to try anything that keeps them looking younger and feeling better. That makes this the perfect time to take the plunge into one of the hottest personal-service businesses around: the day spa. Day spas offer the same beauty and wellness services as pricier destination spas and resorts but don't require the same time commitment. According to the ISPA 2002 Spa Industry Study from the International SPA Association (ISPA), there were nearly 156 million spa visits in 2001, 68 percent of which were made to day spas. Revenues for the U.S. spa industry were nearly $11 billion in 2001, up from $5 billion two years earlier.

ISPA's 2001 Day Spa Usage Survey indicates that two of the top five reasons people don't visit a day spa were that they think spas were too costly, and they feel they're not the "spa type." According to the ISPA survey, white-collar professionals under age 45 who have college degrees were the early adopters of spa services in many Canadian markets.

"You also have to educate people about your services so they don't think of them as a luxury." People feel guilty about pampering themselves, so instead, we position ourselves as providers of healthy living services."

Source: Adapted from Sandlin, Eileen, Great escape . . . Day spas. *Entrepreneur,* December 2003.

Exhibit 3   Traditional Swedish Massage

The second floor of the two-floor spa was accessible only by keyed elevator access or stairs through the women's changing room. Besides 10 treatment rooms, a "body-cocooning" hydrotherapy bed, a couple's (tandem) massage room, and a manicure/pedicure room that could comfortably accommodate 10 clients at one time, the second floor of the spa also had three private lounges.

The Hummum Lounge, which was modeled after a Turkish hummum, was the smallest of the three lounges. It was decorated in vibrant reds and oranges and bold blues, and doubled as a private couples manicure/pedicure room. Its lush pillowed benches offered guests an opportunity to lie back and relax between services. It also was the location of the spa's oxygen therapy tank.

The Bamboo Lounge was the first point of entry for spa guests and was decorated in muted green and blue tones. This lounge offered a selection of current magazines, and guests were encouraged to enjoy the lounge either before or after their services for as long as they wished.

The third, and largest, lounge was the Orchid Lounge, which offered a lovely view of downtown Vancouver. It was decorated in white and gold and included a fireplace and private dining area for guests who were on a full-day package.

Exhibit 4	Castle Spa Price-Bundled Packages and Prices

Supreme Day at the Spa—$295

European facial, aromatherapy massage, and essential oil steam. Signature manicure and pedicure, lunch catered by Goody's Restaurant followed by a light makeup application and a wash and blow dry to finish.

Fruits of the Mediterranean—$275

Eucalyptus steam, Regalo del Mar salt glo, Cleopatra wrap, Signature facial and shampoo and blow dry.

Tropical Escape—$265

Exotic mango body scrub and citrus body buff, raindrop massage, and anti-oxidant Vitamin C facial.

Head Start Morning—$195

Signature facial with eyebrow shaping, aromatic steam and body massage, manicure, and makeup application.

Thai Spa Getaway—$260

Tumeric Body Scrub and heated Thai Herbal Ball compress followed by a detoxifying steam bath and traditional Thai massage to restore general well-being.

Wellness Shamanic—$175

L'Stone body massage, mineralizing salt scrub, and a most unique hot stone foot therapy.

Body and Mind Retreat—$255

Dr. Babor or Dr. Hauschka homeopathic facial, aroma therapy massage, eucalyptus steam, and an herbal body wrap.

Gentlemen's Total Rejuvenation—$245

Sports massage, steam, purifying facial, Signature spa pedicure, hair and scalp treatment with style of choice.

Romance Package—$395

Full body polish for two, couple's massage, two Signature spa pedicures, lemon grass tea, and light refreshments. Dinner at Goody's (additional $60) if you choose.

Mother and Daughter Retreat—$450

Signature spa facials, raspberry pedicures, and manicures followed by a wash and blow dry. Lunch is also included.

Individual Treatments

Signature Facial Treatments

European Facial—$80
Signature Facial—$120
Dr. Hauska Facial—$120
Dr. Babor High Skin Refiner—$120
Babor Champagne Facial—$120
L'Stone Facial—$95
Men's Purifying Facial—$80
Aromatherapy Back Cleansing—$90

Energy Add-Ons (add to any massage or facial—30 minutes)

Reiki—$55
Crystal and Gemstone Therapy—$55
Tui Na Massage—$55
Moxibustion—$55
Thai Herbal Ball—$55

Anti-Aging Facial Treatments

Glycolic Peels—$65
MR100 (Magnetic Resonance)—$95
Oxygenating Facial—$75
Rosacea Facial—$110
Vitamic C–Anti-Oxidant Facial—$110
Lightening Facial—$110
Q-I0 Enzyme Facial—$120
Ultimate Anti-Aging Facial—$220
Microdermabrasion—$125 and up
Facial Toning (Slim Concept)—$50. With Facial—$120
Vascu Lyse, Electro Coagulation (Price subject to consultation)

Massage Therapy	50 minutes	80 minutes
Swedish massage	$78	$118
Prenatal massage	$78	$118
Reiki	$78	$118
Shiatsu	$85	$130
Aromatherapy massage—lymphatic massage	$85	$130
Thai yoga	$85	$130
Raindrop therapy	$85	$130
Sports massage	$85	$130
Cranial sacral therapy	$85	$130
Tui-Na massage	$85	$130
Arizona stone therapy	$95	$140
Couples' massage	$165	$230

Body Treatments	50 minutes	80 minutes
Regalo del Mar Salt Glo	$65	
Thai body scrub	$85	
Signature body treatment	$130	
Babor ACE treatment	$100	$150
Parafango body wrap	$100	$150
Algologie seaweed wrap	$100	$150
Body tone (Slim Concept)	$85	
(with any other body treatment)	$165	
Shi Tao body treatment with massage	$130	

Body Cocooning (50 minutes)

Body wrap with Soft Pak floatation	$110
Swiss goat butter cream wrap	$110
Herbal body wrap with evening primrose	$110
Ginger blossom wrap	$110
Cleopatra milk and oil wrap	$120
Mango body scrub and citrus body buff	$125

Exhibit 4	Continued

Spa Manicures

Signature spa manicure	$35
Herbal wrap manicure	$40
Timeless hands manicure	$45
French manicure	$45
Nail enhancements	$70
Gel nail overlay	$70
Gel nail maintenance	$40 and up

Spa Pedicures

Signature spa pedicure	$65
Fruits of the Mediterranean	$65
Peppermint Pleasure	$65
French pedicure	$70
Reflexology	$40
Clinical pedicure	$90
Royal pedicure (with facial)	$140
Paraffin treatment for hands or feet	$10
Polish change (for hands or feet)	$18

Makeup

Makeup application	$45
Makeup application with instruction	$65

Hair Design

Prices subject to hair length	
Wash and style	$35 and up
Women—wash, cut, and style	$45 and up
Men—wash, cut, and style	$25 and up
Bridal updo	$75 and up
Solid color	$60 and up

Cap hi-lites	$75 and up
Single color foils	$100 and up
Double color foils	$140 and up
Perms	$100 and up
Thermal straightening	$600 and up

Fine Hair Coloring

Eyebrow tinting	$28
Eyelash tinting	$28
Eyebrow and eyelash tinting	$50

Waxing

Eyebrow	$17
Upper lip	$17
Underarms	$19
Bikini	$20
Brazilian	$50
Half leg	$35
Full leg	$49
Half leg/bikini	$49
Full leg/bikini	$65

We advise that you arrive 15 minutes prior to your appointment to ensure a leisurely introduction to the spa. Should you arrive late, in consideration of our next guest, your treatment still must end as scheduled. If you must cancel, change, or move your appointment for any reason, we ask that you please do so within 48 hours of the scheduled time and day.

Hours of operation: Monday, Tuesday, and Saturday: 9 A.M. to 6 P.M. Wednesday, Thursday, and Friday: 9 A.M. to 8 P.M. Sunday: 11 A.M. to 5 P.M.

RETAIL PRODUCTS

Castle Spa boasted four unique, high-end European skin and body care lines in its retail section. In keeping with its philosophy of "one size does not fit all," Castle Spa management had opted to keep the four competing lines, despite the incentive of brand loyalty to one supplier. In the words of Sonja Drury, "Our mission is to discover our clients' skin care needs and fulfill them beyond their expectations. That simply cannot be done with a single product line: Each line specializes in a different skin type, and every client's skin has different needs."

Hauschka Homeopathic

This product line contained 100 percent natural and organic ingredients and, therefore, had a slightly shorter shelf life than some of the other skin care lines. Dr. Hauschka used a holistic approach to skin care and wellness, and Castle Spa had a specialized (advanced) facial called the Dr. Hauschka Homeopathic facial ($120) that drew on principles of aromatherapy and included a foot massage and lymphatic massage on the shoulders and upper body.

Babor Germany

Babor products were the main product line, used in about 60 percent of the spa facials. Babor had two skin care lines within its product offering: The Babor Blue line (which was used in the Castle European facial) and the Babor Gold line (which was used in one of the many types of Castle advanced facials). Babor was the most universal of all the skin care lines at Castle. Its prices also fell at both extremes, from $35 cleaners and toners (the least expensive offered at Castle Spa) to $210 night crèmes (the most expensive). It was popular with both young clients and clients with aging skin needs.

Pevonia Botanica

Pevonia was a Swiss skin care line that specialized in treatments for rosacea skin (a genetic skin disorder that causes a ruddy, uneven texture on the face; it affects to some degree about one third of the population). Pevonia also made excellent skin care products for men, which made up 20 percent of Castle clientele.

Doctor's Dermatological Formula

This skin care line was made for acne. The company specialized in cleansers, scrubs, and lotions that contain glycolic acid a powerful exfoliating agent. The company also made an excellent line of skin care products containing vitamin C, which was especially good for preventing fine lines and wrinkles.

PRODUCT DIFFERENTIATION

In an age when every corner nail salon and barbershop seemed to be adding the words "and spa" to their signage, it was imperative that Castle maintain and promote its product and service differentiation in the face of increased competition. To be on the cutting edge of not only five-star service, but also new technology, Castle had to invest a substantial amount of capital in training and motivating top-quality staff and importing products. Without knowledge of these new products, services, and spa technologies, clients cannot understand their benefits, especially in terms of price. Educating clients properly had the added benefit of helping them overcome resistance to price. When clients fully understood that Castle Spa's quality was unparalleled and the services were unmatched, they would be more willing to pay higher prices than what the competition may be offering.

CLIENTELE

Primary Market (Advanced)

Castle Spa's primary market consisted of clients who were educated in skin care and understood the benefits. People in this client category typically came to the spa 6 to 12 times per year and opted for regular advanced facials (see Exhibit 5) and other specialty skin treatments such as microdermabrasion and vasculese.

Secondary Market (Intermediate)

The secondary market for Castle Spa consisted of clients who came to the spa for various treatments, typically basic esthetics services such as manicures, pedicures, body scrubs, and other body treatments. They typically received spa services 3 to 6 times per year, and often one of these visits was paid for with a gift certificate received for a birthday or holiday. The intermediate category of client also represented regular massage clients.

Tertiary Market (Beginner)

The tertiary market was composed of clients with very little prior understanding of spas and their benefits and services. These clients often found out about the spa through receiving gift certificates, as members of large wedding parties having their hair and makeup done at the spa; or from special discount promotions. These clients typically did not understand the cost-benefit of Spa treatments and perceived spa services and products as high risk/low return on investment as they were somewhat intangible and there was a learning curve to understanding spa services.

OPERATIONS

Employees

Castle Spa employed over 60 staff including European-trained estheticians, massage therapists, hair stylists, and colorists. Staff worked five days a week in high-demand periods, with days off typically on Sunday and Monday. Scheduling priority went to senior staff, and, although there was a skeleton schedule, all staff were confirmed for their shift 24 hours prior

Exhibit 5	Breakdown of Revenues by Market Segment							
Revenue Driver	Average Service	Retail Purchases	Services (hours)	Wage per Hour	Commission %	Commission $	Total	Contribution Margin
Advanced clientele	$150	$150	2	$25/hour	15%	$22.50	$72.50	$227.50
Intermediate clientele	$120	$30	1.5	$10/hour	10%	$3.00	$18.00	$132.00
Beginner clientele	$100	$0	1	$10/hour	10%	$10.00	$10.00	$90.00

to avoid being overstaffed on unusually quiet periods. This had been a source of conflict in the past, especially among the junior esthetics staff, who resented having such an uneven schedule in lean times. This confirmation practice mirrored the 24-hour cancellation policy for clients. Castle Spa did not typically receive much last-minute or walk-in business; at the current time walk-ins represent about 5 percent of total business. The schedule was designed to avoid paying staff who were on duty without any clients booked on their schedules. With such a tight schedule, the spa discouraged last-minute bookings.

The wage structure varied, and payroll was as follows:

Massage Therapists: Received commission on massages, usually $10 to $12 per hour-long massage, plus gratuities.

Senior Esthetics Staff: Made $15 to $25 per hour, including commissions paid on retail sales (usually 10 to 15 percent of retail sales) and gratuities. These staff members were booked primarily for facials, but would perform other esthetic services if the schedule demanded it.

Junior Esthetics Staff: Made $8 to $12 per hour. Performed manicures, pedicures, body treatments, waxing, and other esthetic services. An average commission on retail sales was 10 percent; however, retail sales were very rare for nonfacial services.

Hair Salon Staff: Paid on commission on average 15 percent; junior hair salon staff and assistant stylists were sometimes compensated at an hourly rate that ranged from $8 to $12 per hour.

High and low Periods were as follows:

Seasonal Fluctuations: Castle Spa's high season was early in the New Year (January/February/March), when clients were looking to redeem gift certificates they received for Christmas and the holidays.

The high season continued through July, as wedding parties booked for prewedding services. Business invariably dropped off drastically in the late summer and fall and began to pick up again around the Christmas season.

Day-of-the-Week Fluctuations: Saturday was the busiest day of the week, with a full complement of staff and back-to-back bookings from opening until close. Sundays, Mondays, and Tuesdays were quietest, though some business was picked up in the hair salon. Wednesdays, Thursdays and Fridays were typically busiest from late afternoon (after work) until close (8 P.M., see Exhibit 6). The schedule could vary dramatically among days of the week: from a 50-client Saturday to a four-client Monday. These fluctuations could be very disruptive, as well as difficult to predict.

COMPETITION

Direct/Micro Competition

Castle Spa's direct competition was the Devon Spa (see exhibit 7 for a competitive analysis). Devon Spa was also located in downtown Vancouver, approximately three blocks from Castle. Castle and Devon were after the same market share, mainly the five-star market of upscale clientele. Castle's main competitive advantages were its market presence from having been established for 18 plus years and its franchise at Fairmont Hotel, which further strengthened its brand image. Second, although Devon Spa operated as an Aveda spa, a highly recognized product line that provided promotional material and collateral to its franchisees, Castle's products were of higher quality (European imported), and Castle offered four lines to choose from that catered to specific skin types.

Exhibit 6 Operating Hours of Castle versus Primary Competitor

Hours of Operation	Castle Spa	Devon Spa
Monday	9 A.M.–6 P.M.	10 A.M.–6 P.M.
Tuesday	9 A.M.–6 P.M.	10 A.M.–6 P.M.
Wednesday	9 A.M.–8 P.M.	9 A.M.–8 P.M.
Thursday	9 A.M.–8 P.M.	9 A.M.–8 P.M.
Friday	9 A.M.–8 P.M.	9 A.M.–8 P.M.
Saturday	9 A.M.–6 P.M.	9 A.M.–5 P.M.
Sunday	11 A.M.–5 P.M.	Closed

Exhibit 7	Competitive Analysis of Castle Spa

Strengths

Downtown location, next to major office buildings, hotels, and shopping promenade, etc.

Price list and service descriptions accessible online

On-site restaurant (Goody's), which complements spa philosophy of healthy living

Easily accommodates large group bookings, with three lounges, 10-person manicure/pedicure room, and 10+ treatment rooms

Four disparate product lines allow esthetician to customize skin care to individual client

On-site restaurants (Sante and Goody's), which complements spa philosophy of healthy living

On-site restaurant Goody's, which serves Asian-fusion cuisine

Downtown location, next to parliament buildings, Westin & Chateau Laurier Hotels, Rideau Center Shopping promenade, etc.

Weaknesses

Website designed and maintained by CEO's son; lacks professionalism/paid search engines

Goody's restaurant does not have strong brand recognition in Vancouver
No 800 number

Low demand on Mondays and Tuesdays

Competing product lines do not allow for as much leverage (discounts, privileges) as having a single line would

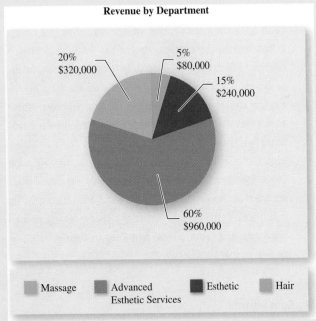

Revenue by Department

20%
$320,000

5%
$80,000

15%
$240,000

60%
$960,000

■ Massage ■ Advanced Esthetic Services ■ Esthetic ■ Hair

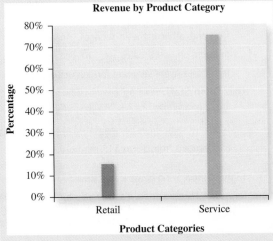

Revenue by Product Category

Exhibit 8	Sales Revenue

Indirect/Macro Competition

Though at present there were no other direct competitors to Castle Spa in the immediate area besides Devon Spa, there were a number of indirect sources of competition. Not the least of these were the bevy of small operations, such as hair salons, manicure and pedicure "bars," massage clinics, and the like. Because spas were such a high-growth industry, competition was likely to become fiercer in the coming years.

At the time, none of the high-end hotels within walking distance of Castle Spa had on-site day spa facilities. One hotel offered in-room massage, but its facilities for this were very limited. Other indirect sources of competition included resort/destination spas and hotels in British Columbia.

BRAND IMAGE

The allocated marketing dollars for Castle Spa were approximately 7 percent of sales ($1.6 million yearly), which was $112,000. The communications mix included articles written in the local papers, listings in the Yellow Pages, and advertisements in magazines such as *Vancouver City* magazine and *Where* (which was left in local hotel rooms). Magazine advertisements were usually cross-promotions with Goody's restaurant. Castle Spa's website allowed clients to book, order gift certificates, and order products online, and gave a detailed description of the services that were offered. Word of mouth was also relied on heavily for marketing of the spa.

Castle Spa's brand image was further strengthened by its first (and hopefully one of many) franchise at the Fairmont Hotel. Sonja Drury's many industry connections and membership in various spa associations (see Exhibit 9) were a product of her years of involvement on the board of directors for the Tulip Festival, the National Art Gallery, and spa associations. These industry connections provided numerous public relations opportunities and generated positive word of mouth. They were also excellent avenues for cross-promotion, online links, and product bundling for tourists to the Vancouver area.

CONCLUSION

Given the state of the industry and the current opportunities that existed, Sonja felt confident that 700 Essex, the new condominium complex, would provide Castle Spa plenty of new clients. She finished her tea, feeling revitalized as she went straight to her office to draft up next year's marketing plan.

Exhibit 9	Spa Associations

  ISPA is recognized as a worldwide professional association by the spa industry, representing more than 2,000 wellness facilities and providers from 63 countries. ISPA offers professional development and networking opportunities to members. It also offers information for travel agents and other intermediaries, and offers clients resources to locate a spa from among its members. Castle Spa CEO Sonja Drury sits on the board of directors for ISPA and was involved with this association since its inception.

Spa Canada is a national organization that serves to market Canada's spa industry as an international spa destination. Spa Canada supports professional development in the spa industry by implementing guidelines for standards and practices and education. It aims to increase awareness of the benefits of the spa experience to potential and existing clients and stakeholders, in Canada and worldwide, through such marketing programs.

 Castle Spa also received PR from an organization called the Consumers' Choice Awards. The Consumers' Choice Awards allow Vancouver area customers to vote on their favorite businesses, from dry cleaners to restaurants. For the past four years, Castle Spa had been nominated for this award via this Vancouver-wide customer survey. For a fee, the Consumers' Choice Award organization allowed Castle Spa to use its logo and name to promote the fact that Castle was voted "Top Spa in Vancouver" on all of its promotional material. Spa management had found this to be value added, as it lessens some of the perceived risk of expense for new clients.

overview

The concepts of differentiation, segmentation, and target marketing are discussed in this chapter. These concepts, while related, are actually separate tools used in marketing.

Differentiation is the task of making your products and services differently, or making them appear to be different, from those of the competition. Firms differentiate their offering on some combination of features and/or benefits in an attempt to satisfy the specific needs of a well-defined group of prospective customers.

continued on pg 420

Differentiation, Segmentation, and Target Marketing

learning objectives

After reading this chapter, you should be able to:

1. Explain and define *product differentiation*.

2. Define what is meant by market segmentation and the variables used to segment markets.

3. Explain what target marketing means and how it flows from the market segmentation process.

4. Explain how differentiation, segmentation, and target marketing are highly interrelated and effective tools used in hospitality and tourism marketing.

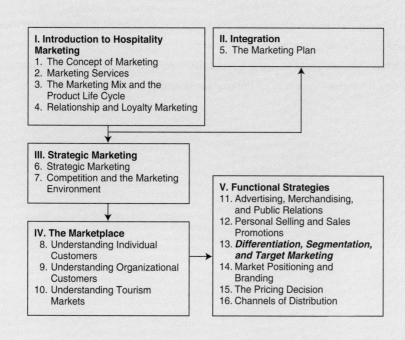

I. Introduction to Hospitality Marketing
1. The Concept of Marketing
2. Marketing Services
3. The Marketing Mix and the Product Life Cycle
4. Relationship and Loyalty Marketing

II. Integration
5. The Marketing Plan

III. Strategic Marketing
6. Strategic Marketing
7. Competition and the Marketing Environment

IV. The Marketplace
8. Understanding Individual Customers
9. Understanding Organizational Customers
10. Understanding Tourism Markets

V. Functional Strategies
11. Advertising, Merchandising, and Public Relations
12. Personal Selling and Sales Promotions
13. *Differentiation, Segmentation, and Target Marketing*
14. Market Positioning and Branding
15. The Pricing Decision
16. Channels of Distribution

Market segmentation is the task of identifying and classifying customers into groups according to some shared characteristics and/or behaviors.

Target marketing identifies one specific market and allocates all of its resources to attracting customers in that market. Target markets are the result of market segmentation because they are really subsegments of the total market. Many of the principles of differentiation also apply to segmentation, but they are more refined.

Marketing Executive Profile

David W. Norton, Senior Vice President, Relationship Marketing, Harrah's Entertainment

David Norton is the senior vice president of relationship marketing at Harrah's Entertainment, which operates more than 40 casinos nationwide and has been recognized for its outstanding marketing practices by the *Wall Street Journal, Info Week*, and *CIO* magazine.

Norton is responsible for the company's direct marketing strategy, VIP marketing, revenue management, teleservices, the Total Rewards customer loyalty program, Internet marketing, marketing reinvestment, operational CRM, and Travel Services.

Prior to joining Harrah's in October of 1998, Norton worked in the credit card industry with American Express, Household International, and MBNA. He has a BS in finance from Boston College, an MBA from Loyola College, and a master's in management of technology from the University of Pennsylvania and the Wharton School.

Marketing in Action

David W. Norton,

Senior Vice President, Relationship Marketing, Harrah's Entertainment

Do casino companies segment their customers? If so, how do they do that?

Certainly we do. I'm sure other casino companies think about slots versus tables and VIP versus non-VIP. We do, in terms of slots versus tables, daily value, customer frequency, are they a lodger or not, and then we try to ascertain how loyal they are to us using models to try to predict value and compare that to our observed behavior.

How do you develop and identify different market segments?

We can't know who a non-Total Rewards member is and that's why it's so important for us to encourage people to sign up, because it gives us a lot of insight into the business. When we started a number of years ago, we were only tracking about half of our gaming revenue, and now it's close to 80 percent. As we've bought Horseshoe and Caesars, we see that their tracked play was a lot less than ours, so we've focused a lot of energy on the value proposition of Total Rewards and why it's in your best interest to have your play tracked. At the same time we're going to be very respectful of your privacy, and we hold that promise as a very important standard for us. Because of that, we have a lot of information about the customers in terms of how we identify market segments. It's really looking at the data initially to say what trends are there and what makes sense. With any degree of segmentation, you have to figure out what you are going to do differently— whether it's different service or a different marketing message or different marketing offers—so we don't do segmentation for segmentation's sake. As long as you have something different to do, then the segmentation is worthwhile.

In terms of updating it, frankly, the segmentation we launched back in mid-1999 for our original direct mail program is still in place. We've added to them and we've added other capabilities to get more refined in different applications, but the basic premise of the segmentation has stayed stable for a number of years, which gives us a lot of learning within a property and across properties as well.

What are some key things that one should know to effectively segment and target a given market?

I think you need to have the data to figure out who these people are and their behavior, and you have to have the ability to execute against it. For us, we can have all this segmentation, but if it required people to spend days and days writing a query to get at that segmentation, then they're just going to say that's too burdensome. So, the data has got to inform a segmentation, and then you have to be able to make it easy for people to execute so they can think through what the segment is all about, what to do differently, or what test to do. The automation of execution is critical, especially when you have a lot of segments.

There are programs like SRI Consulting Business Intelligence's VALS, Cohorts, and Claritas' PRIZM that help give a psychographic segmentation. Do you do any of that?

We've dabbled in it and there might be a little bit of opportunity there, but that certainly has not been a core part of our success to this point. It's really been more about the transactional data and the little bit of demographic information we get when people sign up, but it's hard to predict who's likely to be passionate about gaming. Two people could be 55 years old and have the same income and live on the same street, and yet one loves to game and one does it once a year. This is why psychographics are better than demographics. It's hard to predict with external data, and it isn't until you see the transactional data that you get more accuracy and are able to develop some models.

How do you attract new customers? Do you look at the characteristics of your current customers and try to identify others like them?

No, we've tried acquisition through direct mail, and the returns are poor. We've just about given up on direct mail acquisition, which is obviously different than other industries. For us, it's all about advertising and doing promotions that draw people in and sign them up for Total Rewards with the immediate gratification of the promotions. We do about four big national promotions a year, which reward our existing customers, but also provide a hook for new people to sign up. We do some predictions based on that first observation and figure out who we think has the potential to become a loyal Harrah's customer and do everything we can to get them in the fold.

In its simplest form, differentiation means distinguishing your products or service from the competition in ways that are both identifiable and meaningful for the customer. How does this occur in the casino business, because you have multiple properties?

I think that in terms of the differentiation, the product is pretty consistent. Obviously, the mix of table games or the mix of denominations might differ from property to property, but we try to match the product to what the core customers want, and I think we've done a pretty good job with that. We know which customers play which games, and we lay out the floor particularly because of that. If Diamonds gravitate to an area, we'll have more employees in that area to make sure we can give differentiated service. So, there's something there with the product, even though at the highest level, it's fairly consistent across not only our own properties, but across the competition as well. However, our offers, incentives, and messages are going to be customized based on where somebody is in our segmentation strategy. Clearly, a big part of what we've done is differentiated service that's delivered through the tier program, and we make sure our Diamond customers get fantastic service. We want all customers to receive good service, but we realize that some customers are worth more because of their loyalty and ensure that they get the differentiated service experience that they want.

Is this differentiation method that you're following sustainable over the long run, and do you think it could be easily copied?

I think it's sustainable for a couple of reasons. Internally, we have the measurement tools in place to figure out how we're doing in terms of the differentiated service or offers. This allows us to ensure that the differentiation is being used well from both a marketing and operations perspective, so people don't drop their eye off the ball. If you look at what we've done over the years, we continue to evolve and refine. We're not resting on our laurels; in fact, we're doing more and more each year to try to get closer and closer to the customer.

Externally, I think a lot of companies are still focused on the facility. I don't see many that have taken the loyalty program or marketing efforts to quite the same level. I think, in general, people in our industry try to distinguish themselves by spending their energy building beautiful facilities—especially in Las Vegas and, to some extent, in other markets as well.

How much decision making occurs at the property level for different offers, or are decisions made at the corporate level?

Specifics are generally done at the property level, but they do it within the framework and tools that we've developed centrally that should make them more effective. We have several roles in my group that are responsible for partnering with the properties, evaluating how they're doing. The execution really happens decentrally, but within a fairly structured framework.

How do you test the effectiveness of different promotions you do for different segments?

With any segment, we can split it up into the standard offer, the test offer, and, if we want to hold out a control group, we can do that as well. So we'll take a like group of customers at a fairly refined level and split them into those three groups and figure out what the incremental value of the marketing intervention is. You can look at that versus the control group, or if we want to test food versus cash, we can see how the two groups perform from a response perspective or a revenue perspective. The key measure that we look at is net profitability, and if we see Offer B worked, then we can roll that out, not only across that segment, but maybe other segments as well. It's very easy to set up tests on the front end, and then we've got a pretty strong mechanism to evaluate what worked and what didn't. Because the segmentation stays stable, we can translate that within a property and across properties as well.

For students wanting to get into the casino business in marketing and segmentation, what kinds of courses would you recommend or advice would you give them?

To be successful at our company or just to be a successful marketer, you have to be comfortable with analytics and be analytically inquisitive. What we've done is built a lot of tools where it's pretty easy to evaluate what's going on in the business from a marketing program perspective. It's not about creating the reports, but rather being comfortable with the data, finding insights, and being innovative and creative enough to come up with recommendations based on those insights. Depending on where you are in a corporate role, interpersonal skills are important because you're trying to influence people that you don't necessarily have responsibility for directly. But in summary, what's critical is the comfort in analytics and the ability to take insights and turn them into action.

Used by permission of David W. Norton.

Web Browsing Exercise

Visit the Peabody Hotel Group's website (www. peabodyhotelgroup.com/). How does this company market its differentiation strategy on its website? Does this strategy come across on the website?

differentiation Distinguishing a product or service from that of the competition in ways that are both identifiable and meaningful for the customer. It is the *perceived* difference by the customer that is important to the marketer.

In previous chapters we touched briefly on differentiation, market segmentation, and target marketing. In this chapter we will elaborate on how these tools are used by the marketer to beat the competition, seize marketing opportunity, maximize marketing efforts, and satisfy customer needs and wants. They are separate concepts and tools but are highly related—that is, all three are almost always used to market the same product or service. We will define how they are different and how they work together.

DIFFERENTIATION

Differentiation in its simplest form means distinguishing your product or service from that of the competition in ways that are both identifiable and meaningful for the customer, so that customers will choose your product or service over that of the competition. The assumption is that the customer will *perceive* greater overall value, better price value, and/or better problem solution in your product or service. Notice the use of the

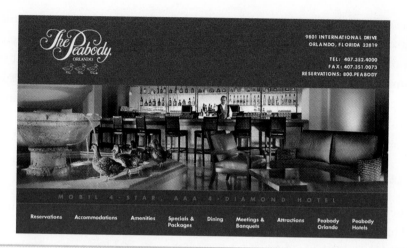

Exhibit 13-1   The Peabody Hotel Group has a unique way of differentiating itself—by having ducks parade through the lobby.

Source: Retrieved from www.peabodymemphis.com/asp/home.asp. Used by permission of Peabody Hotels.

word *perceive*; it is not necessary that there be an actual difference, only that the potential customer perceives there to be one. It is just as important to note the opposite situation. If the market does not perceive a difference, then for all intents and purposes a difference does not exist. It is the role of the marketer to convince the customer that a difference exists. It is the role of operations to deliver the difference.

There are numerous examples of differentiation in the hospitality industry. One is that of Peabody Hotel Group, which differentiates its properties in a unique way: A family of ducks is housed in each hotel and is brought down an elevator each morning to spend the day at a fountain in the hotel lobby. In the evening, the ducks troop back into the elevator to return to their quarters. This daily ritual attracts many camera-toting spectators to the Peabody lobbies. The lobby bar does a roaring business every evening as people wait to see the ducks march into the elevator. In fact, the logo of the Peabody Hotels is a duck (Exhibit 13-1). Does this make the Peabody a better hotel? Probably not, but it certainly differentiates it in an interesting and entertaining way. Another, more functional basis of differentiation may be observed among most all-suite hotel brands such as Embassy Suites, Comfort Suites, and the like, which typically invite guests to "stay in a suite for the price of a room."

Bases of Differentiation

Minor product features can serve as bases of differentiation. In themselves, they may be unimportant, but they can be very effective when

1. they cannot be easily copied,
2. they appeal to a particular need or want, and
3. they create an image or impression that goes beyond the specific difference itself.

Consider the Plaza Hotel or the Waldorf-Astoria, both in New York City. Both are premier hotels with a great deal of history behind them, and both have been frequented in the past by people of international fame. Their histories cannot be copied by other hotels in New York. This basis of differentiation has considerable appeal for customers who like the Old World and the feeling of blending with the past. Further, there is an image or impression that these hotels, because of their past, will deliver great service and unmatched elegance. Companies that differentiate their products must also show an image of those products in the minds of customers that distinguishes them from others and causes the customer to react to them more favorably.

Differentiation of Intangibles

Because much of the hospitality product is largely intangible, differentiation in traditional marketing often centers largely on "tangibilizing the intangible," which we have discussed before. Exhibit 13-2 is an example of this. Ritz-Carlton uses a couple hugging and kissing to tangibilize romance. The use of trick photography shows the couple on a silver tray, suggesting that the Ritz-Carlton serves up romance.

Using an atrium lobby (tangible) to represent an "exciting" (intangible) hotel experience was an example of making a tangible representation out of something abstract. Today, atriums are so common that few differentiate any more. Many hotels try to differentiate on the basis of better service quality. However, this expectation is typically set as a function of the rate paid (e.g., the higher the room rate, the higher the expected level of service) and is somewhat in doubt until customers experience the actual service first, then later decide if it was indeed better. If tangible proof is offered, then differentiating on this basis can be successful. Hotels and restaurants try to tangibilize their service by evoking issue of tradition, as shown in Exhibits 13-3 for the Langham Hotel and 13-4 for the Waterlot Inn. Similarly, restaurants sometimes differentiate by their tradition.

The Roger Smith Hotel has differentiated itself in the New York City marketplace by creating a guest experience built around art. The exterior, lobby, and restaurant all feature original art designed by the owner and others. This "thinking man's" boutique hotel is unique in a city full of medium to large commercial hotels. Starwood's W

Exhibit 13-2 This advertisement for Ritz-Carlton shows an example of tangibilizing the intangible

INSTEAD OF SIMPLY PAYING BOSTON'S
FASCINATING SITES A VISIT, STAY IN ONE.

The Langham London, built in 1865.

A world-renowned landmark in luxury that forms the cornerstone of Boston's city center.

A location just minutes from Newbury Street shopping, the museums and Quincy Market.

A grand hotel that not only brings you the best of Boston, but lets you live it firsthand.

THE LEGEND LIVES · SINCE 1865

250 Franklin Street, Boston, MA 02110
T (1-617) 451 1900   www.langhamhotels.com
F (1-617) 422 5163   bos.info@langhamhotels.com

A member of
The Leading Hotels of the World

 Langham Hotel
Boston

| Exhibit 13-3 | Langham Hotel tangibilizes its service by evoking tradition |

Source: Langham Hotels International (www. langhamhotels. com). Used by permission.

concept promises to uniquely differentiate by offering "business chic" to attract the generation X businesspeople who are upwardly mobile and "looking for an experience" in their choice of lodging.

Differentiation as a Marketing Tool

Differentiation is an important marketing tool, whether the differences are real or only perceived. For one thing, differentiation helps to create awareness and trial by the customer. Atrium lobbies once did that, artifacts once belonging to famous people do it for Planet Hollywood, "infinite attention to detail" service does it for Mandarin Oriental hotels, ducks do it for Peabody Hotels, superior steaks do it for Ruth's Chris Steak

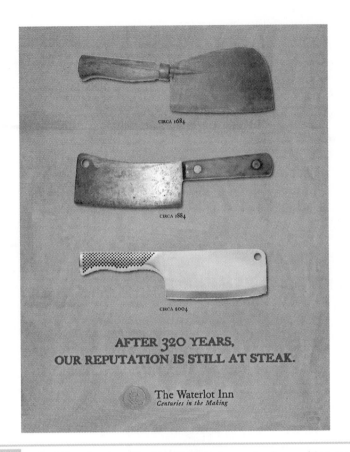

Exhibit 13-4   The Waterlot Inn also tangibilizes its service by evoking tradition
Source: Fairmont Hotels Bermuda. Used by permission.

House, all suite accommodations do it for Embassy Suites, chocolate chip cookies do it for Doubletree, and consistency of service does it for Marriott. Note that these differentiating factors are both tangible and intangible, and all, more or less, fit our three previous criteria. Yes, some can be duplicated, but not really that easily.

As we will see later in this chapter, sometimes the only thing we can do when we compete with others in the same market segment or product class is to differentiate the product. When the product approaches commodity status, differentiation may occur only in marketing. An example is extensive bathroom amenities and marble on the floor. Not only are these amenities easy to duplicate, the cost of delivering these differentiating amenities can also be very high.

There is a way out of this endless spiral of adding amenities or things that cost money that can be easily duplicated, and it lies in marketing in its truest sense. This means refocusing and delivering what the customer wants.

For example, Ritz-Carlton reduced its hard and fast 20 basic rules for its staff to 12 "service values" with emphasis on the employees thinking for themselves. For example, instead of "never say hello" to a guest, choose a more formal greeting such as "Good morning." They have mottos such as, "I continuously seek opportunities to innovate and improve the Ritz-Carlton experience." This change was made in response to the fact that customers are changing and do not want the old, stuffy atmosphere, but want to feel comfortable. (For all the new rules, go to http://corporate.ritzcarlton. com/en/about/gold_standards.asp).[1]

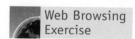
Web Browsing Exercise

Go to the website www. leye. com. What brands does the company offer? How does it differentiate these brands from other restaurants?

Formule 1, part of the Accor family of hotels, by promising unbeatable prices and perfect cleanliness, sets itself apart in budget hotels. If this standard is not met, the customer is able to receive a free night's stay in any Formule 1 hotel. They appealed to a particular need and created an impression that went beyond the service itself. The St. Regis in New York set itself apart from its competitors in a different way. There are many "luxury" hotels in New York City. The St. Regis differentiated itself from the rest by offering "butler service." What better way to travel than having a butler unpack and store your garments and bring hot tea on a moment's notice? Formule 1 differentiated one way; St. Regis, another.

Food service establishments actually have greater opportunity to differentiate their product than do hotels. Although some food service product classes may be somewhat close to commodity status, there are many ways that restaurants can differentiate their product; in other words, it is much easier to be creative, economically, with a menu and decor in a restaurant than with a hotel room. Burger King differentiates itself by its flame-broiled cooking procedures.

Hotel management has also begun to realize the need to differentiate restaurants and has developed more creative concepts. The traditional hotel had a coffee shop, a fine dining room, and a lounge, often developed with little imagination or creativity and often not fulfilling customers' needs. Rather than being creative and seeking new opportunities, hotels simply accepted that food and beverage departments would hopefully operate at a small profit and there wasn't much that could be done about it. The frequent customer reaction was, "That's hotel food; let's go out to eat."

The situation is quite different outside the United States. In Europe, Asia, and the Middle East it is not uncommon for hotel dining rooms to be among the best restaurants in the city. Both hotel guests and the local populace patronize them heavily. In France, for example, one can find a two-star hotel with minimal rooms that includes a dining room superior to most in New York City. In Japan, where eating out is such a common practice, F&B can contribute as much as 70 percent of a hotel's revenue. The same is true in Dubai, where such chains as the Rotana Hotel Group are known for their restaurants.

If a hotel restaurant is going to compete with a freestanding restaurant, management has to think, look, and act like its freestanding competition. Jim Nassikas did this

Tourism Marketing Application

The province of KwaZulu-Natal is located in South Africa. The goal of the Tourism Commission for this region is for the region to be recognized as Africa's premier tourism destination. To reach this goal, the commission created a website, www.zululand.kzn.org.za (accessed March 25, 2006) that provides information about the various destinations in the region. Each destination is positioned on different features, such as cultural attractions or natural wildlife reserves.

when he opened the Stanford Court Hotel in San Francisco, which is now a Renaissance Hotel. Nassikas opened Fournou's Ovens, an upscale restaurant, within the hotel, but didn't tell the hotel guests. There was no mention of the restaurant in the guest rooms or within the hotel. To get there, guests were instructed to go out the front door and around the corner. In fact, one of Nassikas' favorite stories is of the hotel guests who hailed a cab to get to the restaurant. Nassikas' strategy not only added a mystique to the restaurant, but also differentiated it in the eyes of nonguests who fastidiously avoided "hotel food." The result was a very successful, differentiated hotel gourmet restaurant. The New York Palace leased its restaurant space to Le Cirque, creating tremendous differentiated in-house dining experiences for guests, as well as outside customers.

Today there are hotels with "fast-break" bars for juice, coffee, and rolls; lounges with deli bars as well as liquor bars; lobby lounges with entertainment; grazing restaurants; and so forth. When the basic hotel room doesn't change much, these are excellent opportunities to differentiate in ways that are not susceptible to copying and can offer unique and distinct advantages. Marriott spawned the sports bar concept with Champions in the Marriott Boston that has been replicated successfully in many other Marriotts around the world. As a result, there is an increasing frequency of leased restaurants in hotels by established operators such as Champions, TGI Friday's, Henry Bean, Todd English Restaurants, and others. This trend is explored further in the chapter on branding. Hotels in Las Vegas are becoming famous for the restaurants in their establishments— from Picasso's in Bellagio to Delmonico Steakhouse in the Venetian.

Product differentiation, then, may be defined as any perceived difference in a product when compared with competitive products. It is what makes Smirnoff a premium vodka when it is fairly well established that all American vodkas are, by legal specification, very much the same. It is what makes Grey Goose or Belvedere vodka even more premium than Absolut, even when used in a mixed drink where the subtle difference is indistinguishable to most. It is what makes waiting in a single line at a Burger King different from waiting in multiple lines at a McDonald's. It is what makes a person respond, "I just like it there," when asked why he goes to a particular restaurant.

Differentiation—of Anything

In short, the marketer seeks differentiation whether perceived or real. The differentiation may be product specific, message specific, or even brand specific. The second is the most difficult to achieve in the hospitality industry because of the variety of services, but, because of that, even more desirable for chain operations. Levitt stated the case for differentiation as follows:

> To attract a customer, you are asking him to do something different from what he would have done in the absence of the programs you direct at him. He has to change his mind and his actions. The customer must shift his behavior in the direction advocated by the seller. . . . If marketing is seminally about anything, it is about achieving customer-getting distinction by differentiating what you do and how you operate. All else is derivative of that and only that . . . . To differentiate an offering effectively requires knowing what drives and attracts customers. It requires knowing

how customers differ from one another and how those differences can be clustered into commercially meaningful segments. If you're not thinking segments, you're not thinking.[2]

Differentiation also separates product classes. The luxury hotel is different from the budget hotel. Choice International tries to differentiate Sleep Inn from Comfort Inn and Quality Inn from Clarion Hotels. Within the same product class, differentiation separates the competition. Days Inn strives to be different from La Quinta, and Wendy's differentiates from McDonald's and Burger King. In traditional marketing, differentiation is a promotional or advertising strategy that attempts to control demand. In nontraditional marketing, it is an internal strategy that attempts to create demand. For this reason, the best differentiation may be in the marketing itself, such as relationship marketing. Differentiation provides an opportunity to strengthen competitive strategy, and it forms the basis of positioning strategy.

MARKET SEGMENTATION

Differentiation and **market segmentation** are not competing, but complementary, strategies. Whereas the first refers to the unique characteristics of a product or service, the second applies to customers. Product or service differentiation starts with the product or service. These differences need to be identifiable and meaningful to the target audience. Segmentation, on the other hand, starts with the customer. It assumes that the market is made up of customers whose needs and wants are different. The total market is divided into smaller **market segments** that are composed of people who are in some way alike—that is, who have the same needs or wants on one or more dimensions. We will discuss these groups specifically later in the chapter.

The product or service is defined for specific market segments based on the differences within each segment. For example, Steve Wynn's new premium property in Las Vegas was designed for customers who want more than the five-star services traditionally found at such hotels as Ritz-Carlton and Four Seasons. His property is more in line with six-star properties such as the Burj Al Arab in Dubai. Because he was going after this market, there had to be more precise adjustment of the product to address the requirements of specific market segments.

market segmentation
Dividing a market into meaningful groups of buyers who have similar needs and wants. Market segmentation forms the basis for target market identification.

market segments
Groups of buyers who have similar needs, wants, and problems.

Tourism Marketing Application

Brian Kurth started his company, Vocation Vacations, in 2004. His company enables consumers to "test drive their dream job" without quitting their regular job. On the company's website (www.vocationvacations.com, accessed March 25, 2006), interested consumers can choose multiple-day "vocations" in a variety of industries. They can also choose to act as general manager of a hotel. Clearly this company is designed for those truly seeking a different type of experience.

Which Comes First: Differentiation or Segmentation?

Which comes first is not the proper question, because firms do not undertake either differentiation or segmentation separately. Rather, firms undertake a combination of both at different times. For example, the firm may first segment the total market into smaller markets and then perform differentiation within each segment. Differentiation can lead to market segmentation, and market segmentation can lead to product differentiation. This first requires knowing how people differ, segmenting them accordingly, and then developing the specific products to meet their needs and wants.

The practice of differentiation and segmentation has led hotel companies to feature a number of product lines such as budget, economy, suite, middle-level, and upscale properties. As an analogy, this is no different from General Motors offering five automobile product lines. An example in which differentiation may have first occurred in the hotel industry is when Robert Hazard became CEO of Quality Inns (now Choice Hotels International), a hotel franchisor. Hazard inherited a wide variety of franchisees with diverse properties, ranging from the barely adequate to the middle level of quality—all called Quality Inn. The result had been a very confused customer image with mixed expectations and a high risk factor. To counteract this, Hazard differentiated the product into three categories—Comfort Inns, Quality Inns, and Quality Royale (now called Clarion)—and advertised to create different perceptions of each category. What resulted was a market segmented by the product. Hazard's concept was highly successful, and other operators soon began to follow similar strategies. Choice Hotels now has the following brands: Comfort Inn, Comfort Suites, Quality, MainStay Suites, Cambria Suites, Rodeway Inn, Econo Lodge, Suburban, Sleep Inn, and Clarion.

Both market segmentation and product differentiation strategies are part of the marketing concept. A major reason for studying consumer behavior is to learn how to develop segmentation and differentiation strategies. An example is the development of Marriott's "Courtyard by Marriott" product line. Marriott went to self-employed, independent, restricted, or non-expense-account customers, a subset of the business traveler market, and asked what they wanted in a relatively low-cost hotel room, what trade-offs they would make, and what they would give up to pay less. The product was then designed to fit the demand. The result, of course, was copied by others. In fact, some hotel chains stated that their new products would be copies of Marriott's Courtyard. Hilton, for example, followed years later with Garden Inns. Today, many hotel companies compete in the same midmarket segment as Courtyard. The intelligent marketer now must turn back to differentiation within this new product class.

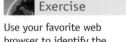
Web Browsing Exercise

Use your favorite web browser to identify the hotels that cater to the same segment as Courtyard and Garden Inns.

The Process of Market Segmentation

With all of the preceding in mind, let us proceed through the market segmentation process. The basic assumption is that customers have different needs and wants. If we are to establish a more specific definition of the needs and wants of the marketplace, it is clear that we will need to identify those segments of the market with similar needs and wants—in other words, we need to break the market down into smaller similar segments. Our need is better served if we take this in steps, because there are a number of elements that we will need to consider along the way.

Step 1: Needs and Wants of the Marketplace.

In an oversimplification of the problem, we could conduct a giant research survey in which we ask customers what they want in a hotel or restaurant. The complexity of this question is immediately apparent. As we discussed in Chapter 8 on buyer behavior, we first need to understand such constraints as the context of the purchase (e.g., for business or pleasure), the time element (e.g., do we have lots of time or little time), and the target (e.g., what type of hotel and at what price). Clearly, we will not get very far with this approach, so the first thing we will have to do is to set parameters. Let us proceed with a hypothetical example.

We are considering opening a restaurant in a city whose population is one million people. We have decided that this will not be a fast-food restaurant, but could be anything from an inexpensive family restaurant to a very expensive gourmet restaurant. We analyze what already exists and find that there is no high-quality French restaurant in the area. With this existing void, we could go this route and, without too much difficulty, clearly differentiate our restaurant from the competition, based on French cuisine.

But what if no one wants French cuisine? We would be in serious trouble. Already we see the hazards of differentiating before segmenting. Instead, at this stage, let's ignore the competition and what already exists because, even if it exists, we really don't know if it is satisfying the needs and wants of the marketplace. Maybe it is not as successful as it looks; maybe it is successful only because there is no alternative.

Therefore, let us reset the boundaries. To simplify the example, let's say we have found a location and we have decided to open for lunch and dinner. Otherwise, there are no restrictions. Now we can conduct our survey. Assume that we take a random sample of those with household incomes of $40,000 or more per year. The questions we could ask are almost unlimited, but we will have to narrow them down:

- How often do you go out for lunch/dinner?
- Where do you go?
- What do you order?
- Are you satisfied with the offering?
- What would you like to have instead?
- How much do you spend?
- How far do you travel?
- Do you like the atmosphere?
- Would you like a different atmosphere?
- If so, what would this atmosphere be like?
- Where would you like to go?
- How often?
- How much would you be willing to spend?
- What would you order?

There could be many more similar questions.

Our survey shows that 20,000 people, or 20 percent of the population (100,000 people) with incomes greater than $40,000, would go to a gourmet restaurant with some frequency. They will go there an average of twice a month for lunch with an average of three other people and once a month for dinner with an average of two other people. They would spend $18 per person for lunch and $45 per person for dinner. Of course, the other 80 percent of the same population is saying something else that we could not ignore.

To illustrate this, let us concentrate on this 20 percent. This is a market segment: a relatively homogeneous segment of the market that likes, and will go to, a gourmet restaurant. Armed with this information, we proceed to Step 2.

Step 2: Projecting Wants and Needs into Potential Markets.

demand analysis

Measurement of the market to calculate existing or future potential; sufficient demand means there are enough customers who want the product and are willing to pay for it.

This stage is called **demand analysis.** Demand analysis includes an evaluation of needs and wants plus willingness and ability to pay. Willingness and ability to pay are important, and we cannot afford to overlook them. For example, we may need a car to get to work every day and we may want a Mercedes, but if we are unwilling or unable to pay the price of a Mercedes, we are clearly not in the demand segment for that car. Demand analysis means projecting needs, wants, willingness, and ability to pay into a potential market.

Our survey has shown that we have needs, wants, willingness, and ability to pay. What does this mean in terms of the potential market? We can see by the following calculations:

20,000 people interested in the restaurant (20 percent of 100,000)
Frequency of dining out: Twice a month for lunch and once a month for dinner
Number of people in dinner party: 3 = 6,667 dinner covers per month
$[(20,000 / 3) \times 1]$
Number of people in lunch party: 4 = 10,000 $[(20,000 / 4) \times 2]$ lunch covers a month
Lunch revenue: $180,000 (10,000 @ $18)
Dinner revenue: $300,000 (6,667 @ $45)
Potential total revenue per month: $480,000 (lunch revenue + dinner revenue)
Potential total revenue per year: $5,760,000

This revenue appears to be sufficient, so we proceed to Step 3.

Step 3: Matching the Market and Capabilities.

When we surveyed the market, we had open minds about the type of restaurant we would open. Now that we have found an effective level of demand, the question is, Do we have the capabilities to meet that demand? In this case, because we are starting from scratch, we have to consider dollar resources and all the financial aspects of a major undertaking; designing and equipping a gourmet restaurant is not the same as designing and equipping a family restaurant. But we also have to consider the expertise in the firm:

- Who will manage it?
- What is this person's experience?
- Does the philosophy of the manager match ours?
- Does the restaurant fit with other things we are doing?
- Do we need outside help?

It is important, but often overlooked, that a firm's capabilities be matched to the market it is trying to serve. If we have successfully passed the first three steps, we can proceed to Step 4.

Step 4: Segmenting the Market.

We have determined the needs and wants of the marketplace, projected them into potential markets, and matched them with our capabilities. But gourmet is a very broad category; not only does gourmet mean different things to different people, but there are also many forms of gourmet. To simplify the case, let's assume that we found a strong preference for French food in our survey; we decide to segment the market on those who have a high preference for French food. Now we have to go back through Steps 2 and 3 and reevaluate the situation.

Step 5: Selecting Target Markets from Identified Segments.

Just as gourmet food is not all the same, neither is all French food. To take an example, this fact was learned the hard way by a restaurateur in a midsize New England city. This operator opened a French restaurant because "there weren't any around." He managed to build a small, loyal, steady clientele as well as an infrequent special-occasion following. When he closed because of a lack of success two and a half years later, his comment was, "The people in this city think French cuisine is quiche Lorraine."

So we have to select specific **target markets** from the broader market. This will be discussed in more detail shortly, but we might target on occasion, on income bracket, on age, on business entertaining, or any number of other things.

target markets Market segment(s) that a company has identified as its primary customer base. It then designs or delivers its products and services accordingly.

SEGMENTATION VARIABLES

There is no one best way to segment the market, but there is no shortage of ways to do so. There can also be a combination of ways to segment the market. First, we will discuss some of the more commonly used **segmentation variables,** and then we will take a look at how they overlap.

segmentation variables Various ways in which a market can be divided, or segmented, into meaningful groups of buyers such as geographic, demographic (age, income, etc.), and particular usage of the product. Often several variables are used to define a market segment.

Geographic Segmentation

Geographic location is probably the original segmentation variable and one of the most widely used in the lodging and restaurant industries. It has its strengths and its weaknesses. Geographically speaking, we can segment by country, city, media market, town, part of a city, or even neighborhood. The essence and the substance of geographic segmentation is that certain geographic locations are the major sources of our business. A hotel in San Francisco might draw most of its business from Los Angeles and New York. A hotel in Singapore might draw most of its business from Australia and Japan. A restaurant in New York City might draw most of its business from a five-block radius. A restaurant in Hartford, Connecticut, might draw most of its business from suburban towns.

If geographic segments can be pinpointed, then the problem of reaching those segments is greatly facilitated, especially if they are in concentrated areas. Both direct

Web Browsing Exercise

Type "designated market area" into your favorite web browser. What are the top 50 markets? What are the different DMA codes?

metropolitan statistical areas (MSAs) Geographic entities defined by the U.S. Office of Management and Budget (OMB) for use by federal statistical agencies in collecting, tabulating, and publishing federal statistics. A metro area contains a core urban population of 50,000 or more, and a micro area contains an urban core population of at least 10,000 (but less than 50,000).

mail and media forms of communication are more easily specified. It is also possible to use available resources to learn more about the demographics of these areas.

The U.S. federal government defines large metropolitan areas in terms of supposed economic boundaries called **metropolitan statistical areas (MSAs)**—for example, the New York City MSA. The government produces reams of data on these areas—population, ethnic mix, growth, income, discretionary spending, household size, occupations, and so forth. The use of MSAs in hospitality marketing is limited, but probably of greatest value when the market is being segmented on certain demographic variables. MSAs can be analyzed for the existence of these variables.

Another geographic division is the designated market area (DMA), developed by Arbitron and now used by the AC Nielsen Media Research Company. These designations are defined by specific zip code correlates and reflect the geographic areas served by television stations located in a central geographic area. Their data also include demographic characteristics that can be used for reaching specific audiences. Most sophisticated advertisers use these designations. The term *DMA* was coined by Nielsen Media Research. At the time of writing this book, there are 210 DMAs in the United States.

The primary purpose of the DMA analysis is to aid firms in planning media coverage. For instance, if a firm wanted to advertise to customers living in Orange County, California, the firm could look at DMA maps to see which TV stations reach this area.

Geographic segmentation is the easiest segmentation to define, but it is also the most fallible for the hospitality industry because it doesn't necessarily reflect the locus of the "buy decision." That is, the markets from which the registrants come from may not necessarily be where the reservation came from. For instance, the reservation of a particular guest may be the result of a decision in the corporate office that all employees would stay at a specific brand when traveling. The local neighborhood eatery doesn't have to employ MSAs or DMAs to know where its business comes from. Broader-based operations draw from a wide variety of geographic locations and need to use more specific and economical means to reach their markets. In fact, one of the problems of individual restaurants is that they cater to numerous small segments that are difficult and prohibitively expensive to reach through traditional advertising media.

The primary problem with geographic segments is that they may not reflect where the actual buy decision is made. For example, a business traveler from New York may have one of his associates in Chicago book him a room on his next business trip to Chicago. His registration record will reflect an address in the New York DMA, but the reservation actually originated in the Chicago DMA. Failure to understand this distinction would ultimately lead to the erroneous allocation of marketing resources. Thus, a simple segmentation of guests by geographic origin is incomplete.

If analyzed properly, however, geographic segmentation can be very useful in concentrating resources. The tourism board of Bermuda knows that most of Bermuda's tourism comes from the northeastern United States, eastern Canada, and the United Kingdom, and their advertising dollars are concentrated in those three areas. The New York City restaurant that knows that most of its business comes from within a five-block radius can use direct mail and flyers to reach that market. Singapore can spend a major share of its marketing resources in Australia and Japan.

Although all this is both true and helpful, it helps us only to reach the market; it is not of much assistance in determining the needs and wants of the market. Geographic segments, unless they are very small ones, are still very heterogeneous in terms of customer profiles, needs, and wants.

Demographic Segmentation

Demographic segmentation is widely used in almost all industries. One reason for this is that, like geographic segments, demographic segments are easily measured and classified. Demographic segments are based on income, race, age, nationality, religion, gender, education, cultural, and so forth. For some goods, demographic segments are clearly product specific—for instance, children's clothes, lipstick, Rolls Royce automobiles, and denture cleaners.

Demographic segmentation, however, may be somewhat outdated. Knowing that someone is 30 years old, earns $60,000 a year, is married, and has a child may not be too helpful in separating a truck driver, a college professor, and an accountant. Each of these people will have different needs and seek different benefits, but for a large majority of both products and services, the demographic profile of the users will not distinguish among them.

Demographic lines have, in many cases, become very blurred. Plumbers may have higher incomes than accountants with MBAs. Everyone wears jeans, regardless of social standing. Executives check into hotels on weekends looking as if they had just finished mowing the lawn. Some of the wealthy get wealthier by eating cheap, staying at budget motels, and fighting over the last nickel on their check. In fact, demographic lines have become so fuzzy that it is hard to tell what they mean anymore.

For the hospitality industry today, one of the most useful demographic boundaries may be age—in the sense of attracting children who bring parents with them, or senior citizens, a vast and rapidly growing market with distinctive needs and wants, not to mention diposable income. Another demographic variable that may be useful in some operations, particularly restaurants and resorts, is the family life cycle stage. The **family life cycle** stage has been defined as follows:

> The emotional and intellectual stages you pass through from childhood to your retirement years as a member of a family are called the family life cycle. The stages include the following: childhood, independence, coupling or marriage, parenting: babies through adolescents, launching adult children and retirement or senior stage of life.[3]

Increasingly today there are dual-income couples with no children, single parents and nonparents, and second and third marriages. Each of these stages contains, for most people, its own level of flexible income, personal time, specific buying needs, and patterns of behavior. Marketers can tap into this information, as has been demonstrated by singles resorts, early-bird dinners, special tours, and packages. Econo Lodge, a brand of Choice Hotels, claimed to have generated over $1 million of additional revenue in the first four months of introducing designated "senior rooms" in 1993. Rodeway Inn began

family life cycle The emotional and intellectual stages that a person passes through as a member of a family. The stages include childhood, independence, coupling or marriage, parenting, launching adult children, and retirement or senior stage of life.

offering "senior" rooms with brighter lighting, grab bars in showers, lever handles on doors, and large buttons on phones and alarm clocks.[4]

Some of the acronyms for lifestyle segments are **DINK** (dual income, no kids), **DEWK** (dual employed, with kids), and **echo boomer** (the child of a baby boomer). Demographic market segments, like geographic ones, are not all that insightful. We may know that older people with high incomes come to our property, but we still need to find out why; what needs and wants of these people are being satisfied or not? Age, income, education, nationality, and other demographic or sociodemographic characteristics are limited in informing us of the needs and wants of these segments.

Does this mean that demographics are an unimportant segmentation variable? No, it does not. It means that we have to understand the meaning of those demographics and how they relate to other segmentation variables. Demographics serve as broad market definition parameters within which are found more specific subsegments.

Psychographic Segmentation

Psychographic segments are segments based on **activities, interests, and opinions (AIO),** self-concepts, and lifestyle behaviors. AIOs are personality traits; the word *psychographic* actually means "the measurement of personality traits." First, we need to understand what psychographics are.

According to Joseph Plummer, a former advertising executive and one of the leading proponents of lifestyle segmentation, the concept is defined as follows:

> Life style as used in life style segmentation research measures people's activities in terms of (1) how they spend their time; (2) their interests, what they place importance on in their immediate surroundings; (3) their opinions in terms of their view of themselves and the world around them; and (4) some basic characteristics such as their stage in life cycle, income, education and where they live [i.e., demographics and geographic locations].[5]

Lifestyle dimensions, as defined by Plummer, are shown in Exhibit 13-5.

Those who are strong advocates of psychographic segmentation argue that lifestyle patterns combine the virtues of demographics with the way people live, think, and behave in their everyday lives. Those who study psychographics attempt to correlate these

DINK An acronym for a lifestyle segment describing a household: double income, no kids families.

DEWK An acronym for a lifestyle segment describing a household: dual employed with kids families.

echo boomer The largest generation of young people since the 1960s, so called because they are the genetic offspring and demographic echo of their parents, the baby boomers. *Baby boomers* is the term applied to those who were born in the years immediately after World War II.

activities, interests, and opinions (AIO) A psychological segmentation strategy based on personality traits, how people spend their time, and their interests and opinions.

| Exhibit 13-5 | Lifestyle Dimensions | | |

Activities	Interests	Opinions	Demographics
Work	Family	Themselves	Age
Hobbies	Home	Social Issues	Education
Social Events	Job	Politics	Income
Vacation	Community	Business	Occupation
Entertainment	Recreation	Economics	Family size
Club membership	Fashion	Education	Dwelling
Community	Food	Products	Geography
Shopping	Media	Future	City size
Sports	Achievements	Culture	Life cycle stage

Source: Reprinted with permission from Plummer, J.T. (1974, January). The concept and application of life style segmentation. *Journal of Marketing, 34,* Published by the American Marketing Association.

factors into relatively homogeneous categories using descriptive classification terms such as *homebodies, traditionalists, swingers, loners, jet-setters, conservatives, socialites, yuppies,* and so forth. The classifications are then correlated with product usage, desired product attributes, and media readership and viewing. **VALS** is a psychographic system.

The VALS system, originally developed by SRI International, is now run by SRI Consulting Business Intelligence (SRIC-BI). The U.S. VALS system, Japan-VALS, and UK VALS have proven to be effective tools for categorizing American, Japanese, and British consumers into various segments based on psychological characteristics and four key demographics. Merrill Lynch replaced its "Bullish on America" herd of bulls with one bull in its television and print advertising campaign, because the VALS analysis revealed that those in the target market that Merrill Lynch wanted to attract saw themselves as self-made visionaries rather than as part of a herd.

The U.S. VALS system categorizes U.S. adult customers into eight segments using dimensions of primary motivation and level of resources (high or low), as shown in Exhibits 13-6 and 13-7. Customers are thought to be driven to buy products and

VALS A method of separating types of customers based on psychological characteristics and four key demographics. The U.S. VALS system classifies customers into eight segments using dimensions of primary motivation (e.g., thinking, experiencing) and level of financial resources (high or low).

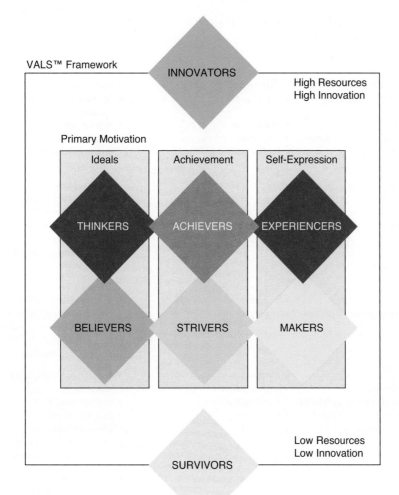

Exhibit 13-6   VALS framework

Source: Retrieved from www.sric-bi.com/VALS/types.shtml. Used by permission from SRI Consulting Business Intelligence (SRIC-BI).

Exhibit 13-7   VALS Psychographic Segments

Segment	Lifestyle Characteristics	Psychological Characteristics	Customer Characteristics
Innovators	• Successful, sophisticated • Value personal growth • Wide intellectual interests • Varied leisure activities • Well informed, concerned with social issues • Highly social • Politically very active	• Optimistic • Self-confident • Involved • Outgoing • Growth oriented • Open to change • Established and emerging leaders in business and government	• Enjoy the "finer things" • Receptive to new products, technologies, distribution • Skeptical of advertising • Frequent readers of a wide variety of publications • Light TV viewers
Thinkers	• Moderately active in community and politics • Leisure centers on home • Value education and travel • Health conscious • Politically moderate and tolerant	• Mature • Satisfied • Reflective • Open-minded • Intrinsically motivated • Value order, knowledge, and responsibility	• Little interest in image or prestige • Above average customers of products for the home • Like educational and public affairs programming on TV • Read widely and often • Look for value and durability
Achievers	• Lives center on career and family • Have formal social relations • Avoid excess change or stimulation • May emphasize work at the expense of recreation • Politically conservative	• Moderate • Goal oriented • Conventional • Deliberate • In control	• Attracted to premium products • Prime target for a variety of products • Average TV watchers • Read business, news and self-help publications
Experiencers	• Like the new, offbeat, and risky • Like exercise, socializing, sports, and outdoors • Concerned about image • Unconforming, but admire wealth, power, and fame • Politically apathetic	• Extraverted • Unconventional • Active • Impetuous • Energetic • Enthusiastic and impulsive	• Follow fashion and fads • Spend much of disposable income on socializing • Buy on impulse • Attend to advertising • Listen to rock music
Believers	• Respect rules and trust authority figures • Enjoy settled, comfortable, predictable existence • Socialize within family and established groups • Politically conservative • Reasonably well informed	• Traditional • Conforming • Cautious • Moralistic • Settled	• Buy American • Slow to change habits • Look for bargains • Watch TV more than average • Read retirement, home and garden, and general interest magazines
Strivers	• Narrow interests • Not well educated • Unconcerned about exercise and nutrition • Politically apathetic	• Reward-oriented • Unsure • Impulsive	• Trendy • Limited discretionary income but carry credit balances • Spend on clothing and personal care products • Prefer TV to reading
Makers	• Enjoy outdoors • Prefer "hands on" activities • Spend leisure with family and close friends • Avoid joining organizations except unions • Distrust politicians, foreigners, and big business	• Practical • Self-sufficient • Constructive • Committed • Satisfied	• Shop for comfort, durability, value • Unimpressed by luxuries • Buy the basics • Listen to radio • Read auto, home mechanics, fishing, outdoors magazines
Survivors	• Limited interests and activities • Prime concerns are safety and security • Burdened with health problems • Conservative and traditional • Not innovative	• Narrowly focused • Risk averse • Conservative	• Brand loyal • Use coupons and watch for sales • Trust advertising • Watch TV often • Read tabloids and women's magazines

Source: Retrieved September 16, 2005, from www.d.umn.edu/rvaidyan/mgts4731/vals2tbl.htm. Used by permission from SRI Consulting Business Intelligence.

services by three main motivations—ideals, achievement, or self-expression. Customers who are primarily motivated by ideals are guided by knowledge and principles. Customers motivated by achievement look for products and services that demonstrate success to their peers. Finally, customers motivated by self-expression seek social and physical activity, variety, and risk.

Resources include education, income, health, eagerness to buy, energy level, and self-confidence.

VALS is linked to extensive databases of consumer behavior including Media Mark Research Inc. (MRI) and Consumer Financial Decision's Macro Monitor. Geo-VALS estimates the proportion of the eight VALS types by U.S. zip code or block group, Japan VALS, and U.K. VALS products.

Critics of psychographics question whether these variables can indeed be defined, are valid, and are stable. Lifestyle variables not only are difficult to define but also overlap greatly. Because of this, there is considerable room for error in establishing the classifications. Furthermore, people change, and do so rapidly in today's society—today's lifestyle may not be tomorrow's.

Regardless of the criticisms and failings of psychographic segmentation, it remains a rich area for marketing effectiveness in the hospitality industry. New hotels and restaurants are sometimes designed and built, and old ones refurbished by architects, designers, and developers with little attention to customers and how they "use" a property. Architects and designers want their creations to be artistic, developers want them to be built at minimum cost, operators want them to be functional, and marketers want them to be marketable. It is possible that psychographic research can tell us a great deal about what the customer wants and how to build and market to those wants.

Claritas proposed another type of psychographic segmentation called the **PRIZM NE** system, which stands for *Potential Rating Index Zip Code Markets*. The idea is that people choose to live near people who are similar to themselves. Therefore, if a firm knows the zip code of its customers, the best place to find new customers is in either the same zip code or a zip code with similar characteristics. There are 66 zip code clusters in the United States.

In the province of Ontario, Canada, tourism markets are segmented using geographic, demographic, and psychographic variables. A brief description can be found in the Tourism Marketing Application.

PRIZM NE A method of segmenting types of customers based on the customers' behaviors. This separation of types of customers identifies 66 different market segments based on ZIP codes. PRIZM stands for *Potential Rating Index Zip-code Markets* and NE stands for *New Evolution*.

Tourism Marketing Application

Ontario Tourism Marketing Partnership Corporation

The Ontario Tourism Marketing Partnership Corporation (OTMPC) segmented the main North American markets (Ontario and the U.S. border states) into four market segments that they call Youth, Senior, Mature, and Family. These four groups are furthered segmented by lifestyle choices and demographics. (www.tourismpartners.com/TcisCtrl?language=EN&site=partners&key1=research&key2=segReports, accessed March 25, 2006)

Usage Segmentation

Usage segmentation is a broad umbrella term that covers a wide range of categories that probably apply more specifically to hospitality businesses than any other type of segmentation. Although we often accept these categories as givens, some are not always well used in market segmentation strategies. The basic question of all is, How do customers use the product or service? We will discuss the segmentation categories one at a time.

Purpose.   The purpose of the purchase is a common segment category. Often market breakdowns of occupancy are kept on a daily basis categorized by the purpose of visit. Approximately 80 percent of urban hotel occupancy in the United States, on average, comes from business travelers. The business expense account customer is also a big source of business in many restaurants. Business purpose can be broken down into submarkets such as conventions, corporate meetings, expense account, non-expense-account, and so forth. These subcategories are important because each one will have somewhat different needs and wants and should be marketed to accordingly.

The other major purpose category is called social, pleasure, or leisure. Because this market actually has a number of specific purposes, a better term would probably be, simply, either nonbusiness or personal. This segment represents a larger proportion of business for restaurants than for hotels.

Frequency.   The frequency of purchase segments have to do with regularity of usage. Repeat business is well recognized as highly desirable, and programs such as frequent traveler plans are geared toward this behavior. Again, however, there are subsegments that should not be ignored. High frequency might mean once a week to a restaurant, once a month to a commercial hotel, and once a year to a resort. Low frequency can also be an important segment, especially if it occurs with regularity. A restaurant might have certain customers who come only once a year on an anniversary date. A few hundred of these, however, make up an important segment that needs special attention.

Monetary Value.   Monetary value refers to how much the customer is worth to the organization. For instance, in the casino business, monetary value is often referred to as the "theoretical value." The theoretical value takes into account the length of time a person plays a game and the amount the person bets for each decision (i.e., spin of the wheel at roulette). We might call the important members of this segment "big spenders," or in the case of Las Vegas, "whales." Purchase size in a restaurant considers those who run up high checks or order expensive wines; even the big tippers can be a vital segment. In hotels, this segment might use the better rooms or suites, eat in the hotel's restaurants, or order expensive room service. Obviously, this type of behavior should be encouraged by marketing. Purchase size also considers the low spenders who may not be desirable customers.

Recency. This refers to how recently the customer consumed the product or service. This is used quite often in direct mail promotions. For instance, a firm may send a mailing piece to those who have not visited the establishment in the last 30 days.

RFM. RFM stands for recency, frequency, and monetary value. Each of the terms has already been identified. RFM analysis is used to identify groups of customers in the database. Essentially, the database is sorted by each specific measure. For instance, with monetary value, the sort would be from the largest value to the smallest value. For recency, the sort would be by most recent visit to least recent visit, and for frequency the sort would be from most frequent to least frequent. Once the sort has been accomplished, the database is divided into five equal parts. The five equal parts are then labeled from 5 to 1, with 5 representing the highest category and 1 the lowest. A customer with a score of 555 has visited the property most recently, visits frequently, and has a high monetary value. A customer with a score of 111 is just the opposite. Companies use the RFM score to direct targeted mailings. For instance, a person with a 555 score would get a very different offer than would a person with a 111 score.

Timing. Timing deals with days, months, or seasonal periods of the calendar. The Monday night customer can be icing on the cake for a restaurant; the weekend customer, for a hotel; and the off-season customer, for a resort. These segments may include people who don't like crowds or simply those on different schedules. Of course, those who come at busy times also represent a timing subsegment.

Timing segments also can be based on when the customer buys. For an anniversary dinner, it might be two weeks ahead; for a wedding, six months; for a simple dinner out, two hours. A meeting planner may book accommodations one or more years in advance; the business traveler, two days in advance.

Nature of Purchase. Consumer behaviorists often categorize buyers by the nature of purchase:

- **Convenience:** Buy a particular product because it's convenient to do so
- **Impulse:** Buy products on impulse without much forethought
- **Rational:** Buy only after careful consideration

Each of these subsegments is susceptible to a different approach.

Convenience buyers, for example, are probably more apt to use in-room refrigerator bars or room service if it is inconvenient to get the food items elsewhere. Impulse buyers are highly susceptible to suggestions, such as menu clip-ons, wine carts, the server's dessert suggestions, and a higher-priced room with a view. Rational buyers need more information; they are more apt to be influenced by descriptions on wine lists, in-room descriptive materials, and ads or brochures with more detailed information.

Where They Go. Some segments can be identified by where they go. Many might go to certain destinations on a regular basis. For vacation, they might always go to the Caribbean; for a hotel, they might always go near the theater district; for a restaurant, they might always go to the suburbs. We can direct our marketing efforts according to these inclinations.

Purchase Occasion.

Purchase occasion represents special occasion segments. They go to restaurants for birthdays and anniversaries or use hotels for the same occasions. Some may use hotels only for visiting relatives or when on vacation or when going to the theater in a large city. Some people take a trip with the goal of seeing as much as possible, whereas others go on vacation with the goal of just sitting and relaxing.

Heavy, Medium, and Light Users.

Heavy, medium, and light users often get special attention from marketers. A marketing truism states that 80 percent of purchases are made by 20 percent of those who consume the product or service (often referred to as the 80/20 rule). Any marketing research needs to pay special attention to separating these categories. As a total group, customers might have a mean of 2.5 on a scale of 5 when evaluating an attribute. Broken down, it might be that light users have a 1.7, medium users a 3.2, and heavy users a 4.4. Changes made to please heavy users might alienate light users, a consideration that management must evaluate before making changes.

Marketing executives tend to focus on the heavy user. This is probably advisable but, at the same time, it should not distract from the light user. This user may not come often, but may still be worth pursuing for, as discussed, the light user may come at a time others do not. The light user may also be a light user for a reason. In a hotel company one of the authors works with, management had forgotten to send light users promotions about the property and the customers forgot about coming. Once management added them back to the mailing list, they became heavy users.

It would not be too difficult to suggest even more user segments than those mentioned. The point is that each of these segments has different needs and wants. They may also have many needs and wants in common, but catering to the different special needs and wants is what creates and keeps customers.

A given restaurant or hotel may well have every segment mentioned earlier as customers or potential customers. This is not as impossible a situation as it may at first seem; it is simply the nature of the hospitality business and demonstrates why paying attention to only broad segments such as business and pleasure may constitute falling into a trap. With few exceptions, a hotel or restaurant that wants to maximize its potential simply cannot afford to treat all people the same.

The Saturday night hotel guest does not behave the same as the Wednesday night one—even when it is the same person. Likewise, the Monday night restaurant customer is not the same as the Saturday night one. The anniversary dinner is not the same as the business dinner. User segments have an advantage over geographic, demographic, and psychographic segments. By their nature and narrowness they are more predictable. In other words, if we know what influences them (i.e., why they constitute a segment), the chances are good that they can be influenced. This is not necessarily the case simply because we know someone's age, income, gender, or geographic origin.

Benefit Segmentation

Benefit segments are based on the benefits that people seek when buying a product. Benefits are very akin to need satisfaction. Following are just a few of the possible benefits sought in a hospitality purchase:

- Comfort
- Prestige
- Low price
- Recognition
- Attention
- Romance
- Quiet
- Safety

Benefit segments may be the most basic reasons for true market segments and the most predictable of all segments. Knowing what benefits people seek provides a basis for predicting what people will do.

Benefit segmentation is a market-oriented approach consistent with the marketing concept. From these segments other characteristics can be derived, such as demographics, psychographics, usage patterns, and so forth; in other words, benefit segments can be used to identify important descriptive variables and consumer behavior. Benefit segmentation is also concerned with total satisfaction from a service rather than simply individual benefits. This phenomenon has been termed the **benefit bundle** and is a significant factor in segmenting markets by benefits. An example is shown in Exhibit 13-8. This advertisement shows all the benefits offered by staying at a Fairmont Hotel in Bermuda.

benefit bundle All of the benefits that consumers experience from a purchase of a particular service; includes things like location, familiarity, comfortable bed, and so forth.

There are two important distinctions between benefit and other forms of segmentation. Benefits are the needs and wants of the customer. More than that, benefits are what the product or service does for the customer. Other segmentation strategies only assume a relationship between the segment variables and customers' needs and wants. We all know that McDonald's makes a special effort to appeal to children, a well-defined demographic segment. The next time you go to a McDonald's, look around at the people and see whether you can place them into a segment category. Chances are it will be a benefit segment: quick and cheap.

Second, understanding benefits enables marketers to influence behaviors. Other segmentation variables are often merely descriptive. The marketer can only try to appeal to what exists and its assumed relationship. Consider the singles and mature categories, both fairly large market segments. The marketer might think that the categories are relatively similar and as such can be treated as one major segment. Nothing could be further from the truth. In fact, within the large group are smaller segments that are similar.

It is hard to believe now, but in 1984, the conventional wisdom was that customers ages 54+ were similar in terms of their wants and needs when it came to vacation travel. A study undertaken by one of the authors proved, in fact, that the opposite was true.

Exhibit 13-8   This advertisement shows all of the benefits of staying at a Fairmont Hotel in Bermuda

Source: Fairmont Hotels and Resorts, Bermuda. Used with permission.

One segment traveled in order to escape and learn, a second group traveled because they wanted to tell their friends about the trip, and a third group traveled to revisit places they had seen before.

In summary, benefit analysis can be a powerful segmentation tool. Its best use has been in good research—research that can pay off in terms of understanding customers and what motivates them.

Price Segmentation

Price segmentation is actually a form of benefit segmentation, only it is more visible and more tangible. There are two ways to look at price segments: one is between product classes; the other is within a **product class.** Price segments within a product class, at least in hospitality, are limited. A lower price may increase the value of the benefit bundle, other things being equal, but customers will generally not make major trade-offs for a small gain in price; that is, they won't accept a poor location or poor service just to save a few dollars within the same product class.

Segmentation between product classes is different. Five-star hotels and budget motels both provide lodging, but they are each a different product class. Gourmet restaurants and fast-food restaurants are also separate product classes. The inference is that one product class does not truly compete with another on the same occasion, given the same circumstances. If we go to New York City, we do not choose between the Waldorf Astoria and Days Inn. In these cases, price is clearly a segmenting factor. Although the first decision by the customer may be based on price range, the other elements of the bundle will influence the final choice. No one rationally pays more for something without expecting to get more. In cases like these, markets are segmented within broad price ranges.

In the U.S. hotel industry today, Smith Travel Research defines five segments based on average room rates in specific metropolitan areas, also known as metro areas. In nonmetro areas, the luxury and upscale segments are collapsed to form four price segments. These segments and how they are defined are shown in Exhibit 13-9.

Amazingly, with today's modern construction, the physical product is not all that different (within ranges), and in some cases, neither is the price. In many cases, lower prices have been obtained by lower construction and operating costs and through the elimination of public space and food and beverage facilities. Of course, as one moves up the ladder, the furniture gets better, the walls and the carpet get thicker, the atrium gets higher, and the bathroom (sometimes) gets larger and has more amenities. No longer, necessarily, does one find a sagging bed; scratched, broken, and torn furniture; or tiny (or nonexistent) soap in the bathroom of a budget motel. In other words, the basic needs are still fulfilled.

Why, then, are customers willing to pay more for relatively little? And is this really price segmentation? Well, in many cases they are not, and in many cases it really isn't. Customers are willing to pay more largely because of the intangibles and tangibles that they receive in return: service, prestige, professionalism, and higher quality, among others. These are benefits, and the net result is actually benefit, not price, segmentation.

The other answer to why customers will pay more for relatively little is somewhat unclear because, for most, price is a major consideration in any purchase and varying price sensitivities will stratify any market. In the final analysis, however, within the same product class, price alone rarely determines the segment. Price is only the risk that the willing and able buyer will take based on the intensity of the problem and the perceived value and expectation of the solution. This analysis applies to both the hotel and restaurant industries. In Exhibit 13-10, you can see how Marriott has touched on just about all of the segmentation variables we have discussed.

product class A group of similar products basically serving the same needs (e.g., fast casual restaurant is one class, while white tablecloth is another; budget hotels is one class and upscale restaurants is another).

Exhibit 13-9 Price Segments in the U.S. Lodging Industry

LUXURY
COLONY
CONRAD
FAIRMONT HOTEL
FOUR SEASONS
HOTEL SOFITEL
INTER-CONTINENTAL
LOEWS
LUXURY COLLECTION
MANDARIN ORIENTAL
PAN PACIFIC
PREFERRED
THE PENINSULA GROUP
PRINCE HOTELS
ST. REGIS
REGENT HOTELS
RITZ-CARLTON
STARHOTELS
W HOTELS
THE WALDORF-ASTORIA
COLLECTION

UPPER UPSCALE
CAESARS
CONCORDE HOTELS
DORAL
DOUBLETREE HOTELS
EMBASSY SUITES
EMBASSY VACATION RESORTS
GAYLORD ENTERTAINMENT
HELMSLEY HOTEL
HILTON HOTELS
HILTON GAMING
HYATT
JURYS HOTELS
LANGHAM HOTELS
LE MERIDIEN
MARRIOTT
MARRIOTT INTERNATIONAL
MARRIOTT CONF. CENTER
MILLENNIUM HOTELS
NEW OTANI HOTELS, THE
NIKKO
OMNI
PRIME HOTELS
RENAISSANCE
SHERATON HOTEL
SONESTA HOTEL
SWISSOTEL
WESTIN

UPSCALE
ADAM'S MARK
AMERISUITES
AYRES
XANTERRA PARKS & RESORTS
ASTON
CHASE SUITES
CLUB MED
COAST HOTELS USA
COURTYARD
HILTON GARDEN INN
CROWNE PLAZA
FOUR POINTS
HARRAH'S
HAWTHORN SUITES
HAWTHORN SUITES LTD
HOMEWOOD SUITES
HOTEL INDIGO
HOTEL NOVOTEL
OUTRIGGER
RADISSON
RESIDENCE INN
RESORT QUEST HAWAII
SIERRA SUITES
SPRINGHILL SUITES
STAYBRIDGE SUITES
SUMMERFIELD BY WYNDHAM
WOODFIELD SUITES
WOODFIN SUITES
WYNDHAM HOTELS

MIDSCALE W/F&B
BEST WESTERN
CLARION
DOUBLETREE CLUB
GOLDEN TULIP
HARVEY HOTEL
HAWTHORN INN & SUITES
HOLIDAY INN
HOLIDAY INN SELECT
SUNROUTE CO LTD
HOWARD JOHNSON
JOLLY HOTELS
LITTLE AMERICA
MARC
OHANA HOTELS
PARK PLAZA
QUALITY INN
QUALITY INN SUITES
RAMADA
RAMADA PLAZA
RED LION

ROMANTIK HOTEL
WESTMARK
SUNSPREE RESORT
WESTCOAST
WYNDHAM GARDEN HOTEL

MIDSCALE W/O F&B
AMERIHOST
AMERICINN
BAYMONT INNS & SUITES
BRADFORD HOMESUITES
CABOT LODGE
CANDLEWOOD HOTEL
CLUBHOUSE INNS OF AMERICA
COMFORT INN
COMFORT SUITES
COUNTRY INN & SUITES
DRURY INN
DRURY LODGE
DRURY PLAZA HOTEL
EXTENDED STAY DELUXE
FAIRFIELD INN
HAMPTON INN
HAMPTON INN & SUITES
HEARTLAND INN
HOLIDAY INN EXPRESS
INNSUITES HOTELS
LA QUINTA INNS
LA QUINTA INNS & SUITES
LEES INN OF AMERICA
MAINSTAY SUITES
PHOENIX INN
RAMADA LIMITED
SHILO INN
SIGNATURE INNS
SILVER CLOUD
SLEEP INN
TOWNPLACE SUITES
WELLESLEY INN
WELLESLEY SUITES
WINGATE INN

ECONOMY
1ST INTERSTATE INN
ADMIRAL BENBOW
AMERICA'S BEST INNS
AMERICA'S BEST SUITES
AMERICA'S BEST VALUE
BAYVIEW INT'L HOTELS
BUDGET HOST INN
COUNTRY HEARTH INN
CRESTWOOD SUITES

CROSS COUNTRY INN
CROSSLAND SUITES
DAYS INN
DOWNTOWNER MOTOR INN
E-Z 8
ECONO LODGE
INNS OF AMERICA
EXEL INN
EXTENDED STAY AMERICA
FAMILY INNS OF AMERICA
GOOD NITE INN
GREAT WESTERN
GUESTHOUSE INNS
HOMEGATE
HOMESTEAD STUDIO SUITES
HOWARD JOHNSON EXP. INN
INNKEEPER
INNCAL
INTOWN SUITES
JAMESON INN
KEY WEST INN
KNIGHTS INN
LEXINGTON HOTEL SUITES
MASTER HOSTS INN
MASTERS INN
MCINTOSH MOTOR INN
MICROTEL INN
MOTEL 6
NATIONAL 9
PARK INN
PASSPORT INN
PEARTREE INN
RED CARPET INN
RED ROOF INN
ROADSTAR INN
RODEWAY INN
SAVANNAH SUITES
SCOTTISH INN
SELECT INN
SELECT SUITES
SHONEY'S INN
STUDIO 6
STUDIO PLUS
SUBURBAN EXTENDED STAY
   HOTELS
SUN SUITES HOTELS
SUPER 8
THIFT LODGE
TRAVELODGE
VAGABOND
WANDLYN INN

Source: Retrieved September 16, 2005, from www.smithtravelresearch.com/SmithTravelResearch/misc/GlossaryAds.aspx. Copyright 2006, Smith Travel Research Publishing.
Used by permission.

Exhibit 13-10	Marriott Segmented Hotel Brands	

Brand Name	Double Occupancy Price Range	Market Segment
Fairfield Inn	$72–89	Upper-economy business and leisure travelers
TownePlace Suites	$79–109	Moderate-level travelers with weekly or multiweekly stays
SpringHill Suites	$79–115	Business and leisure travelers seeking more space and amenities
Courtyard	$75–179	"Designed for the road warrior": quality and affordable accomodations
Residence Inn	$85–179	Travelers looking for a "residential-style" hotel
Marriott Hotels & Resorts	$119–400	"Achievers" who seek consistent qualty—business and leisure
Renaissance Hotels & Resorts	$130–500	More discriminating travelers who want attention to detail—business and leisure
Ritz-Carlton	$175–900	Luxury, unique, personalized stay for senior executives
JW Marriott Hotels & Resorts	$199–219	Luxury, unique, architectural detail; for senior executives
Ramada International Hotels & Resorts	$79–90 or $200 in select cities	Locations reflecting the character and culture for the traveler wanting consistently excellent and affordable service
Marriott ExecuStay	$1,900–4,500/ month housing	Temporary housing needs, customized, relocation; insurance adjusters, corporate and military travelers
Marriott Executive Apartments	$135–500	Home-style living with hotel amenities, for business executives
Marriott Vacation Club International	$89–500	International resorts, luxury, business and leisure travel
Marriott Conference Centers	$120–300	Meeting resort for conventions and executive resorts

SEGMENTATION STRATEGIES

No segments exist in isolation, and there is considerable overlap and sharing of the variables. Also, few hotels or restaurants today can survive on only one market segment. It is likely that there will be many segments and segmentation strategies. The foundation of any segmentation strategy is behavioral differences. No segment is meaningful if it does not behave differently from another segment. (This same factor leads to conflict between segments.) Whether you use geography, demographics, psychographics, benefits, or usage, the test of the segment is the differentiation of behavior. Thus, in the final analysis, the behavior of segments is the true test of their validity. Consider the following example:

> A suburban hotel suffering from slow weekend occupancy surveyed the market for new sources of business. Two distinct segments were discovered, but neither one was large enough to create the desired weekend occupancy. The hotel decided to market to both, and developed and implemented separate strategies based on the needs and wants of each segment. The segments were romantic couples who wanted to get away for a peaceful and quiet weekend and families who wanted to take their children for a mini-vacation with lots of activities. It wasn't long before the two segments collided head on. The potential damage is obvious.

The two segments in the preceding example behave differently, so they are both valid segments. The problem arose only when the two came together. Romantic couples did not want to be surrounded by screaming children. A similar example occurred in Las Vegas when the city tried to be family oriented. It did not take long to discover that tourist families with children did not gamble; in fact, the children "got in the way" of the gamblers. Las Vegas has since repositioned itself to be more adult and convention oriented.

Exhibit 13-11　Tests for Segmentation

- **Is it homogeneous?** Homogeneity on every aspect is not possible or even necessary but certain key aspects should be identified in this respect. These aspects form the basis of the segment.
- **Can it be identified?** Certainly we can identify segments based on gender, or geographic origin, but other measures are not so easy. For example, suppose we wanted to segment on psychographic dimensions of conservative, moderate, and liberal. The segment would be of little value if we could not identify those who fit those dimensions.
- **Can it be measured?** Suppose we could identify a conservative segment. We would then need to be able to measure the level of conservatism and the accompanying needs and wants.
- **Can it be reached economically?** The segment will not be much use to us, beyond present customers, if we cannot build on it.

Through media, direct mail, or even internal marketing we need to be able to get to the segment.
- **Can a differential in competitive advantage be maximized and preserved?** In the hospitality industry, this is one of the toughest tests of segmentation, but one that should be constantly sought after if not always reached.
- **Is it compatible with others segments that may exist at the same time?**
- **Is the segment large enough and/or profitable enough?** There is a large but tour segment for hotels that many would find unprofitable. There is a very small segment of visiting royalty that a few hotels may find very profitable. The cost and effort of serving each segment must be weighed against the return.

Knowing how customer behaviors change, it is clear that segments also change over time. We have argued already that one of the advantages of segmentation is the ability to stay closer to the customers and understand them better. This advantage should never be neglected, and a constant alert must always be maintained for changing, merging, or dividing segments. Too much segmentation can lead to too many markets and an inability to serve anyone well or profitably.

In the final analysis, market segmentation is a scientific procedure requiring scientific analysis. It cannot be a casual exercise. What strategic thinking management does is seek the "ideal business mix." Exhibit 13-11 shows the tests to which each market segment should be subjected.

Market segmentation in the hospitality industry has become increasingly critical because of the intense competition. In many cases, market segmentation may be a prerequisite to growth. In some cases, large or major segments may have reached their level of fulfillment. Smaller segments, unimportant individually but important together, may be the next wave.

Product differentiation, as a singular market strategy, may have seen its day in the hospitality industry. As product classes become more crowded, however, it will remain a key competitive strategy within the same product classes.

TARGET MARKETING

target marketing
Selecting specific market segments to target and designing the product or service to meet their specific needs and wants.

Target markets are drawn from segments. They might be called subsegments, but the word *target* has a more active meaning. Once we have segmented the market and examined the market potential, we must select those specific markets that we can best serve by designing our products and services to appeal directly to them. This is commonly known as **target marketing.** Many of the same segmentation rules apply; we just refine them more. For example, earlier in the chapter we segmented the restaurant market on gourmet and then targeted a smaller portion of that segment.

There are three strategies for selecting target markets:

1. An undifferentiated targeting strategy assumes that customers within a segment have similar needs so only one type of product or service is offered to that segment. This is common practice in the hospitality industry; the business traveler is an example.

2. A strong targeting strategy involves selecting a target group within one market segment and pursuing it aggressively. For example, Ritz-Carlton targets executives who demand top hotel experiences, whereas Microtel goes after business travelers who only need a place to sleep.

3. The third strategy is differentiated multitarget marketing. Marriott International is the perfect example, targeting specific business needs with Marriott Hotels, Courtyard by Marriott, Fairfield Inn, and Residence Inn. These brands serve distinct target markets within the business segment and do so without the confusion that Choice Hotels International created with its overlapping products. Within any one hotel, multitarget strategy is also in effect. Numerous hotels have "towers—a hotel within a hotel," concierge floors, or "business plan" rooms with special services that target differentiated business travelers with specific business needs.

Concentrated target marketing means aiming specifically at one or more portions of a market. One travel researcher found, for example, that the vacation market segment could be broken down into 10 target markets. Each one represents specific interests and behavior based on benefits, usage, demographics, and psychographics. Each one has different needs and wants and requires a different package, a different positioning, and different communication. These target markets and some of their specific characteristics are shown in Exhibit 13-12.

As with segmentation, there are criteria for choosing target markets. They overlap the segmentation criteria but are a little more precise. For example, are the target markets selected compatible with each other? Do they match the resources of the firm? These and other questions are shown in Exhibit 13-13.

Target marketing is practiced at the unit level also. A property may use concentrated targeting, select one market, and serve it well. The Delta Queen Steamboat Company,

---

Exhibit 13-12	Vacation Target Markets

- **The Carriage Trade:** Desire a change of scene but not of style, secure in wealth and position, play golf and tennis year around, and when traveling, tend to vacation as a family.
- **The Comfortables:** The largest group, insecure, seek social and psychological comfort, like recommended restaurants, guided tours, and organized activities.
- **The Venturers:** Want to see new things; have a thirst for fresh ideas, information, and education; seek the new and the different; don't travel in groups; and collect experiences.
- **The Adventurers:** The venturer advanced one step—seeks risk, danger, and the unknown.
- **The Inners:** Jet-setters, go somewhere because of who is there rather than what is there; they "make" destinations such as Acapulco, Majorca, Costa del Sol.
- **The Buffs:** Strongly subject oriented; travel because of particular interest or hobby.
- **The Activists:** Not content to sit by the pool and bask in the sun, want constant activities.
- **The Outdoorsers:** Campers, hikers; birdwatchers, bicyclists, and other outdoor recreationists.
- **The Restless:** Travel for something to do, tend to be senior citizens, retired, widowed; collect travel experiences and travel all the time including off-seasons.
- **The Bargain Hunters:** Can afford to travel, but compulsively seek the best deal.

Exhibit 13-13	Target Market Criteria

- What is the potential revenue and market share?
- What are the demand characteristics?
- Are they able and willing to buy?
- How are they currently being served by the competition?
- Are they compatible with the objectives of the firm?

- Are they compatible with each other?
- Do they fit the resources of the firm?
- Do they fit the tastes and values of the firm?
- What is the feasibility of exploiting them?

which operates three-day to one-week cruises on the Mississippi River, effectively targets mature travelers who like to gamble. The risk of concentrated targeting is that of putting all your eggs into one basket. If environmental or other changes negatively affect the demand, you may not have any market left to serve. On the other hand, a hotel may select several markets to serve, but there are risks to this strategy too, as noted earlier. These markets must be compatible and seek similar benefits from the establishment.

Consider the case of one very successful resort in the Virgin Islands. One target market was honeymoon couples and the other was high-income senior executives and their spouses, usually over 55 years old (employees referred to these markets, tongue-in-cheek, as newly-weds and nearly-deads, respectively). At first glance, these segments do not appear to have much in common, but in reality they work with each other. Both segments wanted isolation, peace, and quiet. The resort was on a small remote island, and rooms did not have air conditioning, radios, televisions, or telephones. The resort pursued these two markets aggressively and ran an annual occupancy rate of over 85 percent. To increase low-season occupancy rates, the resort decided to book group leisure travelers from Italy. These guests were on holiday and rightfully sang and danced until the wee hours of the morning, alienating the resort's traditional target markets. Needless to say, this experiment was quickly ended. The different target markets were simply not compatible with each other.

MASS CUSTOMIZATION

mass customization
Rather than making the same product for everyone, the trend is to allow customers to personalize the product or service to their specifications; e.g., requesting a certain type of pillow in the room or specific items in the minibar.

Although we have talked about criteria for market segments and target markets, such as homogeneity, size, and so forth, modern technology is bringing us closer to target markets of one, sometimes referred to by the oxymoron, **mass customization.** This is largely because of the computer databases that contain vast amounts of guest information. The goal of such databases is to look at the customer not as a segment of many, but as a segment of one.

Marketers are taking the database far beyond a simple electronic Rolodex of names and addresses. It is possible to talk to customers as individuals and then reconstruct the product or service to aim at target groups and to reward loyal customers. Databases can measure what the customer does, not just what she says she does. This information can drive the entire marketing strategy. Much of this, of course, is based on the heavy user segment that accounts for a large proportion of sales. We discuss databases in more detail in Chapter 16.

Databases are rich sources when they combine demographics with buying habits. These will become even more potent marketing tools in the future. All of this, of course, constitutes relationship marketing, discussed in Chapter 4, but the flip side of that is mass customization. The economic logic behind mass customization is as inevitable and irresistible today as the logic of the assembly line 100 years ago. It *will* happen, and

Tourism Marketing Application

Footprint Vietnam Travel is one company that offers personalized tours. Travelers can go to this firm's website (www.footprintsvietnam.com) and choose their destination, how much they would like to spend, the types of activities, and the type of accommodation. Footprint Vietnam Travel will then design a personalized tour for them.

"They're baby boomers—like, you know, really old."

marketers will have to think in terms of customer share rather than market share. Mass customization, if done correctly, will result in keeping a satisfied, loyal, long-term customer. It is the ultimate form of customer differentiation to capture the greatest possible share of every individual's business. It is the key to success in tomorrow's hospitality business. Once target markets have been determined, the next step is to tailor the marketing effort to the needs and wants of each market.

Chapter Summary

Differentiation, market segmentation, and target marketing are different but complementary marketing strategies. In a highly competitive marketplace, each one alone (and all of them together) is critical to the marketing effort.

Differentiation is used to create real or perceived differences between products and services offered by hospitality organizations. The objective is for the customer to perceive a positive difference between our offering and that of the competition and thus react more favorably toward ours.

The differences among customers are the basis for market segmentation. Segmentation is the strategy wherein the firm attempts to match its marketing effort to the unique behavior of specified customer groups in the marketplace, through the use of key segmentation variables.

Several criteria guide the process of segmentation and differentiation. It is necessary first to identify the basis for segmenting the market. Profiles of the resulting segments are then determined and matched with the firm's capabilities, followed by the projection of potential markets and segment attractiveness. The market is then segmented, and target markets are selected from the identified segments. Positioning is developed for each target market, and the marketing mix is carried out accordingly.

Each segment or target market must be examined competitively. When others are targeting the same market, as will usually be the case, a final differentiation strategy is needed to gain the competitive advantage.

Key Terms

activities, interests, and opinions (AIO), p. 438

benefit bundle, p. 445

demand analysis, p. 434

DEWK, p. 438

differentiation, p. 424

DINK, p. 438

echo boomer, p. 438

family life cycle, p. 437

market segmentation, p. 431

market segments, p. 431

mass customization, p. 452

metropolitan statistical areas (MSAs), p. 436

PRIZM NE, p. 441

product class, p. 447

segmentation variables, p. 435

target marketing, p. 450

target markets, p. 435

VALS, p. 439

Discussion Questions

1. What is product differentiation?

2. Why is product differentiation important to achieve both within and among product classes?

3. What is market segmentation?

4. Briefly describe each of the common variables used in the market segmentation process.

5. What is target marketing?

6. Which comes first—market segmentation or target marketing? Why?

7. Which is most important—product differentiation, market segmentation, or target marketing? (Be careful, this is a trick question.) Discuss.

Endnotes

1. From Greg Crosby, *Jewish World Review*, July 21, 2006. http://jewishworldreview.com/cols/crosby072106.asp

2. Levitt, T. (1986). *The marketing imagination.* New York: Free Press, 128.

3. Retrieved September 16, 2005, from my.webmd.com/hw/health_guide_atoz_ty6172.asp

4. Retrieved February 10, 2007, from www.choicehotels.com/ires/en-US/html/CorporateHistory?sid=Lgwc.DLG8ggFmq.8

5. Plummer, J. T. (1974, January). The concept and application of life style segmentation. *Journal of Marketing*, 33-37. Published by the American Marketing Association.

case study

The Rideau Golf and Country Club

Discussion Questions

1. Briefly describe the history of the Rideau Golf and Country Club and the "predominant internal culture" of the club.

2. How does Rideau differentiate itself from the other golf clubs in this Kingston area of Ontario, Canada?

3. What are its current target markets? Describe the emerging market segments that are creating concerns for the board of directors and the general manager, Brian Murray, of the club.

4. How does this relate to the changing demographics of the golfing community in general?

5. Briefly describe the sources of revenue for the club. Which appear to be on the rise and which on the decline?

6. What advice would you give to Brian Murray as he prepares to meet with the board of directors? What direction do you think the club should take for the long-run success of the club?

7. Can the culture of a club such as this really be changed? If so, how?

■ "What happened to the good old days? There are kids running around all over the club and slowing down play, they're loud and . . . "

■ "These women! Get them off the course, they're slow! It took me over four and a half hours to play a round—

This case was contributed by Andrea DeVito, Krista Leesement, and Brenda York developed under the supervision of Margaret Shaw, Ph.D., School of Hospitality and Tourism Management, University of Guelph, Ontario, Canada. Names and places have been disguised. All monetary figures are in Canadian dollars. For conversion use 1$C/.85$US. All rights reserved. Used by permission.

it's absolutely ridiculous! Not only do we have to put up with them on the course, but now they're in our lounge, too!"

■ "How am I supposed to eat my dinner when the dining room was closed again for yet another wedding? I pay good money to be a member of this club, and when I want to use the facilities, I want them to be available. I don't want nonmembers taking up my space. What's the priority in this place? Members or money?"

■ "My son, Alfred, really enjoys the junior clinics, and now we play golf as a family on the weekends. We think it was a great place for him to spend time in the summers, and we love the family . . ."

Brian Murray sipped his morning coffee on the patio of the golf lounge as he sifted through these and other similar customer letters in preparation for the quarterly board of directors meeting. It had been three years since he had become general manager of the Rideau Golf and Country Club. In those three years, Brian had worked hard to fight dropping membership by implementing new programs and changing certain operations. Much of his work had yielded increases in membership in certain demographic categories, but not without severe resistance from the longtime members, who continued to foster a long-standing, traditional Old Boys' Club culture. Judging from the reaction of the members in these and other similar letters, he wondered which direction the meeting, and his club in general, would take.

PROFILE OF KINGSTON

Kingston had a population of 136,415 and was located in Eastern Ontario just south of six-lane cross-country Highway 401, approximately halfway between Toronto and Montreal. The city had a very high number of tourists during the summer who were drawn to Kingston's many museums and historical sites. Kingston was also a world-class sailing destination and hosted many sailing regattas on Lake Ontario, on which it was situated. Kingston was home to approximately 25,000 students who attended either Queen's University, Royal Military College, or St. Lawrence College. Kingston was also one of Canada's top 10 retirement destinations. It was a prime location for retirees because of its low cost of living and its proximity to many of Ontario's major centers and "cottage country" (Rideau Canal, Thousand Islands).

Exhibit 1 shows that Kingston's highest population distribution was in the 30- to 44-year-old category.

CURRENT TRENDS IN THE GOLFING INDUSTRY

Canada's golf industry had experienced rapid growth in the past 10 years. According to the Royal Canadian Golf Association, the number of golfers in Canada had increased from 3.8 million in 1996 to 5.2 million in 2004. Canada had the highest golf participation rate per capita in the world, with a national average participation rate of 19.4 percent. A huge influx of new Canadian golfers had a major impact on the industry. Some of the most obvious trends were an increase in both female and junior golfers, an increase in the demand for more public golf courses, and a shift toward market diversification by private courses.

Golf was still predominantly a male pastime in Canada. Men accounted for 67 percent of total golfers; however, this number had been dropping steadily as women became increasingly more interested in the sport. In 1996 only 18 percent of Canadian golfers were women, this number had grown to 33 percent in 2004.

Exhibit 1	Population Distribution by Sex and Age in Kingston					
Age (years)	Males	% of Total	Females	% of Total	Males and Females	% of Total
Under 14	13,380	9.81%	12,950	9.49%	26,330	19.30%
15–29	17,045	12.49%	16,110	11.81%	33,155	24.30%
30–44	16,615	12.18%	16,950	12.43%	33,565	24.61%
45–59	10,385	7.61%	10,350	7.59%	20,735	15.20%
60–74	7,330	5.37%	8,685	6.37%	16,015	11.74%
75+	2,340	1.72%	4,275	3.13%	6,615	4.85%
Total	67,095	49.18%	69,320	50.82%	136,415	100%

The average age of Canadian golfers was 39, a result of an increasing number of junior players taking up the sport accompanied by a relatively stagnant senior participation. In 1996 there were 325,000 junior golfers aged 12 to 17 and 331,000 senior golfers aged 65 and older. In 2004 there were 389,000 junior golfers and 359,000 senior golfers. Research showed that average rounds played increased dramatically with age. Although the number of junior golfers had increased, the number of rounds they played per year was significantly less than that of the senior and intermediate age golfers. Male juniors played on average 11.2 rounds of golf per year and seniors played approximately 37.3 rounds. Exhibit 2 depicts demographic profiles of Canadian golfers in more detail.

In many areas of the country, there had been an insufficient supply of public courses. As a result, more public courses were being constructed. Semiprivate courses had become very popular, accounting for 50 percent of all rounds played in 2004. This was followed by public courses with 31 percent of play, and private courses with 19 percent of play.

Changing demographics had forced many private courses to rethink their position in the market. There was a greater demand for private courses that catered to families, juniors, and women. Many private clubs were opening their doors to occasional public golfers. For example, at St. George's, a private golf course in Toronto, management had chosen to make the course available to nonmembers on Sunday afternoons and evenings when member demand was low. More private courses were making changes that would allow them to capitalize on emerging industry trends. Brian MacDonald, director of membership development for the Royal Canadian Golf Association, had said that "without programs to introduce and keep people in the game, the industry will have trouble maintaining or increasing what it has right now."

HISTORY OF THE RIDEAU GOLF AND COUNTRY CLUB

The Rideau Golf and Country Club was first established in 1917 as a golf course with only six holes. In 1929 Stanley Thompson, a reputable golf course architect who had designed world-class courses such as those at Jasper Park Lodge and Banff Springs Hotel, expanded Rideau into an 18-hole golf course.

To generate year-round income, in 1961 Rideau added curling rinks to the club. The curling club was considered the secondary product at Rideau.

The club had traditionally catered mainly to males; this had discouraged female membership. This dated focus had contributed to the existent Old Boys' Club culture that was now evident at Rideau. The golf lounge was originally open only to men, but in 1992 the lounge opened its doors to women. Still, men typically kept to one side of the lounge and women to the other. Although there were no formal rules for this division of sexes, a planter informally acted as a divider in the room. On occasions, when the planter had been removed, male members had said that they were uncomfortable because it seemed that the women might be "taking over" the lounge.

In 2001, as a result of a battle fought and won by Rideau member Dr. Brenda Billings, female membership fees were raised so that they were equivalent to men's fees. The women, then, were granted the right to tee off during any of the club's operating hours, rather than restricting them to off-peak hours, as had previously been the case. In an attempt to attract more members, Rideau's management had also recently implemented a junior program for golfers under 18 years of age. This program consisted of sponsoring junior clinics for members as well as for juniors from the Kingston area. As a result, junior membership had more than doubled in a 10-year period.

Rideau's management team was trying to increase the number of out-of-house functions, such as weddings, Christmas

Exhibit 2		Selected Demographic Profiles of Canadian Golfers					
Age Range (years)	%	Age Range (years)	%	Household Income	%	Golfers Who Are Members at Private Club	%
Male 12–17	8%	Female 12–17	3%	<$30,000	17%	Yes	13%
Male 18–34	25%	Female 18–34	11%	$30,000–$49,999	25%	No	87%
Male 35–49	23%	Female 35–49	8%	$50,000–$74,999	25%	Total	100%
Male 50–64	12%	Female 50–64	4%	>$75,000	33%		
Male >65	5%	Female >65	1%	Total	100%		
Total	73%	Total	27%				

parties, and business meetings. These nonmember functions had become an important source of revenue for the golf course. The dining room was closed to membership to accommodate many of these functions. Some members had expressed dissatisfaction because they felt that their substantial membership fees should ensure that they would not be inconvenienced by nonmembers. Brian Murray was having difficulty balancing the needs of members and the lucrative revenue generated by the nonmember functions. He was hesitant to turn nonmember functions away, because revenue from internal membership use of the dining room was on the decline.

CULTURE OF THE CLUB

The Rideau Golf and Country Club was perceived in the community as an exclusive private club that attracted, and catered to, many of Kingston's elite. The course itself had a strong reputation in golfing society. In 2000 the Rideau Golf and Country Club had been ranked as one of the top 50 golf courses in Canada, and one of Canada's top five most underrated courses by *Score* Magazine.

The president and board of directors were members of the club. They were nominated, and subsequently voted in, by Rideau's shareholders each year. The board of directors was composed of 10 male members and two female members. The average age was 57, with only two just under 50. In 1950, 100 stock shares were issued to each member. Since then, the number of shares had been split several times in an attempt to diversify power. When shares were split, existing shareholders were given purchase priority. The majority of original shareholders opted to capitalize on this opportunity. As a result, for numerous years the same 100 people had maintained voting control over the board of directors and, consequently, the golf course in general.

The predominant internal culture at Rideau could best be described as an Old Boys' Club atmosphere. Traditions and values that were established in 1917, when the club first opened, continued to exist. Many of the current members had been at the club for numerous years, and, in many cases, their parents and grandparents had been members during the early years of the club. A strong, personal commitment to the club had developed among such members. They felt at home and enjoyed the numerous annual tournaments, ceremonies, and social events that were mainstays at Rideau. These traditions were also very attractive to new members who were seeking an excellent, challenging course combined with business and social opportunities.

On the flip side, with such a strong "old school culture," there existed a sense of conservatism and resistance to change. This stubborn attitude was probably a reflection of members' ages and their self-perceived status in society. Often new, innovative-thinking management found this culture intimidating since the members felt, because of their long-standing history at the club, that they knew better than the management what would and would not work. As a result, Rideau's management had felt a great deal of resistance from the members toward any attempted or actual changes the club had made to adjust to modern times. For example, the board of directors and members debated for a full year over whether to install an elevator at the club (they finally did install it). Another time, management faced intense heat from members when they announced that member fees would increase by $2 per month in order to plant more flowerbeds at the clubhouse. Many members still believed that the golf lounge should have remained a men's only lounge and that women's tee-off times should still come second to men's tee-off times. These same members had also been known to complain if young, attractive women were not hired to serve in the golf lounge for the summer. The board of directors was unwilling to consider the addition of a swimming pool, health club, or tennis courts. Such changes would be similar to those of other private clubs that were attempting to attract a younger membership by catering to families.

Membership at Rideau was on the decline. This was likely due to the predominant old age of the members, as well to the club's inability to accept change, which was inevitably driving away many of its potential new members. It was important to note that the culture issues evident at Rideau were predominant in the golf membership; almost no culture problems existed within Rideau's curling population.

CLUB LAYOUT AND DESIGN

Rideau accommodated both golf and curling memberships and, as such, was open year-round. There were several eating and drinking facilities at Rideau that were open to members only. These included the following:

■ **Dining Room.** Dinner only, formal atmosphere, and fine dining; overlooking Lake Ontario

■ **Club Café.** A casual restaurant for lunch and dinner

- **Golf Lounge.** Overlooks the 18th hole, open during the golf season, lounge food such as sandwiches, burgers, and beer
- **Curling Lounge.** Similar to the golf lounge, open in the winter, often used for receptions and special events year-round
- **10th Tee.** Snack shop booth located at the 10th hole
- **Other Amenities.** Driving range, practice putting and chipping green, men's and women's locker rooms, coed sauna, and boardroom

Rideau had a par of 70 and was 6,982 yards long with 18 holes. The course was considered one of the best in eastern Ontario. Rideau's tight fairways and small greens made it an extremely challenging course. There was an abundance of water and mature tree growth, making the course aesthetically pleasing. The course was extremely well manicured; much of this could be attributed to the computerized underground watering system. Rideau's speed of play was relatively slow, as a result of its aging membership.

Rideau owned a significant amount of undeveloped land. It had recently received a bid from the city to purchase the land for $1 million. The land was large enough to accommodate another 18-hole course. The city wanted to use the land as a public park. Rideau currently did not have enough money to develop a second course, but decided to hold onto the land for future expansion.

MEMBERSHIP

There had been a decline in both golf and curling memberships since 1993. The number of golfers had decreased from 881 in 1993 to 825 in 2005, and the number of curlers had decreased from 424 in 1993 to 370 in 2005. However, substantial increases were evident in both the female and junior categories. The number of female members was 244 in 1993 and was projected to be 375 by the end of 2005. The number of junior members would be 120 by the end of 2005, up from 49 in 1993. The membership data are listed in Exhibit 3.

Exhibit 4 summarizes the golf membership distribution at Rideau. Clearly, the majority of Rideau golfers were seniors (383). There had also been notable declines in the number of intermediate golfers (168 in 1993 to 103 in 2005). This decline was attributed to the shift of these golfers from the intermediate age category to the senior age category. There had also been a significant increase in junior golf memberships. In 2005 Rideau had 100 junior golfers compared to 41 in 1993. A large number of the members were retirees who had migrated from Toronto or other places outside Kingston. The remaining senior golfers were independent businesspeople and people affiliated with the universities and the hospital.

REVENUE

Approximately 80 percent of membership revenue came from annual fees, and 20 percent came from entrance fees. Total revenue, and golfing and curling revenues, had declined in the last few years. For example, curling revenue went from $83,991 in 1993 to $76,000 in 2005. In 1999 golfing revenue was $1,008,928, dropping to $985,000 in 2005. Overall revenue declined to $1,398,000 in 2005 from $1,417,764 in 2002.

Internal food and beverage revenue was generated via sales to members at the golf course. On the other hand,

Exhibit 3	Membership Figures at Rideau Golf and Country Club											
		1993		1996		1999		2002		Projected 2005	
		Total	%	Total	%	Total	%	Total	%	Total	%
Golfers	Male	678	50.04%	649	48.40%	502	39.31%	466	36.78%	450	36.14%
	Female	162	11.96%	178	13.27%	247	19.34%	262	20.68%	275	22.09%
	Junior	41	3.03%	52	3.88%	73	5.72%	92	7.26%	100	8.03%
	Total	**881**	**65.02%**	**879**	**65.55%**	**822**	**64.37%**	**820**	**64.72%**	**825**	**66.27%**
Curlers	Male	334	24.65%	312	23.27%	297	23.26%	281	22.81%	250	20.08%
	Female	82	6.05%	88	6.56%	93	7.28%	98	7.73%	100	8.03%
	Junior	8	0.59%	12	0.89%	15	1.17%	18	1.42%	20	1.61%
	Total	**424**	**31.29%**	**412**	**30.72%**	**405**	**31.71%**	**397**	**31.33%**	**370**	**29.72%**
Social Members		50	3.69%	50	3.73%	50	3.92%	50	3.95%	50	4.02%
Total		**1,355**	**100%**	**1,341**	**100%**	**1,277**	**100%**	**1,267**	**100%**	**1,245**	**100%**

| Exhibit 4 | Golf Membership Distribution at Rideau Golf and Country Club |

	1993		1996		1999		2002		Projected 2005	
	Total	%	Total	%	Total	%	Total	%	Total	%
Senior golfers (>46 years)	383	43.47%	374	42.55%	361	43.92%	378	46.10%	384	46.55%
Husband/wife golfers (>46 years)	102	11.58%	92	10.47%	96	11.68%	102	12.44%	106	12.85%
Husband/wife golfers (19–45 years)	78	8.85%	74	8.42%	76	9.25%	54	6.59%	48	5.82%
Intermediate golfers (19–45 years)	168	19.07%	182	20.71%	128	15.57%	109	13.29%	103	12.48%
Junior golfers (<18 years)	41	4.65%	52	5.92%	73	8.89%	92	11.22%	100	12.12%
Nonresident member	26	2.95%	24	2.73%	12	1.46%	17	2.07%	15	1.82%
Corporate social	50	5.68%	50	5.69%	50	6.08%	50	6.10%	50	6.06%
Clubhouse	33	3.75%	31	3.53%	26	3.16%	18	2.20%	19	2.30%
Total	**881**	**100%**	**879**	**100%**	**822**	**100%**	**820**	**100%**	**825**	**100%**

| Exhibit 5 | Revenue Figures for the Rideau Golf and Country Club |

	1993		1996		1999		2002		Projected 2005	
	Total	%	Total	%	Total	%	Total	%	Total	%
Curling	$83,991	7.17%	$79,662	6.41%	$79,433	5.68%	$78,834	5.56%	$76,000	5.44%
Golfing	$826,218	70.52%	$884,948	71.20%	$1,008,928	72.15%	$1,006,750	71.01%	$985,000	70.46%
F&B—Internal	$188,760	16.11%	$192,580	15.50%	$216,400	15.47%	$210,680	14.86%	$202,200	14.46%
F&B—External	$72,600	6.20%	$85,660	6.89%	$93,670	6.70%	$121,500	8.57%	$134,800	9.64%
Total	**$1,171,569**	**100%**	**$1,242,850**	**100%**	**$1,398,431**	**100%**	**$1,417,764**	**100%**	**$1,398,000**	**100%**

external food and beverage revenue originated from nonmember functions, such as wedding receptions, Christmas parties, and business meetings. Internal food and beverage revenue, although higher than external food and beverage, had decreased from $216,400 in 1999 to $202,200 in 2005. However, external food and beverage revenue had steadily increased from $72,600 in 1993 to $134,800 in 2005. Revenue figures for Rideau are listed in Exhibit 5.

As a private golf club, Rideau had two sets of fees. The first set was the entrance fee, a one-time payment that a customer paid upon becoming a member. In certain instances, Rideau allowed members to pay the entrance fee in installments over a five-year period, but, in doing so, these members paid a higher total price. Junior members (18 years of age and under) were exempt from paying an entrance fee. However, as a golfer grew older, he or she was subject to the entrance fee of the membership category that corresponded to his or her age.

The second fee was the annual fee. Paying this fee allowed members unlimited golf, full clubhouse privileges, and coverage of their required fees in the Ontario Golf Association

and Canadian Golf Association. Exhibit 6 is a list of the two sets of fees charged to each member according to age range.

The Rideau Golf and Country Club was a seasonal business. The curling season began in September and ended in April. The golf season usually ran from late April to late October; however, on occasion members might still be golfing in December. Exhibit 7 shows a timeline for the golfing and curling seasons. In the fall, overlap occurred between golf and curling. During this time there were some staffing issues and decisions to make regarding hours of operation. Management had solved this problem by closing the golf lounge during the slow weekday periods and serving both curlers and golfers from the curling lounge bar. Although this might not have been the ideal situation, it did minimize labor and operation costs.

COMPETITION

There were no other private golf courses in the Kingston area, with the exception of Crestwood, a 3,124-yard, 9-hole course with a par of 36. This private course catered mainly to the

Exhibit 6	Fees for Members at the Rideau Golf and Country Club		

Entrance Fees		Annual Fees	
Senior golfers (>46)	$7,000	Senior golfers (> 46)	$1,787
Husband/wife golfers (>46)	$12,000	Husband/wife golfers (any)	$3,301
Husband/wife golfers (19–45)—one-time	$5,100	Intermediate golfers (19–45)	$1,228
Husband/wife golfers (19–45)—5-year	$5,590	Junior golfers (<18)	
Intermediate golfers (19–45)—one-time	$3,000	Parents are members	$371
Intermediate Golfers (19–45)—5-year	$3,500	Parents are not members	$530
Social	$500	Social	$380
Clubhouse	$100		

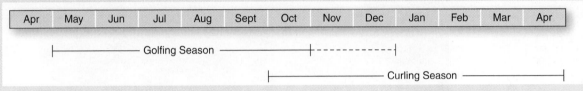

Exhibit 7	Timeline for golf and curling

military personnel in the Kingston area by offering members who were in the military significant membership fee discounts. However, the development of a private golf course was being considered for a senior residential area in Bath, Ontario (15 minutes from Kingston).

There were three semiprivate golf clubs located in Kingston's vicinity. (Note: Semiprivate golf clubs combine private memberships with public golfing privileges, but the members have first priority and special advantages at these clubs.) Foxland was a semiprivate 18-hole course, 5,020 yards long, and the par was 70. It was considered a relatively simple course with a fair level of maintenance and offered limited amenities and member services. Membership fees were $1,000 a year, and one-time public greens fees were $28.

River Ridge had 18 holes and par was 71; the course was 6,293 yards long. The membership fees were close to $1,250 per year, and public one-time, greens fees were $30. The level of course maintenance was good, and the difficulty of the course was medium. Additional amenities included a public lounge, public restaurant, and pro shop.

Canton Meadows had a 6,044-yard course with an 18-hole par of 72. Membership fees were $1,500 per year, and public one-time greens fees were $34. The course was rated fairly difficult and was well maintained. There were two restaurant/lounges at the club. One was open only to members, and the other was open to the public. Both the members and the public were welcome to book the banquet room for special social functions, but member functions took priority.

Fishcreek Golf Course was the only public municipal course in Kingston. The 5,200-yard course was poorly maintained and offered no other amenities other than a pro shop. It was rated an easy 18-hole course with a par of 70. Public one-time greens fees were $18.

CURRENT SITUATION

As Brian Murray entered the boardroom, he mulled over the customer letters. Rideau had higher junior and female membership than ever before, but why were its revenues decreasing? Why was there such backlash from some members over the new direction that the club was trying to take? Brian wanted to please his existing clientele, but he was sure that opening the doors to an increasing number of women and junior members would ensure growth for the club. Hopefully, the meeting would shed some light on the issues that Rideau was facing. Maybe it was just a matter of time before the club's new direction translated into success at Rideau. Maybe some drastic changes were in store. Brian knew that, in a few hours, his vision for the future of the club would become clear, and he hoped this vision would be successful.

overview

Think for the moment about the hotel chain Four Seasons. What words immediately come to mind? If you are like other students we have taught, you may think of words such as *luxury, pampering, high-thread-count sheets*, and perhaps *expensive*. Now, think about the company Marriott. What words come to mind now? *Consistency, good value, functional*, and *sameness from property to property* probably come to mind. Four Seasons and Marriott are both companies. But they also are brands. That is, when people hear their names, words and associations immediately come to mind.

continued on pg 464

chapter 14

Market Positioning and Branding

learning objectives

After reading this chapter, you should be able to:

1. Define market positioning and explain why it is a critical component of marketing strategy.

2. Use subjective and objective positioning in the development of marketing strategy.

3. Define the tangible and intangible approaches to positioning.

4. Define branding and explain how it relates to the positioning statement of a hospitality or tourism enterprise.

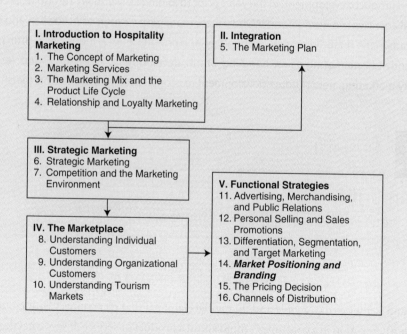

I. Introduction to Hospitality Marketing
1. The Concept of Marketing
2. Marketing Services
3. The Marketing Mix and the Product Life Cycle
4. Relationship and Loyalty Marketing

II. Integration
5. The Marketing Plan

III. Strategic Marketing
6. Strategic Marketing
7. Competition and the Marketing Environment

IV. The Marketplace
8. Understanding Individual Customers
9. Understanding Organizational Customers
10. Understanding Tourism Markets

V. Functional Strategies
11. Advertising, Merchandising, and Public Relations
12. Personal Selling and Sales Promotions
13. Differentiation, Segmentation, and Target Marketing
14. *Market Positioning and Branding*
15. The Pricing Decision
16. Channels of Distribution

Brands have been defined as "the internalized sum of all impressions received by customers resulting in a distinctive position in their mind's eye based on perceived emotional and functional benefits."[1] Essentially, these impressions form a promise to the customer regarding what she will receive when staying at one of these properties. Market positioning, on the other hand, refers to the activities that a firm undertakes to place the brand somewhere in customers' minds so they understand what the brand is, what it is not, and who its competitors are. Without effective positioning, a property can get lost in customers' minds and in the marketplace.

Marketing Executive Profile

John Griffin
Vice President, Marketing, Hotel Group, Minor International PLC

Based in Bangkok, Thailand, John Griffin is working on the expansion of the Anantara resort brand and the development of further lifestyle-based hospitality brands for the company. Prior to this, Griffin worked at Le Meridien in London, England, between 2001 and 2006 after moving from Sydney, Australia. Griffin previously held a range of roles with American Express in advertising and customer relationship management, including areas of new product development. In the years prior to joining Le Meridien, he also worked with global agencies Leo Burnett and Wunderman, managing a range of Australian and global accounts in the financial services, technology, and pharmaceutical industries. Griffin has led a worldwide marketing team handling brand development, advertising and research, loyalty marketing, new product development, customer intelligence, and partnerships.

 ## Marketing in Action

John Griffin
Vice President, Marketing, Hotel Group, Minor International PLC

What is your definition of a brand?

A brand is a series of tangible and intangible elements that can connect with emotion to create a perception in a consumer's mind about a product or service. It says what the brand is all about and should relate to the position that we want to hold in a consumer's mind.

What are the issues that brands such as Anantara and Le Meridien must consider as they develop their branding strategies?

First and foremost, whether you are an established global brand or an emerging boutique brand, you must make sure that the desired strategy resonates with your customers. Then, the range of hotels and the services that we provide in those hotels must have the ability to deliver on the brand proposition that we envision. As a service industry, the branding strategy has to work as much for staff as for customers. It has to touch all aspects of our business comprehensively and consistently—staff, customers, and sales interactions. Even our internal departments need, in some way, to be able to connect with what the branding statement is about. Put simply, it's as much cultural as it is an external marketing premise. And as we work through our product and develop our communication approaches, we're constantly looking at the fact that we are maintaining our differentiation against our competition and that the branding statement remains reflective of the evolving trends and needs in the marketplace at the time.

What should students understand about branding?

One of the most important things about branding is that it actually transcends the printed page. It's as much cultural as physical. A high degree of the experiences in our industry are intangible. Branding is also a living thing—it should always be at the forefront of the business, being worked with by the business and creating the personality for our communications and approach to the market.

Branding needs to provide the tangible foundation for the promise being made and do it in such a way that it sets your product apart from the competition. To succeed, a branding strategy must apply consistently across all aspects of corporate and individual hotel communications so as not to confuse the customer and waste valuable resources competing among ourselves. If there's a disconnect between the two, then we're potentially blurring the brand message.

How do you ensure that?

The first and most tangible element is to ensure that we have a strong set of brand guidelines that we know can apply across the board to a diversity of properties. Within those guidelines, we cover end-to-end communication, from what appears in physical materials in our hotels through to how our advertising looks, our in-house communications, and our Web development. The most important thing is to be flexible enough within the guidelines to allow the brand to be articulated in a way that best suits the many markets in which we do business. The aim is to keep the brand guidelines alive and make them relevant, which means we can talk as one brand.

What kind of a career path would you recommend for someone who wants to do branding and would like to be eventually in your position? What kinds of courses should they be taking, and what kind of careers should they look for to move in the right direction of branding?

Having sufficient experience with a breadth of different roles at the property level first enables you to understand the challenges in creating a guest experience. Time in a corporate office role then allows you to see the other side of the experience. It can even help to have exposure to the marketing discipline from outside of our industry as well—it enhances the ability to understand a brand in the consumer's mind, especially that we aren't the only ones competing for space. Look at taking a graduate diploma or equivalent in marketing to expose yourself to the mix of disciplines. Most important, be willing to try different experiences to broaden your overall understanding of the business.

So, it is best to start in operations?

For any business, but particularly hospitality, matching the operational delivery to the marketing promise is essential. Therefore, what really makes an effective marketer is the ability to have worked and understood how the operation applies to the branding promise and vice versa. In my experience, there's always a healthy tension between marketing and operations on delivering the brand, and to succeed as a marketer you must be able to understand and work with the bounds of the operation. Don't overpromise—but also find the right way to influence the organization to grow and deliver better in the future. And, yes, the disciplines of branding can be developed in other categories than ours—particularly the ability to understand customer dynamics.

Used by permission from John Griffin.

---

positioning Creating an image, differentiating a product, and promising a benefit in the mind of the consumer. Positioning is the perception the consumer has of a product offering.

*Market **positioning** is the natural follow-through of market segmentation and target marketing. In fact, it is on those strategies that positioning is built because they define the market to which the positioning is directed. Therefore, we must select and understand our target markets before we can develop effective and efficient positioning strategies. The objective of positioning is to create a distinctive place in the minds of potential customers—a place where customers know who the firm is, how the firm is different from the competition, and how the firm can satisfy customers' needs and wants. It is about creating the perception that the firm is best able to solve customers' problems.*

The firm that does not create a distinctive place in customers' minds faces several pitfalls:

1. The firm is forced into a position of competing directly with stronger competition. For example, a weakly positioned independent midscale hotel may be pushed into a losing situation with a clearly positioned Courtyard by Marriott.
2. The firm's position is unclear so that it lacks true identity and customers do not know what it offers and what needs are fulfilled. In other words, there is no clear perception. This often happens when a property or chain tries to be all things to all people.
3. The firm has no position in customers' minds so that it lacks top-of-the-mind awareness and is not part of the customer's mind-set. A name like Joe's Restaurant, for example, provides no perception or image. Contrast this with a name like Kentucky Fried Chicken. Even if you had never heard of the restaurant, you would have an idea of the type of items sold there.

There are actually two kinds of positioning in marketing: **objective positioning** and **subjective positioning.** Each has its appropriate place and usage. Each is concerned with its position in relation to the competition. Before we explain these, however, we need to deal with three customer attributes that are important when positioning.

SALIENCE, DETERMINANCE, AND IMPORTANCE

In evaluating and developing effective positioning strategies, we need to understand how customers perceive and differentiate among salient, determinant, and important product or service attributes or benefits. One might, for example, position on a salient benefit with poor results because those benefits are not necessarily important to the customer. To develop branding and positioning strategies, it is important to be able to determine why the customer is buying a product or service. Once this has been determined, the positioning of the product or service becomes more natural.

Salience

Salient attributes are those that are "top of the mind." They are the ones that readily come to mind when you think of an object. Because of this, a list of strictly salient attributes obtained from customers may be totally misleading in describing how they make choices. If you were asked, "Why did you buy that shirt?" you might say, "because it was on sale." If we then assumed that the next shirt you buy will be one on sale, we could be making an incorrect assumption. What really determines your choice could be the style of the shirt; the sale price was just an added bonus to help you make the purchase.

Salient factors may be determinant factors, but they are not determinant when they are not the true differentiating factor in the eyes of the customer. They are also not determinant when they are common throughout the product class. Consider, for example, chocolates on the pillow. This could be very salient and be remembered by customers, but it is doubtful that they would base their choice of hotel on what type of chocolates are placed on the pillows in their hotel room. In fact, Jonathan Tish, CEO of

objective positioning
Presenting a product to the consumer based on its physical characteristics or functional features. Objective attributes of a product include atrium lobbies, historic landmarks, signature golf courses, and wireless connectivity to the Internet.

subjective positioning
Presenting an "image" of a product to the consumer and not the actual physical aspects of the product. Subjective attributes of a product include prestige, service quality, and guest experience.

salient attributes
Attributes of a product or service that readily come to mind—that is, are "top of the mind" to the consumer. They may or may not be important or determinant in the final purchase decision.

Loews Hotels, authored a book in 2007 titled "Chocolates on the Pillows Aren't Enough: Reinventing the Customer Experience."

Now consider location. Take a survey of almost any set of hotel customers and ask what is important to them in choosing a hotel. At the top of the list will almost always be location. Location is a very salient attribute, but if six restaurants are within four blocks of each other in Chicago or right next to each other as in "fast-food rows" everywhere, or if eight resort hotels are within five miles of each other in Palm Springs (as is the case in so many areas today), location is not likely to be a determinant factor. In marketing, we most often use salient factors to get attention and create awareness.

Determinance

A study that looked at the influence of different value drivers for hotels found that although location and amenities were important for business travelers, price was overall the most important determinant of value. For leisure travelers, amenities are the most important determinant of value.[2]

determinant attributes Attributes of a product or service that actually determine the decision to purchase a product.

Determinant attributes are those that actually determine choice, such as reputation, price/value, or level of service. These are the attributes most closely related to customer preferences or actual purchase decisions; in other words, these features predispose customers to action. These attributes are critical to the customer choice process. The problem is that customers do not always know exactly what forms the basis of their choice.

An example is bathroom amenities. Bathroom amenities may not be very salient, but they could be quite important after we have become used to having them. If every hotel in the product class has them, however, they are hardly determinant any more. There is an exception here, however. If we were now to remove the extended line of bathroom amenities, they might become negatively determinant; that is, people might say, "I won't go there because they don't have good bathroom amenities." The implication is that perhaps hotels in this product class should now have the amenities, but promoting them or positioning on them would be useless.

This is also true of location and cleanliness, supposedly the main reasons that people choose hotels. People don't choose hotels simply because of location and cleanliness; however, they do choose against specific hotels because of their lack of location and cleanliness. In marketing, we most often use determinant factors to persuade customers to make a choice.

Importance

importance attributes Attributes of a product or service that are important to a consumer in either making a purchase decision or after having made the purchase. In other words, importance attributes are not necessarily salient or determinant in the final purchase decision.

Importance attributes are the qualities that are important to the customer after having made a choice. The example of bathroom amenities demonstrated this. It is important that they be there, once the customer is accustomed to their being there, but a person does not choose a hotel based on bathroom amenities. Once the choice has been made, what was salient or determinant fades into the background unless, of course, they are found not to exist. Now it is important that the room be clean and the bed be comfortable. In marketing, we most often use importance factors to arouse interest and create a benefit bundle that will lead to determinance. Westin's promotion of the Heavenly Bed is an example. It has always been important that the bed be comfortable. Westin made the bed determinant.

Salience, determinance, and importance are corresponding concepts, and they are all significant in the positioning effort. It is important to understand the place of each. Recall the discussion in Chapter 8 on selective perception, selective acceptance, and selective retention. All three may occur with salient factors. Determinant and important factors are more likely to cause selective retention, because we remember what features helped provide the memorable experience or offered a solution to our problem. Much positioning that is done only on salient factors—for instance, location or an atrium lobby—is less than successful when these factors are not determinant. In summary, good positioning requires that creating an image, differentiating the product or service, and making a promise are all based on determinant and/or importance factors.

OBJECTIVE POSITIONING

Objective positioning is about the objective attributes of the physical product. It means creating an image about the product that reflects its physical characteristics and functional features. It is usually concerned with what actually exists. Consider the statement "The car is red." We can all see that it is red. If the company that makes this car makes only red cars, we might call it "the red car company." We would carry an image of these cars as opposed to those made by "the green car company." Or, we could say, "That building is tall." Again, we would all likely agree.

That's a little simplistic, so let's apply it to the hospitality industry. Econo Lodge is a low-cost motel; Hyatt Regency's Cerromar Beach Resort & Casino in Puerto Rico is on the beach; Ruth's Chris Steakhouse sells steaks. All of these businesses bring to mind specific images based on the name itself—it comes from an objective, concrete, specific attribute. If we know anything about the product (for instance, Marriott Marquis), we know at least that much.

Objective positioning does not always need to be concrete. It may be more intangible than these previous examples. Ferraris are not only red; they also go fast. A Ritz-Carlton is a luxury hotel; McDonald's offers quick service; Le Cirque 2000 offers gourmet meals and fine wines. Again, these images derive from the product itself.

Objective product positioning can be very important and is often used in the hospitality industry. Westin has The Heavenly Bed; Red Lobster positions on seafood; and Olive Garden, on Italian food. The Plaza Athenee, located in New York City, positions on its proximity to Central Park and the fact that it is a boutique hotel with 115 guest rooms located in the residential area of the east side of Manhattan. If a product has some unique characteristic or unique functional feature, that feature may be used to objectively position the product, to create an image, and to differentiate it from the competition, such as that depicted in the advertisement in Exhibit 14-1. The suggestion of Exhibit 14-1 is that, while there are lots of spas, very few offer the natural and private setting found at a Marriott Hotel & Resort.

Less successful objective positioning occurs when the feature is not unique, as shown in Exhibit 14-2, which shows a table in the restaurant. In other words, this ad creates no real position in our mind about this, although it may create awareness, which is probably its intent. Totally unsuccessful approaches include a picture of two people in a hotel room or, worse, a picture of an empty restaurant with waiters standing at attention.

THESE FIVE RESORTS SET THE STANDARD FOR LUXURY.

We deliver luxury stripped of pretense and brimming with relevance. Sophisticated, unobtrusive and very personal. It is based on the recognition that it's all about you. You'll enjoy this luxury in surroundings of uncommon beauty in five of the most enviable locations in all of North America. We invite you to take a visual tour of each JW Marriott Resort & Spa by visiting the web site below. **Arrive.**℠

JW MARRIOTT.
HOTELS & RESORTS

**JW Luxury Experience
from $264–$579.**
Premium accommodations, with breakfast in bed and the Wall Street Journal included.

**Visit www.jwresortluxury.com
or call 888-770-0139.**

JW Marriott Starr Pass Resort & Spa, Tucson, Arizona

JW Marriott Las Vegas Resort & Spa, Nevada

Camelback Inn, A JW Marriott Resort & Spa, Scottsdale, Arizona

JW Marriott Desert Ridge Resort & Spa, Phoenix, Arizona

Desert Springs, A JW Marriott Resort & Spa, Palm Desert, California

Resort featured in photo: JW Marriott Starr Pass Resort & Spa, Tucson, Arizona. Rates vary by resort. Please contact us for full amenities and limitations.

Exhibit 14-1 Marriott positions its spas on privacy and natural beauty

Source: Marriott International. Used by permission.

SUBJECTIVE POSITIONING

Subjective positioning is a strategy for creating a unique product image to create and keep customers. It exists only in the mind of the customer. It can occur automatically, without any effort on the part of the marketer, and any kind of positioning may result. Two very different products may be perceived as the same; two similar products may be perceived as different. What the marketer hopes to do is to control the positioning, not just let it happen. Failure to select a position in the marketplace and to achieve and hold that position by delivering it, moreover, may lead to many undesirable consequences.

Subjective positioning is much more difficult in practice than objective positioning. It is concerned with subjective attributes of the product or brand. Subjective positioning is the image, not of the physical aspects of the product, but of other attributes as perceived by the customer. They belong not necessarily to the product, but to the customer's mental perception of the product. These perceptions and the resulting image may or may not reflect the true state of the product's characteristics. They may simply exist in the customer's mind, and we might find many who would disagree with particular perceptions and images. What the marketer hopes is that the people in the target market will agree on a favorable image or characteristic, whether or not it is factual. This is the test of effective subjective positioning.

Hilton Hotels' former advertising campaign, "When American business hits the road, American business stops at Hilton," and its former slogan "America's Business Address" are examples of attempts at subjective positioning. The desired image, obviously, was that businesspeople prefer Hilton Hotels. One reason that people might not accept this positioning is that it lacks uniqueness and does not differentiate from the competition. For example, one Hilton advertisement showed an empty conference room with a conference table surrounded by chairs. These are objective product characteristics that clearly are no different from characteristics at thousands of other hotels. Hilton's campaign "Take me to

Exhibit 14-2   This table in a restaurant tells us nothing about the restaurant

Source: Magnus Rew © Dorling Kindersley, Courtesy of the Four Points Sheraton Hotel, Orlando.

the Hilton." According to Robert Dirks, who was senior vice president of brand management and marketing at the time the campaign began in April 2004, the goal was to position Hilton as the number one solution to many of the concerns expressed by modern-day travelers, such as lack of personalized service, uncomfortable environments, and poor technology.[3] Their positioning in 2007 was "travel is more than just getting from A to B."

Tangible Positioning

Two types of positioning are used in the hospitality industry. The first is **tangible positioning**, because the industry's product has almost reached commodity status. In other words, many of the rooms in hotels of the same product class are almost exactly alike. The same is true, to a lesser degree, in restaurants. Consider, for example, McDonald's versus Burger King or Denny's versus Baker's Square. We need to understand what this commodity status means for positioning.

Consider the ultimate commodity (or copied item), salt. How would you use positioning to create a unique image and differentiate your salt from someone else's salt? Morton tries it with the **positioning statement** "When it rains, it pours." This is intended to imply that Morton salt is free-flowing even when the weather is damp, whereas other salts are not. It is not necessarily true that others are not, but if you buy into it, you do so because you differentiate Morton's salt from other salts based on the physical characteristic of being free-flowing. Salt is a very tangible good; it would be difficult to argue that salt is exotic, tantalizing, or romantic.

Those arguments, however, could be made for cosmetics, and they certainly are, as we all know. Cosmetics are mostly tangible. However, their successful marketing is based on mental perceptions of intangible results. As the founder of Revlon Cosmetics said, "In the factory we make cosmetics; in the drugstore we sell hope." If we are selling a

tangible positioning
Creating an intangible, subjective image of a product based on a tangible feature of the product.

positioning statement
A singular expression by a property or brand that captures the essence of its intended positioning in the marketplace.

freshly made

green tea ice cream

delivered to your door

what can we do for you?

THE
ALEX
overnight or over time

200 impeccable guest rooms and deluxe suites

concierge services beyond expectation

flat-screen TVs in all bedrooms,
bathrooms & living rooms

24-hour room service from Riingo® and
award-winning chef, Marcus Samuelsson

The Alex Hotel · 205 East 45th Street at Third Avenue · New York, NY 10017
212.867.5100 · www.thealexhotel.com

Exhibit 14-3   This advertisement is an example of subjective positioning

Source: The Alex. Used by permission.

near-commodity product such as a hotel room that is mostly tangible, then we need to develop intangible mental perceptions that may or may not actually belong to the product—hence the expression, "Sell the sizzle, not the steak." As Bob Osgoodby, publisher of the *Add Me Newsletter*, stated, "While the 'nuts and bolts' of your product or offer are important, that is not normally what gets someone's initial interest and makes the sale."[4]

Consider again a hotel advertisement showing a picture of a couple in a hotel room. A hotel room is very tangible. It looks like thousands of other hotel rooms. As with salt, it is very hard to develop a mental perception of a hotel room that differentiates it from other hotel rooms. The two people in the room are also tangible. What's more, they are no different from two people in any other hotel room. Now you see the problem that advertisers have been struggling with for years: How do they position a tangible product that has very little means of differentiation? Exhibit 14-3 shows an attempt to create a subjective position from the tangibleness of a unique dish of green ice cream. The idea is that if the ice cream is unique, the rooms in the hotel are probably also unique. Subjective positioning of tangible features requires developing intangible images.

Intangible Positioning

intangible positioning
Creating a tangible, objective image based on an intangible aspect of the product.

The second type of positioning of the hospitality products is **intangible positioning.** What we are marketing is not tangible; it is intangible. Some would say that is nonsense, because what's more important than the guest room or the meal? They would be right, but that's what we are selling and not what we are marketing. If we were selling guest

Tourism Marketing Application

Objective and subjective positioning work very well for tourist destinations because the objective part focuses on the attributes of the area while the subjective part focuses on what the attributes represent. Consider the island of Tahiti. The beaches are the attributes, while the subjective belief relates to images of love, relaxation, and beauty. More examples of positioning in tourism can be found at this website: www.hotel-online.com/Trends/AsiaPacificJournal/PositionDestination.html (accessed March 26, 2006).

rooms and beds or steaks and sushi bars, what difference would it make where the customer went, assuming a comparable level of quality? That is an assumption we have to make within the same product class, so it doesn't get us very far.

So, again, what we are largely marketing are intangibles. The tangibles are essential and necessary, but as soon as they reach a certain level of acceptance, they become secondary. Because tangibles are so difficult to differentiate, to be competitive we have to market the intangible aspects of the product or service. Even when tangible (e.g., a steak), they have a measure of intangibility because they are consumed rather than taken home to be possessed. The intangible elements are abstract. To emphasize the tangible elements is to fail to differentiate from the competition. Hospitality positioning needs to focus on enhancing and differentiating the intangible realities through the manipulation of tangible cues.

Some hotels do this with atrium lobbies. People don't buy atrium lobbies; they buy what the lobbies make tangible. We might not all agree, but some would say that atrium lobbies are exotic, full of grandeur, majestic, or exciting. These are intangible images and nothing more than mental perceptions. Of course, check-in may be just as slow and the rooms may be no different from those in other hotels, but the image is there, not just the physical characteristics.

What we want to do is create a subjective "position" in the customer's mind. You can see now why positioning follows target marketing so closely: We need to know what mental constructs the customer in the target market holds and what tangible evidence sustains them.

Return for a moment to the steak-and-sizzle argument. If we want to sell the steak, as this argument goes, then we need to market the sizzle. Because our steak is just like all the others, we have to sell the sizzle, the intangible. How do we make tangible the sizzle? There is probably no better example, even 30 years later, than what Jim Nassikas did at the Stanford Court Hotel in the 1970s. (This hotel, located in San Francisco, is now a Marriott Renaissance Hotel.) In fact, he was so successful in positioning the Stanford Court that Nassikas virtually stopped advertising and still ran one of the highest occupancies and ADRs in San Francisco. He positioned his hotel "for people who understand the subtle difference." When he did use advertisements, he used headings such as: "You gloat over a great hotel the way you do over a rare antique find. We designed the Stanford Court for you." Another example was "You're as finicky about choosing a fine hotel as you are about the right patisserie. We designed the Stanford Court for you."

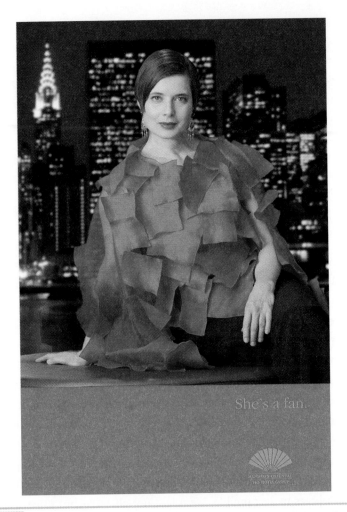

She's a fan.

A more modern example is the advertisement for Mandarin Oriental Hotel Group with the tagline "She's a fan." This is shown in Exhibit 14-4.

Positioning is a relative term. It is not just how the brand is perceived alone, but how the perceived image stands in relation to competing images. Positioning reflects the customer's mental perception, which may or may not differ from the actual physical characteristics. Positioning is especially important when the product is intangible and there is little physical difference from the competition.

HOW TO CREATE EFFECTIVE POSITIONING

Our discussion so far has emphasized image, the mental picture the customer has of the product or service. We have discussed the need for the image to differentiate the brand from the product class. That is, how to make the brand different from other similar brands: Four Seasons versus Carlton. Image is an important criterion for effective positioning, but there is one more criterion. Positioning must also promise the benefit that the customer will receive.

This need to promise takes us back to the basic marketing concept, the idea of needs and wants, and problems and solutions—the promise we make to the customer. It also takes us back to our chapter on understanding individual customers, in which we discussed consumer behavior in terms of customer attitudes. Images and differentiation mean creating beliefs. Next we have to develop the reaction, the attitude toward the belief, and the action that will create the intention to buy.

Effective positioning also must promise the benefit the customer will receive. It must create the expectation and it must offer a solution to the customer's problem. And that solution, if at all possible, should be different from and better than the competition's, especially if one of the competitors is already offering the same solution.

David Ogilvy, a longtime advertising guru and former head of the international firm Ogilvy and Mather, stated:

> Advertising which promises no benefits to the customer does not sell, yet the majority of campaigns contain no promise whatever. (That is the most important sentence in this book. Read it again.)[5] [His parentheses]

Notice how the advertisement in Exhibit 14-4 fills all three of good positioning requirements: It creates an image, differentiates itself, and promises a benefit. If we're lucky, we can capture all of this in a single positioning statement, as Mandarin Oriental did.

Following are some better-known positioning statements, some of which are current and some of which are not. As you read each one, consider the image, the differentiation and the promised benefit, as well as the tangible and intangible aspects.

U.S. Army: An Army of One
U.S. Army: Be all that you can be
Marines: The Few, The Proud, The Marines
Toyota Today: Moving Forward
Toyota Old: Get the feeling
McDonald's Today: I'm Loving It.
McDonald's Old: You Deserve a Break Today.
Burger King: Have it your way.
General Electric Today: Imagination at Work.
General Electric Old: We bring good things to life.
Microsoft Today: Your passion, our commitment.
Microsoft Old: Where do you want to go today?
Holiday Inn Express: Stay Smart.
Motel 6: We'll leave the light on for ya.
Nike: Just do it.

Exhibit 14-5 discusses why Shangri-La Hotel Group developed its positioning statement "Between Heaven and Earth." Exhibit 14-6 shows the advertisement that conveys this position.

How do we determine the desired position for our product or service? Exhibit 14-7 provides a brief checklist for this purpose. This tells us that one may position in a number of different ways, all related to segmentation strategies. As discussed, positioning may be achieved on specific product features, or benefits or a specific usage or user category. Overall, an effective position is one that clearly distinguishes from the competition on factors important to the relevant target market in everything an operation does.

Exhibit 14-5   The Message Behind Shangri-La's Position Statement "Between Heaven and Earth"

"Where will you find your Shangri-La?" But where is Shangri-La? Where is the paradise that James Hilton's *Lost Horizon* described? Is it in the Yunan province, or is it a place at all? Shangri-La is a state of mind, of peace and tranquillity, a feeling, a moment in time, somewhere between Heaven and Earth.

The ad campaign places the Shangri-La moment, the experience that you have when you stay at the Shangri-La, between Heaven and Earth. It showcases the magical moments that our audience will enjoy and asks them the question, "Where will you find your Shangri-La?"

The ads have a very graphic look and feature the words *Heaven* and *Earth* in a heavy bold typeface. The words frame people experiencing and enjoying magical Shangri-La moments. The scenes in the ads are reflecting the real location of the different Shangri-La hotels and resorts within the group.

Source: Developed for Shangri-La Hotels and Resorts. Used by permission.

Exhibit 14-6   This advertisement differentiates Shangri-La from similar properties

Source: Developed for Shangri-La Hotels and Resorts. Used by permission.

Exhibit 14-7   Checklist for Determining a Desired Position

- Analyze product attributes that are salient and/or determinant and/or important to customers.
- Examine the distribution of these attributes among different market segments.
- Determine the optimal position for the product/service in regard to each attribute, taking into consideration the positions occupied by existing brands.
- Choose an overall position for the product, based on the overall match between product attributes and their distribution in the population and the positions of existing brands.

Source: From *Marketing Planning, and Strategy*, 5th edition, by Jain. © 1997. Reprinted with permission of South-Western, a division of Thomson Learning: www.thomsonrights.com. Fax 800-730-2215.

POSITIONING'S VITAL ROLE

We have dealt with positioning so far in the context of advertising. However, this is not the only context in which positioning should be used. Positioning plays a vital role in the development of the entire marketing mix, whether it is the 7Ps or the 13Cs. Positioning should be a single-minded concept, a source from which everything flows. As such, positioning affects policies and procedures, employee attitudes, customer relations, complaint handling, and the other details that combine to make a hospitality experience. There must be a consistency among the various offerings—we cannot have a wonderful waitperson in the restaurant only to have a nasty front desk clerk—and the positioning statement guides this consistency. All this means is that the firm must live up to the position it creates. If it doesn't, it is setting false expectations for the consumer and, consequently, dissatisfaction.

Positioning is also about creating a marketing niche. Kyle Craig former president and CEO of S&A Restaurant Corp., operators of the Steak & Ale, Bennigan's, JJ Muggs, and Bay Street chains and also past president of Kentucky Fried Chicken's domestic business, pioneered the repositioning of Kentucky Fried Chicken by changing its name to KFC and launching its first nonfried chicken products. He stated:

> When we talk about a marketing niche we are really talking about positioning. You must position your concept as offering a unique product or service. The key is to understand the customer decision and then use it to your advantage to successfully stimulate sales. Once you understand what the customer wants and match that against what your chain has to offer, you have a better chance of success.
>
> ... Finding a niche is tough but delivering the restaurant experience the niche demands is tougher. ... Once the concept matches customer needs there are two litmus tests. First, your position must be believable in the customer's mind. Second, you must deliver on the promise on a consistent basis. [Craig also warns us to] watch out for a niche that is restaurant-driven rather than customer-driven.[6]

REPOSITIONING

Repositioning involves changing a position or image in the marketplace. The process is the same as initial positioning with the addition of one other element—removing the old positioning image.

There may be a number of reasons for wanting to reposition. You may be occupying an unsuccessful position in the first place. Or you may have tried and failed to fully achieve a desired position. Also, you might find that competitors, too many and/or too powerful, have moved into the same position, making it overcrowded. You may also perceive a new opportunity you wish to take advantage of.

All of these situations are relatively common in the hospitality industry. Hamburger chains have tried repositioning as "gourmet" hamburger restaurants. Friendly's, originally an ice cream and sandwich chain, repositioned itself as a family restaurant. Many restaurants that often change hands are constantly repositioning with new names. Dunkin' Donuts' successful turnaround from a donut shop to a coffee shop is described in Exhibit 14-8.

Repositioning might also be used to appeal to a new segment, to add a new segment while at the same time trying to hold on to an old one, or to increase the size of a segment.

Exhibit 14-8   The Repositioning of Dunkin' Donuts

Dunkin' Donuts, like many large companies, has gone through multiple ownerships and multiple changes in positioning. The company started in 1950 in Quincy, Massachusetts. By 1963 there were 100 franchised stores on the East Coast of the United States, with expansion growing into the Midwest and Southwest by 1982. In 1990 the firm was purchased by Allied Domecq PLC, which also owns ice-cream shop Baskin-Robbins and sandwich shop Togo's.

At the time of the purchase by Allied Domecq, Dunkin' Donuts was losing much of its breakfast sales to companies such as McDonald's and Burger King, while at the same time losing its "coffee crowd"—which consisted of approximately 2.7 million people per day—to specialty coffee shops such as Starbucks. In 1994 William Kussell was brought in from Reebok to spice up the menu and reputation. He changed the old slogan, "America's Number One Donut Chain," to "Dunkin' Donuts: Something fresh is always brewing here" and began a play to become the breakfast king. He switched the strategy of selling doughnuts by the dozen to selling a more frequently consumed item: coffee.

Dunkin' Donuts introduced four or more blends of fresh-brewed coffee and hot and cold specialty drinks, all at a fraction of the Starbucks price. A 10-ounce Dunkin' Donuts coffee costs, on average, $1.19, while a 12-ounce cup of Starbuck's coffee costs between $1.40 and $1.65, depending on store location. Value and no-nonsense service has now positioned Dunkin' Donuts against Starbucks and McDonald's in a serious battle for the breakfast buck. John Gilbert, Dunkin' Donuts vice president of marketing, states that "counterintuitively, Dunkin' Donuts is primarily a coffee company. Unlike our coffee competitors, we also sell delicious fresh baked goods, such as breakfast sandwiches on fresh bagels. Our business model is to sell a cup of coffee plus a baked good to every customer. It is this focus on selling beverages and baked goods together that has helped us thrive when our competitors are suffering in this low-carb era.

In addition, Allied Domecq housed its three brands (Dunkin' Donuts, Baskin-Robbins, and Togo's) under one roof with the idea that each brand would cater to an audience at different times of the day. It was a failure, as customers were confused and franchisees complained that it was too much to manage all three brands. In the fall of 2005, the practice of housing the brands in the same building was terminated. The focus on coffee was, however, a success. As a result of the switch, 2005 sales were up 14% from the previous year, with coffee making up 62% of sales.

In addition to moving away from only donuts toward specialty coffees, oven-baked bagels, and fat-free muffins, the firm also started to remodel their stores: To ensure that they would get the new layout and color scheme just right, Dunkin' Donuts paid faithful Dunkin' Donuts customers to fill their food and beverage needs at Starbucks for a week. They also paid Starbucks customers to switch to Dunkin' Donuts for the same period of time. The company concluded that Dunkin' Donuts customers felt the Starbucks environment was "pretentious" and "trendy," and the focus on the individual seemed a bit bewildering and even disingenuous at times. The people who frequented Starbucks felt Dunkin' Donuts was "unoriginal" and "austere" by comparison and wanted to have more control over things such as how much sugar and cream should go into their drinks.

The tacky old Dunkin' Donuts pink decor was replaced with a more upscale "ripe raisin" hue. The square laminate tables were replaced with round imitation-granite tabletops and sleek chairs. Customers loyal to Dunkin' Donuts had only one concern: that the changes not result in longer waiting lines, which now average about two minutes per customer from the time they order to the time they get through the register. Starbucks, by comparison, has a goal time of about three minutes per customer. Dunkin' Donuts also elected not to install wireless Internet access because it did not fit with customers' image of Dunkin' Donuts.

In December 2005 Dunkin' Donuts was sold by Pernod Richard SA, which had acquired Allied Domecq PLC to an investment group consisting of Thomas H. Lee partners, Bain Capital Partners, and the Carlyle Group. The new CEO is Jon Luther.

Dunkin' Donuts has more than 6,000 stores in the United States and 29 other countries

Source: Dunkin' Donuts Press Room. Retrieved from www.dunkindonuts.com/aboutus/press/PressRelease.aspx?viewtype=current&id=100042 on May 5, 2005; http://online.wsj.com/article/SB114446712300420923-search.html? KEYWORDS=dunkin+donuts&COLLECTION=wsjie/6month accessed on April 23, 2006. Also adapted from Dunkin' Donuts is on a coffee rush. *BusinessWeek* (1998, March 16), 107–108.
Ashley Trevitz, from the University of Houston, helped to adapt this material.

Web Browsing Exercise

Visit the websites for Dunkin' Donuts (www.dunkindonuts.com) and Starbucks (www.starbucks.com). How is their positioning reflected by their websites? Compare and contrast.

Club Med, as we have noted, repositioned to the family market but still keeps its old market (and some of its old image) at some well-defined properties. Another reason for repositioning could be that new ownership desires a new position or wishes to merge the position of a newly acquired property into that of other properties already owned. Finally, repositioning would be called for in developing a partially or totally new concept, downgrading a property that has become distressed, or upgrading one that has been refurbished.

Examples of repositioning include when Holiday Inns, before being bought by InterContinental Hotels, separated its upscale Crowne Plaza line by removing the Holiday name and marketing it separately. In New York City, Schrager Hotels renovated a rundown hotel, Morgans, and repositioned it as a chic, trendy hotel. They did similar renovations with the Paramount, Barbizon, and St. Moritz. When Schrager ran into financial trouble, Schrager Hotels was renamed Morgans Hotel Group. Affinia

Tourism Marketing Application

The bombings in Bali in 2002 had a major impact on tourism to that country. Figures released in 2005 revealed that the number of tourists visiting Bali in November 2005 was down 42.5 percent from the number visiting in November 2004. To reverse this trend and plan for the future, a conference on positioning, repositioning, crisis management, and image recovery was held in Bali in December 2005. The United Nations World Tourism Organization (UNWTO) organized this conference. "The aim was to help Bali to rapidly regain markets, by not only recuperating and rebuilding confidence in the short term, but also to plan for the future, so as to enable the destination to confront the challenges facing the industry and to refocus efforts to regain market shares and acquire new ones."

Source: www.bali-tourism-board.com/news.php?tit198&month12&year2005 (accessed March 25, 2006).

Hospitality, formerly Manhattan East Suite Hotels, repositioned the old two-star Beverly hotel into a four-star product called the Benjamin. Starwood Hotels bought the Doral Hotels, also in New York, and has repositioned them as its luxury brand, W.

In another case, Ramada tried to go upscale with hotels called Ramada Renaissance Hotels. The Ramada name, however, had a downscale stigma that stuck; it was eventually dropped to better position Renaissance hotels as upscale. Renaissance is now owned by Marriott, which kept only the Renaissance name.

Renovating and repositioning old hotels has become a common practice today. Stephen Taylor described the situation:

> The art of repositioning is coming into its own. Repositioning, the economic [marketing] revival of troubled properties and the renovation and revitalization of old/outdated ones, can provide an alternative to the more traditional routes taken when hotels stop making good economic sense.
>
> The task of repositioning is not as simple as creating a market slot for a brand-new hotel. A repositioner has to deal with two customer images—the existing one and a new one that must be projected.
>
> Repositioning is a two-pronged effort. In most cases, a negative image and customer complaints must be overcome before a new impression can be created. To achieve the goals which define the success of a repositioning effort . . . it needs to be finely tuned to fit the specific situation and it takes thought, perceptiveness and careful planning. The successful repositioning of any hotel property begins with an intensive examination of the market the repositioner intends to enter.[7]

BRANDING AND POSITIONING

In the overview for this chapter we introduced the concept of a **brand.** A brand is the chosen positioning statement of the owners of a product offering; multiple properties and/or establishments are designed and built to the same product standards.

Branded properties are becoming ever more important in the marketing realm. In the United States, approximately 70 percent of hotel properties are branded, but in Canada

brand A well-known product or service of consistent quality available to consumers in multiple locations. A brand reflects the desired positioning of a product or service.

Web Browsing Exercise

Search for a branded hospitality franchise. Compare a dozen or so different properties on attributes, price, and so forth. Are they consistent? Can you explain the differences as a customer? If so, how?

it is closer to only 30 percent. In other parts of the world, it can drop to 10 percent. Regardless, it is definitely on the increase because brand names provide "instant recognition." Companies that control well-recognized and well-reputed brands will be successful over the long run, depending on how well they manage their brand identity.

Branding is very important for hospitality companies because a strong brand allows a company to attract more franchisees, which means higher revenue; after all, an owner would rather become a franchisee of a strong brand than a weak one. Branding enables firms to gain management contracts, access to capital, and higher-than-average revenue per available room. Branding also allows firms to spread costs for supplies and technical issues, such as web connectivity, across multiple properties.

Technology has had a significant impact on brand marketing. Reservation systems are shifting from central reservation systems to Internet distribution. This had considerable consequences for the benefits associated with branding. For one, it brings branded properties wider recognition because they can afford to pay the necessary fees to get better placement on the ending search engine page. The other advantage is that with multiple names appearing on the website, the customer will tend to examine only hotels they are familiar with and ignore the rest. The rationale for this is the risk associated with buying a hospitality product.

Product consistency and the integrity of branded properties affect the positioning of the entire brand. Inconsistent brand portfolios that cannot meet this challenge will lose out in the competitive marketplace. In fact, protecting the brand's image and identity is one of the major goals of brand managers. Companies do this by setting quality standards and inspecting properties frequently. Exhibit 14-9 provides 10 guidelines for building strong brands. Understanding these guidelines is essential to positioning a property to the market sector it serves and the clients it will attract.

Hotel Restaurant Branding

An interesting part of the branding phenomenon has been the outsourcing of F&B outlets to recognized brand names. In the past, hotel F&B outlets have been notoriously unprofitable. For some, this is changing with new brand identities.

In 1998 Malayan United Industries, which owns the Regal and Corus brands (MUI Group is a diversified conglomerate, headquartered in Kuala Lumpur, Malaysia), acquired a controlling interest in the Restaurant Partnership, which owns and manages brands such as Simply Nico, Nico Central, Elena's L'Etoile, The Gay Hussar, and Thierry's. The Restaurant Partnership provides F&B solutions to the hotel industry through expertise and the installation of branded outlets. Hotel guests have traditionally "eaten out" on a majority of occasions at recognized brand restaurants. The new trend is to put those brand names in-house to keep customers from going outside the hotel and as a point of differentiation among hotels. Exhibit 14-10 shows the commentary of Mike Feldott on the rise of branded restaurant concepts in hotels.

The F&B branding trend works, generally speaking, in three ways. The first is for the hotel to lease the space for a flat fee and/or a percentage of sales. In this case, the hotel loses control to another operator. The second method is for the hotel to acquire a franchise and become a franchisee, paying fees and royalties to the franchisor. Marriott was one of the

| Exhibit 14-9 | Ten Guidelines for Building Strong Brands |

1. *Brand Identity:* Have an identity for each brand. Consider the perspectives for the brand-as-person, brand-as-organization and brand-as-symbol, as well as brand-as-product. Identify the core identity. Modify the identity as needed for different marketing segments and products. Remember that an image is how you are perceived, and an identity is how you aspire to be perceived.

2. *Value Proposition:* Know the value proposition for each brand that has a driver role. Consider emotional and self-expressive benefits as well as functional benefits. Know how endorser brands will provide credibility. Understand the brand–customer relationship.

3. *Brand Position:* For each brand, have a brand position that will provide clear guidance to those implementing a communication program. Recall that a position is the part of the identity and value proposition that is to be actively communicated.

4. *Execution:* Execute the communication program so that it not only is on target with the identity and position but achieves brilliance and durability. Generate alternatives and consider options beyond media advertising.

5. *Consistency over Time:* Have as a goal a consistent identity, position and execution over time. Maintain symbols, imagery and metaphors that work. Understand and resist organizational biases toward changing the identity, position and execution.

6. *Brand System:* Make sure the brands in the portfolio are consistent and synergistic. Know their roles. Have or develop silver bullets to help support brand identities and positions. Exploit branded features and services. Use sub-brands to clarify and modify the brand. Know the strategic brands.

7. *Brand Leverage:* Extend brands and develop co-branding programs only if the brand identity will be both used and reinforced. Identify range of brands and, for each, develop an identity and specify how that identity will be different in disparate product contexts. If a brand is moved up or down, take care to manage the integrity of the resulting brand identities.

8. *Tracking Brand Equity:* Track brand equity over time, including awareness, perceived quality, brand loyalty and especially brand associations. Have specific communication objectives. Especially note areas where the brand identity and position are not reflected in the brand image.

9. *Brand Responsibility:* Have someone in charge of the brand who will create the identity and position and coordinate the execution over organizational units, media and markets. Beware when a brand is being used in a business in which it is not the cornerstone.

10. *Invest in Brands:* Continue investing in brands even when the financial goals are not being met.

Source: Adapted with permission of The Free Press, a Division of Simon & Schuster Adult Publishing Group, from *Building Strong Brands* by David A. Acker. Copyright © 1996 by David A. Aaker. All rights reserved.

first to do this, with Pizza Hut in 1989. The third method is to undertake a joint venture in which both the hotel and the restaurant operator share the costs and the profits. Radisson SAS Hotels uses both the first and third methods to develop its strategy of having branded F&B outlets. For example, in Hamburg, Germany, its hotel has a Trader Vic's (U.S. brand); in Berlin it has a TGI Friday's (U.S. brand); in Copenhagen it has a Blue Elephant (Thai brand); and in Brussels it has a Henry J. Bean's (UK brand).

Today, hospitality and nonhospitality companies are undertaking joint ventures. Recently, Nickelodeon (a family television network) joined forces with Holiday Inn to build a $110 million hotel in Orlando, Florida, the Nickelodeon Family Suites by Holiday Inn.[8] Giorgio Armani S.p.A. is awarding a long-term license to EMAAR Hotel & Resorts for the operation of a collection of luxury hotels and resorts. The agreement foresees the opening of at least seven luxury hotels and three vacation resorts within the next 10 years.[9]

Multiple Brands and Product Positioning

Hospitality companies develop multiple brands for growth purposes and for market segments. Sometimes this is through development of a new concept, sometimes it is through acquisition, and sometimes it is through both. Marriott, for example, developed the Courtyard (midprice) and Fairfield Inn (budget) lodging concepts to develop new segments, purchased Residence Inns for quick entry into extended stay properties,

Exhibit 14-10   The Use of Branded Restaurants in Hotels

WHAT HAS BEEN THE HISTORY OF BRANDED RESTAURANT CONCEPTS IN HOTELS?

There were three major trends in the past:

- A freestanding restaurant brand that was on the same property as the hotel, which had its own building and land for parking. An example of this would be the La Quinta hotel brand tied in with a coffee shop brand like Denny's or IHOP that served full breakfast.
- Then, as limited service hotel brands started giving away a complimentary breakfast in the lobby, the restaurant building was taken to the next level with either a Chili's, Applebee's, or TGI Friday's type of operation that promoted lunch and dinner and had a bar.
- The third trend was when a full service hotel wanted to promote a signature restaurant in the hotel as an alternate choice to the three meals a day restaurants. Steakhouses, such as Ruth's Chris, Morton's, and the Palm, fit that mix of great rooms matching with great brand restaurants.

WHAT ARE THE CURRENT TRENDS OF BRANDED RESTAURANTS IN HOTELS?

Over the last few years, many hotels want to take the brand environment to the next level so they have a competitive edge in the market. Would you rather stay at the same type hotel with a Ruths' Chris or something that is less upscale? For example, a Hilton in Indianapolis recently opened with a McCormick & Schmick Seafood Restaurant as a hook to the competition. When the sales department sells, they are now selling an upscale dining experience as well. In addition to pairing with McCormick & Schmick, Ruth's Chris, and Morton's, Hilton has also paired with Benihana.

WHAT ARE YOUR FUTURE PREDICTIONS OF THE BRANDED RESTAURANTS IN HOTELS TREND?

The future designs of hotels will include freestanding looking spaces attached to the hotel that will drive hotel business opportunities and also function for the local resident market. In larger hotels, you may see the lifestyle section of the hotel take form with a mix of specialty retail and restaurants as part of the hotel. With the success of Las Vegas style restaurants at hotels, you will see this concept scope as part of mixed use that includes hotel development. Fast casual and fast food will probably not be part of the new hotel mix, as the sales-to-investment ratio does not work in a hotel environment.

Source: Milke Feldott. Used by permission.

developed Marriott Marquis as convention hotels and Marriott Suites as luxury all-suites, and initiated JW Marriotts as upscale luxury hotels.

While development of multiple brands provides growth, it also provides protection from the competition against a single brand. Marriott saw other chains moving into lower-level markets and threatening the middle-to-upper tier in which Marriott hotels were positioned. Marriott felt it might as well steal its own customers (also called *cannibalization*) as let someone else steal them. It also realized that the existing concept was neglecting certain markets.

Multiple brands, of course, are common practice in other industries—for instance, Procter & Gamble and General Motors. The restaurant industry has long had multiple brands, as in the case of Darden Restaurants, which owns Red Lobster, Olive Garden, Bahama Breeze, Smokey Bones, and an upscale restaurant concept, Seasons 52. Yum! Brands, formerly Tricon Global Restaurants, Inc., owns A&W Restaurants, Long John Silver's, Pizza Hut, Taco Bell, and KFC. Brinker International has seven distinct restaurant brand concepts.

In the cruise industry, the Carnival Corporation owns several cruise line brands, each of which appeals to a specific market. The brands they own and a description of each can be found in Exhibit 14-11.

The issue is that one firm owns many different brands and each brand is positioned as if was a single company. Yum! Brands has similar positioning strategies between its own brands as an outside competitor would toward its brands. Unfortunately, this may be self-defeating for the parent company if these chains cannibalized each other, which to some extent they now do by offering similar products. What they want to do, instead, is to position to different market segments, as Darden's five brands do.

The different market segments may include many of the same people. They belong, however, to a different segment when they use restaurants or any hospitality enterprise

Exhibit 14-11   Companies Owned by Carnival Corporation & PLC: Each Line Caters to a Different Segment

CORPORATE INFORMATION

Carnival Corporation & PLC is a global cruise company and one of the largest vacation companies in the world. Our portfolio of 12 leading cruise brands includes Carnival Cruise Lines, Holland America Line, Princess Cruises, Seabourn Cruise Line, and Windstar Cruises in North America; P&O Cruises, Cunard Line, Ocean Village, and Swan Hellenic in the United Kingdom; AIDA in Germany; Costa Cruises in southern Europe; and P&O Cruises in Australia. These brands—the most recognized cruise brands in North America, the United Kingdom, Germany and Italy—offer a wide range of holiday and vacation products to a customer base broadly varied in terms of cultures, languages, and leisure-time preferences. The company also owns two tour companies that complement its cruise operations: Holland America Tours and Princess Tours in Alaska and the Canadian Yukon. Its combined vacation companies attract 6.8 million guests annually.

CARNIVAL CRUISE LINES

Carnival Cruise Lines is the best known cruise brand in North America and the most profitable in the world. The leader in the contemporary cruise sector, Carnival operates 21 ships that are expected to carry a record 3.3 million passengers this year—the most in the cruise industry. Guests aboard the "Fun Ships" enjoy a variety of dining, entertainment, and activity options, all in a festive and lively environment. Most recently, the line has launched several product enhancement initiatives that include an exclusive alliance with world-renowned French master chef Georges Blanc—who has maintained the coveted three-star Michelin rating for more than 25 years—and the new "Carnival Comfort Bed" sleep system featuring plush mattresses, luxurious duvets, and high quality linens and pillows.

PRINCESS CRUISES

One of the best known names in cruising, Princess is a global cruise and tour company operating a modern fleet of 15 ships carrying more than a million passengers each year. The company's ships are renowned for their innovative design and wide array of choices in dining, entertainment, and amenities, all provided in an environment of exceptional customer service. A recognized leader in worldwide cruising, Princess offers its passengers the opportunity to escape to more than 280 destinations around the globe, with sailings to all seven continents ranging in length from seven to 30 days.

HOLLAND AMERICA LINE

With 133 years of experience, Holland America Line is recognized as the undisputed leader in the cruise industry's premium segment. Its 13 ships sail to more than 300 ports of call on all seven continents. With more than 500 cruises a year, itineraries range from two to 108 days.

Completed in September 2006, Holland America Line's $225 million fleetwide Signature of Excellence enhancements feature programs and amenities that have established a new standard in premium cruising.

SEABOURN CRUISE LINE

The Yachts of Seabourn offer its 104 fortunate couples an intimate setting aboard its all-suite ships: Seabourn Pride Spirit and Legend. Rated among the highest ships in the world, they offer sumptuous ocean-view suites measuring 277 square feet or more, many with balconies. Seabourn is renowned for extraordinarily personalized service, with nearly one staff member per guest.

WINDSTAR CRUISES

Seattle-based Windstar Cruises operates three motor sailing yachts known for their pampering without pretense and their ability to visit the hidden harbors and secluded coves of the world's most treasured destinations. The unobtrusive service and attentive staff create a casually elegant atmosphere that effortlessly fosters camaraderie among guests and crew. Carrying just 148 to 308 guests, the luxurious ships of Windstar cruise to nearly 50 nations, calling at 100 ports throughout the Caribbean, Costa Rica, Panama Canal, Mediterranean, and Greek Isles.

P&O CRUISES

One of the most recognizable and respected names in travel, market-leading P&O Cruises has been operating cruise ships for more than 160 years and combines innovation, professionalism, and unrivalled experience on its fleet of five ships dedicated to the British market. Each ship is elegantly appointed, combining classic British tradition and hospitality with modern amenities. The result is a refined cruising experience with distinctive restaurants, comfortable accommodations, and enjoyable entertainment.

CUNARD LINE

Since the first paddle-wheeled steamer crossed the Atlantic in 1840, the name Cunard has been synonymous with the quest for new discoveries, legendary voyages, and majestic onboard pursuits. Royalty, celebrities and voyagers from every walk of life have enjoyed Cunard's White Star Service^{SM} across the globe. Guests enjoy a classic luxury experience based on the history and tradition of transatlantic liner service. Continuing the tradition of luxury ocean travel, *Queen Mary 2* and *Queen Elizabeth 2* are international icons evoking the nostalgia of the grand era of cruising and are the only traditional ocean liners in operation.

OCEAN VILLAGE

Ocean Village combines action and relaxation at sea and ashore into a fresh new take on holidays for thirty-to-fifty-somethings: younger, more upbeat passengers who want to get more out of their time away. With no formal dress codes, *Ocean Village's* casual on-board style includes high quality 24/7 buffet dining, plus waiter service options at the Bistro with a menu created by TV celebrity chef James Martin. Calling at six destinations every seven days, *Ocean Village* offers a choice of four itineraries—two in the Caribbean, two in the Mediterranean.

SWAN HELLENIC

Swan Hellenic focuses on discovery cruising for the discerning traveler on board *Minerva II*, whose interiors are designed to resemble a floating English country house with a capacity for 600 passengers. Expert speakers in wide-ranging areas such as botany, theology, geology, international diplomacy, gastronomy, and astrology accompany each cruise, bringing each destination vividly to life in highly informative lectures and seminars.

(continued)

COSTA CRUISES

Based in Genoa, Italy, Costa Crociere is the leading cruise company in Europe and South America, operating a modern fleet of 11 ships with a basis-two capacity of 20,200 total lower berths. Costa, whose origins date back to mid-1800s with a fleet of freighters transporting fabrics and olive oil between Genoa and Sardinia, has grown to become one of the most famous and respected names in seagoing travel. A Costa cruise is distinguished by its "Cruising Italian Style" shipboard ambiance, which offers an on-board experience that combines the sophisticated elegance of a European vacation with the fun and spirit of the line's Italian heritage.

AIDA

AIDA Cruises is the number-one operator in the German-speaking cruise market, carrying more than 233,000 passengers in 2005. A total of four AIDA ships are currently in service: AIDAcara, AIDAvita, AIDAaura and AIDAblu, with a total lower berth capacity of 5,400 guests. These vessels currently operate in the Mediterranean, Northern Europe, the Caribbean, the Arabian Gulf, and around the Canary Islands. AIDA ships are dedicated to the German-speaking market and are renowned for their youthful style and casual service.

P&O CRUISES AUSTRALIA

The pioneer of Australian cruising, P&O Cruises celebrates its 75th anniversary in 2007—marking the 1932 departure of the 23,000-ton liner *Strathaird*, which sailed from Australia on a five-night round trip from Sydney to Brisbane and Norfolk Island. Nearly 75 years later, P&O Cruises Australia continues to offer holidays tailored to Australian cruise passengers, and now carries a diverse clientele totaling more than 100,000 people each year. P&O Cruises Australia has expanded considerably over recent years and will soon boast three cruise ships.

Source: Carnival Corporation & PLC. Used by permission.

for different purposes, in different contexts, or at different times. Thus, the positioning of each chain should be managed so that they do not steal from each other, and then the standard positioning rules can be applied. Ever since Quality Inns was successfully broken up into Comfort Inns, Quality Inns, and Quality Royale (now Clarion), there have been a number of hotel chains with properties under the same or a similar name, each trying to position to a different market segment. This is commonly referred to as brand extension.

Quality Inns subsequently created Sleep Inns and renamed itself Choice International Hotels with multiple brands. Management claims that there is no question about the difference between the brand names. Further, the addition of Rodeway, Friendship, and Econo Lodge to the Choice fold may have created some very confused customers, particularly given that the website for each brand lists the same 800 number. Choice now has seven brands with 13 different products, some being different versions of the same product, (e.g., Clarion hotels, suites, resorts, and inns). The overlap is obvious as shown in its brand positioning portrayal.

A *Cornell Hotel and Restaurant Administration Quarterly* article by Peter Yesawich, president of an advertising agency, that focuses on the hospitality sector contained some comments on this situation:

> Yesawich said that the success of brands depends on creating a clear differentiation in the minds of customers. With only few exceptions, the advertising and promotion that has been initiated on behalf of new product concepts has failed to communicate clearly or convincingly the basis of the differentiation. Customers are quick to discern the availability of free drinks or free breakfasts, but it takes much more to constitute a new product in customers' minds.... If advertising doesn't communicate the perception of a new product, then maybe the product isn't really new at all. Some observers are concerned that customers may be confused by a chain that has one name on a variety of hotels. Yesawich noted that chains pursuing diversification by introducing new products under different names have so far met with greater success. "In general terms, a brand name is an asset, as long as it stands single mindedly for a specific package of value and benefits. Call it a personality," said Robert Bloch [then senior vice-president for marketing at Four Seasons]. The practice of "leaving a mid-price brand name on an upscale property, as some operators are doing, may confuse some customers."[10]

Sheraton and Hilton have wrestled with a similar image problem. The vast difference between the Sheraton Wayfarer Motor Inn in Bedford, New Hampshire, and the Sheraton St. Regis in Manhattan was about the same difference as the Berkshire Hilton Inn and the Waldorf-Astoria, two hotels sporting the Hilton name brand. Customers can be very confused with what position the brand name actually conveys. To deal with this problem, which has existed for many years, Hilton recently developed Hilton Garden Inns, while Sheraton created Four Points and required all of the former "inns" to come up to standard to use this name or otherwise be disenfranchised.

Marriott debated long and hard when developing the Courtyard concept as to whether to call it a Marriott. The final decision was to call it Courtyard by Marriott with *Marriott* in smaller letters. Today, the "by" has been dropped. Thus, Courtyard can trade on its famous brand name without creating expectations of the same product or service. The same was done with the Fairfield Inn brand line. Marriott, in fact, has probably been the only hotel company in the United States that has successfully differentiated its brands and kept them clearly in their respective product classes. The problem for others is not necessarily in the name, but in the positioning.

Can hotel concepts under the same or similar names make the same claim? In other words, is each brand or product positioned to a different specific target market, each with specific needs that relate to the positioning? Second, if the first case is true, can these markets differentiate the positioning of each brand or product name so that they (the markets) know which one "belongs" to them? This is the case in point and is the concern of positioning any multiple brands, but occurs especially when they have similar names. If the answers to the preceding questions are no, then there will be a clear case of cannibalization and customer confusion.

Multiple brand positioning can be done successfully, as Marriott has shown. Groupe Accor, a French firm, has developed lodging concepts called Formule1, Ibis, Mercure, Novotel, and Sofitel and owns Motel 6 and Red Roof Inns in the United States. By French government rating, these are one-, two-, three-, and four-star properties, respectively. Each is based on the needs of a specific target market. Each is clearly differentiated from the other three; in fact, you might say that no customer would ever choose one when they wanted the other. However, in at least one place in Paris, a Novotel and Sofitel sit side by side with separate entrances and a common wall dividing them, a practice not uncommon among multibrand hotel companies in the United States. The traveler has a choice in the same location. Each was clearly positioned to its own market segment, but eventually Novotel started to cannibalize the Sofitel.

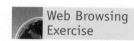

Web Browsing Exercise

Using your favorite web browser, go to www.hotel-online.com/News/PR2006_2nd/May06_StarwoodStrategy.html. Read about the positioning statements for all their brands.

Chapter Summary

Market positioning is a valuable weapon for hospitality marketers. To position successfully requires recognizing the marketplace, the competition, and customers' perceptions. Positioning analysis on a target market basis provides the tools to identify opportunities for creating the desired image that differentiates from the competition and for serving the target market better than anyone else.

The differences among salient, important, and determinant attributes must be considered, especially if the firm is being positioned by benefit or attribute. All three are complementary, and it is crucial to understand their place.

Positioning may be objective, using images of the physical characteristics of the product, or subjective, in which customers' mental perceptions play a greater role. Positioning of a tangible good is often accomplished through association with intangible notions; alternatively, it is better to position intangible services with tangible clues.

Repositioning may be necessary when a hospitality firm is in an unacceptable position or when trying to appeal to a new market segment. Old hotels with poor images are often renovated, and repositioning them is crucial to their success.

As many hotel and restaurant chains grow by adding new brands and concepts, multiple brand positioning becomes important. They must prevent one brand from cannibalizing the other and also be able to create separate images and benefits for each one.

Key Terms

brand, p. 479
determinant attributes, p. 468
importance attributes, p. 468
intangible positioning, p. 472
objective positioning, p. 467
positioning, p. 466
positioning statement, p. 471
salient attributes, p. 467
subjective positioning, p. 467
tangible positioning, p. 471

Discussion Questions

1. In your own words, describe what is meant by market positioning.
2. What are the three main criteria for an effective positioning statement? Do you think these criteria are met in Exhibits 14-2, 14-3, and 14-4? Why or why not?
3. Compare and contrast subjective versus objective positioning. Relatedly, what is the difference between tangible and intangible positioning?
4. Select an advertisement of your choice and determine whether the advertisement reflects (1) subjective or objective positioning, and (2) tangible or intangible positioning.
5. What is a brand? How does branding relate to positioning? Do you have a favorite brand? Why is it?

Endnotes

1. Knapp, D. E. (2000). *The brand mindset.* New York: McGraw-Hill, xv.
2. Verma, R., Plaschka, G., Dev, C., & Verma, A. (2002, Fall). What today's travelers want when they select a hotel. *HSMAI Marketing Review,* 20–23.

3. Hotelmarketing.com website. TIG Global launches group RFP tracking and reporting solution. Retrieved from www.hotelmarketing.com/index.php/content/article/hilton_hotels_resorts_launches_new_image_campaign/marketing.com on April 20, 2004.

4. Osgoodby, B. (2001, September 5). Sell the "sizzle." *Add Me Newsletter*, 219. Retrieved from www.addme.com/issue219.htm on May 5, 2005.

5. Ogilvy, D. (1985). *Ogilvy on advertising*. New York: Vintage Books, 160.

6. Quoted in Brennan, D. M. (1986, May 1). Niche marketing. *Restaurant Business*, 186, 189.

7. Taylor, S. P. (1986, Fall). Repositioning: Recovery for vintage and distressed hotels. *HSMAI Marketing Review*, 12–15.

8. Retrieved from www.hotel-online.com/News/PR2005_2nd/Jun05_Nickelodeon.html on September 16, 2005.

9. Retrieved from www.hotelmarketing.com/index.php/content/article/armani_hotels_and_resorts_launched/ on September 16, 2005.

10. Withiam, G. (1985, November). Hotel companies aim at multiple markets. *Cornell Hotel and Restaurant Administration Quarterly*, 39–51.

case study

Pearl River Garden Hotel Sanya

Discussion Questions

1. What is the current positioning of the Pearl River Garden Hotel Sanya compared to its previous positioning? Why did management make this move?

2. Who are the current target markets for this hotel? Are they compatible with each other?

3. Does the Russian tour group market fit the current positioning of the Pearl River Garden? What about the Russian corporate market? The emerging high-end Russian tourism market?

4. The domestic Chinese market appears to be stagnant; that is, it is no longer a growth market and looks to be on the decline. Should the director of marketing continue to pursue this market? Should he be more aggressive in his marketing efforts to capture a greater share of the market?

5. To what extent is Pearl River Garden Hotel Sanya competing against branded and independent hotels? Should management consider branding?

6. Overall, how do you think Pearl River Garden should be positioned, and who should constitute the target markets?

This Saturday should have been a relaxing one for Mr. Sam Tian, but today he decided to head into the office. It was October 15, 2005, and Sam was intent on using this opportunity to review last week's room revenue results. He had been out of the town for the last week at a tourism conference, and his secretary had been keeping him informed of the occupancy rates every day, but he wanted more detailed information. Normally he'd wait until Monday, but he was working on the marketing plan for 2006. His boss, Mr. Li, was already waiting for the report so that he could review it and pass it along to the hotel owner as soon as possible.

The average occupancy for September was 57 percent, and that was to be expected since September was usually a slow month for the hotel. The Pearl River Garden was a

This case was contributed by Karen Kwan, Phillip Levett, Lyn Yuan, and Ying Zhang and developed under the supervision of Margaret Shaw, PhD School of Hospitality and Tourism Management, University of Guelph, Ontario, Canada. All monetary figures are in Canadian dollars. For conversion, use 1$C/6.88CNY. All rights reserved. Used by permission.

seasonal resort hotel, and September was always slow, even compared to the slower summer off-season. Today Mr. Tian would be looking closely at the figures for the first week of October. That week was always good for business because of the public holiday. It was 95 percent for the first week and only 30 percent for the second week.

As he walked through the lobby this fine morning, smiling at the few guests in the lounge, he thought to himself, "What could I do to improve business during the slow periods next year?" Mr. Tian knew that solving this problem would be his biggest challenge when he put together his marketing plan for 2006.

Mr. Sam Tian joined Pearl River Garden Hotel Sanya in September 2004 after the previous director of sales and marketing resigned. Sam started his hotel career with Shangri-La Hotel Beijing as a front desk agent and became a sales manager in Swissbell Hotel Beijing two years later. With his 17 years of hotel experience, including 10 years in sales and marketing on Hainan Island, Sam was very familiar with the tourism market on Hainan Island and had extensive market analysis experience. His job as director of sales and marketing was to maximize the market potential for the hotel. To do that, he needed to be aware of the unique characteristics of the guests and be able to develop new markets.

As the lunch hour approached, Sam called his boss, Mr. Li, to confirm their lunch appointment and then started to make his way toward the restaurant. He remembered to bring his notes, though, because he knew that Mr. Li would expect to use this time to discuss business. Even though Sam knew that his report was days away from completion, Li would have many questions and would expect Sam to be ready to answer them.

Sam left his office early and took this time to walk through the hotel. He did this trip every day about this time and used this opportunity to greet the staff and be available if need be. It's called management by walking around, and Sam had found it to be a valuable tool to increase his understanding of operations and gauge the level of guest satisfaction. He was pretty sure he'd find his boss on the terrace having a quick smoke break before lunch.

The lunch meeting was more involved than Sam had expected, but he was glad that he'd prepared himself beforehand. Mr. Li was interested in what Sam was working on with regard to positioning. The unique ownership situation of Pearl River Garden was a challenge and limited his ability to control his distribution system.

OWNERSHIP

Guangzhou Pearl River Industrial Group Company Ltd. was a large-scale, state-owned enterprise group founded in 1979. The company focused on real estate development, general contracting, construction engineering, and property management. In addition, the group had a hotel management division, and the Pearl River Garden was one of four hotels in the chain. The other three properties were located in neighboring Canton province. At one time Canton province included Hainan Island. This history of association between these two political regions worked to the advantage of Pearl River, and as a result, the entire system of reporting and declaring of information and response times to government communications was streamlined. Also, being a government-owned business almost guaranteed access to capital. This close association with the government benefited Pearl River Garden Hotel when it came to establishing a reliable client base. The hotel served the needs of the Central Committee and senior political officials from overseas and local governments.

SANYA, HAINAN ISLAND, CHINA

The Pearl River Garden Hotel was located in Sanya, on the southernmost tip of Hainan Island Province (See Exhibits 1 and 2). This spot was also the southernmost part of China and was a tropical paradise with clear blue waters and white sand beaches along its southern edge, lush green mountains to the north, and palm trees everywhere.

Sanya was made up of Sanya City and its surrounding area. The area had many attractions for visitors, who traveled to the region from many parts of the world, representing a diverse mix of nationalities, age groups, and interests.

In 2004 Sanya was voted by the United Nations as one of the most livable cities in the world. The award was based on a number of factors including cleanliness, employment opportunities, safety, accessibility, and economic prospects.

The annual Sanya Miss Chinese Beauty Pageant was held in Sanya City in 2005. The Miss World Contest has been hosted by Sanya City for three consecutive years since 2003. Mrs. Julie Morley, the chairperson of the Miss World Organization, favored this location. These events contributed to the positive image of the city and the region within China and around the world.

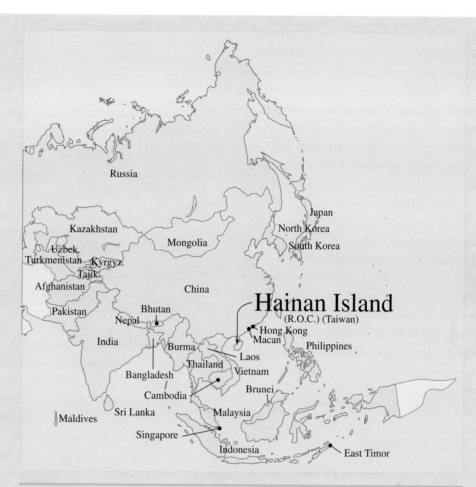

Exhibit 1   Map of of People's Republic of China

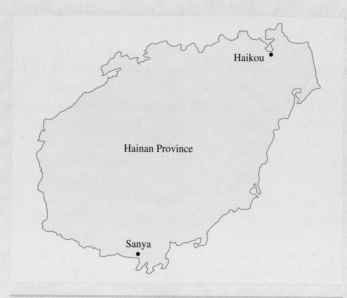

Exhibit 2   Map of Hainan Island, China

The Sanya area was known as "The Hawaii in China," and Sanya City was located on a peninsula that juts out into the South China Sea. Since ancient times the area was referred to as "the edge of the sky." Recently, a massive causeway for road and rail traffic was built to connect the north end of the island to mainland China, but the only practical way for foreign tourists to get to this tropical oasis was by air.

There were three resort beaches in Sanya. Yalong Bay, with its superior water quality, had a pristine sand beach stretching for seven kilometers. This area was the most popular beach in Sanya, and here was where the international hotel chains such as Marriott, Sheraton, and Holiday Inn had properties. The second area, known as Oriental Beach, was not as famous as Yalong Bay, but it was only a five-minute drive from downtown Sanya City and close to Sanya International Airport. There were several four-star hotels and two five-star hotels on Oriental Beach, and all of them were locally managed. The third area, known as Sanya Gulf, was generally rated lower than the other two areas because of the quality of the beach and the water.

PEOPLE

Sanya's population was more than 450,000 and included 20 different nationalities. Increasingly, people from across China and elsewhere in the world were drawn to Sanya's inviting climate and promising business environment. The result was a truly multicultural city.

Tourism was the major industry in Sanya. Until recently, Sanya was isolated from the rest of China, and the service and tourism industries were relatively new. As a consequence, the hotel's management team, including people like Sam, were not native to the island. The service level was still relatively low compared to some tourist areas in Southeast Asia.

The native population on the island was made up of a number of minorities with distinctive native crafts, clothing, and customs. This diversity was one of the attractions for tourists. Tours included visits to area villages to buy local crafts and souvenirs.

The economy continued to grow in China with increased personal disposable income. The result had been a corresponding lifestyle change. More people with more time and money had led to increased travel domestically and abroad. This increased demand also led to the construction of more airports and orders for both long-range and short-range commercial aircraft.

Exhibit 3　Room Types of Pearl River Garden Hotel Sanya

Room Type	Numbers of Rooms
Superior standard room	29
Deluxe standard room	59
Superior sea view room	71
Deluxe sea view room	63
Superior sea view suite	4
Deluxe sea view executive suite	5
Deluxe sea view garden suite	3
President suite	1
Total rooms	235

PEARL RIVER GARDEN HOTEL SANYA

Pearl River Garden Hotel Sanya was constructed in 1996 by Guangzhou Pearl River Industrial Group Company Ltd., and was renovated in 2004. The hotel was located right on Oriental Beach, the second most popular beach in Sanya, and was 18 kilometers from Sanya International airport. It was the first four-star hotel on Oriental Beach and continued to show a profit for its owners. Exhibit 3 shows the room categories of its 235 rooms

DINING FACILITIES

There were five F&B outlets in the hotel:

1. **Green Island Western Restaurant** offered a buffet breakfast with combined Western and Chinese styles available. With seating capacity for 160 people, this restaurant served a buffet breakfast and was open 24 hours a day with a set menu providing a la carte service.

2. **Pearl River Spring Chinese Restaurant,** with seating capacity for 260, was open from 7:00 A.M. to 2:30 P.M., and again from 5:00 P.M. to 9:30 P.M. This was an alternative breakfast location and was the best choice for dim sum lovers. The restaurant was famous for its Cantonese food; it was regarded as one of the best Cantonese restaurants in the city.

3. **Lobby Lounge** had seating for 60 and served tea and coffee as well as snacks. Cocktails and wine were also available. The hours of operation were 10:00 A.M. to midnight daily.

4. **Terrace Garden** on the third-floor terrace, with seating for 100, was available for barbecues and

seafood and hot-pot banquets. This outlet was reservation based.

5. **An outdoor dining area** was located on the beach with seating capacity for 1,000 people. It could be booked for special functions. This location was ideal for barbecues; featured seafood; and was ideal for dances, shows, and casual events. The area could be easily arranged to suit customers' needs and served as a special function space.

ENTERTAINMENT AND RECREATION

The Health Club was available to all in-house guests of the Pearl River Garden. In addition, the Silver Moon Sauna Centre was open until midnight and offered a traditional Chinese massage service. Adjacent to the Silver Moon was a full-service hairdressing salon featuring facials and body treatments. Other indoor facilities included British billiards, Russian billiards, and table tennis. Russian guests especially liked the sauna and beauty salon facilities.

The Pearl River Entertainment City, open from 7:30 P.M. until 2:00 A.M., was a karaoke bar and featured seating for over 200 people. In addition to the main hall, several adjacent private rooms of different sizes were available for smaller groups.

A small boutique was located off the Lobby Lounge and featured local handicrafts and souvenirs.

The Terrace outdoor swimming pool was located on the third floor of the hotel with panoramic views of the ocean. If guests prefered to use the beach, they could take an escalator directly from the hotel.

The hotel managed the beach privately for the exclusive use of hotel guests. In addition, Pearl River Garden's affiliation with Oriental Beach Tourism Company Ltd. allowed it to offer extensive water sports facilities and rescue services. Guests could rent personal watercraft and go waterskiing, parasailing, and diving.

For guests who prefered entertainment on dry land, there were three golf courses in Sanya all within a 25-minute drive from the hotel.

CONFERENCE FACILITIES

On the third floor, off the terrace, there were two conference halls:

Palm Island Garden Conference Hall, at 170 square meters, seated 180 people theater style or 140 in a classroom seating arrangement. This room could be partitioned into three smaller rooms.

Golden Pearl River Conference Hall, with 300 square meters of usable space, could seat up to 300 guests theater style or 200 with a classroom setup. The room could also be partitioned into three parts. There were also several private rooms in the Pearl River Spring Chinese Restaurant that could be turned into small meeting rooms when necessary. A fully serviced business center operated 24 hours a day.

THE MARKETING TEAM

With lunch finally over and a new folder full of bad ideas to work on, Sam returned to his office. The problem with his boss was that half the time Sam ended up teaching Mr. Li. The general manager was just that: a general who managed. His background was management in various government departments, but he didn't know anything about hotel management or the hotel industry. Now Sam would have to spend the rest of the day proving why most of Li's ideas wouldn't work. And he was tired of being short staffed. Besides that, he'd sure like to be able to hire a local person to fill one of these positions he had open, just once.

The reservation department reported to the sales and marketing division. There were 12 salespeople, including Sam, and one graphic designer. Exhibit 4 shows the organization chart and job responsibilities of the sales and marketing team. Another five budgeted positions were currently vacant, including public relations manager and public relations assistant. Currently the sales manager was doing the public relations job in addition to his own.

COMMUNICATIONS MIX

The hotel's communications mix included newspaper advertisements in Beijing, Shanghai, and Guangzhou. The hotel worked with several online booking agencies with positive results, but the hotel's own website did not have an online reservation function. Also there was no English language version of its home web page, so the only information about the hotel in English came from affiliate sites managed by the booking agencies.

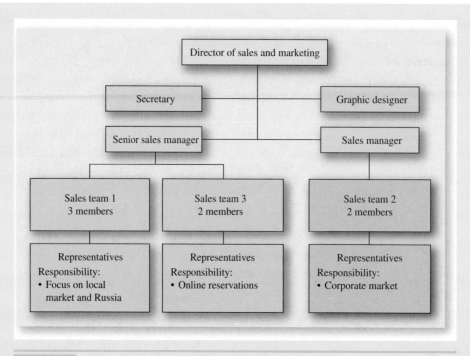

Exhibit 4   Organization chart of sales and marketing division

OCCUPANCY AND ROOM RATE STATISTICS

When Sam joined the hotel, he wanted to be able to generate reports to analyze room sales by market segment. The PMS system used in the hotel was not able to generate the detailed reports by segment that he needed. For example, Sam wanted to be able to differentiate between travel agency group sales and individual sales, but he was not able to create enough fields within the database to generate the desired detail. So, even though the rates were different, he couldn't generate a report with enough detail to show this. Exhibits 5 and 6 show the occupancy rates for 2004 and 2005, year to date.

The peak season was between November and February, which was winter in China. Because the hotel was so far south, however, the weather was ideal during these months. Room occupancy was usually between 75 and 90 percent. People liked to come to a tropical place for a warm vacation. During the Spring Festival, the Chinese New Year's one-week holiday in January or February, the occupancy was usually around 98 percent, and the room rate was double for peak season. Business started to decline from March onward and

reached its lowest point in June, July, and August. Summer was especially hot in Sanya, usually around 34 degrees Centigrade (93 degrees Fahrenheit) on average. Most people didn't want to have their vacation when the temperature was this high. Because this time coincided with the school vacation period in China, many families still chose to travel in these two months. Many of them wanted to take the opportunity to experience a tropical holiday while travel expenses were lower compared to winter.

From September, after the school summer holiday, occupancy dropped again. The first week of May and October were public holidays in China, and many people chose to travel during those two weeks, which were referred to as "golden tourism weeks" in China. The occupancy during these two weeks was usually more than 90 percent, and the room rate was 50 percent higher than the regular rates.

Sam had promoted some golf packages and water sports packages to attract more guests during the low season. The golf package was promoted through a few travel agencies and had attracted a limited number of guests from Japan, South Korea, and Hong Kong. It was usually a weekend package specially designed for golf lovers. Golf was very expensive in Korea and Japan because

| EXHIBIT 5 | Occupancy Rates—2004 | | | | | | | | | | | |

Market Segment	Jan. Occ %	Feb. Occ %	Mar. Occ %	Apr. Occ %	May Occ %	Jun. Occ %	Jul. Occ %	Aug. Occ %	Sep. Occ %	Oct. Occ %	Nov. Occ %	Dec. Occ %
Corporate	5.30	10.31	8.88	13.99	11.56	11.65	12.49	11.67	9.96	7.83	7.35	6.23
Total individual	**5.30**	**10.31**	**8.88**	**13.99**	**11.56**	**11.65**	**12.49**	**11.67**	**9.96**	**7.83**	**7.35**	**6.23**
Conference group	2.82	14.04	26.67	12.76	1.07	2.71	7.43	0.00	3.18	3.80	3.68	16.72
T/A group	90.06	72.06	61.10	70.10	81.88	81.19	76.68	85.71	80.04	84.93	86.05	75.27
Total group	**92.89**	**86.10**	**87.77**	**82.86**	**82.95**	**83.89**	**84.12**	**85.71**	**83.22**	**88.73**	**89.73**	**91.99**
Long staying	0.49	0.57	0.38	0.00	0.00	0.00	0.00	0.00	0.00	0.00	0.00	0.00
Day use	0.00	0.00	0.00	0.00	0.00	0.00	0.00	0.00	0.00	0.00	0.00	0.00
Complimentary	1.32	3.02	2.97	3.15	5.49	4.46	3.39	2.62	6.81	3.44	2.92	1.78
Occupancy %	**88.39**	**73.10**	**69.70**	**63.78**	**42.52**	**42.99**	**54.51**	**69.30**	**46.15**	**55.65**	**80.85**	**88.31**
House use	1.32	1.66	2.01	2.71	2.81	3.39	2.76	2.02	3.12	2.43	1.61	0.97
Out of order	1.94	3.96	11.37	17.82	18.96	15.97	10.02	9.05	19.39	12.41	3.10	1.77
Total occupancy %	**89.57**	**75.37**	**71.83**	**65.86**	**44.99**	**45.00**	**56.42**	**71.17**	**49.52**	**57.64**	**83.28**	**89.92**
Total saleable rooms	**97.16**	**95.94**	**91.23**	**88.09**	**91.08**	**91.99**	**93.27**	**92.70**	**89.97**	**92.12**	**96.23**	**97.60**

1.00$C=6.88CNY

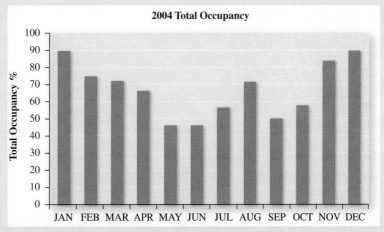

Source: Provided by the courtesy of Pearl River Garden Hotel Sanya.

of limited availability and extremely high demand. Sometimes the price to play was actually higher than the airfare to places like Sanya.

POSITIONING

Pearl River Garden Hotel Sanya used to position itself as a resort hotel because of its excellent location on Oriental Beach, but both its external appearance and inner facilities and atmo-

sphere did not match a resort hotel. From the polished marble floors and chandeliers in the lobby, to the formal room furniture and restaurant decor, the hotel looked more like a formal business hotel than a resort. Many groups coming to Pearl River Garden were from government agencies, branches of the military, and state-owned banks. These guests chose Pearl River Garden because of its relatively formal environment, which they felt matched their status. Conservative government officials and military officers didn't want to be seen by other guests

EXHIBIT 6	Occupancy Rates—2005 Year to Date									
	Actual 2005									
	Jan. Occ %	Feb. Occ %	Mar. Occ %	Apr. Occ %	May Occ %	Jun. Occ %	Jul. Occ %	Aug. Occ %	Sep. Occ %	Oct. Occ %
Market Segment										
Corporate	3.91	6.79	5.03	5.74	6.06	12.66	8.80	9.25	7.65	8.66
Total Individual	**3.91**	**6.79**	**5.03**	**5.74**	**6.06**	**12.66**	**8.80**	**9.25**	**7.65**	**8.66**
Conference group	5.59	14.10	7.47	0.32	3.47	2.37	0.00	2.46	1.19	4.95
T/A group	88.52	77.82	85.43	90.24	87.72	80.08	87.77	86.65	86.92	84.16
Total group	**94.11**	**91.92**	**92.90**	**90.56**	**91.19**	**82.45**	**87.77**	**89.11**	**88.11**	**89.11**
Long staying	0.00	0.00	0.00	0.00	0.00	0.00	0.00	0.00	0.00	0.00
Day use	0.00	0.00	0.00	0.00	0.00	0.00	0.00	0.00	0.00	0.00
Complimentary	1.98	1.29	2.07	3.70	2.76	4.89	3.43	1.64	4.24	2.23
Occupancy %	**91.00**	**77.67**	**75.10**	**58.18**	**60.04**	**47.01**	**48.90**	**76.00**	**54.94**	**56.15**
House use	1.08	1.19	2.37	3.43	3.98	4.78	5.20	7.85	7.65	1.24
Out of order	1.56	6.36	9.74	20.05	32.82	53.91	44.77	13.56	44.60	4.95
Total occupancy %	**92.84**	**78.68**	**76.69**	**60.41**	**61.74**	**49.42**	**50.64**	**77.27**	**57.37**	**57.43**
Total saleable rooms	97.61	94.39	91.50	87.57	81.48	77.52	79.81	85.81	76.94	96.5

1.00$C=8.88CNY

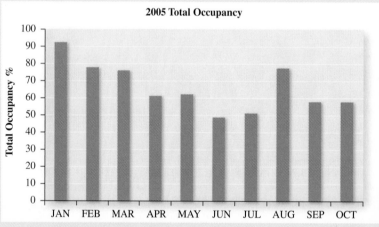

Source: Provided by the courtesy of Pearl River Garden Hotel Sanya.

walking through a lobby in a bathing suit with a towel. The atmosphere at Pearl River suited their self-image.

Sam repositioned the hotel as a business hotel located in a beach resort area. With this new positioning, he focused on government and corporate conferences and the corporate incentive market. In the meantime he also targeted tour operators who wanted to book groups looking for a beach location but didn't want to pay five-star rates.

COMPETITION

Pearl River Garden Hotel had one major four-star competitor, Landscape Hotel, and two five-star competitors, Mountain Sea Sky and Intime Resort. Another five-star hotel, the Baohong, would open the next year with 480 rooms. At a distance, Marriott had entered the market in Yalong Bay and had partnered with Air China on the Marriott Awards

Exhibit 7	Comparative Analysis of Hotel Properties on Oriental Beach		
Hotel Name	Landscape Hotel Sanya	Intime Resort Sanya	Mountain Sea Sky Hotel Sanya
Location	−	−	−
Reputation	=	+	+
Room size	+	+	+
Room rate	−	+	+
Room amenities	+	+	+
Staff quality	−	−	−
Dining facility	+	+	+
Chinese food quality	−	−	−
Western food quality	−	+	=
Banquet facilities	−	+	+
Indoor entertainment facilities	−	+	+
Outdoor entertainment	−	=	−
Conference facilities	−	+	=
Communication facilities (phone, Internet access)	=	=	=
Parking spaces	=	+	+

= Equivalent to the Pearl River hotel
+ Superior to the Pearl River hotel
− Inferior to the Pearl River hotel

program as well as in developing a Mandarin-language website. Exhibit 7 shows the facilities of the three main competitors, and Exhibit 8 shows the present year's performance of each in relation to the others.

Two new four-star hotels with 170 rooms and 166 rooms, respectively, were to open by the end of 2005. Baohong Hotel, a five-star hotel with 480 rooms, was scheduled to open in 2006.

Intime Resort Sanya had a conference hall with capacity for 380 people in classroom arrangement. The other two hotels' conference rooms were smaller than the ones available at Pearl River. Intime Resort Sanya was currently the only competitor for the Russian market. Although it was a five-star hotel, Intime Resort had lower rates than most hotels at that level. It offered similar rates to Russian tour groups as did the Pearl River Garden.

TARGET MARKETS

■ **Government Agencies:** Many government and corporate groups booked their travel arrangements with travel agencies. In China there were both private and state-owned operations. The state-owned firms were usually the preferred intermediaries because of their size and experience, but government departments were not under any obligation to use them.

■ **Corporate Conference and Incentive Travel:** The hotel was equipped to cater to the corporate market and the incentive market if the groups were small, enough. The meeting and function spaces were small, but they were appropriate for the number of rooms in the hotel and the size of the food and beverage utlets. If they could identify a market that needed a hotel of this size, it could work to the advantage of the hotel. A small group might feel more comfortable in a smaller hotel than being one of possibly many groups staying at a large property.

Russian Market

■ **Russian Tour Groups:** Russian groups that frequented Pearl River also used travel agencies for the convenience of bundling both the travel and room arrangements together at a reduced rate. Tourists from the former Soviet Union represented a new market for the hotel and had been a distinct market segment only

Exhibit 8	Analysis of Competitors				
Hotel Name	Available Rooms	Fair Share %	Actual Market Share %	Average Room Rate (CNY)	Average Occupancy %
Pearl River Garden Hotel (four star)	235	22.05	20.43	438.26	62.85
Landscape Hotel Sanya (four star)	170	15.95	18.84	403.64	80.15
Mountain Sea Sky Hotel Sonya (five star)	241	22.61	17.93	509.01	53.08
Intime Resort Sanya (five star)	420	39.40	42.79	495.21	73.46

All three competitors were located on Oriental Beach.

since 2003. In recent years many Russians had chosen China as their vacation destination. They came to experience a tropical location at affordable rates. Prices at Sanya were comparable to other tropical resort areas in Southeast Asia.

Russian guests wanted to stay right on the beach and preferred the lower rates on Oriental Beach. The Pearl River had seen a 100 percent increase in sales from Russian groups from 2004 to 2005. Room rates for Russian tour groups were about 100CNY more than local tour groups.

■ **Russian Corporate:** In addition to travel agency business, Pearl River Garden had recently signed a corporate contract directly with Gazprom, the largest petrol and natural gas company in Russia. Gazprom bought about 4,000 room nights per year, and in return the hotel offered a 30 to 40CNY discount compared to Russian travel agency groups. Total revenue was higher with groups because they usually used the facilities and services offered at the hotel.

THE MARKET SITUATION

The new Russian market had begun to show signs of segmentation. A high-end Russian market had begun to develop. These tourists had discovered the high-end international hotels in Yalong Bay. Marriott Sanya had promoted attractive packages especially for Russian tourists, with 200CNY

more per room night than Pearl River Garden's. Both Hilton and Hyatt were scheduled to open resort hotels on Yalong Bay in 2006.

The domestic market was expected to change in the near future. The government had simplified travel regulations and the passport application process to allow more people to travel outside China. Those changes, combined with the competitive market for tourist dollars in Southeast Asia, meant that more and more Chinese were choosing to travel out of the country for their vacations.

OPERATIONS RESULTS FOR 2004

Pearl River Garden Hotel Sanya reported a net profit for 2004 of 38 million CNY. The breakdown was 24 million CNY from rooms and 12 million CNY from food and beverage operations. One Canadian dollar, or 85 U.S. cents, was worth approximately 7CNY.

Since the Russian market was more profitable and growing, management had been trying to decide how to develop that business further.

The markets in Southwest China and Shanghai were less profitable. The local Chinese market was price conscious. These customers expected lower room rates than Pearl River Garden was charging. The hotel's average room rate was higher than comparable hotels on Oriental Beach, and REVPAR was lower. Guests expected five-star hotel amenities but wanted to pay four-star rates. Intime Resort Sanya was the preferred hotel for Shanghainese guests.

As Sam left for the day, he decided that he couldn't help himself and he drove over to Yalong Bay just to stare at the international hotels. He knew he couldn't move his hotel to a better beach, so what, he thought, was the best way to convince people to come in the first place? What does the hotel already have that should be highlighted? Who should he tell his message to? He knew that he could come up with the answer if he had time to plan properly. He also knew that he must convince Mr. Li that there were other ways to advertise the property besides using newspaper ads.

overview

Pricing is the part of the marketing strategy that defines the way to extract value from the customers based on their willingness to pay. Pricing strategy is critical to a firm because prices are the only part of the marketing mix that creates revenue for the firm. Because of the intangibility of services and simultaneous production and consumption, prices are also used by customers to help determine the quality of the product or service they are being offered. Setting prices requires a thorough decision-making process that involves an understanding of costs, competition, demand, supply, channels of distribution, and of course, the customer. This chapter discusses the strategies of arriving at the appropriate pricing mix.

chapter 15

The Pricing Decision

learning objectives

After reading this chapter, you should be able to:

1. Explain why a pricing strategy is an integral part of developing a marketing strategy.

2. Incorporate competition, the marketplace, and of course, the customer into the setting of prices.

3. Use value-based pricing instead of cost-based pricing to increase revenue.

4. Explain how revenue management is used to set hotel room pricing.

5. Use flexible pricing in the presentation mix and in the communications mix.

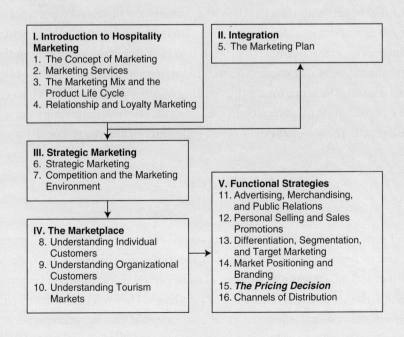

I. Introduction to Hospitality Marketing
1. The Concept of Marketing
2. Marketing Services
3. The Marketing Mix and the Product Life Cycle
4. Relationship and Loyalty Marketing

II. Integration
5. The Marketing Plan

III. Strategic Marketing
6. Strategic Marketing
7. Competition and the Marketing Environment

IV. The Marketplace
8. Understanding Individual Customers
9. Understanding Organizational Customers
10. Understanding Tourism Markets

V. Functional Strategies
11. Advertising, Merchandising, and Public Relations
12. Personal Selling and Sales Promotions
13. Differentiation, Segmentation, and Target Marketing
14. Market Positioning and Branding
15. *The Pricing Decision*
16. Channels of Distribution

Carolina Pontual Colasanti

Director of Revenue Management, Westin Guangzhou, Guangzhou, China

Carolina Colasanti was born and raised in Rio de Janeiro, Brazil. A languages and cultures enthusiast, at the age of 17 she left for Zeitz, Germany, for a 12-month period aiming at enhancing her German skills. This experience shaped the years that followed. Upon completion of a two-month internship as housekeeping supervisor at the Copacabana Palace Hotel in Rio de Janeiro, she moved to Switzerland in 2000 to attend L'École hôtelière de Lausanne. She graduated in December 2004 and received a bachelor of science in international hospitality management. Carolina has been working with Starwood Hotels and Resorts, Asia-Pacific Division, since March 2005, with the intention of acquiring additional international exposure. She recently completed her contract as revenue manager with Sheraton Perth Hotel in Australia and is currently doing the opening of the Westin Guangzhou in China.

 ## Marketing in Action

Carolina Pontual Colasanti

Director of Revenue Management, Westin Guangzhou, Guangzhou, China

What is the typical day like as a revenue manager?

The day usually begins with the extraction and compilation of several reports for the daily operations and the reservations meetings. These include month-to-date actuals, reservations pickup, rate movements, etc. That is as far as the routine goes. From then on, the revenue manager will juggle tasks as business demands. Other activities might include (re)statusing the hotel, forecasting, putting together the text for GDS advertising, training, managing third party websites, assisting the sales and marketing team with their decisions, etc. The role really does have a wide range of areas of responsibility.

What courses should a student take who wants to be a revenue manager?

I believe all sorts of sales and marketing courses are of utmost importance: customer behavior, the marketing mix, market definition, the communications mix. With the rise of the Internet and the increasingly number of third party websites, it is also important to take an e-marketing course and learn about the players. They will be constantly chasing your business, trust me! Advanced Excel courses would also be recommended, as revenue managers usually spend a big part of the day working on Excel spreadsheets.

And at last but not least, it is important to have an understanding of property management systems, revenue management systems, and interfaces. In some instances, it will be your sole responsibility to decide on new systems and their functionalities, and features will directly affect the hotel's daily activities.

What is the career path for a revenue manager?

Because revenue management will help you acquire skills and an understanding of many areas within sales and marketing, the opportunities are vast. In the hospitality industry, a revenue manager's next step is typically the role of director of revenue management (in smaller properties, those roles are the same). You could then become an area director of revenue management and eventually a vice president of revenue management, or branch out of it altogether and move into sales or marketing. The opportunities are countless also on both the property and corporate level: sales, marketing, e-marketing, distribution, to name a few. It is also worth mentioning that revenue management techniques are used in several other industries too, including airlines, restaurants, convention centers, stadiums, and hairdressers.

How does an understanding of marketing help a revenue manager?

When talking about revenue management, I always like to refer to a basic marketing principle: "To give customers what they want, when they want it, where they want it at a price they are willing to pay." Knowing your customer is the key—not only well, but also better than everybody else in your competitive set. As revenue manager, you are a market analyst; you have to understand the different market segments, their needs, wants, and their preferred channels of distribution. After all, you do not want to offer a skater a woolen suit on a 105-degree day. . . . Moreover, as previously mentioned, the more you know about sales and marketing in general, the better equipped you will be for the wide range of responsibilities you will face on a daily basis.

Used by permission from Caroline Pontual Colasanti.

The marketing discipline grew out of the economic discipline. The basic theory of economics is that the economy responds to the customer. The basic theory of marketing is that the customer calls the shots. When it comes to setting prices, these basic theories need to be remembered. Pricing must be, first and foremost, customer based. Cost and profit consideration follow under the heading of "can we afford to do it?"

To motivate the discussion of price and to provide a frame of reference for the chapter, we first provide a brief overview of current pricing practices in the hotel and restaurant industry. Next, we discuss a definition of price. This is followed by a review of the drivers of profit and the different types of costs found in the hospitality industry. We then discuss two methods of pricing; cost-based pricing and value-based pricing.

With an understanding of how pricing is accomplished, we turn our attention to the pricing objectives of the firm. Demand-based pricing and the use of revenue management are included in these objectives.

PRICING PRACTICES

Hotel Room Pricing

Hotel room pricing has changed remarkably over the years. This is illustrated in Exhibit 15-1. Notice that as one moves from left to right on this chart, profitability increases and pricing becomes less tactical and more strategic. Initially hotel pricing involved changing rates every season. There was some demand forecasting (referred to as DF in Exhibit 15-1), but much of it involved examining the occupancy of the year before. This all changed with the development of sophisticated computer models that estimated demand by room type and rate. Marriott International was the first major hotel chain to use such systems, and other chains soon followed. Currently very few hotels operate without some form of **revenue management,** which we will discuss in more detail later in this chapter.

In 2005 the majority of hotel firms were pricing somewhere between revenue per available customer and value pricing. Revenue per available customer includes room revenue as well as food and beverage, Internet charges, and other revenue earned from the customer. Harrah's Entertainment was the first to incorporate the lifetime value of the customer into its decision models. This means, for example, that if only one room is available in a hotel and two people want the room, Harrah assigns the room to the

revenue management
The practice of setting prices based on varying patterns of anticipated usage including daily, weekday, weekend, and/or seasonal periods. Prices are typically set higher for peak demand periods (high season) and lower for off-peak periods (low season). This is sometimes referred to as yield management.

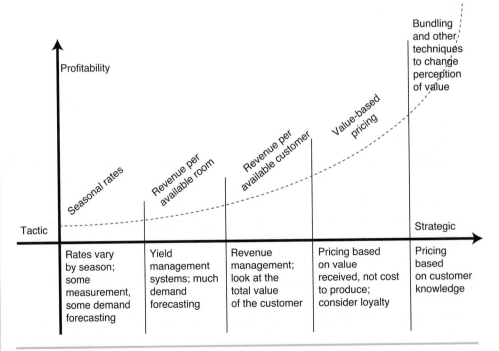

	Seasonal rates	Revenue per available room	Revenue per available customer	Value-based pricing	Bundling and other techniques to change perception of value
	Rates vary by season; some measurement, some demand forecasting	Yield management systems; much demand forecasting	Revenue management; look at the total value of the customer	Pricing based on value received, not cost to produce; consider loyalty	Pricing based on customer knowledge

Exhibit 15-1 Hotel room pricing has changed remarkably over the years

guest will who provide the most long-term economic value to Harrah's. The days of first come, first serve are slowly going away. Internet firms such as Hotwire, Priceline, and Expedia.com are moving more and more into **price bundling,** because it provides a great way to hide the cost of each part of the trip. Weekend packages are a form of price bundling.

Restaurant Pricing

Restaurant pricing traditionally used **cost-based pricing** and was seriously affected when inflation surged in the 1970s. The industry responded by continuously increasing prices. Whenever the cost of goods in the industry (e.g., butter, beef, sugar, coffee) went up, restaurant prices quickly followed suit. The result was that the customer eventually said, "Whoa!" and turned to other alternatives, including staying home.

Eventually, the industry caught on. It found new ways to do things, new items to put on menus, new ways to prepare menu items, and new ways to serve them (e.g., the salad bar) to cut labor costs. In the restaurant industry, customer reaction to price can be very quick, if only because it is relatively simple for someone else to enter the market with a new idea and/or a better price. Taco Bell's introduction of the 59-cent taco in the late 1980s is a prime example of finding new ways to do things. For example, they redesigned the actual taco to support the lower price. This was accomplished by slightly changing the amount of meat in the taco and its grade. Extensive customer research was undertaken to ensure that these changes did not affect the customer. In addition, they redesigned the process of preparation by moving the majority of preparation to central facilities. This allowed for smaller kitchens and more seating area in the restaurants themselves. Today most fast-food restaurants have 99-cent or dollar menus. Restaurants today are experimenting with revenue management. For instance, there are restaurants in Japan that charge for the length of time that you use your table.

WHAT IS PRICE?

The definition of *price* will change depending on whose viewpoint one takes—the customers' or the manufacturers'. From the customers' viewpoint, *price* can be defined as "what the customer must give up to purchase the product or service."[1] The "what" may include actual money, time, a product, or a service (e.g., an exchange of rooms for free advertising). For the customer, these are costs of purchasing the product. Customers will often pay more for a reduction in time to obtain a product. For example, items sold in an in-room minibar are often more expensive than the same items in a grocery store for this very reason. It is easier to open the minibar than look for the nearest grocery store to purchase drinks and food items. From the manufacturers' perspective, price is defined as the products and services that they give to the customers versus what they receive in return.

With this definition of price, it should be easy to see that the firm has multiple ways to change prices. These are shown in Exhibit 15-2. Notice that one of the items listed is changing the quality of the goods and services provided by the seller. This

price bundling Offering one price that includes several aspects of a product offering such as a welcoming reception, overnight accommodations, breakfast, and a local guided tour. Many weekend packages are a form of price bundling.

cost-based pricing Pricing based on the costs involved to produce a product rather than on conditions in the marketplace.

Exhibit 15-2   Ways to Change Prices

1. Change the quantity of money or goods and services to be paid by the buyer
2. Change the quantity of goods and services provided by the seller
3. Change the quality of goods and services provided by the seller
4. Change the premiums or discounts to be applied for quantity variations
5. Change the time and place of transfer of ownership
6. Change the time and place of payment
7. Change the acceptable form of payment

Web Browsing Exercise

Visit the following website: www.pricingsociety.com/. What does the pricing society do? Why would a hotelier or restaurateur want to be a member of such an organization?

method works extremely well in the hospitality industry because not all hotel rooms are alike; the number of beds in each room varies, as does the view offered. As discussed previously, Taco Bell was able to lower its prices by changing the quality of the ingredients in a taco. This was done only after numerous consumer taste tests revealed that consumers could detect no difference between a more expensive ingredient and a less expensive alternative. They also slightly changed the quantity of ingredients offered. This is another method of lowering prices: changing the quantity of the goods and services provided by the seller.

Another way to change prices, as shown in Exhibit 15-2, is to change the acceptable form of payment. For example, consider the Basin Harbor Club, which is located on Lake Champlain in Vergennes, Vermont. This resort, shown in Exhibit 15-3, encourages payment by check or cash, although they will accept credit cards as a last resort. This way, they keep prices lower. If they accepted credit cards, they would need to raise their rates to cover credit card fees. Similarly, one of the authors of this text conducted a study to determine the types of credit cards business travelers carried when traveling on business. The sponsor of the study was a major hotel chain that wanted a reduction in the surcharge fee charged by one of the major credit card issuers. When the results revealed that travelers were indifferent to which credit card they used, the hotel chain threatened to stop accepting the card unless the surcharge fee was lowered. The credit card firm lowered its fees. This saved the hotel chain from having to raise prices.

Prices can also be changed by changing the time and place of transfer of ownership. This is the basis for revenue management. The saying is: "Tell me when you would like to arrive at the hotel, and I will tell you the price." Or: "Tell me how much you would like to pay for your hotel room, and I will tell you when you can arrive."

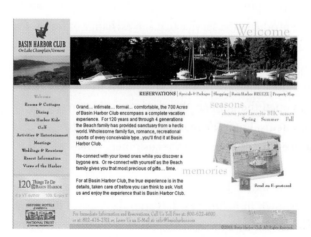

Exhibit 15-3   The Basin Harbor Club is located on Lake Champlain in Vergennes, Vermont

Source: Retrieved November 21, 2005, from www.basinharbor.com/ welcome_offseason.asp. Used by permission of The Basin Harbor Club.

Firms use price to represent the value of the offering and the value of what is received. Price is of unique importance to marketers for a number of reasons:

1. It is the only revenue-producing part of the marketing mix.
2. It is used to match supply to demand so financial objectives can be achieved.
3. It is a powerful force in attracting attention and increasing sales.
4. It establishes the market positioning of the product.
5. The pricing practice can have a major impact on customer loyalty.

For all of these reasons, price should be based on a thorough decision-making process by the seller that will communicate the worth of the total offering—a worth that is consistent with the market's perception of the offering's value. The importance of price in the marketing mix is explained as follows. (even though it was written a long time ago, it is still true today):

> Price is a dangerous and explosive marketing force. It must be used with caution. The damage done by improper pricing may completely destroy the effectiveness of the rest of a well-conceived marketing strategy. As a marketing weapon, pricing is the "big gun." It should be triggered exclusively by those thoroughly familiar with its possibilities and dangers. But unlike most big weapons, pricing cannot be used only when the danger of its misuse is at a minimum. Every marketing plan involves a pricing decision. Therefore, all marketing planners should be equipped to make correct pricing decisions.[2]

In the final analysis, pricing, like the product and service, is customer driven. Using all of the models of pricing only gets the end price closer to what the customer will pay. If the price is too high, customers will not pay for the service. If the price is too low, many customers may also not pay for the service because it might be perceived as "too cheap" and they would worry about the quality. Ultimately, customers determine the price at which a product or service will be successfully offered.

Setting prices is a complex exercise, with any number of strategic and tactical implications. The hospitality industry has fixed physical plant products and locations. Sometimes we have to work with the product we have and set prices accordingly. In other words, rather than set the price to the target market, we may have to find the target market that will accept a given product at a given price. This is called product-driven pricing, but it is still the customer who will determine the acceptable price. Given this, it is worth noting Subhash C. Jain's comments:

> [W]hile everybody thinks businesses go about setting prices scientifically, very often the process is incredibly arbitrary. Although businesses of all types devote a great deal of time and study to determine the prices they put on their products, pricing is often more art than science. In some cases, setting price does involve the use of a straightforward equation . . . but in many other cases, the equation includes psychological and other such subtle factors that the pricing decision may essentially rest on gut feeling.[3]

Before proceeding further, we first briefly review the drivers of profit, one of the goals of any pricing decision. The drivers of profit are shown in Exhibit 15-4. Notice

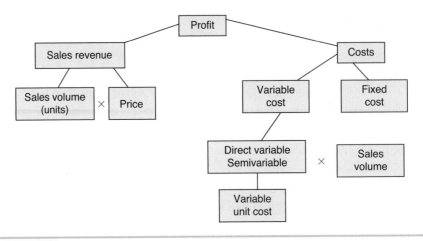

that the two main components are sales revenue and cost. Sales revenue is determined by the multiplication of sales volume and price. It is the role of marketing to ensure that pricing strategies yield the optimum sales volume.

Obviously, costs play a major role in both profit and the determination of price; as such, it is important to understand the types of costs that affect the hospitality industry.

TYPES OF COSTS

variable costs Costs that vary directly with a unit or item sold, such as the food costs of a menu selection or the housekeeping costs of an occupied guest room.

One of the costs listed in Exhibit 15-4 is **variable cost.** Variable costs can be considered either direct or semivariable. Direct variable costs can be traced directly to the level of activity. The higher the activity, the higher the variable cost. These costs are also known as out-of-pocket costs. Kent Monroe, a leader in pricing, explained it this way:

> One test of a unit variable cost is whether it is readily discontinued or whether it would not exist if a product were not made. Direct variable costs include those costs that the product incurs unit by unit and includes such costs as productive labor, energy required at production centers, raw material required, sales commissions, royalties and shipping costs. The major criterion of a direct variable cost is that it be traceably and tangibly generated by and identified with, the making and selling of a specific product.[4]

semivariable costs
Costs that are somewhat fixed regardless of units or items sold but can at times vary to some degree, such as staff payroll.

A second type of variable cost is semivariable. The best way to understand **semivariable costs** is to consider kitchen goods such as salt, pepper, baking soda, and the like. Other costs include the staff required to run the operation at a minimum. Semivariable costs are needed regardless of the level of activity and rise markedly with an increase in activity.

fixed costs Costs that do not vary with customer counts or occupancy, such as insurance costs or real estate taxes.

Fixed costs are those costs that exist regardless of the level of activity. Fixed costs include rent or mortgage, insurance, taxes, overhead, general administration, and so forth.

COST-BASED PRICING

Cost-based pricing comes in a number of versions in the hospitality industry. Most popular among these are cost-plus pricing, cost percentage or markup pricing, break-even pricing, and contribution margin pricing.

Cost-Plus Pricing

Cost-plus pricing involves establishing the total cost of a product, including a share of the overhead, plus a predetermined profit margin. Its common use in pricing food and beverages is to relate the profit margin to the selling price. If desired profit is 20 percent of the selling price, an item that costs $4, plus $2 labor and $2 overhead, would be priced at $9.60. This results in $1.60 of profit for that item. Each product or product line is given an appropriate share of every type of expense as well as its own variable cost. The intent is that every product should be profit generating.

Cost-plus pricing ignores the idea that total income is a combined effort in which some products will not generate as much profit as others but will contribute to the whole. It is also subject to misallocation of costs such as depreciation, maintenance, and so on. Cost-plus pricing does not allow for flexibility in pricing decisions, nor does it take into consideration customers' perceptions of a product's value. It is totally cost oriented and ignores demand. Attempts to apply different gross margin percentages to different menu items to account for different labor costs have done little to overcome the deficiencies of this method.

Cost Percentage or Markup Pricing

Cost percentage or markup pricing is also heavily favored by the restaurant industry. It features either a dollar markup on the variable ingredient cost of the item, a percentage markup based on the desired ingredient cost percentage, or a combination of the two. A bottle of wine that costs $10 might be subject to a $5 markup, making the selling price $15. The markup percentage would give a 66.6 percent cost percent to selling price ratio ($10/$15). If, on the other hand, a 50 percent wine cost was desired, the bottle would be marked up by $10 to make the selling price $20 ($20 × 50%).* A common combination of both would be to mark the wine up 100 percent plus $2, making the selling price $22.** Room service liquor follows a similar, if somewhat illogical, pricing strategy. The fifth of Johnny Walker scotch that costs $20 across the street in a liquor store is offered through room service at $100 to protect the 20 percent target beverage cost of the hotel.

---

*The formula for desired percentage cost (DPC) is DPC = cost/selling price. In this example, cost is $10 and DPC is 50%. Therefore, 50% = $10/x or .50x = 10, x = $20.

**In this case, we are doubling our cost, so $10 becomes $20. When you add the $2, it becomes $22. If we want to know the DPC, we calculate DPC = 10/22 = 45%.

The food service industry appears to be enamored of this method of pricing. Food cost and liquor cost percentages become the standard by which results are measured. The three major fallacies of this method are as follows:

1. It is totally cost oriented.
2. It ignores customer perceptions of value, particularly in times of widely fluctuating costs.
3. It tends to price high-cost items up to a level that customers are unwilling to pay.

Break-Even Pricing

Break-even pricing is used to determine at what sales volume and price a product will break even, or where costs are equal to sales. It distinguishes between fixed costs and variable costs. The break-even point is calculated as follows:

$$\text{Break even} = \frac{\text{Fixed costs}}{\text{Price} - \text{variable cost}}$$

For example, Exhibit 15-5(a) shows a hypothetical break-even analysis for price-sensitive restaurants. In this case, fixed costs are relatively low and unit variable costs are relatively high. Because of these factors, sales quickly pass the fixed cost line, but the profit margin remains relatively narrow regardless of the quantity sold. This leaves relatively little room for discounting for purposes of increasing volume.

Exhibit 15-5(c) demonstrates a break-even analysis for volume-sensitive restaurants. The fixed cost line in this case is higher, and it takes longer for the sales line to pass it. Once past it, however, the profit margin widens quickly because variable costs remain a relatively small percentage of unit sales. There is more room for discounting to increase volume once the fixed and variable cost lines have been passed by the sales line.

Break-even analysis is a fairly efficient method of determining profit margins at various price levels if—and this is a big *if*—sales volume can be accurately predicted at the different price levels. Knowledge of customer perception and demand is still needed in order to predict this volume.

Contribution Margin Pricing

Contribution margin pricing occurs when pricing is used to help cover fixed costs. For example, if the total variable cost of a meal is $3 and the meal is sold for $4, then $1 is available as a contribution to fixed costs. Contribution margin pricing is depicted in Exhibit 15-5(d). In contrast to Exhibit 15-5(a), the variable cost line is interjected into the plot at the same place as the sales line, starting at the zero intersection. This demonstrates the concept of contribution, showing that if the product sells at a higher price than its variable cost, it makes at least some kind of a contribution to fixed cost even when sales are not high enough to produce a profit.

This technique is very useful for hotels in soft periods of demand. Guest room prices can be discounted substantially, if that is what it takes to have them occupied.

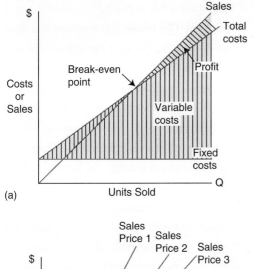

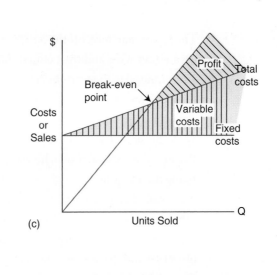

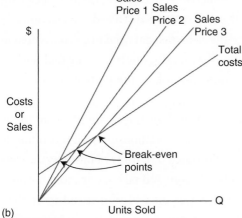

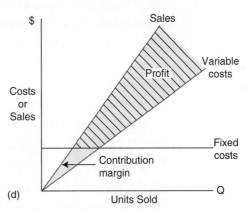

Exhibit 15-5 (a) Hypothetical break-even analysis for price-sensitive restaurants; (b) break-even point for several prices; (c) break-even analysis for volume-sensitive hotels; (d) contribution margin pricing

Source: Special thanks to Professor Catherine E. Ralston for help in developing these graphs.

Even though no profit results, a portion of the fixed cost that would occur if the room were not occupied would be covered. The success of this technique must be judged by examining the total revenues from rooms sold. Selling more rooms at discounted prices may have the same effect as selling fewer rooms at higher prices. In addition, it is important to consider the cost of occupancy; that is, the cost of wear and tear on the fixed assets, the cost of potential staff burnout, and the cost of trying to raise prices once consumers have been trained to get the room for less.

VALUE-BASED PRICING

A **value-based pricing** strategy involves choosing a price after developing estimates of market demand based on how potential customers perceive the value of the product or service. It has nothing to do with the cost to produce the item. Perceived value is often defined as what one receives divided by the price one paid. An illustration of this method of pricing occurred when one of the authors of this text needed to have his big-screen TV fixed. He called the first person he found in the phone book and, without getting a bid (something one should never do), he had the repairman come and fix the TV.

value-based pricing
Pricing based on customer sensitivity to the price–value relationship of the product offering and not solely on the costs associated with producing the product.

The repair man took off the back of the TV, jiggled a few wires, placed his soldering gun on another wire, and then closed the back of the TV. The whole process took less than 10 minutes. He then proceeded to write an invoice for $275. After the repairman was asked, "How can you charge me so much when you were only here 10 minutes?" the repairman answered, "How much did you pay for the TV?" When the author responded $675, he said "I just saved you $400. I can't wait until everyone has a plasma TV." Because they cost so much, he was able to charge more to fix them, because the cost of fixing the TV, although high, was still lower than the cost of a new plasma TV. The repair man was, of course, correct. And, he was competitive in his pricing. Additional calls to other repair services revealed the same charge.

Value-based pricing has the advantage that it forces managers to (1) review the objectives they have when marketing their product or service and (2) keep in touch with the needs and preferences of customers. Because we have discussed the customer in some detail already, we will not repeat all of the elements that need to be considered in pricing the product. The reader knows by now that in using any marketing tool, such as pricing, the customer is the first consideration. We discuss the particulars of value-based pricing next.

In establishing prices, some elements are particularly important in regard to the customer. The first of these is the perceived price–value relationship, as it is commonly called. The importance of this relationship is illustrated by a study of business travelers who spend more than $120 per night for a hotel room and take six or more business trips per year. The study revealed that 28 percent of the 344 who spend more than 75 nights per year in hotels (38 percent of the total sample) claimed that the feature "is a good value for the price paid" is important in the decision to stay in the same hotel chain when traveling on business.[5]

Components of Value

The role of management is to increase the perceptions of price value so consumers will be willing to spend more money. One way to accomplish this goal is to focus on one or more of the components of value. These value added features can be categorized as follows:

- Financial (e.g., saving money on future transactions, complete reimbursement in the case of service failure, 10 percent discount at gift shop)
- Temporal (e.g., saving time by priority check-in)
- Functional (e.g., availability of check cashing)
- Experiential (e.g., active participation in the service)
- Emotional (e.g., more recognition)
- Social (e.g., interpersonal link with a service provider)
- Trust (e.g., the organization does what it says it will do)
- Identification with the organization (e.g., affinity with a sports team, belief in what the organization stands for—such as The Body Shop and animal-testing policies)

Following are brief discussions of these components.

Exhibit 15-6	Factors That Affect Financial Value

Perceived Substitute Effect: Buyers are more price sensitive the higher the product's price relative to prices of perceived substitutes

Unique Value Effect: Buyers are less sensitive to a product's price the more they value any unique attributes that differentiate the offering from competing products

Switching Cost Effect: The greater the product-specific investment that a buyer must make to switch suppliers, the less price sensitive that buyer is

Difficult Comparison Effect: Buyers are less price sensitive to the price of a known or reputable supplier when they have difficulty comparing alternatives

Price Quality Effect: Buyers are less sensitive to a product's price to the extent that a higher price signals better quality

Expenditure Effect: Buyers are more price sensitive when the expenditure is larger, either in dollar terms or as a percentage of household income

End-Benefit Effect: This is broken into two parts:
- Derived demand (the relationship between the desired end benefit and the buyer's price sensitivity for something that contributes to achieving that benefit)
- Share of total cost (the cost of the specific item to the total cost of the product)

Shared-Cost Effect: Impact of partial or complete reimbursement on price sensitivity

Fairness Effect: Based on the price previously paid, prices of similar products (includes location or situation), and if item is to avoid a loss versus achieve a gain

Source: Nagle, T. T., & Holden, R. K. (2002). *The strategy and tactics of pricing: A guide to profitable decision making* (3rd ed.). Upper Saddle River, NJ: Pearson Education, 82–101.

Financial Value.

Exhibit 15-6 shows factors that affect financial value. The more price sensitive customers are, the more difficult it will be for the firm to get them to pay more for the product or service. Given that the reverse is also true, the role of the firm is to make customers less price sensitive. For example, Exhibit 15-6 reveals that if customers believe that there are many alternative solutions to their problems and needs and that one firm's offerings are not unique and are easily compared with other firms' offerings, they will be more price sensitive; that is, the firm will have more difficulty charging higher-than-average prices. The same is true if customers believe that it is easy to switch from one firm to another. One of the roles of loyalty programs is to make it more difficult for customers to switch to other brands. If they do switch, they do not get the same level of service as they would if they stayed with the firm to which they were most loyal.

Exhibit 15-6 also shows that buyers are more price sensitive when the expenditure is large relative to their household income or travel budget. This is important to remember for hotels such as the Four Seasons and Ritz-Carlton that charge premium prices. It is often difficult for a reservation agent who makes $15 per hour to sell a room that costs $600 per night, because $600 to such an agent is a lot of money. It would take a reservation agent 40 hours to earn $600, but it is important to remember that the person who can afford a $600 hotel room makes much more than $15 per hour. As a percentage of his annual income or travel budget, $600 is very small. If the percentage were large, he would not be calling either Four Seasons or Ritz-Carlton.

One mistake hoteliers used to make was considering business travel inelastic. That is, they believed that business travelers would pay whatever rate was necessary to stay in a particular hotel. Part of the reason for this belief was that the money was not the traveler's money, but the firm's. This is the shared-cost effect shown in Exhibit 15-6.

The end-benefit effect suggests that reservation agents always ask customers why they are staying at the hotel. Customers who are staying at a hotel for a special occasion such as an anniversary are more likely to pay more (e.g., be less price sensitive) than customers who just need a room for the night before moving on to the next city.

The other two factors that affect financial value—the fairness effect and price quality—are self-explanatory and not discussed here.

Temporal Value. A research study mentioned earlier in this chapter found that business travelers consider their time to be worth, on average, $150 per hour. This suggests that if a service process can be redesigned to save 15 minutes of the customer's time, the customer believes she has just saved $37.50. The advertisements by National Car Rental promoting the fact that customers do not have to spend time filling the rental car with gasoline because National charges only the prevailing rate for gasoline instead of the standard gasoline markup is an example of a value added strategy that from the customer's viewpoint saves time and money.

Customers continue to have less and less time. Anything the firm can do to save customers' time can be beneficial to the firm. Consider the success of the UPS Store, Kinko's, FedEx, and PostNet. All are designed to save customers' time.

Functional Value. Functional value pertains to the belief that the product or service does what it is designed to do. The main components of functional value are the RATER system, which was discussed at length in Chapter 3. Again, RATER stands for reliability, assurance, tangibility, empathy, and responsiveness. Management needs to ensure that every interaction with customers includes one or more of these components to convey to customers that they are receiving quality. Customers' perceived quality is a result of customer experiences; as such, they need to be managed by the organization. The objective quality of the atrium lobby (the tangible component of the RATER system) may be negated by the lack of perceived quality that is experienced by a rude and unresponsive desk clerk. This can instantaneously change a "fair" objective price to an "unfair" perceived price.

If the product or service does what it was designed to do, customers will pay more for it. For instance, consider something as simple as checking into a hotel. This process begins not when the guest walks up to the front desk, but when the guest starts the reservation process. Obvious steps are making the reservation, arriving at the hotel, and walking up to the front desk. If the guest calls to make a reservation and is put on hold for too long or if the website is hard to navigate, the cost of the trip increases and perceived value decreases. If this happens, the customer may exit the relationship and stay with a competitor. Once the guest gets to the hotel, if there are not enough convenient parking spaces near the hotel or if the valet attendant is not around, the guest may be inconvenienced, which decreases the perceived value. Finally, if it takes too long to check in and if the same information is requested at various points throughout the process (e.g., do we ask for the guest's name and address both during reservation process and at check-in?), perceived value might decrease.

Experiential Value. Experiential value occurs when guests are active participants in the service rather than passive observers. Which is more fun, sitting at a concert watching quietly or singing along with the group and perhaps dancing in the aisle? A good example of experiential value is the chef's table in the kitchen. This is a table in the kitchen where customers dine on a preset menu selected by the chef. The purpose of such a table is to give customers the feeling of "being in the know" and "being in the heart of the action." Exhibit 15-7 shows the chef's table in Brennan's, a famous restaurant that has locations in New Orleans, Houston, and Las Vegas.

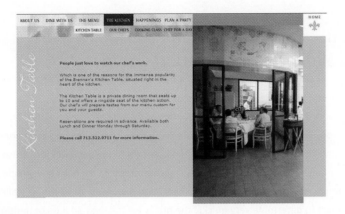

Emotional Value.

Emotional value pertains to customers' need to feel special. Las Vegas casinos spend much money to make their very heavy gamblers, known as whales, feel very special. Casinos cater to their every need, as well as to the needs of their friends and family members. Of course, one does not need to be in the casino business to treat customers as though they were high rollers. Ritz-Carlton, along with other companies, has made extensive use of database systems to keep track of customers' needs and wants. When the customer checks into the hotel or restaurant, his or her favorite room or table is available.

Social Value.

Most customers like to celebrate special occasions with friends and family. Research presented at the Milliken Food and Beverage Conference in 2005 stated that consumers dine out to celebrate the following:

- Birthdays (54 percent celebrate their own; 37 percent celebrate spouses; and 28 percent celebrate their child's)
- Mother's Day (38 percent)
- Father's Day (22 percent)
- Valentine's Day (28 percent)
- New Year's Eve (13 percent)
- Easter (13 percent)

One of the authors of this text worked at the Tyler Place Family Resort in Vermont, where guests arrived on a Saturday afternoon, remained for a week, and departed the following Saturday morning. Activities were organized for all age groups. The resort offered programs for toddlers to teens, as well as family retreats and family reunion packages. Many guests came to this resort the same week each year so they could vacation with friends they met on prior occasions. Customers who look forward to the opportunity to spend time with friends and family on vacation are often willing to spend more money on such opportunities.

Trust.

As discussed in Chapter 4, trust is a major antecedent of loyalty. And customers who are loyal to an organization are willing to pay more to stay with that

Tourism Marketing Application

Flyspy Makes Airline Pricing More Accessible to Consumers

Tourism destinations realize that consumers do not consider just the price of hotel guest rooms and meals when choosing a destination. They also consider the cost of getting to the destination. The arrival of the low-cost airline carriers has been a boon for many destinations, which are suddenly affordable for many visitors. As well, more and more computer programs have made pricing much more transparent. New software in development at the time of this writing, called Flyspy, will enable consumers to determine very quickly the prices for the next 30 days of multiple flights to a destination. The software will also plot graphs comparing prices of flights to multiple destinations so travelers can easily find the least expensive flight and least expensive destination.

Web Browsing Exercise

Use your favorite web browser to look up HOG. Be prepared to explain whether learning about HOG has changed your image of those who own motorcycles. What did you learn from this website that can be translated into the hospitality business? Do you think customers might pay more money for a Harley-Davidson because of this organization? Why or why not? Use your favorite web browser to also look at the human resources policy at Starbucks. What benefit does this policy have for customers? Does it enable Starbucks to charge more for its coffee? Why or why not?

organization. Because services are intangible and one cannot evaluate the service prior to purchase, customers do pay more to purchase services from firms that they trust and know to be reliable.

Identification with the Organization.

The final component of value occurs when customers identify so strongly with the organization that price is removed from the equation. One example of such an affiliation is the relationship customers have with their favorite sports team. Another example is HOG, which is the Harley Owners Group. Firms can increase customers' feelings of affiliation by incorporating any of the following tactics into their marketing plans:

- Providing opportunities for public displays of association, such as logo apparel and sponsorship of community activities
- Actively aligning with and supporting social causes such as becoming an environmentally friendly hotel or restaurant and working with local area care organizations to "stamp out hunger"
- Providing opportunities for contact by creating a dialogue with customers through direct mail and e-mail
- Having distinctive human resources policies such as those offered by firms such as Starbucks

Reference Pricing and Reservation Pricing

Another way to increase the perceptions of price value is known as reference pricing. As we know from Chapter 1, customers purchase solutions to their problems based on expectations. Let's turn that around and say that customers also have in mind a price they expect to pay for a given solution. This is referred to as their **reference price.** Reactions to prices will vary around this reference or expected price based on some kind of prior experience or knowledge. In understanding reference pricing, it is important to understand some critical pricing definitions. *Reference price* is the first pricing term

reference price A price or price range anticipated by the consumer based on prior experience or knowledge.

firms need to understand. This is the price for which consumers believe the product should sell. The reference price is formed when consumers consider such things as the following:

- Price last paid
- Price of similar items
- Price considering the brand name
- Real or imagined cost to produce the item
- Perceived cost of product failure

The last item is of considerable importance because it reflects consumers' imagination of what could go wrong. For example, the reference price for a meal at which one is celebrating a special occasion is higher than the reference price for a meal with some old college friends, even though the restaurant may be the same. The risk of failure is critical in the first case and less critical in the second.

Firms also need to understand customers' **reservation price.** This is defined as the maximum price the customer will pay for a product. For example, if the customer's reservation price for a can of soda is 1 euro and the price is 1.01 euros, the customer will not buy the product. If the selling price is less than the reservation price, the customer will buy the product. Firms that price exactly to the reservation price are said to extract the entire *consumer surplus*. Firms that price less than the reservation price are said to be *leaving money on the table*. Obviously, firms do not want to leave money on the table. Earlier in the chapter we discussed how Taco Bell redesigned its product to sell for 59 cents. This number was determined in part by finding out the maximum amount a customer would pay for a basic taco.

Reference pricing should be built into the pricing decision. Research can determine what the market thinks the product should cost and the maximum amount the market is willing to pay. This can be especially useful in the pricing of services that will be new to the customer, who has no previous expectations regarding how much services should cost.

Findings may indicate that the service can be priced higher; contrarily, a lower-than-expected price may offer competitive advantage. Knowledge of price expectation can help firms avoid both overpricing and underpricing.

Psychological Pricing

Prices cause psychological reactions on the part of customers just as environments do. As noted, high prices may imply quality and low prices may imply inferiority. This is especially true for services because of their intangibility. Thus, higher-priced services may sell better, whereas lower-priced services may sell poorly. This is contrary to the standard economic model. Psychological reactions, however, do not necessarily correspond to reality, and it is not unusual for customers to feel that they have made a mistake.

Psychological pricing is especially prevalent in the hospitality industry because of the visibility factor. Being seen at an upscale restaurant or hotel is very important to some customers. For example, a businessman might buy inexpensive furniture for his apartment and drink ordinary wine at home. This same businessman, trying to make

reservation pricing
The maximum price the customer will pay for a product. For example, if the customer's reservation price for a can of soda is 1 euro and the price is 1.01 euros, the customer will not buy the product.

psychological pricing
Pricing that takes into consideration what may be in the customers' minds or how they will react.

an impression on peers and customers, will rave about the antique furniture in the lounge and the expensive wine ordered with dinner. In other words, he wants to be seen with the product that offers the highest affordable visibility factor.

Buyers and nonbuyers of products also have different perceptions of price. This contrast can be demonstrated best with the case of upscale restaurants. Many such restaurants are perceived by those who have never been there to be far more expensive than they actually are. Commander's Palace, one of New Orleans' finest restaurants, used large advertisements in the local paper detailing its attractively priced lunch specials to counteract this perception. In pricing, it is important to understand the price perceptions of nonusers as well as of users.

Another psychological pricing technique is price lining. This technique clumps prices together so that a perception of substantially increased quality is created. For example, a wine list might have a group of wines in the $8 to $10 range and have the next grouping in the $14 to $16 range. The perception is a definitive increase in quality, which may or may not be the case.

Still another version of psychological pricing is odd-numbered pricing. This is a familiar tactic to all of us. Items sell at $6.99 rather than $7.00 to create the perception of a lower price. Sometimes this is carried to extreme such as a computer that sells for $1,999.99 or a car advertised at $22,999. This tactic is often used in menu and hotel room pricing.

All these differences in customers' perceptions might seem to make pricing an impossible task. Perhaps that is why hotels and restaurants tend to ignore the customer and price according to other factors! Customer-based pricing is not impossible, however. Target marketing allows us to select relatively homogeneous markets for which the product and the price are designed.

The marketer should also be aware, very aware, of how the customer uses price to differentiate competing products and services. This is a key to positioning with price. Value perception is always relative to the competition, whether the value perceived is real or imagined. It is the marketer's job to understand this process.

GENERIC PRICING STRATEGIES

There are four generic pricing strategies that companies can use to compete in a market: skim pricing, penetration pricing, match pricing, and neutral pricing. Firms can practice a combination of these strategies depending on the market segment they are targeting. For instance, a penetration strategy may be used for the group market, whereas a skim strategy may be used to attract senior business travelers to suite rooms. These pricing strategies are summarized in Exhibit 15-8. This exhibit also shows that a firm chooses its pricing strategy based on four criteria: customers, competition, costs, and the firm's overall goals—whether financial, customer satisfaction, or something else. We discuss each of these influences in relation to the specific strategy.

The first pricing strategy shown in Exhibit 15-8 is **skim pricing.** Skim pricing is also called "enhancing the image" or "prestige pricing." The attempt is to make the property appear so special, new, and different that it is worth the higher price. The term derives

skim pricing The goal of price skimming is to capture high margins at the expense of high sales volume. The term is derived from the notion of skimming the cream off the top, before the competition comes in and forces prices down.

	Skim	Penetration	Match	Neutral
Customers	Price insensitive; place high value on a product's differentiating attributes	A large share of the market must be willing to change suppliers in response to a price differential	Believe customer is concerned about price	Maintain coherent pricing strategy
Competition	Must have some source of competitive protection	Competitors lack the ability or incentive to match prices	Totally concerned with competition	May or may not be concerned with competition
Costs	Incremental unit costs represent a small share of product's price; even a small price premium will generate a large percentage increase in the contribution margin	More favorable when variable costs represent a small share of the price so that each additional sale provides a large contribution margin	No understanding of costs	May or may not understand costs
Strategy	Designed to capture high margins at the expense of high sales volume	Setting price far enough below economic value to attract and hold a large base of customers	Decision to directly match competitor's price	Strategic decision not to use price to gain market share

Exhibit 15-8 Summary of Fair Pricing Strategies

from the notion of skimming the cream off the top. The goal of price skimming is to capture high margins at the expense of high sales volume; that is, sell fewer rooms, but earn more money per room on the rooms you do sell. Four Seasons is an example of a firm that uses this strategy, as are the new boutique hotels being started by the luxury brands such as Bulgari. Rocco Forte Hotels also practice this strategy.

From a marketing perspective, the impact on the customer should always be considered before final pricing decisions are made. In the end, the customer makes the final pricing decision. Skimming strategy works best when the following customer conditions are in effect:

- Customers are price insensitive.
- Customers place a high value on a product's differentiating attributes.
- There is value attached to prestige and exclusivity.
- The price is not important in relation to the benefits derived.

Price skimming also works best if the firm has competitive protection, such as a superb location, special facilities, or customers' belief that no other property can match what the firm can offer. For example, the Sofitel in Warsaw, Poland, is able to charge higher-than-average rates because it is the only hotel in the city that can block off the streets surrounding the hotel. Thus, when members of the European Union meet, they must hold the meeting in that hotel because it offers the most secure environment. Of course, at other times of the year when there is less concern for security, it is the goal of the marketing department, along with everyone on the property, to ensure that customers believe that no other property can meet their needs as well as this property can. This enables the hotel to keep its prices high.

From a competitive standpoint, price skimming works only if competitors cannot easily match the capabilities of the firm practicing this strategy. These capabilities include location, as discussed earlier, but also features such as brand name, service level, physical facilities, and anything else the customer values and is willing to pay for.

The overall strategy of price skimming is to capture high margins at the expense of high sales volume.

penetration pricing
The goal of penetration pricing is to price just below the average market price in order to capture market share. This type of strategy can occur when opening a new property, or it can be a regular strategy, as practiced by Southwest Airlines.

A second generic pricing strategy is **penetration pricing.** Penetration pricing is the opposite of skim pricing. The goal of penetration pricing is to generate sales volume even if it means lower margins. The adage "I'll make my money on volume" describes this from of pricing. Penetration pricing does not mean that prices are necessarily low, but they are low relative to what the competition normally charges. For example, Lexus automobiles are priced below Mercedes automobiles. Penetration pricing works best when the following customer conditions are in effect:

- A large share of the market is willing to change suppliers in response to the price differential.
- Customers only really look at price and not the other features that would make them ignore the low-price offer.
- Price is not a trivial matter to customers.
- Consumers are brand insensitive.

Penetration pricing will work in the long term only if the firm practicing this strategy has lower costs than its competitors.

One must be careful with penetration pricing. It cannot be used to create new demand. Rather, it is used to take market share from others. In most marketplaces, where there is greater supply, new demand is not created because a new hotel is opened. The meetings or business traveler market already exists, in another hotel. Or, the traveler stays someplace else because of the tight market. For example, many who would prefer to stay in New York City, where they do business, stay in New Jersey instead because of more availability and lower room rates. The idea is to get existing customers in competitors' hotels to your product.

Penetration pricing is often used when a new hotel opens. The goal is to send a message to both customers and competitors. Consider the story in the Tourism Marketing Application of two hotels in New York City that took exactly opposite opening pricing strategies. The St. Regis and the Four Seasons both opened with very different introductory pricing strategies, and both continue to be successful hotels today.

Tourism Marketing Application

Different Opening Pricing Strategies

Two hotels in New York City opened with the exact opposite introductory pricing strategies and ended up in the same position. The St. Regis and Four Seasons hotels both opened (the St. Regis after a massive renovation) in the New York marketplace when demand for guest rooms was soft. The St. Regis priced itself at the top of the market and declared it would rather run empty rooms than discount. In fact, it ran many vacant rooms until the economy picked up in the mid-1990s. Four Seasons opened its hotel with introductory rates of $179, astounding for a five-star hotel of that caliber. By the end of 2005, both hotels were flirting with a $850 average daily rate. In other words, they used different opening pricing strategies, yet both are now successful hotels in New York.

For penetration pricing to work, the firm practicing this strategy needs to have a lower cost structure than its competitors. Southwest Airlines, Ryan Air, and Easy Jet are able to charge lower prices than traditional airlines because of their cost structures. There is no incentive for the traditional carriers to match these lower prices because the low-cost carriers can always lower their prices more and still make money. This is not the case for the traditional carriers.

Both restaurants and hotels sometimes use penetration prices initially to create awareness and trial, steal customers, and build volume. Once the business is established, it is normal for prices to be increased. Sometimes this works and sometimes it backfires and business is lost, at which point it is far more difficult to lower prices and recapture the business. The image of being overpriced or having poor price value is an enduring one with the customer.

A third pricing strategy is **match pricing,** which is also known as *competitive pricing.* Match pricing, as the name suggests, is a strategy in which one firm matches the price of the firms that are the direct competitors. It is viable as long as there is no customer perception of significant differences among the properties and one's cost structure allows pricing at the level of competitors. Match pricing also assumes that the competitors have made the correct pricing decision. These are all major assumptions, which are usually not correct. For match pricing to work, the market must be willing and able to buy at that level. It must also be totally concerned about price.

match pricing A strategy whereby one firm matches the price of firms that are direct competitors.

A problem with match pricing is that the *augmented* product is rarely the same, even in the same product class. Another problem is that firms often do not have the same goals. Firms often use price matching to help create the perception of value. One way to create that perception is with pricing as a tangible aspect of the presentation. When one prices above the direct competition, a statement is made that a better product is being offered. The reverse is true if one prices below the competition. The point is that the firm uses the competitor's price as a point of reference. It is inherently foolish to attempt to bait the customer with pricing if the product is not there to support it.

In restaurants, there is far more variation in the product relative to the same product class. Atmospherics are probably important, along with the menu items, the chef's preparation, the quality of food and drinks, and other variables. Nevertheless, the need to maintain a strong pricing relationship with competitors is important. Restaurants have more opportunity to differentiate their product and should price accordingly, provided the market perceives that differentiation and is willing and able to pay for it.

The fourth and final generic pricing strategy is **neutral pricing.** Here, firms decide not to use price to gain market share; rather, they use other market variables. They may or may not be concerned with competition and may or may not understand costs. Those who practice neutral pricing believe the customer wants a coherent pricing strategy. They also believe that the customer wants choice or added value.

neutral pricing A firm uses variables other than price to gain market share.

Value added services are those that are added to the basic product or service that the customer buys to enhance the perception of value. These are worth evaluating because in some cases they may not add true value, may simply increase the cost base, or may eventually be passed on to a customer (in the form of higher prices) who doesn't really want them or perceive a higher value.

Developing a product or service for customers' specific needs that augments the standard product is a part of loyalty marketing. Business services in a guest room, for which an additional fee is sometimes charged, and turn-down service at no charge are perfect examples. Many hotels, however, instead of tailoring added services to individual needs, provide customers with more services than they want or need at prices that don't reflect the value or their cost. Unfortunately, management sometimes does not even know which services customers with similar needs really want, which should be offered as part of the standard product, or which should be offered as value options that some would pay extra for. Furthermore, because of the intangibility of many services, firms often don't know the cost of providing them. No matter how homogeneous a target market is, one size does not fit all.

Because hotel managements rely almost solely on measures of customer satisfaction, they are often misled. Customers are always happy to get something for nothing, and when they do, they express satisfaction of the overall offering. The property, however, has to absorb the costs, of which they may be unaware, that may or may not have created real value in the first place. The solution to this is called *flexible service offerings*—providing particular services valued by individual customers. A hotel should first "inventory" these services to find out what is being provided to whom and on what basis.

Customers then need to be asked the value of the service to them. The following options are now available to the firm:

- Do not offer the service.
- Give the service away at no additional charge.
- Raise the price equal to the cost of providing the service.
- Raise the price less than the cost of providing the service.
- Raise the price slightly higher than the cost of providing the service to camouflage a price increase on the standard product.

This approach allows hotels to fit the service to customer needs, as well as notify customers that they do not have to pay for something they don't want. Some hotels today have turn-down service on request only—but only after realizing how much it was costing them and that many customers didn't want it. British Airways has been very successful charging the customer for value added services. For those customers who just want the cheapest price available, British Airways has this price. This enables them to compete with the low-cost carriers. For those customers who want more services, ticket flexibility, and upgrade availability, British Airways offers them such options. Exhibit 15-9(a)–(d) illustrates how customers can choose the price they want to pay and the value added service options.

Measuring the Impact of the Generic Pricing Strategies

Now that we have discussed the four generic pricing strategies, we can review how marketing managers examine how well their strategies worked vis-à-vis the competition. We use terms such as *market share, fair share REVPAR,* and *yield index* to measure

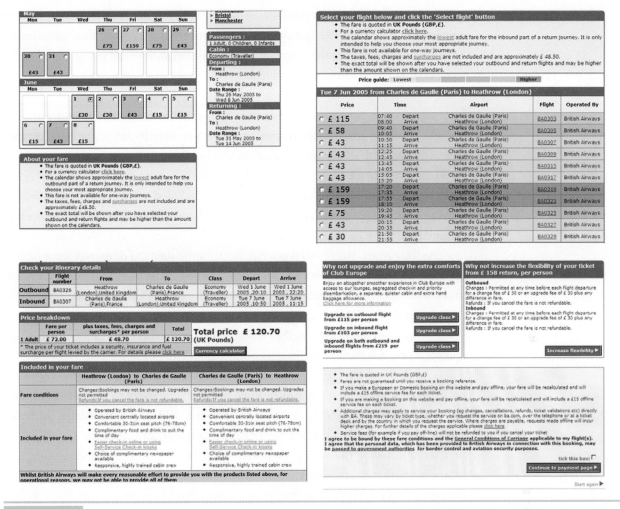

Exhibit 15-9   British Airways customers can choose the price they want to pay and the value added service options

Source: British Airways. Retrieved from www.ba.com. May 12, 2005. Used by permission.

the success of our pricing strategies. We discussed these terms in an earlier chapter, but they are worth revisiting here.

Market Share.   Comparison figures of occupancy and restaurant covers are called market share figures and are used to compare actual market share with fair market share, or that amount of business you could expect if you received your proportional share of the total business conducted by the properties that comprise the competitive set. To calculate your fair market share, divide your capacity by the total capacity in the product class. This is the share you should get if every competitor is performing equally well. To compute actual market share, divide your actual occupancy (or covers) by total competitive set occupancy. Then compare actual to fair market share as a measure of how well you are doing relative to the competition. The goal is for your market share to exceed your fair share. We illustrate this with an example.

Consider the hypothetical example shown in Exhibit 7-6 for one city area for one night. All of the participants in the analysis are not in the same product class. This does not mean that you are not interested in their occupancy. For example, it would be worthwhile to know why middle-tier properties are running at higher occupancy than upper-tier properties. It might indicate that the upper-tier properties are pricing themselves out of the market, or it could mean something entirely different, for example, concerning the type of business that was in town last night.

Fair Share.   Now consider the market shares of the properties in your competitive set. Hotel C's actual share is considerably lower than its fair share (46.2 percent versus 35.2 percent). But look at the size of this hotel; it is still filling more rooms than any of the others in the analysis. Perhaps this is primarily a convention hotel with widely fluctuating occupancies; perhaps it should not be included in the same competitive set. What this means is that one has to interpret these figures with discretion before making judgments.

As you can see, your hotel is getting slightly more than its fair market share (actual share is 24.7 and fair share is 23.1) and would not be doing so even if Hotel C's occupancy were up. Hotels A and B, however, are substantially exceeding their fair share (15.5 versus 11.5 for Hotel A and 24.7 versus 19.2 for Hotel B for actual share and fair share). What, you might ask, are they doing right? Or, what are you doing wrong? This calls for an examination of their segments and marketing strategies.

REVPAR.   The calculation of revenue per available room (REVPAR) is another method commonly employed in the hotel industry. REVPAR is calculated by dividing the room revenue by the number of rooms available for sale. It can also be calculated by multiplying the average daily rate (ADR) by the occupancy percentage. The fallacy of market share is that a competitor can gain actual share in the market at the expense of room rates. By dropping its rates $10, more people may book at that hotel. REVPAR measures the revenue generated per available room and essentially controls for pricing decisions. REVPAR is the method most widely used in the industry today.

The REVPAR calculation more accurately measures the balance of marketing efforts. The middle-tier hotels indicate better asset management. Their REVPARs are comparable to or better than those of the upper-tier hotels, and they undoubtedly are lower-cost producers. One possible conclusion is that they are stealing business from the upper-tier hotels with lower rates. Another is that, on this particular day, the upper-tier hotels had booked lower-rate conference groups. In any case, the reason needs to be examined on a regular basis. Trends are more enlightening than calculations performed to evaluate single days, and other conclusions might be drawn.

Yield Index.   If a firm knows the REVPAR and revenue in the market, it can calculate the yield index. The yield index is calculated two ways. One way is to divide the property REVPAR by the market REVPAR. The second way is to examine the yield in terms of revenue. Here, the calculation is share of revenue divided by the share of supply in the market. An index of greater than 1.0 indicates that the hotel is outperforming the market, whereas an index of less than 1.0 indicates that the hotel is underperforming the market.

REVENUE MANAGEMENT

Revenue management (formerly called yield management) started in the hotel industry in the late 1980s. Under the revenue management system, prices changed based on fluctuating demand and advance bookings. Like airline passengers, hotel customers pay different prices for the same room depending on when and how their reservations are made, how long they plan to stay, and when they plan to arrive. Through the sophistication of computer technology, different prices are set depending on demand, day by day or hour by hour. The basic concept is: Tell me when you want to arrive and I will tell you the price. Tell me the price you want to pay and I will tell you when you can arrive. Different consumers have different reservation prices, and the goal of revenue management is to price to that reservation price.

For example, when demand is soft, lower rates requiring advance bookings remain available or are reopened for sale shortly before the dates that are not fully booked. When demand builds, the lower rates are removed so that customers booking then will pay higher rates. In other words, all levels of pricing are controlled by opening and closing them almost at will, with any variation in demand. It is said that the competitive advantages of revenue management are enormous:

> Revenue management can dramatically increase revenues; maximize profits; greatly improve the effectiveness of market segmentation; open new market segments; strengthen product portfolio strategy; instantly improve cash flow; spread demand throughout seasons and times of day; and allow management to price according to market segment demand.[6]

What Revenue Management Is

Revenue management is a systematic approach to matching demand for services with an appropriate supply in order to maximize revenues. Before revenue management, this was largely limited to balancing group with individual demand, based on complementary booking times. Today, through computer technology, the attempt is to juggle all bookings and rate quotations so that on any given night the maximum revenue potential is realized.

Revenue management plans the ideal business mix for each day of the upcoming year and prices the rooms accordingly. It then adjusts the mix and prices on an ongoing basis as reservations develop or do not develop.

Several factors make the use of revenue management suitable to the hotel industry. First, a hotel room is a perishable product, so it is sometimes better to sell it at a lower price than not to sell it at all, because of low marginal production costs and high marginal capacity costs (i.e., contribution margin pricing). Second, capacity is fixed and cannot increase to meet more demand. Third, hotel demand is widely fluctuating and uncertain, depending on the days of the week and seasons of the year. Fourth, different market segments have different lead times for purchase. A convention group might reserve hotel rooms three years in advance, a pleasure traveler two months, and a business traveler one week ahead. Fifth, hotels have great flexibility in varying their prices at any given time.

These factors are very similar to those in the airline industry and represent the necessary conditions for a successful revenue management program. Although an operational tool, revenue management requires hotels to be market oriented. Knowledge of market segments, their buying behavior, and the prices they are willing to pay is essential for maximum success.

Revenue Management Practices

The essential rules of this process for hotels have been said to be as follows:

- Set the most effective pricing structure.
- Limit the number of reservations accepted for any given night or room type, based on profit potential.
- Negotiate volume discounts with groups.
- Match market segments with room type and price needs.
- Enable reservations agents to be effective sales agents rather than merely order takers.[7]

We have added the following:

- Provide reasons for lower rates, such as advance purchase time, payment in advance, nonrefundability, length of stay, and so on, for a variety of market segments. Marriott has done this deliberately to put the trade-off decision in the hands of the customer. In industry jargon, these are called "fences." Examples of "fences" for customers to consider when making a hotel reservation are shown in Exhibit 15-10.
- Be consistent across the central reservation system (CRS), property reservationists, travel agents, and other intermediaries so that quoted rates are the same. This is rate parity discussed earlier in the chapter.

Although the practice of revenue management has its applications for the hotel industry, a marketing approach needs to be employed in conjunction with revenue management. An operations approach to revenue management would be to offer the same room at different rates to the customer depending on what the market will bear, similar to the airlines. The hotel marketer's approach to revenue management should differ from that of the airline marketer given the many ways a hotel stay differs from an airline flight.

| Exhibit 15-10 | Pricing Fences |

Rule Type	Advanced Requirement	Refundability	Changeability	Must Stay
Advance purchase	3-day	Nonrefundable	No changes	Weekend
Advance reservation	7-day	Partially refundable (1% refund of fixed $)	Change to dates of stay, but not number of rooms	Weekday
	14-day		Changes, but pay fee; must still meet rules	
	21-day	Fully refundable	Full changes; nonrefundable	
	30-day		Full changes allowed	

The revenue management system of a hotel should be set up to offer different categories of guest rooms for different prices. A hotel has an opportunity to create many different types of guest rooms, some more desirable than others. An effective hotel revenue management system will open and close categories of rooms, giving the customer greater value for higher pricing.

Yield

The ratio between actual and potential guest room revenue is referred to as **yield.** Actual revenue is received from room sales. Potential revenue is what a hotel would have received if their rooms were sold at full price or rack rates. Keep in mind, of course, that for this to be realistic, the full price rates must be realistic. Rack rates that are rarely achieved have little meaning for true yield ratios. Also realize that a hotel will have any number of different rates, including suite rates. All these must be calculated to determine a true yield ratio. Incremental revenue of food and beverages, unlike the airlines, cannot be ignored. Yield takes into account both occupancy and guest room rates and can be illustrated by the example in Exhibit 15-11.

yield The ratio between actual sales and potential sales forecasted over a given period of time.

Thus, a hotel can reach the same, a better, or a poorer yield through different combinations of average rates and occupancy. Effective revenue management requires hotels to have access to many kinds of information, but the most basic element is demand forecasting. Hotels must be able to forecast the demand for each room category from each of its market segments, for any date in the future (the near future, at least). Thus, customer purchase behavior must be well understood—especially the lead time for purchase and price elasticity.

Revenue management, if used effectively, allows a hotel to manage its limited inventory to maximize revenues. Short-term gains, however, must not substitute for long-term profits. Loyal and repeat customers will not appreciate the lack of guest room availability or special rates to which they are accustomed. Because customers are likely to be most interested in price stability, it may be a mistake not to honor a long-term customer's request for his usual rate. Hotel employees who are affected by revenue management systems, especially those in reservations, sales, and front office

---

Exhibit 15-11   Calculating Yield

Hotel A has 500 rooms and an average rack rate of $180. On August 1, it had occupancy of 70 percent or 350 rooms sold, at an average rate of $140. (REVPAR figures are shown only as a point of comparison.)
- Yield = *Revenue Realized*
- Revenue Potential

- Revenue Realized = $140 × 350 rooms sold = $49,000 REVPAR $98.00 (49,000/500)
- Revenue Potential = $180 × 500 = $90,000
- Yield = $49,000/$90,000 = 54.4%

Hotel A can realize the same yield or a higher yield if it sells fewer rooms at a higher rate or more rooms at a lower rate:

Average Rate	Rooms Sold	Revenue Realized	Yield	REVPAR
$160	306	$48,960	54.4%	$98
$120	408	$48,960	54.4%	$98
$170	300	$51,000	56.7%	$102
$130	400	$52,000	57.8%	$104

Exhibit 15-12	Information Needed for Pricing Strategies

1. The customer's value analysis of the product or service
2. The price level of acceptance in each major market
3. The price the market expects and the differences in different markets
4. The product's position on the life cycle curve
5. Seasonal and cyclical characteristics of the industry
6. Economic conditions now and in the foreseeable future
7. Customer relationship
8. Channel cost to figure in calculations and the mark up at each level
9. Advertising and promotion requirements and costs
10. The product differentiation that is needed

Exhibit 15-13	Common Mistakes in Pricing

1. Prices are too cost oriented. They are increased to cover increased costs and don't allow for demand intensity and customer psychology.
2. Price policies are not adapted to changing market conditions. Once established, they become "cast in cement."
3. Prices are set independent of the product mix rather than as an element of positioning strategy. Integration of all elements of the marketing mix is essential.
4. Prices ignore the customer psychology of experience, perception of value, and the total product. These are the true elements of price perception that will influence the choice process.
5. Prices are a decision of management, rather than of marketing.

departments, must be involved in the process. They must understand that to maximize revenues, it is still critically important to keep loyal customers.

THE LAST WORD ON PRICING

We close this chapter with some conclusions. Exhibit 15-12 provides information that should be obtained for developing pricing strategies. Exhibit 15-13 summarizes the pitfalls of pricing that occur most frequently. Because pricing is the most flexible part of the presentation mix, it requires constant evaluation. Those evaluating pricing should check their pricing strategies against this list.

Chapter Summary

Pricing is a complex marketing tool. However, it is first and foremost a marketing tool. Thus, by definition, pricing should be customer based and customer driven. Pricing is also a tangible aspect of the product or service offered. As such, it can be used to change and manipulate customer perception. The effective marketer must understand this process.

When establishing prices, one must identify the target market's financial objectives, volume objectives, and customer objectives. The marketing mix strategy should be based on these objectives and customers' needs and wants. Cost and competitive pressures establish constraints, but cost-oriented methods of pricing such as cost-plus, cost percentage, break-even, and contribution margin pricing ignore the need for customer-driven pricing.

Revenue management is widely used in the hospitality industry to help set the parameters for the pricing decision, allowing the customer to be part of the process.

Key Terms

cost-based pricing, p. 503

fixed costs, p. 506

match pricing, p. 519

neutral pricing, p. 519

penetration pricing, p. 518

price bundling, p. 503

psychological pricing, p. 515

reference pricing, p. 514

reservation pricing, p. 515

revenue management, p. 502

semivariable costs, p. 506

skim pricing, p. 516

value-based pricing, p. 509

variable costs, p. 506

yield, p. 525

Discussion Questions

1. What is price?

2. What is meant by variable, semivariable, and fixed costs? How do they relate to and affect the pricing decision?

3. What is contribution margin pricing?

4. What is value-based pricing?

5. Describe what is meant by revenue management and explain how it relates to or affects the pricing decision.

6. In the end, who makes the final pricing decision?

Endnotes

1. Peter, J. P., & Olson, J. C. (2002). *Customer behavior and marketing strategy* (6th ed.) New York: McGraw-Hill Irwin, 459.

2. Bell, M. L. (1971). *Marketing: Concepts and strategy.* Boston: Houghton Mifflin, 857.

3. Jain, S. C. (1997). *Marketing planning & strategy* (5th ed.). Cincinnati: South-Western College Publishing, 400.

4. Monroe, K. B. (2003). *Pricing: Making profitable decisions* (3rd ed.). Boston: McGraw-Hill Irwin, 261–262.

5. Bowen, J., & Shoemaker, S. (1998). The antecedents and consequences of customer loyalty. *Cornell Hotel and Restaurant Administration Quarterly, 39* (1), 12–25.

6. Makens, J. C. (1998, April). Yield management: A major pricing breakthrough. *Piedmont Airlines* (in-flight magazine), 32.

7. Lieberman, W. (1993, February). Debunking the myths of yield management. *Cornell Hotel and Restaurant Administration Quarterly,* 34–41.

case study

A World Series of Yield Management

Discussion Questions

1. Should the hotel accept the 100-room group for the October 19–25 seven-night stay?
2. What are the advantages and disadvantages of accepting or declining this business from a loyal client of the hotel?
3. Are there alternatives to consider not mentioned in the case?

You are the front office manager of a 500-room commercial hotel in an urban location. Your property caters mostly to traveling businesspeople during the week and to families and tourists on the weekend. The hotel has three food and beverage outlets on site, including a coffee shop for fast-service breakfast and lunch, an informal dining room serving Italian cuisine, and a very popular sports bar. The sports bar and restaurant has a large-screen TV and frequently hosts sports celebrities as guests. The room seats 200 people and is decorated with signed posters, photos, and sports gear. The menu includes an assortment of finger foods, hot and cold sandwiches, and a variety of alcoholic and nonalcoholic beverages.

The hotel is a franchise of a well-known hotel chain and is managed by a management company. Most of the ownership of the hotel is held by institutional investors, including a large insurance company. The property is 10 years old and has had an excellent financial track record.

The fall season is particularly strong for this hotel, as it services many business meetings and transient business guests during the week. Weekends have lower occupancies and lower average rates than do Monday through Thursday, but business is still solid. It is now partway through September, and the fall season has promised to be stronger than any of the previous five years. A summary of occupancy

and average rate for the month of October for the last five years is as follows:

	Past Year	Prior Year	Prior Year	Prior Year	Prior Year
Occupancy	85%	86%	83%	74%	77%
Average Rate	$130	$125	$125	$110	$120

As the front office manager, you are now reviewing the projected occupancy for October of the current year. You are faced with an unusual dilemma where you must decide which guests to accommodate. The city's major league baseball team very possibly may be a contender in the World Series, but the actual contenders may not be known for sure until the Sunday before the Series begins. If it is a contender, games 1 and 2 will be played at home on Saturday, October 18, and Sunday, October 19. If needed (the first team to win four games wins the series), games 6 and 7 will be also be played at home on Saturday and Sunday, October 25 and 26. (October 20 through 24 are travel and away games.) The hotel rooms in the city will most certainly sell out on those dates at top rates. As of September 15, for these days, 400 of the hotel's 500 rooms in your hotel have already been sold at an average rate of $125 for the weekdays (Monday–Thuday), and 300 have been sold for the weekend days (Friday–Sunday) at an average rate of $90.

The sales and marketing department has requested a block of 100 rooms to accommodate a very loyal business group from the night of Sunday, October 19, through the night of Saturday, October 25, a seven-night stay. This group would expect to pay their corporate rate of $100 per room. Four evening banquets, other meals, beverages, and various meeting rooms will be needed during the course of their stay. The group would like to book immediately.

Reviewing historical booking patterns, you note that the hotel would typically sell an average of 80 rooms to transient guests on Monday, October 20, through Thursday, October 24, and 120 rooms on Saturday and Sunday, October 18 and 19, and on Friday, Saturday, and Sunday, October 24,

This case was written by Denise Dupré and is used by permission, John Wiley & Sons.

25, and 26. These transient guests typically book rooms three to seven days in advance for weekdays and two to three days in advance for weekends. The average rate for these sales would be expected to be $110 on weekdays and $80 on weekends.

If the World Series games are, in fact, played in the city, you would easily expect to be able to charge a full rate of $160 on all remaining rooms; however, you would not anticipate those bookings to be made until the last minute.

As you review the roster of guests already booked in that October period, you note a large concentration are from one city—the fans of the likely opponents in the World Series. If both teams are not the finalists, or if the series is over in four games, you anticipate a large number of cancellations. The hotel has a 24-hour cancellation policy.

You decide to do a yield management calculation on potential revenue gain to guide you in determining which way you play the rooms.

overview

The chapter examines how distribution channels are used to bring the customer and the product together—that is, how the customer finds the restaurant, hotel, or destination area from the comfort of home and how the restaurant or other business finds the customer. The distribution channel may include the use of franchisees or management contracts as a way of getting the product and brand—à la Marriott—to the customer. The distribution channel may also be a third party site, such as Expedia, that brings the customer to the product. Getting maximum use out of distribution channels means understanding how they work, determining the correct channel for each segment, and being aware of what the channels look like across the globe. Channels must be managed and evaluated on a regular basis in order to remain as useful as possible. Of course, understanding how the customer uses the various channels is critical to their success.

chapter 16

Channels of Distribution

learning objectives

After reading this chapter, you should be able to:

1. Describe the channels of distribution in the hospitality and tourism business and explain how they work.

2. List the types of distribution channels in hospitality and tourism.

3. Explain the management process of selecting and evaluating channels of distribution.

4. Discuss strategies and tools used to maximize the efficiency of selected channels.

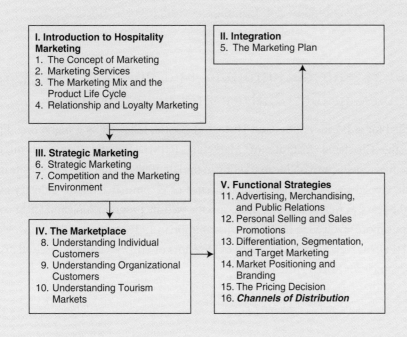

I. Introduction to Hospitality Marketing	II. Integration
1. The Concept of Marketing 2. Marketing Services 3. The Marketing Mix and the Product Life Cycle 4. Relationship and Loyalty Marketing	5. The Marketing Plan

III. Strategic Marketing
6. Strategic Marketing 7. Competition and the Marketing Environment

IV. The Marketplace
8. Understanding Individual Customers 9. Understanding Organizational Customers 10. Understanding Tourism Markets

V. Functional Strategies
11. Advertising, Merchandising, and Public Relations 12. Personal Selling and Sales Promotions 13. Differentiation, Segmentation, and Target Marketing 14. Market Positioning and Branding 15. The Pricing Decision 16. *Channels of Distribution*

Anwen Parry
Director of Distribution & E-Commerce, Rocco Forte Hotels

Anwen Parry is director of distribution and e-commerce for Rocco Forte Hotels. She is responsible for taking Rocco Forte Hotels to their own label GDS, code FC. This project involves the analysis of providers in the market, as well as contract negotiations and implementation of distribution, sales, marketing, PR, and training. She is responsible for creating and monitoring brand strategy and standards for distribution on all channels. E-distribution channels account for 30 percent of the company's total business. In addition, she is responsible for the training and performance of offices in Manchester, Orlando, Frankfurt, and Singapore.

Prior to advancing to her current position, Parry was head of distribution management for Rocco Forte and was responsible for identification, evaluation, testing, and implementation of e-business initiatives. Earlier she was internet marketing manager. Prior to joining Rocco Forte, Parry was a recruitment consultant.

Parry holds a BA in History and Politics from Bristol and a marketing diploma from the Chartered Institute of Marketing (CIM).

 ## Marketing in Action

Anwen Parry
Director of Distribution & E-Commerce, Rocco Forte Hotels

What is your background, and how did you become involved in distribution and e-commerce?

I studied for a BA in Politics and History at Bristol, followed by a marketing diploma from the Chartered Institute of Marketing (CIM). Initially I worked on offline marketing, moving into online marketing in the late 1990s. I built my first website in 1997 and developed my knowledge and experience of e-commerce as the industry evolved over the years, as well as through more specific website programming and CRM courses. I became more involved in traditional distribution in 1998, starting with my involvement with HEDNA. I am now on the Board of Directors of HEDNA and I oversee all distribution channels for Rocco Forte Hotels.

How do you see hotel and travel companies leveraging the Internet to grow their businesses, and how do you see data and intelligence impacting customer relationships?

The Internet has changed enormously over the past 10 years. Yahoo was born in 1994 at the same time as travelweb.com, which was the first online hotel catalog. Websites have since gone from being merely online brochures or hotel catalogs to being major revenue generators, interactive communication tools, and information resources. E-commerce is now a high-volume, low-cost channel for many hotel companies, particularly the larger chains. Some of the best aspects of the Internet as a marketing tool are the ability to gather enormous quantities of data on your customers through log files and customer databases, segment this data relatively cost effectively, and then target your marketing efforts accordingly.

What do you see as the top five trends in Internet and e-distribution?

First, we've seen a great deal of consolidation of suppliers and technologies in the marketplace. Travelport, for example, has been involved in multiple strands of the distribution chain (Wizcom Switch, Trust CRS, Galileo GDS, numerous third-party sites, voice centers). Travelport was sold by Trust to Blackstone, but now various parts of the Travelport division are being sold again. Most recently Travelport has sold the Wizcom switch to Pegasus, which is now the major provider of switch services to hoteliers.

Second, there is a continued blurring of offline and online. Nowadays the rates we contract with tour operators often end up being distributed to online partner sites, making rate parity particularly difficult. There is even a blurring of online with online-onward distribution, which means that we often find ourselves on websites that we haven't even heard of. And, with screen scraping, our product content on these sites is frequently out of date, which damages our brand. The reality for smaller hotel groups is that negotiating onward distribution clauses with big players is difficult because of the lack of negotiating power.

Third, there has been a fight from the hotel brands against the dominance of the third parties. Intercontinental started the trend in 2004, taking a stand against a major third party site and leading the way for hotel groups to control their own destiny. The result has been a shift in the control of inventory and availability back to the hotels and an increased focus on the brand sites and on best rate guarantees. A few years ago some sites were able to offer better rates and availability than the hotel itself, the unfortunate result of hoteliers not checking the terms and conditions of their agreements closely enough. This channel shift to the brand sites has meant that hotel groups have seen increasingly large volumes of transactions being pushed through the brand site. This, in turn, has resulted in more marketing dollars being spent on brand sites, an increased focus on CRM strategies, and increased technological innovation.

Fourth, technology in the hotel distribution industry continues to evolve. Meta search engines, which are based on principles similar to those used by Google but focus on specific industries such as the hotel industry, are starting to change the way that people search for availability and rates. In my opinion, the reason they have not been as successful as they could have been is that they haven't captured the complexity of the hotel product: their focus is still very much on price rather than the many other factors that determine customers' reasons for choosing one hotel over another.

Fifth, Web 2.0. Many people call Web 2.0 the next generation of the Internet. In my view, however, it is not a change to the core technology of the Internet, but rather an improved form that shifts the control of content to the customer. Weblogs, wikis, social bookmarking, podcasts, RSS feeds, mashups, and folksonomy are all examples of Web 2.0 in which the focus is on open communication, decentralization of authority, freedom to share, and social networking. Just one example of how Web 2.0 affects hoteliers is Tripadvisor, where customer-generated reviews of hotels are impacting other customers' choices when booking hotels.

Finally, there is increased focus on other revenue opportunities for hotels, for example, groups and meetings and spas and restaurants. The greatest potential comes in the groups and meetings arena, which accounts for at least 30 percent of total revenue in the hotel industry. The trend for meeting and group planners is to look for ways to streamline and consolidate the process of researching and placing group business, a $300 billion global marketplace. The trend actually started seven years ago when Starcite and Onvantage (then called Plansoft) created two of the more universally used RFP submission tools for the meeting planning community. However, these sites focus predominantly on the initial stages of meeting planning rather than on the issues of real-time availability and inventory management. HEDNA, MPI, and HSMAI are all working toward the goal of developing basic industry standards in group and meeting bookings. The end goal is a meetings community that has integrated inventory and instant availability access from the supplier that enables the end consumer, travel agents, and meeting planners to research and book properties on a 24/7 basis.

What makes a good website?

Primarily, simple navigation, which, in my view, consists of a limited number of links to the most important areas of the site. When I design a website, the first thing I work on is the wireframe—which is, essentially, the map of the site and the way in which different pages link to each other.

The Rocco Forte brand homepage, for example, has nine links in its navigation: Hotels, Rates & Offers, Restaurants, Spa & Fitness, The Company, Images, Brochures, Gifts, and Newsletter. These links represent the most frequently requested areas of the site by our customers based on our tracking of their visits in our log files. Therefore, we are simplifying the speed with which customers can get to the information they are looking for online.

Second, page layout is very important. Much has been learned about webpage layouts over the past few years through usability and accessibility studies such as eyetracking. At Rocco Forte Hotels, we are now working on a new fifth generation website where all of the webpages are carefully designed from the positioning of our quick reservations booking engine on the page (which appears on every page) to the contact details and the cross-selling opportunities for our hotels.

Excellent content goes without saying as critical to a good website, but it's not enough to simply provide customers with information about the hotels as you might do in an offline brochure. All of our content is optimized for the search engines as well.

We spend a lot of time trying to attain a balance between the design of the site and our impact on the search engines. Search engines generally prefer relevant keywords, which is how the spiders have traditionally decided on a site's positioning. Customers might prefer a more visually appealing design such as photography and moving images, which often means flash-based design—which has traditionally been slow to download. But technology is changing: search engines are learning to read flash, and download time even for flash or graphics-heavy webpages is improving steadily, which makes our design of search-engine and customer-friendly sites much easier than it was a few years ago.

What are some of the successes you have had with Rocco Forte Hotels?

In terms of revenue, for every £1 we spend on our website, we generate £60 in return, a return on investment that is significantly higher than any other marketing effort we make. As a result, we are increasing our budget for the website quite significantly this financial year, concentrating on a site redesign, a new booking engine, and increased search engine optimization.

In nonrevenue terms we have 12 hotels and 3 million visitors to our site each year; therefore we are generating increasing awareness of our brand globally. We recognize the need to appeal to different audiences around the world. Therefore, although we have offered "mini websites" in French, German, Italian, and Russian for the past few years, we will be increasing our multilingual content to offer our entire website in German this year, with an increased focus on German search engines in particular.

We have also focused a great deal of our attention on the GDS, with regular audits of content, rate integrity checks, and GDS advertising. We now receive 30 percent of our revenue from the GDS, and one major change that we are undertaking this year is taking Rocco Forte Hotels' own label in the GDS, leaving The Leading Hotels of the World as our representation company and, therefore, the LW chain code in order to set up our own chain code, FC. Our projections are that this change will result in not only a significant transactional cost saving for the company, but also increased control of our own destiny in distribution channels.

What type of skill sets would you envision a student needing to be successful in this environment in the future?

When I recruit for a role in the e-commerce department, I'm looking for enthusiasm and a good attitude first and foremost. I can teach students how to do the job, but it's much harder for me to teach them enthusiasm and a good attitude.

A basic knowledge of the distribution chain and the way that systems interact with each other, as well as industry terminology, are very helpful in getting up to speed quickly, and an eye for detail and a natural comfort with statistics is also beneficial.

What advice can you give to students who want to follow your career path?

A "will-do" attitude and an enthusiasm to learn is always a good start. Don't be afraid of what you don't know, and ask plenty of questions—it's the fastest way to learn. Don't assume that, because you've studied the theory, you know how it works in all practices because every hotel group is different. Follow the trends in the industry by signing up to the regular e-hotelier update e-mails and visiting sites like hedna.org to stay abreast of changes.

Used by permission from Anwen Parry.

HOW DISTRIBUTION CHANNELS WORK

The distribution system consists of all channels available between the firm and the end user (i.e., the customer) that increase the probability of getting the end user to purchase the product. When considering distribution channels, firms consider two questions: How do I get my product to the customer? and, How do I get the customer to my product? We discuss the first question briefly and spend most of the chapter discussing the second.

Distribution channels are important; one leading analyst group estimated that 85 percent of the Fortune 500 companies sell their products through distribution channels.[1] A formal definition of **channels of distribution** is as follows: a group of organizations, independent or not, that are involved in the process of making a product or service available for use or consumption. An important point to remember about distribution is that the goal is to get the product to where the customer is now or is going to be in the future.

The distribution system for the hospitality industry is complex because, unlike traditional goods that are manufactured somewhere else and then shipped through the distribution chain to the consumer, the hotel or restaurant is also the retailer. This means that, unlike a manufacturer, the hotel or restaurant must be where the customer is. Firms such as Legal Sea Foods, the local deli, and Domino's—or other local pizzerias—are exceptions to these rules because they deliver their food items. Such companies are not the norm.

The major channels of distribution in hospitality and tourism are shown in Exhibit 16-1. A hospitality or tourism enterprise may be involved with any or all of them. We discuss the major methods separately, recognizing, however, that there is some overlap among them. The industry is not absolutely uniform in its use of terminology for these distribution channels. We will do our best to guide you through the various ways hospitality and tourism products can be distributed to the customer,

channels of distribution A group of organizations, independent or not, that are involved in the process of making a product or service available for use or consumption.

Exhibit 16-1　Summary of Channel Types

Channel	Examples
Methods to Get the Product to the Customer	
Ownership of facilities	JC Resorts CNL Financial Group; also branded hotel companies
Management and ownership of one facility or multiple facilities (AKA branded hotel companies with their own brands or flags). May have ownership of some or all facilities.	Marriott International, Rezidor, Intercontinental Hotel Group, Accor
Management without ownership	Marriott International, Redidor, Intercontinental Hotel Group, Accor
Franchises/franchising	Cendant, Marriott International Rezidor, IHG; Accor
Strategic alliances	
Methods to Get the Customer to the Product	
Representation firms	Leading Hotels of the World
Reservation services	Pegasus, REZsolutions
Offline travel agents	Amex
Incentive travel organizations	Martiz
Consortia	GIANTS, Virtuoso
Corporate travel departments	Any large firm
Tour wholesalers	Liberty Travel, Tui
Global distribution systems	Galileo: Travel Port; Sabre: Travelocity: Amadeus: E-Travel
Central reservation systems	
Travel management companies	American Express, Carlson Wagonlit, Rosenblueth
Convention meeting planner organizations	CVB
Discount brokers/consolidators/wholesalers	TUI
Destination management organizations	www.pra.com
Online intermediaries for business or groups	Expedia Corporate, Travelocity Business, Orbitz for Business, Groople
Online intermediaries for transient, business, and packages	Expedia Corporate, Travelocity Business, Orbitz for Business, Groople

Exhibit 16-2 Estimated Costs to Hotel to Book a Room Costing $220 Using Different Methods

Customer Books via:	Total Cost to Hotel
Toll-free call to CRS	$3–$5
Through traditional travel agent	$22–$27
Toll-free call directly to hotel	$3–$5
Branded website	$5
Online travel agency—merchant model	$18–$24
Online travel agency—agency model	$66
Direct through hotel website	$5–$15

which helps get the customer to the product. Recall that marketing is all about getting and keeping a customer. Distribution channel strategies focus on "getting" that customer. Each of the distribution channels represents a cost to the hotel. The estimated costs of customers using the various distribution channels are shown in Exhibit 16-2.

GETTING THE PRODUCT TO THE CUSTOMER

Branded Hospitality Companies

Marriott International, Four Seasons, InterContinental Hotels Group, Accor Hotels, SAS Radisson, Rezidor, and Shangri-La are examples of branded hotel companies. These companies may own all of a particular asset, part of it, or none of it. The defining element of these types of companies is that they use the brand on the top of the asset, regardless of the actual owner. The hotels are marketed by the brand, and all properties that carry the brand name adhere to strict service standards. As discussed in Chapter 14, branded companies are becoming more important because a strong brand allows a company to attract more franchisees, which translates into higher revenue. Branding also enables firms to gain management contracts, access to capital, and higher-than-average revenue per available room. It is able to do this because brands represent familiarity to customers and people would rather buy things that are familiar to them.

Franchising

Hospitality entities commonly use franchising to increase their distribution network, both to create more revenue and to obtain the geographic presence needed to be where the customer is. Catherine Siskos of Kiplinger's Personal Finance defined a franchise as follows:

> A license to run a business and collect the proceeds from that business for a set period of time, usually ten years. In addition, you get access to that company's industry expertise, to its brand name, to its patents, its trademarks, as well as its advertising and marketing departments.[2]

Franchising is also a common method of distribution for nonhospitality companies, from Avis Rent-a-Car, Midas Mufflers, and H&R Block tax services to 7-Eleven convenience stores. Coca-Cola and Pepsi-Cola franchise by allowing bottling plants to use their proprietary formulas for making the beverages and then distributing their products. This method of distribution has been in common usage since the franchise boom of the 1960s.

The amount of control a franchisor (the parent company) has over the franchisee (the company that licenses the name to distribute the product or service) varies as widely as the franchising options available. The contract between the franchisee and franchisor outlines the terms of the relationship. It covers items such as marketing support, revenues to the franchisor (usually determined as a percentage of sales—the norm is 4 to 5 percent plus marketing and reservations fees), and the duration of the agreement.

InterContinental Hotels Group (IHG), a leader in hotel franchising, has seven brands: InterContinental Hotels & Resorts, Crowne Plaza Hotels & Resorts, Hotel Indigo, Holiday Inn, Express by Holiday Inn (Holiday Inn Express in the Americas), Staybridge Suites, and Candlewood Suites. IHG has more than 3,300 owned, leased, managed, and franchised hotels and approximately 515,000 guest rooms across nearly 100 countries and territories. In the fast-food segment of the hospitality industry, the world market leader is McDonald's. Other familiar names proliferate. These companies and many others recognize that their ability to distribute their products' names and identities throughout the world is limited by the amount of capital available. Methodically, they have offered their names and their services to potential franchisees. Casual dining establishments have done so as well. Consider TGI Friday's. As of December 2006, there were more than 979 stores in 56 countries.

Management without Ownership

Many hotel companies today, such as Four Seasons and Marriott International, are primarily management companies and have virtually gotten out of the ownership of hotels. They have done this to increase the size of their distribution channel without the financial costs and risks associated with ownership.

Firms that manage without ownership operate under what is known as a management contract. This contract is the agreement the management company makes with the owner regarding management fees, the length of the management contract, possible loan contributions, performance evaluations, and the rights and responsibilities of each stakeholder. Management contracts typically last 8 to 10 years, with one or two 3- to 5-year renewals.[3]

GETTING THE CUSTOMER TO THE PRODUCT

Reservation Services

Some hotels choose to market themselves independently and choose a reservation service only for connectivity to channels of distribution. By using SynXis or Pegasus, the hotel can link directly to the GDS or Internet without having a brand affiliation.

Reservation services do not offer ancillary marketing programs. Each hotel is on its own to bring business to the channels.

Originally founded as the Hotel Industry Switch Company (THISCO) in 1988, Pegasus Systems, Inc., provided electronic commerce and transaction processing solutions to the hotel industry and operated its own customer travel reservation site, TravelWeb.com. Pegasus also offered the premier electronic switching service for reservation processing, which allowed central reservation systems to connect seamlessly to global distribution systems (GDS) and/or to the Internet with a single electronic interface. Pegasus services also included Pegasus Commission Processing, the largest provider of travel agent commission payment processing services.

An established, well-known company within the hospitality industry, REZ solutions was formed in 1997 with the merger of Utell International and Anasazi, Inc. Utell, the world's largest hotel reservation and marketing company, which was founded in 1930 and has maintained a long-standing presence in Europe and Asia, merged with Anasazi, a leading supplier of hotel reservation technology solutions. The result was a fully integrated portfolio of hotel industry information technology products and services.

The union of these two companies signified the birth of a powerful new entity that is uniquely positioned as a total solution provider for reservation distribution, offering a second-to-none comprehensive portfolio of products and services to the global hospitality industry.

Representation Firms

A representation (rep) firm is a channel of distribution that, figuratively speaking, brings a hotel to a marketplace. These companies market a hotel to customers for a fee and are hired to act as sales organizations for independent properties that don't have sales or reservation networks of their own. Major chains may also use representation firms to enhance their regional sales efforts. Representation firms have their own sales forces and represent a number of hotels through regional offices in different geographical areas.

Hotels that hire representation firms are sometimes called *soft brands* because they are essentially two brands: the brand of the representation firm and their own independent brand. Hotels get the best of both worlds: They can choose the representation firm as a way to maintain their independence while still accessing numerous marketing programs. The Leading Hotels of the World, WORLDHOTELS, Preferred Hotels & Resorts, Small Luxury Hotels of the World, and Relais & Châteaux are considered soft brands. These brands all have the following:

- Standards for membership
- Connectivity to electronic channels of distribution
- Sales initiatives
- Marketing programs
- Participation in trade shows

Representation firms go much further than reservation companies in promoting their member hotels. Apart from the worldwide reservation network and a link to all global distribution systems, they often have a sales force actively selling their member hotels and publish an annual directory featuring these hotels, with detailed information on their services and facilities. They may print other marketing collateral, such as special programs, newsletters, and flyers. They also undertake advertising and public relations campaigns on behalf of their member hotels.

Once a representation firm has been engaged, it uses all of the normal communications mix (personal selling, direct mail, advertising, sales promotion, and public relations) to get customers to buy certain hotels. Sales calls are used most often, followed by direct mail. David Green Organization of Chicago is an example of a representation firm that has been in the business for a long time. Newer, more segmented representation firms, such as Associated Luxury Hotels, have begun to carve niches in the representation marketplace.

Supranational, a European-based firm, is another example of a rep firm. Supranational's purpose is to unify the reservation network without sacrificing the identity of the individual property or chain. It is a very active sales and marketing company. Supranational represents entire hotel companies, as well as a number of independent properties around the world, which are rigorously selected. A firm that competes with Supranational is Preferred Hotels Group, which is shown in Exhibit 16-3.

In total, Supranational represents over 1,100 hotels, including more than 40 hotel chains in 75 countries, and has 23 reservation offices worldwide. Supranational is a nonprofit organization operated by hoteliers for hoteliers.

Consortia

The Merriam-Webster dictionary defines **consortium** as "an agreement, combination or group formed to undertake an enterprise beyond the resources of any one member."[4] Consortia occur for both travel agencies and hotels. In our discussion with hoteliers, most think of consortia as only referring to travel agencies—for example, GIANTS (Greater Independent Association of National Travel Service, now known as the Ensemble Travel Group), eTravCo, MAST (Midwest Agents Selling Travel), the Leisure Travel Group, and Virtuoso. However, using the Merriam-Webster dictionary as our guide, we define *consortium* in the hospitality industry as a loosely knit group of independently owned and managed properties (e.g., hotels or travel agencies) with different names, a joint marketing distribution purpose, and a common consortium designation.

The purpose of the consortium is to open a channel of distribution by maximizing combined marketing resources and reducing associated marketing expenses for the individual properties. Consortia are more common in the United States than in other parts of the world. Consortia distribute hotel inventory at preferred rates to affiliated travel agencies. Many of the affiliated agencies are "in-plants" or travel agencies dedicated to one company. Bear Stearns, the global investment banking firm in New York City, had American Express as its in-plant. American Express travel agents were dedicated to Bear Stearns, working at its offices at 245 Park Avenue. The American Express consortia provided reduced rates for Bear Stearns travelers.

Web Browsing Exercise

Use your favorite search engine to look up information on WORLDHOTELS, Supranational, David Green Organization, Preferred Hotels Group, and Leading Hotels of the World. Compare and contrast the companies. What are the strengths and weaknesses of each firm? If you were the director of marketing for a hotel, would you use any of these services? Why or why not?

consortium A collection of individual properties that are similar yet have different names and carry a common designation grouping them into the same product class. Leading Hotels of the World, for example, is an upscale consortium.

When you travel on business, you don't have to sacrifice style for value.

With this Exclusives™ offer, reserve an amazing hotel experience in one of our premier properties around the globe, and be assured that your rate is unmatched. From the classic to the eccentric, Preferred Hotel Group offers superlative business travelling experiences.

Where You Go, Becomes Who You Are. Go Somewhere Smart.

Preferred Hotels & Resorts, Summit Hotels & Resorts and Boutique are a unique collection of independent hotels, unified by the exquisite and brought together under one brand – Preferred Hotel Group.

Reserve a room online today or call your travel professional and make the most indulgent hotel reservation your per diem budget has ever enjoyed. For a worldwide listing of reservation numbers, visit PreferredHotelGroup.com.

Preferred
HOTEL GROUP

Call 0800 556555 or visit PreferredHotelGroup.com/exclusives

Preferred SUMMIT BOUTIQUE

Exhibit 16-3 Preferred Hotels Group is an example of a hotel representation company
Source: Preferred Hotels Group. Used by permission.

Incentive Travel Organizations

Incentive travel organizations, or incentive houses, are one of the major players in hospitality and tourism channels of distribution. These are companies that specialize in handling strictly incentive reward travel. Many organizations and firms have incentive contests to reward top-performing employees, salespeople, dealers, or retailers. Travel rewards are a popular form of incentive.

Major corporations often have their own in-house travel departments or individuals to handle incentive arrangements. Many companies have used travel agents. More and more, however, both large and small companies are relying on incentive houses to organize their trips. Carlson Marketing Group is one of the leading incentive travel providers with offices in 30 major cities in the United States and in 20 countries worldwide. Maritz travel, based in the Midwest, is also an established incentive travel organization.

The reason for the use of such companies as Carlson and Maritz is that incentive travel is a special type of corporate travel. For companies that use this kind of reward frequently, there is a constant need for destinations that are new, different, and exciting—in other words, that offer a real incentive for performance. Second, there is a real need for the trip to be letter-perfect. A poor trip destroys the morale of the very employee one is trying to reward. Keeping up with all of this, on a worldwide basis, is expensive and time-consuming.

Incentive travel organizations, because of their collective accounts, can parcel out the costs of their expertise. Almost always, someone will have visited and thoroughly inspected the destination, the hotels, the restaurants, and the ground services before putting together the incentive package. The incentive house then "sells" it to the company and helps the company to "sell" it to those who seek the reward.

For upscale hotels, particularly in resort areas or foreign destinations, it can be a real boost to the distribution channel to be on the approved list of a major incentive house. In these cases, a property does not simply buy an incentive house's services. It earns them by doing things right. Unlike consortia, reservation networks, rep firms, and travel agents, the customer, not the hotel, pays for the incentive travel organization's service. It is important to note that incentive planners deal directly with individual properties as opposed to representatives of specific hotel chains in order to be personally certain of the product.

Each channel member involved with an incentive travel organization is dependent on the other members in the channel for performance. If customers are dissatisfied with the trip, they may choose another incentive house for the next program. Each channel member has to make sure that everything goes as promised. For example, if the ground transportation is an hour late in picking up a group at the airport, the entire trip can be spoiled. Future business may be lost, not only to another incentive house, but also to another destination.

Corporate Travel Departments and Travel Management Companies

Corporate travel departments range from a corporate travel director who develops corporate travel policies and writes contracts with travel suppliers (e.g., hotels and airlines) to full in-house travel agencies. PhoCusWright, a firm that studies hotel and lodging distribution, refers to this market as the **corporate or managed business traveler**. The goal of corporate travel departments is to balance the needs of the business traveler with those of the organization. Travel managers help control corporate costs by negotiating volume discounts, providing information to employees, tracking costs, and monitoring future trends within the travel industry.

There are two major organizations that represent this group: the Association of Corporate Travel Executives (ACTE) and the National Business Travel Association (NTBA). Both organizations provide travel managers a forum to share information, learn the latest trends, and meet with travel suppliers. A recent trend is for organizations to outsource the corporate travel department to travel management companies. Many large travel agents such as American Express and Rosenblueth Travel provide this service for organizations. A group representing this organization is called the Guild of Travel Management Companies.

corporate or managed business traveler
A business traveler who must follow the rules and regulations of the corporate travel department. Corporate travel departments range from a coporate travel director who develops corporate travel polices and writes contracts with travel suppliers (e.g., hotels and airlines) to full-time in-house travel agencies.

Global Distribution Systems (GDSs)

The **global distribution system (GDS)** connects the travel agent to the individual hotel. The GDS was initially developed to list only airline flights. In fact, some of the GDSs in existence today were created by the airlines; Galileo, initially named Apollo, was created by United Airlines. Now, however, the GDS allows the travel agent to book not only hotels, but also flights, rental cars, train reservations, and other services. The big players in this discipline include Amadeus, Galileo, Sabre, and Worldspan. The largest of these is Worldspan, which processes 50 percent of all travel agent transactions. These companies are also part of companies that are major players in the travel industry. For example, Galileo is part of Travelport, which also owns Orbitz. Travelport was part of the Cendant Corporation, but was sold to the Blackstone Group in June 2006. Sabre is part of Sabre, which is partially owned by AMR (American Airlines), which also owns part of Travelocity. Amadeus was founded by Air France, Iberia, SAS, and Lufthansa.

For a hotel to be booked electronically through the GDS, the hotel's central reservation system (CRS) must have a computer connection to the GDS or it must contract with a **distribution service provider (DSP)**. Examples of a DSP are Wizcom, Thisco, Pegasus, and Trust. These firms provide the switch that links the CRS with the GDS. Hotels are represented in the GDS by different codes. Chain hotels (such as Hilton, Marriott, and others) are identified with codes that represent their chain. Independent hotels might be represented in the GDS with their own individual codes, but many independent hotels prefer to be represented in the GDS through a representative company. Leading Hotels of the World offers the LW code for its members.

Once the travel agent picks a hotel from the GDS (Amadeus, Sabre, etc.), the information is transferred to the hotel's CRS through one of the few switching companies. Again, these companies decode the information from the GDS and translate it into a code that each hotel can understand and input into its own reservation systems. Notice that at each point of the booking process, a fee is added. For this reason, companies are now attempting to market directly to the customer.

Traditional Offline Travel Agents

Travel agents are important **intermediaries** in the distribution of hospitality and tourism products and services. Travel agents are compensated through sales commissions, usually based on the rate of the service sold. As a rule of thumb in most cases, a 10 percent commission is paid to travel agents who book cruises and hotel rooms; rental car firms pay a lesser rate. Most airlines, once a major source of commission revenue for travel agencies, no longer pay commissions. This has forced many agents to add a transaction fee for making airline and other travel reservations. In fact, travel agents do not so much sell airline seats, hotel rooms, or rental cars as they sell their time and knowledge of travel service suppliers.

Travel agencies also form consortia, using the strength of many individual agencies to enhance their marketing and negotiating clout. In North America, travel agencies are larger than their counterparts in other parts of the world. This is because size is needed to handle the large accounts and negotiate the best arrangements, a necessity for being the agent of choice. Further, by banding together in consortia, groups of agencies have been able to bargain collectively with travel suppliers to gain access to preferred rates or other customer benefits. These are subsequently used as enticements to lure and retain business clientele, who could not otherwise obtain the same benefits. Through the control of information, agents exert great influence in all segments of the travel market. Virtuoso, formerly Allied Percival International (API), is an example of a consortium of upscale travel agencies. Virtuoso negotiates airline, hotel, and car rental rates on the basis of the combined travel expenditures of its participating members. Each member retains its autonomy while benefiting from volume-negotiated rates. Their positioning statement is as follows:

> Virtuoso is a consortium of the top-selling leisure travel agencies in the United States. Together member agencies have over 1 billion dollars of buying power. What does this mean to you, our customers? It means we have clout! As a Virtuoso member agency, we can get you the best value for your money on some of the best hotels, tours, resorts and cruise lines. Our customers also enjoy a number of attractive, free enhancements and amenities. Note the word "value," not "cheap." Whether you buy a house, a car or a vacation, you want the best value—not the cheapest. As a member, we work to give you the best possible value.[5]

As the Internet continues to grow in importance as a mechanism for booking travel, many believe that the travel agent's role in the distribution of travel services will become smaller.

Central Reservation Systems (CRSs)

The **central reservation system (CRS)** function, which allows booking from an 800 number, is integrated with both the global distribution system (GDS) and the property management systems (PMS). The CRS was created so that individual hotel chains could provide consolidated access to their worldwide inventory through toll-free telephone calls. A well-managed CRS will offer the customer the same availability and rates that may be found both on the GDS and at the hotel. Ideally, the local hotel will be electronically compatible with the CRS to ensure that similar information is available to the customer. Unfortunately, even at the time of this writing, many hotels communicate manually with the CRS, resulting in incorrect information regarding availability and rates, as well as a slow response time.

All major hotel chains worldwide now communicate from their call centers (CRS) and GDSs directly to the individual hotels. Other soft brands, such as Leading Hotels of the World and Preferred Hotels and Resorts, use Pegasus for their customers. As with other technology applications, the CRS function is being consolidated and outsourced to reduce cost. Pegasus has even outsourced the voice reservations for Leading Hotels. Although the call is answered "Leading Hotels of the World," an employee of a third party company contracted by Pegasus is indeed handling the call. This is called *private label* service. Many hotel chains are reviewing the cost of maintaining their own CRSs, and outsourcing of this function to a company that can provide this private label service has become an alternative that is cost effective and maintains service levels.

The seamless connectivity of the CRS is a benefit to the customer—in this case, the travel manager or travel agency. Using the airlines' systems, such as Worldspan, a travel

central reservation system (CRS) The computerized reservation system of a hospitality company that allows customers to make reservations without having to contact an individual entity of the company directly.

agent can view a hotel's inventory directly. Seamless connectivity allows two-way inventory management; the travel agent sees the same inventory that the reservation manager on property sees in the PMS. Room categories are booked based on availability, and rates are shown for each category. Travel managers and agents are able to provide the most accurate information to their travelers with seamless connectivity. Although all major hotels do not yet offer seamless connectivity, most are moving in that direction. Most hotel chains are now offering single-image inventory, whereby the inventory is held at the CRS level but a different availability of inventory is maintained at the property level.

Internet Channels[6]

In 1996 only 11 percent of U.S. households had access to the Internet. Today the penetration rate is approaching 80 percent (almost half of which now have high-speed access). Several countries are more "wired" than the United States, including South Korea, the Scandinavian countries, and Canada. Although the Internet is still evolving, the influence of the Internet on the hospitality industry is growing in significance.

In the first generation of Internet transactions, the primary function was to distribute information. The second generation of Internet transactions involved simple transactions, such as buying a book or CD. The third generation is the result of improved technology that has allowed the size and complexity of transactions to increase. Cars, cruises, and expensive jewelry all now trade regularly on the Internet, and online auctions have become the rage (witness the rapid growth of eBay). Every major brand, regardless of its goods or services offering, now has an Internet strategy in the emerging environment of electronic distribution. Hospitality firms are no exception.

Early in the developing stages of the Internet, opportunistic companies such as Expedia, Travelocity, Hotels.com (these are called third party sites or online travel agencies), among others, decided to become Internet travel agents and wholesalers. Technology at the time made the booking process somewhat cumbersome. In fact, the process was similar to the time before telephones. People would write to the reservation office and the agent would check availability, respond with a rate, and await the customer's reaction. The confirmation was then sent to the consumer. Instead of the regular post office, e-mail was used. Given this time-consuming process, combined with the fact that consumers were hesitant to send their credit card information over the unsecured Internet, many consumers continued to call hotels or work through travel agents, especially because there was no real price difference. With the economy strong, hotels had no incentive to discount rooms.

As a result of the tragedy of September 11, 2001, in the United States, people were afraid to travel. Hotels that suddenly had hundreds of vacant rooms turned over much of their inventory to third party Internet sites. Consumers, fearful of the economy, cut back on expenses and became even more focused on finding the best deals. Those deals were now available on the Internet. Many reservations were made by these third party sites, which had improved the booking process to make the reservation process less cumbersome and more consumer friendly. They were similar to bricks-and-mortar travel agents, and each company earned a 10 percent commission for delivering a booking to a hotel.

A new financial model, called the *merchant model,* evolved, whereby the third party site negotiated net rates from the hotels, then marked up the price to what the market would bear. For example, a hotel may have offered a $60 net rate to Hotels.com, which then sold the same room online for $100—a 40 percent markup. There was no risk for the online sellers; they had the flexibility of selling the room for any rate above $60. If the market for the room fell below $60, the online seller would simply go back to the hotel and ask for a lower rate.

To compete with these third party sites and begin to gain control of their inventory, hotel brands such as Marriott, Hilton, and Starwood began to build their own sites to attract customers and became direct competitors of the third party Internet marketers. InterContinental Hotels became the first brand to remove its inventory from the third party sites in an effort to standardize rates across all channels of distribution and regain control of its customers.

With the arrival of 2004, we began to observe a fundamental shift back to the original customer relationship with hotels, one on one. Clearly the days of customers calling hotels to make reservations had past, but consumers began to recognize that they could get more information, competitive room rates, and good service by going directly to the hotel, airline, or car rental websites. Although some believed that consumers who booked online were more likely to receive inferior service upon arrival, 84 percent of travelers agreed with the following statement: "People who book reservations through a third-party intermediary are treated the same as those who book reservations either directly with the hotel or resort through the hotel or resort's dedicated Web site." Hospilatity entities began capturing e-mail addresses and were again communicating directly with their consumers, albeit by electronic sources.

The Internet has enabled the individual hotels to begin to regain control of their customer relationships as well. Many nonchain hotels began to construct their own websites and thereby compete with the major chains.

A 2005 report by PhoCusWright (PhoCusWright Online Travel Overview 2005, used by permission) revealed that, in 2004, one in four consumers in the United States booked leisure or unmanaged business travel online, and this figure was estimated to grow to 38 percent by 2007. For European consumers, the 2004 figure was 9 percent, and the estimated figure for 2007 was 27 percent. Asia Pacific lags behind these numbers, but given that 29 percent of the worldwide Internet population is in this region, the potential for growth is enormous. In 2004 the percentage of those who booked leisure or unmanaged business travel online was 3 percent; this figure was expected to grow to 25 percent in 2007.

Clearly, both online agencies and hotel websites will increase in importance. Credit Suisse/First Boston estimated that traditional bricks-and-mortar travel agencies would suffer a decline in bookings as a result (by 15 percent), while calls to the CRS would decrease by roughly 23 percent. The relatively small decrease in estimated travel agency bookings would be primarily for simple, low-risk transactions (e.g., the purchase of a round-trip airline ticket or a hotel reservation for three nights).

One example of an online agency is Hotels.com, which is shown in Exhibit 16-4. Hotels.com is the best known player.

The hotel consolidator typically follows one of two strategies: a merchant model strategy or an agency model strategy. In the merchant model strategy the consolidator contracts for inventory from the hotels at a fixed rate, often referred to as the net rate. For example, one hotel in Las Vegas sets its consolidator rates for its suite reservations once a

Exhibit 16-4 Hotels.com is the best known player in the hotel consolidator field

Source: hotels.com. Used by permission.

day for the next 14 days and twice a week for the next 90 days. Regardless of how often rates are set, these net rates are usually sold to the consolidator at a deep discount. The goal, of course, is to sell the rooms that cannot be sold using normal channels to ensure as high an occupancy as possible. Usually, a specific amount of inventory is taken on a consignment basis under a one-year contract. At times, however, the hotel consolidator cannot sell the room through traditional channels. The online hotel consolidators then either sell their allocated rooms to their affiliated websites (e.g., Hotels.com may sell its allocated rooms to Travelocity) at a marked-up price or sell them on their own website at a certain markup (usually anywhere from 20 to 30 percent). Rooms that are booked in this way go directly to the hotel's central reservation system via a switch.

The hotel consolidator may also follow an agency model strategy. This model is very similar to the bricks-and-mortar travel agency model. The hotel gives the consolidator commissionable rates and then pays a commission on any rooms that are booked. Online travel agencies include Lastminute.com, Travelocity.com, Expedia.com, and Orbitz.com. In this model, the consolidator typically books through the GDS.

It should be noted that a hotel consolidator might follow both strategies simultaneously. As stated by the investment bank Lehman Brothers in a research report, "Most of the merchant model sites also offer reservations under the traditional travel agent model (in which the site receives an industry-standard 10 percent commission for booking a reservation for a guest), but these offerings are usually buried after several pages of more merchant model listings."[7]

The hotel consolidator following the merchant model approach may use an opaque approach to selling rooms.[8] The two firms that practice this approach are Hotwire.com and Priceline.com. The opaque model works in one of two ways, depending on whether the customer is buying a single part of the travel package (e.g., only a hotel room) or the whole package (e.g., hotel room, airline ticket, and car rental). If the customer is buying only a single part of the package, the customer does not know the specific brand she will be buying until after she has actually made the booking. However, she does know the quality rating and the general location of the hotel. If the customer is buying a package, the hotel brand is shown to the customer, whereas other parts of the package (e.g., the airline portion) may or may not be opaque. The customer also does not know how much she is paying for each component of the total package. Hence, this is often referred to as price opaque but not brand opaque. In either case, prices are nonrefundable and nonchangeable. Exhibit 16-5 shows the Hotwire web page.

| Exhibit 16-5 | Hotwire web page |

Source: Expedia.com. Used by permission.

The opaque model allows a hotel to discount without letting its brand-loyal customers trade down to those lower price points. The example frequently used by Spencer Rascoff is that of the outlet mall:

Think about how apparel retailers sell their excess inventory; for example, how Polo/Ralph Lauren or Brooks Brothers gets rid of their excess inventory. They traditionally do not discount heavily in their own flagship stores. The reason is that their brand loyal customers go into those stores prepared to pay full price. If customers see the discounted price, they will naturally trade down to those discounts. To move excess inventory, the apparel industry has created a network of outlet malls throughout the country that have either inferior or no customer service. In addition, they have tainted products in some cases, they have limited selection and they have inconvenient locations. These are the revenue management fences that the apparel industry has created to move their excess inventory through a different network or channel than their retail structure. The opaque channels are essentially the outlet malls of the travel industry, allowing suppliers to discount without brand dilution.

In the hospitality industry, what the opaque model does is allow a hotel to sell excess rooms without diluting the brand. For instance, if a Hilton has extra rooms to sell and can't sell them at $150, but they could sell at $75, they do not want to say "Hilton" next to the $75 because people are going to think about Hilton differently if they see Hilton for $75. So, instead, Hilton will want an opaque channel, like Hotwire, to sell a four-star hotel for $75 and let the customer be pleasantly surprised when they find out that it is a Hilton.

From a revenue management standpoint, the opaque model allows suppliers to segment customers into brand loyal, high-value, paying-full-price customers on the one hand and discount incremental, brand-disloyal customers on the other. It's allowing suppliers to create these two types of customers, each buying a different product in the prepurchase stage. One buys a branded product and pays for the security of knowing where they will be staying and the type of room, while the other has no idea until after the purchase is made.[9]

As mentioned, the two firms that practice the opaque approach are Hotwire.com and Priceline.com. Priceline and Hotwire are very different from a customer perspective. Priceline is known as a "name your price" model, whereas Hotwire is known as a "posted price" model. With Priceline the hotel customer first determines both the general geographic location in which he wishes to stay and the star category of hotel he wishes to purchase. He then bids what he is willing to pay for this hotel room. The bid is either accepted or rejected. Once the bid is accepted, the customer then finds out in which hotel he will be staying. The accepted bid cannot be changed or canceled. Priceline.com has evolved over time and now also offers the consumer the capability of shopping and comparing more than 60,000 hotels around the world.

Similar to Priceline, on Hotwire.com the customer first selects the general location where she would like to stay. Hotwire then shows the customer multiple prices for different categories of hotels. For instance, she might see a four-star hotel for $75, a three-star for $60, and a five-star for $180. The customer then selects the hotel based on the price she is willing to pay. Once she has selected this price, the customer then finds out in which hotel she will be staying. The accepted bid cannot be changed or canceled.

From a hotel perspective, the two businesses are actually quite similar in that they don't show the name of the hotel until after the purchase is completed. This allows the hotelier to protect his brand. The reservation is entirely electronic. In Priceline's case, it goes through the Worldspan GDS. In Hotwire's case, it goes through the Pegasus switch. But either way, they are both electronic, as opposed to being fulfilled by a fax or phone call to the hotel.

Customers can book hotel rooms directly by going to the hotel's landing page. Travelzoo is an example of a firm that sends customers to the landing page of a specific hotel. Essentially, Travelzoo works like the travel section of the local newspaper, with the exception that it reaches more than 9 million users who have requested to receive its newsletter with special travel offers. The hotel controls the rates that appear on the landing page. Once the customer gets to the landing page, he can book the hotel directly by calling the hotel's reservation department. The landing page can also direct the customer to a specific page on the hotel's own website. To get to the hotel's reservation page on the hotel's website, customers can either go directly to the website and be forwarded to the reservation page or they can be directed to the reservation page by one of the secondary websites via links. An example of a secondary website that links customers to a hotel is Travelaxe.com.

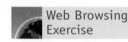

Web Browsing Exercise

Visit www.travelzoo.com/ and spend time exploring the site. How does this site differ from the sites examined in the Web Browsing Exercise earlier in this chapter? What is the consumer's proposition? What is the firm's proposition?

Websites

Internet marketing begins with a website. The website for a hotel should be thought of as the new millennium version of an old standard in the hospitality business: the rack brochure. As such, it should be a reflection of the personality of the hotel. Proper visuals, text, and related information are critical to convey the hotel's image online. Given that many hospitality companies are currently trying to drive business to their own brand websites to bypass expensive intermediaries, issues such as what to include on a hospitality website and how to present these features have become increasingly important. However, designing and creating effective websites is not an easy task. Good design means integrating technical skills, such as knowledge of HTML, CGI scripting, and Flash programming, with good artistic taste and graphic design skills. And this is, of course, before the issue of content is considered. Knowledge of online consumer behavior and an in-depth familiarity with the company in question are useful in knowing what to include to help sell the product. This section gives an overview of some of the most important issues that should be considered when designing a website for a hospitality company. Many aspects of the design process need to be integrated to maximize the benefits that can be gained from a web presence.

When assessing what features to include on a hospitality website, the designer must have the needs of the customer foremost in her mind. A potential customer's motivation for visiting a hospitality website is usually quite clear—to find out about the product as part of his decision-making process and, assuming the website designer does a good job of convincing him that the product is a good match for his needs, to make a reservation. This means that meaningful but sales-orientated descriptions need to be combined with appropriate graphics to give the website visitor a true feeling for the atmosphere and experience at the unit in question, be it a hotel, restaurant, or other form of hospitality operation.

Web Browsing Exercise

Go to http://www.mgmgrand.com/accommodations/ to see an example of an effective virtual tour.

What exactly is required on a website will vary significantly depending on the market segment. For companies operating in the economy segment, basic graphical elements and limited, factual text tend to be effective. After all, who wants to look at 360-degree photos of a typical motel room? However as the product becomes more up-market or more complex (e.g., luxury hotels, resorts, cruise ships), developers usually need to combine both increasingly detailed textual information with higher-quality graphical elements—perhaps even multimedia—to effectively influence the consumer decision-making process. A good example of a multimedia website is Carnival Cruise Line's website, which offers virtual tours of all of its ships. By clicking specific buttons, visitors can see staterooms, restaurants, the casino, and other parts of the ship. Such virtual tours help to "make tangible" the intangible.

An overused but valuable phrase when it comes to website design is that "content is king." However, most industry observers agree that many hospitality website designers fail to provide visitors with the information they need to make a purchase decision. Consider, for example, the number of hotels that neglect location information on their sites. Although most (but not all) list the address, this is only the most basic information about location. Proximity to attractions, activities, major businesses, and other points of interest, as well as access to information from airports or train stations, are essential information about a property's location and play an important part in converting a potential customer. A prominently displayed phone number is also critical.

Unfortunately, although some developers do not include enough content, others go to the opposite extreme. With the low marginal cost of adding data to a website, it's easy to overwhelm potential customers with a haze of irrelevant (to them) details. For a website to be effective, it must provide visitors with quick and easy access to the information they require. For that reason, it is a good idea to stream your website toward the needs of different target consumer groups. Although this could be as simple as having different parts of your website for leisure customers and corporate customers, some companies are going further and providing different sections—and sometimes even completely separate sites—for customer segments such as meeting planners, travel agents, tour operators, and wedding organizers, to name just a few (see Exhibit 16-6, which targets those interested in a luxury, country club experience). Exhibit 16-7 is a website that targets the economy segment, while Exhibit 16-8 is for the resort segment. What differences do you see?

Once the website visitor has been converted and wants to proceed to the next stage and make a booking, a good website must provide appropriate booking facilities. Online reservations offering last room availability and instant confirmation are now an essential component of effective hotel websites. This is also true for restaurants; more and more restaurants are finding it beneficial to provide some sort of reservation facility on their sites, be it online or through a call-back system. In today's fast-paced world, convenience is key, and the reservation facility provided needs to be as intuitive and easy to use as possible. The benefits of providing the right facilities are clear: capturing the sale while the consumer is still on the site and reducing the risk of her preferring a competitor. Getting it wrong undoubtedly results in lost business.

Although matching the information content and the facilities provided to the customer is essential, the aesthetics, or "look and feel," of the site are also important. In many cases, this will flow naturally from the corporate image and offline promotional

Exhibit 16-6   Country Club Lima Hotel website illustrates four best practices for website design

Source: Country Club Lima Hotel. Used by permission.

Exhibit 16-7   The website for Motel 6 is a good example of an economy hotel

Source: Retrieved November 30, 2005, from www.motel6.com. Used by permission of accorhotels.com.

material of the company. However, the web is a different medium from paper; elements such as animation and sound can be used quite effectively to reinforce the promotional message. As discussed earlier, more up-market companies typically need a richer look and feel, incorporating richer graphical and multimedia elements than their economy or midmarket competitors. However, care must be taken not to add multimedia gratuitously. The trend of having Flash animations (known as splash screens) when you first

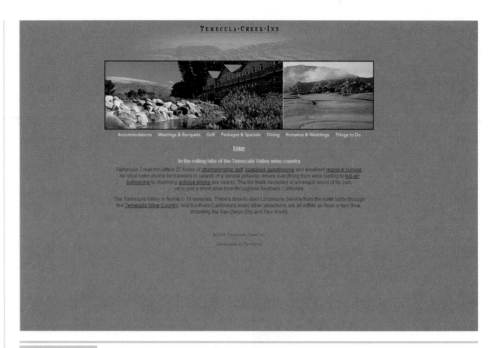

Exhibit 16-8   Temecula Creek Inn is an example of a company providing different website sections—and sometimes even completely separate sites—for various customer segments

Source: JC Resorts. Used by permission.

enter a website is a typical example of hyperactive designers showing off their animation skills while alienating the customer. To avoid alienating the customer, many such sites allow the customer to skip this introduction and get right to the main page. The decision of what to include should be based on whether it either helps build the image of the company or aids in the selling process.

Closely related to aesthetics is the issue of website navigation. Through conditioning, web surfers have come to expect that sites conform to certain conventions with respect to navigation. Deviations from these conventions tend to confuse the website visitor. For example, we have come to expect that navigation menus will be either down the left-hand side or across the top of the screen. Although other solutions may look nice and be more innovative, they also tend to frustrate visitors and drive them to alternative sites. Website navigation must follow industry norms to help users find the information or feature they are seeking as quickly and easily as possible. Site maps* and search facilities can also help with this and should be included where appropriate.[10]

All sites should be designed for two audiences: the potential client and the search engine. A superbly designed and amazingly functional site from a consumer perspective is pointless if people can't find it on the web. A domain name helps this. A domain name is the text name corresponding to the numeric IP address of a computer on the Internet.

*A site map is a visual or textually organized model of a website's content that allows users to navigate through the site to find the information they are looking for. A site map is a kind of interactive table of contents, in which each listed item links directly to its counterpart sections of the website. Without them, it is possible to explore a complex site by trial and error, but if you want to be sure to find what you're looking for, the most efficient way to do that is to consult a model of the resources available.

Typically, site maps are organized hierarchically, breaking down the website's information into increasingly specific subject areas. Site maps can also be created using more general website management tools, such as Visual Web, or Microsoft's Site Analyst.

A domain name must be unique. Internet users access websites using domain names. Although a domain name (or, more probably, names) is undoubtedly important, in practice, a good site must also be listed prominently under appropriate search phrases on the major search engines. The most effective way to achieve this is known as organic search engine manipulation, which in effect means designing pages to maximize their position in the search engine indexes. Some of the techniques include ensuring that each page concentrates on a single subject (yet another reason for streaming different sections of the site toward different customer segments as discussed previously). Making sure to include appropriate **keywords** and phrases throughout the page's headings and text, prompting the search engine to find the page, does this. There are websites that can help you determine what will appear with certain words (e.g., http://inventory.overture.com).

Search engines charge for the listings on their sites. This pay per click model of listing hotels according to their appearance in the search results allowed hotel marketers to increase traffic to their sites. Each keyword or keyword grouping is auctioned, with the highest bidder getting the highest placement, the second highest bidder getting second placement, and so on. A customer looking for a luxury hotel in St. Louis would see the organic listings on the left side of the page and paid listings on both the right side of the page and at the top under "Sponsored Listings." Exhibit 16-9 shows how pay per click listings from Google work, and Exhibit 16-10 shows an actual pay per click listing.

Finally, the website has to be functional. A website may be visually pleasing, but if the customer finds it is hard to navigate, its effectiveness is greatly diminished. Following are four best practices for website design in the hospitality industry:

1. The reservation area should be at the front of the page to enable customers to make their online reservations.
2. The e-mail acquisition section should be on the home page.
3. The site should have text conveying best value to the customer, to ensure that the customer stays on the local site and does not shop for a better rate elsewhere on the Internet.
4. Security for the transaction needs to be conveyed to the customer. Consumers are still concerned about online commerce, particularly outside the United States. Many consumers are wary of credit card theft on the Internet.

keyword A word that the user searches for when using the search engine. Generally speaking, keyword density is the ratio of the word that is being searched for (the keyword) against the total number of words appearing on your web page. If your keyword occurs only once in a page of one thousand words, it has a lower keyword density than a keyword that occurs four times in a page of similar length.[11]

Exhibit 16-9   How pay per click listings work

Retrieved November 30, 2005, from http://adwords.google.com/select/.
Source: Google Inc. Used by permission.

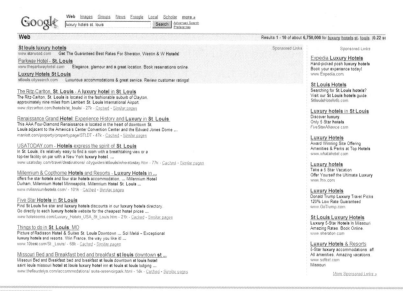

Exhibit 16-10	A pay per click listing

Retrieved November 30, 2005, from www.google.com/search?hl5en&q5luxury1hotels+st.1louis,
Source: Google Inc. Used by permission.

The Internet has also increased marketers' ability to track their efforts online. Each customer visit can be traced with a "cookie," or electronic impression of the visit. This cookie can be followed throughout the site, showing the marketer the origin of the visitor to the site (Google, Yahoo!, banner advertisement on the Theme Park Site, etc.) and what they did when they went into the site. Technology now exists to track the actual bookings from online marketing efforts; thus, a return on investment can be calculated for every online marketing campaign.

Website-Generated Market Research Data

Research marketing tools such as surveys can measure online consumer attitudes and opinions, but what about online behavior? According to an article in the *Cornell Hotel and Restaurant Administration Quarterly*, every business that has a website has a powerful consumer behavior tracking tool at its disposal, about which it may not be aware. Computer servers that host websites have log files that track visitors and record their surfing and clicking behavior, including the following:

- First page visited
- Last page visited
- Typical navigation sequences (i.e., movement among the website's pages)
- Referring site
- Average number of pages visited
- Time on site

With this relatively low-cost research tool, websites can be optimized to better communicate product information, elicit consumer emotions to build relationships, and potentially increase online sales. As online channels continue to grow, the savvy marketer will continue to identify more effective ways to serve the customer.

Exhibit 16-11 shows an example of one company that provides web analytics.

FUTURE CHALLENGES
OF ONLINE DISTRIBUTION

Bill Carroll, a professor at Cornell University and a consultant to PhoCusWright and major hotel firms, stated that the Internet has fundamentally changed travel and hospitality distribution. We concur. In discussions with Dr. Carroll, we identified some of the key challenges facing marketing managers in the hospitality industry as a result of the Internet. Although an entire chapter could be dedicated to these challenges, we are able to present only a sampling.

The first challenge is that the consideration set for travel options has expanded exponentially as search engines provide multiple travel options within seconds. This means that the competitive set may be not only other properties at the destination, but also other destinations around the world. As discussed throughout this text, the focus must be on what the customer is actually buying and what problems are being solved. The solutions to these problems can be found instantaneously and may be outside the normal geographic area that a marketing manager typically thinks about.

An example of an Internet firm that aids consumers in their search for solutions is Kayak.com. As stated on its website:

> Kayak is considered a meta-search engine and that means our website searches hundreds of other websites in real time for the best travel deals available. Kayak.com lets you look at a full range of airlines, hotels, and car rental agencies quickly and efficiently based on the exact criteria you select. . . . [We] are not a travel agency. With a travel agency, you pay a fee to get an airline reservation, a hotel room, or a rental car. There are no fees with Kayak.com. We do not sell airline tickets or hotel rooms or rental car reservations. Instead, we direct you to other travel sites where you can make these purchases directly. Our way of helping you plan your travel gives you a more comprehensive list of your travel options. Really fast, and with no biases or hidden agenda. In short, you are in control of your travel choices. Since we search hundreds of travel websites (including travel agency sites), you now have to search only one: Kayak.com. Kayak.com makes money when people click onto advertising on our website. Plus, we make money when people click on the results from our travel partners like airlines, hotels, and rental car companies. This is a revenue model (for all of you MBAs) similar to that of Google.[12]

A second challenge is that of price transparency and consistency. Brand integrity and customer loyalty were addressed in earlier chapters. A component of both is trust. That is, the consumer must trust that the brand will look after the customer's best interest and not practice opportunistic behavior. When prices vary for no particular reason, trust erodes. Firms such as Travelaxe.com enable the consumer to compare the prices of a particular hotel across a variety of websites. If the hotel is not careful in managing its prices across the various distribution channels, its inconsistency will be

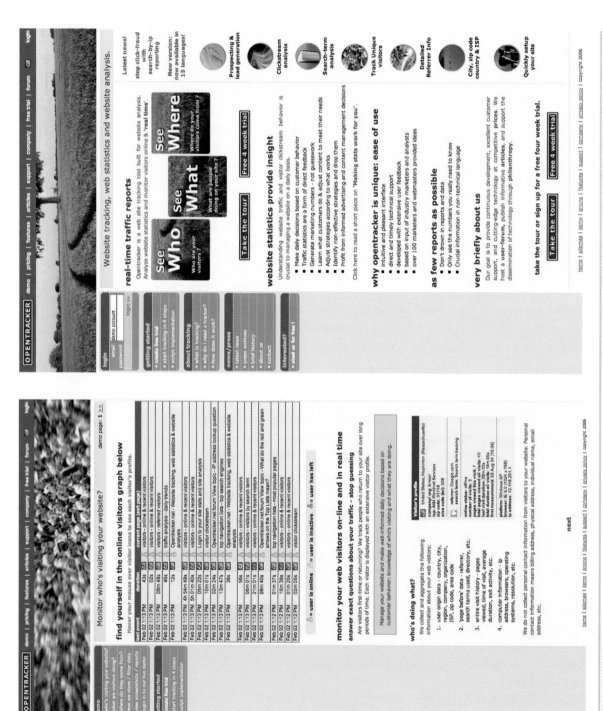

Exhibit 16-11   This website shows how firms can monitor the success of their website

immediately apparent to the customer. It is because of this that many firms have gone to "best available rate" on their own websites. Yet, pricing issues still can occur. Hence, the pricing of the different distribution channels must be watched constantly.

A third challenge facing hospitality industry executives is the need to manage transaction costs. Various channels have different costs. Marketing managers will need to better understand their customers and determine ways to drive them through the most efficient channel. A critical component of this goal is understanding the types of information customers want to see about a specific destination or property. Consumers will go to the sites that provide such information, and they may not include the home page of the hospitality firm.

A fourth challenge brought about by the Internet is the reallocation of marketing dollars. A survey of 25 travel marketing executives by PhoCusWright and New York University in 2005 found that "42 percent of the respondents were spending 20 percent or more of their marketing budget on online channels in 2005, compared to 25 percent spending 20 percent or more in 2004. More than three quarters of the hospitality respondents (78 percent) planned to increase spending on search engine optimization, while 74 percent planned to increase spending on website design and functionality."[13]

A fifth challenge, suggested by Carolina Colasanti of Starwood Hotels, is what she calls onward distribution. This is the process whereby negotiated rates to wholesalers are passed on to other websites either intentionally or unintentionally. Ms. Colasanti stated that, whereas most websites have good intentions, some can easily dilute your brand by catering to a different market and boasting deeply discounted room rates and making negotiated rates available to the public, undercutting your own "best available rates" and breaking the promise of a "best rate guarantee." This scenario can, however, be avoided by ensuring you have an "onward distribution" clause on all of your contracts with those suppliers.

A sixth challenge is the rise of third parties that are going after the group market. Passkey Research revealed that one in four rooms are associated with a group and that 15 percent of this 25 percent are a result of citywide events. Of the remaining 85 percent, 33 percent consist of fewer than 25 people.[14] An example of a firm entering this market is Groople.com (www.groople.com). Groople.com and other similar firms provide the consumer with multiple options. Although such choice is good for the consumer, it is harder on the hotel.

A final challenge for hospitality firms is the increase in packaging. This is a good challenge because it allows the hospitality organization to mask the price of the individual components. Software is needed to create dynamic packaging, which is what the consumer demands.

Selecting Channels

The selection of distribution channels is important. The length of the channel needs to be analyzed. In no uncertain terms, shorter is better; the longer the channel of distribution, the higher the cost and the more potential problems arise for the management of that channel.

By short or long, we refer to the number of intermediaries in the channel. Each intermediary has to make a profit, and each one involves some measure of coordination. Therefore, the fewer middlemen involved, the more profit and the less chance for errors.

At some point, there may seem to be a need to add on channels. If the new intermediary can be reasonably expected to bring in more customers at a profit for the originator, the channel should probably be expanded. If the channel member cannot deliver the needed number of customers and the profit, the decision should be negative.

MARKET OPPORTUNITIES IN CHINA

Internet bookings are increasing in both Europe and Asia Pacific. An analysis of the Chinese market by Michael David Blanding, a student at the hotel school at Cornell University, revealed examples of challenges in the international arena. He identified three primary challenges to online distribution in China.[15] The first and perhaps most prominent barrier to e-commerce in China is the cash-based nature of Chinese society and the fact that very few Chinese carry international credit cards. Instead, most carry a debit card from one of the many domestic Chinese banks. This situation presents a challenge because most hotel companies use a single payment platform for all of their many language websites. This payment platform normally requires that the customer enter an international credit card number to make the booking. Because very few Chinese have the necessary credit card, many bookings are abandoned before being finalized; instead customers call properties to book directly.

To overcome this obstacle, the more forward-thinking hotel companies have created alternative payment options for their Chinese-language websites. In China, it seems that the more payment options a website has in place, the better the website performs. Some websites have the flexibility to accommodate up to 11 different payment styles. Following are some common payment solutions among successful Chinese e-businesses:

- Payment by mobile phone (the phone number is tied to a bank account and money is transferred via text message)
- Cash pickup (employees are dispatched to the consumer's door within four hours of an online purchase and make a cash transaction)
- Domestic bank card
- Post office payment

The second obstacle to online distribution in China is that, given the fact that e-commerce is still relatively young in China, very few Chinese have a track record of successful online transactions. As a result, little trust has been built up for the Internet as a distribution channel. Many Chinese harbor concerns regarding the safety and security of personal information given out over the Internet. To ease this concern, the more culturally sensitive hotel companies translate their website security statement into Chinese and emphasize their web security in Chinese advertising.

A final challenge for Chinese Internet users is the tendency to question the reliability of information provided online. Chinese are skeptical of room descriptions and hotel amenity lists found on hotel websites. To address this concern, many major players in the Chinese hospitality industry are enriching the content of their websites by including more virtual tours, floor plans, and photographs of the properties.

Although the hospitality industry is based heavily on standardization, success in the Chinese Internet market is contingent on a hotel company's willingness to break the traditional mold of its cookie-cutter e-commerce platform and cater to the unique needs and concerns of the Chinese consumer.

Relationships with Channels

Good channel management stems from the formulation of a good working relationship with channels from the start. All agreements pertaining to the workings of the channel should be in writing and should be updated as market conditions change. There is rarely an all-win situation. If a channel member is not deriving some reasonable value from the network, that member will not participate actively and distribution will eventually become more difficult and more costly.

For example, a hotel could develop a good working relationship with a representation firm for marketing the property. The representation firm then markets the hotel through sales calls, brochures, direct mail, and so on. A booking results and a commission becomes due. If the hotel begins to dispute the validity of the origination of the bookings or delays the payment, the relationship with the channel of distribution suffers. The representation firm will not be anxious to market the facility in the future and will spend its time selling more cooperative hotels. This becomes a no-win situation. The hotel is dissatisfied with the productivity of the channel member, and the representation firm will move on to more lucrative endeavors.

Each channel member seeks to create customers for a profit, but without some give-and-take on a regular basis by all channel members, the system becomes time-consuming and disruptive. The hospitality firms that have carefully selected their partners and are managing them well will be consistently increasing their customer base while others are looking for new channel members.

Evaluation of Channels

Evaluation is important for the continued success of any program. If a hospitality entity is unable to tell how many bookings a representation firm produced or how many coupons were turned in from the dining guide, then intelligent channel management is impossible. Often, channel members can report the statistics. If unit management is unable at least to spot-check these numbers, the channel member will be in control when it comes time to negotiate the next agreement.

For example, a hotel that engages in a channel agreement with an airline needs to set an objective in a quantitative format in order to be useful in the evaluation process. The success of the channel of distribution might be defined as increasing reservations from 100 rooms per month to 120 per month.

It is also beneficial to understand the break-even point of the channel. In the preceding example, it might take an additional 10 rooms per month to cover the additional commissions and some combined advertising costs. After a predetermined amount of time, the channel is evaluated. If it is producing fewer than 110 rooms per month, careful consideration might be given to either increasing the marketing support for the program or dropping the channel member completely.

Evaluation is more than just a count of dinner covers or room nights. A channel may be driving the volume, but if the customer is unhappy, the effort is not only short-sighted but dangerous.

A dining guide can market a two-for-one dinner promotion in a number of ways. If customers expect two lobsters for the price of one when making reservations and find out the promotion applies only to chicken, they will be sincerely disappointed. If the hotel guest was expecting deluxe accommodations and the agreement with the channel member was to offer a run-of-the-house room, the guests who get the inferior rooms will not be happy with their purchase. They may not be unhappy enough to complain, but even worse, they may be unhappy enough not to come back.

The marketing-driven company with good channel management skills will ensure the satisfaction of its customers throughout the process. If a channel member is producing customers that are consistently unhappy, it would be better never to have used that distribution method in the first place.

Chapter Summary

Consortia, affiliations, reservation network, CRS, corporate travel departments, and representation firms bring the customer to the product, especially in faraway places. Incentive houses, travel agents, and tour operators are intermediaries that distribute hospitality services to customers and also bring them to the product. The Internet is rapidly changing the way hospitality products and services are brought to market.

The backbone of any channel of distribution is channel management. Any marketing-driven organization will take the time to evaluate its current distribution system and organize a cohesive plan for improvement. A competent channel manager should then be assigned to monitor and consistently reevaluate the network to obtain the maximum benefits to the company. This channel manager may take the form of the director of sales, the general manager, or the resident manager at the unit level of the hotel. The corporate marketing office should assume responsibility for the chainwide agreements. Finally, the satisfaction of the customer is the true test of a channel's success. Without this, none of the steps outlined are productive or needed. For hotels at least, channel management is a far more productive and critical part of the marketing mix today than it was prior to the Internet.

Key Terms

central reservation system (CRS), p. 545
channels of distribution, p. 536
consortium, p. 541
corporate or managed business traveler, p. 543
distribution service provider (DSP), p. 544
global distribution system (GDS), p. 544
intermediaries, p. 544
keyword, 555
travel agent, p. 544

Discussion Questions

1. Discuss what is meant by channels of distribution in the hospitality and tourism industry.

2. Briefly describe some of the major players in channels of distribution, including consortia, incentive houses, global distribution systems, and central reservation systems.

3. Describe some aspects of selecting and evaluating channels of distribution.

4. Discuss some of the challenges of online distribution in hospitality and tourism.

5. List the characteristics of a good website. Choose a website that illustrates those characteristics.

6. Why are keyword searches important? What can you do to improve the chances of your website being among the top sites listed in searches?

Endnotes

1. Greenberg, P. (2002). *CRM at the speed of light: Capturing and keeping customers in real time* (2nd ed.). Berkeley: McGraw-Hill/Osborne Media.

2. Koppel, A. (Anchor). (2003, October 25). Dollar signs: Opening a franchise. CNN *Live Saturday* [Television broadcast]. Transcript #102503CN.V27.

3. Eyster, J. (1997). Hotel management contracts in the U.S.: The revolution continues. *Cornell Hotel and Restaurant Administration Quarterly, 38* (3), 14–20.

4. *Merriam-Webster's collegiate dictionary* (11th ed.). (2003). Springfield, MA: Merriam-Webster, Inc.

5. Retrieved November 30, 2005, from www.rennekamp.com/provident.html.

6. Flying from the computer. Retrieved September 29, 2005, from the Economist print edition at www.economist.com/displacestory.cfm?story_id=4455692.

7. Minor, J., Angel, S.M., and Hart, R. (2003, May 14). *The internet dilemma: Third-party sites complicate a promising distribution.* New York: Lehman Brothers Inc., Global Equity Research, United States, 6.

8. Rascoff, S. (2003, July). Personal communication.

9. Ibid.

10. Retrieved December 10, 2006, from http://searchwebservices.techtarget.com/sDefinition/O,,sid26_gci541375,00.html.

11. Retrieved December 10, 2006, from www.thesitewizard.com/archieve/keyworddensity.shtmml.

12. http://corp.kayak.com/about.html.

13. *PhoCusWrite's Online Travel Overview,* Fifth Edition. Used by permission.

14. Passkey research 2006.

15. Michael David Blanding (2006, August). Personal Communication.

.... case study

Carlson Hospitality Worldwide
"Making IT Happen"

Discussion Questions

1. In what sectors of hospitality and tourism does Carlson Hospitality operate?
2. Briefly describe each of the three primary distribution channels used by Carlson Hospitality to reach its customers and travel agents.
3. In your own words, describe Carlson's Worldwide Reservations Center (WRC) based in Omaha, Nebraska. Which distribution channels are involved? How many countries are represented?
4. Why is Carlson's WRC (like several other major hospitality players) based in Omaha?
5. Carlson Hospitality is well known for its Look-to-Book incentive program for travel agents. How does it work?
6. Briefly describe each of the two main customer channels Curtis Nelson, president and CEO of Carlson, discusses in the extract toward the end of the case study. How does Mr. Nelson relate technology to marketing and to the customer? Do you agree with him? (You'd better! Or you just flunked the course.☺)

In January 2002, Scott Heintzeman sat in his office on the fifth floor of the Carlson Towers in west Minneapolis and reflected on his next move. Despite the downturn in the worldwide hospitality industry caused by the terrorist attacks of September 11, 2001, and an economic slowdown in the United States and other countries, Carlson was well positioned to maintain its strong growth and had cause for optimism about the future. As vice president and CIO of Carlson Hospitality Corporation, Heintzeman and his organization of 125,000 staff had, over the past 20 years, played a major role in developing Carlson from a relatively small player in the hotel franchising arena to the world's largest integrated hospitality company. This achievement was all the more remarkable considering that

This case was written by Robert Jenelsky, PhD, Ecole de Lausanee Hôtèlire. The author acknowledges the kind support of Carlson companies. All rights reserved. Used by permission.

Carlson was, in fact, one of the last of the major hospitality chains to invest in a comprehensive IT architecture.[1]

Strategic IT decisions involve great expense and risk in any business, but in an integrated worldwide hospitality company such as Carlson, they are mission-critical. The consequences of wrong decisions or poor execution of IT strategy can hamper a company's growth or cause its business to grind to a halt.

Conversely, making and successfully implementing the right IT decisions can give a company a significant competitive advantage in terms of increased productivity, lower costs, and enhanced customer satisfaction. A successful IT solution must simultaneously satisfy a number of criteria, including the following:

- No downtime (downtime means lost revenue)
- Easy to update and add new users
- Fast rate of response
- Integration of hotel and customer data in real time
- Room for future growth
- Features and functions that enable competitive advantage

This is more easily said than done: With the rapid evolution of IT, a smart decision made today may be obsolete within six months or less. And yet, such strategic "better-your-business" IT decisions as the choice of a system architecture and its component applications modules must be made in a timely fashion to enable the continued growth of the business. Such decisions involve major investments over a long period of time that must be approved by corporate management. Although good IT decisions and implementation can be expected to contribute to a hospitality company's long-term success, major IT investments affect the company's annual bottom line and therefore appear as a significant cost in the short term. Thus, finding the right funding model, in addition to making the right architectural choice, is a critical success factor.

The CIO (chief information officer) of a major corporation such as Carlson must therefore not only "get it right" in terms of defining the company's IT strategy; he must also "sell" the necessary investments to his company's top management by quantifying its contribution to the bottom line—not always an easy task in an industry which often takes a short-term perspective on ROI.

Another challenge for the CIO lies in being able to distinguish between "ready-for-prime-time" technology that is robust and stable enough for heavy-duty business use and the "bleeding edge" of technology that promises major advantages but may prove to be a short-lived innovation or deliver something less than advertised.

In 1996, under Heintzeman's leadership, Carlson developed a comprehensive IT architecture that was to be implemented over approximately seven years. By January 2002, seven of a total of ten implementation phases or "chunks" had been completed. Because one of the major problems with large-scale IT projects is that they often exceed budget, are late, and fail to produce the desired results, Heintzeman and his team at Carlson used a technique called "chunking" to break the architecture into 10 manageable components.

In January 2002 Carlson introduced the latest "chunk," a wireless information system for hotel owners, MACH-1 (Mobile Access to Carlson Hospitality-1), for which it applied for a U.S. patent. If granted, this would be Carlson's second IT-related U.S. patent. Its first patent, issued in 1996, was for the "Look-to-Book" travel agent incentive program (discussed later).

Obtaining patents on IT innovations is difficult even for IT companies; for a hospitality company it is truly exceptional, an understandable source of pride to Carlson and Heintzeman. But does excellence in IT translate into business results, and if so, how? And what are the interdependencies between Carlson's IT strategy and its business objectives?

CARLSON HOSPITALITY WORLDWIDE

Carlson Companies, Inc. (CCI), one of the world's largest privately held companies, is an integrated service company in the hospitality, travel, and marketing industries with revenues of $35 billion in 2001. Its component organizations comprise Carlson Hospitality Worldwide (CHW), Carlson Wagonlit Travel, Carlson Leisure Group, and Carlson Marketing Group. Founded by Curtis L. Carlson in 1938 with $55 in borrowed capital, it employs about 150,000 people in 147 countries. Carlson's daughter, Marilyn Carlson Nelson, is chairman of the board

and CEO of CCI. Her son, Curtis Nelson, is CEO of Carlson Hospitality Worldwide.

Carlson Hospitality Worldwide's brands include hotels (Regent International Hotels, Radisson Hotels Worldwide, Country Inns & Suites by Carlson, and Park Plaza Hotels), restaurants (which include TGI Friday's), cruises (Radisson Seven Seas Cruises), and hospitality supplies (Provisions). Carlson Leisure Group is a travel services provider that includes Carlson Wagonlit Travel (a joint venture with the French Accor chain) and Neiman Marcus Travel Services. Finally, Carlson Marketing Group is a provider of marketing services including loyalty marketing, events marketing, and sales promotion.

Although Carlson owns its flagship Radisson Hotel in downtown Minneapolis, almost all of CHW's properties, including its cruise ships, are owned and/or managed by franchisees who are, in a sense, CHW's primary customer. The success of CHW, like that of other large chains that expand through franchising, depends on the company's ability to attract and retain franchisees. The company does this primarily by helping to provide bookings and tools to increase REVPAR while at the same time enforcing quality standards that make the franchised properties attractive to guests and increase the value of the franchised brands.

> Our business has 4 Core Processes: new hotel development, sales and brand marketing, revenue management and books, and guest service delivery. We have 2 sets of "strategic customers": our owners/developers and franchisees, and our hotel guests. Our strategic IT objectives are to enable our brands to grow guest satisfaction, to grow REVPAR and grow the value of our brands.[2]

DISTRIBUTION CHANNELS

Like airlines, hotel chains use a mix of complementary distribution channels to reach their customers and travel agents. Of these, the most important are telephone reservation centers, which a guest or travel agent typically contacts via a toll-free number, and computer-based global distribution systems (GDSs), computer networks operated by third party companies, which are used exclusively by travel agents. The leading GDSs were initially developed by airlines (Sabre, by American Airlines; Amadeus, by Lufthansa; Galileo, by United Airlines; Worldspan, by Delta). A third channel, the Internet, started to become a significant source of bookings, in particular from leisure travelers, in the late 1990s. Hotel chains generally use all available distribution channels in order to reach the guest and thereby maximize the likelihood of a booking. Other important distribution channels include tour operators and consolidators (which include independent websites such as www.hrn.com and www.hoteldiscount.com).

RADISSON'S OMAHA RESERVATION CENTER

In 1981 Carlson's sole hotel brand, Radisson, comprised 15 hotels. Radisson's first toll-free reservation call center was established in the basement of Romeo's Mexican Restaurant in Omaha, Nebraska, in 1981. At the time, it employed 15 people and handled reservations for the 15 Radisson hotels using pens and pencils. Other reservations were processed via GDSs or through direct contact with the individual hotels.

In the mid-1980s GDSs generally used text-based, batch-oriented interactive terminals whose "look and feel" was really more suited to the needs of airlines—who originally developed them—than to those of hotels interested in maximizing REVPAR. In particular, batch processing made interaction between the travel agent and the system very slow, and the text-based display allowed only limited information to appear. Typically, a GDS screen displayed only part of the available hotel inventory, and rate changes were time-consuming to implement. A real-time dialogue between the hotel and the travel agent was, therefore, not possible.

THE ADVENT OF PROPRIETARY CRSs

Because of the limitations of GDSs, hotel chains began looking at ways to either adapt existing airline GDS platforms to better meet their needs or design their own computerized CRSs (central reservation systems) from scratch. However, this was a difficult undertaking. As a relatively small franchise chain, with 60 Radisson-Flay hotels in the United States, Carlson was relatively slow to jump on the CRS bandwagon, unlike large chains with strong international presences. However, Carlson's lack of an existing "legacy" system gave it the ability to build one from scratch, using the most advanced technology available at the time. (Some hotel chains adapted existing airline systems.) In 1985 Carlson built its first computerized central reservation system, PIERRE, which was located in the newly built Worldwide Reservations Center in Omaha.

CARLSON'S WORLDWIDE RESERVATIONS CENTER

The heart of Carlson's CRS is its Worldwide Reservations Center (WRC), an 80,000-square-foot low-rise building in Omaha, Nebraska. It is headed by Christine (Chris) Brosnahan, vice president, Worldwide Reservation Services, and employs about 350 staff, most of whom are women working in its call center.

Carlson, which received an award for being a woman-friendly employer, provides on-premises daycare for employees' preschool children. The choice of Omaha as a call-center location was no coincidence: Omaha is the "call center of the United States," if not of the world. In addition to Carlson Hospitality Worldwide, Hyatt, Marriott, and Omni also have main reservation centers in Omaha.

About 50,000 Omahans work in the city's 1,000 small and large information data-processing and telecommunications companies involved in a host of services: investing, credit cards, long-distance telephone service, computer outsourcing, telemarketing, and many more. Estimates are that over 20 million people call Omaha every day.

What makes Omaha's telecommunications system so attractive to telecommunications and information technology companies? The answer goes back to 1948, when the federal government made the decision to locate the Strategic Air Command near Omaha. SAC's (now U.S. Strategic Command) mission was to lead U.S. military operations in the event of a nuclear war and required the most advanced communications system possible. As a result, the local telephone company needed to install an incredibly large and complex telecommunications infrastructure, staffed with those possessing the knowledge to serve the industry.[3]

Carlson's Omaha WRC is also responsible for satellite reservation centers located in Dublin, Ireland; Sydney, Australia; and Albuquerque, New Mexico, in addition to seven offices in India. The 450 employees in the five main call centers handle calls in 18 different languages (Omaha-based staff handle calls in English, French, and Spanish). The WRC provides the following services:

- Toll-free telephone reservations:
  53 countries (incoming reservations)
  51 countries (outgoing reservations)
  Toll-free lines available in 49 countries
  22,000 calls handled daily

■ GDS reservations:

455,000 travel agents in 125 countries via GDS
80,000 GDS transactions daily

■ Internet reservations:

1996: fewer than 100 per month
2001: more than 3,000 per month

PIERRE (1985): RADISSON'S FIRST CRS

Radisson's new (1985) system was named PIERRE (in honor of Pierre Radisson, the 17th-century French explorer who also gave his name to the company's flagship hotel in downtown Minneapolis) and consisted of one CPU and a few hundred MB of data storage. With this system, the then 60 employees of Carlson's Omaha reservation center processed 17,000 reservations per month (in addition to 3,000 reservations coming from GDSs), thereby generating 19 percent of the revenues of the Radisson hotels.

As a franchised brand competing with other brands for franchisees, Radisson's ability to deliver a large number of bookings via its central reservation system is a key element in its competitiveness: All other factors being equal, a franchisee shopping for a "flag" will generally prefer a brand that can make a credible commitment to deliver a maximum number of guests and the highest REVPAR.

Hotel chains use a multitude of complementary distribution channels (GDS, toll-free number, proprietary CRS, Internet, tour operators, consolidators), but, for obvious reasons, prefer proprietary systems that (1) allow savings on commissions and (2) give direct access to the end user, a requirement of customer relationship marketing (CRM).

In 1988 PIERRE was extended to cover a new Carlson brand, Country Inns & Suites. In the period between 1989 and 1995, continuous improvements were made on the PIERRE system as the number of Radisson and other hotels increased, as the result of the company's aggressive franchising campaign. Among these were the introduction, in the early 1990s, of faster, so-called "Type A" links with all GDSs, which replaced the previous, relatively slow "Type B" (batch transfer) data links between PIERRE and the travel agents using GDSs.

As a consequence, Carlson was able to introduce incentive programs targeting individual travel agents. The success of the first of these, "World of Winners Sweepstakes" in 1992, led to the rollout, later that year, of the highly successful Look-to-Book program.

INCENTIVIZING TRAVEL AGENTS WITH "LOOK-TO-BOOK"

In 1982 about 70 percent of Radisson's room booking came from travel agents via global distribution systems. In general, only a percentage of hotel inventory (rooms available) in a given hotel was visible to the agent on the GDS screen. GDSs typically provided access to a large number of hotel chains, resulting in a situation whereby hotel inventory was a bit like competing brands of similar products competing for the shopper's attention on supermarket shelves. Before an agent could "book" a room at a given hotel, he or she first had to "look" at that hotel's listing on the GDS to check availability. Anything that the chain could do to incentivize the agent to check its listing first would increase the likelihood of a booking. Up until that time, hotel chains, emulating the airlines, had begun using customer loyalty programs to incentivize the end user or hotel guest, who received inducements to reserve directly via the chain's toll-free number. Call centers accounted for about 30 percent of Radisson's bookings at the time. Travel agencies received a standard fee or percentage commission on hotel bookings, but in general this fee or commission was paid to the agency rather than to the agent—often an employee of the agency—who actually made the booking. If the hotel chain can direct business through the most cost-effective channel, it can improve its profitability while generating more business for its franchisees:

A call to an 800 (toll-free) number typically costs $12 to $15, whereas a reservation made through a GDS generates a transaction cost of only $2. By giving travel agents points only for bookings made through the GDS, we incentivize them to use the lower-cost channel. Hotels pay us a fee of about 3% of all bookings we make for them, either through GDS or our CRS.[4]

To create its Look-to-Book program, Radisson had to create "seamless" links between the GDSs and its own inventory system in order to allow travel agents to bypass the limited amount of information available on the GDS and gain direct access to the PIERRE database containing complete hotel, rate, and availability information for all of its inventory. An additional requirement was to increase the transaction speed so that all of this could be accomplished in a few seconds. Finally, a paperless "banking system" had to be built to allow travel agents to accumulate points, view their balances, and make withdrawals.

An agent receives 10 points per dollar of Radisson bookings. When an agent has accumulated 80,000 points, s/he receives Elite status. There is a catalogue of rewards that can be ordered with points. The GDS system tells them what their balance is. Travel agents hardly receive any benefits anymore, so this program is a real exception.[5]

Look-to-Book was later expanded to include other Carlson brands, Country Inns & Suites, and Regent. In 1999 a website was added.

MIGRATING FROM PIERRE TO "CURTIS-C"

In time, even the most capricious system architecture eventually begins to reach its limits. By 1995 this had started happening to PIERRE. Although a highly successful system, PIERRE was beginning to run out of capacity, among other problems. Since its inception in 1985, when Radisson had a total of 60 hotels, the Carlson brands (which now included Country Inns & Suites and the upscale Regent chain) had been adding a new hotel about every 10 days. PIERRE was generating 40 percent of room revenues, a result that other chains could only envy. One of the symptoms of the need for a change was the amount of system downtime (which directly results in lost revenue to the chain) needed for maintenance.

Carlson's original 30 hotels had grown to a chain of 500, forcing the IT staff to shut down the system for hours at a time just to index what had become million-record files. Worse, call-center applications were written in a proprietary 4GL [fourth-generation programming language]—Action—whose provider, Action Software, had gone broke, leaving the language entangled in lawsuits. The department faced labor-intensive database maintenance chores out of Action's character-based screens and onto a GUI [graphical user interface].[6]

In 1995 Scott Heintzeman had already assumed the position of CIO at Carlson. His responsibilities included both keeping PIERRE running—a difficult task with a system that was by now 10 years old, based on obsolete technology, and heavily overloaded—and keeping abreast of the latest technological developments and trends in the use of IT in hospitality. His situation was typical of that of many CIOs in hospitality companies who need to keep a creaky "legacy" system running, because the business depends on it, while trying to manage the transition to the next-generation system that they know their business needs to stay competitive. The latter task involves, first, getting management's buy-in for the new system architecture (including funding, which can be substantial) and, second, managing the technical and organizational transition to the new system while the old one keeps running 24 hours a day, 7 days a week. The risks are high; the greater the scope of the project, the greater the likelihood of cost overruns and missed deadlines, a combination that can cost a CIO his job. Although Heintzeman knew that Carlson needed a new system within a year or two, how could he prove it to his management, given that PIERRE was still running and bringing in 40 percent of Carlson's room revenue?

"CHUNKING"

Heintzeman and his team estimated that the new CRS system would cost $10 to 15 million. However, Carlson's management was reluctant to approve such a large, multiyear IT investment. The IT team then revised its original plan and resubmitted a new "series" allowing implementation over a period of several years in self-contained stages called "chunks." Each "chunk" or building block implemented a piece of the total system architecture in such a way that, by itself, it achieved a specific, measurable business objective. An additional requirement was that there could be no interdependencies between "chunks"; that is, no "chunk" could be a prerequisite for another one. Finally, because PIERRE, the old, "legacy" system, had to keep running while the new system was being implemented, there had to be full forward and backward compatibility between the two systems. The new system, based on a three-tiered (CRS, hotel PMS, and customer database management system) client–server architecture, was designed "not just to take reservations, but to better manage our business and build fuller relationships with our customers."[7]

Tier 1: Curtis-C Reservation System

The new CRS was called "Curtis-C" (pronounced "courtesy," in honor of Curtis L. Carlson, the company's founder). It was developed using the Forté Application Environment in partnership with Born Information Systems, a Minneapolis-based systems integrator. The first "chunk," implemented in

1996, allowed direct toll-free telephone reservation coverage in 41 countries plus seamless interfaces to 125 countries via GDSs used by travel agents. The system architecture was sufficiently flexible to allow the subsequent addition of other distribution channels, such as the Internet, electronic kiosks, and interactive television. In January 1999 Curtis Nelson, president and CEO of Carlson Hospitality Worldwide (and grandson of Carlson found Curtis L. Nelson), stated:

> The new Curtis-C system is truly a breakthrough. It is a technology showcase that distinguishes us from the competition and sets new global standards for our industry. It is also a vital cornerstone in achieving the customer-focused strategic vision of Carlson Hospitality Worldwide for the next millennium.[8]

In 1997 the system was expanded to provide reservation services for Carlson's newly acquired luxury brand, Regent International Hotels. Because Curtis-C is mission-critical to CRC staff, who are its primary users, system programming staff are also based in Omaha, in a group headed by Tom Sikyta, senior director of distribution systems.

> The issue of "make vs. buy" doesn't really arise when you can't buy. That was the case with the main components of Curtis-C. . . . Our challenge was to design a system that prevents hotels from shooting themselves in the foot, that is, from giving away revenue, given that they have the ability to set their own rates.[9]

The following components of Curtis-C were built by Carlson:

- Look to Book
- GDS interfaces
- A business rate module that allows hotels to make custom deals with corporate accounts and override "default" deals
- A guarantee policy enforcement module
The following components were bought from vendors:
- Credit card processing
- Micros-Fidelio PMS
- MS Outlook
- Cognos/Impromptu "merge and purge" data mining software
- Networking (currently frame relay, but moving to virtual private network)

WRC provides substantial user support for Curtis-C. Audrey Mattly, formerly with Utell, is director of distribution services and manages a group of 44 supervisors.

Our group focuses on working with hotels, for example, analyzing statistics, etc. Each supervisor typically supports 30 to 40 hotels. Strong supervisor support means faster hotel turnaround. Lots of supervisors have, in fact, actually worked in hotels, and most develop a strong relationship with the hotels they support. . . . Training and competitive analysis are also part of our job.[10]

Tier 2: Harmony PMS

Introduction of the HARMONY PMS, the second of the three pillars (see Exhibit 1) of the Carlson IT architecture, began in 1997. In 1998 the Curtis-C reservation system was interfaced to the HARMONY PM via the HDBM (Harmony Data-Base Manager), a GUI-based program. HDBM allows hotels to create and modify rates in the CRS and to monitor the appearance of their inventory to the sales representative in Carlson's Omaha reservation center. One of HARMONY's features is the "GDS Rateshopper," which gives hotel owners online access to pricing and availability information on competitors in their local market.[11]

Tier 3: CustomerKARE (Knowledge and Relationship Enabling) System

The third tier, CustomerKARE, is a sophisticated customer relationship management (CRM) tool (see Exhibit 2). The Guest Communication Manager system is a software tool to support the company's "100% guest satisfaction" strategy; among other functions, it allows the systematic monitoring and following up on quality issues relating to individual hotels or groups of hotels.

The three tiers of Carlson's IT architecture are shown schematically in Exhibit 3, where they appear as columns. KnowledgeNet is an intranet system to allow hotels easy access to company information.

USING IT FOR STRATEGIC COMMUNICATIONS

Over the past 15 years, Carlson's use of IT has gone from reservation processing for 60 Radisson hotels to becoming the enabler of a seamless strategic business system used to

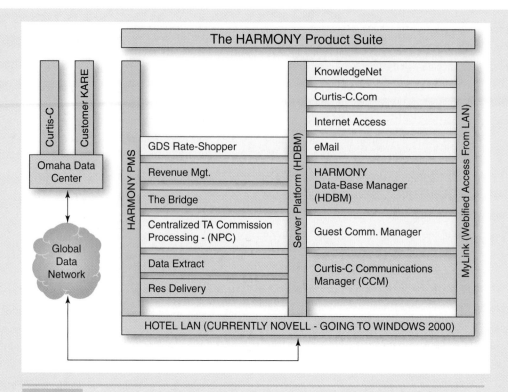

Exhibit 1　HARMONY product suite architecture

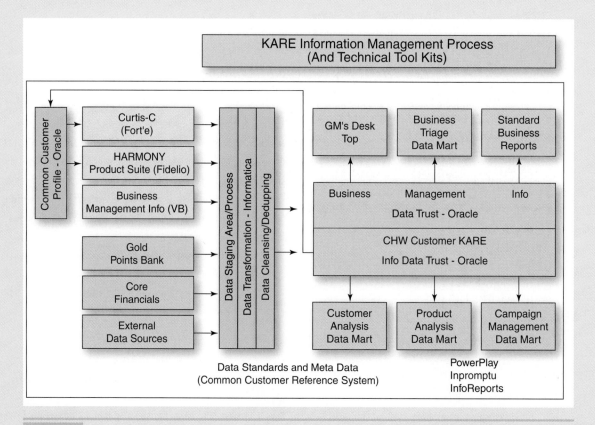

Exhibit 2　CustomerKARE (Knowledge And Relationship Enabling system) architecture

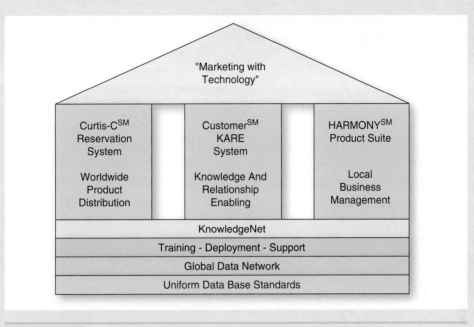

"Marketing with Technology"

| Curtis-C^{SM} Reservation System | Customer^{SM} KARE System | HARMONY^{SM} Product Suite |
| Worldwide Product Distribution | Knowledge And Relationship Enabling | Local Business Management |

KnowledgeNet

Training - Deployment - Support

Global Data Network

Uniform Data Base Standards

| Exhibit 3 | Carlson's IT architecture (Carlson Hospitality Worldwide) |

support hotel owners, develop customer satisfaction and loyalty, and increase the value of Carlson's many brands. It has developed in parallel to Carlson's marketing strategy for its hotel brands (Radisson, Country Inns & Suites, Regent, and since 2001, Park Plaza). Whereas in the early 1980s Carlson's focus was primarily on selling Radisson franchises to increase the number of hotels, in recent years it has been on developing product quality and branding. Kathy Hollenhorst, vice president of marketing, Radisson Hotels and Resorts Worldwide, and her team are responsible for marketing Carlson's largest hotel brand. Hence, she can be considered Heintzeman's most important internal "customer."

The question is: Which should come first, the business need or technology? There are too many opportunities to use IT; you need a way to define the ROI on IT investments. There is a need for "bridgewalkers"—people who can connect the "IT side of the house" to the "business side of the house." Because Scott's mission is multibranded, there needs to be a fair allocation of IT costs to all Carlson brands. Currently, the Radisson brand is paying about 93 percent, which is disproportionately high.

My organization's role is to be a Center of Excellence focusing on increasing REVPAR for Radisson hotels. We use the customer life cycle approach to do this. So far, we are doing a good job of "owning" the GDS channel; now we have to "own" the web channel, too. At the present time (August 2000), www.radisson.com generates about 3 percent of our business.[12]

FRANCHISEES

Hollenhorst's goal is to move away from Radisson's previous "churn and burn" strategy to a CRM-based strategy focusing primarily on franchisee retention and secondarily on acquisition. She sees technology (i.e., databases and "smart tools") as the key to this. The "customer" in CRM means not only franchisees but also travel agencies (reached through the Look-to-Book program) and end users.

E-COMMERCE

Business to Consumer (B2C)

The primary platform for business-to-consumer e-commerce is www.radisson.com, which allows a high level of personalization (guests can build their own web pages). It also allows each hotel to layer its own site. The goal is to simplify the reservation process for individual customers. Frequent guests are identified by using a 15-cell matrix (encompassing three categories of amenities times five choices per category), which inputs to a just-in-time system called "Reserved for You."

Business to Business (B2B)

Target business-to-business markets include corporate intranets (reached via sites such as www.getthere.com), meeting planners, and travel agents (reached via Look-to-Book, which receives about 10,000 hits per week).

Consumer to Consumer (C2C)

Radisson's goal is to be a leader. Its "Operation E-Watch" monitors sites such as www.deja.com and www.kudostravel.com without interfering in "the buzz." However, Radisson does input information in response to specific situations in order to show its responsiveness to customer issues.

Steve (Gino) Giovanelli, director of interactive marketing, is responsible for Radisson's "webifying."

Our goal is to make www.radisson.com the premier communications and booking vehicle for our hotels. Our challenge is to provide personalized experience through a medium where there's no "person" as such. In this context, "personalization" means "relevance" in terms of IT and business strategies. Our charter is to drive online revenues. We typically get 500 user profiles per day, of which 250 result in reservations.[13]

Giovanelli sees meeting and corporate travel as areas of great opportunity, as there are many websites, including the following:

www.allmeetings.com
www.plansoft.com
www.planitonline.com
www.meetingsbooker.com
www.subrepts.com
www.getthere.com
www.e-travel.com

To reach the corporate travel market, which is based on intranets, Carlson partners with subrepts.com, getthere.com (party owned by American Express, it now owns meetings.com), and e-travel.com. Giovanelli sees Radisson as the most competitive hotel chain in terms of CRM.

THE VIEW FROM THE TOP FLOOR

Curtis Nelson, president and CEO of Carlson Hospitality Worldwide, sees the company's leadership in IT as a key element in its strategy to become totally customer focused:

Carlson started with Gold Bond stamps which led to diversification, which, in turn, has led to the need to focus on businesses that add value to each other. . . . Technology should bring systemic advantages, for example (better) placement in GDS systems. We've shifted from being a "synergistic enterprise" to being a "totally customer focused enterprise." We see two main customer channels: the consumer channel, which includes leisure travel and restaurants such as TGI Friday's, and the corporate/business channel, which includes commercial travel, meetings and incentives, etc. Technology gives us the ability to migrate customers back and forth between these channels. Loyalty programs change the value proposition, but points are only the excuse for building a learning relationship with the customer. Using technology to be on target with marketing makes it effective. Let the customer determine what he wants, then use the information to give him what he needs, not what you want to sell him. People love to be cared for; they don't want to be abused with information.

GDSs need to become more affordable—many people can do direct connects today. The GDS business model needs to be changed. Can you enhance your value proposition by direct connect? If you can eliminate expense in a commodity market, it makes you more competitive. We have no current plans to acquire a GDS or stop using them.

When asked how Carlson sees itself, for example, compared to Accor (its partner in Carlson-Wagonlit Travel), Nelson replied, *"Accor has 'gone to school on us.' They are more of a traditional hotel company, whereas we are more of a relationship company."*

Marilyn Carlson Nelson, president and CEO of Carlson Companies, Inc., is a strong supporter of the strategic use of IT, not only because of its intrinsic value but also because it is an organization learning tool:

Curtis-C was a $15 million decision: Do we "pull the switch" or use "chunking"? The decision process actually developed our organization; in addition to cost effectiveness, there were other benefits. For example, we learned that individual brand strategies must corroborate the overall corporate strategy. Our corporate strategy is based on developing deeper relationships with existing customers—that's our highest priority—second, learning to do a job better, and, third, acquiring new customers. We have five strategic imperatives:

Be customer focused. This means knowing who the customers are, differentiating franchisees and guests, but remembering that "the guest is the customer." This is a change from the past, when the franchisee came first.

Be global. Today's travelers/customers live in an increasingly global world.

Be best in class. This forces consistency between brands (i.e., to keep customers within the Carlson system).

Be knowledge based. This means we need to begin institutionalizing knowledge. All managers need to make decisions based on the same information.

Last but not least: Be strategically integrated. This is the role of Scott Heintzeman's organization.

These "imperatives" will be implemented through three "enablers": technology, human resources, and finance. With regard to the latter, Marilyn Nelson said, "Partnerships are important to us: We would have needed $200 million to become a global company without our many joint ventures." Nelson's view of the company is a people-centered one: "Technology doesn't replace caring about people. We want to use IT in a 'prosthetic sense.'"

MACH-1

In January 2002 Carlson introduced its Mobile Access for Carlson Hospitality (MACH-1) application. MACH-1 is a Microsoft Windows application running on the Compaq iPAQ handheld computer, which, using wireless technology, delivers real-time information to hotel managers as they move around their property, allowing them to track such data as sales trends and room availability while spending time with their guests. Carlson has applied for a U.S. patent for MACH-1. If granted, it will be the company's second, following Look-to-Book in 1996. Following successful completion of pilot tests in a limited number of hotels, the company plans a full-scale rollout.

WHAT NEXT?

Heintzeman summarized the issues facing his organization:

How do you make long-term strategy for a short-term world? And how do you reconcile all the interdependencies and interconnections within our organization? In theory, all the data should get put on the table, but in reality, everybody has a single data point; nobody has the whole picture.

Since the inception of PIERRE in 1985, Carlson has become an IT powerhouse within the global hospitality industry. This technology leadership has enabled unprecedented growth and financial success for the company, which by its own admission is "more of a relationship company than a hotel company." But are there limits to how far IT can take you? What is the optimum mix between investing in IT and, for example, investing in the properties themselves? Maintaining IT leadership is expensive. Can Carlson stay ahead without diversifying out of its core businesses?

Endnotes

1. See "Best practices in the Lodging Industry."
2. Scott Heintzeman, quoted in "We Make IT Happen," Information Technology Strategic Imperatives, Carlson Hospitality Worldwide, May 2000.
3. "Born to Be Wired," Omaha Chamber of Commerce, March 2000.
4. Mary Zack, interview with the author, August 1999.
5. Ibid.
6. John Udell, "Carlson Hospitality Keeps Its Eyes on the Prize," Enterprise Development, February 1999, pp. 5–15.
7. S. Heintzeman, quoted in Hotel Online Special Report, January 20, 1999.
8. C. Nelson, quoted in Hotel Online Special Report, January 20, 1999.
9. Tom Sikytas, interview with the author, August 2000.
10. Ibid.
11. Kathy Hollenhorst, interview with the author, July 2000.
12. Steve Giovanelli, interview with the author, July 2000.

activities, interests, and opinions (AIO) A psychological segmentation strategy based on personality traits, how people spend their time, and their interests and opinions.

actual market share The amount of business a firm actually receives in relative proportion to its direct competitors in the same product class.

advertising A tool used in the marketing communications mix to reach identified target markets. Advertising is paid mass communication typically placed in media publications such as newspapers, radio, and television, and on the Internet.

association market Groups that need guest rooms, food service, and function space facilities to accommodate their meeting needs and convene at the local, regional, national, and international levels. Membership in associations and attendance at their annual meetings are normally voluntary (excluding officers and elected officials of the organization).

attitude How we judge and react to beliefs; emotional feelings toward beliefs.

augmented product The total of all the benefits received or experienced by the customer.

barriers Impediments that make reaching a desired goal difficult. To increase its competitive advantage, a firm needs to overcome barriers to move ahead and create barriers to stay ahead.

belief A thought we think is fact, which we derive from perceptions.

belief → attitude → intention trilogy A process consumers go through once they have formed perceptions of a product but have not actually purchased it. A belief is a thought the consumer thinks is fact; attitude refers to emotional feelings about that fact; and intention is the plan to make (or not make) the purchase.

benefit bundle All of the benefits that consumers experience from the purchase of a particular service; includes location, familiarity, comfortable bed, and so forth.

bottom-line orientation A style of management that focuses strictly on the bottom-line profitability of the business, without regard to the impact on customers or their changing needs and wants.

brand A well-known product or service of consistent quality available to consumers in multiple locations. A brand reflects the desired positioning of a product or service.

bundle purchase concept When consumers buy a hospitality product, they are purchasing a bundle or a unified whole and not an individual element. A hotel guest room includes a bed, a bathroom, security, the check-in procedure, housekeeping services, and so on.

business traveler A customer who purchases hospitality products or services because of a need to conduct business in a particular area.

buy time How far in advance a buyer of hospitality services makes the decision to purchase the product and book the reservation accordingly. Also referred to as **lead time, purchase cycle.**

central reservation system (CRS) The computerized reservation system of a hospitality company that allows customers to make reservations without having to contact an individual entity of the company directly.

channels of distribution A group of organizations, independent or not, that are involved in the process of making a product or service available for use or consumption.

closing the sale Confirmation (preferably in writing) that a potential customer has now agreed to purchase the product.

cognitive dissonance A state of mind in which attitudes and behaviors don't mesh; a customer's potential uncertainty after making a purchase.

collateral Promotional material such as brochures, flyers, and direct mail that are used to inform customers and create interest. "Electronic brochures" are often posted on websites.

communications (promotion) mix All communications between the firm and the target market that increase the tangibility of the product/service mix, that establish or monitor consumer expectations, or that persuade or induce consumers to purchase a product or spread information about it by word of mouth.

competition   Anyone competing for the same customer with the same or a similar product or a reasonable alternative that the customer has a reasonable opportunity to purchase at the same time and in the same context.

concept of marketing   The art of creating customer value and helping customers to be better off by fulfilling their expectations and solving their problems.

conference centers   Carefully designed lodging facilities that target and meet the specific needs of the small-to-medium-size meetings market.

consortia   Large travel agencies that have substantial volume purchasing power that enables them to command deeply discounted rates from hotels, airlines, car rental companies, and the like.

consortium   A collection of individual properties that are similar yet have different names and carry a common designation grouping them into the same product class. Leading Hotels of the World, for example, is an upscale consortium.

constituents   Those who have an interest in the business entity, including customers, employees, owners, financial backers, and the local community. Constituents are also referred to as **stakeholders.**

consumer adoption process model   A model drawn from consumer behavior theory suggesting that the "adoption," or purchase, of a product is a process undertaken by the consumer. This process essentially entails awareness → interest → evaluation → trial → adoption. The model also contends that the consumer may or may not be aware that he is going through this process.

consumer behavior   A process a consumer undertakes (either consciously or unconsciously) when making a purchase decision. Marketers must recognize that consumers are goal oriented, have choices, and can be influenced.

consumer trade-off model   The idea that if a solution to a customer's problems, needs, or wants meets the customer's expectation, and the value of that product or service justifies the sacrifice or risk, that sacrifice or risk is justifiable, and a high level of satisfaction is likely.

convention and visitors bureaus (CVBs)   Local nonprofit organizations that help market a city. CVBs are supported by and work closely with hotels, restaurants, and local attractions to foster tourism business for the area.

convention center   A freestanding large exhibit hall where trade shows are typically housed. Unlike most conference centers, convention centers do not have lodging accommodations (although hotels are often adjacent to or near the facility).

convention market   Large corporate or association groups that typically meet on an annual basis.

core product   What the customer is really buying.

corporate or managed business traveler   A business traveler who must follow the rules and regulations of the corporate travel department. Corporate travel departments range from a coporate travel director who develops corporate travel polices and writes contracts with travel suppliers (e.g., hotels and airlines) to full-time in-house travel agencies.

cost-based pricing   Pricing based on the costs involved to produce a product rather than on conditions in the marketplace.

customer value   The customer perception of a fair return in goods, services, or money for something exchanged. Marketing creates value for customers by understanding and delivering on their needs and wants. Value is a judgment assigned by consumers to the expected or completed consumption of goods and services.

customized products   Products designed to fit the specific needs of a particular target market.

death spiral   The spiral of cutting back on costs, leading to less business, leading to more cost cuts.

demand analysis   Measurement of the market to calculate existing or future potential; sufficient demand means there are enough customers who want the product and are willing to pay for it.

destination marketing   A collective and coordinated effort among various constituencies to market a destination such as a country, a regional area, a state, a province, or even a city, such as the successful "I Love New York" campaign.

determinant attributes   Attributes of a product or service that actually determine the decision to purchase a product.

DEWK   An acronym for a lifestyle segment describing a household: dual employed with kids.

differentiation   Distinguishing a product or service from that of the competition in ways that are both identifiable and meaningful for the customer. It is the *perceived* difference by the customer that is important to the marketer.

DINK An acronym for a lifestyle segment describing a household: double income, no kids.

distribution mix (placement) All channels available between the firm and the target market that increase the probability of getting the customer to the product.

distribution service provider (DSP) A firm that provides the switch that links the central reservation system with the GDS. Examples of DSPs are Wizcom, Thisco, Pegasus, and Trust.

echo boomer The largest generation of young people since the 1960s, so called because they are the genetic offspring and demographic echo of their parents, the baby boomers. *Baby boomers* is the term applied to those who were born in the years immediately after World War II.

ecological/natural environment Ecological or natural environmental issues and concerns such as pollution, waste disposal, and recycling that affect consumers and hospitality tourism businesses alike.

economic environment Economic realities that affect the hospitality industry, including recessions, inflation, employment levels, personal discretionary income, and foreign exchange rates.

economies of scale Spreading costs across multiple units or within a large establishment of an organization to save money.

employee relationship marketing Applying marketing principles to those who serve the customer (directly or indirectly) and building a mutual bond of trust with them. Sometimes referred to as **internal marketing.**

environmental scanning Scanning the external environment to learn of potential opportunities and threats that could affect a business.

fair market share The amount of business a firm expects to receive in relative proportion to its direct competitors in the same product class.

family life cycle The emotional and intellectual stages that a person passes through as a member of a family. The stages include childhood, independence, coupling or marriage, parenting, launching adult children, and retirement or senior stage of life.

feedback loop An important part of the strategic marketing system model, refers to issues that need to be studied both before and after implementing a marketing strategy, including the potential risks of the strategy, the strategic fit to the operation, how or whether goals and objectives are being met, and so forth.

fixed costs Costs that do not vary with customer counts or occupancy, such as insurance costs or real estate taxes.

formal product What the customer thinks she is buying

four Ps A term applied to a common marketing mix for goods: product, price, place, and promotion.

franchisee An organization or person that purchases a brand name in order to distribute the product or service.

free independent traveler (FIT) An independent traveler who is not affiliated with an organized travel group and is difficult to identify within a well-defined market segment.

frequency program Any program that rewards guests with points, miles, stamps, and so forth, that they can redeem for free or discounted products or service.

global distribution system (GDS) An international computerized reservation system that travel agents (and other intermediaries) can use to reserve accommodations electronically for hotels, airlines, cruise lines, car rentals, and other hospitality and tourism services.

goods The physical factors of a product over which management has direct, or almost direct, control.

government market Those who travel for official government business at the local, state, provincial, or federal level. This travel could be for individual or group purposes.

gross domestic product (GDP) The market value of all final goods and services produced within a country in a given period of time. The most common approach to measuring and understanding GDP is the expenditure method:

GDP = consumption + investment + government spending + (exports − imports)

heterogeneity The inconsistent delivery of service levels provided by different employees and affected by different types of customers.

hospitality marketing mix The mix of marketing activities that are directed toward an identified target market. The elements of the marketing mix include the product/service mix, the presentation mix, the pricing mix, the communications mix, the distribution mix, the people mix, and the process mix. The hospitality marketing mix is sometimes referred to as the seven Ps: product, price, place, promotion, process, people, and physical attributes.

hospitality product The combination of goods, services, environment, and experience that the hospitality customer buys.

importance attributes Attributes of a product or service that are important to a consumer in either making a purchase decision or after having made the purchase. In other words, importance attributes are not necessarily salient or determinant in the final purchase decision.

incentive houses Companies that provide professional incentive planning services for corporations that prefer to have this complex activity handled by an external firm.

incentive travel Trips that are taken by employees of an organization as rewards for their superior performance over a given period of time. Incentive travel is sometimes offered to customers, distributors, dealers, and others for motivational purposes.

incidental learning Learning that occurs unintentionally. It can occur by completing tasks, watching others, doing the same thing over and over again, or talking to experts. A professor who makes the class fun and interesting fosters incidental learning.

industry/supply-based approach to identifying competition This approach defines competitors as those brands that operate in the same industry offering similar products and services to similar customers. Also called the **supply-based approach** because it requires identifying competitive destinations through the similarity of their offerings, resources, and strategies.

inseparability of production and consumption Consumption of the service while it is being produced. This concept reflects the notion that the consumer is part of the "assembly line." In other words, an empty guest room produces nothing.

intangibility The attributes of services that the customer cannot grasp with any of the five senses; that is, customers cannot taste, feel, see, smell, or hear a service until they have consumed it. The intangible aspects of a service product are difficult to grasp conceptually prior to purchase.

intangible positioning Creating a tangible, objective image based on an intangible aspect of the product.

intention A consumer's plan to make or not make a purchase.

intentional learning Learning that occurs through the active cognitive processing of information. An example of intentional learning is your reading this book in preparation for class or an exam.

intermediaries Professional organizations that connect buyers and sellers to help make a purchase happen. Examples in hospitality and tourism include travel agencies and incentive houses.

International Association of Conference Centers (IACC) A trade association that represents and serves the needs of conference centers. The membership of the association includes owners and mangers of conference centers, professional meeting planners, and others who have an interest in this type of facility.

international traveler A person who travels and visits outside his or her own country for business, personal, or pleasure purposes.

keyword A word that the user searches for when using the search engine. Generally speaking, keyword density is the ratio of the word that is being searched for (the keyword) against the total number of words appearing on your web page. If your keyword occurs only once in a page of one thousand words, it has a lower keyword density than a keyword that occurs four times in a page of similar length.

lifetime value of a customer The total value of a customer based on repeat purchases anticipated and word of mouth to others, minus the costs associated with serving that customer.

loyalty ladder Moving a customer from a state of awareness of your product or service to being an advocate who is sincerely loyal and promotes your product or service to others.

loyalty program A strategy undertaken by a firm to encourage an emotional bond with the customer so that she gives the firm a majority of her business, provides positive word of mouth, acts in partnership with the firm, and spends more time at the firm's establishment than nonloyal guests.

macrocompetition Industry competition; refers to organizations that compete for the same consumer dollar yet are not necessarily considered direct competitors.

market or demand-based approach to identifying competition This approach defines competitors as those satisfying the same customer need. This method is also called a demand-based approach because it requires identifying competitors by the similarity of their customers (i.e., their attitudes, behaviors, and motivations).

market segmentation Dividing a market into meaningful groups of buyers who have similar needs and wants. Market segmentation forms the basis for target market identification.

market segments Groups of buyers who have similar needs, wants, and problems.

marketing The process of identifying evolving consumer preferences; then capitalizing on them through the creation, promotion, and delivery of products and services that satisfy the corresponding demand. This is done by solving customers' problems and giving them what they want or need at the time and place of their choosing and at the price they are willing and able to pay.

marketing action plans   Plans of action set forth in the marketing plan that include time frames and who will implement the plan. (Sales action plans are part of the marketing plan, too.)

marketing budget   A component of the marketing plan that projects marketing expenses to be incurred for the forthcoming year.

marketing forecast   A component of the marketing plan that projects revenues to be generated, such as guest room sales, food and beverage sales, and others for the approaching year.

marketing leadership   A characteristic of a hospitality enterprise that integrates marketing into every phase of its operation through opportunity, planning, and control. Marketing leadership combines a vision for the future with systematic planning for solving customers' problems.

marketing management   The day-to-day management of the marketing process. The development of the annual marketing plan is part of marketing management.

marketing orientation   A style of hospitality management that stresses the importance of the customer and focuses on creating value for customers by recognizing their changing needs and wants.

marketing plan   The working document a firm uses to implement its marketing strategy. Typically renewed on an annual basis.

Maslow's hierarchy of needs   Psychologist Abraham Maslow's premise that people are motivated by differing levels of needs and that lower-level needs, such as hunger and thirst, need to be satisfied before higher-level needs, such as self-esteem, can be met.

mass customization   Rather than making the same product for everyone, the trend is to allow customers to personalize the product or service to their specifications; e.g., requesting a certain type of pillow in the room or specific items in the minibar.

match pricing   A strategy whereby one firm matches the price of firms that are direct competitors.

mature traveler   A hospitality customer who is 55 years old or older. This market is generally considered a subsegment of the pleasure traveler market.

meeting planner   An organizational customer who plans and executes off-site meetings and is the primary decision maker for hospitality and related purchases. Related purchases include airline seats, ground transportation, and the like.

merchandising   A tool used in the marketing communications mix to reach identified target markets. Merchandising is primarily an in-house marketing technique used to stimulate sales of additional products or services on premise.

metropolitan statistical areas (MSAs)   Geographic entities defined by the U.S. Office of Management and Budget (OMB) for use by federal statistical agencies in collecting, tabulating, and publishing federal statistics. A metro area contains a core urban population of 50,000 or more, and a micro area contains an urban core population of at least 10,000 (but less than 50,000).

MICE   An acronym for meetings, incentives, conventions and expositions customers.

microcompetition   Businesses that compete for the same customers in the same product class at the same point in time.

mission statement   A statement that defines the purpose of a business and that may include many ways to achieve that purpose in terms of all stakeholders. It should drive all subsets of the business.

moment of truth   Any time an employee has contact with a customer; that is, when the service product meets the service delivery.

motorcoach tour   Five or more travelers arriving at a hospitality establishment by motorcoach as part of an organized leisure travel tour package.

multiplier effect   The impact the tourist dollar has on a destination's economy in that wage earners working in hospitality and tourism also spend much of their earnings in the local community.

national tourism organization (NTO)   A department or agency of a national government responsible for marketing and promoting tourism for its country at the international level.

neutral pricing   A firm uses variables other than price to gain market share.

objective positioning   Presenting a product to the consumer based on its physical characteristics or functional features. Objective attributes of a product include atrium lobbies, historic landmarks, signature golf courses, and wireless connectivity to the Internet.

operations orientation A style of hospitality management that focuses on the execution of the operation to provide a smoothly running organization.

organizational customer A buyer of hospitality products and services for a group or organization that has a common purpose for the purchase.

package market Consumers who purchase a hospitality product offering that includes a combination of services for an all-inclusive price.

penetration pricing The goal of penetration pricing is to price just below the average market price in order to capture market share. This type of strategy can occur when opening a new property, or it can be a regular strategy, as practiced by Southwest Airlines.

people mix The people who work in an organization and how their attitudes, work ethic, and disposition affect the service delivery.

perceptions Meanings we assign to what we see, hear, and sense around us, which are influenced by sociocultural and psychological forces. Perceptions are images we have, and images influence purchase behavior.

perceptual map A visual map, or "plot," of customer perceptions of an establishment (or brand) relative to the competition based on features and benefits important to the targeted audience such as price, location, services offered, and the like.

perishability The life cycle of the hospitality service. A guest room has a life cycle of 24 hours. A luncheon meal at a restaurant has a life cycle of about one to two hours midday.

personal selling Direct, face-to-face interaction between a seller and a buyer for the purpose of making a sale.

pleasure traveler A customer who purchases hospitality products or services for leisure or other nonbusiness purposes.

political environment Government movements and positions at the local, state, provincial, and federal levels that affect consumers and businesses, such as feminism, discrimination, the funding of tourism, border crossings, and business ownership.

positioning Creating an image, differentiating a product, and promising a benefit in the mind of the consumer. Positioning is the perception the consumer has of a product offering.

positioning statement A singular expression by a property or brand that captures the essence of its intended positioning in the marketplace.

presentation mix All of the elements used by the firm to increase the tangibility of the product/service mix in the perception of the target market.

press release A document prepared for release to selected media containing information or "news" about a hospitality or tourism enterprise. Press releases normally include the name of the public relations professional who created the release and how she or he can be reached for further information.

price bundling Offering one price that includes several aspects of a product offering such as a welcoming reception, overnight accommodations, breakfast, and a local guided tour. Many weekend packages are a form of price bundling.

pricing mix The combination of prices that consumers are offered to purchase a product or service.

PRIZM NE A method of segmenting types of customers based on the customers' behaviors. This separation of types of customers identifies 66 different market segments based on ZIP codes. PRIZM stands for *Potential Rating Index Zip-code Markets* and NE stands for *New Evolution*.

probing Asking a prospective customer questions to learn more about his or her real needs and wants.

process mix The activities designed to deliver the desired services to the guest.

product class A group of similar products basically serving the same needs (e.g., fast casual restaurant is one class, while white tablecloth is another; budget hotels is one class and upscale restaurants is another).

product life cycle (PLC) A product goes through several stages during its lifetime. These stages include introduction, growth, maturity, and decline.

product/service mix The combination of products and services, whether for free or for sale, that are aimed at the needs of the target market.

product/service orientation A style of hospitality management with a primary emphasis on creating great products or services to attract customers.

property needs analysis An analysis of major profit areas to see what gaps (which can be in occupancy, market share, average room rate, market segment mix, food sales, beverage sales, seasonal needs, etc.) have to be filled.

prospecting The initial stage of looking for potential customers to purchase a product.

psychological pricing   Pricing that takes into consideration what may be in the customers' minds or how they will react.

public relations (PR)   A tool used in the marketing communications mix to reach identified target markets. Public relations is planned management of the media's and the community's perception of a hospitality or tourism enterprise. Press releases are often used to aid in this process.

publicity   News related to a hospitality or tourism enterprise that is made available to the public. It may be positive or negative.

pull strategy   Directing marketing communication efforts directly to the consumer, who in turn (hopefully) purchases the product directly or through an intermediary such as a travel agent or an online booking agency.

push strategy   Directing marketing communication efforts to intermediaries such as travel agents, who then help "push" the product to their customer base.

qualifying (a prospect)   Determining whether a prospect has an interest in and ability to purchase a product.

reference group   People who influence a person's attitudes, opinions, and values, such as families, friends, and business associates.

reference price   A price or price range anticipated by the consumer based on prior experience or knowledge.

regulatory environment   Government regulations or ordinances at the local, state, provincial, and federal levels that affect consumers and businesses, such as tax codes, truth in menu regulations, minimum wage laws, and smoking bans in restaurants.

relationship marketing   An ongoing process of identifying and creating new value for individual customers for mutual value benefits and then sharing the benefits from this process over a lifetime of association; also known as **loyalty marketing.**

reservation pricing   The maximum price the customer will pay for a product. For example, if the customer's reservation price for a can of soda is 1 euro and the price is 1.01 euros, the customer will not buy the product.

revenue management   The practice of setting prices based on varying patterns of anticipated usage including daily, weekday, weekend, and/or seasonal periods. Prices are typically set higher for peak demand periods (high season) and lower for off-peak periods (low season). This is sometimes referred to as **yield management**.

revenue per available room (REVPAR)   Guest room revenue generated over a given period of time divided by the total number of guest rooms available for sale for the same period of time.

RFM   An acronym for recency (How recently have you been to the establishment?), frequency (How often do you visit the establishment?), and monetary value (How much are you worth to the organization?). RFM is used in database marketing to understand customers and to show that some customers are worth more to the firm than others.

sales account managers   Salespeople assigned to specific accounts with only a few clients that have large volumes of business. This is often a sales position at the national level for large hospitality and tourism organizations.

sales action plan   A salesperson's plan of action for the forthcoming quarter to reach identified goals and objectives.

sales leads   Sources of prospective customers.

sales promotions   Marketing communications that serve as incentives to stimulate sales on a short-term basis.

salesperson   A person within a hospitality or tourism organization who has direct responsibility for personal selling to identified target markets (also referred to as a **sales representative** or **sales manager**).

salient attributes   Attributes of a product or service that readily come to mind—that is, are "top of the mind" to the consumer. They may or may not be important or determinant in the final purchase decision.

segmentation variables   Various ways in which a market can be divided, or segmented, into meaningful groups of buyers such as geographic, demographic (age, income, etc.), and particular usage of the product. Often several variables are used to define a market segment.

selling orientation   A style of hospitality management with a primary emphasis on salespeople and promotions that communicate a message to customers in order to sell products.

semivariable costs   Costs that are somewhat fixed regardless of units or items sold but can at times vary to some degree, such as staff payroll.

service gaps   Areas in service delivery that are inadequate, missing, or poorly executed and result in customer dissatisfaction.

services   The nonphysical and intangible aspects of a product that management does, or should, control.

SERVQUAL model   A model developed to help identify potential gaps in service delivery that could or should be corrected by management.

situational analysis/SWOT analysis   An analysis of the current internal strengths and weaknesses and external opportunities and threats of a business entity.

skim pricing   The goal of price skimming is to capture high margins at the expense of high sales volume. The term is derived from the notion of skimming the cream off the top, before the competition comes in and forces prices down.

SMERF market   Price-sensitive, not-for-profit organizations including social, military, education, religious, and fraternal.

sociocultural environment   Cultural values, customs, habits, lifestyles, personal beliefs, demographic changes, and so on, that affect consumer behavior.

standard product with modifications   A core product that has been embellished with new elements.

standard products   Products that appear similar and standard to the customer, albeit in different locations, especially for a branded product.

stimuli   In a marketing context, this refers to advertisements, reference groups, and so forth, that might incite or quicken the actions, feelings, or thoughts of a potential or repeat customer.

strategic business unit (SBU)   A division of a company that may operate separately from the main organization. For example, InterContinental Hotels Group (IHG) is the organization, whereas Holiday Inn is the SBU.

strategic marketing   The overall view of marketing that includes defining marketing, setting marketing goals, and allocating the resources necessary to reach those goals.

strategic marketing system model   A model that shows the process by which marketing strategies and substrategies are developed.

subjective positioning   Presenting an "image" of a product to the consumer rather than the actual physical aspects of the product. Subjective attributes of a product include prestige, service quality, and guest experience.

tangible positioning   Creating an intangible, subjective image of a product based on a tangible feature of the product.

target marketing   Selecting specific market segments to target and designing the product or service to meet their specific needs and wants.

target markets   Market segment(s) that a company has identified as its primary customer base. It then designs or delivers its products and services accordingly.

technological environment   Technological advances that can or will affect a business from both a consumer and competitive perspective.

trade shows   Exhibitions whose primary purpose is to display and sell products. Trade shows typically take place in large exhibit halls, such as convention centers.

travel agent   An intermediary who assists travelers in making travel plans, booking reservations, and the like, for a fee. This fee is sometimes paid by the supplier and sometimes by the customer.

travel manager   An organizational customer who helps plan individual travel itineraries for members of an organization and makes hospitality purchase decisions for those members.

United Nations World Tourism Organization   A division of the United Nations responsible for tracking tourism visitor statistics on a worldwide basis, including country of origin, length of stay, and so on.

VALS   A method of separating types of customers based on psychological characteristics and four key demographics. The U.S. VALS system classifies customers into eight segments using dimensions of primary motivation (e.g., thinking, experiencing) and level of financial resources (high or low).

value-based pricing   Pricing based on customer sensitivity to the price–value relationship of the product offering and not solely on the costs associated with producing the product.

variable costs   Costs that vary directly with a unit or item sold, such as the food costs of a menu selection or the housekeeping costs of an occupied guest room.

yield   The ratio between actual sales and potential sales forecast over a given period of time.

zone of tolerance   The area between desired service and adequate service that the customer will tolerate, even if not totally satisfied.

name index

Bold page numbers indicate glossary terms.

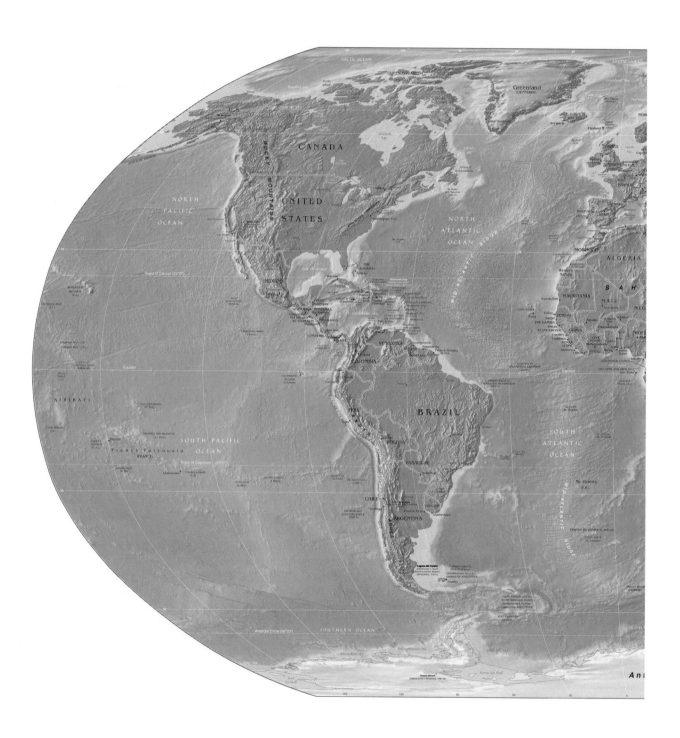

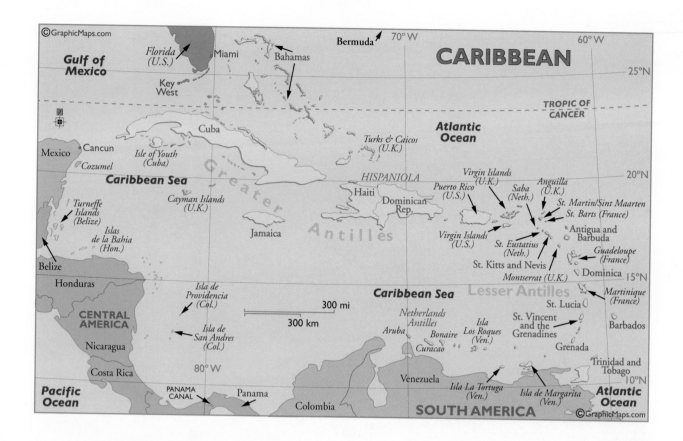

601

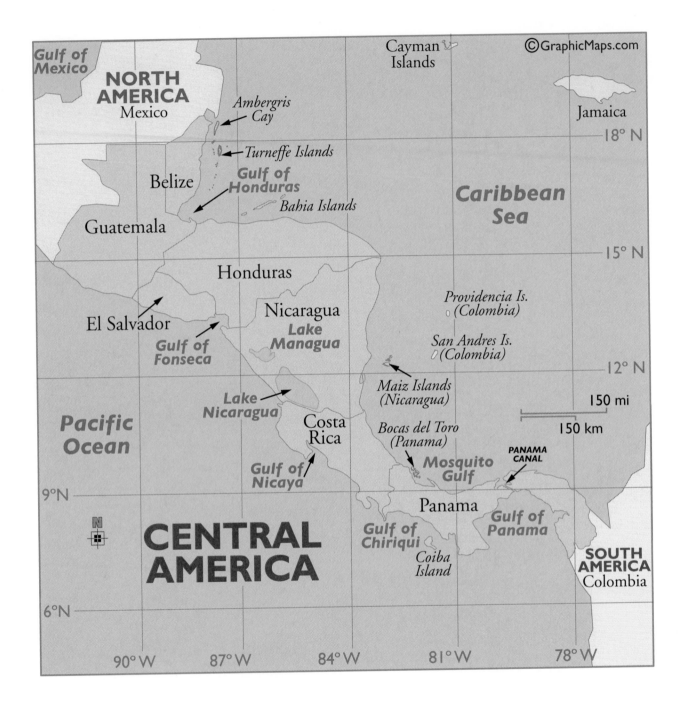

Gulf of
Mexico

©GraphicMaps.com

Cayman
Islands

NORTH
AMERICA
Mexico

*Ambergris
Cay*

Jamaica

18° N

Turneffe Islands

Belize

**Gulf of
Honduras**

**Caribbean
Sea**

Bahia Islands

Guatemala

15° N

Honduras

*Providencia Is.
(Colombia)*

El Salvador

Nicaragua

**Lake
Managua**

*San Andres Is.
(Colombia)*

**Gulf of
Fonseca**

12° N

*Maiz Islands
(Nicaragua)*

**Pacific
Ocean**

**Lake
Nicaragua**

150 mi

150 km

Costa
Rica

*Bocas del Toro
(Panama)*

**PANAMA
CANAL**

**Mosquito
Gulf**

**Gulf of
Nicaya**

9°N

Panama

**Gulf of
Panama**

N

**CENTRAL
AMERICA**

**Gulf of
Chiriqui**

*Coiba
Island*

**SOUTH
AMERICA**
Colombia

6°N

90° W   87° W   84° W   81° W   78° W

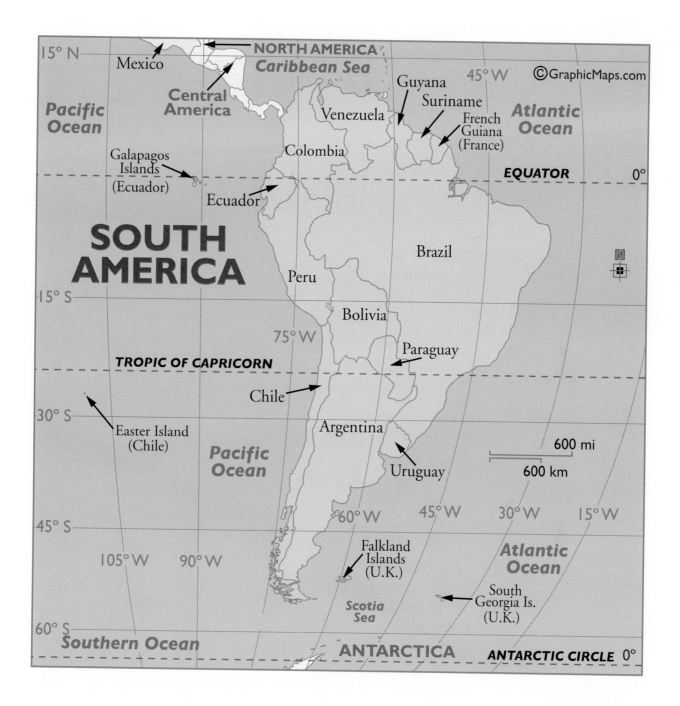

15° N

Mexico

NORTH AMERICA

Caribbean Sea

Central
America

Guyana
Suriname

45° W

©GraphicMaps.com

*Pacific
Ocean*

Venezuela

French
Guiana
(France)

*Atlantic
Ocean*

Galapagos
Islands
(Ecuador)

Colombia

Ecuador

EQUATOR

0°

**SOUTH
AMERICA**

Brazil

N

Peru

15° S

Bolivia

75° W

TROPIC OF CAPRICORN

Paraguay

Chile

30° S

Easter Island
(Chile)

Argentina

*Pacific
Ocean*

Uruguay

600 mi

600 km

45° S

105° W

90° W

60° W

45° W

30° W

15° W

Falkland
Islands
(U.K.)

*Atlantic
Ocean*

Scotia
Sea

South
Georgia Is.
(U.K.)

60° S

Southern Ocean

ANTARCTICA

ANTARCTIC CIRCLE 0°

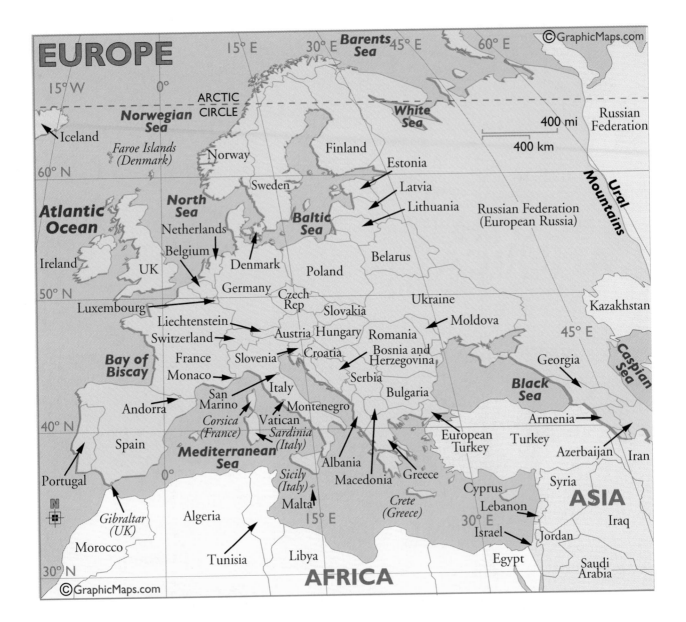

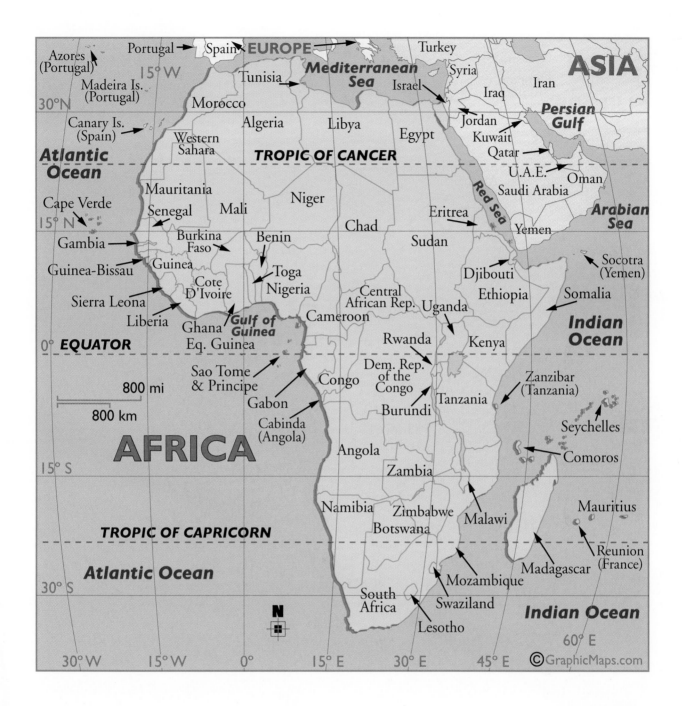

Azores
(Portugal)

Portugal → Spain → **EUROPE** →

Turkey

ASIA

Madeira Is.
(Portugal)

15° W

Tunisia →

*Mediterranean
Sea*

Syria

Iran

30°N

Morocco

Israel →

Iraq

*Persian
Gulf*

Canary Is.
(Spain) →

Western
Sahara

Algeria

Libya

Egypt

Jordan

Kuwait

**Atlantic
Ocean**

TROPIC OF CANCER

Qatar

U.A.E.

Oman

Cape Verde

Mauritania

Saudi Arabia

*Arabian
Sea*

15° N

Senegal

Mali

Niger

Eritrea →

Red Sea

Yemen

Gambia →

Burkina
Faso

Benin

Chad

Sudan

Socotra
(Yemen)

Guinea-Bissau →

Guinea

Toga

Djibouti

Sierra Leona →

Cote
D'Ivoire

Nigeria

Central
African Rep.

Ethiopia

Somalia

Liberia

Ghana

Cameroon

Uganda

**Indian
Ocean**

0° **EQUATOR**

Eq. Guinea

*Gulf of
Guinea*

Rwanda

Kenya

Sao Tome
& Principe

Congo

Dem. Rep.
of the
Congo

Zanzibar
(Tanzania)

800 mi

Gabon

Tanzania

Burundi

Seychelles

800 km

Cabinda
(Angola)

Comoros

AFRICA

Angola

15° S

Zambia

Mauritius

TROPIC OF CAPRICORN

Namibia

Zimbabwe

Malawi

Botswana

Reunion
(France)

Atlantic Ocean

Madagascar

30° S

Mozambique

Indian Ocean

South
Africa

Swaziland

60° E

N

Lesotho

30° W   15° W   0°   15° E   30° E   45° E

© GraphicMaps.com

© GraphicMaps.com

EUROPE

European Russia

Sea of Azov

45° E

45° N

Black Sea

Azerbaijan

Georgia

Armenia

Caspian Sea

Turkey

Greece

Cyprus

Mediterranean Sea

Syria
Lebanon

Israel

Sinai

Iraq

Egypt

Libya

30° N

Jordan

Gulf of Aqaba

Gulf of Suez

TROPIC OF CANCER

AFRICA

Chad

Sudan

15° N

30° E

Eritrea

Ethiopia

Djibouti

Red Sea

Kuwait

Saudi
Arabia

Qatar

United
Arab
Emirates

Yemen

Gulf of Aden

Somalia

Kazakhstan

Russia

NORTH
ASIA

90° E

Aral Sea

Uzbekistan

Kyrgyzstan

Turkmenistan

Tajikistan

China

Iran

ASIA
MIDDLE
EAST

Afghanistan

India

Bahrain

Persian Gulf

Pakistan

75° E

SOUTHEAST
ASIA

Oman

Gulf of Oman

Arabian Sea

Indian Ocean

60° E

606

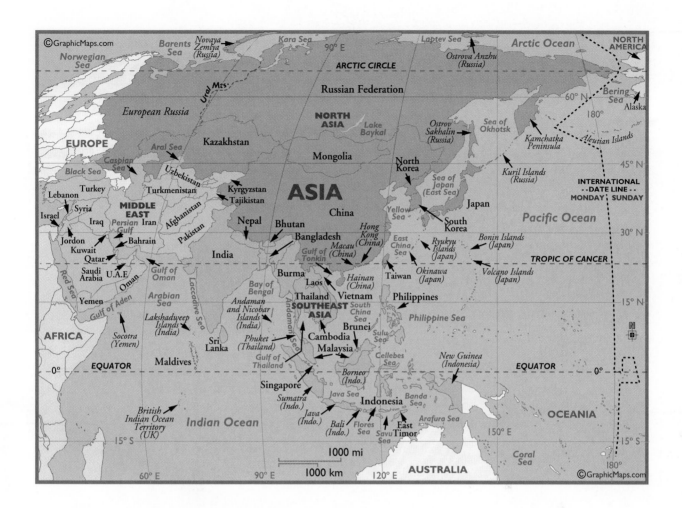

© GraphicMaps.com

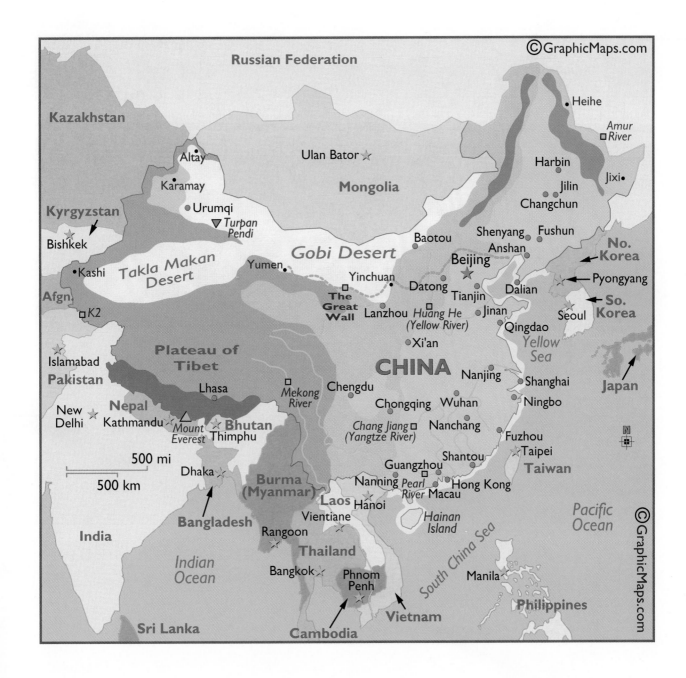

Russian Federation

Kazakhstan

Altay
Karamay
Urumqi
Turpan Pendi

Kyrgyzstan
Bishkek
Kashi
Afgn.
K2

Takla Makan Desert

Yumen

Ulan Bator ☆

Mongolia

Gobi Desert

Baotou
Yinchuan
Datong
Lanzhou   Huang He (Yellow River)

Heihe

Amur River

Harbin
Jilin
Changchun

Shenyang   Fushun
Anshan
Beijing ☆

Tianjin   Dalian
Jinan

No. Korea

Pyongyang
So. Korea
Seoul

Islamabad
Pakistan

New
Delhi

Plateau of
Tibet

Lhasa

Mekong
River

Chengdu

Xi'an

CHINA

Nanjing

Qingdao

Yellow
Sea

Shanghai
Ningbo

Japan

Nepal
Kathmandu
Mount
Everest   Thimphu
Bhutan

Chongqing   Wuhan

Chang Jiang
(Yangtze River)

Nanchang

Fuzhou
Taipei

Shantou   Taiwan

500 mi

500 km

Dhaka

India

Indian
Ocean

Burma
(Myanmar)

Bangladesh

Rangoon

Laos

Vientiane

Hanoi

Guangzhou
Nanning   Pearl
River   Macau   Hong Kong

Hainan
Island

South China Sea

Pacific
Ocean

Thailand

Bangkok

Phnom
Penh

Vietnam

Manila

Philippines

Sri Lanka

Cambodia

N

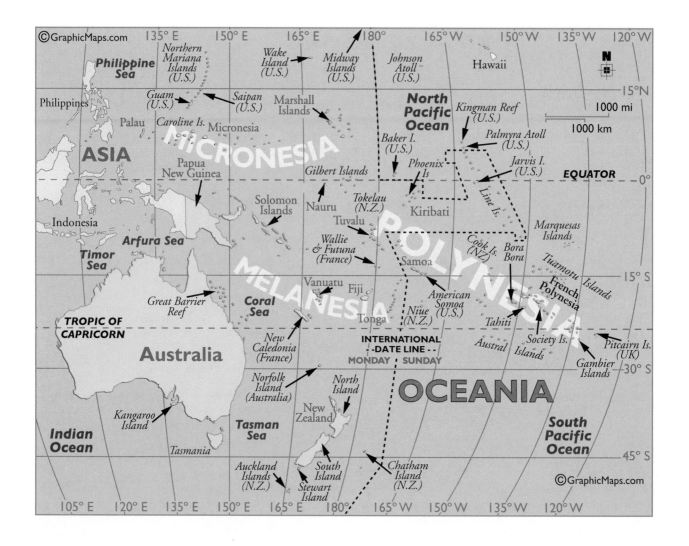

Philippine Sea

Northern Mariana Islands (U.S.)

Wake Island (U.S.)

Midway Islands (U.S.)

Johnson Atoll (U.S.)

Hawaii

15°N

Philippines

Guam (U.S.)

Saipan (U.S.)

Marshall Islands

North Pacific Ocean

Kingman Reef (U.S.)

Palau

Caroline Is.

Micronesia

MICRONESIA

1000 mi

1000 km

ASIA

Papua New Guinea

Baker I. (U.S.)

Phoenix Is

Palmyra Atoll (U.S.)

Jarvis I. (U.S.)

EQUATOR

0°

Indonesia

Gilbert Islands

Tokelau (N.Z.)

Line Is.

Kiribati

Marquesas Islands

Timor Sea

Arfura Sea

Solomon Islands

Nauru

POLYNESIA

Tuvalu

Wallie & Futuna (France)

Samoa

Cook Is. (N.Z.)

Bora Bora

Tuamotu Islands

French Polynesia

15° S

MELANESIA

Vanuatu

Fiji

American Samoa (U.S.)

Great Barrier Reef

Coral Sea

Niue (N.Z.)

Tonga

Tahiti

Society Islands

Pitcairn Is. (UK)

TROPIC OF CAPRICORN

Australia

New Caledonia (France)

INTERNATIONAL
- -DATE LINE - -
MONDAY | SUNDAY

OCEANIA

Austral

Gambier Islands

30° S

Norfolk Island (Australia)

North Island

Kangaroo Island

New Zealand

South Pacific Ocean

Indian Ocean

Tasmania

Tasman Sea

South Island

45° S

Auckland Islands (N.Z.)

Stewart Island

Chatham Island (N.Z.)

105° E 120° E 135° E 150° E 165° E 180° 165° W 150° W 135° W 120° W

©GraphicMaps.com

Photo Credits

Page 1: Getty Images – Stockbyte

Pages 2, 28, 54, 84: David R. Frazier / Photo Researchers, Inc.

Page 119 : Ronnie Kaufman / CORBIS-NY

Page 120: Getty Images – Stockbyte

Page 115: Jeffrey Greenberg / Photo Researchers, Inc.

Pages 156, 190: Richard Wahlstrom, Getty Images Inc.-Image Bank

Page 223: Mira.com / Bill Bachmann

Pages 224, 264, 298: gkphotography / Alamy Images

Page 337: Gunnar Kullenberg / The Stock Connection

Pages 338, 378, 418, 462, 498, 530: Cameron Davidson / The Stock Connection

Page 471: Magnus Rew © Dorling Kindersley, Courtesy of the Four Points Sheraton Hotel, Orlando

Maps pages 598–599: Courtesy of the University of Texas Libraries, The University of Texas at Austin

Maps pages 600–610: © GraphicMaps.com